Effects of Atmospheric Pollutants on Forests, Wetlands and Agricultural Ecosystems

NATO ASI Series

Advanced Science Institutes Series

A series presenting the results of activities sponsored by the NATO Science Committee, which aims at the dissemination of advanced scientific and technological knowledge, with a view to strengthening links between scientific communities.

The Series is published by an international board of publishers in conjunction with the NATO Scientific Affairs Division

A	Life Sciences	Plenum Publishing Corporation
B	Physics	London and New York
C	Mathematical and Physical Sciences	D. Reidel Publishing Company Dordrecht, Boston, Lancaster and Tokyo
D	Behavioural and Social Sciences	Martinus Nijhoff Publishers Boston, The Hague, Dordrecht and Lancaster
E	Applied Sciences	
F	Computer and Systems Sciences	Springer-Verlag Berlin Heidelberg New York
G	Ecological Sciences	London Paris Tokyo
H	Cell Biology	

Series G: Ecological Sciences Vol. 16

Effects of Atmospheric Pollutants on Forests, Wetlands and Agricultural Ecosystems

Edited by

T.C. Hutchinson and K.M. Meema

Institute for Environmental Studies, University of Toronto
Toronto, Ontario, M5S 1A4, Canada

Springer-Verlag
Berlin Heidelberg New York London Paris Tokyo
Published in cooperation with NATO Scientific Affairs Division

Proceedings of the NATO Advanced Research Workshop on Effects of Acidic
Deposition on Forests, Wetlands and Agricultural Ecosystems held at Toronto,
Canada, May 12–17, 1985

ISBN-13:987-3-642-70876-3   e-ISBN-13:987-3-642-70874-9
DOI: 10.1007/987-3-642-70874-9

Library of Congress Cataloging in Publication Data. NATO Advanced Research Workshop on Effects of Acidic Deposition of Forests, Wetlands, and Agricultural Ecosystems (1985 : Toronto, Ont.) Effects of atmospheric pollutants on forests, wetlands, and agricultural ecosystems. (NATO ASI series. Series G, Ecological sciences ; vol. 16) "Proceedings of the NATO Advanced Research Workshop on Effects of Acidic Deposition on Forests, Wetlands, and Agricultural Ecosystems, held at Toronto, Canada, May 12–17, 1985"—T.p. verso. Includes indexes. 1. Plants Effect of air pollution on—Congresses. 2. Plants, Effect of acid deposition on—Congresses. 3. Forest ecology—Congresses. 4. Wetland ecology—Congresses. 5. Agricultural ecology—Congresses. 6. Air—Pollution—Environmental aspects—Congresses. 7. Acid deposition—Environmental aspects—Congresses. I. Hutchinson, T.C. (Thomas C.), 1939-. II. Meema, K.M. III. North Atlantic Treaty Organization. Scientific Affairs Division. IV. Series: NATO ASI series. Series G, Ecological sciences ; vol. 16. QK751.N38 1985   574.5'264   87-20640
ISBN-13:987-3-642-70876-3 (U.S.)

This work is subject to copyright. All rights are reserved, whether the whole or part of the material is concerned, specifically the rights of translation, reprinting, re-use of illustrations, recitation, broadcasting, reproduction on microfilms or in other ways, and storage in data banks. Duplication of this publication or parts thereof is only permitted under the provisions of the German Copyright Law of September 9, 1965, in its version of June 24, 1985, and a copyright fee must always be paid. Violations fall under the prosecution act of the German Copyright Law.

© Springer-Verlag Berlin Heidelberg
Softcover reprint of the hardcover 1st edition 1987

2131/3140-543210

TABLE OF CONTENTS

Acknowledgements..	ix
Participants of the Workshop..................................	xi
Workshop Summary...	xv
Hutchinson TC	

PART I: GEOGRAPHICAL OVERVIEWS OF FOREST DECLINE

Development and cause of novel forest decline in Germany..	1
Prinz B, Krause GHM and Jung K-D	
Results of studies on forest decline in northwest Germany...	25
Matzner E and Ulrich B	
Forest damage in Switzerland, Austria and adjacent parts of France and Italy in 1984.......................	43
Bucher JB	
Air pollution and forest damage in Norway................	59
Tveite B	
Forest decline - the view from Britain....................	69
Binns WO, Redfern DB and Rennolls K	
Deterioration of red spruce in the northern appalachian mountains...	83
Johnson AH	
Sugar maple decline in Ontario..........................	101
McLaughlin DL, Linzon SN, Dimma DE and McIlveen WD	

PART II: RESPONSES OF PLANTS TO ACIDIC DEPOSITION

Deposition and forest canopy interactions of airborne nitrate..	117
Lindberg SE, Lovett GM and Meiwes K-J	
Responses of herbaceous and woody plants to the dry deposition of SO_2 and NO_2.............................	131
Mansfield TA, Whitmore ME, Pande PC and Freer-Smith PH	
Plant cuticle as a barrier to acid rain penetration.......	145
Berg VS	
The response of plant reproductive processes to acidic rain and other air pollutants.............................	155
Cox R	
Consequences of cloud water deposition on vegetation at high elevation...	171
Unsworth MH and Crossley A	
Acid deposition, nutrient imbalance and tree decline: a commentary..	189
Tomlinson GH	

PART III: OTHER STRESSES INVOLVED IN FOREST DECLINE

Response of trees to drought.................................. 201
   Tyree MT, Flanagan LB and Adamson N

Responses of American forests to photochemical
oxidants.. 217
   McBride JR and Miller PR

Ozone toxicity - is there more than one mechanism of
action?... 229
   Pell EJ

Ethylene: a possible factor in the response of plants
to air pollution and acid precipitation....................... 241
   Reid DM

Responses of forests in decline to experimental
fertilization... 255
   Zoettl HW

Decline as a phenomenon in forests: pathological and
ecological considerations..................................... 267
   Manion PD

PART IV: DENDROCHRONOLOGICAL STUDIES

The use and limitations of dendrochronology in studying
effects of air pollution on forests........................... 277
   Cook ER

Dendroecological analysis of acidic deposition effects
on forest productivity.. 291
   LeBlanc DC, Raynal DJ and White EH

Trace metal uptake and accumulation in trees as
affected by environmental pollution........................... 307
   Baes III CF and McLaughlin SB

PART V: SOIL ACIDIFICATION AND POSSIBLE TOXICITY

Air pollution and soil acidification.......................... 321
   Abrahamsen G

A discussion of changes in soil acidity due to natural
processes and acid deposition................................. 333
   Johnson DW

Soil acidification and metal solubility in forests of
southern Sweden... 347
   Tyler G, Berggren D, Bergkvist B, Falkengren-Grerup U,
   Folkeson L and Ruhling A

Differences in aluminum mobilization in spodosols in
New Hampshire (USA) and in the Netherlands as a
result of acid deposition..................................... 361
   Mulder J and van Breemen N

Limits of cation leaching of weakly podzolized forest
soils: an empirical evaluation................................ 377
   Morrison IK and Foster NW

Effects of heavy metals and aluminum on the root
physiology of spruce (Picea abies Karst.) seedlings....... 387
  Godbold DL, Tischner R and Huttermann A

Aluminum toxicity in forest tree seedlings............... 401
  Eldhuset T, Göransson A and Ingestad T

PART VI: BRYOPHYTE AND LICHEN STUDIES

The effect of simulated acid rain on boreal forest
floor feather moss and lichen species.................... 411
  Hutchinson TC, Scott M, Soto C and Dixon M

Annual absorption of gaseous air pollutants by mosses
and vascular plants in diverse habitats.................. 427
  Winner WE and Atkinson CJ

Effects of quantitative and qualitative changes in air
pollution on the ecological and geographical
performance of lichens................................... 439
  Seaward MRD

PART VII: AGRICULTURAL RESPONSES

Crop responses to ozone-sulphur dioxide mixtures......... 451
  Ormrod DP, Deveau JL, Allen OB and Beckerson DW

Stomate-dependent and stomate-independent uptake of
NO_x and effects on photosynthesis, respiration
and transpiration of potted plants....................... 463
  Saxe H

The effects of acid rain, alone and in combination
with gaseous pollutants, on growth and yield of
crop plants.. 481
  Shriner DS and Johnson JW Jr

PART VIII: WETLANDS

The natural and anthropogenic acidification of
peatlands.. 493
  Gorham E, Janssens JA, Wheeler GA and Glaser PH

Interactions of Sphagnum with water and air.............. 513
  Clymo RS

Sources of alkalinity in Precambrian shield watersheds
under natural conditions and after fire or
acidification.. 531
  Bayley SE and Schindler DW

Responses to acidic deposition in ombrotrophic mires
in the U.K... 549
  Lee JA, Press MC, Woodin S, Ferguson P

The stratigraphic record of atmospheric loading of
metals at the ombrotrophic Big Heath bog, Mt. Desert
Island, Maine, U.S.A..................................... 561
  Norton SA

Proton cycling in bogs: geographic variation in northeastern North America.................................. 577
   Urban NR, Eisenreich SJ, Gorham, E

PART IX: ECOSYSTEMS AS A SOURCE OF SULPHUR EMISSIONS

Potential sulphur gas emissions from a tropical rainforest and an southern Appalachian deciduous forest.... 599
   Haines B, Black M, Fail J Jr, McHargue L and Howell G

PART X: CAUSE/EFFECT RELATIONS FOR PLANT DAMAGE

Perspectives on establishing the relationships between acidic deposition and vegetation responses............... 611
   Evans LS and Lewin KF

PART XI: MEETING SUMMARIES - Group Reports

Forest Decline.. 619
   Rapporteur: Johnson A

Forest Soils.. 625
   Rapporteur: Foster N

Persectives of Air Pollution Effects on Crop Plants........ 627
   Rapporteur: Evans LS

Wetlands.. 631
   Rapporteur: Gorham E

Author Index.. 637
Species Index... 639
Subject Index... 643

ACKNOWLEDGEMENTS

The editors are grateful for the financial assistance provided for the NATO Advanced Research Workshop by the Eco-Science panel of NATO which enabled the Toronto meeting to be mounted. Invaluable financial support was also received from Environment Canada, the Ontario Hydro and the Department of Forestry, University of Toronto, in Canada, and from the Electric Power Research Institute (EPRI) in the United States.

The programme was developed with the able assistance of Eville Gorham of the University of Minnesota, Gunnar Abrahamsen of the Norwegian Institute for Soil Research at Aas, and with advice from Drs. Ellis Cowling of North Carolina State University, Lance Evans, of Manhattan College, New York, and George Krause of Landesanstalt fur Immissionsschutz des Landes Nordrein-Westfalen in West Germany.

We also acknowledge the direct and indirect assistance of the staff of the Institute for Environmental Studies at the University of Toronto at all stages in the planning and running of the meeting.

Naomi Adamson acted as an assistant in the running of the actual meeting, and the staff of the Guild Inn, Toronto, provided a suitable atmosphere for an international group of 40 scientists to discuss and argue the details of effects of atmospheric pollutants on terrestrial ecosystems.

We also thank the late Dr. M. di Lullo and Dr. C. Sinclair, both NATO Science Directors, for their patience in the lengthy scientific review process we felt necessary for publication of a volume on this important topic. We also gratefully acknowledge critical comments and scientific advice provided by the scientific reviewers of individual chapters.

T.C. Hutchinson and K.M. Meema

PARTICIPANTS IN THE NATO ADVANCED RESEARCH WORKSHOP,
TORONTO, 12-17 MAY, 1985

Abrahamsen, G.
Norwegian Forest Research
 Inst.
Postbox 61, 1432 As-NLH
Norway

Balsillie, D.
Director of Air Resources
 Branch
880 Bay St., 4th Floor
Toronto, Ontario
Canada, M5S 1Z8

Bayley, S.
Dept. of Botany
University of Manitoba
Winnipeg, Manitoba
Canada, R3T 2N6

Berg, V.S.
Biology Dept.
University of Northern Iowa
Cedar Falls,
Iowa 50514, U.S.A.

Binns, W.O.
Research & Development Div.
Alice Holt Lodge
Wrecclesham, Near Farnham
Surrey GU10 4LH, UK

Bondietti, E.
Environmental Science
 Division, Oak Ridge
National Lab., P.O. Box X
Oak Ridge, TN 37830, U.S.A.

Brydges, T.
Ontario Ministry of
 the Environment
A.P.I.O.S.,
135 St. Clair Ave. W.
Suite 100
Toronto, Ontario
Canada, M4V 1P5

Bucher, J.B.
Swiss Federal Institute
 of Forestry Research
CH-8952 Birmensdorf
Switzerland

Clymo, R.
Queen Mary College
University of London
Mile End Rd.
London, E1 4NS, UK

Cook, E.R.
Lamont-Doherty Geological
 Observatory of Columbia Univ.
Palisades, N.Y. 10964, U.S.A.

Cox, R.M.
Maritimes Forest Research
 Centre
P.O. Box 4000, Fredericton
New Brunswick, Canada

Curtis, C.R.
Dept. Plant Pathology
Ohio State University
201 Kottman Hall
2021 Coffey Rd. Columbus
Ohio 43210-1087, U.S.A.

Evans, L.S.
Manhattan College, Plant
 Morphogenesis Laboratory
Manhattan College Parkway
Riverdale, N.Y. 10471, U.S.A.

Fayle, D.
Ontario Ministry of Natural
 Resources,
Whitney Block, Rm. 3631
99 Wellesley St. W.
Toronto, Ontario
Canada M7A 1W3

Foster, N.
Great Lakes Forest Research
 Centre, C.F.S.
P.O. Box 490, Sault Ste. Marie
Ontario, Canada, P6A 5M7

Godbold, D.
Institut fur Forstbotanik
Busgenweg 2
D-3400 Gottingen
West Germany

Gorham, E.
Dept. of Ecology and Behavioral Biology, 107 Zoology
University of Minnesota
318 Church St. S.E.
Minneapolis, MN 55455, U.S.A.

Haines, B.L.
University of Georgia
Botany Dept.
Athens, Georgia 30602
U.S.A.

Havas, M.
Institute for
 Environmental Studies
University of Toronto,
Toronto, Ontario, Canada
M5S 1A4

Hutchinson, T.C.
Institute for Environmental
 Studies and Department
 of Botany, University of
Toronto, Toronto,
Ontario, Canada, M5S 1A4

Ingestad, T.
University of Agricultural
 Sciences
S-750 07 Uppsala, Sweden

Jacobson, J.S.
Environmental Biology Program
Boyce Thompson Institute
Tower Road, Ithaca,
NY 14853, U.S.A.

Jeglum, J.
Canadian Forest Service
Box 490, Sault Ste. Marie
Ontario, Canada, P6A 5M7

Johnson, A.H.
Dept. of Geology
University of Pennsylvania
240 South 33rd. St.
Philadelphia, PA 19104, U.S.A.

Johnson, D.
Environmental Science Division
Oak Ridge National Laboratory
P.O. Box X, Oak Ridge,
TN 37830, U.S.A.

Kerekes, J.J.
Canadian Wildlife Service
Atlantic Region, Biology Dept.
Dalhousie University, Halifax
Nova Scotia, Canada, B3H 4J1

LaPointe, L.
Ministre de l'Environment
3900 rue Marley
Sainte-Foy, Quebec
Canada, G1X 4E4

Lee, J.A.
Dept. of Botany
University of Manchester
Manchester M13 9PL, UK

Legge, A.H.
Kananaskis Centre for
 Environmental Research
University of Calgary
2500 University Dr. N.W.
Calgary, Alberta
Canada, T2N 1N4

Lindberg, S.E.
Oak Ridge National Laboratory
P.O. Box X
Oak Ridge, TN 37830, U.S.A.

Linzon, S.N.
Ministry of the Environment
Air Resources Branch
880 Bay St., Toronto
Ontario, Canada, M5S 1Z8

Mansfield, T.A.
Dept. of Biological Sciences
University of Lancaster
Lancaster LA1 4YQ
Lancashire, United Kingdom

Manion, P.D.
Suny College Environmental
 Science and Forestry
Dept. of Forest Botany
 and Pathology
Syracuse, NY 13210, U.S.A.

Martin, H.
Environment Canada, Atmospheric
 Environment Service (L.L.O.)
4905 Dufferin St., Downsview
Ontario, Canada, M3H 5T4

Matzner, E.
Institut fur Bodenkunde und
 Waldernahrung, Universität
 Göttingen,
Busgenweg 2, D-34 Göttingen
West Germany

McBride, J.
Dept. of Forestry
University of California
Berkley, California, U.S.A.

McLaughlin, S.B.
Environmental Science Division
Oak Ridges National Laboratory
Oak Ridge, TN 37831, U.S.A.

Meema, K.M.
Institute for Environmental
 Studies, University of
 Toronto, Toronto
Ontario, Canada, M5S 1A4

Morrison, I.
Canadian Forest Service
Great Lakes Research Centre
P.O. Box 490, Sault Ste Marie
Ontario, Canada, P6A 5M7

Norton, S.A.
Dept. of Geological Sciences,
University of Maine at Orano
110 Boardman Hall,
Orano, Maine 04469, U.S.A.

Ormrod, D.P.
Dept. Horticultural Science
University of Guelph
Guelph, Ontario
Canada, N1G 2W1

Pell, E.J.
Dept. of Plant Pathology,
Pennsylvania State University
211 Buckhout Lab.
University Park
PA 16802, U.S.A.

Pitelka, L.
E.P.R.I., P.O. Box 10412
Palo Alta, CA 94303, U.S.A.

Prinz, B.
Landesanstalt fur Imission-
 sschutz des Landes NRW
Wallneyerstrasse 6
4300 Essen 1, West Germany

Raynal, D.J.
State University of New York
College Env. Sci & Forestry
Syracuse, NY 13210, U.S.A.

Reid, D.
Dept. Biology
University of Calgary
2500 University Dr. NW
Calgary, Alberta
Canada, T2N 1N4

Robitaille, L.
Ministry of Energy & Resources
2700 rue Einstein, Sainte-Foy
Quebec, Canada, G1P 3W8

Saxe, H.
Royal Veterinary & Agricultural
 University, Dept. of Plant
 Physiology & Anatomy
Thorvaldsenvej 40, DK-1871
Copenhagen V, Denmark

Schindler, D.
Freshwater Inst., Canadian
 Dept. Fisheries and Oceans
501 University Cres.
Winnipeg, Manitoba
Canada, R3T 2N6

Seaward, M.R.D.
Dept. of Biology
University of Bradford
Bradford BD7 1DP
Yorkshire, UK

Shriner, D.
Oak Ridges National Laboratory
P.O. Box X, Oak Ridge
TN 37830, U.S.A.

Tomlinson, G.F.
Domtar Inc./Research Centre
Senneville, Quebec
Canada, H9X 3L7

Torrenueva, A.L.
Environmental Studies & Assess-
 -ment Dept., Ontario Hydro
700 University Ave., Toronto
Ontario, Canada M5G 1X6

Tveite, B.
Norwegian Forest Research Inst.
Post Box 61, 1432 As-NLH
Norway

Tyler, G.
Vaxtekologiska Intitutionen
O. Vallgatan 14
S-223 61 Lund, Sweden

Tyree, M.
Dept. of Botany,
University of Toronto,
Toronto, Ontario
Canada, M4S 1A1

Unsworth, M.H.
Inst. of Terrestrial Ecology
Edinburgh Research Station,
Bush State, Penicuik
Midlothian, EH26 OQB, UK

Urban, N.
Dept. of Civil and
 Mineral Engineering,
122 Civil & Mineral Eng. Bldg.
500 Pillsbury Dr. S.E.
MN 55455-0220, U.S.A.

van Breemen, N.
Agri. Uni. of Wageningen
Dept. of Soil Sci. & Geology
P.O. Box 37, 6700 AA
 Wageningen, The Netherlands

Winner, W.
Dept. of Plant Pathology
 & Physiology, Virginia
Polytechnic Inst. & State Uni.
2314 Terra Bella State
Blacksberg, Virginia 24061,
U.S.A.

Zoettl, H.W.
Institut fur Bodenkunde
 und Waldernahrungslehre
Albert-Ludwigs-Universitat
D-7800 Freiburg i.Br.
West Germany

WORKSHOP SUMMARY

T.C. Hutchinson

The NATO Advanced Research Workshop detailed in this volume was held in Toronto, Canada, in 1985. The purpose of the Workshop was to provide a "state of the art" report on our knowledge of the sensitivities and responses of forests, wetlands and crops to airborne pollutants. Approximately 40 scientific experts from nine countries participated. Most participants were actively involved in research concerning the effects of air pollutants on natural or agro-ecosystems. These pollutants included acidic deposition, heavy metal particulates, sulphur dioxide, ozone, nitrogen oxides, acid fogs and mixtures of these. Also invited were experts on various types of ecosystem stresses, physiological mechanisms pertinent to acid deposition, and other areas that were felt by the director to be of direct relevance, including: effects of ethylene on vegetation, the physiology of drought in trees, the nature and role of plant cuticles as barriers to acid rain penetration, the use of dendrochronological techniques in reconstructing the time of onset and the subsequent progression of growth declines, the ability of soils to naturally generate acidity, the role of Sphagnum moss in natural peatland acidity, the use of lichens as indicators of changing air quality, and the magnitude of natural emissions of reduced sulphur gases from tropical rainforests and temperate deciduous forests.

The Workshop included a series of invited presentations and subsequent group discussions. These presentations were designed to allow syntheses of our present knowledge as well as detailed questioning and discussion. In addition, participants divided themselves amongst four smaller working groups which met frequently throughout the five-day meeting. The objective of the working groups was to provide written statements of consensus of what is known, what is unknown, and what needs to be known in the following areas:

(1) the responses of forests to airborne pollutants, with particular consideration of the possible role of air pollutants in the problem of forest decline and tree dieback now occurring widely in (a) Europe, most notably in Germany, Austria, Switzerland, Poland, Czechoslovakia and France, (b) uplands of eastern U.S.A., and (c) most recently, sugar maple-dominated forests in Quebec and adjacent areas of the USA, and in Ontario and New Brunswick, Canada;

(2) the responses of soils and soil microbial populations to atmospheric inputs, especially acidic pollutants, as well as the nature and chemistry of soil acidification processes;

(3) the responses of crop plants and agro-ecosystems to air pollution, especially gaseous pollutants and acid rain;

(4) the responses of wetlands and peatlands to atmospheric pollutants, with special consideration of the sensitivity of wetland

species to changes in the acidity and chemical status of wet and dry deposition.

The findings of the four working groups are given in the concluding section of this volume.

Geographical Overviews of Forest Decline

In order to provide a perspective on the extent and severity of the forest decline problem, the opening presentations dealt with a country-by-country evaluation of the status of forest decline in Europe, followed by the examples of red spruce decline in the Northern Appalachians of eastern U.S.A., and the recent and alarming decline and dieback of sugar maple forests in Quebec, and, to a lesser extent, Ontario in Canada.

The recent decline problems were first reported from W. Germany in the late 1970s. While decline of Norway spruce was initially reported in the Hartz Mountains in the north, forested areas of the Black Forest, Bavaria and in the Erzgebirge mountains to the east were all reported to show symptoms of decline by the early 1980s. This decline was initially reported from forests on acidic, nutrient-poor soils, especially at high altitudes, and with symptoms of chlorotic older needles, followed by premature shedding of these needles, with a consequent canopy thinning. Several other coniferous species in addition to Norway spruce have been affected, as have deciduous species, especially European beech. Other reported symptoms include the following: a clumping of upper branches in a characteristic "witches broom" form, a tendency for branches to hang downwards, and chlorosis most prevalent on needles exposed to the sun. The yellowing of needles starts at the needle tip and moves towards the base. Foliar analyses have often shown magnesium deficiency in affected trees. Potassium deficiency has also been reported. While the symptoms occur initially in older trees, eventually all age classes are affected, and trees at the edge of forests are the first to be affected. The trees die rather rapidly, i.e., within 2-3 years from the onset of initial symptoms. The areas in Germany most severely affected are often far from urban-industrial sources of pollutants.

In contrast to the new forest decline problem in Germany, Austria and Switzerland, the present forest damage occurring widely in Poland, Czechoslovakia, and East Germany seems to relate directly to proximity to urban-industrial pollution sources, especially point sources of sulphur dioxide, and therefore, appears to be the direct effect of gaseous phytotoxicity. Chlorotic banding of needles characteristic of SO_2 fumigations is found in these regions.

The forest decline working group compared the symptoms exhibited in the West German forests with those of the extensive red spruce decline occurring at high altitudes in the Adirondacks and northern Appalachian mountains of the northeastern U.S.A. It was felt that the symptoms of decline were different to those in Europe. Mortality is still largely confined to mid- to upper slopes. Death of twigs and needles occurs on the top of the crown and at the ends of the branches, while browning of needle tips is common in the spring and late winter. The affected trees generally have pale foliage. Examination of radial increment growth shows an abrupt reduction in growth in all age classes about 20 years ago.

The occurrence of sugar maple forest decline in Quebec, Canada, was first reported in 1981. As in European beech, affected trees die back from the tips of the upper branches. The leaves on these branches are dwarfed, pale and yellowish in mid-summer, often developing premature, red autumn colouration. The decline initially affects mature trees and those on uplands on nutrient-poor soils, particularly acidic soils. Eventually younger trees are also affected. Trees on poorly drained sites are vulnerable as well. Interestingly, L. Robitaille reported at the workshop that American beech, as well as several other deciduous species in the maple forest such as black ash and yellow birch have been affected. Foliar analyses show low concentrations of calcium and magnesium in the leaves of declining trees. The spread of this decline, like that in Germany, has been precipitously rapid and the maple syrup industry is in jeopardy.

McLaughlin et al. reported on their Ontario, Canada, studies up to May 1985. Their role has been to respond to the numerous requests for help and investigation from members of the Ontario Maple Syrup Producers Association who had reported problems since 1982. Foliar and soil analyses, evaluations of trees marked in permanent plots, examination of roots and shoots for pathogens, etc., were reported. Since the 1985 Workshop, the problem has become more extreme in Quebec, and is reported from 6 northern U.S.A. states and from New Brunswick and Ontario in Canada. Recent data show the declining tree foliage to be significantly lower in Ca, Mg, N and P, both in Ontario (Hutchinson and Kinch, unpublished) and in Quebec (Hendershott, unpublished). Similarities to the deciduous forest problems of Europe seem to be emerging, though in perhaps a repeat of societal reaction, there are many scientists who feel the declines are perfectly normal events which occur from time to time, and only that in the Eastern Townships of Quebec is actually a cause for alarm. A number of commercial maple syrup producers have gone out of business. Ominously perhaps, other hardwood species are also being reported as in decline, but to a lesser extent.

Amongst other regional reports, Tveite noted that the relative vitality of Scots pine and Norway spruce in Norway did not relate to gradients of acidic deposition or gaseous pollutants. However, reports of forest decline similar to that in Germany have been made from southwestern Sweden. Binns reported that in the United Kingdom forest decline of the German forest type has not been reported, though it should be noted that many of the forest plantations of northern England and southern Scotland are relatively young, i.e., < 40 years old. Their relative youth may be a key factor in the absence of decline symptoms which elsewhere seem always to be first seen, and most severe, in mature trees.

Other Stresses Involved in Forest Decline

The role of climate, especially in triggering the German forest decline and that of red spruce in the eastern U.S.A., was considered in detail. Cook and Johnson both indicated, using dendrochronological techniques, that the decline appeared to start in the mid-1960s, though visible foliar symptoms were not reported until 10-12 years later. Johnson reported that on mountains in Vermont, New York State and New Hampshire the onset of tree decline followed two well-marked climatic events, especially the occurrence of prolonged and widespread drought. He also noted a positive correlation between the extent of decline and altitude. Cook noted additional declines in annual increment growth above that of annual climatic fluctuations which seemed to be unique to the last 20 years. Increased air pollution, including ozone and acid fog exposures, over this time could be factors in this. Johnson sounded a cautionary note by his observations that synchronous declines of red spruce occurred between 1871 and 1885 in eastern North America. Since major droughts may have been triggers of these hundred-year events, we cannot be confident in ascribing a key role to air pollution in the red spruce decline of the last 20 years when a similar effect on growth occurred in the last century in much 'cleaner' air. Indeed, Lebanc, Raynal and White concluded from careful dendrology studies of plantation conifers exposed to a range of acid deposition intensities that the annual fluctuation and pattern in growth could not be ascribed to these acidic inputs.

In further support of climate as a factor in forest decline, Tyree et al. emphasized the similarities between the response of trees to droughts and the symptoms seen in many of the declining hardwoods. They discussed the detrimental effects on canopies and on terminal branches of loss of turgor in the fine branches of the vascular tissues.

In the German forest situation, large areas of decline are rather remote from major industial pollution sources, especially of sulphur dioxide and sulphate deposition. This fact led to much discussion and debate at the workshop, especially as to the role of ozone and acid fogs in forest decline. A consensus did emerge that, while the 1975-1976 widespread droughts may have been a trigger (or pre-disposing factor), air pollution in one form or another seems to be much the most likely new factor causing the new decline. Though the alternative hypotheses which have emerged from Germany have tended to emphasize differences, it seems that they really have much in common. These hypotheses include the role of airborne pollutants in accelerating foliar leaching of nutrients, especially of bases from leaves, the predisposition of acidic, nutrient-poor soils to be less able to maintain the supply of bases into the roots, the added toxic effect of soluble aluminum on fine root systems, increases of aluminum in the soil solution in surface soils under acidifying conditions, and an agreement that both wet and dry acid deposition could be involved in the foliar leaching and in soil acidification. The possible role of acid fogs bathing the trees at higher altitudes and accelerating foliar leaching was also widely accepted for European and North American higher altitude decline areas. The specific role of ozone was discussed at length and some agreement was reached that photooxidants are amongst the few pollutants which have increased in concentration in the affected forest over the past 20 years, though a good historical data base is lacking. Nitrogen oxides emanating from automobiles and from high temperature combustion are also believed to play a synergistic role with acid fogs, sulphur dioxide and ozone.

The special severity of the forest problems near autobahns was considered. Both Reid and Hutchinson suggested that organic gases and vapours could be involved, especially ethylene and hydrocarbons. Ethylene is a particularly interesting 'candidate' gas as it is a natural, ubiquitous component of both urban and rural air. It is produced from burning vegetation, automobile exhausts, city incinerators, and places of incomplete combustion. It is also a natural microbial product in forest soils and in plants. The reason we felt it merited more detailed consideration is that it is a natural plant hormone, and the work of Abeles (1973, see chapter by D. Reid) indicates that many of the rather unusual symptoms shown by some of the trees in declining forests in Germany and elsewhere are rather like those induced by ethylene exposure.

McBride and Miller discussed the large impacts that oxidants, including ozone, have had on the forests of the Pacific southwest of the United States. For example, in zones exposed to 0.08 to 1.2 ppm O_3 as an hourly average, as many as 33% of the ponderosa pine trees died, while only 6.9% died in zones where the hourly average of ozone exposure was < 0.08 ppm. Though the oxidant levels are still high, the death of trees has slowed or ceased in many areas, after the loss of the approximately 15% ozone-sensitive trees. Similar, although less severe, effects have been reported for jeffrey pine and for epiphytic lichen species. The overall effect is a change in the species composition of the forest. McBride reported that in the eastern U.S.A. ozone levels in the summer months in high altitude forests are now sufficient to cause detectable damage to foliage of white pine and act synergistically with SO_2 to enhance the problem. The ozone levels in Germany were not felt by the North American ozone experts to be sufficient per se to cause the degree of severity of forest decline observed.

Mansfield reported that in comparative experiments using crops and woody species, simultaneous exposures to SO_2 and NO_2 often produce results that would not be predicted from the individual effects of the two gases. Grasses and cereals were able to recover from initial, severe suppressions of growth, but in woody plants such recovery is limited. Support for this idea is seen in the area around the Sudbury smelters, in Canada, where surviving trees are still stunted and suppressed, though SO_2 levels and metal deposition was greatly reduced in 1972. In birch, Mansfield reported that low concentrations of SO_2-NO_2 mixtures stimulated growth while increasing levels caused severe growth inhibitions.

Lindberg et al. provided a critical re-appraisal of the efficiency of bulk collectors in accurately measuring dry deposition. For nitrate, they estimated that the unmeasured flux exceeds that measured as bulk deposition by factors of 1.3 to 6. The unmeasured fraction was calculated to be dry deposition of nitric acid vapour or fog and cloud droplets.

Unsworth emphasized the unique stresses which high altitude forests face from air pollutants. These stresses include increased deposition of photochemically-generated pollutants such as ozone and nitric acid vapour, the increased quantity of rain and snow at higher altitudes as compared with forests at lower altitudes, the precipitation also often being polluted, and the effective capture by vegetation of wind-driven cloud water. This cloud water is often considerably more acidic than rain due to condensation on the aerosol water droplet surface. Very acidic mountain mist and fogs have been measured in the Adirondacks, in Scotland and in California.

The possible role of an enhanced nitrogen 'fertilizer' input into the forests was also discussed. Evans and A. Johnson suggested that an enhanced nitrogen availability may ameliorate shoot growth and may prolong the growth period into autumn, such that normal winter hardening, or acclimation, may be curtailed or be incomplete. Such unhardened growth may then be especially susceptible to frosts and to winter kill factors.

Cox reported interesting experiments on the effects of acidic droplets on pollen placed on the stigma of various tree species. Pollen germination and subsequent pollen tube growth varied with species, but was distinctly pH-dependent. The conifer pollen performed best under the more acidic conditions, while sugar maple and yellow birch pollen was more susceptible to low pH. Amongst the conifers, white pine (Pinus strobus) was more sensitive than other species tested. Actual seed set in the field from hand-pollinated trees was reduced at pHs currently recorded from daily rain event samples in eastern Canada.

The role of pathogens, especially fungal pathogens, but also rickettsia viruses, bacteria, and insect pathogens in forest decline is one that is receiving increased attention. A variety of fungal pathogens, particularly Armillaria mellea are commonly isolated from declining trees. Whether or not the role of fungal pathogens is as a primary cause or as a secondary consequence of decline was debated and requires evaluation for each circumstance. Manion suggested in his presentation that forest declines are nothing new, many species have been affected over the years, and that most often no specific cause is found. He emphasized the pioneering work of Sinclair and Houston (see Manion, this volume) in present-day decline models which emphasize the spiral of events and interactions designated as stress-triggered diseases by which healthy trees become affected and altered by stresses, with invasion by fungal pathogens in the later stages. These pathogens may, in fact, be only weakly pathogenic. He also commented on the decline problem occurring in some Pacific forests, especially in Hawaii, studied by Mueller-Dumbois (1983) who has described a synchronous cohort of trees which decline and die together. These ideas led to lively debate. There seems little disagreement that both the red spruce decline in North

America and the more general forest decline in central Europe are the result of complex factors and interactions, with an ultimate role for fungal pathogens, beetles, etc. not proven. The nutritional status of these trees seems to be affected which in itself may alter both root mycorrhizal associations and root pathogen invasion.

Responses of Soils to Airborne Pollutants

Zoettl reported on his analyses of soils and foliage of healthy and declining trees is south-west Germany. For Norway spruce the analyses of needles from declining trees most often indicated a deficiency in magnesium. On soils high in calcium, he reports foliar deficiencies in potassium, while on a calcareous soils, the yellow chlorotic foliage appeared due to manganese deficiency. His fertilizer trials in the field on these various soils provided very promising positive responses to additions of these deficient elements. Zoettl demonstrated photographically dramatic re-greening responses of selected trees in his trials. It is worth noting that, since the 1985 workshop, the German Federal government has allocated 140 million DM for fertilizer applications to declining forests.

Tomlinson supported the contention that soil nutrient deficiency is a key factor in the declines with examples from the eastern U.S.A. of foliar deficiencies of Mg, K and Ca at particular locations. He ascribed this to accelerated soil acidification due to acidic inputs from polluted air masses.

The famous Solling experiments of Ulrich, Mayer, Matzner and others in the Hartz mountains first revealed forest decline problems in Germany and the Ulrich hypotheses have greatly stimulated research both in Germany and elsewhere. A key component discussed at the workshop was the increase in aluminum in soil solution in the re-wetting periods following droughts at the time of strong acid pulses.

In the soils' presentations, much of the attention focused on natural sources of soil acidity, on the potential and actual impacts of acidic deposition on these processes, and on soil chemistry. Abrahamsen and D. Johnson reported the occurrence of natural acidification of soils by fast-growing forests, especially coniferous ones, due to ion exchange with the roots during nutrient uptake and to the production of organic acids during litter decomposition. Johnson reported that increases in soluble aluminum have been noted at some sites without concomitant reductions in exchangeable base cations, perhaps due to dissolution of interlayer Al^{+3} in the clay minerals. Mulder and van Breeman studied aluminum mobilization in soils from the Netherlands and from Hubbard Brook, U.S.A. They found that strongly acidic atmospheric inputs are buffered in spodosols, dependent upon the base status of the soil. Those soils rich in silica and poor in basic cations buffer acidity primarily through dissolution of aluminum. Spodosols richer in bases neutralize acidity primarily by the solubilization of basic cations.

While reports on soil acidification due to acidic deposition are very limited, the theoretical reasons advanced to suggest that soils are extremely well buffered against acidification appear contradicted in practice by the re-sampling of soil profiles at 30-50 year intervals reported by Tyler (this volume) and by Hallbäcken and Tamm (1986). An historical perspective on changes in soil pH changes is rare because of lack of samples taken at identical locations at reasonably long time intervals and with well-described measurement techniques. However, G. Tyler et al. presented compelling evidence that in southern Sweden a substantial acidification of forest soils has occurred during the last decades. This acidification has occurred in both coniferous and in deciduous forests, but is the most severe in spruce stands. The lower pH has increased the solubility of several elements in the soil, including magnesium, aluminum, cadmium and zinc. High concentrations of these in the soil water may already be detrimental to root systems, and may be absorbed and translocated internally in the plants to sites where they cause physiological stresses.

The possible roles of heavy metals, aluminum and nutrient imbalance were also emphasised by Matzner and Ulrich, and by Godbold et al. The latter state that "Comparison of the metal concentrations shown to inhibit root elongation in nutrient solutions to those found in the soil solution from the humus layer taken from under trees showing the symptoms of decline suggests that the levels of Hg, Pb and Zn are sufficiently high to influence root growth." Ingestad reported on experiments his group have carried out in Sweden using novel culture conditions. They determined in comparative studies that Norway spuce appears to be the species most sensitive to aluminum, and Scots pine to be the most tolerant. However, they caution that in the soils they used in the field the aluminum levels should not be sufficiently high to be a serious hazard for Norway spruce. Interestingly, Hutchinson, Bozic and Munoz-Vega (1986) report red spruce to be the most sensitive to elevated aluminum levels in the soil solution in comparisons of five North American coniferous species tested as seedlings. Thus, Norway spruce and red spruce have been shown to be especially sensitive to elevated aluminum levels in experiments and they also seem to be the species showing the earliest and most widespread declines. The role of aluminum in the European forest decline was first postulated by B. Ulrich in 1980 and has been suggested as a factor in red spruce by T. Siccama and A. Johnson and in sugar maple by Hutchinson.

The substantial deposition of potentially toxic elements now occurring in many of the declining forests may lead to an accumulation of these elements in the soil, with the possible consequence of inhibition of important microbial processes. Baes and McLaughlin reported on metal levels in tree rings of several species. In those collected near industrial point sources, the increased aerial deposition and soil burdens of the metals was reflected in the wood analysis. However, they felt that at other sites where lower concentrations were involved, the correlations were not always good, nor could lateral movement of elements be ruled out.

Bryophyte and Lichen Studies

Lichens have been used more than any other group of plants and animals as indicators of air quality. Species differ markedly in their sensitivity to such gaseous pollutants as SO_2, O_3, NO_x and F. The dearth of lichen species around most major conurbations has been remarked upon for at least 60 years. Interestingly, Seaward has been able to show the expansion of certain extremely pollution-tolerant species, e.g., Lecanora conizaeiodes, into urban areas where they have replaced more sensitive species on tree bark. The distribution and occurrence of lichen species has been mapped in some detail in Europe, revealing a recent recovery and re-invasion of some urban centres in northern England where air quality has now improved, e.g. Leeds, Manchester, Bradford. The devastation of lichen floras in the forests of Poland, Czechoslovakia and East Germany appears to be the direct consequence of industrial emissions, especially of SO_2, as is the associated death of forest trees in these regions.

Winner and Atkinson reported on their comparative and experimental findings of SO_2 absorption by mosses versus vascular plants. During the course of a year, mosses absorbed more SO_2 than the vascular plants, with absorption relating especially to periods of wetting and drying cycles. The differences between mosses and vascular species could be as much as 400-fold in any one habitat, due largely to the absence of a cuticle in the mosses. In contrast, much of the sulphur in the vascular plants, even in polluted regions, was found to enter the plants from the soil.

In experiments reported by Hutchinson et al., plots in a mature jack pine forest in the Canadian boreal forest region were sprayed from 1981 to 1985 during the growing season with simulated acid rain of a range of pHs. The feather moss Pleurozium schreberi, which together with Hylocomium splendens covers many thousands of square kilometres in the boreal forest areas of Europe, North America, and Asia, was found to be very sensitive to sprays of pH 3.5 or less. At pH 2.5, the moss populations were killed within two years, and significant death and loss of photosynthetic capacity occurred at pH 3.5. Depletion of calcium and magnesium was noted even in filaments not showing visible signs of damage at pH 3.5. The forest floor lichens Cladina stellaris, C. rangiferina and C. mitis were also adversely affected by the sprays but to a lesser extent than the feather mosses. Strongly acidic rain events can be expected to have deleterious effects on these lower plants and may already be acting as a selective force in their populations. Indeed, some specific morphological and anatomical abnormalities were noted in C. rangiferina, which have the potential for use as pollution warning "markers."

Responses of Agro-ecosystems

A number of studies were described which dealt with the responses of crop plants to acid rain sprays, both in the greenhouse and in open-top fumigation chambers, and to ozone, nitrogen oxides and sulphur dioxide, as well as to gaseous - acid rain mixtures. In the studies of Shriner and Johnson, a number of crops differed quite markedly in their sensitivity to foliar acid rain sprays. Radish showed foliar injuries at spray pHs of 3.3, while Davis soybeans were not affected. Sprays of pH 2.3 applied to only the foliage of tomato caused foliar lesions, while sprays applied to the soil alone without foliar contact actually enhanced growth. Sprays of pH 3.3 in combination with ozone at various levels caused damage of an additive nature. Shriner and Johnson suggest that the "net effect of acid rain on vegetation growth may be the result of offsetting hydrogen ion-induced negative effects and nitrate- and sulphate-induced nutritional effects. If so, the nutrient status of vegetation is an important determinant of the direction and magnitude of response to acid rain." The group summary report on air pollution - crop interactions supports the contention that detrimental effects of acid rain on crops is many times lower than that of ozone and for North America, and the dominant air pollution problem for crops is ozone and oxidant exposures. This appears also to be the case for North American forests in general, aside from the specific problem areas of decline.

Ormrod et al. reported on a most useful way of using factorial experiments to generate crop response of 'surfaces' for several crops to O_3 and SO_2 and combinations. They are also working on sequential exposures to O_3 and SO_2 which may be a much more realistic approach since the coincidence of high SO_2 and O_3 is an unlikely one compared with pollution episodes dominated by one or the other.

Berg described some of the basic biophysics and physiological properties of cuticles in vascular plants and epicuticular waxes. The movement of substances through the cuticle is controlled by the conductance of the cuticle to the substance and the ease with which rain or fog droplets wet the surface. The presence of crystalline epicuticular waxes on the surface of the cuticle, as in cabbage, greatly reduces the potential for exchange of ions and molecules between drop and leaf by decreasing contact between the drop and the leaf surface. While erosion of cuticular waxes by acidic droplets theoretically should not occur, SEM micrographs shown by Hutchinson, Caporn and Adams (paper not in volume) of the leaf surfaces of cabbage sprayed with solutions of pH 3.2 and 2.8 showed an apparent loss of and structural change in the epicuticular waxes reminiscent of solubilization of wax in an organic solvent. This paper on the structural aspects of leaf surface responses also reported the occurrence of crystals high in Ca and S content, probably as $CaSO_4$, on the surface of leaves treated with pH 3.0 solutions of sulphuric acid and when sprayed with a 2:1 sulphuric:nitric mixture. The occurrence of foliar leaching of bases and of cellular collapse of epidermal cells was discussed at the meeting. Species appear to differ widely in the ability of their leaves to neutralize acidic droplets. The sensitivity of crops to acid rain-induced foliar lesions seems greater than for most tree species and, indeed, for most non-crop plants. Part of this difference may relate to man's deliberate selection for crops with thin cuticles associated with rapid growth.

Responses of Wetlands to Airborne Pollutants

The wetlands sessions brought together experts from Europe and North America for probably the most detailed consideration of sensitivity of wetland and peatland systems to aerial inputs of pollutants yet undertaken. There is a sense that these systems have been neglected in the overall acid rain concerns of the past 15 years, perhaps on the assumption that many wetlands are already acidic and are comprised of acid-tolerant vegetation. Indeed, Sphagnum bogs generate a great deal of acidity themselves so might be considered tolerant to aerial inputs of inorganic acids. However, a number of historical lines of evidence suggest that wetlands are by no means impervious to the effects of aerial pollution, including acid rain.

Gorham et al. provided a wide-ranging review of the distribution and characteristics of wetlands. They noted the importance of wetlands as transitional ecosystems between terrestrial and aquatic systems. They suggest that fens with weakly acid surface waters and low alkalinity (with 40 μeq.l^{-1}) are vulnerable to acidic inputs and can be invaded by Sphagnum. Acid deposition to peatlands is largely neutralized by plant uptake and by microbial reduction of associated nitrate and sulphate. They also emphasized (a) the sensitivity of Sphagnum species to air pollutants, and (b) the occurrence of natural invasions of Sphagnum and associated species on moist deglaciated soils, riverine flood plains and even truly upland soils in the process of paludification.

The special and dominant role of Sphagnum in the peat-accumulating wetlands was described by Clymo. He discussed the formation of polyuronic acids by Sphagnum with release of H^+ ions into the bog waters. Much of the zonation and differences in Sphagnum species composition of these bogs relates to their differential sensitivity to high pH, high Ca and to moderate concentrations of σ-phosphate, NO_3^- and NH_4^+. They are also sensitive to moderate concentrations of HSO_3^-. Clymo supported the idea that the disappearance and destruction of Sphagnum bogs in much of northern England was due to their sensitivity to air pollution, especially to the toxic sulphite ion generated from SO_2 emissions. Lee et al. noted for uplands of northern England that Sphagnum species have virtually disappeared from the southern Pennines, probably due to higher SO_2 levels. Lee presented evidence that as industrial activity declines, nitrogen deposition is of growing importance, and interestingly, the activity of the nitrate reductase enzyme system in the Sphagnum increases in response to nitrate deposition. This is reminiscent of reports of increases in superoxide dismutase activity in poplar cuttings exposed to sulphur dioxide, and of higher activities in SO_2-tolerant clones.

In a study reported by Urban et al., the hydrogen and proton cycling in Marcell bog in Minnesota was examined. This is in an area of limited acid rain input, but subject to storms coming off the prairies as well as from the industrial mid-west of the United States. They found that production of organic acids was the dominant source of acidity, which buffered the bog water at pH 4.0. The sequestering of elements in the peat was also a significant source of acidity due to ion exchange, while the weathering of the atmospherically-deposited dustfall was a significant source of alkalinity. Overall, they reported that net acidity, generated as a result of biological uptake, gave high values in maritime bogs and lower values in mid-continental bogs. They also found that bogs have a large capacity for sulphate reduction. This is an increasingly important source of alkalinity as rates of sulphate deposition increase. In eastern North America, more than 60-90% of annual sulphate deposition is retained as reduced sulphur in bogs. Clymo estimated in his report that 1-2% of the earth's surface is covered in wetlands. Thus, the acid-generating potential of wetlands and their sensitivity to inputs of strong inorganic acids and gaseous pollutants are of considerable significance.

Bayley and Schindler took a rather different approach in that they analysed the chemical and hydrological record for three small watersheds in an area of NW Ontario subject to only mildly acidic rain (pH 4.9 average). They found the watersheds generated little alkalinity and did not export significant quantities. Perhaps surprisingly in view of various published reports and the general conception held by the scientific community, forest fires did not increase the alkalinity yields of terrestrial watersheds, but rather caused a small increase in export of acidity due to higher releases of strong acid anions. This finding allows one to speculate that in the much more polluted peatlands and watersheds of Europe, there is a potential for a long-term and significant discharge of acidity. Bayley and Schindler also reported that a two-year experimental acidification of a small wetland at 10-times normal deposition rates had caused little change in the acidity of outflow from the wetland, with almost complete retention of sulphate and nitrate in the watershed.

In a final report on the wetlands, Norton discussed the historical records held in peat cores due to past inputs of atmospheric dusts, pollutants, acid rains, etc. He discussed the use of ^{210}Pb techniques for determining mobilization and migration of elements in the peat, a concern vital to interpretation of peat core elemental analyses. He showed that net accumulation rates of lead and zinc in peat roughly mimic those observed in sediments of nearby lakes. However, considerable zinc moves out of the system, and lateral migration in hummock-hollow peatlands was also shown. It seems, however, that peatlands, like estuaries, may be major retention sites for the organic and inorganic pollutants aerially deposited over long periods of time and that destruction of these peatlands by industrial emissions or removal of surface peats and fuels can lead to major exports of acidity, heavy metals etc. from the watershed.

Conclusions

Overall, the Research Institute achieved its major goal of bringing together many of the leading scientists concerned with the effects of atmospheric pollutants on natural and agricultural ecosystems. It broke new ground in its attention to wetlands. It was agreed that serious problems of forest dieback and decline have appeared within the past 10 years in Europe and in eastern North America affecting an increasing number of both coniferous and deciduous species. In many of the affected areas the symptoms are not readily ascribed to the older types of sulphur dioxide fumigations. Delegates agreed that air pollutants were very probably involved in most of the major forest declines recently appearing. These pollutants include both wet and dry acidic deposition, acidic fogs, ozone, nitrogen oxides and possibly organic hydrocarbons such as the gaseous plant hormone, ethylene. Symptoms differ between the red spruce decline in North America and the German forest decline of Norway spruce. Ozone was recognized as the major economic pollutant affecting crops and forests in North America. The role of nutrient depletion and forest nutrition as a whole appears central to the occurrences of decline in the German, Austrian, and Swiss forests, to the problems of sugar maple in Canada and the USA and perhaps to the red spruce problems. Delegates recognized that tree declines are not new but felt the magnitude of multi-species effects seem in the present situation to merit the greatest concern by government, industry and the academic communities, and that complex interactions seem to be involved, requiring interdisciplinary approaches. Both pathogens and major climatic perturbations may be involved at least as interacting factors, and that soil chemistry and microbial health need always to be considered in such studies.

REFERENCES

Abeles FB (1973) Ethylene in plant biology. Academic Press, New York

Hallbäcken L, Tamm CO (1986) Changes in soil acidity from 1927 to 1982-1984 in a forest area of south-west Sweden. Scand J For Res 1: 219-232

Hutchinson TC, Bozic L, Munoz-Vega G (1986) Responses of five species of conifer seedlings to aluminum stress. Water Air Soi Pollut 31: 283-294

Mueller-Dombois D (1983) Canopy dieback and successional processes in Pacific forests. Pacific Science 37: 317-325

DEVELOPMENT AND CAUSES OF NOVEL FOREST DECLINE IN GERMANY

B. Prinz, G.H.M. Krause and K.-D. Jung

Landesanstalt fur Immissionsschutz des Landes Nordrhein-Westfalen (LIS), Wallneyer Strasse 6, 4300 Essen 1, FRG

ABSTRACT

A clear distinction must be made between the decline and the death of certain forest species and forest areas, which we prefer to call "novel forest decline" ('neuartige Waldschäden'), and the classical form of "smoke-damage", caused mainly by sulphur dioxide. The new type of forest decline was first recognized in Southern Germany during the early to middle seventies. It is remarkable that this phenomenon has been concentrated on areas previously considered to be clean air regions. Other areas outside Southern Germany followed with significant delay. Forests at high altitudes have been especially endangered. The typical symptoms of novel forest decline are not comparable to those caused by classical air pollutants such as sulphur dioxide, nitrogen oxides, chloride, fluoride, and heavy metals. The most common symptom is chlorosis of needles exposed to sunlight, which is correlated with deficiency of nutrients, mainly magnesium.

The explanations of the various causes of this new kind of forest decline have undergone remarkable changes. Up to 1982 the terms "Waldsterben" and "acid rain" were used more or less as synonyms, and acid rain was considered to have caused acidification of soil, with mobilization of soluble aluminum ions affecting the root system. Thus trees become more vulnerable to additional stress factors, including other air pollutants. An alternative hypothesis introduced at the end of 1982 considered the effect of ozone and other photooxidants on the upper parts of the tree as the primary cause. Changes in membrane permeability resulting in an enhanced nutrient leaching, as well as reduced assimilate production affecting fine root growth, are major steps in a spiral of continuous decreasing vitality. This new explanation shifted the emphasis from sulphur dioxide to nitrogen oxides as the emissions of primary importance.

The following observations must be taken into consideration when arguing for or against competing hypotheses.

1. Against sulphur dioxide <u>alone</u>: In some of the most severely damaged areas ambient air concentrations of sulphur dioxide are extremely low with corresponding low sulphur contents in needles.
2. Against acid rain <u>alone</u>: Lichens proved to be sensitive to acid rain but show luxuriant growth, particularly in the most damaged areas.
3. Against influence of soil <u>alone</u>, including soil acidification: No clear correlation between degree of damage and soil

condition or type of bedrock could be found on a broad scale. Severe damage is sometimes also occurring in calcareous soil.

The following lists experimental evidence to support the hypothesis that the combined effect of ozone and acid rain or fog is the primary factor in forest decline:

(1) inhibition of photosynthesis by ozone; (2) impairment of root development by ozone; (3) disturbance of allocation and translocation of assimilates by ozone; (4) seasonal variation of chlorophyll content and supply of essential nutrients to the needles; (5) enhanced leaching dependent on ozone-concentration, pH of rain or fog, previous damage of needles and age of needles; (6) relation between nutrient deficiency and increased photosensitivity; (7) disturbance by ozone of reactions controlled by phytohormones; (8) preference of occurrence of damage on soil with low nutrient status.

These arguments are the most appropriate for fir and spruce. Beech and pine (in this case _Pinus nigra_, var. _austriaca_ (Hoess/Neum.)) may be severely damaged by the direct impact of ozone alone, without further interactions with soil, rain or fog. Although much progress has been made by identifying several possible causes of the novel forest decline, much more work is needed to understand the complex interactions of these causal variables.

INTRODUCTION

The explanation of the novel forest decline in the Federal Republic of Germany has a short but exciting history. Just two and a half years ago the situation very much resembled the time of the inquisition when alternative beliefs were officially not permitted. As expressed in more detail elsewhere (Prinz 1983), science, politics, and administration came to something like a pact, allowing only one cause for the explanation of forest decline and supressing each dissenter. Today the discussion has become much broader: each opinion is carefully listened to, and as in former times of science, the proof of an assumption is what really counts and not the assumption itself.

This proof must fulfill the following conditions:

(1) all the specific symptoms of the damage must be related to the causal factor(s) in question; (2) the temporal development of the damage must parallel temporal occurrence of the causal factor in question, including the consideration of accumulation effects; and (3) the spatial distribution of the damage must largely coincide with the spatial distribution of the causal factor in question.

Following this concept, the phenomena of the novel forest decline ("neuartige Waldschäden") in the FRG will be described. The term "novel forest decline" was officially introduced in order to separate this new kind of forest decline from the classical form

of air pollution injury or "smoke damage" widely distributed in
the past and mainly caused by sulphur dioxide. Secondly, the
situation of air pollution in the most heavily damaged areas is
described, and from this a hypothesis of causality is derived.
Finally, the experimental results that support this hypothesis
are represented.

DEVELOPMENT AND DISTRIBUTION OF FOREST DECLINE

Chronology

Reports from Bavaria state that fir has been noticeably diseased
since the beginning of the 1970s. This disease, known as "fir
dieback," has increased alarmingly since 1976-77 and is still
prevalent today. Since the autumn of 1980 similar disease
symptoms to those in fir have been observed in spruce, and the
extent has increased dramatically in 1982 and 1983. Reports from
the state of Baden-Württemberg indicate that since the middle of
the 1970s fir has been affected by "fir dieback" followed by
decline of spruce beginning in 1980. Especially in the Black
Forest, the intensity of damage and the areas in which trees are
affected have increased rapidly since the autumn of 1981. Both
in Bavaria and in the Black Forest douglas fir and individual
deciduous species (mainly beech, sycamore, mountain ash and
alder) have recently become affected as well.

The current state of injury has been reported by BML (1984).
Table 1 shows forest injury in 1983 and 1984 by tree species.
In 1982 only about 8% of total forested area was reported as
injured, while 34% was affected in 1983 and 50% in 1984. However
this table must be interpreted with caution. The method of
survey has changed slightly during the three years, and it is not
certain that damage of biotic or climatic origin was excluded
from the record of "decline" trees. Since surveys were undertaken as early as June in all three years, and since the summer
of 1983 was extremely dry and hot, it must be taken into
consideration that some damage in 1984 is the consequence of
drought stress in 1983.

Distribution

Figures 1a and 1b show the spatial distribution of all injury
classes and of all tree species for 1983 and 1984 in the 58
"growing districts" of the Federal Republic of Germany that are
more or less homogeneous with respect to climatologic and
orographic features. The measurement unit applied in these maps
is the percentage of an area with damaged forest stands related
to the total forest area in the respective growing district. The
most striking result is the high percentage of damaged forest
stands in the Fichtelgebirge near the border of the German
Democratic Republic and the CSSR, followed by parts of the Upper
Palatine, including the chalk mountains of the Frankonian Jura,
the Bavarian Forest and the Black Forest. Generally, damage

Table 1. Forest injury 1983 and 1984 by tree species on the basis of inventories by the Ministry of Food, Agriculture and Forestry, Bonn. BML (1984).

Tree Species	Area of Injury (all classes)			
	million hectares		percentage of species' area	
	1983	1984	1983	1984
Spruce	1.195	1.477	41	51
Pine	0.641	0.866	44	59
Fir	0.135	0.152	75	87
Beech	0.326	0.631	26	50
Oak	0.090	0.269	15	43
Others	0.161	0.303	17	31
Total	2.549	3.698	34	50

concentrates, with some remarkable exceptions, on the southern region in the Federal Republic of Germany, as well as on the higher altitudes of the "subalpine mountains" in other parts of this country.

It is important to note that the recent forest decline started outside the Black Forest and the Bavarian Forest, in the other parts of the Federal Republic of Germany, with a time lag of roughly 1 to 1 1/2 years, i.e., in autumn 1982. In the Bavarian and Black Forests, the decline occurred for the first time on the west slopes above an average altitude of 800 to 900 m, and in the Northern Black Forest, above 400 to 500 m, although lower areas were also affected, beginning in 1982. The same cycle was repeated in the other parts of the Federal Republic of Germany where, in 1983, the damaged areas extended for the first time also down to the bordering plains, after the notable appearance of the decline on the higher altitudes in 1982.

Stand, Environment and Injury

The injury affects spruce and fir of virtually all age categories, although with certain differences in symptomatology. In nearly all forest stands symptoms appear preferentially on dominant trees. Furthermore, the trees at the edges of stands are more frequently affected than those within stands. Among the edge trees, a clear distinction can generally be made between their outward sides, which are injured most, and their inward sides, giving the impression of a "directional attack." The injury seems also to occur preferentially on the inner edges of small glades or other clearings resulting from turns and forks of paths and roads. The impression arises that the causal agent(s) can attack more effectively here than under the protection of a closed canopy. That is, those trees that are in some way exposed to "light" and "air", are most severely affected (Prinz et al. 1982).

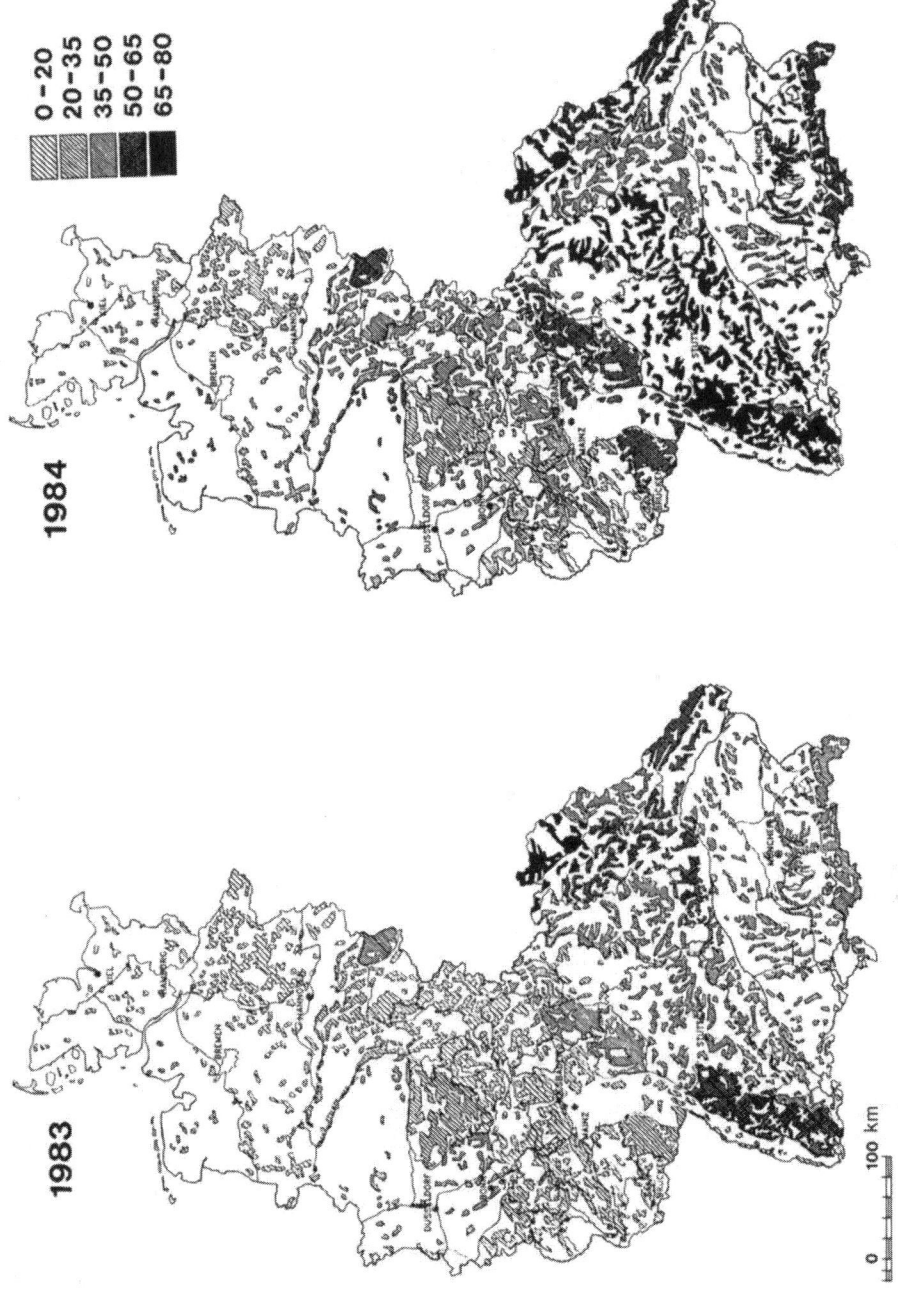

Figure 1. Spatial distribution of forest decline for all tree species and injury classes in % of total forested areas in the Federal Republic of Germany. (a) 1983 (b) 1984.

General Symptoms

Silver Fir and Norway Spruce: On silver fir (Abies alba Mill) and Norway spruce (Picea abies (L.)) needles first show lightening in colour to chlorotic green, followed by a yellow-green speckled appearance and finally intense yellow discolouration. Oldest needles drop, leaving a green fringe of the youngest needles which eventually will start yellowing in autumn. In silver fir, needle discolouration is not as predominant as in Norway spruce, since silver fir loses impaired needles much more readily than does Norway spruce, where several age classes of discoloured needles may exist, so that the affected tree has an overall bright yellow appearance for a long time. The phenomenon of discolouration and needle loss starts from the base of the crown, extending to the top and spreads from the inner to the outer branches. Especially in silver fir, trees show premature reduction in height increment, leading to formation of the "stork's nest" appearance, which is normal in old trees but not in younger trees. In both species the fine root system and number of normal mycorrhizal rootlets are reduced. In the above-ground part of the trees, adventitious buds often develop in an increased manner.

One symptom always observed is that the upper side of the needles shows much more yellowing than the lower side. Furthermore, a shading or shielding effect is observed, in that the upper, sun-exposed parts of the branches are yellow while the ones immediately below are more or less green. In younger trees the yellow appearance predominates often with concurrent normal growth of the terminal shoot, while older trees very soon lose most of their needles. On some, obviously genetically different trees of Norway spruce, the partly needle-less twigs of second order are pendant. This is often called "silver tinsel effect," which is considered by some people, however, to be a normal growth habit of Norway spruce.

Other Species: The symptoms of the other tree species are not as distinct. In common beech (Fagus sylvatica L.) leaves start turning yellow or yellow to brown in the middle of the summer leading to premature leaf drop. This leaf drop, however, begins at the shoot tips, unlike normal leaf drop that begins in autumn. Serrated or unsmooth margins of the leaves caused by wrinkling of tissue between the veins are also typical.

Foliar Nutrients and Soil Conditions

Fig. 2 and Fig. 3 demonstrate that the yellowing of Norway spruce needles, a distinct symptom of the novel forest decline, is accompanied by a pronounced deficiency of magnesium and calcium. This deficiency varies for each element depending on needle age (Prinz et al. 1982).

An elemental analysis of injured and uninjured needles of spruce and fir from southern Black Forest (Table 2) shows that the classical air pollutants cannot be considered a main cause of forest decline, particularly because in yellow needles sulphur

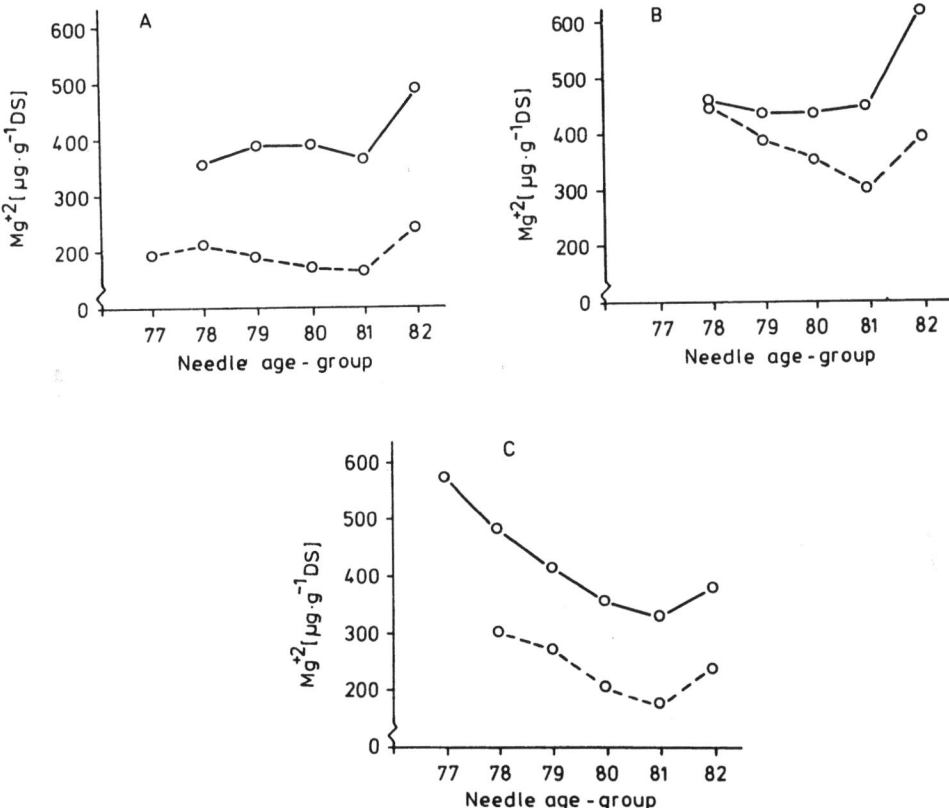

Fig. 2. Magnesium content in needles of Norway spruce as a function of needle age-group and discolouration (green needles: o———o, yellow needles: o - - - o). Samples were taken at Neuhof, Southern Black Forest (A), Egge mountains, slightly injured tree (B), and Egge mountains, severely injured tree (C). The Egge mountains are situated in the eastern part of Northrhine-Westphalia.

content, essential for nutrition, is so low that it reaches the deficiency range (Krause et al. 1983b, Prinz 1983).

In general it can be concluded from many reports in Germany that injury to conifers occurs in spite of differences in bedrock (including limestone) and is observed on soils with high and low buffering capacity and in a broad pH-range. This point is emphasized in two official reports of the Bavarian government (Anonymous 1983a,b). Soil scientists in Baden-Württemberg (Kenk et al. 1984) state from their own experiences that difficulties in supply of nutrients alone never have been the cause of broadly scattered decline or dieback of forests and therefore soil

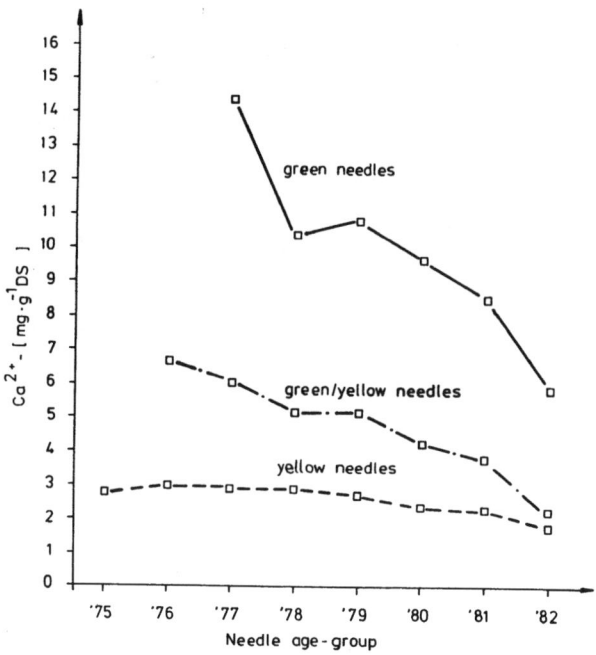

Fig. 3. Calcium content in needles of Norway spruce as a function of needle age-group and discolouration. Samples were taken at Kalbelescheuer, Southern Black Forest.

Table 2. Content of sulphur, fluoride, chloride, zinc, lead, and cadmium ($\mu g \cdot g^{-1}$ DS) of 10 year old Norway spruce and silver fir trees from Southern Black Forest (Neuhof, forest district Staufen).

Element/ Colour of needle		Spruce	Needle Age			Fir		
	1979	1980	1981	1982	1979	1980	1981	1982
S green	1280	890	950	1350	1765	1550	1335	1110
yellow	550	510	500	570	760	585	680	490
F green	3.6	4.0	3.0	3.2	8.4	5.1	3.9	2.9
yellow	4.8	2.9	3.0	2.6	4.7	5.5	4.3	2.4
Cl green	710	970	780	580	600	350	585	410
yellow	500	530	670	620	370	460	310	340
Zn green	83	64	75	72	82	68	61	38
yellow	58	47	48	43	43	39	32	24
Pb green	5.7	6.0	5.5	5.9	4.3	4.4	2.6	5.4
yellow	5.8	4.0	5.0	5.3	5.2	7.2	5.7	5.0
Cd green	0.8	0.4	0.4	0.9	0.4	0.5	0.5	0.5
yellow	0.3	0.6	0.3	0.4	0.3	0.4	0.7	0.5

nutrients may have only a minor role in the development of forest decline in the Federal Republic of Germany.

Distribution of Bark Lichens

Prolific lichen growth has been observed in all damaged forest stands in Bavaria and Baden-Württemberg having suitable climatological and orographic conditions. The forestry administration in Baden-Württemberg regards the abundant lichen population as a secondary result of defoliation and an indication of injury. Within a stand, branches of injured trees have more lichens than healthy trees. Furthermore, tree trunks show extremely prolific lichen growth in the most heavily damaged stands. This is particularly true for the lichen species Hypogymnia physodes. Many lichens have been used as sensitive indicators of air pollutants such as sulphur dioxide, hydrogen chloride, hydrogen fluoride and heavy metals in the bio-effects monitoring system in Northrhine-Westfalia. One of these is H. physodes, which is described in the U.K. as moderately tolerant of SO_2, but in studies in Germany where transplants were made into the Ruhr Valley, the lichens died within a few weeks at the most SO_2-polluted sites. (Schönbeck 1969).

AIR POLLUTION

It is essential to consider air pollution in the areas that are remote from urban and industrial aggregations and where the recent forest decline has been first observed.

Sulphur Dioxide

According to official estimations (Sartorius 1984), the emission of sulphur dioxide has decreased since 1970 in the Federal Republic of Germany while the emission of nitrogen oxides increased by roughly 50% between 1966 and 1982. A continuing upward trend of the emission of nitrogen oxides is indicated also by the development of the concentration of nitrogen oxides in the ambient air within the last decades at different stations in the Federal Republic of Germany, including the so-called clean air stations outside of industrial and urban aggregations. The upward trend on the side of ambient air pollution is, however, not nearly as distinct as on the emission side because a great proportion of the emitted nitrogen oxides is transformed from nitrogen oxides to nitrate before reaching the acceptors.

In Bavaria, a comprehensive comparison was made between the degree of damage and the level of sulphur dioxide. From this it was determined that the concentration of sulphur dioxide in the Fichtelgebirge (northeastern Bavaria, bordering the CSSR) is indeed very high, with 70 $\mu g.m^{-3}$ as an annual average and single peak values on a half hour basis of more than 1,500 $\mu g.m^{-3}$. On the other hand, the concentration in the most severely damaged parts of the southern Bavarian Forest as well as of the southern Black Forest was extremely low, with 10 to 20 $\mu g.m^{-3}$, and less as an annual mean (Obländer et al. 1983). In other words, major areas exhibiting forest dieback in Germany do not experience high exposures to sulphur dioxide.

Ozone

Ozone is the only component in areas of higher altitudes and/or remote from industrial and urban aggregations that tends to occur at higher concentrations. Oblander and Hauss (1985) showed that the mean annual concentration of ozone correlates positively with altitude at different stations in Baden-Würtemburg. Fricke (1983) shows that, especially at the South German mountain stations, the general upward trend of ozone concentration during recent years is apparent. Still more striking is the increase of ozone within the whole troposphere, an increase from 1981 to 1982 which was as much as in the two 7-year periods before, and an increase also in winter, independent from global radiation (Attmannspacher et al. 1984).

The overall increase of ozone, even during winter, was also shown by Feister and Warmbt (1984). According to their investigations, during the last 25 years a steadily increasing ozone concentration could be found on the island of Rügen in the Baltic Sea (German Democratic Republic), which was highest in the months of May and August but which also occurred during the winter months. For the monitoring station in the subalpine mountains of the Erzgebirge, Warmbt (1981) found an increase of annual mean ozone concentration from 35 $\mu g.m^{-3}$ in 1955 to 65 $\mu g.m^{-3}$ in 1980. This is of special interest because air pollution in the Erzgebirge has been regarded almost exclusively to be caused by sulphur dioxide. In Fig. 4, examples of long lasting ozone records by Feister and Warmbt as well as by Attmannspacher are combined. In the Bavarian Forest (Brotjacklriegel), ozone showed an upward trend between 1980 and 1983 but the situation in the Black Forest (Schauinsland) seems to be more stable.

Since August 1983, two monitoring stations have operated in damaged forest stands in the Eggegebirge and Eifel in Northrhine-Westphalia. During the first year of monitoring, the ozone concentration was higher by a factor of 2.0 and 2.4, respectively, than in the Ruhr area. The concentration of sulphur dioxide reached 40 and 30% of that in the Ruhr area; but there were remarkably high peak values of 500 $\mu g.m^{-3}$ in the Eggegebirge and 800 $\mu g.m^{-3}$ in the Eifel and even more than 1,000 $\mu g.m^{-3}$ during a period of extremely high air stability in January 1985 (Pleffer and Buck 1985). Of course, it is not possible to prove that in all stands with forest decline the ozone concentration has changed in strict synchrony to the degree of damage.

Acidity in Rain and Fog

It can generally be stated that during the period between 1950 and 1970, acidity in rain went up from an average level above pH 5.6 to a level below pH 5.6 in central Europe due to an increase in sulphur dioxide emission and a simultaneous decrease of

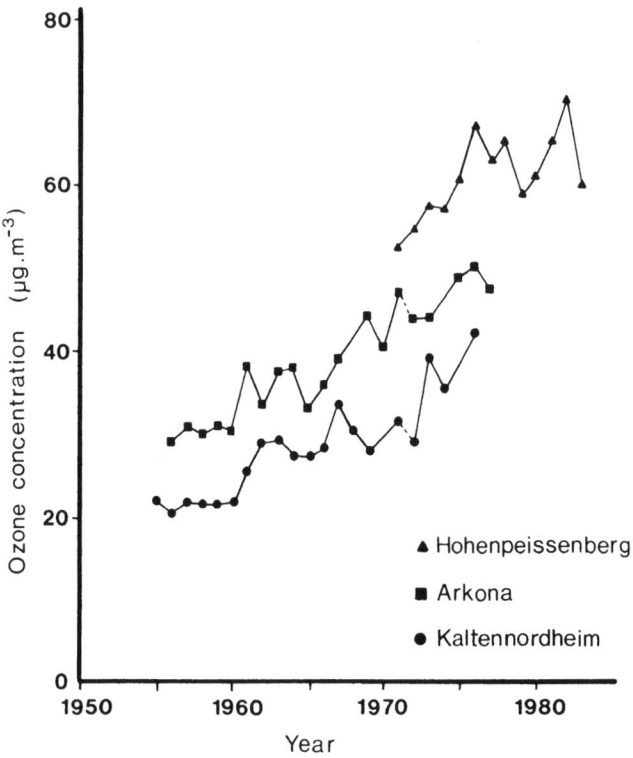

Fig. 4. Trends in mean annual ozone concentration at three rural monitoring stations. Arkona is situated on the Baltic coast of the Democratic Republic of Germany at an altitude of 42 m. The locations of Kaltennordheim (485 m a.s.l.) and Hohenpeissenberg (975 m a.s.l.) are subalpine and alpine sites in the German Democratic Republic and the Federal Republic of Germany, respectively. (From Ashmore et al. 1985).

basically reacting dust emission (Prinz 1984). Since then, the acidity has remained more or less constant. The rain is least acidic in industrial regions such as Duisburg/Northrhine-Westphalia, with a range of pH between 4.5 and 5.0 (measured as bulk deposition), while south of the neighbouring city of Essen, with relatively low dust load, the acidity is equal to remote areas such as Eggegebirge, with a range of pH between 4.0 and 4.5.

The most definite trend in rain acidity is the substitution of sulphate by nitrate in the rain samples (Guicherit and van den Hout 1982). The pH does not seem to have changed significantly during this time. The trend of pH is even upward now in some damaged forest stands, for example in the southern Black Forest (Schauinsland).

In Bavaria, a relationship between forest decline and H^+ deposition could not be found (Rudolph 1983; Reiter 1983). In the

Black Forest, a slight correlation between H^+ deposition under canopy and degree of forest decline may exist (Evers 1984).

The acidity in fog is a different phenomenon. The subalpine areas are characterized by long exposures to fog and, on the whole, fog is characterized by a much greater buffering capacity than rain. So even with the same pH, fog may be much more harmful than rain, especially in connection with nutrient leaching (Prinz et al. 1982).

Pollutant Deposition as a Function of Stand Structure

It is probable that not all micro-spatial variations in the degree of damage can be attributed to corresponding variations in the concentrations of air pollutants. It is, therefore, necessary to understand the influence of such factors as height above ground and wind speed on pollutant deposition. Monitoring the flux, i.e., the product of concentration multiplied by wind velocity, is thought to be more relevant for the uptake of air pollutants by plant than the concentration alone. This parameter is called "Immissionsrate," and it can be measured with an exposed wet surface. The so-called IRMA device has been developed especially for this purpose by Luckat (1972).

The flux of sulphur and fluoride was monitored within a corridor between the northeastern border of the Ruhr area and the lee-side of the Weser mountains with IRMA sensors and at 1.50 m and 25.0 m height on a broadcasting tower in the Teutoburger Forest. It was concluded that the flux is significantly higher in the mountains than in the plain next to the Teutoburger Forest and remote from the Ruhr area. The flux at 25.0 m height is 5 times higher than that just above ground at the same site (Schwela and Radermacher 1985). This last observation demonstrates that forest stands on ridges of hills and on stands opened by previous damage or by silvicultural activity are much more endangered by air pollutants than are protected stands with closed canopies such as on slopes of hills or mountains or especially in valleys.

CAUSES OF THE NOVEL FOREST DECLINE IN THE FEDERAL REPUBLIC OF GERMANY AND THEIR EXPERIMENTAL PROOFS

Fir and Spruce

In the following sections, the hypothesis that ozone in combination with (acid) rain and fog plays a key role in the novel forest decline in fir and spruce stands is explained. The mode of action is demonstrated in Fig. 5. The broken lines are considered of perhaps minor importance because they are not yet sufficiently proved, although they are possible in principle. We think that the feedback mechanism of the decay of fine roots (and perhaps mycorrhiza), caused by an insufficient supply of assimilates, is of major importance in that tree decline increases even if the level of air pollution remains constant. In other words, the rate of change of some variable within a

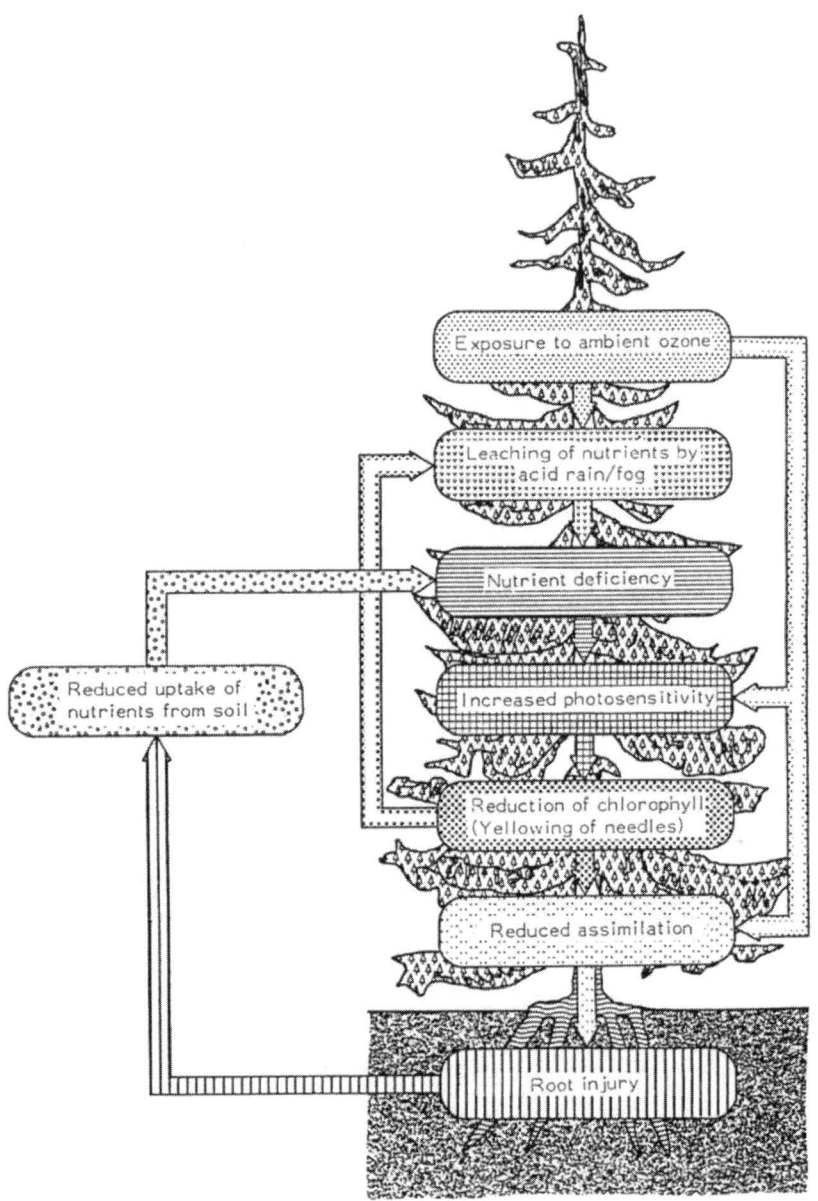

Fig. 5. Diagram of injury development of coniferous trees induced by combined exposure to ambient ozone and acid fog and rain. According to this scheme, root development is disturbed mainly by the attack of air pollutants on the upper part of the tree, although nutrient deficiency in the soil may be an important contributing factor. The broken lines represent chains of effects that are either of minor importance or not yet sufficiently proved. (Modified from Krause et al. 1983a.)

biological system, and not necessarily the absolute level of the variable, is the significant factor.

The experimental results that speak in favour of our hypothesis are the following:

Leaching Effect: At the beginning of our investigations we fumigated 5-year-old spruce with ozone concentrations of 150 µg.m^{-3}, 300 µg.m^{-3} and 500 µg.m^{-3}. After four weeks of treatment, twigs of different age classes were cut off and washed separately in artificial acid rain for up to 23 hours. The results determined for the magnesium efflux in the washing solution are represented in Fig. 6. From this it can be derived that the single twigs apparently did not reflect the influence of ozone in a representative manner. On the other hand, the age of the needles had a definite influence on the amount of efflux. This suggests that leaching of older needles in a natural stand can be assumed to be much higher than is found for younger trees in an experiment. Furthermore, during the washing procedure the conductivity of the solution was lowered and the pH went up. This means that, at least in acid solution, the needles exchange metal-ions with low mobility in the solution for protons with high mobility.

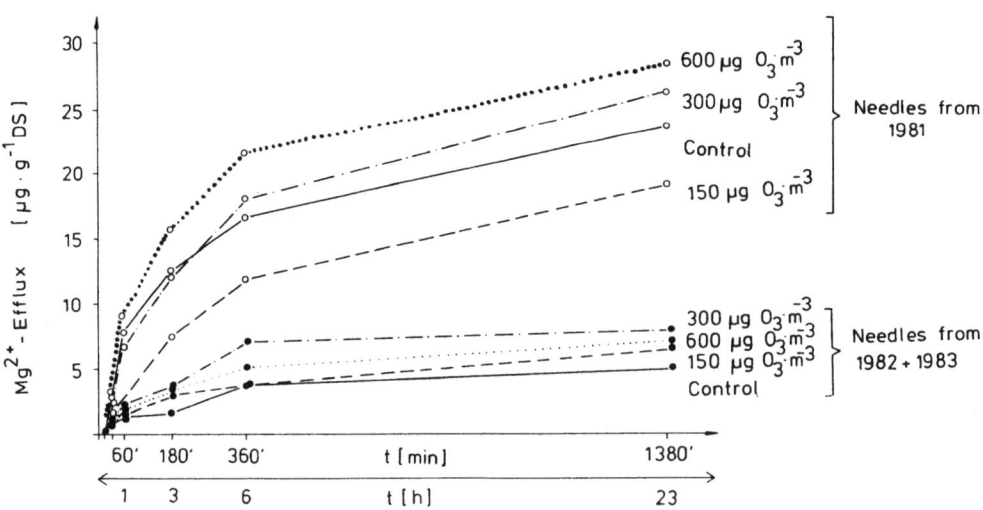

Fig. 6. Magnesium efflux (µg.g^{-1} DS) after ozone fumigation from differently aged needles of Norway spruce. For further details of the applied method see text. While the influence of needle age is quite obvious, the specific method applied here has not proved representative for the effect of different zone concentration on leaching (Krause et al. 1983a).

Within a second experiment described in more detail by Krause et al. (1983a,b) and Krause et al. (1985), whole trees were intermittently treated with ozone (0, 200 and 600 µg.m^{-3}) and artificial acid fog (pH 3.5). The solution applied was analyzed both before and after each fogging event for efflux of magnesium, calcium, potassium, zinc, ammonium, sulphate, nitrate, and phosphate. In an additional experiment, the ozone concentration was held constant and the pH varied. With the exceptions of phosphate and chloride, the efflux of all elements was enhanced by ozone. The pH had an influence only on the efflux of cations. Examples of the results of this leaching experiment are represented in Fig. 7. For more details see Krause et al. (1985). From Fig. 5 and the experiment described previously, the conclusion can be drawn that the efflux of cations is dependent on ozone concentration, pH, previous damage of needles, and needle age. This conclusion is at variance with the findings of Skeffington and Roberts (1985), who reported in their experiments on pine trees that ozone seemed to be without influence on leaching.

<u>Deficiency of Nutrients and Photosensitivity</u>: Four-year-old spruce trees were grown on soil with high or low nutrient supply to examine the relation between nutrient deficiency and photosensitivity.

Half of these treatments were exposed to light intensity of 12 - 22 (average 15) kilolux and half to 28 - 36 (average 30) kilolux (Krause et al. 1985; Prinz et al. 1985). It was shown that under conditions of poor nutrient supply the full light exposure induced a dramatic decay of chlorophyll with severe chlorosis, while the trees on the well-fertilized soil reacted rather moderately to light (Table 3).

Table 3. Change in chlorophyll content (mg.g^{-1} fresh weight) in needles (1983) of spruce grown in soil with high and low nutrient content under a light regime of 30 and 15 kilolux for 12 h per day over 30 weeks.

Soil	Light	Chlorophyll content [mg.g^{-1} fresh weight]	No. of Observations
Low nutrient content	30	0.62 ± 0.13	15
	15	1.20 ± 0.28	16
High nutrient content	30	1.93 ± 0.44	17
	15	2.29 ± 0.46	15

In a second experiment, the rate of apparent photosynthesis of "light" and "shaded" trees on poor soil was reduced by 50% in the "light trees." On the basis of chlorophyll-content, however, the efficiency of assimilation was equal or even somewhat higher in

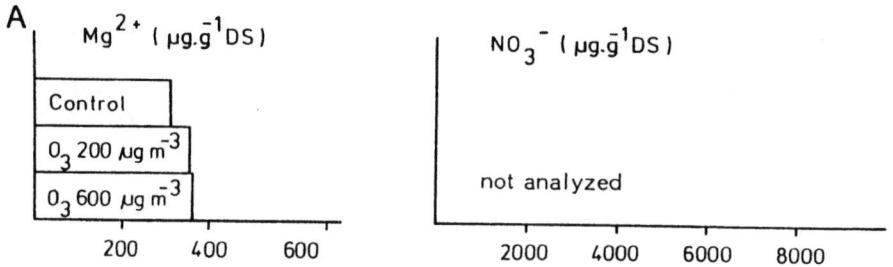

Exp. 1: Healthy trees from a tree nursery, well fertilized soil.

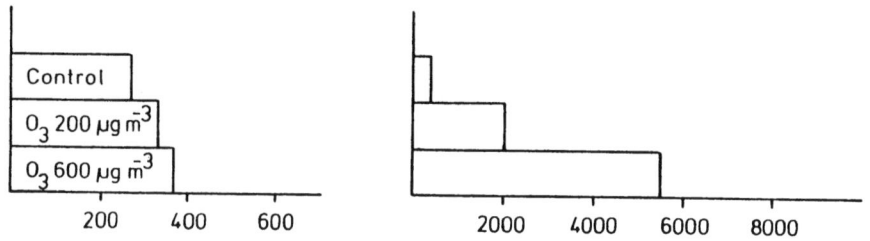

Exp. 2: Healthy trees from a tree nursery, soil taken from the Black forest.

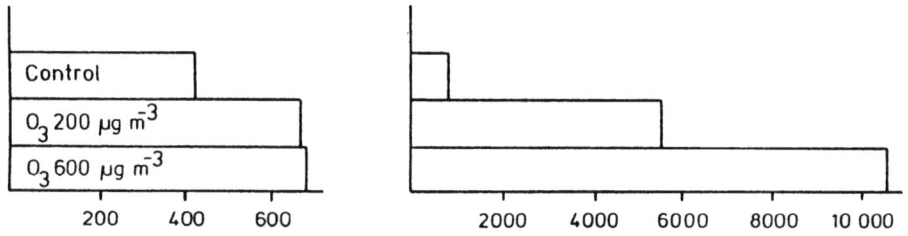

Exp. 3: Predamaged trees and soil taken from the Black forest.

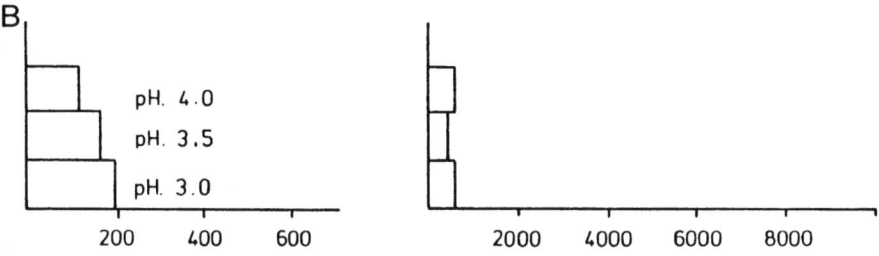

Healthy trees from a tree nursery, unfertilized soil.

Fig. 7. Combined effect of ozone and simulated acid fog on leaching of Mg^{2+} and NO_3^- from Norway spruce. In (A), the ozone concentration was varied (0, 200, and 600 µg.m^{-3}) and the pH held constant. In (B), the ozone concentration was held constant (200 µg.m^{-3}) and the pH varied (pH 4.0, 3.5 and 3.0).

"light trees" than in "shaded trees." This may be caused by the formation of special light chloroplasts (Lichtenthaler et al. 1982).

In this experiment, light-induced chlorosis was not caused by magnesium but possibly nitrogen deficiency. One question remaining to be solved concerns the mechanism of chlorophyll decay. The two principle ways are photooxidation of already existing chlorophyll, perhaps via reduction of accessary pigments like β-carotenes, and reduction in the formation rate of new chlorophyll. According to Powles (1984), interactions exist between frost stress and photooxidation of chlorophyll. This may be the reason that chlorosis is often found to be extremely high during winter.

Inhibition of Photosynthesis: Four year old spruce seedlings were fumigated in the LIS with an average concentration of 450 ug $O_3.m^{-3}$. After 3 weeks, the rate of photosynthesis of treated plants was significantly lower than the photosynthesis of control plants, reaching a difference of 25 - 30% after 5 weeks. Since visible symptoms such as mottling were barely detectable, it can be concluded that under the influence of ozone physiological reactions start earlier than visible injury (Krause and Prinz 1986).

Impairment of Root System by Ozone Exposure: According to Fig. 5 we consider the impairment of root development by lack of assimilates as a very important step in the development of the novel forest decline. Some evidence for this assumption is provided by Blum and Tingey (1977). Recent experiments with pine seedlings (Pinus eliottii Engelm.) used the relatively low concentration of 150 µg $O_3.m^{-3}$ for seven hours per day, which is equal to less than 50 µg $O_3.m^{-3}$ as a daily average (Tingey, 1985). There was a reduction in dry substance of the roots by 34 - 38%, compared with a reduction of the above ground parts of the plant by 16 - 17%. The exposure period was 112 days.

Matzner et al. (1984) argue that, contrary to our assumption, impairment of root development by direct attack of air pollutants on needles and leaves is more or less irrelevant. They cite Meyer (1984), who found a sufficient starch content in injured roots. Meyer's findings support the idea that disadvantageous soil conditions, rather than insufficient assimilate allocation due to disturbed physiology in the above-ground parts of the tree, causes root damage.

In recent investigations, however, Meyer (1985) kept spruce seedlings under full-light and shaded-light (8% of day light) conditions. In the latter case, he found a dramatic reduction in fine root growth, although the above ground part of the trees exhibited no visible reaction. He states, therefore, that the starch content of older roots is not affected by a reduced assimilation while the development of new roots is inhibited dramatically.

Disturbance of Allocation and Translocation of Assimilates: There
are several investigations demonstrating the disturbance of
allocation and translocation of assimilates under the influence
of ozone. This is of special importance for conifers because the
needles of different age classes have specific functions as
sources and sinks of assimilates. According to Clark (1961),
from the month of June on, 36% of assimilates are provided by the
youngest needles of white spruce (Picea glauca (Moench) Voss.)
and the remaining 64% is due to photosynthesis in the remaining
six age classes. For balsam fir (Abies balsamea (L.) Mill.) the
figures are 42% and 58%, respectively. The importance of the
older needles, graduated according to age, is to store assimilates for bud growth in next year and to prevent exaggerated and,
therefore, harmful storage of assimilates in the youngest
needles.

McLaughlin et al. (1982) investigated allocation in healthy white
pine trees (Pinus strobus L.) grown in the neighbourhood of Oak
Ridge, Tennessee. In June, assimilates were translocated from
the older to the youngest needles and in November vice versa,
from the youngest to the older needles. In obviously injured and
ozone-stressed trees, this allocation scheme was severely
disturbed as shown by C^{14}-method. The number of hours exceeding
150 µg $O_3.m^{-3}$ at the investigation site was much lower than in
the heavily damaged areas in East-Bavaria and the Black Forest.
McLaughlin et al. (1982) concluded that such stressed trees
appear to be particularly susceptible to further decline from
additional stresses of either biotic or abiotic origin and lose
capacity to respond to favourable years for growth.

These observations show the importance of the attack of air
pollutants on the above ground parts of the plant in connection
with disturbances of allocation and translocation of assimilates.
This mechanism must, therefore, be taken into consideration when
discussing the causes of the novel forest decline, especially the
decay of the fine root system as an important symptom.

Seasonal Variation of the Chlorophyll Content and Nutrient Supply
in the Needles: Mies and Zöttl (1985) have studied the
chlorophyll content and the content of essential nutrients in
discoloured and healthy needles of Norway Spruce (Picea abies
(L.) Karst.) over the seasons of the year. They found that
colour differentiation started in the previous year's needles in
May at the time of bud break and development of new needles.
Chlorophyll content, as well as content of some essential
nutrients, including magnesium, then decreased in the yellow
needles but remained more or less constant in healthy needles.

This result is very much in agreement with the observation of
Prinz et al. (1982) that normally only the older needles turn
yellow. The content of magnesium - a very mobile element - is
highest in the oldest needles, due to exposure and accumulation
time, and in the youngest needles, due to transport of magnesium
into these needles with greatest metabolic activity. This means
that magnesium deficiency and discolouration are not necessarily
the result of preliminary leaching. They may be the consequence

of complicated internal reactions and turn-overs during the course of the growing season in such trees, which may be weakened by a shortage of nutrients due to leaching caused by ozone exposure in combination with acid fog/rain. This continuing impairment may have a long period of development before first symptoms become visible.

It is clear from this statement that soil conditions are a very important "contributing" factor. The probability of damage of single trees depends very much on the magnesium content in soil if the comparison is made on microspatial base, i.e., between neighbouring trees (LIS, unpublished results). This conclusion is also in accord with the observation of Zech and Popp (1983) and ourselves that damage could be reversed by applying magnesium sulphate via needles. Within a couple of weeks the formerly discoloured needles became dark green. Overall, we suggest that factors such as ozone concentration and exposure time, as well as the intensity of fog exposure, have a dominant role in the development of specific symptoms of novel forest decline.

Disturbance of Reactions Controlled by Phytohormones: Greenhouse fumigation experiments with Norway spruce trees (Picea abies) with average ozone concentration of 450 $\mu g.m^{-3}$ were started at the end of October when trees were dormant. Under these conditions all trees of the control group developed new shoots between the fifth and eighth week, while all ozone-fumigated trees even after the 11th week, were without any sign of budding. This gives rise to the question of whether ozone also influences the production or decay of some specific phytohormones in the tree. The involvement of phytohormones, such as abscisic acid, in the development of the novel forest decline has been suggested earlier by Schutt (Schutt et al. 1982). It is important to examine this behaviour more extensively in the future.

General Conclusion on Damage of Fir and Spruce: To begin with the negative statements, we think that the following exclusions should be kept in mind when the causes for the novel forest decline are discussed.

a) We think that sulphur dioxide by itself should be excluded as a cause of the novel forest decline, since no correlation between degree of damage and ambient concentration of this air pollutant can be found. Moreover, needles of heavily injured trees in Southern Germany frequently show a very low sulphur content (600 μg $S.g^{-1}$ DS).

b) "Acid rain" by itself should be excluded because, according to our experiments, lichens proved to be much more sensitive to acid fog than fir and spruce (Prinz et al. 1982; Krause et al. 1983a). On the other hand they are observed to grow luxuriously at sites in the Black Forest in which trees are dying.

c) The soil, including soil acidification, by itself should be excluded, since all types of soil and bedrocks are involved, even in some cases, calcareous ones.

d) Even ozone by itself should be excluded with regard to spruce and fir because these species have proved to be rather resistant (Prinz et al. 1982; Krause et al. 1983a, 1985) and the typical symptoms of mottling induced by ozone exposure are not identical to the type of discolouration found in nature.

As a general, positive conclusion, it is obvious that novel forest decline is very much related to nutrient supply, a point emphasized at an early stage by Rehfuess (1981) and Rehfuess et al. (1982). On the other hand, it is well known that trees may grow even on extremely poor soils if adapted to these conditions from the beginning. The most relevant point in the development of the novel forest decline, with its specific symptomatology, seems to be an imbalance of source-sink-relation between crown and root system. Starting with leaching of nutrients and possible direct attack of ozone on the above ground part of the tree, the root system is affected by a shortage of assimilates. Further uptake of nutrients is increasingly lowered, which works like a spiral whenever started. Other factors, including soil and climate, may contribute to this process. For example, major steps in the development of the novel forest decline occurred after the dry and hot summers in 1976, 1980, 1982 and 1983, while after two relatively wet years (1984, 1985) a certain stabilization or even recovery has taken place. The cause may be a reduced uptake of nutrients during the dry summers or an increased transpiration rate interacting with ozone, which also showed high ambient concentrations in these summers.

A contributing element in the development of damaged forests is the enhanced infestation by insects, such as bark beetles, or infection by fungi such as Lophodermium picea or Rhizosphaera kalkhoffii. The real danger from these secondary pathogens is the epidemic spreading after the primary impact of air pollutants.

Beech and Pine

In contrast to spruce and fir, beech and pine are thought to be sensitive enough to be damaged directly by ambient ozone concentrations without further interaction with soil, fog or rain.

Fumigations in the greenhouse with ozone at 150 $\mu g.m^{-3}$ and artificial light of 35 kilolux caused distinct bronzing of the leaves of common beech (Fagus sylvatica L.) after six weeks of exposure. Shaded leaf parts remained green. As with symptoms in natural stands the tissue between the veins became malformed so that the leaves had a wrinkled appearance. The bronzing originated from small spots between the first order veins following treatment with ozone at 300 $\mu g.m^{-3}$ and 35 kilolux. Damage soon changed to intercostal necrosis. At 15 kilolux, the intensity of damage was much less severe (Krause et al. 1983b).

Beech trees were fumigated with ozone from February to October 1984 in chambers with 70 air changes/hour standing under open air condition at the field-station in Essen-Kettwig. These chambers are supplied with charcoal-filtered air and represent otherwise "normal" climatic conditions. Two treatments were used. In the first, ozone concentration was 150 µg.m^{-3} during 5 hours per day mixed with sulphur dioxide at 100 µg.m^{-3} continuously. In the second treatment, 150 µg O_3.m^{-3} and 60 µg SO_2.m^{-3} were applied continuously plus an episode of 500 µg SO_2.m^{-3} for 2 hours, every fortnight. The two treatments were considered to be typical of an "industrial area" and a "clean air area," respectively.

Preliminary leaf drop occurred in beech in the ozone-treatment, but did not occur in the control- or SO_2-treated plants (Prinz et al. 1985). Since this effect was pronounced under the "clean air area" regime, it is very likely that relatively high ozone concentrations during day and night, which are most typical for elevated regions with forest decline in German damaged forest stands, have an additional dose-effect in comparison to single peak values of the same concentration.

Austrian Pine (Pinus nigra var. austriaca (Hoess) Neum.) was also very sensitive when fumigated continuously with 150 ug O_3.m^{-3}. The oldest needles turned yellow during the summer, became brown, and finally dropped off. A slight mottling was noticeable on the yellow needles but this was much less distinctive than in spruce needles.

There seems to be a great need for further research to study the influence of climatic conditions on the effect of ozone on beech and pine. We have the feeling that this factor causes a great variation in kind and intensity of the damage. The classification of this point is absolutely necessary to unequivocally interpret the symptoms in natural stands.

CONCLUSION

The evidence mentioned here does not have the quality of axiomatic proofs. But in relation to type, spatial distribution and temporal development of injury, our understanding of the causes of forest decline has at the moment, in our opinion, the highest probability of proof among the alternative hypotheses.

We do not agree with the statement of Ulrich (1982) that the burden of proof is not the task of the one who puts forward a certain hypothesis but the task of the one who denies to accept it. In contrast, we think that it is the responsibility of every scientist to provide his own proofs, carefully and in an unprejudiced manner.

On the other hand, no one would be able to act in a sensible way if he waited for total proof of his judgement. This will never be reached in empirical science. Therefore, the concept of highest probability as proposed here is the only useful concept and, ultimately, the only possible one.

Acknowledgement

The authors would like to thank Dr. Paul Miller, Riverside for helpful comments during preparation of the manuscript.

REFERENCES

Anonymous (1983a) Waldschäden in Bayern (Stand: Sommer 1983), Staatsministerium für Landesentwicklung und Umweltfragen, Bayerisches Staatsministerium für Ernährung, Landwirtschaft und Forsten

Anonymous (1983b) Bayerische Staatsforstverwaltung Information, 1/83. Sonderheft, Waldersterben.

Ashmore M, Bell N, Rutter J (1985) The role of ozone in forest damage in West Germany. Ambio 14: 81-87

Attmannspacher W, Hartmannsgruber R, Lang P (1984) Langzeittendenzen des Ozons der Atmosphäre aufgrund der 1967 gewonnenen Ozonmessreihen am Meteorologischen Observatorium Hohenpeissenberg. Meteorol. Rdsch. 37: 193-199

Bundesministerium für Ernährung, Landwirtschaft und Forsten (BML) (1984) Neuartige Waldschäden in der Bundesrepublik Deutschland. Bericht des Bundesministers für Ernährung, Landwirtschaft und Forsten anlässlich der Waldschadenserhebung 1983. BML, Postfach 14 02 70, 5300 Bonn 1, FRG

Blum U, Tingey DT (1977) A study of the potential ways in which ozone could reduce root growth and nodulation of soybean. Atmos Environ 11: 737-739

Clark J (1961) Photosynthesis and respiration in white spruce and balsam fir. State University of New York, College of Forestry at Syracuse University, Syracuse NY

Evers FH (1984) Personal Communication

Feister U, Warmbt W (1984) Long-term surface ozone increase at Arkona (54.68°N, 13.43°E). In: Proc Quadr Internat Ozone Symposium, 3-7 September 1984, Chalkidiki, Greece

Fricke W (1983) Grössraumige Verteilung und Transport von Ozon und Vorläufern. In: VDI Berichte 500 Saure Niederschläge - Ursachen und Wirkungen - VDI-Verlag GmbH, Düsseldorf p 55

Guicherit P, van den Hout (1982) The global NO_x-cycle. In: Schneider T, Grant L (eds) Air pollution by nitrogen oxides, Elsevier Scientific Pub Co, Amsterdam, p 15-29

Heitefuss F (1985) Personal Communication

Kenk GP, Unfried P, Evers FH, Hildebrand EE (1984) Dungung zur Minderung der neuartigen Waldschäden - Auswertung eines alten Düngeversuchs zu Fichte im Buntsandstein - Odenwald. Forstw Cbl 103: 307-320

Krause GHM, Jung KD, Prinz B (1983a) Neuere Untersuchungen zur Aufklärung immissionsbedingter Waldschäden. VDI-Berichte 500: 257-266

Krause GHM, Prinz B, Jung KD (1983b) Forest effects in West Germany. In: Proceedings of the symposium air pollution and the productivity of the forest October 4 and 5, 1983, Izaak Walton League, Washington DC, p 297-332

Krause GHM, Jung KD, Prinz B (1985) Experimentelle Untersuchungen zur Aufklärung der neuartigen Waldschäden in der Bundesrepublik Deutschland. VDI-Berichte 560: 627-656

Krause GHM, Prinz B (1986) Zur Wirkung von Ozon und saurem Nebel auf phänomenologische und physiologische Parameter an Nadel - und Laubgehölzen im kombinierten Begasungsexperiment. In: KFA-Jülich (ed) Wirkungen von Lu/tverunreinigungen auf Waldbäume und Waldböden. Statusseminar vom 02-04.12.1985 Kern/orschungsanlage Jülich, ISSN 0343-7639, p 208-221

Luckat S (1972) Ein Verfahren zur Bestimmung der Immissionsrate gasförmiger Komponenten. Staub Reinh Luft 32: 484-486

Lichtenthaler HK, Kuhn G, Prenzel U, Buschmann C, Maier D (1982) Adaptation of chloroplast-ultrastructure and of chlorophyll protein levels to high-light growth conditions. Z. Naturforsch 37c: 464-475

Matzner E, Ulrich B, Murach D, Rost-Siebert K (1984) Zur Beteiligung des Bodens am Waldsterben. In: Ulrich B (ed) Berichte des Forschungszentrums Waldökosysteme/Waldsterben, Band 2, p 1

McLaughlin SB, McConathy RK, Duvick D, Mann LK (1982) Effects of chronic air pollution stress on photosynthesis, carbon allocation and growth of white pine trees. Forest Sci. 28: 60-70

Meyer FH (1984) Mykologische Beobachtungen zum Baumsterben. AFZ 39: 212-228

Meyer FH (1985) Einfluss des Stickstoff-Faktors auf den Mykorrhizabesatz von Fichtensamlingen im Humus einer Waldscha densfläche. AFZ 40: 208-217

Mies E, Zöttl HW (1985) Zeitliche Änderung der Chlorophyll- und Elementgehalte in den Nadeln eines gelb-chlorotischen Fichtenbestandes. Forstw Cbl 104: 1-8

Obländer W, Wörth R, König E, Braunger H, Schröter H (1983) Ergebnis und Interpretation von zweijährigen Schwefeldioxid-Immissions-Messungen an Tannenbeobachtungsflächen im Schwarzwald und in angrenzenden Wuchsgebieten. Allg Forst-u.J.-Ztg 154: 175-180

Obländer W, Hanss A (1985) Zwischenbericht über Schadstoffmessun gen in Waldgebieten Baden-Württembergs. LfU-Bericht Nr.9/85

Pfeffer HU, Buck M (1985) Messtechnik und Ergebuisse von Immissionsmessungen in Waldgebieten. VDI-Berichte 560: 127-155

Powles SB (1984) Photoinhibition of photosynthesis induced by visible light. Ann Rev Plant Physiol 35: 15-44

Prinz B, Krause GHM, Stratmann H (1982) Vorläufiger Bericht der Landesanstalt für Immissionsschutz uber Untersuchung zur Aufklärung der Waldschäden in der Bundesrepublik Deutschland. LIS-Bericht 28: 154 p. Landesanstalt für Immissionsschutz des Landes NW, Wallneyer Strasse 6, 4300 Essen 1, FRG

Prinz B (1983) Gedanken zum Stand der Diskussion über die Ursache der Waldschäden in der Bundesrepublik Deutschland. Forst- und Holzwirt 38: 450-468

Prinz B (1984) Ergebnisse und Folgerungen aus dem VDI-Kolloquium in Lindau und der VDI-Studie "Säurehaltige Niederschläge". VDI-Berichte Nr. 495, 39-48

Prinz B, Krause GHM, Jung KD (1985) Untersuchungen der Landesanstalt für Immissionsschutz des Landes NW zur Problematik der Waldschäden. In: Symposium der Schwaben AG, Waldschäden 1985 - Theorie und Praxis auf der Suche nach Antworten 24. - 25.01.1985, Stuttgart-Hohenheim, p 143-194

Rehfuess KE (1981) Über die Wirkung der sauren Niederschläge in Waldökosystemen. Forstw Cbl 100: 363-381

Rehfuess KE, Bosch C, Pfannkuch E (1982) Nutrient imbalances in coniferous stands in Southern Germany. In: Internat. workshop on growth disturbances of forest trees. Jyräskyla/Finnland 10. - 13.10.1982

Reiter R (1983) Basiserarbeitung zum Problem "Waldschäden im Bayerischen Nordalpenraum". Materialien 298. Ed., Bayer. Staatsministerium für Landesentwicklung und Umweltfragen, München

Rudolph E (1983) Funktionale Zusammenhänge zwischen Kenngrössen des Regenwassers und Wirkungskriterium. In: VDI-Berichte Nr. 500 Saure Niederschläge - Ursachen und Wirkungen - VDI-Verlag GmbH, Düsseldorf, p 231

Satorius R (1984) Ursprung, Transport und Umwandlung von Luftverunreinigungen. In: Kolloquium Waldsterben - Diagnose und Therapie, FGN Berlin e.V. 02. - 03.04. 1984

Schönbeck H (1969) Eine Methode zur Erfssung der biologischen Wirkung von Luftverunreinigungen durch transplantieite Flechten. Staub-Reinhaltung de Luft 29: 14-8

Schütt P (1982) Das Krankheitsbild - verschiedene Baumarten, gleiche Symptome. Bild der Wissenschaft 12: 86-101

Schwela D, Radermacher (1985) Untersuchungen zur Belastung durch Luftverunreinigungen in quellfernen Gebieten mittels Bioindikatoren, IRMA- und Staubniederschlagsmessungen. Staub, Reinhalt. Luft 45: 284-287

Skeffington RA, Roberts TM (1985) The effects of ozone and acid mist on scots pine saplings. Oecologia 65: 201-206

Tingey D (1985) Personal Communication

Umweltbundesamt (UBA) (1983) Die zeitliche Entwicklung der überregionalen Pegel von Schwefeldioxid, Stickstoffdioxid, Schwefel im Schwebstaub und der Konzentration von Wasserstoffionen im Niederschlag. Monatsberichte aus dem Messnetz des Umweltbundesamtes. Jg. 7, 1982, Nr. 1.

Ulrich B (1982) Lässt sich Schädigung beweisen? Sonderheft der LOLF. Mitteilungen 1982, Recklinghausen, p 9-11

Warmbt W (1981) Langjährige Messungen des bodennahen Ozons in der DDR. Technik und Umweltschutz, Publ. Nr. 23, Leipzig: DT Verl. für Grundstoffind. 62-77

Zech W, Popp E (1983) Magnesiummangel, einer der Gründe für das Fichten- und Tannensterben in NO-Bayern. Forstw Cbl 102: 50-55

RESULTS OF STUDIES ON FOREST DECLINE IN NORTHWEST GERMANY

E. Matzner* and B. Ulrich

Institute of Soil Science and Forest Nutrition, University of Göttingen, FRG
*Research Center Forest Ecosystems/Forest Decline, University of Göttingen, FRG

ABSTRACT

Forest soil acidification in Germany during the past few decades has been reported in numerous papers. H^+ budgets reveal that acid deposition represents the major input of strong acidity to forest soils and therefore is the main cause of soil acidification under middle Europe deposition conditions.

Important ecological effects of soil acidification are: nutrient losses; reduced biological activity; release of potential toxins to roots; groundwater acidification; and reduced elasticity with respect to natural acidification pulses.

These effects are demonstrated using data from polluted forests of Northwest Germany. Special emphasis is given to the interrelationship between acid deposition and natural acidification pulses and to the implication of soil acidification for root growth and forest decline. Experimental results from Al-toxicity studies, root biomass and root dynamics studies, in addition to root regeneration experiments, are presented.

NEED AND MEANING OF AN ECOSYSTEM-ORIENTED HYPOTHESIS

The "Waldsterben" in central Europe can be viewed as a decline of the forest ecosystem. There are natural and man-made factors influencing the forest ecosystem and possibly playing a role in its decline. There are some scientific constraints which should be kept in mind when discussing forest ecosystem decline. All attempts to explain tree dieback are of hypothetical nature. The rules to deal with hypotheses as laid down in mathematical statistics apply also to hypotheses on forest decline. The testing of hypotheses is subject to two kinds of risks:

    risk I: a wrong hypothesis is accepted
    risk II: a valid hypothesis is rejected

The minimum requirements to accept or reject a hypothesis are defined by the data quantifying the hypotheses.

An absolute minimum requirement is the proof that a specific load or stress actually exists. Regarding large scale forest dieback in various parts of the world, this proof has been furnished for atmospheric loads of SO_2, NO_x, O_3, acidity and the accompanying deposition of nutrients, acids, heavy metals, metalloids, and hydrocarbons.

Trees are components of ecosystems, together with other green plants (primary producers), with secondary producers (mainly decomposer organisms, but also consumers), and with the soil as the environment for roots and decomposers. Any one factor influencing the ecosystem may affect any one of these components. In addition, there may be indirect effects operating between these components:

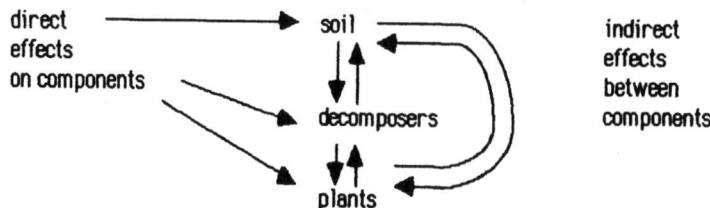

The great danger in dealing with complex systems, such as ecosystems, is the neglect of possible cause/effect relationships. If a well documented effect is only part of a more general syndrome, but is considered as the only cause, it may lead to serious errors in the treatment of the problem. Such errors arise from risk II: the rejection or the neglect of valid hypotheses. This risk may be taken especially if two hypotheses are considered as being alternatives; the acceptance of one hypothesis necessarily excludes the other. In a complex ecosystem such an alternative is a very rare case. It is more likely that an external factor influences the system along different pathways, directly and indirectly as defined above. Neglect of an effect is a rejection of a valid hypothesis without making any test. From a systems point of view, it is as important to search for possible effects so as to quantify effects already described. An ecosystem-oriented hypothesis on forest decline has therefore to take into account all possible effects.

RECENT RESULTS FROM ECOSYSTEM STUDIES

Because of the high filtering activity of the canopy, forest stands are heavily impacted by the deposition of various air pollutants such as heavy metals (Mayer 1981), acids (Ulrich 1983; Matzner 1984b) and organic compounds (Matzner 1984a). The deposition of acidity as well as the effects on soils and through the soil on the growth of the roots will be stressed in this paper in more detail.

Within terrestrial ecosystems, the soil represents the only permanent buffering compartment. Since few forest soils contain carbonates, only alkali and alkali earth silicates are available to neutralize acid deposition in many forest soils. Since alkali and alkali earth silicates are present in high amounts in soils, the rate of buffering and dissolution of these compounds, rather than the amount of silicates available, is the most critical factor in determining acid tolerance.

It has been demonstrated that silicate weathering with subsequent release of nutrient cations is a very slow process, reaching maximum buffering rates of 1-2 kmol $H^+.ha^{-1}.y^{-1}$ (Bache 1982; Mazzarino et al. 1983; Johnson et al. 1981; Fölster 1985). These rates are small in comparison to the rates of emission of acidic compounds, indicating that the neutralization capacity of lime-free soils is exceeded. Acidification of soils to the point of Al solubilization and release is the final consequence. This stage has already been documented in large forested areas in central Europe during the last few decades (Butzke 1981; Grenzius 1984; Evers 1983b; von Zezschwitz 1982).

H^+ budgets for forests show that acid depostion represents the major input of strong acidity to forest soils (Ulrich et al. 1979, Matzner 1984c; Driscoll et al. 1982) especially in central Europe, and therefore must be considered as the major cause of soil acidification.

Ecological consequences of soil acidification are:

(1) reduction of base-saturation of exchangeable cations, with subsequent losses of Ca and Mg from the soil by leaching (e.g., Matzner et al. 1984);
(2) increase of the concentration of potentially toxic ions to roots (H^+, Al^{3+}, Mn^{2+}...) in the soil solution;
(3) reduction of decomposer activity and increase of the organic top layer (Matzner 1985; Ulrich et al. 1980);
(4) reduced elasticity of the soil with respect to seasonal acidification pulses; and
(5) groundwater acidification.

Point 2 and 4 need to be discussed in more detail because of their importance when evaluating the relationship of soil acidity and forest decline.

Increase of the Concentration of Potentially Toxic Ions to Roots

Figure 1 shows the development of pH and Al concentrations of the seepage water under a beech and a spruce stand in the Solling region. Under beech, the Al concentration of the seepage water doubled between the periods 1969-1973 and 1976-1982. A more drastic development was found under spruce, where Al concentrations increased from 2-4 mg.L^{-1} in 1973 to 15-18 mg.L^{-1} in 1977.

The Ca/Al ratio of the soil solution has been found to be a good indicator of Al-toxicity in roots. Rost-Siebert (1983) found reduction in root growth of Norway spruce seedlings in cases in which the Ca/Al ratio was less than 1. In Solling, the Ca/Al ratio of the soil solution at 1 m depth decreased steadily, although it was already low (0.2) at the beginning of the measuring period (Matzner et al. 1983).

Because of the high Al concentrations in seepage water, the Al-budget of the soil is negative, i.e., the soil acts as a source. By setting up a budget for total acidity (H^+ + the cation acids: NH_4^+, Al^{3+} and Mn^{2+}), one can demonstrate that most of the acidity reaching the systems by deposition of H^+ and NH_4^+ is transferred to other systems (deeper soil layers, groundwater) in the form of Al^{3+} and Mn^{2+} ions, themselves strong acids (Ulrich and Matzner 1983). The current output of total acidity under the spruce stand is about twice the input (Matzner et al. 1984). This observation is related to the behaviour of SO_4^{2-} within the soil.

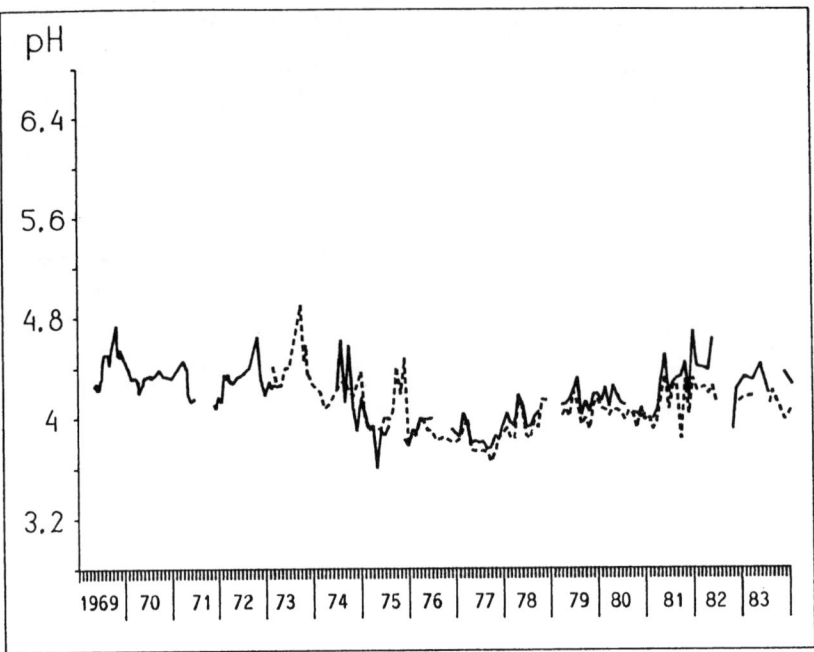
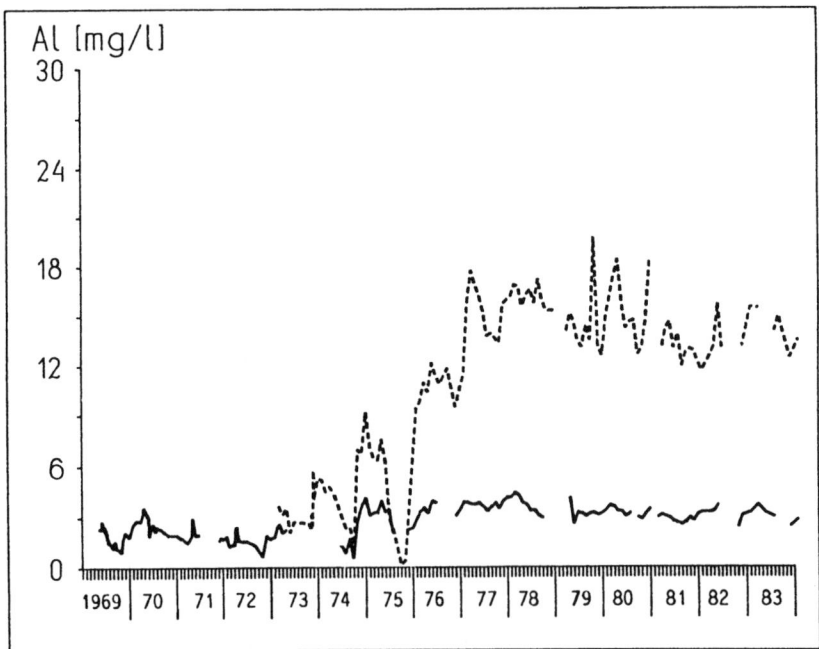

Fig. 1. The pH and Al concentration of seepage water under a beech and a spruce stand in the Solling region. For site description see Ulrich et al. (1979). The soil solutions were extracted by lysimeters at a depth of about 1 m. ---- = spruce; —— = beech

Since SO_4^{2-} is the most important anion (in equivalents) in precipitation, SO_4^{2-} adsorption and precipitation play major roles in determining the chemistry of the soil solution, and in regulating the amount of seepage losses of cations and acids.

The annual budgets of the mineral soil of the spruce stand for Al and S are given in Fig. 2. Positive values indicate that the soil is acting as a sink, and negative values indicate that it is a source. The budgets include: input to the mineral soil from the forest floor, output with seepage water, and plant uptake from within the mineral soil.

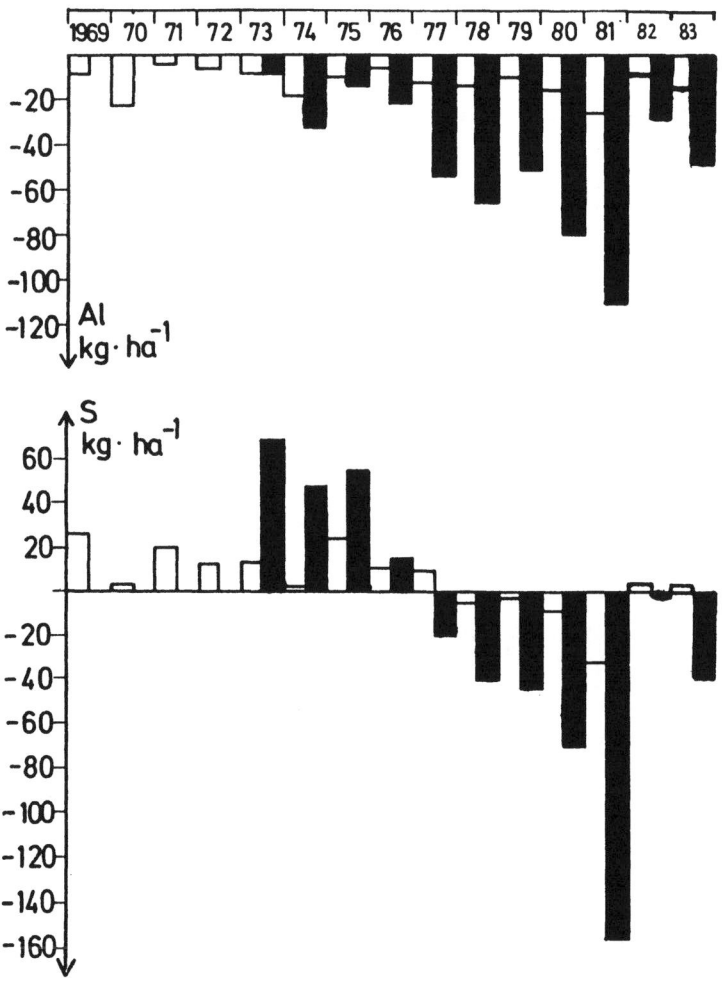

Fig. 2. Change in soil storage of S and Al calculated from the flux balance of the mineral soil.   □ = beech   ■ = spruce

From 1973 to 1975, more than 50% of the S input was retained by the soil. Calculation of chemical equilibria (Prenzel 1983) reveals that the major mechanism of S accumulation in these soils is the formation of $AlOHSO_4$ (jurbanite), probably following adsorption of SO_4, according to the following equation:

$$\text{base} \quad \text{acid} \quad \rightleftharpoons \quad \text{acid} \quad \text{base}$$
$$AlOOH + H_2SO_4 \quad \rightleftharpoons \quad AlOHSO_4 \quad H_2O$$

This reaction leads to the accumulation of acidity in the form of an Al hydroxo sulphate within the soil. Laboratory experiments show that $AlOHSO_4$ may be solubilized as a result of pH changes (Prenzel 1983). Further soil acidification or fluctuations of soil pH (see below) will mobilize $AlOHSO_4$:

$$AlOHSO_4 \quad H^+ \quad \rightleftharpoons \quad Al^{3+} \quad SO_4^{2-} \quad H_2O$$

This seems to have happened after 1976, resulting in increasing net losses of S from the soil, and to high concentrations of Al in seepage water (Fig. 2).

The behaviour of S in this ecosystem is of great importance for predicting water acidification, since S retention was found to be reversible, rapidly releasing the acidity previously stored in the soil.

Al-ions have been known to be potentially toxic to plant roots for about 70 years (Hutchinson 1983). The suggestion that the increasing levels of Al-ions, with toxic effects on tree roots are involved in forest decline was first made by Ulrich et al. (1979) and has been the subject of much debate (e.g., Rehfuess 1981; Zöttl 1983). However, recent results from laboratory experiments have shown root damage to spruce and beech seedlings by H^+, Al^{3+} and Mn^{2+}-ions (Junga 1984; Rost-Siebert 1983; Hüttermann and Ulrich 1984). The concentrations necessary to induce toxicity symptoms in these laboratory experiments were lower than those found in soil solutions of acid soils in the field. Furthermore, experiments with 45-Ca showed that low levels of Al in the soil solution drastically reduced the uptake of Ca (Junga 1984; Hüttermann and Ulrich 1984). A similar effect occurred with Mg (Junga 1984; Evers 1983a). The symptoms of Mg deficiency in spruce needles that have been reported in Germany on acid soils can be explained by antagonistic effects of Al^{2+} and H^+ ions during ion uptake. This conclusion is confirmed by X-ray probe analysis of roots indicating that Mg and Ca deficiency is already present in root cells (Bauch and Schröder 1982) of damaged trees.

Since the effect of Al^{3+} or H^+ on the growth of roots does not only depend on the absolute concentration of Al but also on the concentration of basic cations like Ca and Mg, the Ca/Al or Ca/H^+ ratio is a useful indicator for acid toxicity in tree roots (Ulrich et al. 1984a).

Reduced Elasticity of the Soil with Respect to Seasonal Acidification Pulses

The ion cycle within forest ecosystems may be simplified to two main processes: mineralization and ion uptake. Depending on the cation/anion balance of the process, both may be connected with proton turnover. If the rate of one of these processes exceeds that of the other, net production or consumption of protons may occur in the soil. The production of protons during mineralization is expected when organic N is nitrified (HNO_3 production). If the rate of nitrification exceeds the rate of nitrate uptake, protons are produced in the soil, causing an acidification pulse. The acidification pulse ends as the nitrate is taken up by plants and microorganisms. Acidification pulses can be characterized as seasonal processes related to discouplings of the ion cycle.

The rate of nitrification is influenced by soil temperature as it is a microbial process. Normally, the mean soil temperature is far from optimum for nitrification, and a significant nitrification increase is expected in warm years with higher-than-average soil temperature. Thus, acidification pulses are linked to warm years.

The effect of seasonal acidification pulses on soil organisms and plant roots depends on the ability of the soil to buffer the increasing amount of protons. At this point the interrelationship among acidification pulses, deposition of acidity, and other processes leading to soil acidification (e.g., harvesting) becomes clear, since acid deposition continuously stresses and slowly exhausts the buffering capacity of the soil. A well-buffered soil with a high base saturation is able to buffer acidification pulses by exchanging H^+ for Ca and Mg, without the release of toxic ions like H, Al, Fe and Mn into the soil solution. After the soil has lost this ability, acidification pulses increasingly will lead to toxic conditions in the soil.

Forest decline in central Europe is generally characterized by a significant increase in forest damage following warm and dry years. This phenomenon was first recognized in the case of <u>Abies</u> (silver fir) decline in South Germany, and in other stands in Europe (Schroter 1983). A reduction in the supply of water to trees in the warm and dry years represents an additional stress that alone may cause serious injury to plants. The 1975 and 1976 droughts were in many areas certainly of such magnitude. The drought or water stress is increased if a strong acid pulse has caused root damage, limiting the ability of roots to take up water. The hypothesis of the occurrence and effect of acidification pulses was investigated in the years of 1981-1983 in different forest ecosystems (Matzner and Thoma 1983). Only the results derived from the 100-year old spruce stand in the Solling region are discussed here, since data on the dynamics of the fine soil roots are also available for this stand.

Figure 3 shows the variation in soil solution nitrate concentration and pH in the Solling stand as well as the fine root dynamics in the uppermost soil (Murach 1984).

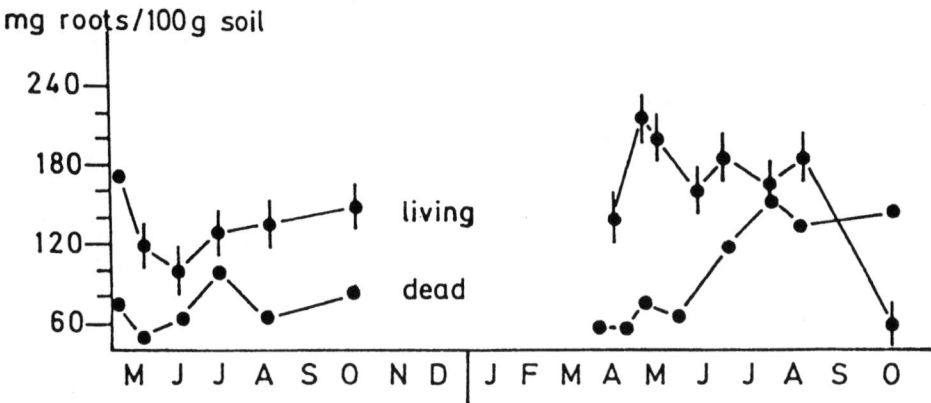
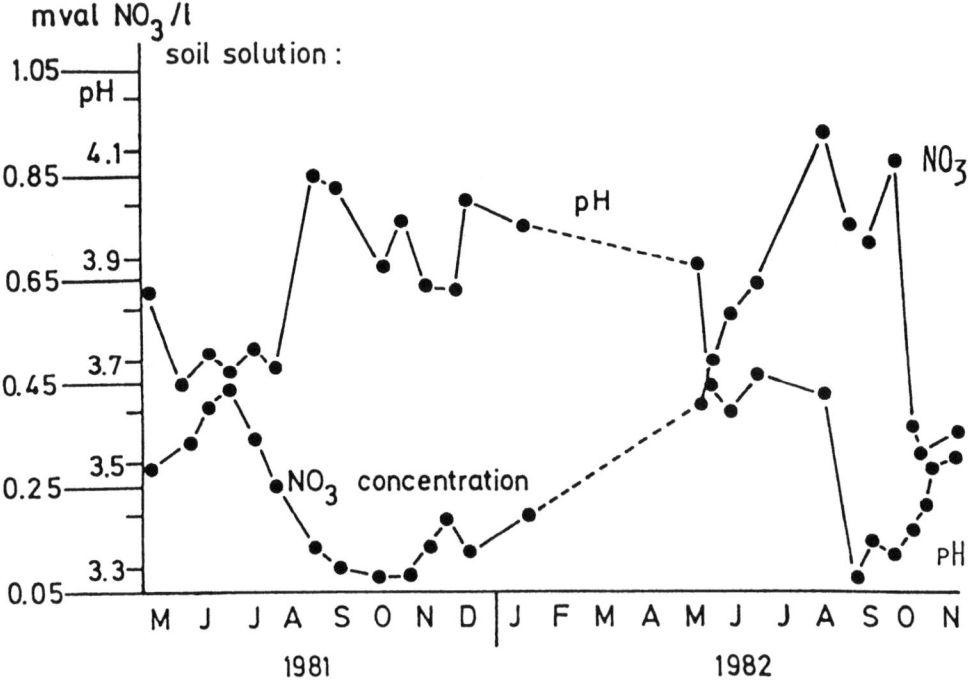

Fig. 3. Variation in soil solution nitrate concentration and pH from the uppermost 10 cm of soil in the Solling stand as a function of time. The data on the fine root biomass are from 0-5 cm.

During the cool and wet year of 1981, the increasing nitrate concentration (nitrification minus uptake) in May caused a decrease of the soil solution pH. The pH values increased in autumn 1981 as a consequence of nitrate uptake (proton consumption). Again in May, the nitrate concentration of the soil solution increased in 1982, accompanied by a decrease in pH. The favourable climatic conditions in 1982 resulted in an accelerated rate of nitrification which far exceeded the rate of nitrate uptake. This acidification pulse resulted in a drastic decrease of the soil solution pH to a level of 3.28 in September 1982. The close relationship between nitrate concentration and pH (the concentrations of H^+ and nitrate fluctuate in equivalent amounts (Matzner and Thoma 1983)) indicates that the upper soil of this stand no longer possesses any fast-reacting buffer characteristics. This is also evident from the fact that the Al concentration only slightly increases during 1982 (Matzner and Thoma 1983).

Laboratory experiments with Norway spruce seedlings (Rost-Siebert 1983) indicated that roots are damaged by H^+ when the pH of the soil solution falls below 3.5. The acidification pulse in 1982, therefore, should have resulted in a serious stress on the roots of the horizon considered. Fluctuations in the amount of living fine root biomass followed the changes in solution chemistry. The amount of living roots declined in May 1981 as the pH of the soil solution dropped, and recovered as the pH increased. This pattern was repeated in May 1982. Due to the strong acidification pulse and the subsequent increased toxicity to acid, the root biomass dropped in October 1982 to the lowest value recorded during the investigation.

Figure 4 shows the pH, and the NO_3^- and Al concentrations during 1981 to 1983 at a depth of 20 cm. The 1983 season was very warm and rather dry. The pH drop of 1982 in the soil at the depth of 20 cm reached a minimum pH of about 3.6. This coincides with the pH drop in 1982 at a depth of 10 cm (Fig. 3). However, in 1983 the NO_3^- concentrations increased to higher levels as compared to 1982 and caused a drastic increase in the Al concentration of the soil solution. These data clearly show two types of acid stress following acidification pulses: H^+-stress in 1982 and Al-stress in 1983.

ROOT GROWTH AND ROOT DISTURBANCE IN DECLINING FOREST STANDS

Disturbance of the root system has been observed when sought in any declining forest stand in West Germany. Symptoms of root disturbance include strong vertical gradients of the fine root biomass (Murach 1984), increased percentage of dead fine roots of damaged stands compared with undamaged ones (see Fig. 5), diseased mycorrhizae (Meyer 1984), abnormal morphology of branching (Blaschke 1981), damage of the endodermis (Hüttermann 1982), and also changes in chemical composition (Bauch and Schröder 1982; Murach 1984).

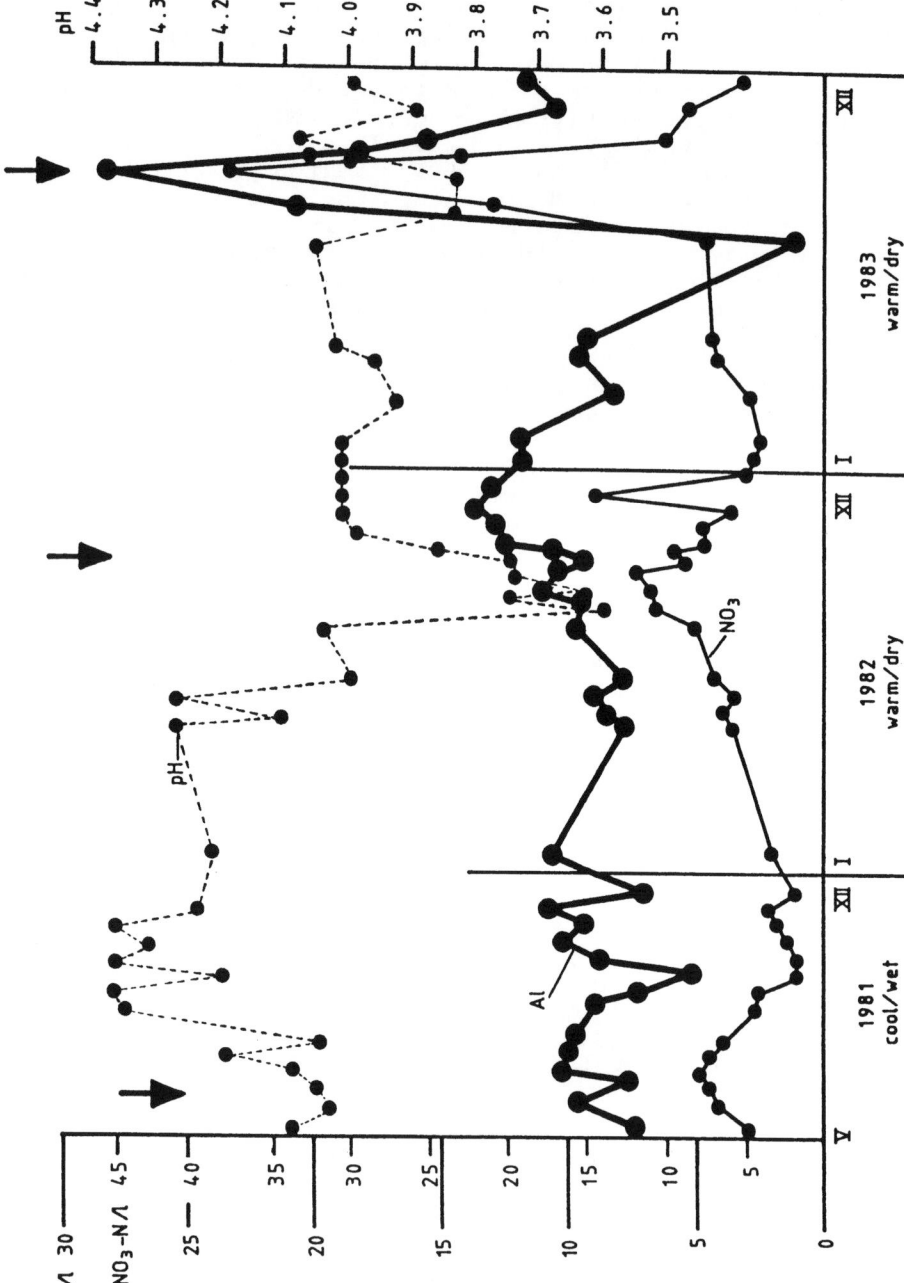

Fig. 4. The pH and NO_3^- and Al concentrations in the soil of the Solling stand at a depth of 20 cm from 1981 to 1983.

These damages place a predisposing stress on the whole tree by reducing water and nutrient uptake. At present two major cause/effect hypotheses are discussed which can explain the root damage. One is the soil-derived acid toxicity hypothesis mentioned above. In contrast, Schütt et al. (1983) and Wentzel (1983), as well as Prinz et al. (1982), Mohr (1983) and Lichtenthaler and Buschmann (1983) see the primary cause of the decline as the direct action of air pollutants on the leaves of the plant; the observed root damage is explained by a reduced supply of assimilates to the root system. However, as far as we know, no experimental data are available from damaged stands to support the direct foliar effect hypothesis.

Measurements of root biomass dynamics conducted in forest stands of North Germany exclude assimilate deficiency as the primary cause of the observed root dieback in these systems. Murach (1984) reported a strong decrease in the living fine root biomass in the top soil of the Solling spruce stand in autumn 1982 and related this finding to the drop in pH following the 1982 acidification pulse (see Fig. 3). Since the biomass of the roots of the deeper soil was constant or increased slightly, Murach excluded assimilate deficiency as a cause of the observed root damage in the top soil. "Living" and "dead" roots were subdivided on the basis of morphological parameters (Murach 1984).

Biomass inventory studies in 2 young (12 year) Norway spruce stands (Schulte-Bisping and Murach 1984) showed that the total fine root biomass of the damaged stand was about three times higher compared with the neighbouring (but sheltered) healthy stand, while the biomass of living fine roots is about the same (Fig. 5). These findings cannot be explained by assimilate deficiency. On the other hand, soil data indicate strong soil acidification under the damaged stand.

In order to test the importance of soil acidity in regulating root growth, an experiment was conducted in 1984. Corings (8, 40 cm depth) were taken and refilled with 3 different soils: (1) an acid soil from the B horizon of the Solling spruce stand (pH $CaCl_2$ = 3.9, base saturation < 4% of CEC), (2) a Solling soil limed to pH 4.8 base saturation ≈40% of CEC, and (3) a mixture of fertilized peat and sand. After the growing season the corings were taken again and the biomass of living and dead fine roots were measured. The stand investigated was a heavily damaged stand of Norway spruce (87 years old, more than 50% needle losses) growing on acid, loess-derived soil, underlain by sandstone.

The stand had an overall fine root biomass of about 2.760 kg.ha^{-1} of which 650 kg were represented by the dead fraction. The results of the in-growth studies are presented in Fig. 6.

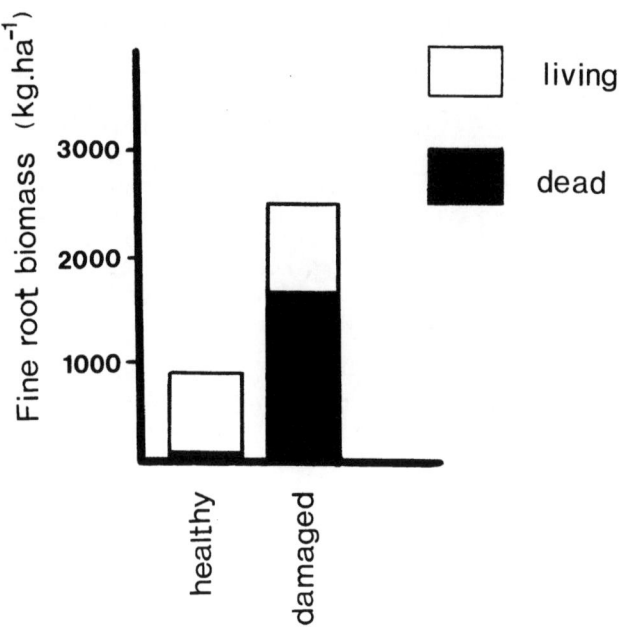

SOIL DATA

Depth (cm)	pH (CaCl$_2$)	
	103 (damaged)	65 (healthy)
0 - 5	2.9	3.4
5 - 15	3.0	3.7
15 - 30	3.3	4.0
> 30	3.4	4.2

Depth (cm)	% H$^+$ of CEC	
	103 (damaged)	65 (healthy)
0 - 5	42	4
5 - 15	37	3
15 - 30	25	0
> 30	17	0

Fig. 5. Fine root biomass of two young (12-year old) Norway spruce forests in northern Germany; one site, 103, is considered to be damaged, while the second site, 65, is considered to be healthy.

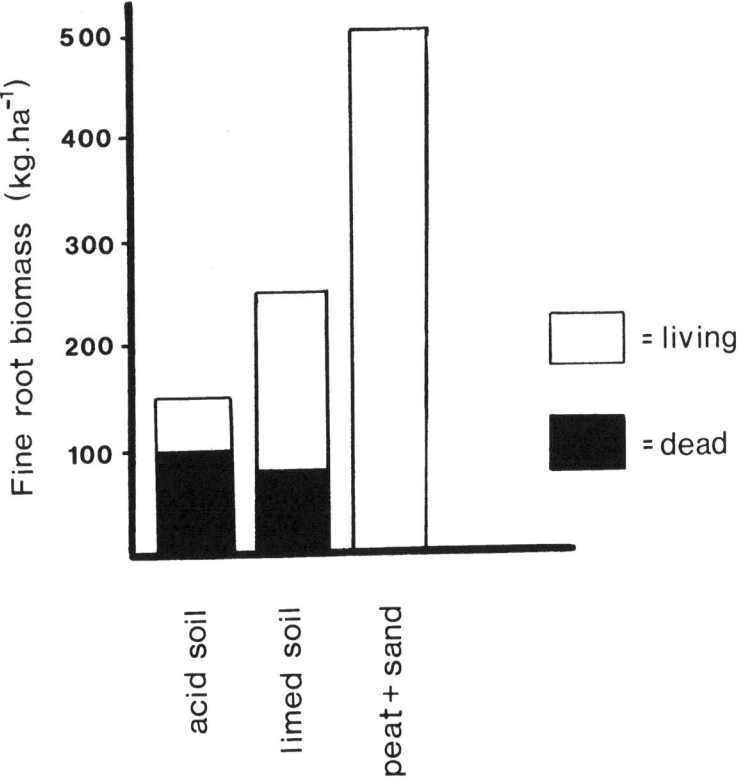

Fig. 6. Fine root biomass of 87-year old heavily-damaged Norway spruce stand in which 3 cores of soil were replaced with (1) acid soil, (2) limed soil, and (3) peat + sand.

Only very few roots grow into the acid soil corings and about 2/3 were found to be dead. The limed treatment increased the total amount of ingrowing fine roots by nearly 100% compared with the acid treatment, with only one-third of the total biomass being dead. The highest fine root biomass was observed in the peat-sand treatment, which reached 505 kg.ha^{-1} with 100% living fine roots. Again these data clearly show the influence of the chemical soil state causing root disturbances and emphasize the importance of the chemical and physical soil conditions for forest decline.

Root dieback not only occurs in damaged mature stands, but is also often observed in tree seedlings (Gehrmann 1984). The deficiency of assimilates as a primary cause of the necrotic root dieback of tree seedlings is improbable, especially since such root damage disappears after liming, and the root damage is already obvious before the cotyledons appear (Gehrmann 1984). In such as case, soil-borne toxicity is probably primarily repsonsible for the root damage.

If the hypothesis of root damage induced by deficiency of assimilates is valid, one would expect a reduction of the fine root biomass before the reduction of the annual wood increment. However, investigations (Ulrich et al. 1984b, in connection with Athari 1983) have shown that this is not the case. Irrespective of the reduction of the wood increment, the trees maintain a fine root biomass of around 3000 kg.ha^{-1} (the fraction of "living" roots decreases with increasing needle loss). Only trees with needle losses above 50% exhibited a reduced fine root biomass. This is not unexpected. It shows that the allocation of photosynthates to a growing (regenerating) root system is preferred. This conclusion is in agreement with earlier observations that under adverse conditions the wood increment is reduced in favour of root growth (Fiedler 1963; Keyes and Grier 1981; Krauss and Wobst 1935; Meyer 1961). The reduction of photosynthesis caused by physiological damage, needle losses, and eventually by a reduced efficiency of the photosynthetic process caused by direct action of air pollution, is a serious stress to the tree. It is, however, not the only one, but accompanied by the stress caused by soil acidity and acidification pulses.

IMPLICATIONS OF THE CHEMICAL SOIL CONDITIONS FOR FOREST DECLINE

Forest decline, which has become evident in many parts of central Europe, is of a complex nature (Schütt 1977). It cannot be attributed to a single stress factor of abiotic or biotic origin.

In order to elucidate the relationship between stress factors and forest declines, Manion (1981) developed a concept which describes forest declines. He also mentions one characteristic phenomenon of forest declines, that is, the disagreement among various researchers on both the "cause" of forest decline, and the relative importance of the specific stress factors involved. Each investigation attempts to find the "cause".

To explain processes of complex origin such as forest decline, Manion distinguishes between three groups of stresses: predisposing, inciting and contributing stresses (see also Manion, this volume).

Predisposing stresses are long-term factors which are relatively static or non-changing, such as climate, soil type, the genetic potential, the age of the tree, and the long-term effects of air pollutants. These factors put a permanent stress on the plant, and weaken it in such a way that other factors may become critical. The air pollutant stress includes, in our opinion, the chronic effects of low concentrations of gaseous pollutants, and of continuing high proton input to forest soils from the atmosphere or from internal production. This leads to nutrient deficiency and to the release of toxic ions in the soil solution.

The second group of stresses is called "incitants". These are of short duration, and may be abiotic or biotic in nature. Examples of incitants are insect defoliators, frost, drought, salt spray, and the short-term effects of high concentrations of air pollutants. They generally produce a drastic injury. Recovery of the plant is impeded by the presence of the predisposing stress. We think that the soil acidification pulses arising within the soil from specific precursor climatic conditions should be added here.

The third group of stresses, called "contributing", may follow the first two. Bark beetles, cancer fungi, root and sap rot fungi are persistent infestations that are often blamed for the condition of the hosts.

The concept given by Manion provides a framework to cover several existing hypotheses on forest dieback in an unconstrained way.

CONCLUSIONS

There is clear evidence from experiments and ecosystem monitoring that soil acidification due to acid deposition and its effects on root growth and nutrient uptake is one of the major factors involved in many of the phenomena collectively called "Waldsterben" in Germany. It is certainly not the only cause (as there is also evidence for other mechanisms), but according to Manion (1981), can be seen as a predisposing stress.

REFERENCES

Athari S (1983) Zuwachsvergleich von Fichten mit unterschiedlich starken Schadsymptomen. Allg Forstzeitsch 653-655

Bache BW (1982) The implications of rock weathering for acid neutralization. In: Ecological effects of acid precipitation. Swedish National Environment Protection Board, Report snv pm 1636

Bauch J, Schröder W (1982) Zellulärer Nachweis einiger Elemente in Feinwurzeln gesunder und erkrankter Tannen und Fitchen. Forstw Cbl 101: 285-294

Blaschke H (1981) Veränderungen der Feinwurzelentwicklung in Weisstannenbeständen. Forstw Cbl 100: 190-195

Butzke H (1981) Versauern unsere Wälder? Erste Ergebnisse der Überprüfung 20 Jahre alter pH-Wert-Messungen in Waldböden Nordrhein-Westfalens. Forst- u. Holzwirt 36: 542-548

Driscoll CT, Likens GE (1982) Hydrogen ion budget of an agrading forested ecosystem. Tellus 34: 283-292

Evers FH (1983a) Ein Versuch zur Aluminium Toxizität bei Fitche. Forst- u. Holzwirt 12: 305-307

Evers FH (1983b) Orientierende Untersuchungen langfristiger Bodenreaktionsänderungen in südwestdeutschen Düngungs-Versuchsflächen. Forst- u. Holzwirt 38: 317-320

Fiedler JH, Hunger W, Zant R (1963) Untersuchungen über die Bodendurchwurzelung der Fichte. Arch f Forstwesen, 1214-1223

Fölster H (1985) Proton consumption rates in holocene and present-day weathering of Acid Forest Soils. In: Drever JI (ed) The chemistry of Weathering, Reidel, Dordrecht, p 197-209

Gehrmann J (1984) Einfluss von Bodenversauerung und Kalkung auf die Entwicklung von Buchenverjüngungen (Fagus sylvatica L.) im Wald Ber d Forsch zentr Waldokosyst/Waldsterben d, Univ Göttingen, Bd 1: 1-213

Grenzius R (1984) Starke Verauerung der Waldboden Berlins. Forstwiss Cbl 103: 131-139

Hüttermann A (1982) Immissionsschäden im Bereich der Wurzeln von Waldbaumen. Mitt d Landesanst f Okologie, Sonderheft: Immissionsbelastungen von Waldokosystemen, Recklinghausen

Hüttermann A, Ulrich B (1984) Solid phase-solution-root interactions in soils subjected to acid deposition. Phil Trans R Soc Lond 305: 353-368

Hutchinson TC (1983) A historical perspective on the role of aluminum in toxicity of acidic soils and lake waters. In: Int Conf Heavy Metals in the Environment, vol 1, Sept 1983, Heidelberg

Johnson NM, Driscoll CT, Eaton JS, Likens GE, McDowell WH (1981) "Acid rain", dissolved aluminum and chemical weathering at the Hubbard Brook Experimental Forest, New Hampshire. Geochim Acta 45: 1421-1437

Junga U (1984) Stierlkultur als Modellsystem zur Untersuchung des Mechanismus der Aluminium-Toxizität bei Fichtenkeimlingen (Picea abies Karst.), Ber d Forsschungszentrums Waldökosyst/Waldsterben d Univ Göttingen, Bd 5: 1-175

Keyes MR, Grier CC (1981) Above- and below-ground net production in 40 year old Douglas fir stands on low and high productivity sites. Can J For Res 11: 559-605

Krauss GA, Wobst W (1935) Über die standörtlichen Ursachen waldbaulicher Schwierigkeiten im vogtländischen Schiefergebirge. Thar Forstl Jahrb 86: 169-246

Lichtenthaler HK, Buschmann C (1984) Beziehungen zwischen Photosynthese und Baumsterben. Allgem Forstzeitschr 15: Heft 1, S 12-16

Manion PD (1981) Tree disease concepts. Prentice Hall Inc., Inglewood Cliffs, NJ, p 399

Manion PD (this volume) Decline as a phenomenon in forests: pathological and ecological considerations.

Matzner E (1984a) Annual rates of deposition of polycyclic aromatic hydrocarbons in different forest ecosystems. Water Air Soil Pollut 21: 425-434

Matzner E (1984b) Deposition und Umsatz chemischer Elemente im Kronenraum von Waldeständen. Ber d Forsch zentr Waldökosyst/Waldsterben d Univ Göttingen, Bd 2: 61-87

Matzner E (1984c) Proton turnover in forest ecosystems. In: Agren GI (ed) State and change of forest ecosystems. Swed Univ Agric Sci, Report nr 13: 303-311

Matzner E (1985) Effect of acid precipitation on soils – The principles demonstrated in two forest ecosystems of North Germany. Symp on Effects of air pollution on forest and water ecosystems, Helsinki, p 47-56

Matzner E, Thoma E (1983) Auswirkungen eines saisonalen Versauerungsschubes im Sommer/Herbst 1982 auf den chemischen Bodenzustand verschiedener Waldökosysteme. Allgem Forstzeitschr, 677-682

Matzner E, Khanna PK, Meiwes KJ, Ulrich B (1983) Effects of fertilization on the fluxes of chemical elements through different forest ecosystems. Plant and Soil 74: 343-358

Matzner E, Khanna PK, Meiwes KJ, Cassens-Sasse E, Bredemeier M, Ulrich B (1984) Ergebnisse der Flüssemessungen in Waldökosystemen. Ber d Forschungszentrum Waldökosysteme/Waldstreben d Univ Göttingen, Bd 2: 29-49

Mayer R (1981) Natürliche und anthropogene Komponenten des Schwermetallhaushalts von Waldökosystemen. Gött Bodenkdl Ber, Bd 70: 1-292

Mazzarino MJ, Heinrichs H, Fölster H (1983) Holocene versus accelerated actual proton consumption in German forest soils. In Ulrich B, Pankrath J (eds) Effects of accumulation of air pollutants in forest ecosystems. Reidel, Dordrecht, Netherlands

Meyer FH (1961) Gefügewandel der Mykorrhiza deutscher Waldbäume als Folge von Abwandlungen des Bodenmilieus. Forschungsergeb z Förderung d forst Erzeugung III, Hiltrup b München, p 24-62

Meyer FH (1984) Mykologische Beobachtungen zum Baumsterben. AFZ, 212-228

Mohr H (1983) Zur Faktorenanalyse des "Baumsterbens" - Bemerkungen eines Pflanzenphysiologen. Allg Forst u. Jagdztg 154: 105-110

Murach D (1984) Die Reaktion von Fichtenfeinwurzeln auf zunehmende Bodenversauerung. Gött Bodenkdl Ber 72: 1-127

Prenzel J (1983) A mechanism for storage and retrieval of acid soils. In: Ulrich B, Pankrath J (eds) Effects of accumulation of air pollutants in forest ecosystems. Reidel, Dordrecht, 157-170

Prinz B, Krause GHM, Stratmann H (1982) Waldschäden in der Bundesrepublik Deutschland. LIS-Berichte 28, Landesanstalt für Immissionschutz, Essen

Rehfuess KE (1981) Über die Wirkungen der sauren Niederschläge in Waldökosystemen. Forstw Cbl 100: 363-381

Rost-Siebert K (1983) Aluminium-Toxizität und -Toleranz an Keimpflanzen von Fitche (Picea abies Karst.) und Buche (Fagus silvatica L.). Allgem Forstzeitschr, 686-689

Schröter H (1983) Krankheitsentwicklung von Tannen und Fichten auf Beobachtungsflächen der FVA in Baden-Württemberg. AFZ 26/27: 648-649

Schütt P (1977) Das Tannensterben. Der Stand unseres Wissens über eine aktuelle und gefährliche Komplex-Krankheit der Weisstanne (Abies alba Mill.). Forstwiss Cbl 96: 177-186

Schütt P, Blaschke H, Hoque E, Koch W, Lang KJ, Schuck HJ (1983) Erste Ergebnisse einer botanischen Inventur des "Fichtensterbens". Forstwiss Cbl 102: 158-166

Schulte-Bisping H, Murach D (1984) Inventur der Biomasse und ausgewählter chemischer Elemente in zwei unterschiedlich stark versauerten Fichtenbeständen im Hils. Ber d Forschungszentrum Waldökosyst/Waldsterben d Univ Göttingen, Bd 2: 207-266

Ulrich B (1983) Soil acidity and its relations to acid deposition. In: Ulrich B, Pankrath J (eds) Effects of accumulation of air pollutants in forest ecosystems. Reidel, Dordrecht, p 33-45

Ulrich B, Matzner E (1983) Abiotische Folgewirkungen der weiträumigen Ausbreitung von Luftverunreinigungen. Forschungsbericht 104 026 15, UBA, Berlin

Ulrich B, Mayer R, Khanna PK (1979) Die Deposition von Luftverunreinigungen und ihre Auswirkungen in Waldökosystem im Solling. Schriften d Forst Fak d Univ Göttingen, Bd 58, Sauerländer-Verlag, Frankfurt/a.Main

Ulrich B, Mayer R, Khanna PK (1980) Chemical changes due to acid precipitation in a loess derived soil in central Europe. Soil Sci 130: 193-199

Ulrich B, Meiwes KJ, König N, Khanna PK (1984a) Untersuchungsverfahren und Kriterien zur Bewertung der Versauerung und ihrer Folgen in Waldböden. Forst u Holzwirt 39: 278-286

Ulrich B, Murach D, Pirouzpanah D (1984b) Beziehungen zwischen Bodenversauerung und Wurzelentwicklung von Fichten mit unterschiedlich starken Schadsymptomen, Forstarchiv 55: 127-134

Wentzel KF (1983) Waldbauliche Erfahrungen zur Erkennung der Immissionswirkungsmechanismen. Allg Forst -u J Ztg 154: 181-185

van Breemen N, Mulder J, Driscoll CT (1983) Acidification and alkalinization of soils. Plant and Soil 75: 283-308

von Zezschwitz E (1982) Akute Bodenversauerung in den Kammlagen des Rothaargebirges. Der Forst- und Holzwirt 37: 275-276

Zöttl HW (1983) Zur Frage der toxischen Wirkung von Aluminium auf Pflanzen. AFZ, 206-208

FOREST DAMAGE IN SWITZERLAND, AUSTRIA, AND ADJACENT PARTS OF
FRANCE AND ITALY IN 1984

J.B. Bucher

Swiss Federal Institute of Forestry Research, CH-8903
Birmensdorf, Switzerland

ABSTRACT

By 1984 that type of forest damage which has no obvious cause, but which is generally attributed to air pollution had reached considerable proportions in Switzerland, Austria, and the adjacent parts of France and Italy. In Switzerland, 34% of all trees were regarded as no longer healthy. In Austria at least one third of the trees investigated were found to be damaged. These figures, however, should not be regarded as universally applicable, since the survey from which they stem focussed on Norway spruce and was not representative of all regions. In the Italian province of South Tyrol, the state of health of the forests, as reflected by only 20% damage in Norway spruce, appeared somewhat better. The forests of the Vosges and the Jura in France, in contrast, seemed to be more severely affected, with only 60% of the trees still healthy. The survey, which employed both single tree evaluation and a grid of sample plots, showed that the phenomenon of forest decline is not limited to Germany alone, but is also affecting neighbouring countries. Although attempts were made in the surrounding countries to apply the German system of evaluation, differences in methods and interpretation necessarily arose, so that the figures for 1984 should be compared with caution. The results so far available do not justify a general condemnation of the idea that air pollution is the cause of the present forest damage. Nevertheless, the possibility that air pollution may well have contributed to this damage is indicated by both the intensity of emission in Switzerland and the results of nationwide surveys of snow cover and the pollutant burden of Norway spruce needles conducted in Switzerland and Austria in 1983. Demands are currently being made that forest damage inventories take criteria of air quality more seriously into account and the influence of emissions be considered from the epidemiological viewpoint. In view of the fact that there is no means of making a simple differentiation between healthy and damaged trees, we recommend that damage be evaluated in terms of foliage loss (5% classes) and that degrees of damage be represented in terms of relative sum frequency.

INTRODUCTION

From 1980 onwards, the state of health of silver firs and spruces on permanent plots in Baden-Württemberg, FRG, deteriorated rapidly, but from no apparent cause (Schröter 1983). The widespread occurrence of similar damage was also reported in Bavaria (Schütt 1982). In 1982 a detailed survey was conducted within the German forest service; the results showed beyond doubt that this type of forest damage was not confined to the southern FRG but was affecting the whole country. In the following year, systematic sampling was used to inventory the occurrence of the phenomenon, which in the meantime had been labelled "Waldsterben" or forest decline, and was generally attributed to air pollution. Comparable symptoms had begun to appear in the neighbouring countries, and forest damage inventories were either implemented or at least deliberated by the relevant bodies. The inventory results for Switzerland, Austria, and the adjacent parts of France and Italy, however, only became comparable in 1984, when each of these countries conducted nationwide inventories on the basis of sample plots, together with single tree evaluation. Consequently, the information presented here is based on inventories of damage of unknown origin. Damage due to known causes, such as windthrow or insect epidemics, is only included to the extent to which it occurred on the actual sampling plots.

FOREST DAMAGE INVENTORIES OF 1984

Switzerland

The 1984 inventory of forest damage in Switzerland was conducted by Schmid-Haas and his co-workers (Anonymous 1984a; Schmid-Haas 1985; Schmid-Haas et al., in press). Criteria of damage assessment, inventory methods, and evaluation were all based on the system applied in the FRG (Anonymous 1984b; Schopfer and Hradetzky 1984), so that in spite of some slight differences in the selection of the sampling plots, the findings in the two countries can be validly compared. The Swiss inventory was based on a systematic random sample grid of 4x4 km, covering the whole country. For reasons of time and organization, the inventory had to be limited to public forests with road access. In view of the location of the possible sampling areas, the tract sampling procedure seemed most appropriate. Each tract of land was divided into 8 areas of 500 m^2, and within each area, every tree of over 20 cm dbh was scrutinized for signs of damage. The primary criterion of single tree evaluation was foliage loss in the crown in relation to the crown bulk of the surrounding healthy trees (100%). Relative foliage loss was basically estimated to 5% accuracy, although for the purposes of analysis it was grouped into 5 discontinuous intervals. In accordance with the system applied in Germany, trees with up to 10% leaf loss were classified as healthy, those with 11-25% leaf loss as slightly affected, and those with 26-60% leaf loss as moderately damaged; a leaf loss of over 60% signified severe damage, and the fifth class comprised completely dead trees. In addition, site conditions and any damage obviously attributable to known biotic or abiotic factors were registered for each tree. The inventory was conducted between the beginning of July and the end of September.

As Table 1 shows, 34% of all trees of over 20 cm dbh in the public, accessible forests of Switzerland must be classified as damaged: 26% are slightly affected, 7% are moderately damaged, and 1% are severely damaged or actually dead. More than 50% of the damaged trees are pines, but over a third of all other conifers are affected as well. Broadleaf trees seem to be somewhat less damaged than conifers, although oak accounts for almost a third of all damaged trees and thus closely rivals spruce. Beech, the most common broadleaf tree in Switzerland, exhibits slightly less damage than oak, even though the percentage of moderately affected individuals is much greater than for oak. The species least damaged so far seem to be maple and ash.

Table 1. Forest damage in Switzerland in 1984. Percentage of trees (weighted according to stem ground projection) per class of damage and mean leaf loss.

Tree species	No. of samples	Damage class and % leaf loss					Mean leaf loss %
		0 0-10%	1 11-25%	2 26-60%	3 >61%	4 dead	
Spruce	11968	65	28	6	0	1	11.2
Fir	3805	61	28	9	1	1	12.7
Pine	1019	47	33	17	2	1	18.2
Larch	958	64	27	7	1	1	11.9
Beech	6095	75	22	3	0	0	8.2
Oak	564	70	29	1	0	0	9.1
Maple	737	83	14	2	1	0	6.1
Ash	663	83	14	3	0	0	7.7
Total all species	26927	65.8	26.3	6.8	0.5	0.7	11.2

Damage occurs over the whole of Switzerland, but there are areas of concentration in the inner alpine valleys and at high altitudes, and damage tends to increase from west to east. Both this trend and the fact that there may be great differences between adjacent tracts can clearly be seen in the case of spruce, the most common tree in Switzerland (Fig. 1).

Schmid-Haas based his analysis of the inventory data solely on the arithmetic mean of foliage loss, since this parameter is more appropriate than the relative frequency of arbitrarily defined class intervals. The degree of damage seems to be correlated with certain site characteristics. In the case of spruce, it is related to relief, being considerably greater on hill crowns, and sometimes also in hollows, than on level sites. Fir exhibits the same trend, although the differences are not statistically significant. Neither beech nor other species display any significant trends. Spruce also shows a marked increase in damage with increasing slope inclination and height above sea level. Further, the degree of injury is related to site quality: the less favourable the site, the greater the damage. Soil, as a site characteristic, seems to have no effect otherwise, or at

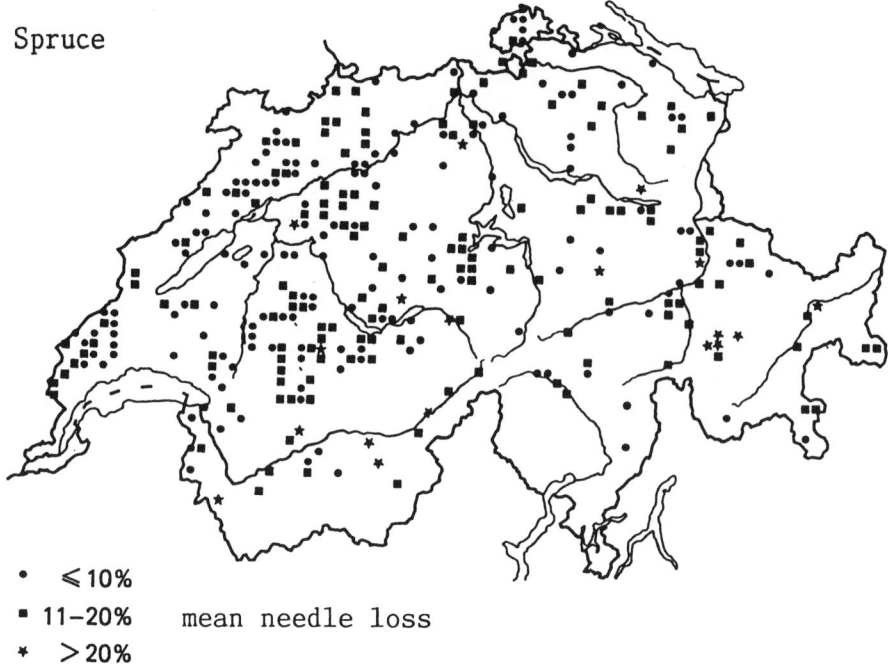

Fig. 1. Mean needle loss in spruces in areas (tracts) in which at least 10 trees were sampled (Schmid-Haas 1985).

least the acidity of the underlying bedrock shows no correlation. On both limestone and acidic rock, injury is approximately the same in all species. Site exposure displays no clear relationship to damage. Correlations between stand characteristics and degree of injury were found, but these were only slight. In pure, single-storied stands, spruce or beech display less damage than in multi-storied ones, and spruces in relatively dense stands are somewhat less affected than those in strongly thinned, open stands. Furthermore, injury is slightly greater in older stands of spruce and silver fir than in younger ones. Stand composition affects only spruce: this species generally exhibits less damage in mixed deciduous-conifer forest than in pure conifer and mixed conifer stands. The degree of injury, however, varies markedly with the position of the individual within the stand. In spruce and beech stands, the greatest damage occurs in the trees at the edge of the stand, and in both species the dominant individuals are more severely affected than the co-dominant ones. In silver fir and pine stands, on the other hand, co-dominant trees display greater injury than those at the margin.

Austria

In Austria, a forest damage inventory was conducted in 1984 as part of an emergency programme. The findings presented here are based on as yet unpublished data of the Federal Institute of Forest Research, Vienna, and were collected by Pollanschütz

(1984, 1985a). The inventory was based on a grid of 4x4 km, expanded or contracted in some locations. In 1984 more than 60% of the approximately 2000 permanent sample plots were surveyed. As the investigation did not include all the federal states, and in some states, comprised only a limited number of district forest inspections, the figures cannot be regarded as representative and should not be extrapolated as indicating the total area of damaged forest (Pollanschütz, pers. comm.). On each of the open sample plots, 30 dominant and predominant trees aged mainly between 60 and 100 years were evaluated. The inventory was generally limited to the major species, spruce, pine, and fir, although in some cases 20 individuals belonging to secondary species were also considered. In contrast to the Swiss inventory, neither implementation nor single tree evaluation was based on the system applied in the FRG (Anonymous 1984b). Foliage loss in the crown, which is usually the most important criterion of evaluation, was recorded not in terms of percentage foliage loss, but on the basis of illustrations representing different classes of damage, namely normal crown condition, slight, medium, and severe foliage loss, and dead. The condition of the tip region, the pattern of needle loss, site characteristics, and damage of known origin were also recorded. The inventory was conducted between mid-July and mid-September.

The initial results of the forest damage inventory of the 1984 emergency programme are shown in Table 2. It should be noted that only six forest district inspections were carried out in Niederösterreich, and only one in Steiermark, and that no figures are given for Vienna or the Vorarlberg. Two-thirds of the trees assessed are classified as healthy, one-third as damaged. Pine seems to be more severely affected than spruce (Strohschneider 1985). The degree of forest damage in the various federal states is best compared on the basis of mean degree of foliage loss in the crown or mean degree of damage per stand or sampling plot (index "5"). According to Pollanschütz (pers. comm.), index 1.2 indicates a normal condition; index 1.2 to 1.5, a transitional state; index 1.5 to 2, slight damage; 2 to 2.8, medium injury; and 2.8 or above, severe injury.

Italy (South Tyrol)

In Italy, a forest damage inventory exists only for the autonomous province of the South Tyrol. This inventory (Minerbi 1985) is based on a sample grid of 8x8 km covering the whole province. On each sample plot at least 30 predominant or solitary trees were assessed. Spruce was selected as the reference species, since it occurs over the whole province and displays symptoms of damage most clearly, but pine and silver fir were also considered as far as possible. Assessment of damage was modelled on the methods applied in Austria (see Pollanschütz 1984), but the German classification of degree of damage was used (Anonymous 1984b). Trees with up to 10% foliage loss were evaluated as healthy, those with 11-25% loss as slightly affected, and those with 26-60% foliage loss as moderately damaged; a loss of 61-99% signified severe injury, and the final class comprised completely dead trees. As in other inventories, site and stand character-

Table 2. Forest damage in Austria in 1984. Percentage of trees per class of damage and degree of foliage loss in the crown or damage per stand or sampling plot (index "5") in the various federal states (Pollanschütz 1985a).

State	Tree species	No. of samples	Class of damage 1	2	3	4	Mean degree of damage (index "5")
Burgenland	Spruce	210[a]	32[a]	41[a]	24[a]	3[a]	1.97[a]
	Pine	1785	37	43	-	20	2.03
Kärnten	Spruce	7619	67	29	4	0.2[a]	1.37
	Pine	1080	47	48	-	5	1.63
Nieder-österreich[b]	Spruce	3010	43	51	6	0.4[a]	1.66
	Pine	2179	52	46	-	2	1.53
Ober-österreich	Spruce	7685	67	31	2	0.1[a]	1.35
Salzburg	Spruce	6985	72	26	2	-	1.30
Steiermark[c]	Spruce	509	90	9	1	-	1.11
	Pine	840	91	9	-	0.5[a]	1.08
Tirol	Spruce	6725	67	24	8	1	1.42
	Fir	772	44	33	17	6	1.42
	Pine	618	77	21	-	2	1.25

[a] not statistically verified
[b] only 6 "district" inspections
[c] only 1 "district" inspection

((According to Pollanschütz (10) defoliation is defined as "degree of crown thinning" in classes, viz. none, slight moderate, severe and complete (dead). These "classes of damage" are approximately comparable with classes of the percentage of needles/leaves loss as follows: 1=0-20%; 3=41-60%; and 4= 60%, according to the usual norms.))

istics as well as injury of known origin were recorded. The inventory was conducted during the summer.

Of the trees assessed, 82% of the spruces but only 75% of the pines were classified as healthy, while for silver fir the figure is as low as 61%. Table 3 includes the data for damage of known origin. Needle loss in spruce can to some extent be attributed to the nationwide epidemic of spruce needle rust (Chrysomyxa sp.) which occurred in the preceding year, and to a lesser extent to needle cast fungi (Lophodermium piceae, Rhizosphaera kalkhoffii). Injury due to honey fungus (Armillaria mellea) and yellowing of new shoots due to the spruce needle miner (Epinotia pygmaeana) were also identified. In silver fir, macroscopic inspection revealed only a few fungal diseases (e.g. Lophodermium macrosporum) or insect invasions (e.g. Mindarus abietinus). In both spruce and silver fir, much mechanical damage due to rock falls or logging activities was found. At around 2%, however, this degree of damage is regarded as normal, and was also recorded in the Swiss inventory (see Schmid-Haas 1985). In pine, in contrast, the extent of damage due to known causes was quite high, at 19%. These causes included needle blight fungus (<u>Lophodermium pinastri</u>), pine shoot beetle (<u>Blastophagus</u> sp.) and mistletoe (<u>Viscum album</u>). Damage of unknown origin was 16% in spruce, 35% in silver fir, and 6% in pine. In terms of spatial distribution, there were tendencies towards more severe damage in the east and greater intensity north of Bolzano.

Table 3. Forest damage in Italy (South Tyrol) in 1984. Percentage of trees per class of damage and mean damage[a] (Minerbi 1985).

Tree species	No. of samples	Damage class and % leaf loss					Mean damage %	Recogn. damage %
		0 0-10%	1 11-25%	2 26-60%	3 >60%	4 dead		
Spruce	3638	81.6	13.8	2.2	0.4	0	5.5	1.9
Fir	153	62.1	27.5	7.2	0.6	0	11.2	2.6
Pine	143	74.8	4.9	0	1.4	0	2.8	18.9

[a]Mean damage $P = \dfrac{(nv)100}{KN}$

n = no. of trees per class
v = class no.
K = class no. -1
N = total no. of trees

France (Alsace, Lorraine, Franche-Comté)

In France the forest damage inventory has so far covered the Vosges in Alsace-Lorraine and the Jura region of Franche-Comté, and includes the public forests managed by the State Forest Office. It is based on a sampling grid of 1x16 km (1 km east-west, 16 km north-south). On each sample plot, 24 trees of over 7.5 cm dbh were assessed in terms of crown condition. The system of evaluation was the same as that used in the FRG (Anonymous 1984b), but the classification was different. Trees with a foliage loss of up to 9% were evaluated as healthy, those with a loss of 10-19% as slightly affected, and those with a loss of 20-59% as severely damaged; a loss of 60-99% signified a moribund state, and one of 100% indicated final death. Site and stand characteristics were also recorded. The inventory was conducted in spring and autumn (Anonymous 1984c). The findings presented here are based on the data for autumn.

In north-eastern France the injury is considerably greater in conifers than in broadleaf trees. In the Vosges and the Jura the difference between the two tree groups is smaller. However, damage is concentrated at certain locations within these regions. Injury is greater on the western slopes of the northern

Vosges than on the eastern slopes, and in the Jura it is most severe around Pontarlier. Table 4 summarizes the condition of the trees evaluated. In the Vosges about 60% of all trees are still healthy, 25% are slightly affected, 13% are severely damaged, about 2% are moribund, and 0.5% are dead. The figures for the Jura are similar, with 65% of the trees still healthy. In both regions, however, there are marked differences between the various tree species.

Table 4. Forest damage in France (Alsace, Lorraine, Franche-Comté) in 1984. Percentage of trees per class of damage (Anonymous 1984c).

Region	Tree species	No. of samples	Damage class and % leaf loss					3+4+5
			1 0-9%	2 10-19%	3 20-59%	4 60-99%	5 100%	
Vosges (Alsace, Lor-raine)	Conifers							
	Spruce	905	51.1	33.1	14.2	1.4	0.2	15.8
	Fir	1914	40.7	33.0	23.1	2.6	0.6	26.3
	Pine	519	46.4	36.4	14.1	2.7	0.4	17.2
	others	107	80.4	14.0	0.9	2.8	1.9	5.6
	Broadleaf							
	Beech	967	87.0	9.8	2.2	0.5	0.5	3.2
	Oak	529	87.5	8.7	3.6	-	0.2	3.8
	others	209	84.5	9.8	3.0	1.7	1.0	5.7
	Total all species	5237	59.6	24.9	13.2	1.8	0.5	15.5
Jura (Franche-Comté)	Conifers							
	Spruce	115	62.6	22.6	13.9	0.9	-	14.8
	Fir	293	41.3	37.2	19.8	1.7	-	21.5
	Pine	6	50.0	33.3	16.7	-	-	(16.7)
	Others	3	33.3	66.7	-	-	-	(0.0)
	Broadleaf							
	Beech	188	82.4	15.5	2.1	-	-	2.1
	Oak	37	78.4	21.6	-	-	-	(0.0)
	others	152	86.9	11.8	1.3	-	-	1.3
	Total all species	794	64.6	24.4	10.2	0.8	-	11.0

AIR POLLUTION AS A POSSIBLE CAUSE OF FOREST DECLINE

In the FRG, even before the first survey of forest damage had been made, a committee of experts reached the conclusion that the nationwide forest damage would not have occurred if there had been no air pollution (Anonymous 1983a). Smith (1984) also regards the widespread air pollution of today as the greatest anthropogenic threat to the forests of the temperate zones.

There is no doubt that the degree of air pollution in Europe has increased, or that there are scarcely any regions with "clean" air left. These facts are generally blamed on the transboundary distribution of air pollutants over large areas. However, measurements, calculations, and, to some extent, the pattern of forest damage, have convinced experts in Switzerland and Austria that "home-made" pollution should not be underestimated and may even be the main cause of the situation (Anonymous 1984d; Donaubauer 1983).

The main features of the air pollution situation in Switzerland may be described as follows: population density is high and most of the country is inhabited; there is intensive industrialization in the secondary sector, which produces relatively little in the way of waste gases; and the energy supply stems almost exclusively from hydro-electric and atomic power stations. As early as 1981 the Federal Office for the Environment evaluated the degree of air pollution in both town and country as questionable (Anonymous 1982). The general situation is shown in Table 5 (Anonymous 1984d).

Table 5. Typical figures for the general intensity of air pollution in representative regions of Switzerland in the 1980s (Anonymous 1984d).

Air pollution Type	Species	Unit	High altitude	Lowland country	Con- urbations	Town centres
Gaseous emission (yearly means)	SO_2	$\mu g.m^{-3}$	2-3	8-12	30-40	50-70
	NO_2	$\mu g.m^{-3}$	2-3	20-30	30-50	60-140
	HC ($-CH_4$)	$\mu gC.m^{-3}$	<5	5-10	50-200	100-500
	O_3	$\mu g.m^{-3}$	60-90	40-70	30-50	20-30
Dry deposition	S	$g.m^{-2}.y^{-1}$	0.5	2	6	10
	N	$g.m^{-2}.y^{-1}$	0.1	1	2	3
Wet deposition	S	$g.m^{-2}.y^{-1}$	1	2	3	3
	N	$g.m^{-2}.y^{-1}$	0.2	0.5	1	1
	H^+	$mg.m^{-2}.y^{-1}$	5	15	70	79
	Na^+	$mg.m^{-2}.y^{-1}$	437	251	98	84
	K^+	$mg.m^{-2}.y^{-1}$	325	178	61	32
	Ca^{2+}	$mg.m^{-2}.y^{-1}$	873	465	466	280
	Mg^{2+}	$mg.m^{-2}.y^{-1}$	87	111	61	42
	NH_4^+	$mg.m^{-2}.y^{-1}$	216	562	527	475
	$NO_3^-(N)$	$mg.m^{-2}.y^{-1}$	150	409	490	560
	$SO_4^{2-}(S)$	$mg.m^{-2}.y^{-1}$	438	818	1188	1074
	Cl^-	$mg.m^{-2}.y^{-1}$	716	808	907	686
	(Acidity)	pH	5.4	4.8	4.26	4.13
	(Volume of rain)	$mm.y^{-1}$	1188	912	1225	1056

Because of the particular conditions in Switzerland regarding emissions, car exhaust fumes, and thus nitrogen oxides, seem to be of special importance. Over the last 25 years, the pH values of precipitation have remained relatively stable, ranging from 4.2 to 4.7, but the concentration of nitrates has steadily increased (Tripet and Wiederkehr 1983). This is also very high in mist and fog, as the recent analysis by Stumm et al. (1985) shows. In 1984, pH values as low as 2.3 were recorded in high fog, and ion concentrations ten to a hundred times greater than those occurring in rain were found. The influence of car exhaust fumes is also reflected in the increasing ozone load, which may reach levels harmful for the forests several times during summer (Bucher 1984). There is no doubt that the peak ozone concentrations of over 250 $\mu g.m^{-3}$ were mainly of anthropogenic origin, as we were recently able to demonstrate through our PAN-measurements (Landolt et al. 1985). In the nursery at our Institute, which is a typical Swiss Mittelland station, the proposed ozone threshold of 80 $\mu g.m^{-3}$ (a 95%-figure for any given month) was continually exceeded throughout the entire vegetation period of 1984 (Bleuler and Bucher 1985).

Measurements of air quality are a valuable aid in assessing the threat which emissions pose to the forest. At the moment, however, there is not enough data to supply definite proof that air pollution is the cause of forest decline. Inventories of forest damage based on purely forestral criteria provide only limited evidence as to the cause of damage. For instance, the study by Schöpfer and Hradetsky (1984) did not include comprehensive data on emissions, and the assumption of air pollution as the cause of forest decline was very much a matter of definition (e.g., exposure to wind = emission damage, damage between 600 and 900 m a.s.l. = ozone damage). Similar difficulties were experienced by Schmid-Haas in analysing the data from the forest damage inventory in Switzerland. He, too, could only find indirect evidence that air pollution is the probable cause of forest decline, by establishing that the damage is not solely due to climate or faulty management (see Schmid-Haas 1985). Last (1983; Last et al. 1984) and others (Bucher 1985) have repeatedly pointed out that air pollutants should be more closely examined as a possible cause of forest damage, and that large-scale epidemiological studies should be conducted. In Switzerland, we carried out such a study in 1983. We distributed questionnaires regarding forest damage of unknown origin among the Forest Service, and also analysed samples of needles from spruces on 840 sites distributed over the whole country for indicator elements reflecting pollution and nutrient supply (Bucher et al. 1984; Landolt et al. 1984; Kaufmann et al. 1984). The limits and possibilities of the methods have been discussed elsewhere (Bucher 1985). Although questionnaires are certainly no substitute for inventories of sample plots, it was interesting to find that the situation in 1983, as recorded in two independent inventories (Mahrer et al. 1984; Schlaepfer et al. 1985) was quite accurately reflected in the returned questionnaires. Comparison of the results with maps of forest damage in 1983, and maps of pollutant distribution based on the needle analysis, showed that the sites with increased needle content of sulphur,

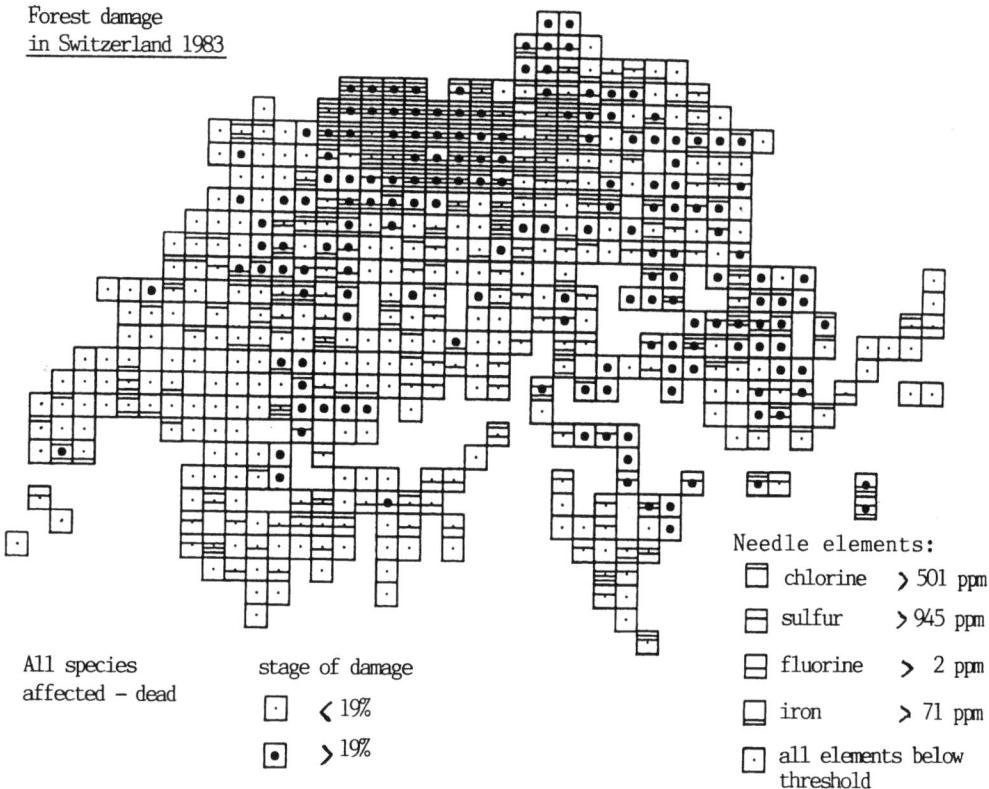

Fig. 2. Assessed forest damage and needle element contents of chlorine, sulphur, fluorine and iron (pollution indicators) in Norway spruce. The figure shows the transformed data per unit area (8x8 km) with emphasis on the highest 33%-quantile of forest damage and elements listed.

chlorine, fluorine and iron were located in the areas with the greatest forest damage (Fig. 2).

A nationwide study of pollutant stress in the forests was also conducted in Austria in 1983. One investigation of the pollutant content of snow cover showed that areas even outside those zones previously known to be polluted were affected, and that considerable concentrations of pollutants occurred in these areas (Stefan 1984a; Anonymous 1983b). The bio-indicator network and the analysis of spruce needles reflected the known areas of pollution and forest damage (see Anonymous 1984e). However, there was no evidence of a severe threat (Stefan 1984b). Since, in Austria, the bioindicator network is permanent and coincides with the grid for forest damage inventories, it will be possible to conduct epidemiological studies on pollution as a cause of forest decline.

DISCUSSION AND CONCLUSIONS

The forest damage inventories in Switzerland, Austria, France and Italy were basically intended to follow the system of damage assessment applied in the FRG. However, questions of organization, as well as differences in interpretation, necessarily resulted in variations in methods of survey and evaluation. The greatest differences in interpretation concerned the distinction between healthy and damaged trees. My own opinion is that the threshold of damage should be set higher than 10% foliage loss. Three years ago I (Bucher 1982) used illustrations dating from 1912 (Balsiger 1912) to demonstrate that individual trees then regarded as local ornaments or as perfect specimens would now be classified as slightly or moderately damaged (Müller, pers. comm.; Fig. 3). Similar opinions are held in Austria (Pollanschütz 1984) and also in Scandinavia (Bengtsson and Lindroth 1984; Abrahamsen, pers. comm.), where the threshold of damage is set at a foliage loss of 20%; and in France it is generally only trees with severe injury that are regarded as actually damaged (Bonneau, pers. comm.; see also Anonymous 1984c). The factor of personal interpretation could be precluded if foliage loss were evaluated in constant, fairly narrow class intervals of 5%, and damage were consistently represented in terms of relative sum frequencies (see Fig. 4). One problem arising all over Europe, but especially in the mountainous countries (Minerbi 1985), is that of the reference tree; in other words, what condition of crown should be taken as healthy (100% foliage). It should also be borne in mind that, when inventories aimed at determining the cause of damage are conducted at any particular time of year, it may be possible to measure the degree of leaf loss, but it is not always possible to identify the biotic factor which may originally have caused the damage (see Roloff 1985). Annual inventories of forest condition, based on a nationwide sampling grid should therefore be complemented by making several surveys per year on permanent sample plots. Finally, it should be emphasized once more that forest damage inventories only provide information on suspected causes insofar as the relevant parameters are assessed (Bucher 1985). In spite of these critical remarks, I still believe that forest damage inventories are a valuable aid in identifying problem areas, and that we have good reason to regard the present pollution situation as a threat to our forests.

Acknowledgements

I would like to express my thanks to Dr. P. Schmid-Haas and F. Mahrer of Birmensdorf, Dr. Pollanschütz and Dr. Stefan of Vienna, and Dr. Bonneau of Nancy for suggestions and data. I am also indebted to Mrs. M.J. Sieber for the English translation of the manuscript.

Fig. 3. "Spruce with ideal crown form" (Balsiger 1912). This tree, photographed in 1912, would today be classified as having 35-40% needle loss (Müller, pers. comm.).

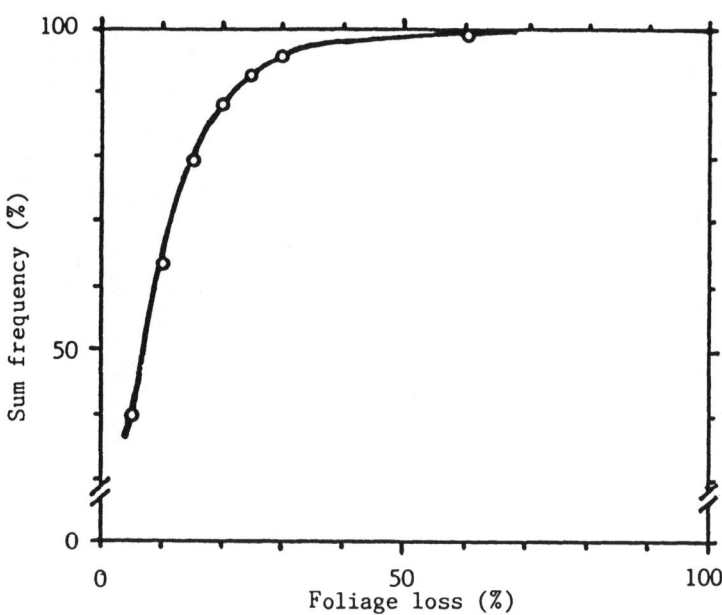

Fig. 4. Forest damage in Switzerland 1984. Relative sum frequency of damage classes (unpublished data; Schmid-Haas 1985)

REFERENCES

Abrahamsen G (1985) Personal communication
Anonymous (1982) Luftbelastung 1981. Messresultate des Nationalen Beobachtungsnetzes für Luftfremdstoffe (NABEL). Schriftenreihe Umweltschutz Nr. 6, Bundesamt für Umweltschutz, Bern, p 25
Anonymous (1983a) Der Rat von Sachverständigen für Umweltfragen, Waldschäden und Luftverunreinigungen. Sondergutachten März 1983. Kohlhammer, Stuttgart/Mainz (1983) p 172
Anonymous (1983b) Die erste bundesweite Untersuchung über die Schadstoffbelastung von Schnee. Allg Forstztg 94: 343
Anonymous (1984a) Ergebnisse der Sanasilva-Waldschadeninventur 1984. Bundesamt für Forstwesen und Eidg. Anstalt für das forstliche Versuchswesen, Bern und Birmensdorf, p 27. Order at: EAFV, Bibliothek, CH-8903 Birmensdorf
Anonymous (1984b) Zur Diagnose und Klassifizierung der neuartigen Waldschäden 1984, Farbbild-Heft der Allg Forstzeitschrift, BLV-Verlagsgesellschaft München (1984), p 24
Anonymous (1984c) Observation de l'état sanitaire des forêts domaniales et communales gérées par l'Office National des Forêts. Resultats 1984. Etude O.N.F. In: Pluies acides. Le dépérissement des forêts attribué à la pollution atmosphérique. Ministère de l'Agriculture, Secrétariat d'État à l'Agriculture et à la Forêt. 78, Rue de Varenne - 75700 Paris
Anonymous (1984d) Waldsterben und Luftverschmutzung. Eidg. Departement des Innern, Bern, p 126

Anonymous (1984e) Waldschäden durch Luftverunreinigungen. Allg Forstztg 95: 229

Balsiger R (1912) Die Plenterwirtschaft als Lichtwuchsbetrieb. Schweiz Z Forstwes 63: 205-214

Bengtsson G, Lindroth S (1984) De första resultaten fran skogsskadeinventeringen 1984. Sveriges Lantbruksuniversitet, Inst f skogstaxering 1984-12-13P 13 (unpublished reproduction)

Bleuler P, Bucher JB (1985) Luftbelastung 1984 im Raume Birmensdorf (ZH). Schweiz Z Forstwes 136: 665-670

Bonneau M (1985) Personal communication

Bucher JB (1982) Waldschäden durch Immissionen? Physiologische Veränderungen und ökotoxikologische Wirkmechanismen. Probleme der Differentialdiagnose. Gottlieb Duttweiler-Institut, CH-8803 Rüschlikon: 91-109

Bucher JB (1984) Bemerkungen zum Waldsterben und Umweltschutz in der Schweiz. Forstw Cbl 103: 16-27

Bucher JB (1985) Luftverunreinigungen und gesunder Wald. Impacts de l'Homme sur la Forêt. Symposium IUFRO, Strasbourg/F 17-22 Sept 1984. INRA, Paris 1985, Les Colloques de l'INRA n° 30

Bucher JB, Kaufmann E, Landolt W (1984) Waldschäden in der Schweiz 1983 (1. Teil). Interpretation der Sanasilva-Umfrage und der Fichtennadelanalysen aus der Sicht des forstlichen Immissionsschutzes. Schweiz Z Forstwes 135: 271-287

Donaubauer E (1983) Immissionsschäden an Oesterreichs Wald. Allg Forstztg 5/83: 115-116

Kaufmann E, Bucher JB, Landolt W, Jud B, Hoffmann CH (1984) Waldschäden in der Schweiz 1983 (3. Teil). Auswertungen zu den Hauptbaumarten Fichte, Tanne und Buche. Schweiz Z Forstwes 135: 817-831

Landolt W, Bucher JB, Kaufmann E (1984) Waldschäden in der Schweiz 1983 (2. Teil). Interpretation der Sanasilva-Umfrage und der Fichtennadelanalysen aus der Sicht der forstlichen Ernährungslehre. Schweiz Z Forstwes 135: 637-653

Landolt W, Joos F, Mächler H (1985) Erste Messungen des PAN-Gehaltes der Luft im Raume Birmensdorf (ZH). Schweiz Z Forstwes 136: 421-426

Last FT (1983) Direct effects of air pollutants, singly and in mixtures, on plant and plant assemblages. In: Ott H, Stangl H (eds) Saure Niederschläge, eine Herausforderung für Europa. Proc Symp, Karlsruhe 19-22 Sept 1983, p 105-126

Last FT, Fowler D, Freer-Smith PH (1984) Die Postulate von Koch und die Luftverschmutzung. Forstw Cbl 103: 28-48

Mahrer F, Brassel P, Stierlin HR (1984) Erste Eergebnisse zum Waldsterben aus dem Schweizerischen Landesforstinventar (LFI) Schweiz Z Forstwes 135: 289-306

Minerbi ST (1985) Ergebnisse der Waldschadenerhebung 1984 in Südtirol. Informationsschrift des Landesforstinspektorates Bozen. LFI 1/85: p 26

Müller E (1985) Personal communication

Pollanschütz J (1984) Aufnahmeschlüssel für die österreichische Waldzustandsinventur nach bundeseinheitlichen Richtlinien (Provisorische Fassung für "Sofortprogramm WZI-1984"). Forstliche Bundesversuchsanstalt Wien (unpublished reproduction)

Pollanschütz J (1985a) Waldzustandsinventur. Uebersicht - Ziele - Grundsätze sowie Ergebnisse der Uebersichtsauswertung 1984. Forstliche Bundesversuchsanstalt Wien (unpublished reproduction)

Pollanschütz J (1985b) Personal communication

Roloff A (1985) Untersuchungen zum vorzeitigen Laubfall und zur Diagnose von Trockenschäden in Buchenbeständen. Allg Forstz 40: 157-160

Schlaepfer R, Mandallaz D, Commarmot B, Günter R, Schmid B (1985) Der Gesundheitszustand des Waldes im Revier Schaffhausen. Zur Methodik und Problematik der Erhebung auf Betriebsebene. Schweiz Z Forstwes 136: 1-18

Schmid-Haas P (1985) Der Gesundheitszustand des Schweizer Waldes 1984. Schweiz Z Forstwes 136: 251-273

Schmid-Haas P, Hämmerli F, Walther G (in press) Die Schweizerische Waldschadeninventur 1984. In: Schriftenreihe der VDI-Kommission Reinhaltung der Luft

Schöpfer W, Hradetzky J (1984) Der Indizienbeweis: Lufverschmutzung massgebliche Ursache der Walderkrankung. Forstw Cbl 103: 231-248

Schröter H (1983) Entwicklung des Gesundheitszustandes von Tannen und Fichten auf Beobachtungsflächen der FVA in Baden-Württemberg. Allg Forst- u J-ztg 154: 123-131

Schütt P (1982) Das Krankheitsbild - verschiedene Baumarten, gleiche Symptome. Bild d Wiss 12/82: 86-101

Smith WH (1984) Auswirkungen von regionalen Luftschadstoffen auf die Wälder in den USA. Forstw Cbl 103: 48-61

Stefan K (1984a) Schadstoffbelastung von Schnee. Ergebnisse der bundesweiten Stichprobenuntersuchungen 1983. Forstliche Bundesversuchsanstalt Wien, p 21 (unpublished reproduction)

Stefan K (1984b) Oesterreichisches Bioindikatornetz 1983. 1. Information für den Dienstgebrauch. Forstliche Bundesversuchsanstalt Wien, p 19 (unpublished reproduction)

Strohschneider M (1985) Regierungsforstdirektorenkonferenz: Erste Ergebnisse der Waldzustandsinventur. Holz-Kurier 5/85: 7

Stumm W, Sigg L, Zobrist J, Johnson A (1985) Der Nebel als Träger konzentrierter Schadstoffe. Neue Zürcher Zeitung 12: 71. Order at: EAWAG, CH-8600 Dübendorf

Tripet I, Wiederkehr P (1983) Etude du problème des précipitations acides en Suisse. Ecole Polytechnique Fédérale de Lausanne, Institut du Génie de l'Environnement, p 77

AIR POLLUTION AND FOREST DAMAGE IN NORWAY

B. Tveite

Norwegian Forest Research Institute, 1432 Aas-NLH, Norway

ABSTRACT

The regional pattern of air pollution in Norway is discussed. Depositions of sulphur and nitrogen are at maximum in the southwestern part of Norway and concentrations of pollutants in air and precipitation decrease northwards.

A study of regional variations in tree vitality (expressed as relative crown density) of Norway spruce and Scots pine showed no relation to the pattern of air pollution. Tree vitality decreased at northern latitudes, at higher altitudes and in older stands. A few permanent observation plots in southeastern Norway indicate no major changes in crown density during the period 1978-1984.

Tree ring studies have given no consistent evidence for effects of long range transported air pollutants on tree growth. Examples of some regional forest injuries, which have generally been related to climatic factors, are given.

Dramatic forest declines in the near future are not expected, mainly because of the low concentrations of gaseous pollutants. Occasionally, however, high ozone concentrations along the southeastern coast need increased attention and study of possible effects.

INTRODUCTION

No convincing evidence exists at present of relations between long-range transported air pollutants and forest damage in Norway. This paper will discuss the general problem of air pollution in Norway and present results from some investigations concerning the issue. Some statements about the probability of dramatic forest declines in the near future will also be given.

Damages from wind, snow, drought, forest insects, and fungi have always been a normal situation in Norwegian forests. Outbreaks of the spruce bark beetle (Ips typographus) killed about 5 million m^3 of timber in the southeastern parts of Norway during the years 1971-1980. The outbreaks were initiated by extensive storm damages in 1969 and extreme drought during the years 1974-1976 (Bakke 1983). Apart from direct damages, remaining trees or stands may have lowered vitality. Symptoms of forest declines in Central Europe are mostly non-specific, i.e., different causes may give similar symptoms. The impact of natural factors on tree vitality is, therefore, most important in a discussion of declines or damages which are thought to be related to long-range transported air pollutants.

DEPOSITION AND CONCENTRATIONS OF AIR POLLUTANTS

Atmospheric pollutants deposited in Norway originate mainly from remote sources in the densely populated and industrialized areas of Central Europe, but also from sources within the country. Approximately 10% of the total deposition of sulphur in Norway originates from local sources (OECD 1979). A general feature is the consistent decrease in concentrations of pollutants both in air and precipitation from Central Europe and northwards (Fig. 1).

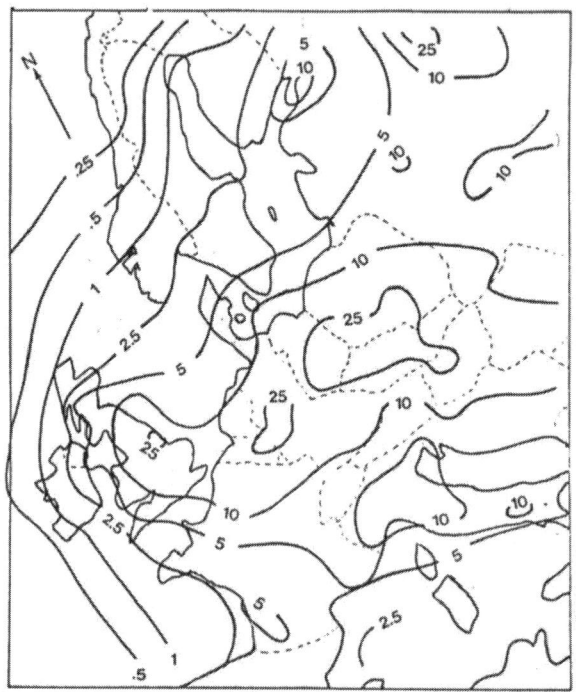

Fig. 1. Calculated annual average SO_2 concentrations ($\mu g\ SO_2\ m^{-3}$) in the atmosphere over Europe in 1974. (From OECD 1979).

Within Norway we have a fairly good picture of precipitation chemistry during the last decades. More than 50 background stations have been in operation during some period as part of national or European networks (Overrein et al. 1980; EMEP 1984). The knowledge about gaseous pollutants is more limited.

Sulphur dioxide concentrations are generally very low in Norway, as compared with Central Europe (Fig. 1). In rural areas average yearly concentrations can be up to 3-4 $\mu g\ SO_2 \cdot m^{-3}$ in the most affected regions (SFT 1983). A regional pattern is difficult to establish because of limited data, but a decreasing south-north gradient is indicated. In the far north, influences from eastern sources are indicated. Maximum daily concentrations have occasionally reached 60 $\mu g.m^{-3}$ in the southwestern part of the country.

Measurements of nitrogen oxides are limited. A one-year study (1982) at a station in southern Norway showed a yearly average of 3.9 $\mu g\ NO_2.m^{-3}$ with a daily maximum of 50 $\mu g.m^{-3}$ (SFT 1983).

Rainfall acidity and concentrations of SO_4, NO_3, and NH_4 all decrease in a consistent pattern northwards (Fig. 2). The pH of precipitation is about 4.2 - 4.3 in the southernmost parts and 4.7 or higher in northern parts apart from the very far north. Episodes with pH of less than 3.2 are nearly non-existent. The concentrations of NH_4-N and NO_3-N are roughly equal and show the same regional pattern.

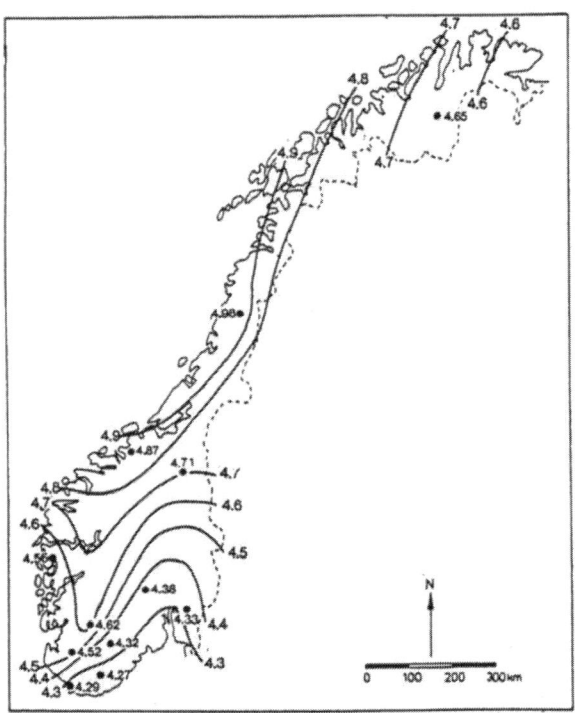

Fig. 2. pH at Norwegian background stations in 1982. (From SFT 1983).

The wet deposition pattern reflects variations both in concentrations and in precipitation amounts. Thus, the deposition is largest in the southwestern part of the country.

The relatively low SO_2 concentrations result in dry deposition, contributing less to the total deposition than in the large emission countries. In south Norway, dry deposition is considered to vary between 15 and 35% of the total deposition (Joranger et al. 1980). The dry deposition of nitrogen compounds may be of the same order (Fowler 1980).

Sulphur and nitrogen deposition (estimated dry deposition included) are similar, varying from 20-25 kg.ha^{-1}.y^{-1} in the southwestern parts to 1-4 kg.ha^{-1}.y^{-1} in the northern half.

Ozone as a potential damaging gasous pollutant has received increasing attention during the last years. Measurements since 1977 have shown maximum hourly concentrations of up to 300 µg.m^{-3} (Grennfelt and Schjoldager 1984). A tentative map has been drawn up to show the probability of maximum hourly concentrations to exceed certain concentrations (Schjoldager 1984). The geographical pattern is similar to the pattern of precipitation acidity with the highest values along the southeastern coastline (Fig. 3).

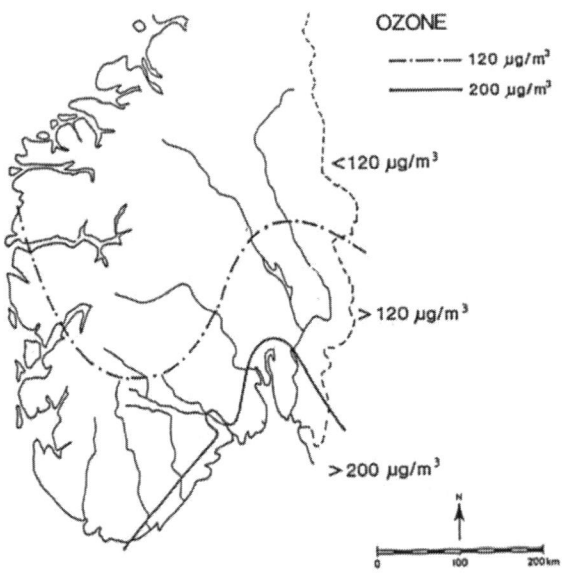

Fig. 3. Areas with an estimated 50% probability of the maximum hourly ozone concentration to exceed 120 µg.m^{-3} and 200 µg.m^{-3} during a summer. (From Schjoldager 1984).

FOREST VITALITY

The alarming reports of forest declines in Central Europe, especially during the last three years, have also caused major public concern in Norway. The concern was further accentuated by reports from southern Sweden in the autumn of 1983 that similar decline symptoms were developing as in Central Europe.

The main symptom of the decline appears to be a general thinning of the crown caused by abnormal needle loss, and in Norway spruce this is often combined with a yellowing of the needles. The symptoms are non-specific, which means that several natural causes are also possible, such as drought, frost, wind and aging effects.

In 1984 an inventory was carried out in connection with the National Forest Survey. The inventory covered more or less six counties in a south-north transect from 58°-65° n.l. and used a systematic clustered sampling scheme. Sample trees of Norway spruce (<u>Picea</u> <u>abies</u>) and Scots pine (<u>Pinus</u> <u>sylvestris</u>) were classified according to relative crown density, trying to judge the actual needle biomass in relation to a fully dense crown (class width 10%). Only the upper half of the crown was used in the spruce classification. Nearly 10,000 trees were classified. In addition to usual sample tree and plot parameters, the tree dominance and position in relation to stand edges were noted.

The data have not been fully evaluated at present. Some general features, however, are clear. First of all, there is a general decrease in tree vitality (as measured by crown density) from south to north (Fig. 4). This is the opposite of the pollution pattern. Within all counties there is a general decrease in vitality with increasing age or development stage (Fig. 5). The

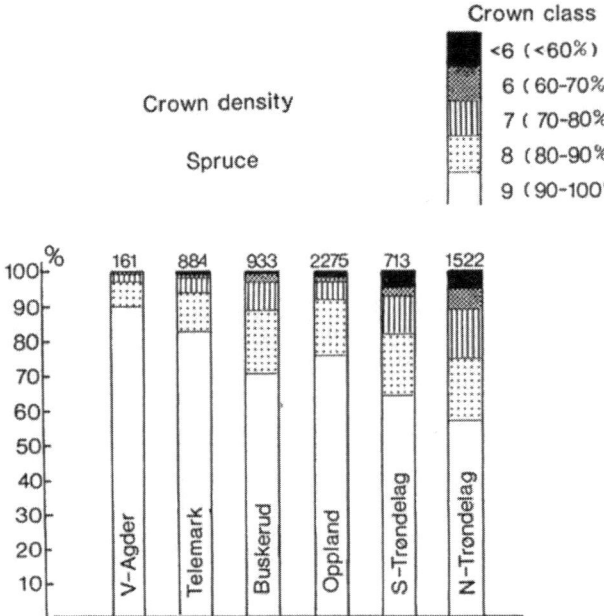

Fig. 4. Crown density distribution of Norway spruce in six counties in a gradient from south (V-Agder) to north (N-Trøndelag). Number of sample trees at the top of each column. Crown class 9 covers trees with fully dense crowns (90-100%). Unpublished data from the National Forest Survey 1984.

data also show a clear tendency to reduced vitality with increasing altitude, especially when approaching the timberline.

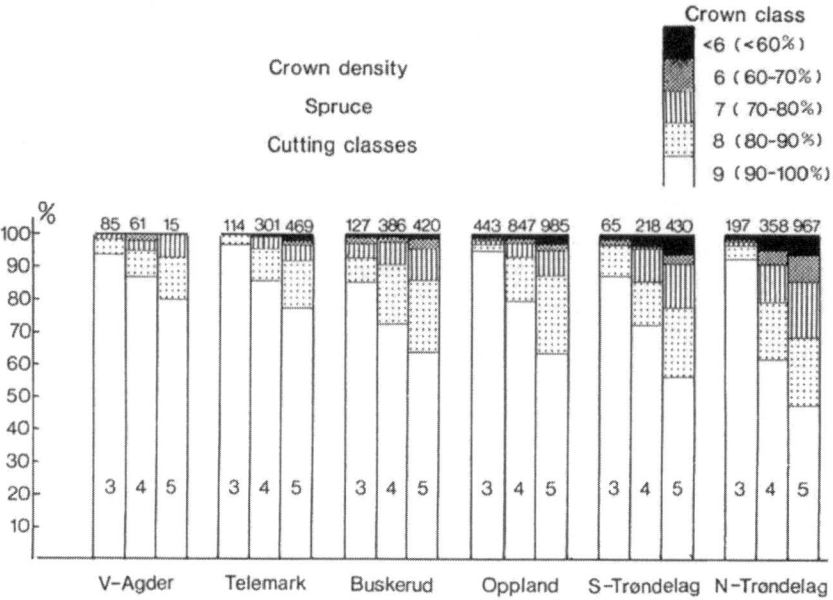

Fig. 5. Crown density distribution of Norway spruce in different cutting classes. Class 3 = young forests, class 4 = middle aged forests and class 5 = mature forests. See Fig. 4 for further explanations. Unpublished data from the National Forest Survey 1984.

Natural factors, such as climatic stress and aging effects, overshadow possible effects of air pollutants on tree vitality. The contribution from different factors remains difficult to assess because little is known about "normal" variations in crown density. According to the classification used in central-European surveys, more than 50% of the standing volume of old spruce is damaged in the northernmost county which was surveyed in 1984 (N-Trøndelag). Air pollution levels are probably not far from background levels in these areas.

A study of several vitality parameters, including crown density, in old spruce forests was undertaken from 1978 to 1981, to look at general vitality. Six permanent observation plots were established in this connection, all in southeastern Norway. The plots have been followed up to 1984. No major changes have occurred during this period. Trees classified as bad in 1978 are still bad. Most dead trees have been killed by spruce bark beetles. Plots with many dead trees have usually many trees with reduced crown density. One explanation may be that stand destruction by bark beetle attacks increases the stress on the remaining trees in the stand. A different explanation may be that stands stressed by drought, for example, are predisposed to beetle attacks.

TREE GROWTH

Tree growth and vitality are generally positively related, especially within the same stand. Tree ring studies of trees classified as less vital according to crown density have often shown that growth reductions probably have occurred before obvious visual decline symptoms (e.g., Kenk 1983).

Extensive tree ring studies in Norway have, however, shown no consistent trends in ring width variations from about 1910 up to 1972-74 that can be related to regional patterns of air pollution (Strand 1980). Nor have any consistent trends been found when sites of different assumed sensitivity to acid input have been compared (Tveite 1975; Abrahamsen et al. 1976; Tveite 1984).

The main problem in such studies is that it is very difficult to exclude alternative explanations of observed differences. These explanations include differences in stand history, differences in normal age trends, and differences in response to weather.

REGIONAL INJURIES

During the last 15 years, regional injuries have occurred which have been concentrated to the southeastern parts of the country. Extreme drought in 1974-1976 with 30-60% of normal summer precipitation caused severe drought injuries over large areas. In 1977 large spruce bark beetle attacks started in the same areas. The attacks declined from 1980 and were insignificant in 1982 (Bakke 1983).

A wide-spread injury to younger spruce in southeastern Norway occurred in the late winter/early spring of 1977, causing needle loss and bark injury of mainly the last year's shoot. The damage was most severe in plantations of southern provenances on high quality sites. The injury was presumed to have been initiated by unfavourable weather in 1976 and triggered by temperature fluctuations in February (Horntvedt and Venn 1980). A similar injury has occurred this spring in mainly the same areas.

In the southwestern parts, local attacks of the fungus Gremeniella abietina ("Scleroderris-cancer") have occurred, causing top dying, especially in young spruce plantations established during the 1960s.

Climatic factors are the most probable causes of these injuries. It is, however, difficult to answer questions about possible influences of regional air pollution. This is especially the case when hypotheses of complicated interactions are presented.

PROBABILITY OF DRAMATIC FOREST DECLINES

The field observations which have been reviewed give no clear evidence of forest decline or forest damage related to regional air pollution. The concern about the forest declines in Central Europe poses, however, the question of whether similar declines will occur in Norway in the near future. The following points speak against dramatic changes, given that the declines are caused by air pollution.

1. The concentrations of gaseous air pollutants do reach such levels in forest areas of Central Europe that direct damage can be expected. The concentrations in Norway are much lower and apart from ozone the probability of vegetation damage seems very small.

2. Central Europe has over a much longer period had substantially higher depositions of air pollutants. It is, therefore, not very likely that the Norwegian forests will be damaged at approximately the same time as in Central Europe.

3. In old mining areas, smelting of sulphur-containing mineral ores has caused forest destructions - most probably because of high SO_2-concentrations. It is reasonable to assume that there has been substantial dry deposition of sulphur in such areas and that increased soil acidification has been the result. Forests have nevertheless normally invaded the areas when the sulphur emissions stopped, apart from areas with high concentrations of heavy metals. The site productivity has probably been reduced, but trees can still grow normally.

4. A number of experiments to study effects of artificial acidification on tree growth have been carried out in the last 10-15 years. Some of the experiments have shown reduced growth when large amounts of sulphur as diluted sulphuric acid have been applied. But only a few experiments using extremely high concentrations have shown damage symptoms somewhat similar to what has been observed in Central Europe.

5. The forest declines in Central Europe are found on both rich and poor soils. If the damages are caused mainly by soil acidification, it is difficult to understand that the damages do also occur on soils rich in calcium and magnesium (Rehfuess 1983).

6. The survey of variations in tree vitality in 1984 in Norway shows no relation to the regional pattern of air pollution.

REFERENCES

Abrahamsen G, Bjor K, Horntvedt R, Tveite B (1976) Effects of acid precipitation on forest and freshwater ecosystems in Norway. SNSF-project, Norway, FR 6/76: 36-63

Bakke A (1983) Host tree and bark beetle interaction during a mass outbreak of Ips typographus in Norway. Z ang Ent 96: 118-125

EMEP (1984) Summary report from the chemical co-ordinating centre for the second phase of EMEP. Co-operation programme for monitoring and evaluation of the long-range transmission of air pollutants in Europe. Norw Inst Air Res, EMEP/CCC Rep 2/84

Fowler D (1980) Wet and dry deposition of sulphur and nitrogen compounds from the atmosphere. In: Hutchinson TC, Havas M (eds) Effects of acid precipitation on terrestrial ecosystems, Plenum Press, New York, London, p 9-27

Grennfelt P, Schjoldager J (1984) Photochemical oxidants in the troposphere: A mounting menace. Ambio 13: 61-67

Horntvedt R, Venn K (1980) Frost injury to Norway spruce during the winter of 1976-1977. Europ J For Pat 10: 71-77

Joranger E, Schaug J, Semb A (1980) Deposition of air pollutants in Norway. In: Drabløs D, Tollan A (eds) Proc int conf ecol impact acid precip, SNSF-project, Norway 1980, p 120-121

Kenk G (1983) Zuwachsuntersuchungen in geschädigten Tannen-Beständen in Baden-Württemberg. (Growth studies in diseased silver fir stands in Baden-Württemberg.) Allg Forstzeitschr 38: 650-652

OECD (1979) The OECD programme on long range transport of air pollutants. Measurements and findings. Second edition. Organization for economic co-operation and development, Paris 1979

Overrein LN, Seip HM, Tollan A (1980) Acid precipitation - effects on forest and fish. Final report of the SNSF-project 1972-1980, SNSF-project, Oslo-Ås

Rehfuess KE (1983) Walderkrankungen und Immissionen-eine Zwischenbilanz. Allg Forstzeitschr 38: 601-610

Schjoldager J (1984) Photochemical oxidants in Norway. A summary. Norw Inst Air Res, 5 p

SFT (1983) Overvåking av langtransportert forurenset luft og nedbør. (Monitoring of long range transported polluted air and precipitation.) Statens forurensningstilsyn. Statlig program for forurensningsovervåking, Rapp 108/83, 228 p

Strand L (1980) Acid precipiation and regional tree ring analyses. SNSF-project, Norway, IR 73/80, 36 p

Tveite B (1975) Sur nedbør - skogproduksjon. Regionale årringundersøkelser. (Acid precipitation - tree growth. Regional tree-ring investigations.) SNSF-project, Norway, TN 11/75, 49 p

Tveite B (1984) Sur nedbør og årringgranskingar. (Acid precipitation and tree ring analyses.) Norsk Skogbr 30(10): 23-25

FOREST DECLINE - THE VIEW FROM BRITAIN

W.O. Binns, D.B. Redfern* and K. Rennolls

Forestry Commission, Alice Holt Lodge, Wrecclesham, Farnham, Surrey, U.K.

* Forestry Commission, Northern Research Station, Roslin, Midlothian, U.K.

ABSTRACT

The new occurrence of Silver fir decline in Central Europe in the 1970s went virtually unnoticed in Britain, but its 'spread' to Norway spruce from about 1980 onwards caused concern. By 1982 it was clear that this was a new kind of damage which could not be explained by known biological or natural climatic agents alone, pollutants in some form apparently being involved. West Germany pioneered survey methods and in June 1984 an FAO/ECE Working Group succeeded in getting general agreement on systems for future surveys.

There is so far no fully satisfactory explanatory hypothesis: direct SO_2 damage and soil acidification are improbable in some affected forests and some of the damage to Norway spruce seems different in kind to the rest.

A survey of Norway and Sitka spruces and Scots pine in Britain, stratified for altitude, rainfall and sulphur deposition rates, was made in winter 1984. The survey had the structure of a factorial design, replication being provided by six geographical zones. Well-grown stands between 30 and 45 years old were assessed from the four margins. The main characteristics assessed were crown density on the five-point West German scale, age to which needles were retained, and needle yellowing.

Because of incomplete coverage for all species of all the combinations of the environmental variables, regression analysis rather than analysis of variance was used, using plot altitude, estimated rainfall and sulphur deposition rates as regressor variables. There was virtually no yellowing reported. Crown density and needle retention were significantly affected by geographical zone. Scots pine retained its needles longer where sulphur deposition was higher. Taking all species together, needles were retained longer at higher altitude and crowns were thinner as rainfall increased. There was a suggestion that Sitka spruce crowns were less dense where sulphur deposition was high. The state of health is regarded as satisfactory and the poor condition of some plots can be explained without invoking pollution damage: no new phenomena were found.

Winter damage in 1983-84 to seven species of conifer in north-western Britain differed in important respects from forest decline, although it was worst on older trees at higher altitudes. There was no needle yellowing, and the damage appeared to be an isolated incident, due most probably to fluctuating temperatures accompanied by strong winds during the winter, though it is impossible to rule out other factors.

Work has been started on open-topped chamber experiments to determine the effects of both ambient and increased concentrations of pollutants using clonal stock of commercial species, which will be tested for sensitivity to pollutants.

HISTORY AND DEVELOPMENT

Silver fir (Abies alba Mill.) has suffered outbreaks of dieback or decline on many occasions in the past, and Wachter (1978), and more recently Kandler (in press), have reviewed the literature. Because silver fir is hardly planted in Britain, British foresters have paid little attention to the species and its problems, and indeed were for some time unaware of the new outbreak of 'Tannensterben' in the 1970s.

Reports of dieback of Norway spruce (Picea abies (L) Karst.) began to reach Britain in 1980 and 1981, and this aroused more interest. A workshop was convened in Göttingen in May 1982 to discuss the problem of forest decline, but no agreement was reached about its cause or extent (Ulrich and Pankrath 1983); the evidence adduced on that occasion did not convince British participants that this was a new problem.

In autumn 1982, the first two authors went to see forest damage in North-Rhine Westphalia, Lower Saxony, Bavaria, and Baden-Württemberg, looking mainly at Norway spruce. A number of instances of decline that we saw were associated with well-known biological agents, but there remained a residue of cases, which appeared, by elimination, to be caused by environmental factors such as climate, air pollution, or soil acidification (Binns and Redfern 1983).

The FAO/ECE Working Group on Air Pollution spawned an ad hoc working group (Impact of Air Pollution on Forests) which met in Geneva in April 1983 and in Freiburg im Breisgau in June 1984. There were representatives at this second meeting from most western European countries and from Bulgaria, Czechoslovakia and Hungary, as well as from Canada. The representatives agreed on criteria for assessment and surveys of forest health, based on the system worked out in West Germany.

The Commission of the EEC in Brussels had tabled a Draft Regulation in June 1983, designed to provide forests in the community with increased protection against fires and acid rain. This met with considerable opposition from a number of Member States and little progress had been made by early 1985. The EEC also organized a meeting in Karlsruhe in September 1983 (Ott and Stangl 1983) and commissioned a special report on the subject which was presented to delegates at that meeting (Anon 1983a).

From 1982 onwards the media in Britain seized with alacrity on the reports of damage to forests, in particular in West Germany, and 'Acid Rain' became the catchword for impending catastrophe. The more extreme claims are now no longer made and there is a more balanced appreciation of both the complexity of the subject and the actual form of the phenomenon in mainland Europe.

Annual surveys of forest damage have been made in Germany since 1982 (Anon 1982, 1983b, 1984). The first, admittedly consisting of a mixture of estimates and assessments, reported some 8% of the forest affected by decline, and the totals rose to 34% in 1983 and 50% in 1984. Until the summer of 1983, damage had been restricted to forests in the mainland mass of Western and Central Europe, with most of it at high altitude. However, in that year there were reports of a similar phenomenon in the southernmost counties of Sweden. The results came from a questionnaire, and interpretation was cautious, but many similarities with German results - worst damage at the higher altitudes, on old trees, and on westerly aspects - suggested that the same unidentified causes were operating there too.

As a consequence of our realization that there was a new problem of forest damage in Western Europe, we decided early in 1984 to undertake a survey of forest health in Britain and announced this at the FAO/ECE meeting in Freiburg referred to above.

CAUSES OF DECLINE

This workshop will be discussing in detail current hypotheses on the causes of forest damage. Both our own observations and evidence from the literature suggest that Norway spruce is suffering from more than one form of damage. For example, the symptoms of the chlorosis seen in the foothills of the Bavarian Alps are quite distinct from those in the Black Forest or Fichtelgebirge (Rehfuess 1985). Although there appears to be no evidence so far that a living organism is responsible for the decline (though see Kandler, in press), the impression of a spreading sickness, creeping slowly down the hills into the valleys, is very strong and engenders a feeling that we have all missed some feature or failed to look at the evidence in the right way.

None of the hypotheses proposed is entirely satisfactory, but some causes (e.g., direct SO_2 damage to foliage, direct damage to roots through soil acidification) seem very unlikely on at least some of the sites. One pollutant which seems to be present in higher concentrations in West Germany than in Britain is ozone (Roberts and Blank 1984). However, there are major differences in climate and forest characteristics in Britain compared with most of those countries where forest decline has recently appeared; Britain is windier, with warmer winters and cooler summers, forests are regarded as old at 50 years, and there is only a small area of forest over 400 m above sea level (Binns 1985).

THE BRITISH SURVEY

The choice of species to survey was relatively easy. Sitka spruce (<u>Picea</u> <u>sitchensis</u> (Bong.) Carr.) was included because of its outstanding importance in British forestry, and Norway spruce, although less and less important, gave a link with the observations being made elsewhere on the Continent. Finally, despite misgivings about difficulties in assessment and interpretation, Scots pine (<u>Pinus</u> <u>sylvestris</u> L.) was included because it is still widely distributed in Britain, of considerable importance to private foresters, and is also native, while both spruces are introduced. Broad-leaved trees were excluded because damage always seems to appear first on conifers. The survey and its results are more fully described by Binns <u>et al</u>. (1985).

Assessment Systems

Although the assessment methods were to be based on the system agreed upon at Freiburg, there are considerable differences in the age distribution and management of British forests compared with those of Germany. With Dr. Hartmann from the Lower Saxony Forest Research Institute in Göttingen advising us, we modified the German system a little. The main change was to assess trees on stand margins only, because the full stocking resulting from regular planting patterns (even when thinned) precluded observation of the crowns from within most stands. In addition, an estimate of the number of years for which the needles were held was included. The two most important characters assessed were the degree of needle yellowing and the crown density, the latter on the five point scale used in Germany. On this scale the percent departure from full crown density (or needle loss) is estimated in five classes: Class 0, 0-10%; Class 1, 11-25%; Class 2, 26-60%; Class 3, 61-99%; Class 4, dead.

Tight restrictions were imposed on the characteristics of the stands to be assessed. For example, stands were to be rejected if edge trees had recently been removed, if the edge had been exposed by windthrow or clear-felling during the last five years, if they were on the forest boundary, if there was significant windthrow, or if there was no access from one or more sides.

Selection of Forests

The distribution of forests in Britain is shown in Fig. 1. This figure is somewhat misleading because it includes all Forestry Commission land, about a quarter of which is at present unplanted and 20 per cent of which will never be planted. Furthermore, it includes only those private forests managed under an agreement with the Forestry Commission, so that a large number of small private forests, as well as some larger ones, are not shown. Moreover, at this scale a number of very small woods do not appear at all. The proportion of conifers and broadleaves, the division between state and private, and the areas in the three countries of Britain are shown in Table 1.

Fig. 2. Land in Great Britain above 800 ft (244 m), solid. Omissions as for Fig. 1. In addition, some very small areas omitted.

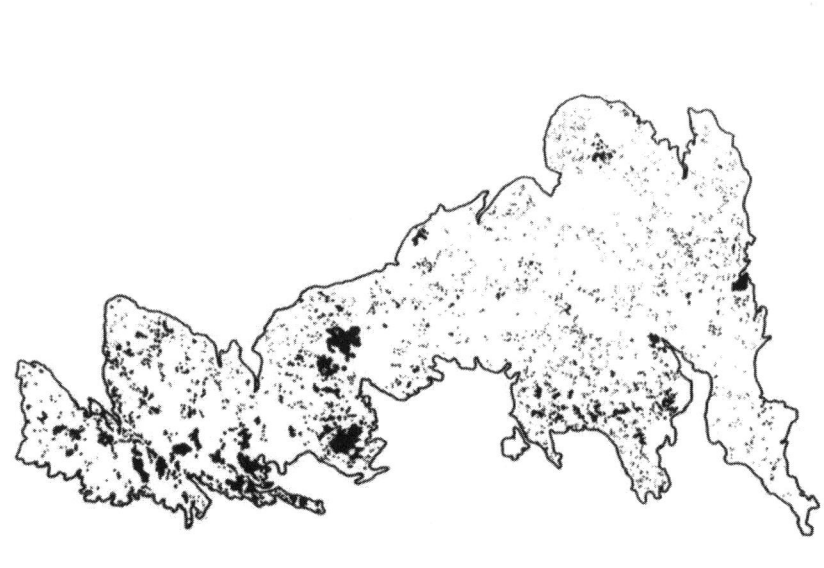

Fig. 1. Forestry in Great Britain 1976, Mainland and Anglesley only. Forestry Commission, all land; private forestry, dedicated and approved woodland only.

Table 1. Forest and woodland (thousand ha) in Great Britain 1984. (Data from the Forestry Commission 1984).

	Coniferous high forest		Broad-leaved high forest		Coppice[1]	Total productive woodland	Unproductive woodland	All woodland
	FC	private	FC	private				
England	206	191	40	381	37	855	90	945
Wales	130	43	6	53	2	234	10	244
Scotland	515	338	4	72	–	929	71	1000
GB	851	572	50	506	39	2018	171	2189

[1] All except 1000 ha in private ownership.

In West Germany assessments are made at the intersecting points of 4 x 4 km grid, but it was considered that in Britain a systematic approach would be both inefficient and wasteful because of the species restriction and the non-uniform distribution of forest. It was also decided, even at this early stage, that we should determine if any new damage found could be linked to any major environmental factors. In Central Europe the damage is usually worst at high altitudes, and there are suggestions that drought may also be a factor in the decline. We therefore stratified the country into land above and below 800 ft (244 m) and into land receiving more or less than 1000 mm average annual rainfall (Figs. 2 and 3). We also wished to stratify for pollution, but information on rainfall acidity, hydrogen ion loading, and nitrate deposition was incomplete for rural areas (Barrett et al. 1984). The only pollutant adequately monitored was sulphur, as shown in Fig. 4; therefore the country was divided into areas receiving more or less than $2 \text{ g S.m}^{-2}.\text{y}^{-1}$, as dry deposition. Britain was divided into six geographical zones and candidate forests were selected in all combinations of the three strata which occurred in the zones.

Forest decline usually appears first on older stands, but we wished to resurvey periodically; therefore, lower and upper limits of 30 and 45 years were set. (It is important to realize that few stands of commercial conifers in Britain are retained beyond 60 years of age and that the faster growing ones are normally felled by 50 years.) It was also decided to restrict the survey to stands which had grown well all their life; therefore, lower limits were set for the three species: General Yield Classes of 8, 10 and 12 for Scots pine, Norway spruce and Sitka spruce, respectively (Hamilton and Christie 1971). Lists of stands satisfying the yield and elevation criteria for the three species were printed out for the candidate forests from the Forestry Commission database, and from these lists the surveyors randomly selected the stands to be examined. Making the final choice was time-consuming and many stands had to be rejected because they did not satisfy one or more of the conditions set. The approximate location of the plots finally chosen is shown in Fig. 5.

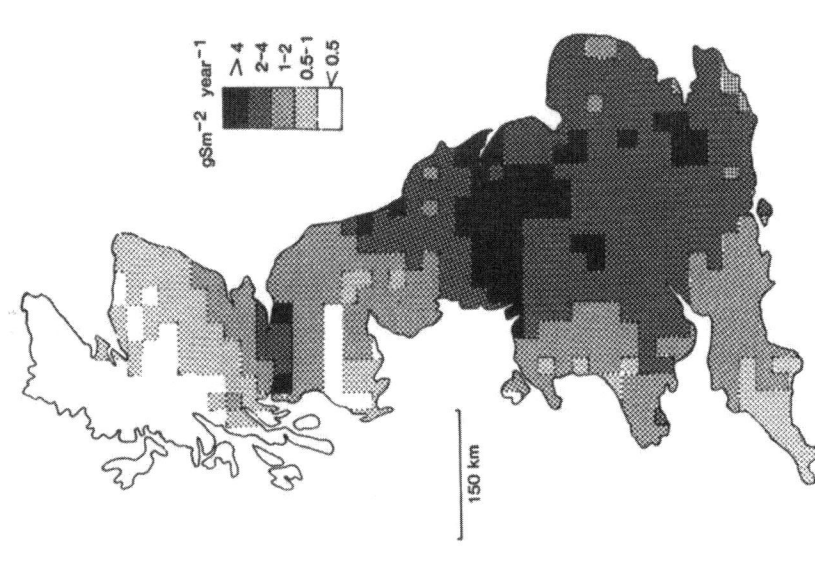

Fig. 4. Dry deposition of sulphur in Great Britain for 20 x 20 km squares. (From Barrett et al. 1984).

Fig. 3. Land in Great Britain receiving more than 1000 mm average annual rainfall, solid. (Simplified from Barrett et al. 1984). Omissions as for Fig. 1.

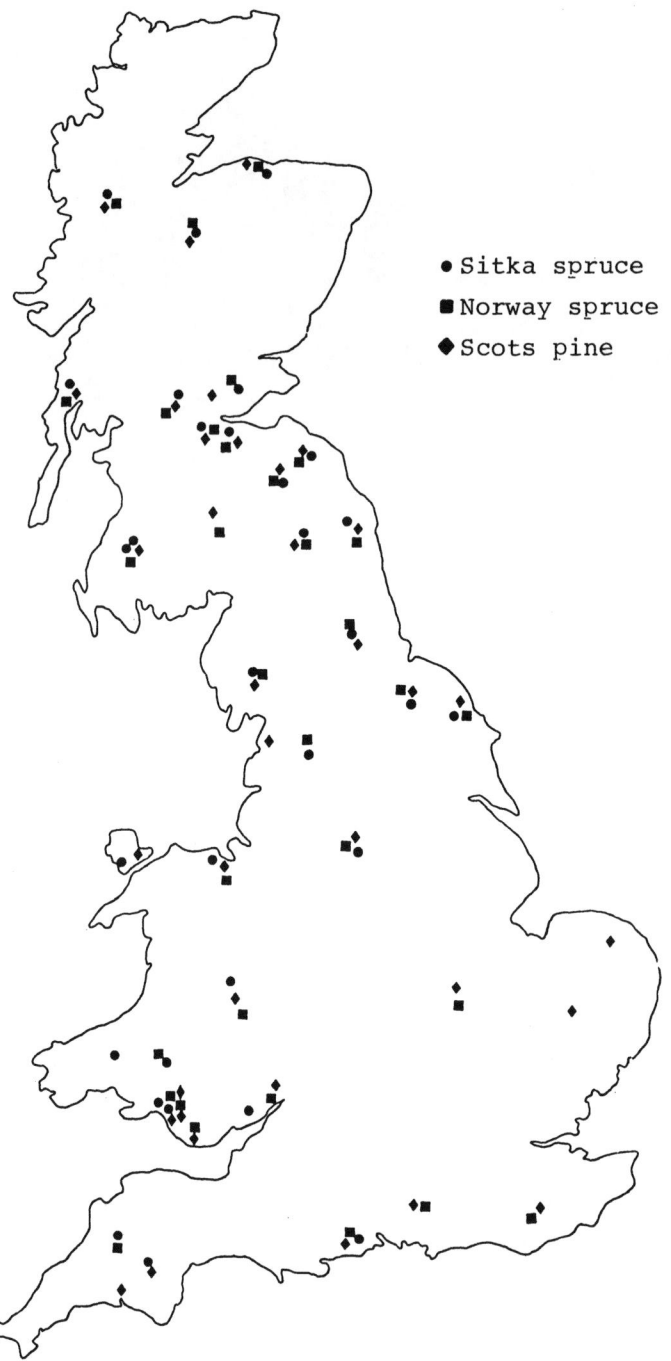

Fig. 5. Approximate location of assessment plots for the 1984 Survey of Forest Heath.

Analysis of Results

The original survey design had taken the form of a 2^3 factorial experiment, the replication being provided by the six regional zones. However, only two of the zones had stands with all eight combinations of variables for all three species. Restricted analyses of variance were performed, but the sensitivity was necessarily reduced and few significant results were detected. The factors had in fact been adopted largely to ensure that the sites spanned a wide range of environmental conditions. The final analysis of the survey was therefore performed by regressing various damage variates upon the regressor variables of altitude, mean total rainfall, mean summer rainfall (both assessed from 1941-1970 averages from the Meteorological Office), and the sulphur deposition value for the 20 km squares within which each plot fell (Fig. 4). This allowed data from each region (treated as a factor in the analysis) to be pooled, providing a much more sensitive analysis of each of the species separately.

Stepwise regression was used, the variates and factors being added sequentially to produce maximal decrease of the sum of the squared residual errors. The signficance of each term added was assessed using an F-test.

The average crown density and needle age for these species are are shown in Table 2.

Table 2. Percentage distribution of sample trees by crown density classes (% departure from full crown density) and average needle life.

	needle life (yrs)	0 0-10%	1 11-25%	2 26-60%	3 61-99%	4 Dead
Sitka spruce	6.2	65.3	28.3	5.8	0.4	0.1
Norway spruce	6.4	71.4	25.9	2.5	0.3	0.2
Scots pine	2.2	48.0	29.5	16.9	5.4	0.2

The needle holding and crown density for both spruces was regarded as satisfactory. Because trees, like other organisms, suffer periodically from pests and diseases, and are also subject to climatic damage, even where pollutants are extremely low, one would always expect to find a proportion of the crops shedding needles prematurely, most often because of damage from wind or from Green spruce aphis (Elatobium abietinum Walker), resulting in rather transparent crowns. Thus, the figures of 28 and 26% in Class 1 for the spruces are not regarded as new or unusual.

The results for Scots pine, however, suggest a poorer state of health. As already mentioned, we were aware that assessment of this species would present problems. This is pointed out in the 1984 West German survey, and the survey of forest health carried out in Sweden in that year does not report results for Scots pine, indicating that consistent interpretation of results has been difficult. Scots pine seldom carries more than three years' needles, and the average of 2.2 during the winter can be regarded as normal. It has a much more open crown than the spruces, which makes it difficult to establish a datum for estimating departure from a full crown, using a subjective system. In addition, in Britain the species has several insects and fungi which can affect crown health, and a number of the stands with less than average needle age and low crown density had been affected by one or more of these agencies. Furthermore, in a windy country there is always some mechanical damage in tree crowns during winter. Thus, although the bare statistics presented for Scots pine crown density might look alarming at first sight, the condition is familiar and causes no concern.

The damaged spruce forests in West Germany commonly show (in addition to other, non-specific symptoms) a characteristic yellowing of older needles which analysis shows to be associated with low magnesium concentrations. None of the stands surveyed showed any appreciable yellowing, and spruces elsewhere in Britain which have yellow needles are found to be suffering from potassium or nitrogen deficiency. Potassium is often deficient in upland organic soils in Britain and in 1982-83, for example, the Forestry Commission treated some 24,000 hectares of forest with PK fertilizer.

The regression analysis showed up some clear regional differences that have not yet been examined further. Other than that, there was only one regressor term which was significant at the 95% level: Scots pine holds its needles longer in the areas of higher sulphur deposition. Whether this is due to a stimulating effect of sulphur on the trees, a depressive effect of sulphur on pests or diseases, or a correlation with some other factor (such as increased nitrogen deposition) it is not possible to say.

There were suggestions that Norway spruce and Scots pine held their needles longer at high altitude and that crown thinning in these species increased with an increase in rainfall, although these effects were not significant at the 95% level, when species were analyzed separately. However, pooling data for all three species made the first of these effects significant at the 95% level, but not the second. Schmidt-Vogt (1983) quotes older evidence showing a strong correlation between altitude and length of needle life for Norway spruce. We would also expect continental species such as Scots pine and Norway spruce to grow better in the lower rainfall parts of Britain, although this would not be expected for Sitka spruce.

WINTER SHOOT DEATH OF CONIFERS IN NORTH-WESTERN BRITAIN, 1983-84

There has been some publicity over winter damage to several species of conifer in northwestern England and northwestern Scotland, first seen in spring and summer 1984. Considerable effort was devoted to mapping the extent of the damage, which took the form of severe foliage browning and shoot death on Sitka spruce, Norway spruce, Scots pine, Lodgepole pine (<u>Pinus</u> <u>contorta</u> Douglas), Douglas fir (<u>Pseudotsuga</u> <u>menziesii</u> (Mirb.) Franco), Noble fir (<u>Abies</u> <u>procera</u> Rehder) and Grand fir (<u>Abies</u> <u>grandis</u> Lindley) (Redfern <u>et al</u>., in press).

The gross appearance of damage varied between species as a result of differences in morphology and in the length of time damaged needles were retained, but detailed examination showed a similar pattern of symptoms on all species.

Damage consisted primarily of shoot death, which occurred during the 1983-84 dormant season; there was no evidence of significant previous injury. Dieback was not confined to 1983 shoots but frequently extended into older internodes, resulting in the death of substantial first and second order branches. The dead shoots were distributed on affected trees in a characteristic manner, being concentrated in a zone of variable extent between the upper few whorls of the tree, which remained healthy in all but extreme cases, and the lower one-half to three-quarters of the crown which also escaped injury. There was marked variation in the severity of damage sustained by adjacent trees and even in the most severely damaged crops a few individuals remained virtually unscathed.

In Sitka spruce, the species most commonly affected, the severity of damage increased both with elevation and crop age and was most severe in crops more than 45 years old and at elevations greater than about 250 m. By contrast, damage was generally rare and insignificant in crops less than 20 years old or below about 200 m.

Damage to species other than spruce occurred at scattered locations throughout Scotland and northern England, but on spruce significant damage appeared to be virtually restricted to western forests from Cumbria to Sutherland, although it was widespread within that area. Slight damage occurred in the east, but at much higher elevations than in the west. Minor damage occurred in crops which probably total several thousand hectares, whereas serious damage was much more restricted and occurred only at scattered locations involving a few hundred hectares.

Although the symptoms shown by affected spruce bore some resemblance to those attributed to 'acid rain' in Germany, they differed in important respects. Most notably, damage in Britain occurred during a single dormant season and was not associated with the striking magnesium deficiency symptoms characteristic of the German syndrome. The evidence strongly suggests that damage on all species was caused by certain weather phenomena during the 1983-84 winter: in particular, alternating periods of mild and cold weather accompanied by strong winds. This conclusion receives some support from the accounts of similar damage on some of the species in previous winters. However, since the precise mechanism of damage remains unknown, the possibility has to be considered that trees were predisposed to climatic injury by other environmental factors.

EXPERIMENTAL WORK

Experiments designed to detect pollution damage to trees in forests, using large open-topped chambers with introduced pollutants as well as filtered and unfiltered air, have been started. Three experiments are envisaged, one in south-eastern England (high O_3 pollution), one in northern England (high SO_2 pollution) and one in eastern Scotland (moderate pollution). The work will concentrate on clones of conifers which will be tested for susceptibility to pollutants before final selection for the experiments, to ensure that they are both representative of the stock used in large-scale forestry and cover a range of sensitivities.

CONCLUSIONS

The 1984 survey of Sitka and Norway spruce and Scots pine in Britain has shown no evidence of any new form of forest damage. Stands in less than perfect health can be accounted for without invoking pollution damage.

The winter damage to seven species of conifer in north-western Britain, however, is new in the form seen on Sitka spruce and exceptionally severe on other species. While weather patterns in the 1983-84 winter are most likely to be responsible, proof is lacking and may well remain so.

Acknowledgements

Members of the Forestry Commission's Field Surveys Branch worked hard under very difficult conditions during winter 1984 to complete the survey on time; their contribution is gratefully acknowledged. We would also like to thank S.C. Gregory and J.E. Pratt for their help with the investigations of winter shoot death of conifers.

REFERENCES

Anon (1982) Waldschäden durch Luftverunreinigung. Schriftenreihe des Bundesministers für Ernährung, Landwirtschaft und Forsten. Reihe A: Angewandte Wissenschaft Heft 273
Anon (1983a) Acid rain: a review of the phenomenon in the EEC and Europe. Environmental Resources Ltd. Graham and Trotman
Anon (1983b). Neuartige Waldschäden in der Bundesrepublik Deutschland. Bericht des Bundesministers für Ernährung, Landwirtschaft und Forsten zur Waldschadenserhebung 1983
Anon (1984) Waldschadenserhebung 1984. Presse- und Informations stelle des Bundesministers für Ernährung, Landwirtschaft und Forsten

Barrett CF, Atkins DHF, Cape JN, Fowler D, Irwin JG, Kallend AS, Martin A, Pitman JI, Scriven RA, Tuck AF (1984) Acid deposition in the United Kingdom. Warren Spring Laboratory, Stevenage, Herts

Binns WO (1985) Are Britain's forests threatened by 'acid rain'? In: Campbell JK (ed) The acid rain inquiry. Proceedings of Scottish Wildlife Trust meeting, Edinburgh, September 1984, p 64-72

Binns WO, Redfern DB (1983) Acid rain and forest decline in W. Germany. Forestry Commission Research and Development Paper 131

Binns WO, Redfern DB, Rennolls K, Betts AJA (1985) Forest health and air pollution: 1984 survey. Forestry Commission Research and Development Paper 142

Forestry Commission (1984) Forestry facts and figures 1983-84. Forestry Commission, Edinburgh

Hamilton GJ, Christie JM (1971) Forest management tables (metric). Forestry Commission Booklet 34

Kandler O (in press) "Waldsterben": Immissions - versus Epidemie-Hypothesen. In: Kortzfleisch (ed) Waldschäden 1985. Kolloquium an der Universität Hohenheim. Oldenburg Verlag

Ott H, Stangl H (eds) (1983) Acid deposition: a challenge for Europe. Proc Karlsruhe Symp 1983. Commission of the European Communities, Directorate General for Science, Research and Development, Brussels

Redfern DB, Gregory SC, Pratt JE, MacAskill GA (in press). Foliage browning and shoot death in Sitka spruce and other conifers in northern Britain during winter 1983/84. European J Forest Path

Rehfuess KE (1985) On the causes of decline of Norway spruce (_Picea abies_ Karst.) in Central Europe. Soil Use Manag 1: 30-32

Roberts TM, Blank L (1984) Effects of air pollution on crops and trees. Proc 51st Ann Conf Nat Soc for Clean Air, Brighton, England

Schmidt-Vogt H (1983) Nadeljahrgangs-Erhebungen zur Beurteilung der Immissionsschäden bei Fichte. Der Forst- und Holzwirt 38: 391

Ulrich B, Pankrath J (eds) (1983) Effects of accumulation of air pollutants in forest ecosystems. Reidel, Dordrecht, p 389

Wachter A (1978) Deutschsprachige Literatur zum Weisstannensterben (1830-1978). Zeitschrift für Pflanzenkrankheiten und Pflanzenschutz 85: 361-381

DETERIORATION OF RED SPRUCE IN THE NORTHERN APPALACHIAN MOUNTAINS

A.H. Johnson

University of Pennsylvania, Philadelphia, Pennsylvania, U.S.A.
19104.

ABSTRACT

A synchronous, large-scale deterioration of red spruce (<u>Picea rubens</u> Sarg.) is quantitatively documented in the high-elevation coniferous forests of northern New York, Vermont and New Hampshire. Tree-ring analyses and field studies indicate that this phenomenon has been in progress for approximately 20-25 years. The beginning of this process coincided with two well-known climatic anomalies. The elevational band in which spruce have deteriorated to the greatest extent (900-1200 m) receives particularly high rates of deposition of acidic substances and trace metals due to inputs of polluted cloud water. A resolution of the cause(s) of red spruce deterioration is made difficult by the co-occurrence of natural and anthropogenic stresses, and the lack of applicable mechanistic models and data.

INTRODUCTION

As awareness of acidic deposition increased over the last decade, a number of possible effects on forest ecosystems were postulated. Theoretical and experimental investigations have delineated the potential for effects of acidic substances on many aspects of plant and forest function, and interactive effects of acidic substances with other pollutants, such as ozone, and with natural stresses are beginning to receive attention. Thus, there is a reasonable amount of evidence of what acidic deposition might do to forests, but as yet those effects have not been clearly discernible in the field.

The finding of widespread mortality of red spruce (<u>Picea rubens</u> Sarg.) in the northern Appalachian Mountains has led to much speculation about whether or not acid deposition is a cause. Because there are many possible pathways by which airborne acidic substances and/or other airborne pollutants could be contributers, there are many avenues for inquiry. At present, there is limited data, and so few of the possible pathways can be confidently eliminated from consideration. As a consequence of the current lack of information on mechanisms, only circumstantial evidence supports claims of acid rain damage to spruce in eastern North America. The suspicion that atmospheric deposition may play a role in the deterioration of red spruce is probably justified by the fact that large scale changes in the forest have taken place in high elevation areas which are receiving airborne heavy metals and acidic substances at rates that are apparently greater than those experienced by all other forested areas in North America (except those near smelters); what is unknown at this time is whether the deposited materials are assimilated benignly by the forest, or whether one or more life-sustaining processes have been altered sufficiently by airborne pollutants to cause mortality.

HIGH-ELEVATION CONIFEROUS FORESTS OF THE APPALACHIANS

Gradients in flora and environmental conditions in the eastern montane forests have been described by numerous investigators (e.g., Siccama 1974; Myers and Bormann 1963; Harries 1966; Adams et al. 1920; Scott and Holway 1969; McIntosh and Hurley 1964; Whittaker 1956). In the Green Mountains (VT), sugar maple (Acer saccharum Marsh) is the most important canopy species (by density and basal area) below 750 m, and balsam fir (Abies balsamea (L.) Mill.) is the most important above 900 m. Red spruce is a minor species below 750 m and above about 1150 m. It is most important in stands of the transition and lower boreal zones (760-1150 m), in which it is often the oldest and largest species. Vegetation patterns are similar, but somewhat more complex in the Catskill, Adirondack and White Mountains due to the fact that they are irregular massifs rather than linear ridges like the Green Mountains (Siccama 1974; Holway et al. 1969; McIntosh and Hurley 1964). Pure, naturally regenerated stands of red spruce in the northern Appalachian mountains are rare. Disturbances from logging, fire and windthrow have a significant effect on vegetative patterns of the high-elevation forests, so that the current mosaic is a complex of disturbance patches superimposed on elevational gradients (i.e., Foster and Reiners, 1983).

Deposition of Acidic Substances and Heavy Metals

The subalpine coniferous forests of the northern Appalachians are above cloud base for considerable portions of the year. The estimated duration of cloud cover varies from approximately 200 hours per year at 600 m to 40% of the year at 1220 m (Siccama 1974; Lovett et al. 1982). Cloud moisture tends to be considerably more acidic than precipitation with average H^+ concentration during the growing season of $288 + 193$ $\mu mol.L^{-1}$ (Lovett et al. 1982). Based on a model of cloud droplet capture and measurements of the chemistry of ten cloud events, Lovett et al. (1982) estimated cloud water inputs of major ions to a subalpine balsam fir forest at 1220 m on Mount Moosilauke (NH) to be considerably greater than bulk precipitation inputs (Table 1). Overall, the total deposition of acidic substances (H^+, NH_4^+, SO_4^{-2}, NO_3^-) in the subalpine balsam fir forest is estimated to be 3 to 6 times greater than in the lower lying hardwood forest at nearby Hubbard Brook, NH (Likens et al. 1977). The estimates of deposition of acidic substances at high elevation exceed the values calculated from low-elevation collectors of the major monitoring networks by several-fold. Values for major ion deposition to the coniferous forest at 1000 m in the northern Green Mountains estimated from rime ice chemistry, cloud chemistry and droplet capture by artificial collectors (Scherbatskoy and Bliss, 1984) are in general agreement with the estimates of Lovett et al. (1982).

Table 1. Annual deposition in bulk precipitation to a northern hardwood forest at Hubbard Brook, NH (Likens et al. 1977) compared to estimated annual deposition by bulk precipitation and cloud droplet capture in a subalpine balsam fir stand on Mt. Moosilauke, NH (Lovett et al. 1982).

ion	Hubbard Brook (northern hardwood forest) (1963-74) bulk precipitation ($kg \cdot ha^{-1} \cdot y^{-1}$)	Mt. Moosilauke (subalpine balsam fir) (1980-81) bulk precipitation ($kg \cdot ha \cdot y^{-1}$)	cloud water	Sum
H^+	1.0	1.5	2.4	3.9
NH_4^+	2.9	4.2	16.3	20.5
Na^+	1.6	1.7	5.8	7.5
K^+	0.9	2.1	3.3	5.4
SO_4^{-2}	38.4	64.8	137.9	202.7
NO_3^-	19.7	23.4	101.5	124.9

As in the case of lake sediments, Pb has been demonstrated to be a useful indicator of atmospheric deposition to forest soils (Reiners et al. 1975; Johnson et al. 1982; Friedland et al. 1984 a,b; Hanson 1980). The amount of Pb in the forest floor increases with elevation. Amounts at 1000 m are approximately 2 g $Pb \cdot m^{-2}$, which is 3 to 4 times higher than the amounts in the forest floor at 600 m (Johnson et al. 1982; Friedland et al. 1984a). It is not clear whether or not to attach biological significance to the current concentrations of trace metals in the forest floor; however, the data are consistent with the contention that subalpine coniferous forests receive considerably higher rates of atmospheric deposition than the adjacent lower altitude hardwood forests.

With the exception of areas near large sources of S, N or metal emissions, no other forests in eastern North America are receiving atmospheric inputs of trace metals and acidic substances at rates as great as those estimated for high elevation coniferous forests of Vermont and New Hampshire. It is important to restate, however, that there is at present no clear evidence which can be used to assess whether we should or should not expect chronic effects on subalpine biota resulting from past or current atmospheric inputs.

Characteristics of Red Spruce in Northern Appalachian Subalpine Forests

Red spruce is abundant in cool, moist climates of the high elevations of the southern and northern Appalachians, in coastal regions from Maine to Nova Scotia, and in interior areas of southern Quebec, northern Vermont, northern New Hampshire, northern Maine, New Brunswick and Nova Scotia (Burgess et al. 1984). Sites with a high moisture-holding capacity are particularly favourable (McIntosh and Hurley 1964). Red spruce older than 300 years are reasonably common in the northern Appalachians (Siccama, 1974; Burgess et al. 1984), and tree ring patterns suggest that some individuals may remain suppressed beneath the canopy for 80 years or more, attesting to the shade tolerance of the species. Those characteristics are important in allowing spruce to compete in the long term with its more vigorous but shorter-lived competitors, balsam fir and white birch (Betula papyrifera var. cordifolia [Marsh] Regel). At present, red spruce in all size classes can show symptoms of disease or injury in the montane forests of New York, Vermont and New Hampshire (Johnson and Siccama 1983; Burgess et al. 1984; Friedland et al. 1984c). Individuals show needle discolouration and death, crown thinning, an altered branching pattern, and the eventual death of branches and individuals. Friedland et al. (1984c) observed that in all size classes, browning of the newest needles was widespread after the winters of 1983-84. They suggest that the most common observation of foliar loss - that it is most noticeable at the outer tips of the branches - could be accounted for by repeated winter damage to new foliage. Typically (but not uniformly), the crowns of mature, declining red spruce appear to die back from the top downward, and from the outside inward (Johnson and Siccama 1983). Severely declining trees generally have five or fewer year classes of needles. Juveniles also show foliar loss at the top of the crown, and in cases where foliage extends to near ground level, the tops may show severe symptoms of foliar loss and repeated winter injury, while the lower-most branches show no injury and retain 10 or more year classes of needles. This has been interpreted as an effect of snowpack protecting the lower-most branches (Friedland et al. 1984c).

Studies of root and butt rot fungi in the northern Appalachians indicate the presence of pathogens which play a secondary role. Early studies (Hadfield 1968; Mook and Eno 1956) indicated infection by Cytospora kunzii, Fomes pini and Armillaria mellea, and the absence of insects and pathogens which might be a primary cause of disease. A recent study of red spruce in hardwood, transition and boreal stands in the White Mountains, Green Mountains, Adirondacks and Catskills showed that declining spruce were infected with Armillaria mellea, and that when stratified by elevation, there was a positive correlation between the number of roots colonized by the fungus and the amount of crown deterioration (Carey et al. 1984). Armillaria infected approximately 25% of the recently dead and severely declining trees in the boreal zone, 55% in the transition forests, and approximately 78% in the hardwood zone, indicating that Armillaria is an agent related to deteriorating red spruce but, owing to its absence in many declining trees, not the main cause of deterioration.

DECLINE, DISEASE AND INJURY

Burgess et al. (1984) have offered an important perspective on the current stature of red spruce in the northern Appalachians. Most commonly, "decline" has been the term used to describe the current deterioration of red spruce (i.e., Johnson and Siccama 1983; Carey et al. 1984; Siccama et al. 1982; Scott et al. 1984). Declines differ from most diseases in that they are not the result of a single causal agent. Manion (1981) has suggested a model for declines whereby three categories of stress may act in sequence to produce mortality. Environmental factors (i.e., poor site quality, airborne pollutants, nutrient deficiencies) may predispose a tree to greater than normal damage from a short term initating stress (i.e., drought, frost damage, insect defoliation, severe air pollution event), which is followed by contributing stresses, often lethal attack from biotic agents (i.e., root pathogens, bark beetles). Because there are several possible stress factors in each of the three categories, declines can be very difficult to resolve into their components. Historically, several tree declines dating back to the 1930s have been documented in North America (Manion 1981; Houston 1981; Mueller-Dombois 1983), often without satisfactory resolution of the causes. Most incidences of decline show a strong relationship with drought conditions (Burgess et al. 1984; Houston 1981) with the presence or absence of several other factors presumably determining the ultimate fate of the affected trees, since not all droughts trigger declines. It is also important to note that large-scale synchronized canopy diebacks have occurred and are occurring in unpolluted areas (Mueller-Dombois 1983).

The characteristic of slowly advancing dieback in red spruce (defined here as loss of foliage and branch death) and the presence of secondary pathogens are consistent with the concept of decline. On the other hand, declines are most commonly associated with mature trees (Manion 1981) and the observation of symptoms in all size classes (Johnson and Siccama 1983; Friedland et al. 1984c) is dissimilar to most other known declines (Burgess et al. 1984). The observation of needle damage and dieback in red spruce in essentially all size classes indicates that injury (short-term effects of a stress) is probably involved at some stage in spruce deterioration. There is no evidence to date that red spruce are deteriorating solely because of a biotic disease (defined here as a long-term interaction of a host with a pathogen), but it is important to note that the present understanding of biotic agents associated with deteriorating red spruce is incomplete.

Deterioration of Red Spruce Populations

Johnson and Siccama (1983) reported on the status of red spruce in northern hardwood, transition and boreal stands in the Appalachians. Data on the extent of crown dieback are shown in Fig. 1. The percentage of spruce which are in an advanced state of deterioration shows a positive correlation with elevation at the northern sites. Mortality and crown dieback at the Appalachian sites south of the Adirondacks are substantially lower.

Table 2 and Fig. 2 summarize data regarding the deterioration of spruce populations during the past two decades in quantitative surveys at Camels Hump, VT and Whiteface Mt., NY. Those data

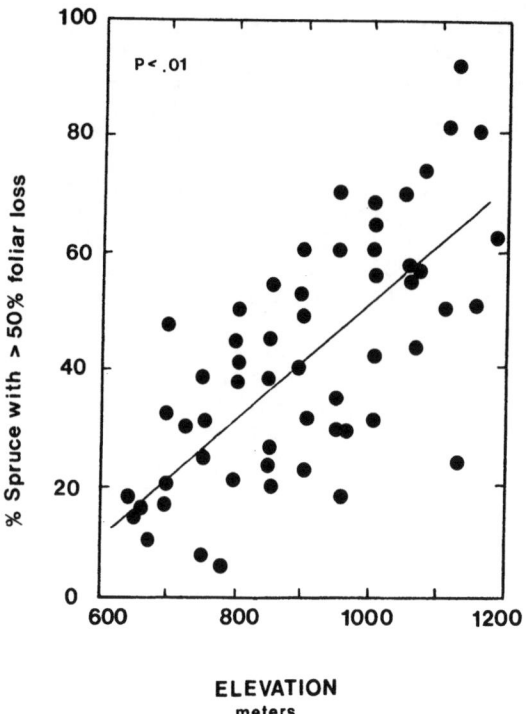

Fig. 1. Populations of dead and severely declining (>50% foliar loss from the live crown) red spruce at systematically sampled sites at Mt. Washington, NH; Whiteface Mt., NY and Mt. Mansfield, VT. Only trees >10 cm dbh reaching into the canopy were sampled (after Johnson and McLaughlin 1986). Data are for 56 transects, which had 15 to 20 spruce on each transect. Thirty transects were on east facing slopes, and 26 were on west facing slopes.

indicate about a 40 to 70% decrease in spruce basal area since the mid-1960s. Basal area decreases of roughly the same proportions occurred in the 2-10 cm dbh and >10 cm dbh size classes. In general, most of the change in total stand basal area is accounted for by the loss of spruce. Increases in white birch in <2 and 2-10 cm dbh classes have been recorded in the Green Mts. and at Whiteface Mt. as expected (Johnson and Siccama 1983), since birch is a vigorous colonizer of disturbed high elevation sites (i.e., Holway et al. 1969; Foster and Reiners 1983). Deteriorating or dead red spruce can occur as scattered individuals in conifer or hardwood stands where spruce is a minor component. Most commonly, patches of dead and deteriorating red spruce are observed in the subalpine coniferous forests in which balsam fir and white birch seedlings or saplings are abundant. In some cases, the patches have enlarged to a hectare or more in size due to blowdown. This secondary disturbance effect probably accounts for the decrease in balsam fir at high elevations on Whiteface Mt. (Scott et al. 1984). Above 900 m, red spruce less than 2 cm dbh decreased in density or frequency at Camels Hump and Whiteface Mt. by 32-79% between the mid-1960s and the later surveys, suggesting a displacement of red spruce by other species as the new canopy forms (Johnson and Siccama 1983; Siccama et al. 1982).

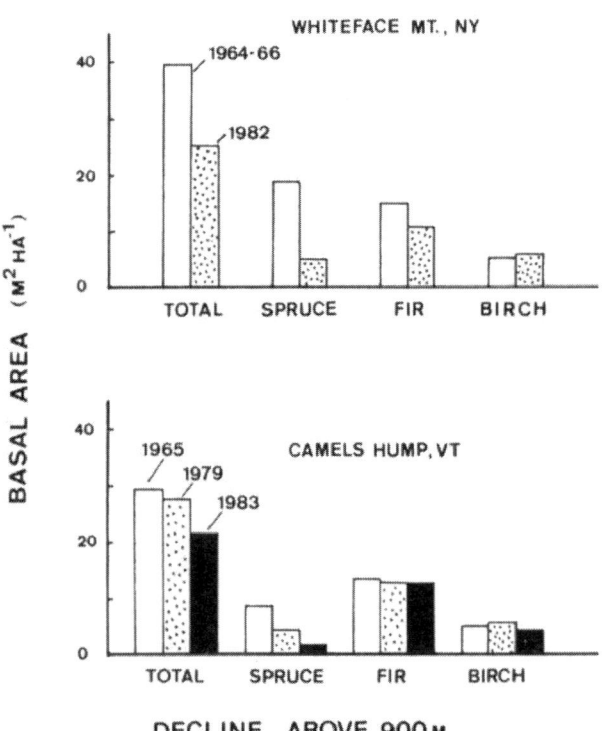

Fig. 2. Changes in live basal area over the past two decades for major species at Camels Hump, VT and at Whiteface Mt., NY. Basal area includes live trees >2 cm dbh at Camels Hump, and >10 cm dbh at Whiteface Mt. Data are for stands above 900 m at Whiteface and above 760 m at Camels Hump. (After Siccama et al. 1982; Scott et al. 1984, Helen Whitney, Univ. VT, pers. comm.)

Initiation of Red Spruce Deterioration

The current deterioration of red spruce in the northern Appalachians appears to have begun sometime between the late 1950s and mid-1960s. This conclusion is based on documentation of patches of dying spruce in Vermont and New Hampshire in the early 1960s and by examination of tree rings. Wheeler (1965), Kelso (1965) and Tegethoff (1964) observed a "general dying of spruce" with symptoms similar to those recently described near Swift River, NH, Wildcat Mt., NH, Greeley Ponds, NH, and at

Hancock Branch, NH. The outbreak of dying red spruce was first reported in 1962 and the area affected had increased by 1964 or 1965. Browning of foliage of red spruce in all stages of deterioration, and browning and mortality in young spruce < 0.6 m tall were noted. The early observers considered the causes to be severe winter conditions combined with drought. Hadfield (1968) and Mook and Eno (1956) studied dead and dying spruce in Vermont and observed several secondary agents, but no primary pathogens or insects. Siccama (1974) observed that spruce in all size classes were suffering unusual mortality (particularly in the hardwood forest) during the period of 1964-1966.

Qualitative examination of annual growth increments of red spruce in the northern Appalachians suggested a series of abnormally narrow rings beginning during the early 1960s and often continuing to 1982 when the trees were cored (Johnson and Siccama 1983). Mean ring widths and ring width indices are shown in Figs. 3-5 for representative transects of the northern study sites. Those data show a marked change in red spruce growth sometime during the early mid-1960s. The same pattern was found for ring width indices for red spruce in hardwood spruce stands in the central Adirondacks (Roman and Raynal 1980) and in old-growth red spruce at Lake Arnold, NY (Cook, this volume).

The tree ring evidence is consistent with the contention that red spruce have been deteriorating for approximately 20-25 years, and that the deterioration has been synchronous across the northern Appalachians. Cook (this volume) has shown that the ring width decline is not solely a function of long-term response to climate, but may be a threshold effect of climatic stress or disease, or perhaps a response to airborne pollutants. Acceptable interpretations of the available field reports and tree ring evidence are that sometime around 1960-65, the balance of environmental factors stressed red spruce beyond its ability to recover completely and a decline ensued, or that continuing or increasing stress to which red spruce is particularly susceptible has caused continuing, widespread injury, mortality, and disintegration of high elevation stands. There is, at present, no indication that balsam fir are declining except as a result of gap formation and the attendant increase in microclimatic stress (i.e., increased susceptibility to wind and winter damage). Tree ring series from balsam fir generally do not show the abrupt change observed in red spruce in the 1960s, except at the front of naturally occurring fir waves which occur at elevations above the spruce-fir zone (Johnson and McLaughlin 1985; Marchand 1984).

SUMMARY OF POSSIBLE CAUSES

The dieback symptoms are currently most pronounced in the sub-alpine coniferous forest, which is an environment subject to stress from wind, cold winter temperatures and nutrient poor soil, and subject to comparatively high rates of acidic and heavy metal deposition. Sorting out the factors responsible for the large-scale deterioration of red spruce is complicated by the range of demonstrated and potential stress factors which must be

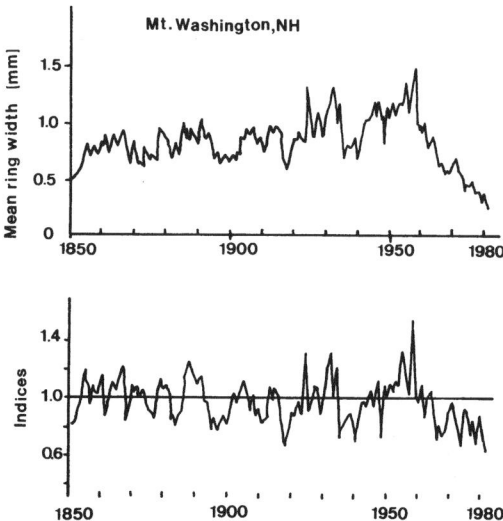

Fig. 3. Average ring width values and average tree ring indices for red spruce on the east slope of Mt. Washington, NH. data based on 2 cores from each of 15 trees at 800-900 m (after Johnson and Mclaughlin 1986). Individual chronologies ranged from 80 to 245 years (x = 144y). Ring width indices are created using the procedures of Blasing et al. 1983. Means are biweight means, weighted means of raw ring widths which minimize the importance of the outliers (Mosteller and Tukey 1977).

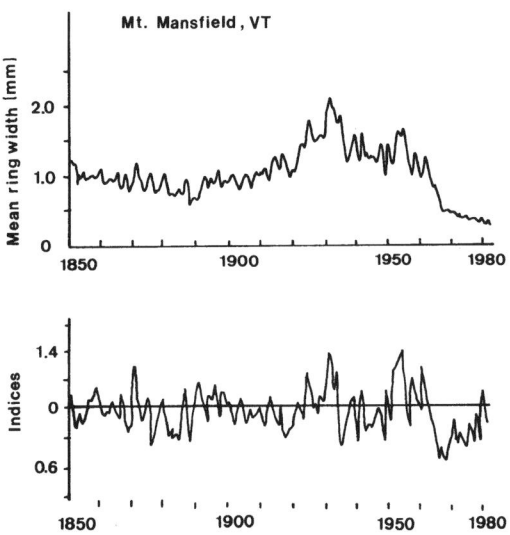

Fig. 4. Average ring-width values and average tree ring indices for red spruce on the east slope of Mt. Mansfield, VT. Data are based on 2 cores from each of 15 trees at 800-900 m. The cause of the release after 1920 is not known, but occurs at low elevation from approximately 80-210 years (x = 133y).

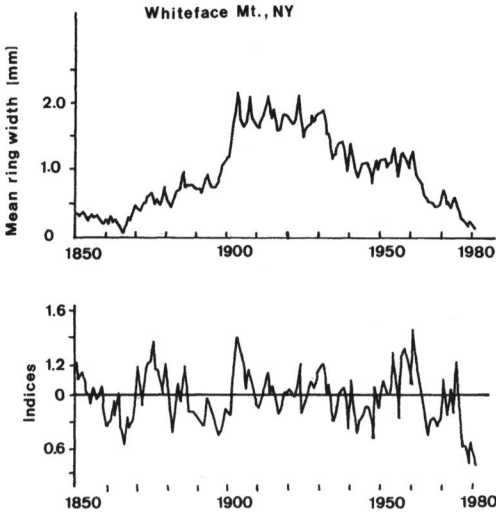

Fig. 5. Average ring width values and average tree ring indices for red spruce on the east slope of Whiteface Mt., NY. Data are based on 2 cores from each of 15 trees at 800-900 m. A strong release is noted following logging of the site in the 1890s (Holway et al. 1969). Length of the chronologies ranged from approx. 80 to 210 years (mean age = 104 years).

considered, the background of mortality due to normal stand dynamics, and the secondary effects of stand destabilization. The break-up of the stands proceeds as a positive feedback system - the gaps increase climatic stress and susceptibility to wind, then blowdown enlarges the disturbance patches, exposing more trees. It appears that the break-up of the deteriorating stands was triggered by rather extensive mortality in spruce, and the observation that red spruce in all size classes now show the development and progression of dieback argues that at least some of the factors causing spruce mortality are still operating. Given the current state of understanding, air pollution is a reasonable candidate for consideration, but no direct evidence linking ambient levels of airborne pollutants to stress in red spruce has yet been offered. On the other hand, reports of widespread, severe and synchronized mortality in red spruce between 1871 and 1885 in eastern North America (Hopkins 1891, 1901) suggest that large-scale, synchronous deterioration of red spruce populations could be unrelated to pollutant deposition, presumably triggered by natural events. The episodes mentioned above occurred within about a 15-year period in West Virginia, New York, New Hampshire, Maine and New Brunswick, and were tentatively attributed to spruce beetle (Dendroctonous sp.) which were probably acting as a secondary agent, following some other stress. Thorough investigation of climatic conditions in the last few decades of the 19th century might provide useful information for comparison with the climatic patterns of the 1960s.

Table 2. Changes in live basal area of red spruce in the Green Mts. and at Whiteface Mt. during the past two decades.

Site	elevation (m)	no. plots[c]	interval	Stand basal area initial/final BA (m²/ha)	Red Spruce initial/final BA (m²/ha)	BA change by dbh class (%) >10 cm	2-10 cm
Camels Hump, VT[a]	883-1158	43/25	1965-79	32.1/27.7	6.84/3.81	-44	-42
Camels Hump, VT	762-853	20/20	1965-79	26.5/25.1	6.62/3.56	-40	-84
Jay Peak, VT[a]	883-1158	32/31	1965-79	30.2/28.1	3.65/3.37	+02	-69
Abraham, Bolton Mt. VT (pooled)	762-853	26/24	1965-79	31.5/34.1	2.82/0.34	-87	-90
Whiteface Mt., NY[b]	900-1200	20/20	1964-82	39.5/24.1	16.5/4.8	-72	-16
Whiteface Mt., NY[b]	500-900	12/12	1964-82	37.6/28.3	9.3/5.4	-43	n.d.

[a] Siccama et al. 1982; Johnson and Siccama 1983
[b] Scott et al. 1984
[c] Siccama, unpublished data

In evaluating the possible primary causes of the current red spruce deterioration, I regard the recognized possibilities as follows:

1. Unlikely: Natural stand dynamics due to aging or successional characteristics.

2. Possible but untested: Drought, biotic disease; repeated or severe winter damage resulting from climate change; pollutant effects; combinations of pollutant and natural stresses.

The pattern of widespread and synchronous mortality, mortality in all size classes, mortality in stands of widely differing basal area, age, composition, and disturbance history argue against aging or competitive pressures as key factors causing mortality. Some of the stands, particularly those on Whiteface Mt., were subject to harvesting late in the 19th century and some of the high elevation areas have been undisturbed for at least 200-300 years by logging or fire (Figs. 3-6), Holway et al. 1969; Foster and Reiners 1983; Johnson and McLaughlin 1985). While the regrowth of spruce-fir stands after harvesting can produce an abrupt decrease in ring width (Meyer 1929), the change in ring width occurs in stands which show no signs of release and are thus not well described by a model developed for even-aged stands. The deterioration of stands which were harvested less than a century ago argues against the contention that the observations of dead spruce are due to the break-up of old stands, and that the decline of stands at Camels Hump apparently began when basal area values were fairly low (Siccama, 1974). Mortality and changes in radial growth characteristics in all size and age classes and in forests with different disturbance histories argue against the idea of "cohort senescence" (mortality of an even-aged canopy) which has been applied to other synchronous episodes of tree death (Mueller-Dombois 1983). Additionally, Foster and Reiners (1983) indicate that the current episode of spruce mortality is a disturbance regime different from that which has produced the current vegetation patterns in disturbance patches in a virgin forest at Mt. Washington, NH.

Widespread drought in the northeastern U.S. in the mid-1960s was particularly prolonged and severe (Cook and Jacoby 1977). As droughts are often associated with declines of other species (Manion 1981; Houston 1981; Burgess et al. 1984), the drought of the 1960s has been forwarded as a possible cause of the current spruce deterioration (Kelso 1965; Tegethoff 1964; Hadfield 1968, Johnson and Siccama 1983). It is difficult to assess adequately the severity of drought stress in high elevation forests. Examination of Palmer Drought Severity Index values (which are, at best, modest representation of high elevation drought) shows that peak drought years were 1964-66 (Johnson and Siccama 1983). The data of Siccama (1974) indicated that there were 40 to 60 cm of rain at Camels Hump during the ice-free season of 1964-66, with a strong gradient of increasing precipitation with increasing elevation. The increase in precipitation with increasing elevation argues against drought as the principal cause, as the degree of deterioration increases with increasing elevation. The persistence of dieback symptoms in smaller size classes in several types of forests (Friedland et al. 1984c) also argues against a drought of 20 years past as being the principal cause of the continuing deterioration. Finally, the observations of patches of dead and dying spruce in 1962 (Kelso 1965; Tegethoff 1964; Wheeler 1965) indicate that mortality was occurring prior to the period of most intense drought.

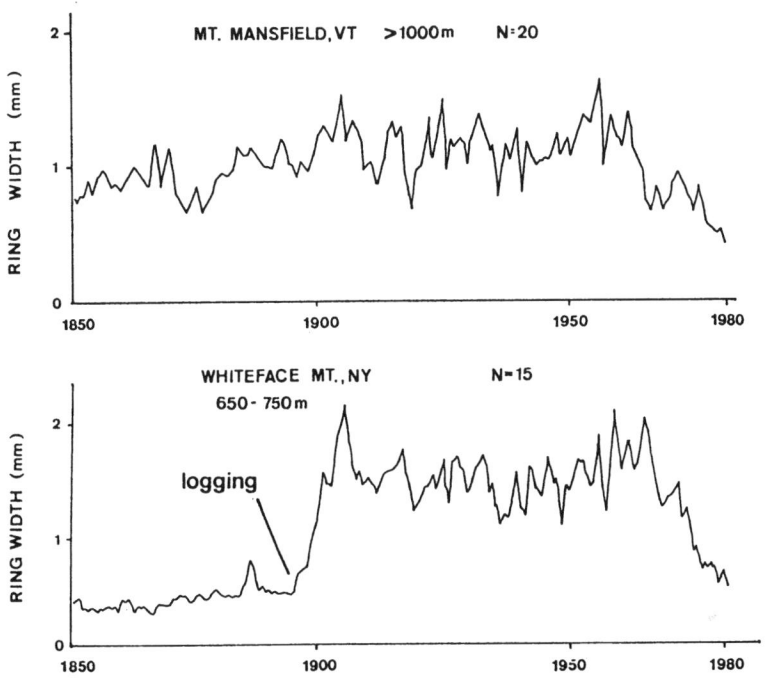

Fig. 6. Average ring-width values for red spruce in a relatively (1983) undisturbed stand at high elevation on Mt. Mansfield, VT and a heavily cut stand at low elevation at Whiteface Mt., NY.

Based on all reported observations over the past 20 years, it appears unlikely that red spruce are deteriorating solely because of a biotic agent. While no primary pathogens have been reported, there have not been any extensive, systematic studies. Further work is needed in this regard.

A second climatic anomaly occurred during the 1960s which could have had a regionwide, adverse influence on red spruce, namely the occurrence of abnormally cold winters from 1962-1971 across the eastern U.S. (Namias 1970). Several observations suggest that such an occurrence could be of consequence. Response function analyses of tree ring chronologies (Conkey 1979; Cook, this volume) suggest an adverse effect of cold winter temperatures on growth, and numerous reports of severe winter injury suggest that red spruce is susceptible to winter damage (Curry and Church 1952; Pomerleau 1962; Friedland et al. 1984c; Burgess et al. 1984). Early observers of the recent deterioration of spruce in New Hampshire observed severe winter damage in stands with dying spruce and speculated that severe winter conditions may have contributed to the mortality (Kelso 1965; Wheeler 1965; Tegethoff 1964). Conkey (1985) and Diaz and Namias (1983) have shown evidence of cool spring temperature in New England during the 1960s and 1970s. The effects of shortened growing seasons and colder than normal winters may be of particular importance to spruce growing near the top of its elevational range (1000-1200 m), where the growing season may be critically short even under average conditions. Diaz and Quayle (1978) showed that the average winter temperature in the 1960s and 1970s was considerably colder (3°F) than the long-term average, and it seems possible that the string of severe winters has been an important factor in the current episode of spruce deterioration.

Hypotheses involving pollution effects are abundant in the popular literature and in the literature of the acid deposition research effort. Acidic substances, heavy metals and gaseous pollutants, alone or in combination, could alter life-maintaining processes in forests. Whether or not they can do so to the extent necessary to cause mortality cannot be adequately evaluated now. Many pathways involving direct effects or soil-mediated effects are possible (EPA 1984; Burgess et al. 1984), but are not yet demonstrated to be of consequence in subalpine forests. A clear assessment awaits the results of experiments designed to test specific hypotheses regarding the effects of ambient levels of pollutants on key biological and biogeochemical processes. In this regard, the initiation of changes in spruce diameter growth and mortality during periods of climatic anomalies suggests that interactions of pollutant stress and water or cold stress should be investigated.

Testing of hypotheses involving interactions of natural and pollution stress will be particularly important since it will probably be very difficult to reconstruct the effect of the 1960s climatic anomalies in a way that will show clearly that climatic factors alone could have triggered the recent deterioration.

REFERENCES

Adams CC, Burns GP, Hankinson TL, Moore B, Taylor N (1920) Plants and animals of Mt. Marcy, New York. Parts I, II, III. Ecology 1: 71-94, 204-233, 274-288
Blasing TJ, Duvick DN, Cook ER (1983) Filtering the effects of competition from ring width series. Tree Ring Bull 43: 19-30
Burgess RL, David MB, Manion PD, Mitchell MJ, Mohen VA, Raynal, DJ, Schaedle M, White EH (1984) Effects of acidic deposition on forest ecosystems in the northeastern United States: an evaluation of current evidence. New York State College of Environmental Science and Forestry, Syracuse, NY
Carey AC, Miller EA, Geballe, GT, Wargo PM, Smith WH, Siccama TG (1984) Armillaria mellea and decline of red spruce. Plant Disease 68: 794-795

Conkey LE (1979) Response of tree ring density to climate in Maine, U.S.A. Tree Ring Bull 19: 19-38

Conkey LE (1985) Red spruce tree ring widths and densities as indicators of past climate in Maine, U.S.A. (unpublished manuscript)

Cook ER, Jacoby Jr GC (1977) Tree-ring-drought relationships in the Hudson Valley, New York. Science 198: 399-401

Cook ER (this volume) The use and limitations of dendrochronology in studying the effects of air pollution on forests.

Curry JR, Church TW (1952) Observations on winter drying of conifers in the Adirondacks. J Forestry 50: 114-116

Diaz HF, Quayle RG (1978) The 1966-77 winter in the contigious U.S. in comparison with past records. Monthly weather review 106: 1393-1421

Diaz HF, Namias J (1983) Association between anomalies of temperature and precipitation in the western U.S. and western northern hemisphere and 700 mb height profiles. J Clim and Appl Met 222: 352-363

EPA (U.S. Environmental Protection Agency) (1984) The acidic deposition phenomenon and its effects. Critical Assessment Review Papers, vol 2, Effects sciences. U.S. EPA, Washington, DC

Foster JR, Reiners WA (1983) Vegetation patterns in a virgin subalpine forest, Crawford Notch, White Mountains, New Hampshire. Bull Torrey Bot Club 110: 141-153

Friedland AJ, Johnson AH, Siccama TG (1984a) Trace metal content of the forest floor in the Green Mountains of Vermont: Spatial and temporal patterns. Water Air Soil Pollut 21: 161-170

Friedland AJ, Johnson AH, Siccama TG, Mader DL (1984b) Trace metal profiles in the forest floor of New England. Soil Sci Soc Am J 49: 4222-4225

Friedland AJ, Gregory RA, Karenlampi L, Johnson AH (1984c) Winter damage to foliage as a factor in Red spruce decline. Can J For Res 14: 963-965

Hadfield JS (1968) Evaluation of diseases of red spruce on the Chamberlin Hill Sale, Rochester Ranger District, Green Mountain National Forest. File Report A-68-8 5230. USDA- Forest Service Northeastern Area, State and Private Forestry Amherst FPC Field Office, Amherst, MA, p 7

Hanson DW (1980) Acid precipitation induced chemical changes in subalpine fir forest organic soil layers. MSc Thesis, The Graduate School, Univ of Maine, Orono

Harries H (1966) Soils and vegetation in the alpine and subalpine belt of the Presidential Range. PhD Thesis, Rutgers Univ, New Brunswick, NJ, p 542

Holway JG, Scott JT, Nicholson S (1969) Vegetation of the Whiteface Mountain region of the Adirondacks. In: Holway JG, Scott JT (eds) Vegetation-environment relations at Whiteface Mt. in the Adirondacks. Report No. 92, Atmospheric Sciences Research Center, State University of New York at Albany, p 1-44

Hopkins AD (1891) Forest and shade tree insects. II. W Virginia Experiment Station Third Ann Report, p 171-180

Hopkins AD (1901) Insect enemies of the spruce in the Northeast. U.S. Dept Agriculture Div of Entomol Bull No 28, new series p 15-29

Houston DB (1981) Stress triggered tree diseases, the diebacks and declines. USDA Forest Service NE Inf. 41-811. Washington, DC, p 36

Johnson AH, Siccama TG, Friedland AJ (1982) Spatial and temporal patterns of lead accumulation in the forest floor in the northeastern United States. J Environ Qual 11: 577-580

Johnson AH, Siccama TG (1983) Acid deposition and forest decline. Environ Sci Technology 17: 294a-305a

Johnson AH, McLaughlin SB (1986) The nature and timing of the deterioration of red spruce in the northern Appalachian Mountains. In: Acid deposition: long term trends. National Academy Press, Washington DC, p 200-230

Kelso EG (1965) Memorandum 5220, 2480 of July 23, 1965. US Forest Service Northern FPC Zone 6 Main Street Amerst, MA

Likens GE, Bormann FH, Pierce RS, Eaton JS, Johnson NM (1977) Biogeochemistry of a forested ecosystem. Springer-Verlag, New York, p 146

Lovett, GM, Reiners WR, Olson RK (1982) Cloud droplet deposition in a subalpine balsam fir forest: hydrological and chemical inputs. Science 218: 1303-1304

Manion PD (1981) Tree disease concepts. Prentice Hall Inc, Englewood Cliffs, NJ, p 399

Marchand PJ (1984) Dendrochronology of a fir wave. Can J For Res 14: 51-56

McIntosh RP, Hurley RT (1964) The spruce-fir forests of the Catskill Mountains. Ecology 45: 314-326

Meyer WH (1929) Yields of second growth spruce and fir in the Northeast. USDA Tech Bull No 142, p 52

Mook PV, Eno HG (1956) Relation of heart rots to mortality of red spruce in the Green Mountain National Forest. USDA Forest Service. Northeastern Forest Experiment Station. Forest Res Note 59, Upper Darby, PA

Mosteller F, Tukey JW (1977) Data analysis and regression: a second course in statistics. Addison-Wesley, Reading, Mass, p 588

Mueller-Dombois D (1983) Tree-group death in North American and Hawaiian forests: a pathological problem or a new problem for vegetation ecology? Phytocoenologia 11: 117-137

Myers O, Bormann FH (1963) Phenotypic variation in <u>Abies balsamea</u> in response to altitudinal and geographic gradients. Ecology 44: 429-436

Namias J (1970) Cliatic anomaly over the United States during the 1960's. Science 170: 741-743

Pomerleau R (1962) Severe winter browning of red spruce in southeastern Quebec. Can Dept For. Biomonthly Prog Report 18:3

Reiners WA, Marks RH, Vitousek PM (1975) Heavy metals in subalpine and alpine soils of New Hampshire. Oikos 26: 264-275

Roman JR, Raynal DJ (1980) Effects of acid precipitation on vegetation. In: Raynal DJ, Leaf AL, Manion PD, Wang CJK (eds) Actual and potential effects of acid precipitation on a forest ecosystem in the Adirondack Mountains. New York State ERDA Report 80-28, Albany, NY, p 4-1-4-63

Scherbatskoy T, Bliss M (1984) Occurrence of acidic rain and cloud water in high elevation ecosystems in the Green Mts. of Vermont. In: Samson PJ (ed) Transactions of the APCA specialty conference: the meteorology of acid deposition. Hartford, CT, Oct 16-19, 1983, p 449-463

Scott JT, Holway JG (1969) Comparison of topographic and vegetation gradients in forests of Whiteface Mountain, New York In: Holway JG, Scott JT (eds) Vegetation-environment relations at Whiteface Mountain in Adirondacks. Report No 92, Atmospheric Sciences Research Center, State Univ of New York, Albany, p 44-88

Scott JT, Siccama TG, Johnson AH, Breisch AR (1984) Decline of Red Spruce in the Adirondacks, New York. Bull Torrey Bot Club 1211: 438-444

Siccama TG (1974) Vegetation, soil and climate on the Green Mountains of Vermont. Ecol Monogr 44: 325-349

Siccama TG, Bliss M, Vogelman HW (1982) Decline of red spruce in the Green Mts of Vermont. Bull Torrey Bot Club 109: 163-168

Tegethoff AC (1964) Memorandum 5220 Sept 25, 1964. White Nat Forest. High elevation spruce mortality. US Forest Service Northern FPC Zone, Amherst, MA, p 2

Wheeler GS (1965) Memoradum 2400-5100 July 1, 1965. White Mt National Forest, Laconia, NH

Whittaker RH (1956) Vegetation of the Great Smoky Mountains. Ecol Monogr 26: 1-80

SUGAR MAPLE DECLINE IN ONTARIO

D.L. McLaughlin, S.N. Linzon, D.E. Dimma and W.D. McIlveen,
Ontario Ministry of the Environment, Air Resources Branch, 880
Bay St., Toronto, Ontario, Canada, M5S 1Z8

ABSTRACT

In response to numerous requests from members of the Ontario Maple Syrup Producers Association, a 20 point program was designed in 1984 to determine the role that acidic precipitation is playing in the decline of sugar maple trees in Ontario. A total of eight comparable sites were chosen for the study, seven of which are in the Muskoka-Parry Sound districts of central Ontario. One site in northwestern Ontario near Thunder Bay serves as a check location, being situated outside the zone of heavy acidic atmospheric deposition.

At each site, permanent observation plots have been established to determine the incidence and degree of tree decline. These plots will be revisited periodically over a number of years to provide a record of changes in tree condition. Adjacent to each plot, healthy trees and trees in various stages of decline were selected for sampling purposes. Foliage, twigs and roots were analyzed for 22 elements including Ca, Mg, Al, Pb, Cd, Mn, B, Zn and K. To assess the sensitivity of soil to acidic precipitation, soil samples collected from a soil pit at each site were analyzed for pH, base saturation, texture and the same chemical elements listed for tree parts.

Acidic atmospheric deposition rates, forest management practices, tree diseases, insect defoliation, site disturbance, tree age, site quality and weather records were investigated at the site locations. Sugar maple trees at each site are being studied also by dendrochronological methods. Increment cores were analyzed chemically by the PIXE multi-element scan method.

Preliminary results from field observations and laboratory examinations showed that the soil in the Muskoka-Parry Sound area is quite acidic with a low cation exchange capacity, low base saturation, and high soluble aluminum content. Lower amounts of nitrogen, sulphur and boron were found in foliage at the tops of declining trees compared to lower portions of the crown. Higher aluminum contents were found in the fine roots of declining trees compared to healthy trees. Root extremities of declining trees were infected by root rot organisms, predominantly _Armillaria mellea_, while roots of healthy trees were not as frequently infected. A forest tent caterpillar outbreak in 1976 and 1977 caused extensive defoliation and tree mortality, and radial growth studies showed that declining trees did not recover whereas healthy trees did recover following the insect attacks. Radial growth in the sugar maple stands generally has shown a reduced rate of growth for the past three decades, with declining trees showing a much greater loss of growth than healthy trees. Further, tapped trees on the same woodlot have displayed poorer growth than untapped trees.

It may be tentatively concluded from the first year's study that acidic precipitation is an additional stress to site, soil, adverse weather, insects, disease and management practices in the decline of sugar maple trees in Ontario.

INTRODUCTION

The decline of sugar maple in a rural area was first reported in Ontario in Grey County in 1952 (Griffin 1965). Symptoms of damage included discoloured foliage and death of young twigs followed by crown dieback. The progression was slow and trees seldom died. These symptoms agreed with those reported by Kessler (1963) in Michigan. The dieback was associated with overexposure due to overcutting in the stands. In addition, on dry sites, grazing by animals may have accelerated tree decline. Recovery occurred in moderately affected trees in both Ontario and Michigan.

Over the years, maple decline has been observed at intervals in different parts of Ontario, with severe outbreaks occurring in 1976 and 1977 in the Parry Sound and Owen Sound districts. In the latter episode, forest tent caterpillar defoliated the forests, and over 50% of the sugar maple trees were killed on 8000 ha and 500 ha in the Parry sound and Owen Sound areas, respectively (Canadian Forestry Service 1980). The insect population collapsed and most of the surviving sugar maple trees recovered.

In the spring of 1984, a number of maple syrup producers from the Muskoka area of Ontario complained to the Ministry of Agriculture and Food about the increase in dieback and mortality of sugar maple trees in the last six years. The producers felt that acidic precipitation may be involved and that continued sugar maple decline could jeopardize the local maple syrup industry.

The Ontario Ministry of the Environment was asked to investigate the role of acidic precipitation in this recent maple decline outbreak in Muskoka. A field study was designed by the Phytotoxicology Section, Air Resources Branch to examine the etiology of the declining sugar maple trees. Questionnaires were given to each woodlot owner to provide a history of stand management. Permanent observation plots were established in seven woodlots in the Muskoka area of south-central Ontario and in one woodlot near Thunder Bay in northwest Ontario. Soil from the woodlots and foliage, twigs and roots from sugar maple trees exhibiting a gradient of decline symptoms were collected for chemical analyses. Increment cores were collected from sampled trees, and discs were taken from felled trees to examine chronological growth patterns. In addition, atmospheric acidic deposition rates, forest management practices, the presence of disease and the history of insect defoliation, site disturbance, tree age, site quality and weather records were investigated at each study site location.

This report briefly describes the field methodology and presents a summary of preliminary results.

STUDY SITES

The study was conducted at eight sites, seven in Muskoka and one near Thunder Bay. Only five of the Muskoka woodlots had a history of management for maple syrup products. The sites were chosen in consultation with the Ontario Maple Syrup Producers Association and the Ontario Ministry of Agriculture and Food to represent a gradient of decline symptoms. Their geographical range extended from just north of Gravenhurst to Sundridge and from Gibson just east of Georgian Bay to Dorset. These sites were located within the Georgian Bay Section of the Great Lakes - St. Lawrence Forest Region, as defined by Rowe (1972). The Thunder Bay site was approximately 30 km NW of the city and was situated within the Quetico Section of the Great Lakes - St. Lawrence Forest Region. All sites were mature hardwood forests and contained sugar maple (_Acer saccharum_ Marsh.) as the dominant tree species. Associated tree species included red maple (_Acer rubrum_ L.), white ash (_Fraxinus americana_ L.), American beech (_Fagus grandifolia_ Ehrh.), black cherry (_Prunus serotina_ Ehrh.), white birch (_Betula papyrifera_ Marsh., yellow birch (_B. alleghaniensis_ Britton) and, occasionally, hemlock (_Tsuga canadensis_ (L.) Carr.).

Ground and aerial observations revealed that maple decline occurred throughout Muskoka with varying degrees of severity. Severe decline occurred in scattered, isolated pockets and more frequently on higher exposed sites. However, this pattern was not consistent. Occasionally, a few severely affected trees were observed distributed among healthy trees in an otherwise unaffected woodlot. On other sites most of the mature sugar maple over several hectares exhibited thin but normally green-coloured crowns. Similar decline symptomatology has been observed in sugar maple stands near Thunder Bay and elsewhere in Ontario on calcareous soil sites.

Atmospheric monitoring in Ontario has determined that the Muskoka region receives about 35 $kg.ha^{-1}.y^{-1}$ total wet sulphate deposition. Acidic deposition in the northwestern (Thunder Bay) section of the province is approximately 8 $kg.ha^{-1}.y^{-1}$. For this reason the Thunder Bay site was originally intended as an "acid rain control" area.

RELATIVE DECLINE INDEX

A detailed numerical rating system was designed to identify symptoms typically attributed to sugar maple decline. The rating system included observations of the number of dead branches, the percentage of the crown displaying undersized and discoloured foliage, and the presence of faults such as cracks, tap holes, cankers, fungal fruiting bodies and other wounds. Basic mensurational data were also recorded. The presence of individual decline symptoms and faults were recorded and assigned a weighted value. The weighted values for each tree were added to obtain a total score for the tree. This score was called a decline index. The healthiest trees have the lowest decline index and, conversely, the decline index increases with deteriorating tree condition. It may be that some of the parameters used in the decline index reflect different causes of tree decline. The index is merely used as a standardized way of assessing relative health.

A 20m x 20m permanent observation plot was established at each of the 8 study sites and every tree greater than 10 cm DBH was rated using the numerical decline index. A tree with a decline index score of less than 15 was considered essentially healthy, a score of 16 to 35 denoted a tree exhibiting light to moderate decline symptoms, and a severely affected tree scored greater than 35.

When the data from the seven Muskoka study sites were combined, 58% of the trees were considered healthy, 20% had light to moderate decline symptoms, and 22% exhibited severe maple decline. The decline gradient was similar for trees from the Thunder Bay site although a slightly higher percentage (68%) were in the healthy category.

There was no apparent relationship between the average decline index of each study site and stand stocking, basal area or total plot biomass. However, the decline index was related positively and significantly with diameter breast height ($t = 2.748$, $p < 0.01$) and can be described with the regression line equation:

Decline Index = 21.719 + 0.112 Diameter Breast Height

Therefore, it would appear that tree size or age was a predispositional factor to tree decline. These data are summarized in Table 1.

At least one half of the trees had one or more trunk faults or wounds. There was a tendency towards a higher decline index on faulted trees. The average decline index of sugar maple trees which were free of trunk faults and had not been tapped for maple syrup production was 20.6. The average decline index of the trees which had not been tapped but had at least one trunk wound was 19.4. Trees which were both tapped and wounded had a mean decline index of 31.9. However, trees which had been tapped but were otherwise fault-free had an average decline index of 33.2.

SOIL CHARACTERISTICS

At each of the eight study sites a soil pit was dug to classify the soil and determine its sensitivity to acidic precipitation as it may relate to tree decline. The site was described according to the guidelines prescribed by the Ontario Institute of Pedology, University of Guelph (1982). The soil was described and classified by the "System of Soil Classification of Canada" (Canadian Department of Agriculture 1978). Soil samples collected from horizons in each pit were air-dried, disaggregated, ground and sieved to two size fractions (10 mesh ASTM, 100 mesh ASTM) and analyzed for the following: pH (both H_2O and $CaCl_2$); exchangeable calcium, magnesium, potassium and aluminum; cation exchange capacity (CEC); base saturation (BS); pyrophosphate and dithionite extractable iron and aluminum; organic carbon; total carbonates; total nitrogen; extractable aluminum and sulphate; heavy metals (copper, nickel, lead and zinc) and texture. A summary of some of these analyses (mineral soil samples) from each study site is found in Table 2.

Table 1. Relationship between diameter class and tree decline index in sugar maple permanent observation plots.

Woodlot Owner	Woodlot Location	Mean Decline Index[a] per Diameter Class[b]					
		10-15 cm	16-20 cm	21-25 cm	26-30 cm	31-35 cm	>35 cm
Miller	Sundridge	18.1	18.7	37.0	37.0	35.8	34.6
Fincham	Burks Falls	12.3	17.7	12.5	21.8	44.0	—
Griffith	Burks Falls	12.0	21.0	12.0	26.6	12.0	19.0
MacLachlan	Huntsville	12.0	12.0	23.3	30.3	29.0	39.2
Boothby	Dorset	14.0	12.0	26.0	26.6	30.3	12.0
Veitch	Utterson	14.2	12.5	17.0	33.2	32.0	40.0
Gibson I.R.	Gibson	17.3	19.5	24.5	19.5	14.2	34.0
Mean, Muskoka Sites		14.3	16.2	21.8	27.9	31.8	34.8
Number of Trees		39	32	24	34	29	22
% of Total Trees		21.7	17.8	13.3	18.9	16.1	12.2
Mt McQuaig	Thunder Bay	10.5	11.0	26.2	33.7	25.3	29.7

[a] Decline Index gradient defined in text; [b] as Diameter Breast Height in cm. Decline Index = 21.719 + 0.112 Diameter Breast Height ($t = 2.748 \leq 0.01$, all sites combined).

The soils in the Muskoka area have developed on coarse textured glacial till material. The major rooting zone within the profiles rarely exceeded 40 cm in depth. Profiles consisted of a thin (less than 10 cm) surface horizon (Ah) rich in organic materials underlain by well-defined mineral horizons. Subsurface horizons (Bf, Bhf) were enriched with iron and aluminum.

All seven Muskoka soil sites were classified as naturally acidic podzols. The mean pH was 4.8 with soil pH values ranging from 4.0 to 5.3. The cation exchange capacities were very low, averaging 1.4 and ranging from 0.4 to 3.2. Base saturation varied from 4.4% to 60%. Soluble Al ($CaCl_2$ extracted) concentrations varied from a low of 5 ppm of a high of 33 ppm. No $CaCO_3$ was detected at any of the Muskoka soil sites. The soil characteristics of the Thunder Bay site were considerably different from the Muskoka sites, thus limiting the use of Thunder Bay as a control site.

Table 2. Soil sensitivity parameters of the mineral soil of selected sites in Muskoka and Thunder Bay.

Woodlot Location	Decline Index	Soil Classification	pH[a]	CaCO$_3$ (%)	CEC[b]	BS (%)	Exch Bases[b]	Sand (%)	SolAl[a] ($\mu g \cdot g^{-1}$)	Sol SO$_4$[a] ($\mu g \cdot g^{-1}$)	Sensitivity Rating[c]
Miller	34.7	Sombric Ferro-humic Podzol	4.4–5.2	ND[d]–ND	0.61–2.6	22–34	0.14–0.88	68–98	5–19	8–24	sensitive
MacLachlan	27.2	Sombric Humo-ferric Podzol	4.9–5.1	ND	0.37–1.5	17–28	0.10–0.29	37–89	7–25	10–18	sensitive
Griffith	26.6	Gleyed Ferro-humic Podzol	4.8–4.9	ND	0.35–1.9	4.4–21	0.04–0.33	65–95	6–33	10–19	sensitive
Gibson I.R.	22.4	Sombric Ferro-humic Podzol	4.2–5.2	ND	0.45–2.3	11–27	0.05–0.33	85–95	7–33	3–19	sensitive
Mt. McQuaig	21.1	Gleyed Eutric Brunisol	6.3–6.9	0.04–0.17	5.7–8.6	97–100	5.7–8.6	44–80	0.08–0.26	4–6	moderately sensitive
Veitch	21.0	Gleyed Ferro-humic Podzol	4.0–5.0	ND	0.80–1.8	27–38	0.22–0.65	67–80	12–20	18–40	sensitive
Boothby	19.6	Sombric Ferro-humic Podzol	4.8–5.3	ND	0.50–2.7	29–60	0.15–1.6	51–80	6–16	7–25	sensitive
Fincham	17.6	Sombric Humo-ferric Podzol	4.8–5.0	ND	0.55–3.2	16–44	0.18–1.42	32–85	8–30	10–27	sensitive

[a] pH – (H$_2$O); Sol Al – CaCl$_2$ extracted; Sol SO$_4$ – water extracted;
[b] in meq.100 g^{-1}
[c] based on base content (the product of CEC and BS; Wang and Coote 1981); [d] ND – Not detected

According to Cowell et al. (1981) and Wang and Coote (1981), podzolic soils contain the lowest amounts of bases and are the most susceptible to acidic precipitation impact (with regard to tree growth) due to base element deficiencies and aluminum toxicity problems. In contrast, the less acidic brunisolic soil (Thunder Bay) may be more sensitive to changes in soil pH due to acidic precipitation. Linzon and Temple (1980) found that soils with low pH values in the Parry Sound area were not pH-altered by acidic precipitation over an 18-year period from 1960 to 1978, whereas a brunisolic soil with an intermediate pH became more acidic.

Soils with pH values greater than 6.5 are in the carbonate buffering range. Soils in the 5.0 to 6.5 pH range are in the silicate buffering range whereby mineral weathering offsets changes in soil solution pH. Soils with pH 4.2 to 5.0 are in the cation exchange buffering range in which clay particles provide active exchange sites. Soils with a pH of less than 4.2 are buffered by aluminum hydroxides (Tomlinson 1983). The Muskoka soils lie primarily within the cation exchange and aluminum buffering range. Due to the coarse texture of the soils, the main source of cation exchange would be organic colloids, the majority of which are located in the thin surface horizon above the active rooting zone. Because of the acidic nature of the Muskoka soils, aluminum will be available to plants. In contrast, the Thunder Bay soil is only slightly acidic and also has a trace amount of carbonates. Carbonates will neutralize incoming acidity and maintain soil pH at a level that prevents aluminum from becoming soluble.

Wang and Coote (1981) classified soil sensitivity (with regard to tree growth) on base content, using the product of cation exchange capacity and base saturation (<6 meq/100 gm exchangeable bases = sensitive, 6-15 meq/100 gm= moderately sensitive, and >15 meq/100 gm = non-sensitive). Based on these criteria, all seven Muskoka sites are classified as "sensitive" and the Thunder Bay site as "moderately-sensitive".

These data suggest that the soil of the Muskoka area could potentially contribute to the mechanisms of tree decline.

TISSUE CHEMICAL CONCENTRATIONS

Foliage, twigs and roots were collected from trees displaying a gradient of decline symptoms from each of the eight study sites. In addition, soil samples were taken from around each of the sampled trees. All samples were analyzed for the following elements; Fe, Mn, P, N, Al, B, Ca, Cd, Cl, Co, Cr, Cu, F, K, Mg, Mo, Na, Ni, Pb, S, V and Zn. The premise was to compare chemical concentrations between healthy and declining tree populations.

Foliar samples were collected in early August, labelled, placed in polyethylene bags and refrigerated until they were processed (not washed and oven-dried). Analysis was performed by either ICAP or AAS, depending on the element string.

Of the 22 elements, only K was significantly lower in foliage from declining trees (0.77% - healthy versus 0.68% - declining, t = 2.60, $p \leq 0.05$). Conversely, in declining tree foliage, N was significantly higher than in the healthy trees (1.90% - healthy versus 2.05% - declining, t = 10.12, $p \leq 0.05$). Calcium and Mg, two elements under study in connection with forest decline in the northeastern USA and Europe, do not appear to be implicated in Muskoka. Neither the Ca or Mg content was significantly different in the two classes of trees. The foliar elemental data were comparable with foliar concentrations reported in the literature. Foliar nutrient concentrations reflect the availability of soil nutrients, and as there was no consistent relationship between low foliar chemical levels and tree decline, it was concluded that site micro- and macro- nutrients were not limiting.

The Muskoka foliar concentrations were also comparable with an extensive chemical data base developed by on-going biogeochemical research conducted in the same vicinity by the Environment Ministry's Acidic Precipitation in Ontario Study (W. Gizyn, pers. comm.).

Additional foliage was collected from healthy and declining trees felled for stem analysis in early September. Leaves were sampled from the very top and very bottom of the crowns. The greatest difference in the four data sets (top versus bottom, healthy versus declining) was observed between the top and bottom of the crown. There was a definite gradient towards lower levels of virtually all elements in the upper crown foliage. The upper leaves most often displayed decline symptoms. Taking into account the variability, the P, Ca, Mg, Pb, Ni, N, B and S levels in healthy trees were substantially lower in upper crown foliage. In addition, N, B and S were considerably lower in the upper crown foliage of declining trees. When the healthy and declining data sets were pooled to provide a sample population suitable for statistical testing, it was observed that P, Ca, N, B, S, Pb and Ni were significantly lower in upper crown foliage at the 5% level and Mg at the 1% level (Table 3).

Morrison (1985) noted a similar gradient from upper to lower crowns in sugar maple and reported comparable foliar concentrations from stands near Sault Ste. Marie in Ontario. It is known from biogeochemical research in Ontario that throughfall and stemflow in hardwood forests are nutrient-enriched relative to ambient precipitation (W. Gizyn, pers. comm.). Foliar leaching by rainfall is assumed to be a naturally occurring and important process in nutrient cycling in the forest ecosystem.

The chemical concentrations of twigs collected from healthy and declining trees, like the foliar analysis, did not implicate site nutrient deficiencies as factors contributing to the observed tree decline. All element levels except K were similar or higher in twigs from declining trees. Iron, P and N were found to be significantly higher ($P \leq 0.05$) in twigs from declining trees than from healthy trees.

Table 3. Foliar chemical concentrations of mature sugar maple trees (both declining and healthy) from the Muskoka Region of Ontario.

Element	Foliar Chemical Concentration[a]		Statistical Significance[b]
	Top of Crown	Bottom of Crown	
Fe	109	95	NS
Mn	1210	1630	NS
P	0.14%	0.18%	5% (t = 3.06)
N	1.92%	2.28%	5% (t = 2.79)
Al	70	60	NS
B	63	81	5% (t = 2.37)
Ca	8240	12000	5% (t = 2.84)
Cd	0.3	0.4	NS
Cl	0.04%	0.05%	NS
Cu	6	7	NS
Cr	1	2	NS
F	< 1	< 1	--
K	0.77%	0.89%	NS
Mg	1010	1680	1% (t = 3.34)
Mo	< 0.5	< 0.5	--
Na	15	12	NS
Ni	1.7	2.4	5% (t = 2.84)
Pb	1.7	2.2	5% (t = 2.73)
S	0.18%	0.22%	5% (t = 2.92)
V	< 1	< 1	--
Zn	27	29	NS

[a] not-washed, oven-dried, ppm (except % where noted), mean of 20 replications; [b] paired t-test.

Soil samples 0-30 cm in depth were collected from within the drip line of each of the sampled trees. Potassium was found to be significantly higher (t = 3.51 P $\leq$ 0.05) in soil around the declining trees. Therefore, soil $\overline{K}$ was not the reason for depressed foliar K levels. There were no statistically significant differences between the healthy and declining soil data sets for the remaining 21 elements.

The root samples were divided into "fine" roots (1 - 3 mm diameter) and "terminal" roots (0.5 - 1 cm diameter). All root samples were washed for 30 minutes in a warm water ultrasonic bath to remove adhering soil particles prior to processing for analysis.

The chemical concentrations were marginally lower in fine roots from declining trees for all elements except K, Al, Na and Ni (Table 4). The K levels in declining fine roots were 38% higher than healthy roots, although a large S.D. negated the significance of this relationship. The Al levels averaged 47% higher in fine roots from declining trees (4000 ppm - declining versus 2730 ppm - healthy, t = 2.41, p $\leq$ 0.05). This Al root relationship in the Muskoka area was consistent with every pair of healthy and declining trees investigated. The Al concentration of terminal roots was substantially lower than the fine roots, suggesting Al is not translocated even short distances along the root in the same concentration that it is absorbed at the root tip.

Many of the fine roots from the declining trees were dead, and root death is consistent with the observed crown decline symptoms. The occurrence of dead roots with elevated Al concentrations in unhealthy maple trees in the Muskoka region may suggest that Al toxicity is involved in the phenomenon of decline. It should be noted that Al concentrations in fine roots from declining trees from the Thunder Bay site were not higher than in healthy trees. At this site, soluble Al levels in soil pits were overall much lower than at Muskoka (Table 2). The soil collected from around the trees was analyzed for total, not soluble, chemical concentrations. There was no relationship between Al levels in fine roots of a particular tree and total soil Al in the soil around that tree.

It has been suggested that the dead root component of the root samples collected from declining trees may passively adsorb available soil ions and that this may account for the elevated Al levels in these samples. This can be discounted since passive adsorption should result in similarly elevated concentrations of light and heavy metals and halide ions. This was not apparent in tree roots from declining root samples. This suggests that Al may be selectively absorbed, although it does not appear to be translocated beyond the fine root system.

Table 4. Root chemical concentrations of healthy and declining sugar maple trees – Muskoka region.

Element	Chemical concentration[a]				Statistical Significance[b]
	Healthy Tree Roots		Declining Tree Roots		
	1-3 mm	0.5-1 cm	1-3 mm	0.5-1 cm	
Fe	785	268	730	350	NS
Mn	232	332	189	302	NS
P	0.60%	0.50%	0.50%	0.40%	5% (t = 2.45)
N	0.77%	0.45%	0.62%	0.53%	NS
Al	2730	2200	4000	1880	5% (t = 2.41)
B	11	11	10	12	NS
Ca	5200	6370	4600	6330	NS
Cd	1.2	0.8	1.0	0.9	NS
Cl	0.01%	0.01%	0.01%	0.01%	—
Cr	1	1	1	1	NS
Cu	10	4	7	3	NS
F	< 1	< 1	< 1	< 1	—
K	0.16%	0.20%	0.22%	0.10%	NS
Mg	890	760	745	840	NS
Mo	< 0.5	< 0.5	< 0.5	< 0.5	—
Na	37	17	58	23	5% (t = 2.93)
Ni	5	3	8	4	NS
Pb	5	2	3	2	NS
S	0.08%	0.03%	0.07%	0.03%	NS
V	2	1	3	1	NS
Zn	69	42	56	40	NS

[a] washed, oven-dried, ppm (except % where noted) mean of 10 replications; [b] paired t-test, mean Healthy vs mean Declining

Root samples from healthy and declining trees were collected for starch analysis. The roots (less than 1 cm diameter) were extracted with ETOH and each sample's colorimetric absorbance was compared with known absorbance versus concentration curve of D-glucose to determine starch content. The root starch concentration of healthy trees averaged 12.7% and ranged from 9 to 18%. Conversely, the root starch concentration of declining sugar maple trees averaged 9.7% and ranged from 3 to 14%. There was a marked tendency towards greatly reduced root starch levels in severely declining trees, although this could not be tested statistically because of the small sample population. Low root starch reserves reduce the tree's vigor which impedes its defence mechanisms and may predispose the tree to secondary pathogenic infection.

In summary, based on total chemical concentrations in the soil collected around the trees and from foliage, nutrient deficiencies were not implicated in the decline of sugar maple in Muskoka. A chemical gradient existed with higher foliar concentrations at the bottom of the crown compared to the top. This could be related to leaching by ambient rainfall. Accelerated foliar leaching by acidic precipitation could not be documented in 1984 because similar sampling of foliage from top to bottom of felled trees was not conducted in Thunder Bay, an area which receives substantially less acidic deposition than Muskoka.

INSECTS AND DISEASE

Armillaria sp. was observed in all Muskoka study sites. The characteristic black shoe-string rhizomorphs of Armillaria were present on approximately 50% of the root samples collected, and occurred more frequently on declining tree roots.

A severe infestation of forest tent caterpillar occurred in the study area in 1976 and 1977. The trees at some of the sites were defoliated twice in one season. Trees stressed by defoliation are predisposed to infection by A. mellea (Wargo and Houston 1974). Parker and Houston (1971) attributed the subsequent death of defoliated trees to A. mellea which invaded the weakened roots.

Although the total growing season (April to September) precipitation subsequent to defoliation was not significantly below normal, the May 1977 rainfall was below the five percentile level. In addition, April 1976 and June 1983 received less than 50% of the 20-year mean rainfall for these months. These drought conditions occurred at critical times for tree growth. The woodlot owners indicated that tree decline began between 1978 and 1981. It would appear that repeated insect defoliation, drought stress and subsequent root infection by A. mellea were contributing factors to the present decline syndrome.

INCREMENTAL GROWTH MEASUREMENTS

Two increment cores were taken at breast height from each of the six sample trees at each of the eight study sites. The cores were mounted, sanded and stained with phloroglucinol to highlight the annual growth rings. The ring widths were observed through a microscope and measured to an accuracy of 0.01 mm on a Bannister Incremental Measuring Machine interfaced with an Apple II microprocessor. Accompanying software modelled after Fayle and MacIver (1983) provided on-line ring-series graphics to continuously compare the measured cores and ensure accurate ring dating. Additional software programs transformed the actual ring width measurements into tree-ring indices as described by Fritts (1966). This is accomplished by fitting a regression line curve to each ring width series and dividing the actual ring width by each yearly value of the fitted curve. The resultant tree-ring index has a mean of approximately 1.00 and a variance which is independent of tree age, position within the stem and mean growth of the tree. The individual tree-ring indices were averaged to generate a sugar maple chronology for each study site, a chronology for healthy and declining tree populations and a Muskoka regional sugar maple chronology.

Indexed ring width and total growing-season rainfall was not significantly related at any of the eight study sites. Although the correlation was better, the relationship was still not statistically significant when total May, June and July rainfall was regressed with indexed ring-widths. These data indicate that rainfall was not a primary limiting factor to sugar maple growth in Muskoka.

Figure 1 is a histogram illustrating 10-year mean index values for the healthy and declining maple tree populations from 1905 to 1984. The tree condition, declining or healthy, is relative to observations made in the summer of 1984. For most of this century the "declining" trees were actually producing wider growth rings than the healthy trees, even though they are of the same species, approximately the same age and inhabit the same site. However, in the last decade, the declining trees averaged considerably less growth than the healthy trees (mean 10-year index value 0.94-declining versus 1.18-healthy). Also, the 10-year mean ring index value for the declining trees has decreased every decade subsequent to peak growth obtained in the period 1945-1954. For example, from 1945 to 1954 growth rings from the declining trees averaged 25.3% wider than those of the healthy trees. From 1955 to 1964, rings from the declining trees averaged 17.6% wider. In the next 10-year period, the mean ring width value was 11.7% wider than that of the healthy trees and from 1975 to 1984 the ring widths from the declining trees were 16.0% narrower. The healthy trees have not experienced a parallel trend in reduced ring growth. In addition, the growth depression associated with the 1976/77 caterpillar defoliation was considerably more severe in the declining trees (Figure 2). Furthermore, the healthy trees recovered from the defoliation, producing much wider rings subsequent to the collapse of the insect population, whereas no parallel growth recovery was apparent in rings from the declining trees.

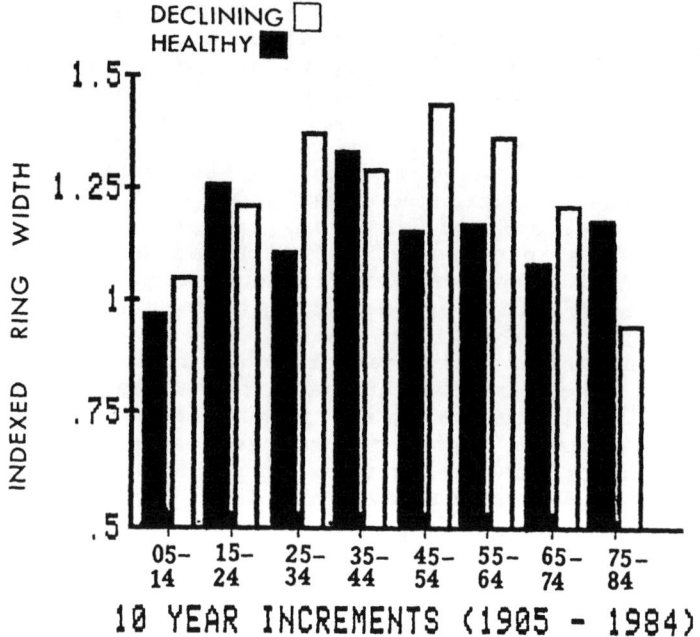

Fig. 1. Incremental growth of healthy and declining sugar maple trees from Muskoka, 1905 to 1984.

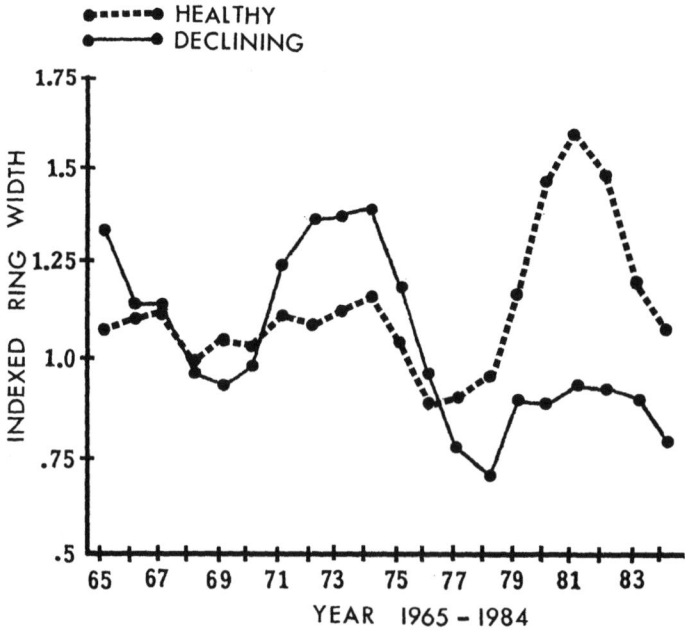

Fig. 2. Annual growth of healthy and declining sugar maple trees from Muskoka, 1965 to 1984. The area suffered an outbreak of forest tent caterpillar from about 1975 to 1978.

These data suggest that the current decline episode in Muskoka was likely initiated by the 1976/77 insect defoliation. Drought periods in the spring of 1976, 1977 and 1983 and infection by *A. mellea* hastened the decline of weakened trees. However, the declining trees appear to have been growing progressively poorer for 20 years previous to the insect epidemic, suggesting that these trees may have been predisposed to decline by physiologic, genetic or environmental factors.

CONCLUSIONS

It may be tentatively concluded from the first year's study of sugar maple decline at Muskoka, that the severe epidemic of forest tent caterpillar in the late 1970s combined with spring droughts in 1976, 1977 and 1983 were prime inciting factors. *Armillaria* root rot, tree age and site management were contributing factors. Some data from this study suggest that acidic precipitation is an additional environmental stress in the Muskoka area.

REFERENCES

Canadian Department of Agriculture (1978) The Canadian system of soil classification. Pub No 1646, Supply and Services Canada, Ottawa, p 164

Canadian Forestry Service (1980) Survey Bulletin. Forest insect and disease conditions in Ontario, summer 1980. Great Lakes Forest Research Centre, Sault Ste Marie, Ont, p 9

Cowell DW, Lucas AE and Rubec CDA (1981) Development of ecological sensitivity rating for acid precipitation impact assessment. Working Paper No. 10, Lands Directorate, Environment Canada, Cat. No. EN13-4/10E, Ottawa

Fayle DCF, MacIver D and Bentley CV (1983) Computer graphing of annual ring widths during measurement. Forestry Chron, Dec: p 291-293

Fritts HC (1966) Growth-rings of trees: Their correlation with climate. Science 154: 973-979

Gizyn W (pers. comm.) Acidic Precipitation in Ontario Study, Ontario Ministry of the Environment

Griffin HD (1965) Maple dieback in Ontario. Forestry Chron 41:295-300

Kessler KJ Jr (1963) Dieback of sugar maple - 1962. Research Note LS-13, Lake States For Expt Sta, USDA

Linzon SN and Temple PJ (1980) Soil re-sampling and pH measurements after an 18-year period in Ontario. Proc Int Conf Ecol Impact Acid Precip, Norway, SNSF Project, p 176-177

Morrison IK (1985) Effect of crown position on foliar concentrations of 11 elements in *Acer saccharum* and *Betula alleghaniensis* trees on a till soil. Can J For Res 15: 179-183

Ontario Institute of Pedology (1982) Field manual for describing soils, 2nd edn. OIP Pub No 82-1, p 38

Parker J and Houston DR (1971) Effects of repeated defoliation on root and root collar extractives of sugar maple trees. For Sci 17: 91-95

Rowe JS (1972) Forest regions of Canada. Dept. of Environment, Canadian Forestry Service, Pub. No. 1300

Tomlinson GH (1983) Dieback of red spruce, acid deposition, and changes in soil nutrient status - A review. In: Ulrich B and Pankrath J (eds) Effects of accumulation of air pollution in forest ecosystems. D. Reidel. Pub Co, p 331-342

Wang C and Coote DR (1981) Sensitivity classification of agricultural land to long-term acidic precipitation in eastern Canada. Land Resources Research Institute Contribution No. 98, Research Branch, Agriculture Canada, p 9

Wargo PM and Houston DR (1974) Infection of defoliated sugar maple trees by Armillaria mellea. Phytopath 64: 817-822

DEPOSITION AND FOREST CANOPY INTERACTIONS OF AIRBORNE NITRATE[*]

S.E. Lindberg, G.M. Lovett, and K-J. Meiwes[*]

Environmental Sciences Division, Oak Ridge National Laboratory, Oak Ridge, Tennessee, U.S.A. 37831

[*]Institute of Soil Science and Forest Nutrition, University of Göttingen, Göttingen, West Germany

ABSTRACT

Preliminary data suggest that atmospheric deposition of nitrate is generally higher to West German forests than to forests in the United States; however, many of these data are based only on bulk deposition. In two detailed studies of deposition processes in high- and low-elevation forests in the eastern United States, we quantified total wet plus dry deposition of nitrate using several approaches. Our data suggest that nitrate deposition by rain, particles, vapours, and cloud droplets significantly exceeds nitrate input measured in bulk deposition; hence, we cannot accurately determine current levels of input to many forests. Using data on the behaviour of deposited nitrate in the canopies of the two forests studied in detail, we estimated the magnitude of nitrate deposition not collected by bulk samplers used routinely in the Solling forest in West Germany. The results suggest that the unmeasured flux exceeds that measured as bulk deposition by factors of 1.3 to 6, depending on different assumptions. Calculation of deposition velocity indicates that this unmeasured fraction could be accounted for by dry deposition of nitric acid vapour or by deposition of fog or cloud droplets.

INTRODUCTION

Historical data indicate that atmospheric emissions of nitrogen oxides have increased in recent years, by 30% in the United States and by 50% in West Germany, and that levels of nitrate in air, deposition, and some soils have also increased (Robinson and Homolya 1983; Lobel and Thiel 1983). Recent evidence suggests a possible relationship between forest decline symptoms and high levels of nitrate deposition (Friedland et al., in press). For example, many symptoms of forest decline occur initially in high-elevation forests that receive elevated nitrate input because of orographic and fog/cloud-water effects (Lovett 1983).

The major sources of nitrogen oxides to the atmosphere are automobiles and power plants in which fossil fuels are combusted at high temperatures. In North America, approximately 50% of these emissions can be attributed to mobile sources such as automobiles (Robinson and Homolya 1983). In West Germany, automobile density is nearly 10 times higher than that in the United States as a whole (Fig. 1). Even in the more populated northeastern United States, automobile density is less than half

---

[*] Copyright of this paper: U. S. Government

of that in West Germany. These figures are reflected in the nitrogen oxide (NO_x) emission densities for each country (Fig. 1). The mean NO_x emission density in West Germany exceeds the average value for the United States by a factor of 6. However, for some smaller areas such as Bavaria, West Germany, and the Appalachian region of the United States, the values are comparable.

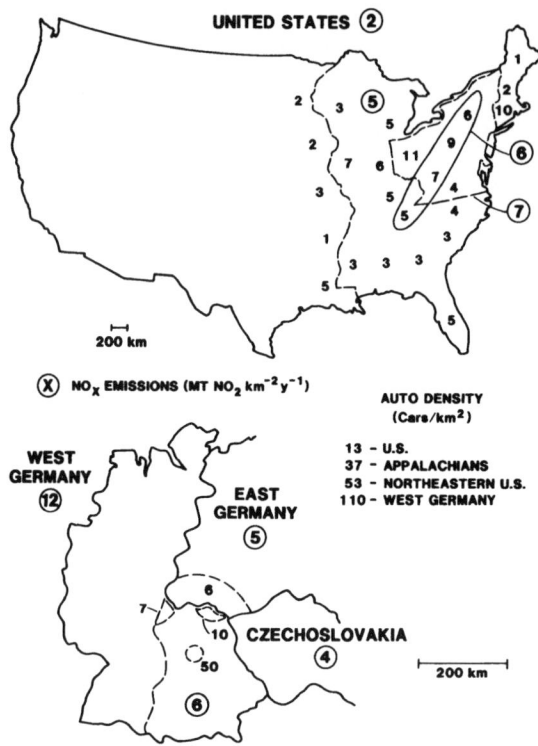

Fig. 1. Automobile densities (tabulated) and NO_x emission densities (values on maps) in West Germany and the United States. Circled numbers on the maps are means on regional or national scales, while uncircled numbers are state or smaller area NO_x emission densities.

Emitted nitrogen oxides are converted to different nitrate species in the atmosphere by the action of several airborne oxidants (NAS 1983). Recent studies indicate that wet deposition of nitrate is becoming increasingly important in the input of acidity to various ecosystems (Galloway and Likens 1981; Brimblecombe and Stedman 1982). However, atmospheric nitrate exists not only in precipitation but also in particle and vapour forms such as $Ca(NO_3)_2$, NH_4NO_3, and HNO_3 (NAS 1983). These can be efficiently scavenged by the high surface area of forest canopies (Hosker and Lindberg 1982).

Nutrient cycling studies in forests frequently estimate the input of airborne nitrate from bulk deposition collected by continuously open containers (NAS 1983). This method underestimates the true input because it is limited to the collection of wet deposition plus some small, but unknown, fraction of the total dry deposition. Hence, we cannot accurately determine the current levels of nitrate input to many forests. In this paper we summarize results of a recent detailed field study of wet and dry deposition of nitrate to a low-elevation deciduous forest in the southeastern United States (Lindberg et al. 1986). We compare these results with those published for a high-elevation conifer forest in the northeastern United States (Lovett et al. 1982), and with those recently summarized for bulk deposition in both high- and low-elevation forests in West Germany (Johnson and Cowling, in press; Matzner et al. 1982).

DEPOSITION MEASUREMENT METHODS

The chemistry, deposition, and canopy interactions of airborne nitrate were studied for 2 years in an oak stand (Quercus prinus, Q. alba) at Walker Branch Watershed, a deciduous forest in eastern Tennessee (340 m MSL). The site is 20 km downwind of a commercial coal-burning power plant (Lindberg and Harriss 1981), but is in a semi-rural area in hilly terrain dominated by forests with canopies of 20 to 25 m height. From 1981 to 1983 we studied the total atmospheric deposition of several plant nutrients and strong acids to the forest canopy, the flux of these materials to the forest floor, and the effects of acid deposition on soil and canopy cycling processes. Results of these studies have been published elsewhere (Lindberg et al. 1986; Johnson et al. 1985; Lovett et al. 1985).

To provide independent estimates of atmospheric input, several methods were employed to collect precipitation, particles, and vapours. We used automatic samplers for wet-only precipitation and throughfall events (Lindberg 1982); open containers for bulk deposition and collars for stemflow (Richter et al. 1983); inert surfaces and leaf extractions for dry deposition of large (>2 µm) particles (Lindberg and Lovett 1985); and filter packs, impactors, and standard analyzers for suspended particles and vapours (Lovett and Lindberg 1984; Huebert 1983; TVA 1982). Total dry deposition was determined from surface fluxes of large particles, extrapolated to the full canopy using a statistical model of throughfall fluxes (Lovett and Lindberg 1984), and from vapour and small particle (<2 µm) air concentrations, using deposition velocities measured on-site or elsewhere (Huebert 1983; Hicks 1984). Deposition velocities were as follows: during the growing season, HNO_3 vapour = 2 $cm.s^{-1}$ and small particles = 0.2 $cm.s^{-1}$; during the dormant season, HNO_3 vapour = 0.5 $cm.s^{-1}$ and small particles = 0.05 $cm.s^{-1}$.

MECHANISMS OF NITRATE DEPOSITION

We reported that dry deposition is a major input process for trace metals and sulphur to the Walker Branch forest (Lindberg and Harriss 1981). Dry deposition is also particularly important for NO_3^-, contributing ~60% of the total annual deposition of 7.6 kg NO_3^--N.ha^{-1}.y^{-1} (Fig. 2). Dry deposition clearly dominates the input during the summer growing season when the canopy is fully developed, providing a significant surface area for interaction with suspended particles and vapours. Wet and dry deposition are comparable during the winter dormant season when the canopy is barren and when the air concentrations of most particle constituents are at a minimum (Lindberg et al. 1986). As a result, the ratio of total summer/winter atmospheric deposition is ~3:1, indicating that nitrate input to this ecosystem peaks during the forest growing season when foliage exposure and vegetation sensitivity to airborne pollutants are at a maximum.

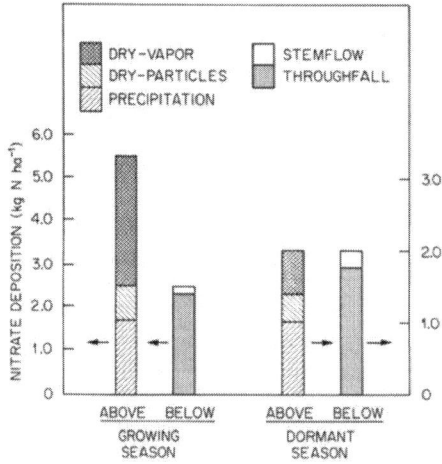

Fig 2. The contribution of several atmospheric deposition and internal transfer processes to the total flux of nitrate to the canopy and to the forest floor at the Walker Branch forest. Note different scales for dormant (November-March) and growing (April-October) seasons.

Most of the airborne NO_3^- at this site exists as HNO_3 vapour, a species not considered in deposition studies until quite recently (Huebert 1983). The mean annual air concentration of total NO_3^- (particle plus vapour forms) above the forest is 1.0 µg N.m^{-3}, ~70% of which is HNO_3 vapour. Nitric acid vapour is a major sink for atmospheric NO_x and is a highly soluble gas that is efficiently scavenged by precipitation (NAS 1983). However, its solubility apparently also results in efficient deposition to plant canopies during dry periods. Measurements at this site suggest that dry deposition of HNO_3 vapour contributes ~50% of the total annual atmospheric flux of NO_3^- to the canopy, making it the single most important nitogen oxide species studied here (Fig. 2). Dry deposition of small-particle NO_3^- is negligible compared to the flux of NO_3^- on large particles. Approximately 70% of particle NO_3^- apparently exists in the large fraction (Lindberg et al. 1986), which accounts for 99% of the particle flux to the canopy and 24% of the total NO_3^- dry deposition. Particle size data for NO_3^- must be interpreted cautiously because of possible sampling artifacts (Appel and Tokiwa 1981).

Dry deposition of NO_x also contributes to the total dry flux of NO_3^-, assuming oxidation of NO_x in the plant canopy. Dry deposition of NO_x at Walker Branch was estimated in a concurrent study (Kelly and Meagher 1985) to be ~2 kg $N.ha^{-1}.y^{-1}$, suggesting an even larger contribution of dry deposition to total input (~70%). However, this study utilized a chemiluminescent NO_x detector, and it is not clear to what extent these measurements of NO_x already include HNO_3 vapour (Kelly and Meagher 1985). Despite the acknowledged uncertainty in determining dry deposition in general (NAS 1983), we can still conclude that dry deposition is probably the dominant mechanism in the airborne flux of NO_3^- to this forest, contributing up to 70% of the total.

CANOPY INTERACTIONS OF DEPOSITED NITRATE

The fate of material deposited on the forest canopy is reflected to some extent in the ion flux in throughfall and stemflow solutions collected below the canopy (Fig. 2). During the winter, when the canopy is leafless, the above- and below-canopy fluxes are comparable, indicating only a moderate influence of the exposed bark surface. However, during the growing season, the canopy has a significant influence on the quantity of nitrate deposition that reaches the forest floor. Absorption by the canopy decreases the flux of atmospheric NO_3^- by ~50%, indicated by a NO_3^- uptake of 3.1 kg $N.ha^{-1}$ during this 7-month period.

Uptake of NO_3^- by the forest canopy is expected for a nitrogen-deficient ecosystem and apparently involves both wet- and dry-deposited nitrate. However, we believe that the generally longer residence time of dry deposition on foliage relative to precipitation favours the uptake of this material, although foliar absorption of NO_3^- from precipitation may also occur to some extent (Lovett and Lindberg 1984). Interestingly, the difference between the atmospheric flux of NO_3^- above and below the canopy during the growing season is nearly equal to the dry deposition of HNO_3 to the canopy during this period (3.0 kg $N.ha^{-1}$). The fate of absorbed NO_3^- in the canopy is not well understood. If nitrate reductase is present, the ion may be assimilated by plant cells. Fertilizer studies suggest this to be the case for some tree species (Eberhardt and Pritchett 1971). Another possibility is incorporation into epiphytes on the leaf surface, with subsequent release of the nitrogen in dissolved or particulate organic matter. Release of organic nitrogen in throughfall from forest canopies is well known (e.g., Carlisle et al. 1966).

DEPOSITION AND CANOPY INTERACTIONS IN OTHER FORESTS

The general conclusion that dry deposition may be a significant mechanism of input to forests is supported by published calculations (Fowler 1980; Shannon 1981); however, we are not aware of any directly comparable field data on nitrate deposition to forests. Dry deposition was identified as being potentially important in the atmospheric flux of nitrogen to forests in southern Sweden by Grennfelt et al. (1980). These authors used published air concentrations and deposition velocities to conclude that dry deposition contributed 40% of the total input of 4.2 kg NO_3^--N.ha^{-1}.y^{-1} and that HNO_3 vapour was the major form of dry-deposited nitrogen. In the United States, Huebert (1983) used micrometeorological methods during a 1-month study to measure short-term HNO_3 vapour dry-deposition velocities to a grassy field. His data yielded an estimate of dry deposition of 0.34 kg NO_3^--N.ha^{-1}.$month^{-1}$, roughly equal to the wet deposition measured nearby. Particle deposition was not determined. Hence, our results suggest a somewhat greater role of dry deposition (64% of total) than do results from these studies.

Atmospheric deposition is most commonly quantified in ecosystem studies by ground-level bulk precipitation collectors that are continuously open to the atmosphere. We can compare our results with deposition estimates from bulk precipitation collected at ground level during a concurrent study in this forest (Johnson et al. 1985). Atmospheric nitrate deposition estimated from bulk precipitation was 2.9 kg NO_3^--N.ha^{-1}.y^{-1}, which is only 38% of the value determined from separate collection of wet and dry deposition (Fig. 2). Our data suggest that bulk deposition significantly underestimates wet plus dry deposition of many ions, particularly those with an important vapour or small-particle component in the atmosphere (e.g., NO_3^-, SO_4^{2-}, and H^+; Lindberg et al. 1984). This comparison suggests that biogeochemical cycling studies in which system inputs are based solely on measurements of bulk precipitation must be interpreted cautiously. Unfortunately, it has generally been the case that no other data are available for forested sites.

In the United States there are published measurements of the regional wet deposition of NO_3^- based on routine monitoring (e.g., NADP 1985). The patterns for 1982 are shown in Fig. 3 in isopleth form. For comparison, the data described above for the Walker Branch forest are given to illustrate the degree to which wet deposition may underestimate total fluxes. Also shown in Fig. 3 are estimates of total NO_3^- deposition by cloud impaction and bulk precipitation to a high-elevation fir forest (Mt. Moosilauke) in the White Mountains of New Hampshire (Lovett et al. 1982), and bulk deposition fluxes of NO_3^- measured at several West German sites experiencing forest decline (compiled from Johnson and Cowling, in press). The regional wet-deposition isopleths suggest that the input of NO_3^- at Walker Branch is between 1.8 and 3 kg N.ha^{-1}.y^{-1} (the actual annual wet deposition is 2.8, Fig. 2). Hence, the use of precipitation chemistry monitoring data alone to estimate NO_3^- input to a forest such as this can lead to an underestimate of over 60%.

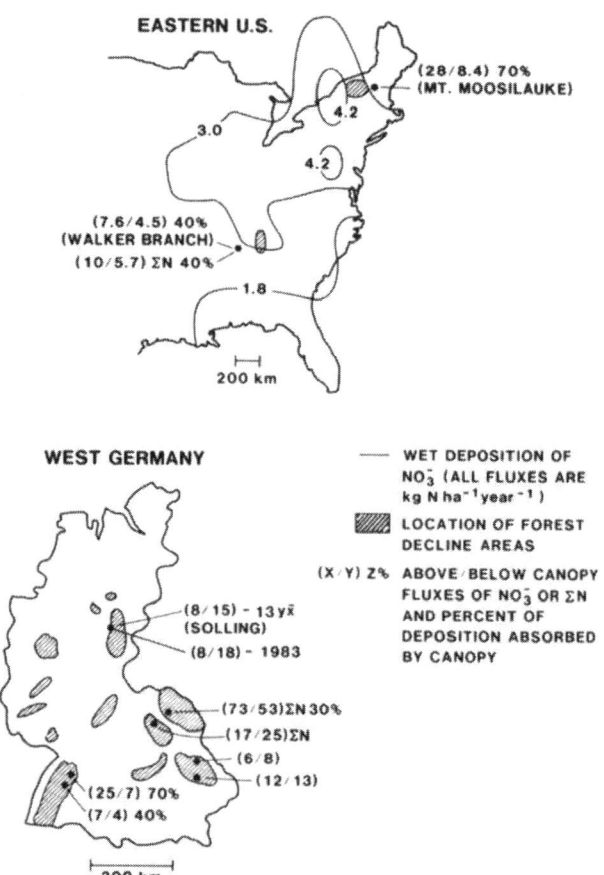

Fig. 3. Recent data on inorganic nitrogen fluxes above and below several forest canopies in the United States and West Germany, and isopleth lines showing wet deposition of NO_3^--N measured by the U.S. National Atmospheric Deposition Program. The values for individual sites are either NO_3^--N (if not labelled) or NO_3^--N plus NH_4^+-N (labelled ΣN). The individual values in the numerator for Walker Branch represent total wet plus dry nitrate deposition, while those for Mt. Moosilauke are bulk plus cloud-water nitrate deposition. All values in West Germany are for bulk deposition. The values in the denominator in each case represent the flux in throughfall plus stemflow. When the difference in fluxes suggests in-canopy uptake (flux above greater than that below), the rate is indicated as a percent of the flux to the canopy.

For the high-elevation forest the potential error is more severe; the isopleths suggest a NO_3^- input of less than 3 kg $N.ha^{-1}.y^{-1}$. However, Lovett et al. (1982) estimated the input of NO_3^- to this forest to be approximately an order of magnitude higher. In high-elevation forests such as this (1220 m above MSL), nitrate deposition is thought to be dominated by impaction of NO_3^--rich cloud droplets onto the canopy, rather than by dry deposition. Cloud droplets, like atmospheric vapours and particles, are scavenged by the forest canopy far more efficiently than they are collected by bulk deposition collectors. Using an estimate that this forest was immersed in clouds 40% of the time, the authors calculated that cloud deposition of NO_3^- was 23 kg NO_3^--N.ha^{-1}.y^{-1}, or 4.3 times the measured bulk deposition (5.3 kg NO_3^--N.$ha^{-1}.y^{-1}$).

The sum of cloud plus bulk deposition was used to estimate the total input of 28 kg NO_3^--N.$ha^{-1}.y^{-1}$ in Fig. 3; hence, this value does not include a separate measure of dry deposition. However, airborne NO_3^- is low at remote locations such as this (Kelly et al. 1985), and is efficiently absorbed by cloud droplets such that its transfer to the canopy probably occurs primarily by cloud impaction. These diverse results for both high- and low-elevation forests in the United States support the conclusion that bulk deposition leads to significant errors in quantifying total NO_3^- input to forests.

The data on nitrogen deposition to West German forests suggest that nitrate input is generally higher than that in the United States (Fig. 3). However, these data are difficult to interpret for many reasons. All of the measurements are based on bulk deposition, some for NO_3^- and some for total inorganic nitrogen $[(NO_3^- + NH_4^+) = \Sigma N]$, and much of the data must be considered as preliminary (Johnson and Cowling, in press). Thus, it is also true for West Germany that we do not know the true deposition of nitrogen to forests, including those forests that show various symptoms of decline.

Although the data are limited, it may be possible to estimate both dry and total deposition of nitrate to West Germany forests, using existing bulk deposition and throughfall data. These estimates rely on certain assumptions about the behaviour of deposited nitrate on foliage, based on measurements of total deposition above and below the canopy. For example, the data from the two studies in the United States indicate significantly higher fluxes of NO_3^- in total deposition above the canopy than those measured in throughfall and stemflow (Fig. 3), indicating NO_3^- uptake by the canopy. The amounts absorbed range from 40% of the total input by the oak canopy at Walker Branch (for both NO_3^- and ΣN) to 70% by the fir canopy in the Northeast. The data discussed above for Walker Branch suggest that the absorbed material may be primarily dry deposition because of its longer residence time on canopy surfaces. If this is the case, the absorbed NO_3^- at Walker Branch represents nearly 65% of the total dry deposition. Thus, over half of the dry-deposited NO_3^- is not accounted for in throughfall plus stemflow.

Data for three of the West German sites indicate similar behaviour, with the amount of NO_3^- or ΣN retained in the canopy ranging from 30 to 70% of the above-canopy deposition. However, the data from four of the West German sites indicate a very different behaviour of deposited NO_3^-; the fluxes below the canopy exceed the bulk deposition fluxes above (or to) the canopy by 1 to 20 kg $N.ha^{-1}.y^{-1}$. Comparing the below-canopy fluxes at the two study sites in the United States with bulk deposition to the canopy reveals this same pattern (flux below greater than that above), with the differences being 1.8 (Walker Branch) and 3.1 (Mt. Moosilauke) kg $N.ha^{-1}.y^{-1}$. Patterns such as these are commonly interpreted as indicating a net removal of NO_3^- from the canopy, the source being either washed-off dry deposition or leaching of foliar NO_3^-.

Assuming foliar leaching of NO_3^- to be negligible, a reasonable assumption for most nitrogen-deficient forests, net removal of NO_3^- can be used as a measure of dry deposition or of some other input process not sampled in bulk deposition (e.g., cloud-water or fog deposition). Such calculations also assume no foliar uptake of dry-deposited NO_3^-, an unreasonable assumption based on much of the data in Fig. 3. However, it may be possible to infer dry-deposition rates to a West German forest using the relationships among bulk deposition, throughfall plus stemflow flux, and dry deposition as measured at the Walker Branch forest. Similarly, the relationships among bulk, throughfall, and cloud deposition at the Mt. Moosilauke site could also be used. These relationships indicate that annual net throughfall (throughfall plus stemflow minus bulk deposition) is ~30% of annual dry deposition for the oak forest at Walker Branch and ~10% of annual cloud input at the fir forest at Mt. Moosilauke. For the purposes of our calculations, we will use a value of 20% and bracket the estimates of the "unmeasured" dry or fog deposition at the West German site by assuming that this value could lie anywhere from 20% (which would indicate a significant in-canopy uptake) to 100% (no uptake).

We used these assumptions to infer dry (or unmeasured) deposition rates from bulk deposition and throughfall data collected in a spruce stand at the Solling forest in West Germany (Fig. 3). This site was chosen because of its long historical record of nitrogen cycling data and the availability of recent air concentration and element flux data, and because these data indicate an increase in the NO_3^- flux below the canopy in recent years (Matzner et al. 1982; Ulrich and Pankrath 1983; Lobel and Thiel 1983; Johnson and Cowling, in press). The air concentration data are particularly useful because they can be used to determine whether or not the dry-deposition estimates are reasonable. This can be done by calculating dry-deposition velocities, V_d, (Chamberlain 1966) from the ratio of the estimated dry-deposition rates divided by the measured air concentrations. These deposition velocities can then be compared with published values for other forests.

The results of our calculations using different assumptions are summarized in Table 1. The estimated "dry"-deposition rates

Table 1. Nitrate dry deposition fluxes and deposition velocities (V_d) calculated from field data collected at the Solling spruce forest. These values have been estimated for the "dry" or unmeasured atmospheric input not collected by bulk deposition samplers in the field.

	FIELD DATA		
Process	Data set	Deposition[a]	Rainfall (cm)
Bulk precipitation (BP)	1983	8.0	103
Throughfall[b] (TF)	1983	18	71
Net (TF-BP)	1983	10	
Bulk precipitation	13-year mean	8.4	98
Throughfall	13-year mean	15.4	72
Net	13-year mean	7.0	

RESULTS OF CALCULATIONS USING 1983 FIELD DATA

Assumptions[c]	Flux[a] Net	Flux[a] "Dry"	Assumed form of dry deposition	Typical mean air concentration ($\mu g \cdot m^{-3}$, as N)	Calculated V_d ($cm \cdot s^{-1}$)
A/C	10	10	NO_x	5	0.6
A/D	10	10	HNO_3	2	1.5
B/C	10	50	NO_x	5	3
B/D	10	50	HNO_3	2	8

[a] kg NO_3^--N·ha^{-1}·y^{-1}
[b] Stemflow is negligible in this forest.
[c] We estimated the "dry" (or unmeasured) deposition velocities needed to account for NO_3^- in the spruce forest net throughfall with the following assumptions:

A. Foliar leaching of NO_3^- is negligible; hence, all net NO_3^- is dry deposition washed from canopy.
B. Foliar leaching is negligible, but net NO_3^- represents only 20% of dry deposition due to canopy uptake (as measured at Walker Branch Watershed, see text).
C. Net NO_3^- results from NO_x deposition.
D. Net NO_3^- results from HNO_3 vapour deposition.

range from 10 to 50 kg NO_3^--N.ha^{-1}.y^{-1}. These values are 2 to 6 times higher than the dry deposition rate measured at Walker Branch, while the larger value is nearly twice the cloud/fog deposition rate measured at Mt. Moosilauke. However, these data suggest that the increased NO_3^- flux below the spruce canopy, assuming no uptake (assumptions A/C and A/D, Table 1), can be readily attributed to dry deposition of either NO_x or HNO_3 because of the reasonable values (actually low for HNO_3) of the deposition velocities needed to account for net NO_3^- in throughfall. Estimates of V_d in the literature (Huebert 1983; Fowler 1980) are on the order of 0.2 to 0.5 cm.s^{-1} for NO_x to pine forests and 2 cm.s^{-1} for HNO_3 to a deciduous forest, both of which have lower canopy surface areas than typical spruce forests. Because the V_d values to the large surface area spruce canopy could theoretically be higher, it is probable that some in-canopy absorption may have reduced the measured net NO_3^- in throughfall, leading to low estimates of "dry" deposition.

Assuming that 80% of the dry deposition was absorbed by the canopy (assumption B), as occurs at Walker Branch, yields unreasonably high values of V_d for NO_x and HNO_3. Two likely explanations of the high values of V_d under case B are that less than 80% of the dry deposition is absorbed in the canopy, and/or that an additional deposition process is occurring which is not sampled by the bulk precipitation collectors. The interception of nitrate-rich fog droplets by the forest canopy is one such process that is characterized by a high deposition velocity (Lovett et al. 1982). Deposition of NO_3^- by fog, HNO_3, and NO_x could theoretically account for the unmeasured deposition in case B. While the results of these calculations are suggestive, the need for more research is clear.

THE NITROGEN DEPOSITION HYPOTHESIS

During a recent exchange of North American and West German scientists (U.S./F.R.G. Exchange) working in the area of air pollution effects on forests (Johnson and Cowling, in press), a hypothesis involving nitrogen deposition was developed in an attempt to explain various symptoms of forest decline. The hypothesis developed during the exchange consists of several subhypotheses; the one regarding deposition is as follows: "Elevated levels of nitrogen deposition to forests in a form available for direct foliar uptake, such as HNO_3 dry deposition, disrupts normal nutrient cycles and physiological processes, resulting in a stress on the tree and on the ecosystem. In combination with existing stresses, this results in certain symptoms found in declining stands."

Intensive studies of nitrogen deposition and cycling on an ecosystem level and detailed physiological research on atmospheric nitrogen uptake and effects under controlled conditions are needed to test these hypotheses. Specific questions which should be addressed include the relative importance of HNO_3 vapour and cloud-water nitrate in the total nitrogen deposition to forests, the extent to which and the forms in which deposited nitrogen is absorbed in the canopy, the extent to which and the process by which foliar-absorbed nitrogen is utilized by trees, and the degree to which forest nitrogen requirement is met (or exceeded) by direct canopy absorption. These answers should be combined with results of physiological experiments to test the general validity of the co-hypotheses that canopy nitrogen, deposited from the atmosphere in excess of forest needs, places an additional stress on the ecosystem through (1) induced nutrient deficiencies; (2) reduced nitrogen uptake by roots (if the foliar uptake pathway is significant); (3) increased NO_3^- leaching from soils (accompanied by nutrient cations and/or aluminum); (4) altered carbon/nitrogen allocation, root/shoot growth, and hormonal imbalances; and (5) increased plant susceptibility to secondary stresses such as pathogen invasion and frost (Johnson and Cowling, in press).

Our data suggest that dry deposition plays a major role in atmosphere/canopy interactions of nitrate in a deciduous forest in the eastern United States. Chronic exposure of the canopy to dry-deposited particle and vapour forms of NO_3^- increases the opportunity for deposition/foliage interactions because of the longer in-canopy residence time of dry deposition relative to precipitation. These interactions can result in both NO_3^- and H^+ ion uptake and loss of nutrient cations (Lovett et al. 1985). We estimate that atmospheric deposition supplies inorganic nitrogen at a rate comparable to 40% of the annual needs of this forest, based on measurements of annual woody increment (nutrient content of new wood tissue produced each year) (Lindberg et al. 1986). If this proportion is raised because of increased industrial and automotive emissions, forests may satisfy increasing portions of their nutrient requirements by assimilation of airborne material, while also being exposed to increasing levels of airborne trace contaminants. The effects of excess nitrogen, hydrogen ion, or trace metal deposition are possibly already being manifested in high-elevation spruce forests in the eastern United States and Europe (Johnson and Siccama 1983).

Acknowledgments

We are particularly grateful to our West German colleagues of the U.S./F.R.G. Exchange who made their preliminary data available for analysis: K. Kreutzer and K. Rehfuess, University of Munich; B. Prinz, Landesanstalt Für Imissionschutz, Essen; B. Ulrich and E. Matzner, University of Göttingen; W. Zech, University of Bayreuth; H. Zöttl, University of Freiburg.

This research was funded as part of the National Acid Precipitation Assessment Program. Publication No. 2514, Environmental Sciences Division, Oak Ridge National Laboratory. Partial assistance was also obtained from the Electric Power Research Institute, and from the U.S. Department of Energy under contract No. DE-AC05-84OR21400 with Martin Marietta Energy Systems, Inc.

REFERENCES

Appel BR, Tokiwa Y (1981) Atmospheric particle NO_3^-, sampling errors due to reactions with particulate and gaseous strong acids. Atmos Environ 15: 1087-1089

Brimblecombe P, Stedman DH (1982) Historical evidence for a dramatic increase in the nitrate component of acid rain. Nature 298: 460-562

Carlisle A, Brown ES, White EJ (1966) The organic matter and nutrient elements in the precipitation beneath a sessile oak canopy. J Ecol 54: 87-98

Chamberlain AC (1966) Transport of <u>Lycopodium</u> spores and other small particles to rough surfaces. Proc R Soc Lond A296: 45-70

Eberhardt PJ, Pritchett WL (1971) Foliar applications of N to slash pine seedlings. Plant Soil 23: 731-740

Fowler D (1980) Removal of sulphur and nitrogen compounds from the atmosphere in rain and by dry deposition. In: Drablos D, Tolan A (eds) Ecological impact of acid precipitation. SNSF Project, Oslo, Norway, p 23

Friedland AJ, Gregory RA, Karenlampi L, Johnson AH (in press) Winter damage to foliage as a factor in red spruce decline. Can J For Res

Galloway JN, Likens GE (1981) Acid precipitation: the importance of nitric acid. Atmos Environ 15: 1081-1085

Grennfelt P, Bengtson C, Skarby L (1980) An estimation of the atmospheric input of acidifying substances to a forest ecosystem. In: Hutchinson TC, Havas M (eds) Effects of acid precipitation on terrestrial ecosystems. Plenum, New York, p 29

Hicks BB (1984) Dry deposition. In: The acidic phenomenon and its effects, vol 1, chapters 1-8, EPA 600/88-83-016A. U.S. Environmental Protection Agency, Washington, DC

Hosker RP, Lindberg SE (1982) Review article: atmospheric deposition and plant assimilation of airborne gases and particles. Atmos Environ 16: 889-910

Huebert BJ (1983) Measurements of the dry-deposition flux of nitric acid vapour to grasslands and forests. In: Pruppacher HR, Semonin RG, Slinn WGN (eds) Precipitation scavenging, dry deposition, and resuspension. Elsevier, New York, p 785

Johnson AH, Cowling EB (in press) Declining forests of the Federal Republic of Germany and the eastern United States: a summary of characteristics and hypotheses, Environ Sci Technol

Johnson AH, Siccama TG (1983) Acid deposition and forest decline. Environ Sci Technol 17: 294A-305A

Johnson DW, Richter DD, Lovett GM, Lindberg SE (1985) The effects of atmospheric deposition on K, Ca, and Mg cycling in two forests. Can J For Res 15: 773-782

Kelly JM, Meagher JF (1985) Nitrogen input-ouput relationships for three forest sites in eastern Tennessee. Tennessee Valley Authority, Air Quality Branch, Muscle Shoals, Alabama

Lindberg SE (1982) Factors influencing the concentrations of trace metals, sulfate, and hydrogen ion in rain. Atmos Environ 16: 1701-1709

Lindberg SE, Harriss RC (1981) The role of atmospheric deposition in an eastern United States deciduous forest. Water Air Soil Pollut 15: 13-31

Lindberg SE, Lovett GM (1985) Field measurements of particle dry deposition rates to inert surfaces and leaves in a forest canopy. Environ Sci Technol 19: 238-244

Lindberg SE, Lovett GM, Richter DD, Johnson DW (1986) Atmospheric depositional canopy interactions of major ions in a forest. Science 231: 141-145

Lobel J, Thiel WR (eds) (1983) Acid precipitation. Verein Deutscher Ingenieure, Dusseldorf, Federal Republic of Germany, p 165-172

Lovett GM (1983) Atmospheric deposition to forests. In: Cronan CS (ed) Forest responses to acidic deposition. Land and Water Resources Center, University of Maine, Orono, Maine, p 7

Lovett GM, Lindberg SE (1984) Dry deposition and canopy exchange in a mixed oak forest determined from analysis of throughfall. J Appl Ecol 21: 1013-1028

Lovett GM, Reiners WA, Olson RK (1982) Cloud droplet deposition in subalpine balsam fir forests: hydrological and chemical inputs. Science 218: 1303-1304

Lovett GM, Lindberg SE, Richter DD, Johnson DW (1985) The effects of acidic deposition on cation leaching from three deciduous forest canopies. Can J For Res 15: 1055-1060

Matzner E, Khanna PK, Meiwes KJ, Lindheim M, Prenzel J, Ulrich B (1982) Elementflusse in Waldokosystemen in Solling-Datendokumentation. Gottingen Bodenkundliche Berichle 71, Universitat Gottingen, Federal Republic of Germany

NADP (National Atmospheric Deposition Program) (1985) Annual data summary of precipitation chemistry in the United States. Lindberg SE, Bowersox V, Bigelow D, Knapp W, Olsen T (eds) National Atmospheric Deposition Program, Colorado State University, Fort Collins, Colorado

NAS (National Academy of Sciences) (1983) Acid deposition-atmospheric processes in eastern North America. National Academy Press, Washington DC

Richter DD, Johnson DW, Todd DE (1983) Atmospheric sulfur deposition, neutralization, and ion leaching in two deciduous forests. J Environ Qual 12: 263-270

Robinson E, Homolya, JB (1983) Natural and anthropogenic emissions. In: The acidic deposition phenomenon and its effects, critical assessment review papers, vol 1, chapter A-2, 600/8-83-016A. U.S. Environmental Protection Agency, Washington, DC

Shannon JD (1981) A model of regional long-term average sulfur atmospheric pollution, surface removal, and net horizontal flux. Atmos Environ 15: 689-701

TVA (Tennessee Valley Authority) (1982) Ambient air quality monitoring system data summary, TVA/ONR/ARP-82/18. TVA (Monitoring Section), Muscle Shoals, Alabama

Ulrich B, Pankrath J (eds) (1983) Effects of accumulation of air pollutants in forest ecosystems. Reidel, Dordrecht, Holland

RESPONSES OF HERBACEOUS AND WOODY PLANTS TO THE DRY DEPOSITION OF SO_2 and NO_2

T.A. Mansfield, M.E. Whitmore, P.C. Pande and P.H. Freer-Smith

Department of Biological Sciences, University of Lancaster, U.K.

ABSTRACT

Simultaneous exposure of plants to SO_2 and NO_2 pollution can produce responses that would not be predicted from the individual effects of the two gases. The action of short, acute doses of SO_2 + NO_2 has not been fully explored, and there are synergistic effects and other dose-response peculiarities, which can at present only be tentatively explained. The synergism beween SO_2 and NO_2 has also been revealed by long-term exposures of herbaceous and woody plants. In the case of grasses and cereals, there is a severe suppression of growth in SO_2 + NO_2 in midwinter, but there can be a remarkable recovery in spring and summer. In woody plants, the capacity for such recovery is limited, and 2-year fumigations revealed enhanced responses in the second year. It has been possible to produce a good dose-response curve for the grass Poa pratensis exposed to mixtures of SO_2 and NO_2 for periods of up to 50 days. Good progress has also been made towards establishing a dose-response relationship for a broad-leaved tree, Betula pendula. In both cases, low concentrations appear to stimulate growth, but growth is dramatically inhibited as concentrations increase. The dry deposition of SO_2 and NO_2 from the concentrations present in rural air in industrialized countries is likely to produce growth stimulation in some circumstances, and inhibition in others. Initial growth stimulations may have long-term deleterious effects; they could, for example, increase sensitivity to frost.

INTRODUCTION

Studies with air pollutants applied singly in controlled fumigations provide the basis of much of our present understanding of the effects of gases, such as SO_2, NO_2 and O_3, on plants. During recent years, however, there have been some major studies involving combinations of pollutants, and the results of these suggest that the known action of individual components may not always be a reliable guide to the responses of plants exposed simultaneously to two or more toxic gases (Runeckles 1984).

Effects of the dry deposition of mixtures of SO_2 and NO_2 are especially relevant to many situations in the field in industrialized countries at higher latitudes, for example in northwest Europe (Heck 1982). SO_2 and NO_x are produced simultaneously by most combustion processes, and from a given process they may be emitted in nearly fixed proportions. Nevertheless, their ratio in the field is variable, and changes in the ratio may be important in determining responses. In this paper most experiments were conducted with equal concentrations of SO_2 and NO_2 on a volume/volume basis. This corresponds to a ratio of 1.39 for $SO_2:NO_2$ in terms of mass per unit volume. For rural locations in the United Kingdom this mass ratio is thought to be about 1.25, and, therefore, the experiments represent a reasonable first approximation to the composition of the atmosphere in the field.

SHORT-TERM EXPOSURES TO SO_2 AND NO_2

Tingey et al. (1971) found that the threshold concentrations for leaf blemish caused by short, acute doses of SO_2 and NO_2 were different when the two gases were applied together. They found that under conditions in which the threshold concentrations for visible injury were about 500 ppb for SO_2 and 2000 ppb for NO_2 applied for 4 hours, mixtures of SO_2 + NO_2 caused damage at concentrations between 50 and 250 ppb. Six plant species were used and these synergistic effects were observed in all of them, although a dose-response relationship was not clear. For example, 150 ppb NO_2 + 100 ppb SO_2 caused 24% injury to the upper surfaces of the two primary leaves of Phaseolus vulgaris, but 150 ppb NO_2 + 250 ppb SO_2 caused only 4% injury. Later studies of effects of pollution on stomata (reviewed by Mansfield and Freer-Smith 1984) offer a possible explanation: lower concentrations of SO_2 may cause stomatal opening and higher concentrations closure. Thus, the dose received by the plant may not simply be related to the gas concentration in the air surrounding the leaves. Amundson and Weinstein (1981) found that less-than-additive leaf blemish on three cultivars of Glycine max was associated with decreased leaf conductances in treatments with SO_2 + NO_2, and Ashenden (1979a) observed short-term increases in transpiration in Phaseolus vulgaris in 100 ppb SO_2 or 100 ppb NO_2 but noted that both pollutants together at the same concentrations caused a reduction in transpiration from the second day onwards.

Thus, it is extremely difficult to predict the damage that may occur in the field due to short-term exposures to SO_2 + NO_2. The responses of stomata to pollutants are so complex (Unsworth and Black 1981) that it is not to be expected that there will always be a simple relationship between response and pollutant concentration in the atmosphere. It is likely that the visible damage induced by these short, acute exposures is related to the amounts of SO_2 and NO_2 taken into the leaves, but further studies are necessary to establish whether or not this is the case.

The nature of the injury reported by Tingey et al. (1971) is of special interest. Leaf blemish in SO_2 + NO_2 treatments often resembled that caused by ozone, and it was thus distinguishable from that due to SO_2 or NO_2 alone. It is possible that symptoms in the field attributed to O_3 could sometimes be due to SO_2 + NO_2. Such confusion would, however, only occur where peaks of SO_2 and NO_2 were coincident, usually in urban areas.

LONG-TERM EXPOSURES TO SO_2 and NO_2

Impact on Herbaceous Plants

Effects on growth: The first study of the effects of prolonged exposure to the two pollutants was by Ashenden and Mansfield (1978). Four grasses (Dactylis glomerata L., Phleum pratense L., Poa pratensis L. and Lolium multiflorum Lam.) were exposed to weekly means of 68 ppb SO_2, 68 ppb NO_2 or 68 ppb of each pollutant, for 20 weeks over winter. (The pollutants were applied at 110 ppb for 103.5 hours a week, with clean air for the remaining time.) Growth was severely inhibited by SO_2 + NO_2 in all four species, and in three of them the effects of the combination of the two pollutants were synergistic. In the case of Phleum pratense, the plants grown in SO_2 + NO_2 weighed only 14% of the controls.

Ashenden (1979b) and Ashenden and Williams (1980) reported detailed studies of growth over winter in the same four grasses. Fumigation with 68 ppb SO_2 or 68 ppb SO_2 + 68 ppb NO_2 caused reductions in dry weight and leaf area which were followed by a decreased production of leaves and tillers. The only species that was markedly affected by 68 ppb NO_2 alone was Poa pratensis, in which dry mass was reduced, although the number of tillers was maintained.

As a result of these early studies, Poa pratensis was selected for further investigation because the time course of its responses to the two pollutants appeared to be of interest. Whitmore and Mansfield (1983) grew this species in concentrations of 100 ppb for 103.5 hours each week, giving weekly means of 62 ppb. The treatments were clean air, SO_2 alone, NO_2 alone, and SO_2 + NO_2. Seedlings were exposed from emergence in the autumn, and harvests were made at regular intervals thereafter. The results of dry weight determinations from one experiment are shown in Fig. 1.

At the time of the final harvest in May, the total dry weight was only 55% of the controls in the treatments with SO_2 and NO_2 separately, and just 26% of the controls after exposure to SO_2 + NO_2. At this stage, therefore, interactive effects of the two pollutants were not apparent. However, the development of statistically significant effects of SO_2 + NO_2 preceded those of the individual pollutants, for example at the harvests in February and March when there were substantial growth reductions in SO_2 + NO_2 but only small (not statistically significant) effects of SO_2 or NO_2 alone. At this stage there were statistically significant interactions between SO_2 and NO_2 ($p<0.01$).

This experiment was repeated over another winter; the fumigation was continued until after the plants had flowered, and the final harvest was in September.

The results of this experiment (Fig. 2) were in broad agreement with those of the previous one, showing substantial growth reductions over winter. The complete recovery of the dry weight

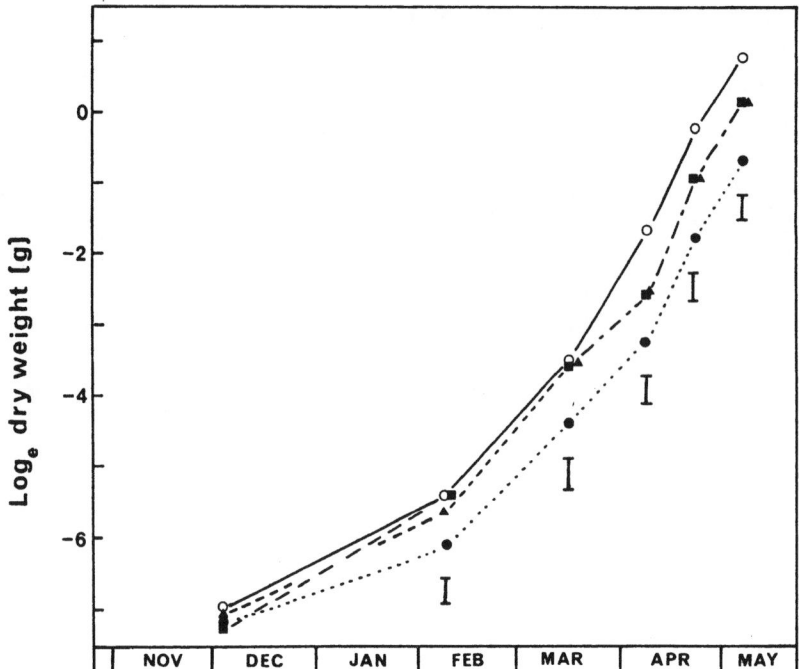

Fig. 1. The growth of *Poa pratensis* cv. Monopoly over winter 1979/80 which was grown from germination in clean air (O) or air containing weekly mean concentrations of 62 ppb NO_2 (■), 62 ppb SO_2 (▲) or 62 ppb NO_2 + 62 ppb SO_2 (●). Values are means of ten replicates except at the first harvest where block treatment measurements only were made. The growth in the NO_2 + SO_2 mixture was significantly lower than in the other treatments ($p<0.05$) from February to May. Redrawn from Whitmore and Mansfield (1983).

of the shoots in the polluted plants by the end of the experiment in the later experiments was unexpected. Complete recovery even occurred in SO_2 + NO_2 treatment, in which there was an 80% reduction in dry weight during March.

The reason for this remarkable recovery cannot be fully explained. Three possible reasons (not mutually exclusive) are:

(1) Reduced growth during the winter was the result of a greater sensitivity of the polluted plants to cold stress (Davison and Bailey 1982; Baker et al. 1982). As the weather became warmer this factor disappeared and the polluted plants recovered quickly.

(2) The smaller size of the plants in SO_2 and SO_2 + NO_2 in spring was sufficient to reduce the initiation of flower primordia. The smaller production of culms then allowed more photosynthates to be utilized for vegetative growth, allowing relative growth rate to recover.

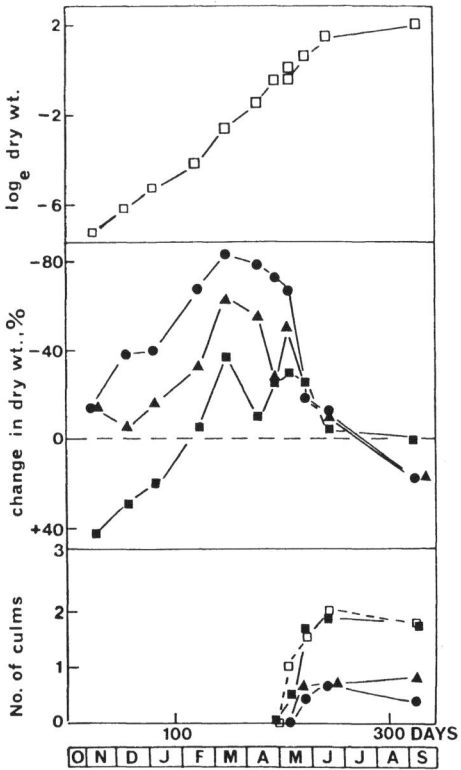

Fig. 2. Effects of exposure to 62 ppb SO_2 (▲), 62 ppb NO_2 (■) or 62 ppb SO_2 + 62 ppb NO_2 (●) on growth and flowering of Poa pratensis. Plants were exposed from germination in October 1980 to September 1981. The upper graph shows dry weight changes for the controls (□), these being for the whole plant until May and thereafter for the shoots only. The number of culms in clean air and in NO_2 was significantly greater ($p<0.05$) than in SO_2 and SO_2 + NO_2. From Whitmore and Freer-Smith (1982).

(3) During summer these levels of SO_2 and NO_2 led to a beneficial foliar uptake of sulphur and nitrogen which enhanced relative growth rate.

We hope to explore the reasons for the marked changes in response in future work. It may be necessary to gain a much more detailed understanding of the physiological and biochemical responses to SO_2 and NO_2 before we can fully explain the results of these experiments.

Effects on yield: If effects of SO_2 and SO_2 + NO_2 similar to those found with Poa pratensis were to occur with cereals, there could be serious reductions in grain production in heavily polluted areas. We have conducted one long-term experiment on winter barley and winter wheat, the results of which show that substantial effects on the crop yield do not necessarily occur, even though the sequence of responses is not dissimilar from that in P. pratensis. Some of the data for wheat (cv. Avalon) are shown in Fig. 3.

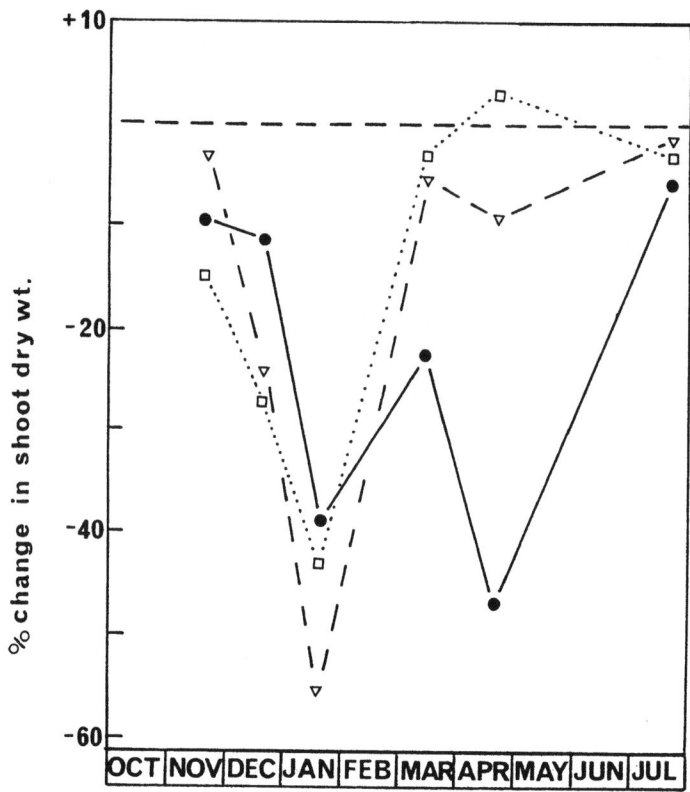

Fig. 3. Percentage changes in shoot dry weight (relative to clean air controls) of winter wheat exposed to weekly mean concentrations of 62 ppb SO_2 (▽), 62 ppb NO_2 (●) and 62 ppb SO_2 + 62 ppb NO_2 (□). Over-winter growth reductions were significant at $p<0.001$ in all treatments (see Pande and Mansfield 1985).

There were substantial reductions in dry weights in all treatments in December and January, but there was a rapid recovery thereafter in the SO_2 and SO_2 + NO_2 treatments. The plants exposed to NO_2 alone recovered somewhat later. These recoveries in dry weight of the shoots did not appear to be at the expense of grain production, which was little affected by the treatments, as shown in Table 1.

The results of a parallel experiment on winter barley (Pande and Mansfield, 1985) were very similar to those for wheat. Grain production in the polluted plants may have been conserved as a result of a reduced diversion of photosynthesis towards new tillers in the polluted plants.

Table 1. Final yield of winter wheat after the early growth reductions shown in Fig. 3.

Treatment*	No. of Spikes	Spike dry wt.(g)	1000-grain wt.(g)
Control	14.20	51.75	53.98
NO_2	13.80	50.21	49.50
SO_2	11.80	50.41	50.59
$SO_2 + NO_2$	12.44	49.93	49.57

* There were no significant differences (at p<0.05) between any of the treatments.

Clearly, there are comparable effects of pollution between grasses and cereals, involving growth depressions in winter and some recovery thereafter. Our experiments suggest that the economic consequences may be small, but this conclusion needs to be approached with great caution until we know more about possible repercussions of the growth responses in winter. There was little evidence of serious frost injury to cereals in these experiments, but it is possible that such injury could occur in plants exposed to pollutants in more severe winter conditions (Baker et al. 1982). In that case, the rate of recovery in early summer would be delayed and greater effects on grain production would be expected.

Dose-response studies: The experiments reported above used concentrations of SO_2 and NO_2 which are only likely to be encountered in heavily polluted areas at the present time. It is important, therefore, to determine whether the lower concentrations found in rural areas are likely to cause similar responses. Attempts to establish dose-response relationships between growth of grasses and SO_2 concentration have not been very successful, probably because of the varying conditions employed by different workers (Bell 1982; Mansfield and Freer-Smith 1981). We have attempted to define dose-response functions more precisely by conducting fumigations under controlled conditions using plants of Poa pratensis of uniform age. Seedlings were exposed from emergence to mixtures of SO_2 + NO_2 containing 40, 70 or 100 ppb of each gas, for periods of 4 to 50 days. The numbers of treatments required meant that experiments had to be performed on several occasions, but environmental conditions varied little. Photon flux densities were between 200 and 250 $\mu mol.m^{-2}.s^{-1}$ for 12 h photoperiods, and temperatures were 17-19°C at night and 22-25°C during the day. The results of the three experiments are shown in Fig. 4.

It was clear from these results that dry weight could be related to dose of $SO_2 + NO_2$, particularly in the longest fumigation. Further studies were made with exposures from 4 to 50 days, and these have been incorporated with those in Fig. 4 to produce a dose-response curve (Fig. 5). Here the effects were expressed in terms of percentage changes with respect to the 'controls', which were exposed to 7 ppb SO$_2$ + 7 ppb NO$_2$ (i.e., partially filtered air).

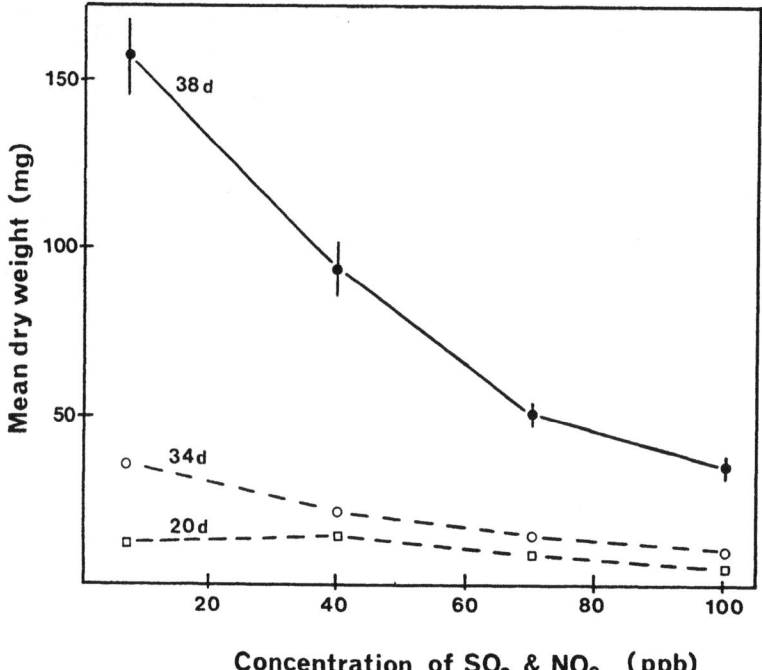

Fig. 4. Results of three experiments of different durations to determine the relationship between the dry weight of <u>Poa pratensis</u> (cv. Monopoly) and concentration of SO$_2$ + NO$_2$. Seedlings were exposed from germination. Concentrations were 40 ppb (●), 70 ppb (O) and 100 ppb (□) applied for 4 to 50 days.

In view of the uncertainties surrounding the relationships between doses of SO$_2$ + NO$_2$ and the responses of plants in previous studies by various authors, the results of these experiments have been most encouraging. We must, however, recognize that the curve in Fig. 5 applies only to one species growing in defined conditions. Deductions should not be made about other species in different situations.

The nature of the curve, with growth stimulations by low doses, then a fairly abrupt transition to inhibitory effects, does not come as a surprise. Stimulatory effects of small doses of SO$_2$, NO$_2$ or their mixture have often been reported in the past. It is clearly of great importance to be able to determine when the transition from growth stimulation to inhibition occurs if we are to make predictions about responses in the field. The curve suggests that maximum stimulation of growth occurs with doses of about 1000 ppb days and that there is a 40% loss of dry weight after 3000 ppb days. These changes in dose could readily occur within the known ranges of the concentrations of SO$_2$ and NO$_2$ in rural air (Fowler and Cape 1982). We must, therefore, expect the prediction of effects in the field to be a very difficult exercise. The construction of curves like that in Fig. 5, for a range of plant species in different conditions, will be very important as a basis of future attempts at predictive modelling.

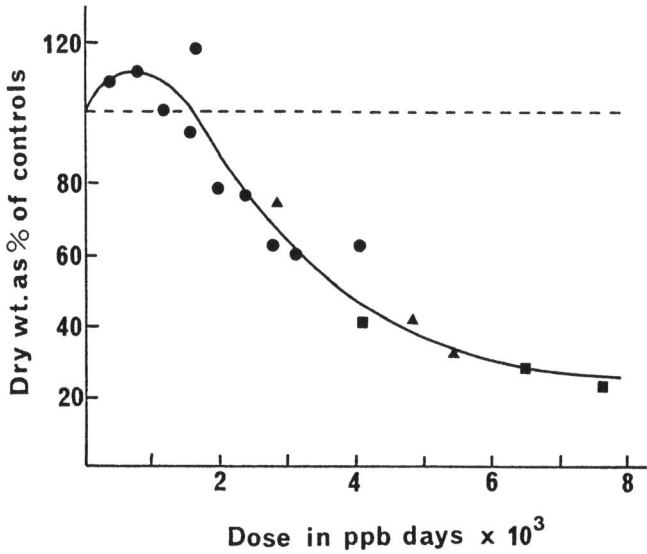

Fig. 5. Dose-response curve showing effects of SO_2 + NO_2 mixtures on growth of <u>Poa pratensis</u>. The results of several experiments under similar conditions are included. Dose is defined as concentration in ppb x days of exposure. Concentrations were 40 ppb (●), 70 ppb (▲) and 100 ppb (■) applied for periods between 4 and 50 days. (Example of calculation: 40 ppb SO_2 + 40 ppb NO_2 for 20 days = 1600 ppb days). From Whitmore (1985).

Impact on Woody Perennials

Effects on growth: Long-term fumigations of woody plants have suggested that the responses to treatments with SO_2, NO_2 and SO_2 + NO_2 are basically similar to those occurring in herbaceous plants (Freer-Smith 1984, 1985). Effects also vary with time of year, but a very important difference from herbaceous plants is that effects during winter are not followed by recovery. Inhibitory effects on a number of growth parameters develop more quickly in SO_2 + NO_2 than in SO_2 alone, and the initial response to NO_2 (during the first spring and summer of exposure) is one of growth stimulation. The nature of the responses of both broad-leaved deciduous and evergreen coniferous species has been found to alter in the second year, following fumigation over winter. In particular, the damage done by SO_2 was enhanced and the stimulatory effects of NO_2 were lost. Our studies of woody plants have shown that there are great differences in response, depending on species, cultivar, provenance and genotype. Some results for <u>Tilia cordata</u> and <u>Betula pubescens</u> are shown in Fig. 6.

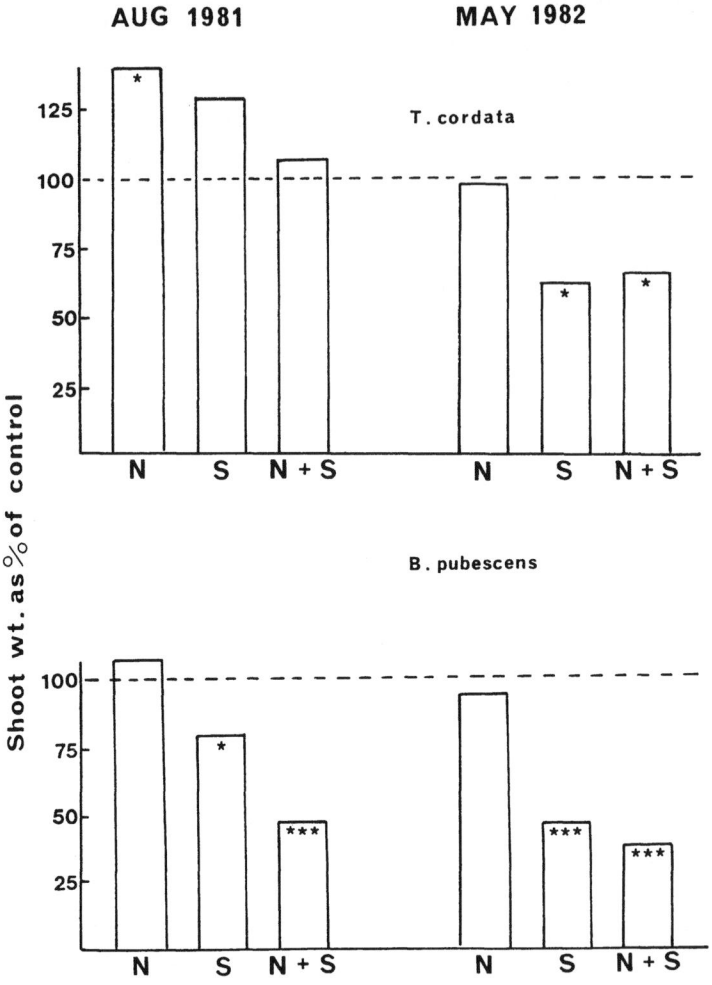

Fig. 6. Effects of 62 ppb NO_2 (N), 62 ppb SO_2 (S) singly and in mixtures (S + N) on the shoot dry weights of two deciduous trees. They were exposed for two years, and the effects shown (% of controls) were in the first growing season of exposure (August 1981) and in the spring of the following year (May 1982). Significant differences: *P<0.05, ***P<0.001

<u>Dose-response studies</u>: In woody plants the dose-responses to treatments with $SO_2 + NO_2$ have the same characteristics as those in herbaceous plants (Freer-Smith 1984, 1985). The data for <u>Betula pendula</u> in Table 2 can be compared directly with those for <u>Poa pratensis</u> in Fig. 5.

Table 2. The effects of mixtures of SO_2 and NO_2 at four concentrations of each gas (30, 50, 70 and 90 ppb) on the dry weights of <u>Betula pendula</u>. The seedlings were grown for 50 days in 14 h photoperiods with a photon flux density of 240 $\mu mol.m^{-2}.s^{-1}$ (photosynthetically active radiation). Temperatures were 19-23°C during the day and 14-16°C at night. There were 14 seedlings per treatment. Doses in ppb days were calculated as in the example in the legend of Fig. 5.

	Mean dry weights (g)			Total dry wt. (g) & percentages of controls	Root: shoot ratio	Dry wt (g) of fallen leaves
	Roots	Stems	Leaves			
Control	0.301	0.216	0.425	0.942(100%)	0.456	0.025
3000 ppb days	0.327	0.250	0.431	1.008(107%)	0.478	0.029
5000 ppb days	0.357	0.212	0.404	0.973(103%)	0.563	0.054
7000 ppb days	0.274	0.194	0.310	0.773(83%)	0.569	0.059
9000 ppb days	0.239	0.140	0.258	0.637(68%)	0.580	0.053
LSD at p=0.05	0.118	0.068	0.105	0.266	0.133	0.021

The responses of woody plants to $SO_2 + NO_2$ are also determined by photoenvironment. Freer-Smith (1985) found a complex interaction between SO_2, NO_2 and the total photon flux received by <u>B. pendula</u>. The data suggest that concentrations of 30-40 ppb may stimulate growth in favourable growing conditions in the field and act in an inhibitory manner when growth is slow.

BIOCHEMICAL EFFECTS OF $SO_2 + NO_2$

A full understanding of the complex responses we have described will depend on a knowledge of events at the sub-cellular level. Wellburn <u>et al</u>. (1981) performed some useful preliminary studies on several grasses taken from the area of long-term fumigations at Lancaster, described earlier in this paper. Within a week after the start of treatment with NO_2, there were significantly higher activities of nitrite reductase than in control plants. However, the presence of SO_2 together with NO_2 appeared to destroy the ability of the plants to respond to NO_2 in this way. These findings were true for <u>Lolium perenne</u>, <u>Dactylis glomerata</u>, <u>Phleum pratense</u> and <u>Poa pratensis</u>. Their implication is that nitrite ions, which are known to be highly toxic, will accumulate in plants treated with $SO_2 + NO_2$ but not NO_2 alone. Wellburn (1985) has recently made significant progress in determining the intracellular levels of critical anions such as nitrite within plant cells, and this is a step towards further understanding of the precise mechanisms of injury.

THE SIGNIFICANCE OF GROWTH STIMULATIONS IN RESPONSE TO SO_2 + NO_2

Attention has usually been focussed on the immediately obvious deleterious effects of SO_2 and NO_2 pollution. Many experimenters have noted, however, that low doses can cause growth stimulations, and these are clearly seen in the dose-response study on Poa pratensis (Fig. 5). In the context of typical fumigation experiments, these effects are often regarded as beneficial because growth is stimulated and because there are clear gains in fresh and dry weight by the end of the treatments. It must, however, be remembered that in such experiments plants are rarely exposed to biotic or abiotic stresses; they are well supplied with water, maintained in favourable temperatures, and are protected from attacks by pathogens.

The fact that activities of nitrate and nitrite reductases increase in leaves exposed to NO_2 demonstrates that its uptake increases the supply of available nitrogen. It is known that increased nitrogen can reduce tolerance of frost, drought and heat stress in various plant species (Levitt 1972). Greater infestations by insect pathogens are documented for plants growing in atmospheres polluted with SO_2 or NO_2 (Port and Thompson 1980; Dohmen et al. 1984).

Marie and Ormrod (1984) found that a treatment with SO_2 + NO_2 insufficient to reduce the dry weight of leaves of tomato led to a marked fall in leaf water content. Whitmore et al. (1985) have noted anatomical changes in leaves of grasses exposed to SO_2 + NO_2. Mesophyll cells were much larger and had thinner walls. Such responses occurred without appreciable changes in dry weight.

The impact of responses of this sort on plants exposed to a variety of stresses in the field needs to be assessed before we can estimate the agricultural and ecological consequences of the dry deposition of SO_2 and NO_2.

Acknowledgements

We are grateful to NERC, the EEC and the Royal Society for financial support.

REFERENCES

Amundson RG, Weinstein LH (1981) Joint action of sulfur dioxide and nitrogen dioxide on foliar injury and stomatal behaviour in soybean. J Environ Qual 10: 104-106

Ashenden TW (1979a) Effects of SO_2 and NO_2 pollution on transpiration in Phaseolus vulgaris. Environ Pollut 18: 45-49

Ashenden TW (1979b) The effects of long-term exposure to SO_2 and NO_2 pollution on the growth of Dactylis glomerata and Poa pratensis. Environ Pollut 18: 249-258

Ashenden TW, Mansfield TA (1978) Extreme pollution sensitivity of grasses when SO_2 and NO_2 are present in the atmosphere together. Nature 273: 142-143

Ashenden TW, Williams IAD (1980) Growth reductions in Lolium multiflorum and Phleum pratense as a result of SO_2 and NO_2 pollution. Environ Pollut (Series A) 21: 131-139

Baker CK, Unsworth MH, Greenwood P (1982) Leaf injury on wheat plants exposed in the field in winter to SO_2. Nature 299: 149-151

Bell JNB (1982) Sulphur dioxide and the growth of grasses. In: Unsworth MH, Ormrod DP (eds) Effects of gaseous pollutants in agriculture and horticulture. Butterworths, London, p 225

Davison AW, Bailey IF (1982) SO_2 pollution reduces the freezing resistance of ryegrass. Nature 297: 400-402

Dohmen GP, McNeill S, Bell JNB (1984) Air pollution increases Aphis fabae pest potential. Nature 307: 52-53

Fowler D, Cape JN (1982) Air pollutants in agriculture and horticulture. In: Unsworth MH, Ormrod DP (eds) Effects of gaseous pollutants in agriculture and horticulture. Butterworths, London, p 3

Freer-Smith PH (1984) The responses of six broadleaved trees during long-term exposure to SO_2 and NO_2. New Phytol 97: 49-61

Freer-Smith PH (1985) The influence of SO_2 and NO_2 on the growth, development and gas exchange of Betula pendula Roth. New Phytol 99: 417-430

Heck WW (1982) Future directions in air pollution research. In: Unsworth MH, Ormrod DP (eds) Effects of gaseous pollutants in agriculture and horticulture, Butterworths, London, p 411

Levitt J (1972) Responses of plants to environmental stress. Academic Press, New York, p 697

Mansfield TA, Freer-Smith PH (1981) Effects of urban air pollution on plant growth. Biol Rev 56: 343-368

Marie BA, Ormrod DP (1984) Tomato plant growth with continuous exposure to sulphur dioxide and nitrogen dioxide. Environ Pollut (Ser A) 33: 257-265

Pande PC, Mansfield TA (1985) Responses of winter barley to SO_2 and NO_2 alone and in combination. Environ Pollut (Ser A) 39: 281-291

Port GR, Thompson JR (1980) Outbreaks of insect herbivores on plants along motorways in the United Kingdom. J Appl Ecol 17: 649-656

Runeckles VC (1984) Impact of air pollution combinations on plants. In: Treshow M (ed) Air pollution and plant life. John Wiley, Chichester, UK, p 239

Tingey DT, Reinert RA, Dunning JA, Heck WW (1971) Vegetation injury from the interactions of NO_2 and SO_2. Phytopathol 61: 1506-1511

Unsworth MH, Black VJ (1981) Stomatal responses to pollutants. In: Jarvis PG, Mansfield TA (eds) Stomatal physiology. Cambridge University Press, p 187

Wellburn AR (1985) Ion chromatographic determination levels of anions in plastids from fumigated and non-fumigated barley seedlings. New Phytol 100: 329-340

Wellburn AR, Higginson C, Robinson D, Walmsley C (1981) Biochemical explanations of more than additive inhibitory effects of low atmospheric levels of sulphur dioxide plus nitrogen dioxide on plants. New Phytol 88: 223-237

Whitmore ME (1985) Relationship between dose of SO_2 and NO_2 mixtures and growth of *Poa pratensis*. New Phytol 99: 545-553

Whitmore ME, Mansfield TA (1983) Effects of long-term exposures to SO_2 and NO_2 on *Poa pratensis* and other grasses. Environ Pollut (Ser A) 31: 217-235

Whitmore ME, Davies WJ, Mansfield TA (1985) Air pollution and leaf growth. In: Baker NR, Davies WJ, Ong C (eds) Control of leaf growth. Cambridge University Press, p 295

PLANT CUTICLE AS A BARRIER TO ACID RAIN PENETRATION

V. S. Berg

Biology Department, University of Northern Iowa, Cedar Falls, Iowa 50614 U.S.A.

ABSTRACT

Exposure of leaf cells to acid precipitation is minimized by the presence of the plant cuticle, a waxy layer that covers the surface of leaves. The movement of substances through the cuticle is controlled by the conductance of the cuticle to the substance, and by the ease with which drops of precipitation wet the surface. When present, crystalline epicuticular waxes on the surface of the cuticle greatly reduce the potential for exchange of ions and molecules between drop and leaf by not allowing contact between the drop and the leaf surface. Some foliar leaching is normal, but increased leaching caused by acid precipitation has been observed both in the laboratory and in the field. Rates of leaching are typically low unless there has been damage to cells. Damage has been observed even in the absence of obvious disruption of the cuticle. Erosion of epicuticular waxes by acidic solutions has been observed for only a few species. Erosion of the cuticle would increase leaching due to the associated increased wetting of the leaf surface, and would alter conditions for microbial populations, including pathogens. For some species under some conditions, direct damage due to elevated acidity may prove important. Under most conditions, however, the direct effects of acidity on foliage are likely to be minor compared with the responses to gaseous pollutants, such as ozone.

INTRODUCTION

Despite numerous reports of acidic precipitation in Europe and North America, there are few reports of extensive damage to foliage caused by this direct exposure to low pH. More common are reports of ionic materials leached from leaves, but the rate of leaching from undamaged tissue is low (Tukey 1970; Cronan and Reiners 1983). Movement of substances through plant cuticle is limited, since the cuticle acts as an effective barrier to ions and polar compounds. The nature of the barrier, and the mechanisms by which it is breached, are the subjects of this paper.

THE STRUCTURE OF THE PLANT CUTICLE

The cuticle is a multi-layered structure covering much of the above-ground surface of plants: leaves, flower parts and uncorticated stems. The cuticle must simultaneously perform several tasks. It must allow photosynthetically-active radiation to pass through to the cells, keep water from evaporating from the leaf, prevent the leaching of materials from the leaf by precipitation, protect the leaf from chemical and physical damage, and repel pathogens. At the same time, it must remain plastic during leaf expansion, while maintaining chemical and mechanical stability.

These criteria are met by the unique structure of the cuticle.

Like other biological structures, cuticles vary from species to species, from individual to individual, and are modified by the conditions under which the organism develops. Since there is considerable variation between species in the thickness and in the pattern of cuticular layers, it is necessary to investigate the particular species of interest. Nonetheless, useful generalizations can be made (Martin and Juniper 1970). An idealized cuticle is shown in cross section in Fig. 1.

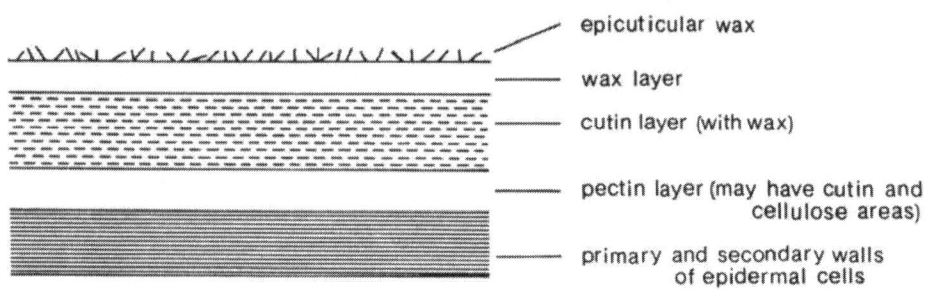

Figure 1. Idealized cuticle in cross section. Epicuticular waxes may be absent. The arrangement of the layers below the wax layer is highly variable between species, as are the thickness and proportions of all the layers. The wax layer is typically 2-10 um thick but may be much thicker or thinner on individual plants or cell types.

A layer of pectin separates the epidermal cell walls from the cuticle proper. The pectin follows the walls of the cells and extends between them. On its inner surface, the cuticle also follows the contour of the cells and may extend down between them. It is possible to isolate cuticles from leaf surfaces at this level with a mixture of pectinases and cellulases.

To the outside of the pectin is a layer of interspersed cellulose, pectin and cutin. The relative proportions and locations of these substances are highly variable. Exterior to this layer lies a region of cutin, a complex cross-linked polyester polymer (Kolattukudy 1980). It is the cutin which gives the cuticle its mechanical integrity and long-term resistance to degradation. There are molecular pores in isolated cutin layers whose permeability to water changes with the water activity of vapour in contact with the membrane (Schönherr and Schmidt 1979).

The cutin layer is overlaid by and infused with wax. The wax forms a continuous layer that provides the final barrier between the plant and the external environment. The term "wax", as it is used for lipids in cuticle, refers rather broadly to a mixture of long-chain compounds, typically saturated, unbranched hydrocarbons and closely related compounds with a single functional group, as well as wax esters. Among the common compounds are n-alkanes, aldehydes, ketones, β-diketones, primary and secondary alcohols and wax esters. They are, as a group, hydrophobic, but the chemical properties of the constituents of the wax layer do not necessarily account for all the differences in the interaction with liquids on the surface. They may form ordered molecular arrangements, as seen by electron microscopy (Jarvis and Wardrop 1974) and temperature and electron spin resonance studies (Schönherr et al. 1979). Compounds with thirty or more carbons frequently occur. The resistance to movement of ions and water across the cuticle is due almost totally to the wax layer (Schönherr 1976). There may be considerable intrusion of cutin strands into the wax layer. Cutin has been reported to extend across the entire depth of the wax to the outer surface (Roberts et al. 1948; Reed and Tukey 1982). These reports must be treated with some caution, as the wax layer is rather plastic. This is useful to a growing plant but may lead to deformation upon sectioning.

On the surface of the cuticle there may be an additional layer of waxes in the form of crystalline epicuticular wax. These are often associated with the presence of the more polar wax compounds, such as alcohols and β-diketones (Tulloch 1976). It is this material which accounts for the glaucous bloom on some plant parts. The epicuticular wax may assume a variety of shapes such as rods or tubes or upright flakes. The type of crystals produced depends on both the chemical composition of the material and the rate and order of delivery to the leaf surface (Jeffree, et al. 1975).

MOVEMENT OF MATERIALS THROUGH THE CUTICLE

Despite recent research on both water movement and ion movement through cuticles, little is known about the mechanisms and pathways by which materials move through them. This is of importance both for leaching of mineral ions from the plant and for penetration of protons from acid rain. Some of the older experimental data supporting unidirectional movement of materials through the cuticle (Yamada et al. 1964) is questionable because of lack of attention to such factors as stirring and wettability. Older reports of open channels through the cuticle (ectodesmata; Franke 1967) have been largely discredited. It is not known how the path of water movement is related to the path of ion movement, but presumably any pathway available to ions is also a pathway of water movement. What has been demonstrated is that ions move through the cuticle according to size and charge (MacFarlane and Berry 1974), suggesting movement through pores lined with polar compounds. The movement of protons through intact cuticle has been conclusively demonstrated (Dreyer et al. 1981). Rates of movement are slow for ionic and polar compounds, but may be much higher for non-polar compounds (Bukovac et al. 1971), including gaseous pollutants such as SO_2 (Lendzian 1984).

For many plants, however, the most important barrier to ion

movement is not the continuous wax layer of the cuticle, but
rather it is the epicuticular wax layer. If present, this layer
reduces the contact between drops of precipitation and the leaf
surface (Fig. 2). In general, glossy leaves, which lack epicuticular wax, are much more wettable than glaucous or dull leaves.
The conductance of a cuticle to any dissolved material depends on
the area contacted, so wettability is critical (Leece and
Kenworthy 1972). Once contact between drop and leaf is established, the conductance of the cuticle to the material in question becomes important (Fig. 3). The reduction in contact area
afforded by epicuticular wax is almost totally a physical phenomenon, as the crystals hold the water drops above the continuous
wax layer. Trichomes or other irregularities, which are themselves covered with cuticle, serve a similar role in reducing
contact between leaf and water drops.

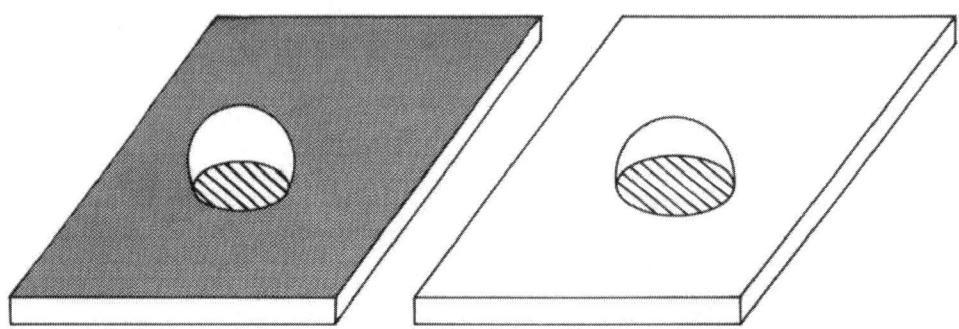

Figure 2. Effect of wettability of leaves on area of leaf
contacted by drops of precipitation. The surface of the leaf on
the left has a relatively low wettability, as is typical of
leaves with epicuticular wax. Only a small area of the drop is
in contact with the surface. The leaf on the right has a higher
wettability, as is normally the case for glossy leaves. A larger
area of the leaf surface is in contact with the drop.

It is now thought that water may move through the continuous wax
layer either through hydratable molecular pores, or by diffusion
through the wax. There is evidence for both mechanisms, in
different species, from investigations with isolated cuticles
(MacFarlane and Berry 1974; Schönherr and Schmidt 1979). There
is additional evidence of the existence of hydratable pores from
work with excised compound leaves of Berberis aquifolium (Seymour
1980). It is possible that small ions also move, along with
water molecules, through these hydratable pores. The mere
existence of these pores remains speculative, as they cannot be
seen with electron microscopy. Nor would we expect to be able to
see them, as they should not be "open" except when hydrated. It
is thought that the pores are lined with the more hydrophilic
lipids of the cuticular wax, and are continuous, though they may
be anastomosing and might follow tortuous paths. The more hydrophilic molecules would naturally aggregate in the process of wax
layer formation, as this arrangement represents a lower energy
configuration. When hydrated, these pores would enlarge, and

there would be an elastic or plastic deformation of the surrounding molecules. If the pore model accounts for ion movement as well as for water, then the nature of the pores (number, size, chemical species) would be important in determining how easily material is leached through the cuticle. In fact, the chemistry and arrangement of the cuticle is more important than thickness per se, although a thicker cuticle of a given type will obviously provide more resistance to the movement of materials than an equivalent thin one.

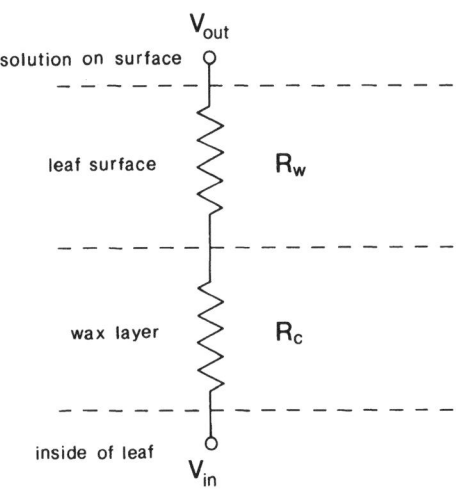

Figure 3. Resistance network equivalent to the leaching and penetration pathway through cuticle. V_{out} is equivalent to the concentration of material in solution on the leaf surface; V_{in} is equivalent to the concentration of the material inside the cuticle, including that on the ion exchange sites. R_w is equivalent to the wettability of the leaf surface; R_c is equivalent to the resistance of the cuticle to movement of materials.

The inner layers of the cuticle provide relatively little resistance to the movement of ionic and polar materials (Dreyer et al. 1981). The cutin is also much less resistant to the diffusion of materials than the wax layer associated with it (Haas and Schönherr 1979), so that any strands of cutin that intrude into the wax layer probably serve as channels for the movement of water, ions and polar compounds. The inner layers of the cuticle, and the cell walls themselves, do however have an important role in leaching. These areas offer ion exchange sites where basic cations (Ca^{2+}, Na^+, etc.) from the cells or the xylem stream can be exchanged for protons from the outside environment.

The degree to which ion exchange occurs farther toward the outside of the cuticle is unknown. In experimental systems involving continuous leaching, rates of ion loss from the leaves were highest at the beginning of the leaching period, and then were reduced. Total quantities of ions leached in several days were sometimes as high as five times the amount in the leaf, indicating resupply from the rest of the plant (Stenlid 1958).

Stomata provide an obvious route by which materials might move into or out of the leaf, but in fact these openings are of little importance unless wetting agents have been applied to the leaf. The edges of the guard cells are covered with a layer of cuticle which typically forms a ledge. This ledge, due to its small radius of curvature, does not permit water to enter the stomate (Martin and Juniper 1970). This is fortunate, as wetting the stomata prevents gas exchange. Keeping stomata dry is a major role of the cuticle. Many plants, especially conifers, have epicuticular wax covering the stomatal surfaces, greatly reducing wettability.

POTENTIAL BIOLOGICAL SIGNIFICANCE

Leaching of mineral ions from plants may be accelerated by acid rain through ion exchange involving cations from the plant and protons from the rain. This process may partially neutralize acidic droplets on the leaf surface (Adams and Hutchinson 1984). For plants with intact cuticles, variations in rates of leaching are probably determined principally by the wettability of the leaf, then by the conductance of the cuticle to the ions, rather than by the availability of ions for exchange. If there is damage to the cuticle, destruction of cells by acid rain may ensue. There is an intermediate case in which cuticles are thin due to rapid leaf expansion or for special organs such as secretory trichomes or hydathodes, which must pass materials through the cuticle (or an opening in it) to the outside. Here the potential for both leaching and damage increases. The amount of damage would in any case depend on the susceptibility of the cells as well as the intercellular pH.

Mechanical or chemical damage to the cuticle increases the rate of ion movement, either by thinning the wax layer or removing the epicuticular wax. Even rainfall can erode the epicuticular wax layer of some plants and have dramatic effects on the wettability of the cuticle (Martin and Juniper 1970). Surfactant materials are used in pesticide sprays to increase wetting and therefore penetration of applied chemicals. The residue increases wetting of the cuticle by natural precipitation as well. Exudates from either plants or insects similarly increase wettability of the leaf surfaces. Direct chemical damage to the wax layer of the cuticle is not common. In fact, one of the traditional methods for isolating intact cuticles from leaves is to soak them in strong acid (Schönherr and Huber 1977). The cutin and the waxes remain structurally intact after this treatment, as most of the compounds in both are resistant to acid attack. If strong acid penetrates the cuticle, however, it erodes the underlying epidermal cells which support the cuticle, and mechanical damage can then be expected. It is also possible that some SO_2, NO_2 and NO pass through the cuticle in non-ionic form (Paparozzi and Tukey

1983; Lendzian 1984), causing damage which may, in turn, affect the cuticle through loss of mechanical support.

SUPPORTING DATA

There are data to support many of these predictions. Laboratory studies have shown differences in the sensitivity of plant species to acid rain applied to the leaves. Damage to leaves, fruits and petals has been demonstrated for a variety of species, chiefly at pH values below 4.0 (Evans et al. 1977a; Evans et al. 1977b; Wood and Borman 1977; Evans et al. 1978; Evans and Curry 1979; Evans et al. 1981; Lee et al. 1981; Forsline et al. 1983; Proctor 1983; Paparozzi and Tukey 1983). Collapsed cells have been seen under the cuticle, with the cuticle intact, supporting the idea that disruption of the cuticle is secondary to elimination of its support (Evans et al. 1977a; Paparozzi and Tukey 1983; Adams et al. 1984). The fact that most of the damage has been seen at pH values below 4.0 (often well below) explains why relatively little acid-induced direct foliage damage has been reported in the field despite widespread occurrence of acid precipitation.

Acid damage to leaf surfaces has been shown to be more common with specific cell types or leaf ages. Paparozzi and Tukey (1983) found that for birch, an acid-sensitive species, most lesions were on or between minor veins. Very young leaves were least sensitive, perhaps due to the reduced wettability associated with the high density of trichomes. The most sensitive leaves were the oldest. There was no difference in thickness of the cuticle after acid rain treatment, although in the most severe cases there were holes through the cuticle, always extending to injured epidermal or palisade cells. For kidney beans, flakes of epicuticular wax were present over some injured areas. Evans et al. (1978) found that recently expanded leaves of poplar were most sensitive, and that most damage occurred near vascular strands and near stomata. For the especially sensitive Brassica oleracea variety Stokes Viking Golden Acre, erosion of the epicuticular wax by acidic drops has been demonstrated (Hutchinson et al., this volume) but this is not the usual case. Much more work will be necessary for us to understand the relative importance of cuticle chemistry, cuticle structure and leaf structure, all of which change with leaf age, in determining the extent and location of damage.

There have been several reports of increases in leaching associated with acid precipitation (Wood and Bormann 1975; Tamm and Cowling 1977; Richter et al. 1983; Scherbatskoy and Klein 1983). These have been used in modelling effects of acid precipitation (Chen et al. 1983). Increased leaching has been observed at pH levels where no visible foliar damage was seen, suggesting that the predicted ion exchange was taking place. The level of leaching by a deciduous forest canopy was considerably greater than that required for neutralization, however, indicating that ion exchange due to acid precipitation is not the dominant source of leaching of cations (Richter et al. 1983). The most dramatic increases in leaching were seen at pH levels associated with damaged tissues, usually about pH 3.3 or below. Changes in rates of leaching at pH levels not associated with tissue damage were much smaller than leaching rates seen at normal rainwater pH. One possible consequence of increased leaching may be neutralization of acidic drops by rapid efflux of cations as shown in leaves of Artemesia tilesii, an acid-tolerant arctic species (Adams and Hutchinson 1984).

A further reason to be concerned by acid rain damage to the cuticle is that the cuticle is critical in preventing pathogen entry into leaves. Acid damage which results in the physical breakdown of the cuticle will leave the plant vulnerable to attack by pathogens which are normally unable to penetrate cuticle.

A reduction in yield due to simulated acid rain has been demonstrated for several field crops (Lee et al. 1981; Evans et al. 1982). Except for reductions in marketability due to damaged leaves (Lee et al. 1981), yield typically was not well correlated with the visible damage to the leaves (Evans et al. 1982; Proctor 1983; Forsline et al. 1983). Yield reduction of up to 7% in soybean occurred in apparently healthy plants with no foliar lesions.

SUMMARY

Direct acid damage to foliage occurs, but seldom at pH values above 4.0 due to the presence of the intact plant cuticle. The cuticle does permit the movement of protons into the leaf and mineral ions out of the leaf, but only at low rates. The primary barriers to ion movement are the epicuticular waxes, which prevent wetting by precipitation, and the wax layer, which limits ion movement. It appears that most damage to the cuticle is a secondary effect associated with the loss of support from injured underlying cells. Leaching of ionic materials from the leaves is promoted by acid precipitation, but the effect is small unless the underlying cells are damaged. Minor foliage damage does not correlate well with yield reduction.

Damage to plant cells caused by acid precipitation is seldom observed in the field. By reducing wettability and limiting penetration and leaching, the cuticle normally protects plant surfaces against injury caused by acid precipitation.

REFERENCES

Adams CM, Hutchinson TC (1984) A comparison of the ability of leaf surfaces of three species to neutralize acidic rain drops. New Phytol 97: 463-478
Adams CM, Dengler NG, Hutchinson TC (1984) Acid rain effects on foliar histology of Artemesia tilesii. Can J Bot 62: 463-474
Bukovac MJ, Sargent JA, Powell RG, Blackman GE (1971) Studies on foliar penetration. VIII. Effect of chlorination on the movement of phenoxyacetic and benzoic acids through cuticle isolated from the fruits of Lycopersicon esculentum L. J Exp Bot 22: 598-612

Chen CW, Hudson RJM, Gherini SA, Dean JD, Goldstein RA (1983) Acid rain model: canopy module. J Env Eng 109: 585-603

Cronan CS, Reiners WA (1983) Canopy processing of acidic precipitation by conifers and hardwood forests in New England. Oecologia 59: 216-223

Dreyer SA, Seymour VA, Cleland RE (1981) Low proton conductance of plant cuticles and its relevance to the acid-growth theory. Plant Physiol 68: 664-667

Evans LS, Curry TM (1979) Differential responses of plant foliage to simulated acid rain. Amer J Bot 66: 953-962

Evans LS, Gmur NF, Da Costa F (1977a) Leaf surface and histological pertubations of leaves of Phaseolus vulgaris and Helianthus annuus after exposure to simulated acid rain. Amer J Bot 64: 903-913

Evans LS, Gmur NF, Kelsch JJ (1977b) Pertubations of upper leaf surface structures by simulated acid rain. Env Exp Bot 17: 145-149

Evans LS, Gmur NF, Da Costa F (1978) Foliar response of six clones of hybrid poplar. Phytopath 68: 847-856

Evans LS, Curry TM, Lewin KF (1981) Responses of leaves of Phaseolus vulgaris L. to simulated acid rain. New Phytol 88: 403-420

Evans LS, Lewin KE, Cunningham EA, Patti MJ (1982) Effects of simulated acidic rain on yields of field-grown crops. New Phytol 91: 429-441

Franke W (1967) Ectodesmata and foliar absorption. Amer J Bot 48: 683-691

Forsline PL, Musselman RC, Kender WJ, Dee RJ (1983) Effects of acid rain on apple tree productivity and fruit quality. J Amer Soc Hort Sci 108: 70-74

Haas K, Schonherr J (1979) Composition of soluble cuticular lipids and water permeability of cuticular membranes from Citrus leaves. Planta 146: 399-403

Hutchinson TC, Adams CM, Caporn SJM (this volume) Neutralization of acidic droplets on leaf surfaces.

Jarvis LR, Wardrop AB (1974) The development of the cuticle in Phormium tenax. Planta 119: 101-112

Jeffree CE, Baker EA, Holloway PJ (1975) Ultrastructure and recrystallization of plant epicuticular waxes. New Phytol 75: 539-549

Kolattukudy PE (1980) Biopolyester membranes of plants: cutin and suberin. Science 208: 990-1000

Lee JJ, Neely GE, Perringan SC, Grothaus LC (1981) Effect of simulated sulfuric acid rain on yield, growth and foliar injury of several crops. Env Exp Bot 21: 171-185

Leece DR, Kenworthy AL (1972) Influence of epicuticular waxes on foliar absorption of nitrate ions by apricot leaf disks. Aust J Biol Science 25: 641-643

Lendzian KJ (1984) Permeability of plant cuticles to gaseous air pollutants. In: Kojoil MJ, Whatley FR (eds) Gaseous air pollutants and plant metabolism. Butterworths, London, p 77-81

MacFarlane JC, Berry WL (1974) Cation penetration through isolated leaf cuticles. Plant Physiol 53: 723-727

Martin JT, Juniper BE (1970) The cuticle of plants. St. Martin's Press, New York, p 347

Paparozzi ET, Tukey HB Jr. (1983) Developmental and anatomical changes in leaves of yellow birch and red kidney bean exposed to simulated acid precipitation. J Amer Soc Hort Sci 108: 890-898

Proctor JT (1983) Effect of simulated sulfuric acid rain on apple tree foliage, nutrient content, yield and fruit quality. Env Exp Bot 23: 167-174

Reed DW, Tukey HB Jr. (1982) Light intensity and temperature effects on epicuticular wax morphology and internal cuticle ultrastructure of carnation and Brussel sprout leaf cuticles. J Amer Soc Hort Sci 107: 417-420

Richter DD, Johnson DW, Todd DE (1983) Atmospheric sulfur deposition, neutralization, and ion leaching in two deciduous forest ecosystems. J Env Qual 12: 263-270

Roberts EA, Southwick MD, Palmiter DH (1948) A microchemical examination of McIntosh apple leaves showing relationship of cell wall constituents to penetration of spray solutions. Plant Physiol 23: 557-559

Scherbatskoy T, Klein RM (1983) Response of spruce and birch foliage to leaching by acidic mists. J Env Qual 12: 189-195

Schonherr J (1976) Water permeability of isolated cuticular membranes: the effect of cuticular waxes on diffusion of water. Planta 131: 159-164

Schonherr J, Huber R (1977) Plant cuticles are polyelectrolytes with isoelectric points around three. Plant Physiol 59: 145-150

Schönherr J, Schmidt HW (1979) Water permeability of plant cuticles. Dependence of permeability coefficients of cuticular transpiration on vapor pressure saturation deficit. Planta 144: 391-400

Schönherr J, Eckl K, Gruler H (1979) Water permeability of plant cuticles: the effect of temperature on diffusion of water. Planta 147: 21-26

Seymour VA (1980) Leaf cuticle: an investigation of some physical and chemical properties derived from a study of Berberis. PhD Thesis, University of Washington, Seattle

Stenlid G (1958) Salt losses and redistribution of salts in higher plants. In: Ruhland W (ed) Encycl Plant Physiol 4: 615-637

Tamm CO, Cowling EB (1977) Acidic precipitation and forest vegetation. Water Air Soil Poll 7: 503-511

Tukey HB Jr (1970) The leaching of substances from plants. Ann Rev Plant Physiol 21: 305-324

Tulloch AP (1976) Chemistry of waxes of higher plants. In: Kolattukudy PE (ed) Chemistry and biochemistry of natural waxes. Elsevier, Amsterdam, Oxford, London. p 235-287

Wood T, Borman FH (1975) Increases in foliar leaching caused by acidification of an artificial mist. Ambio 4: 169-171

Yamada Y, Wittwer SH, Bukovac MJ (1964) Penetration of ions through isolated cuticles. Plant Physiol 39: 28-32

THE RESPONSE OF PLANT REPRODUCTIVE PROCESSES TO ACIDIC RAIN AND OTHER AIR POLLUTANTS

R.M. Cox

Canadian Forestry Service - Maritimes, P.O. Box 4000,
Fredericton, New Brunswick, Canada, E3B 5P7

ABSTRACT

In vitro and in vivo experiments to investigate the effects of acidic and trace element components of polluted rain on pollen germination and tube growth are described. The sensitivity to pH of pollen in vitro was related to the species' position in the canopy, whether it was a conifer or deciduous, and to the habitat in which the species normally occurred, i.e., boreal forest or calcareous woodland. Most pollen assayed revealed significant inhibitions of germination and germ tube growth in response to pHs currently recorded for daily rain samples collected in eastern Canada. The in vitro effects of trace elements in combination with pH on pollen function are discussed in relation to cation stimulation and synergistic interactions. In vivo inhibition of pollen of evening primrose, Oenothera parviflora, by simulated acid rain is compared with similar inhibition reported in response to SO_2 fumigations which have reduced seed set in some species. Given a degree of overlap in gene expression of tolerance in pollen and the plant from which it came, some implications of pollen mortality brought on by air pollution during pollination are discussed.

INTRODUCTION

Pollen germination and germ tube growth are known to be among the more sensitive botanical indicators of atmospheric pollution (Feder 1968; Stanley and Linskens 1974; Feder 1981). Atmospheric pollutants may affect the vulnerable reproductive processes at the time of pollination, either by directly reducing pollen viability and pollen tube growth or by affecting the chemical environment of the stigmatic surface. These effects may reduce stigma receptivity and change pollen stigma interactions which, in turn, may affect the quality and quantity of seed stocks.

Studies on conifers located at different distances from urban and industrial areas (pollution sources) have shown marked reduction in cone dimensions, seed weight and viability, together with reduced pollen viability (Pelz 1963; Podzorov 1965; Mrkva 1969; Antipov 1970). Furthermore, these effects may occur at pollution levels lower than that required for foliar injury (Houston and Dochinger 1977).

Ozone (O_3) is a widespread atmospheric contaminant, and at low concentrations is known to have inhibiting effects on pollen germination and pollen tube growth of tobacco and corn (Feder 1968; Mumford et al. 1972). Correlations between pollen sensitivity and varietal foliar sensitivity to O_3 in tobacco, petunia, and tomato have been demonstrated by Feder and Sullivan (1969) and Feder (1981). However, Benoit et al. (1982) found no such correlation between foliar and pollen sensitivity in eastern white pine (Pinus strobus L.) but they did determine that moist pollen was more sensitive to O_3 than was dry pollen.

The inhibitory effects of sulphur dioxide (SO_2) on pollen viability have been known since the pioneering work of Sabachnikoff (1912). Dopp (1931) assayed the SO_2 sensitivities of pollen from many species in different humidities and clearly demonstrated that the toxicity of SO_2 increased under high relative humidities (RH).

The acidic conditions caused by SO_2 fumigations under wet or high RH conditions were assumed by Dopp (1931), who suggested that the increased effects of the fumigations on in vitro germination over the effects on in vivo germination were due to the extra buffering of the stigmatic surface. The acidity of sulphur dioxide solutions is due to the equilibrium:

$$SO_2 \cdot xH_2O \longleftrightarrow HSO_3^- (aq.) + H_3O^+ + (x-2)H_2O$$

(Powell and Timms 1974).

The role of acidity in the toxicity of SO_2 fumigation to pollen under wet or high RH conditions is not well documented. Karnosky and Stairs (1974), however, reported that the pH of an agar/sucrose gel pollen culture media decreased from pH 7.0 to pH 5.0 after a 4-h fumigation with 1.4 ppm SO_2. They noted that the depression in pH severely inhibits pollen germination and tube growth of the Populus deltoides Marsh. pollen tested. The H^+ ions present in such assays are, therefore, implicated as an important component in SO_2 toxicity to pollen. This view was also supported more recently by Murdy (1979) and DuBay and Murdy (1983a,b).

Many species have been investigated in regard to effects of SO_2 on pollen under wet conditions or high relative humidities in vitro (reviewed by Cox 1983b). Murdy (1979), who exposed flowering plants from different populations of Lepidium virginicum L. to in vivo SO_2 fumigations under conditions of high RH (> 90%), found an increase in fruit sterility. Furthermore, the degree of experimentally-induced sterility related to ambient air SO_2 concentrations of the place of origin of the L. virginicum populations. This suggests an evolution of SO_2 tolerance by the reproductive system of exposed plants in SO_2 polluted areas.

DuBay and Murdy (1983a), however, were unable to increase sterility in L. virginicum under SO_2 fumigation, even though a 50% reduction in in vivo pollen germination was achieved. This may have been due to their selection of plants with good background self-fertility or autogamy.

Another study (Murdy and Ragsdale 1980) of SO_2 suggested that in in vitro effects of SO_2 pollen germination and germ tube growth under high RH conditions may emulate in vivo effects on stigmas. The resultant reduced germination and initial tube growth of pollen observed in vivo may reduce seed set. Such an in vivo reduction of pollen germination after fumigations with SO_2 at the time of pollination was shown to reduce seed set when exposed at 90% RH in Geranium carolinianum (DuBay and Murdy, 1983b).

Although acid inhibition of pollen activity has been known at least since the work of Dopp (1931), other investigations have also noted the importance of acidity in pollen germination and/or tube growth since that time (Cooper 1939; Brewbaker and Majumder 1961; Kratky et al. 1974). In addition Iwanami et al. (1978) considered pollen inviability in beehives to be due to the acidity of royal jelly.

Masaru et al. (1980) however, were finally to verify the importance of H^+ ion inhibition of pollen tube growth by assaying Camellia pollen in three inorganic acids and their corresponding ammonium salts. The Camellia pollen was markedly inhibited by pH 3.2. This begins to suggest a possible effect of regional acid precipitation on plant reproduction.

The above observations, together with the descriptions of pollen responses to trace metals (Masaru et al. 1980; Arvind and Malik 1976; Strickland and Chaney 1979), prompted investigations into the effects of a range of acidities and trace metal concentrations on in vitro pollen germinations of a wide range of Canadian forest species. The effects of soil differences at the site of pollen collection on in vitro pollen response to pH, together with in vivo effects of simulated acid rain in stigma receptivity, were also considered.

IN VITRO ASSAYS

Sensitivity to pH

To determine the response of pollen to pH, pollen of a wide range of forest species was collected from various individuals of potentially sensitive species (Cox 1983b). The pollen was assayed as described by Cox (1983a). After appropriate transformations (Cox 1983a) LD_{50} dosages of pH were computed for the pollen germination data (Fig. 1) of the 12 species which were all significantly influenced by pH treatments ($p < 0.01$). Two thirds of the species had pollen that was inhibited in germination ($\alpha = 0.05$) by pH 3.0 (1000 µEq $H^+ \cdot L^{-1}$) whereas few effects were observed at pH 4.6 (25 µEq $H^+ \cdot L^{-1}$) (Cox 1983a). The LD_{50} dosage is the computed dosage of pH that would produce a 50% probability of response, i.e., failure to germinate. This toxicological approach is used to describe and rank the broad sensitivity groups (Fig. 1). Broad-leaved canopy tree species were the most sensitive, with LD_{50} dosages of pH 3.95 to 3.63. The open habitat herb Oenothera parviflora L. and white pine, Pinus strobus L. were also sensitive, the latter having an LD_{50} dosage of pH 3.6. Next in sensitivity were the species of the understory and ground flora that had LD_{50} dosages of pH 3.58 (Maianthemum canadense Desf.), to pH 3.14 (bush honeysuckle, Diervilla lonicera Mill.). These are followed in sensitivity by three conifers with LD_{50} dosages that range from pH 3.19 for jack pine, Pinus banksiana Lamb. to 2.94 for hemlock, Tsuga canadensis (L.) Carr.

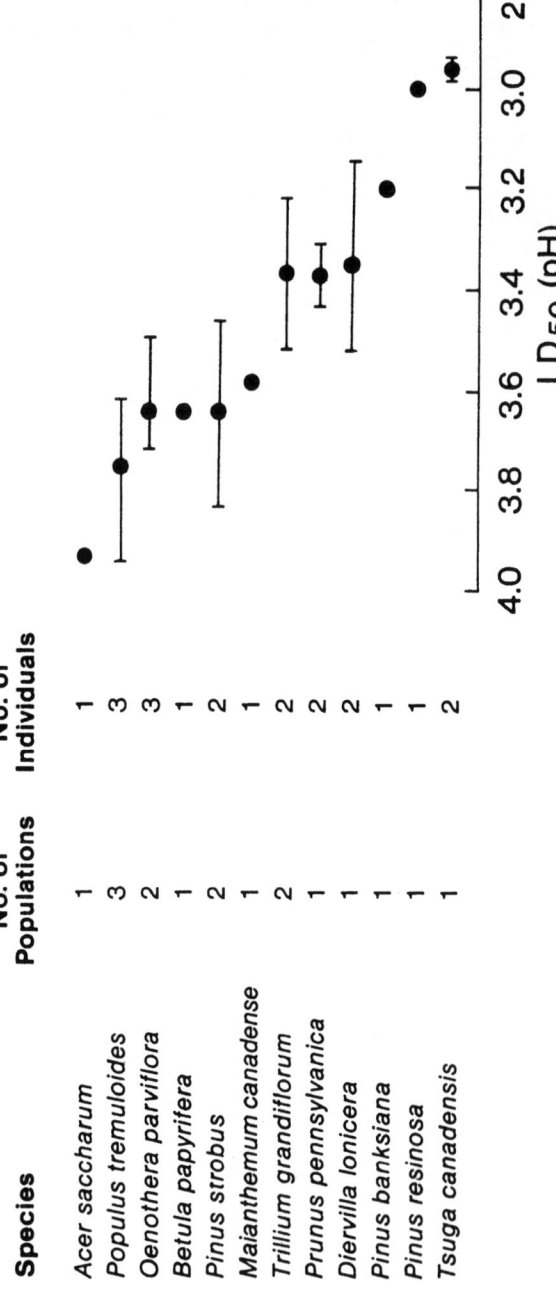

Fig. 1. The means and ranges of LD_{50}s for germination responses of pollen sampled from individuals of various forest species cultured at various pHs.

Sensitivity to Combinations of Different pHs and Trace Elements

Bioassays of Trillium grandiflorum pollen using factorial combinations of Cd and pH have been reported by Cox (1983b). Although the cadmium concentrations were in excess of those expected in rain (Galloway et al. 1982), the 1.0 µM (112 µg.L^{-1}) concentration did approach the maximum concentration expected to accumulate on vegetative surfaces from both dry and wet deposition (Lindberg et al. 1981). At lower concentrations of the metal in solution at pH 3.6, a stimulation of germination and tube growth occurred. Generally the inhibitory effect of low pH outweighed the Cd effects. However at 50 µM Cd at pH ≤ 3.6 synergistic concentration/pH interactions caused significant reductions in both germination and tube growth.

Copper levels recorded in rural wet deposition worldwide has a median value of 0.4 µg.L^{-1} with a maximum value of 150 µg.L^{-1} (Galloway et al. 1982). Plant surfaces may be exposed to concentrations many times in excess of that present in wet deposition alone (Lindberg and Harris 1981). To determine the sensitivity of a wide range of pollens to combinations of various pHs and copper, pollen from 11 species was assayed in factorial combination of 50, 100, 200, and 400 µg.L^{-1} as chloride with pH 5.6, 4.6, 3.6, and 2.6 (solutions adjusted with dilute H_2SO_4). These bioassays were carried out in a random block design of 5 blocks of 50 µL standing drop cultures each containing approximately 80-120 pollen grains. Pollen of Pinus strobus, Pinus resinosa, and Pinus banksiana was obtained from air dried strobili collected in late June, while Tsuga canadensis pollen was collected in late May. These pollens were air dried and then stored over silica gel for about 1 month before being stored in Vacutainer tubes at 4.0 K Pa pressure (vacuum) at -15°C. Pollen from the three pines was assayed at the same time in three different assays while Tsuga pollen was assayed later. The stored conifer pollen was allowed to hydrate for 1.5-2 h on glass slides in a moist chamber before inoculation of liquid cultures of a modified Brewbaker media (Brewbaker and Majumder 1961) containing 20% w/v sucrose with 15 µg.L^{-1} mycostatin and chloramphenicol. The cultures were incubated at 24°C ± 1 for 69.5 and 67 h for the pines and Tsuga pollens, respectively. After incubation 50% v/v acetic acid was added. The pollen was then examined under the microscope for percentage germination and tube growth.

Pollen of the deciduous trees Betula papyrifera, Betula alleghaniensis, and Populus tremuloides was obtained from excised branches from a woodlot at Maple Tree Improvement Institute, Ontario, in early April. This pollen was air dried and stored for two weeks over silica gel before it was assayed as above, except no antibiotics were used. The paper and yellow birch pollen cultures contained 10% sucrose whereas the aspen cultures contained 5% sucrose and were incubated for 4.0, 3.25, and 4.5 h, respectively.

Acer saccharum pollen was obtained from newly-opened flowers of a protandrous individual on an excised branch collected the previous day, May 3, 1983, from the University of Toronto woodlot at Maple, Ontario. This pollen was assayed fresh using the same design described above except the sucrose concentration of the culture media was 15% w/v and was incubated for 4.4 h at 26°C after which the cultures were scored in the usual way. Trillium grandiflorum and Diervilla lonicera pollen was collected from a single individual from populations at Maple on May 14 and July 8, respectively, and was assayed the next day in 10% sucrose at 28°C ± 1 for 5.5 and 5.0 h, respectively. The pollen of Oenothera parviflora, was collected from an individual growing at Erindale Ontario and was assayed fresh in 10% sucrose at 24°C for 4.0 h.

The in vitro sensitivity of the germination response of pollen sampled from individuals of 11 species tested in a pH/Cu factorial experiment is shown in Table 1. The species are ordered according to their relative range in pH sensitivity, i.e., those pHs that first caused significant inhibition ($p < 0.01$) for all pollen tested. While all species showed a response of pollen to lowered pH, not all were sensitive to the Cu exposures provided ($0-0.4$ mg.L^{-1}).

Table 1. Summary of effects of pH (2.6 - 5.6) and copper (0 - 0.4 mg.L^{-1}) on in vivo pollen germination (arcsin) of 11 forest plant species.

Species	pH effects		Cu effects		
	Overall sign.	Threshold of inhibition ($\alpha = 0.05$)	Overall sign.	pH of Cu effect ($\alpha = 0.05$)	Response
Populus tremuloides	**	5.6 - 4.6	**	4.6	Inhibition
Oenothera parviflora	***	5.6 - 3.6	N/S	4.6	Stimulation
Acer saccharum	***	4.6 - 3.6	***	5.6 - 4.6	Stimulation Inhibition
Betula alleghaniensis	***	4.6 - 3.6	***	4.6	Inhibition
Betula papyrifera	***	4.6 - 3.6	N/S	None	None
Diervilla lonicera	***	4.6 - 3.6	***	5.6 - 4.6	Mixed
Trillium grandiflorum	***	4.6 - 3.6	*	5.6	Stimulation
Pinus strobus	***	4.6 - 2.6	N/S	None	None
Tsuga canadensis	***	4.6 - 2.6	N/S	5.6	Stimulation
Pinus resinosa	***	3.6 - 2.6	N/S	None	None
Pinus banksiana	***	3.6 - 2.6	N/S	5.6	Mixed

Overall significance: * = $p < 0.05$; ** = $p < 0.01$; *** = $p < 0.001$

Significant effects ($p < 0.05$) of copper were noted for five species, the pHs at which these effects were noted were 5.6 and 4.6. Not all effects of copper on pollen germination were inhibitory, in fact the pollen germination of O. parviflora, T. grandiflorum, and T. canadensis was stimulated by copper at pH 5.6-4.6. The pollen of A. saccharum was stimulated by low levels of copper but inhibited by the higher dosages. D. lonicera pollen also showed a mixed response to copper, whereas P. tremuloides and B. alleghaniensis pollen showed significant inhibitions by copper at pH 4.6.

The above thresholds of response were obtained from assay conditions that significantly affect pollen responses. A more precise estimate can be obtained from a dose response regression model. The probit-procedure (SAS 1982) was used to determine the dosage that will cause a 50% probability of response (failure to germinate) (LD_{50}) which could be computed only if two or more points on any given response curve significantly differed from the 100% and 0% response. When all or nothing responses to assay conditions are experienced, no estimate of the LD_{50} can be computed because a lack of critical dosages (close to the theoretical LD_{50}). This occurred with the pollen of A. saccharum and B. alleghaniensis to pH in these pH/Cu assays. These two species must be considered to have pollen responses very sensitive to a change in pH from pH 4.6-4.3 at all copper concentrations. The LD_{50} pH values of the conifer pollen were signficantly lower than those computed for the broad-leaved species ($p < 0.001$ for a Mann-Whitney v test).

Using the pollen germination data, LD_{50} dosages for pH were computed (Fig. 2). Here the pH 5.6 no copper control treatment was used to estimate the natural response rate, this enabled the examination of changes in the LD_{50} due to additions of the copper. These LD_{50} dosages for pH are plotted against copper concentration in Fig. 2. Changes in sensitivity in both directions were apparent due to addition of copper. P. tremuloides pollen became more sensitive to acidity by the addition of 0.2 ppm Cu to the medium, whereas pollen of P. resinosa became more tolerant to acidity when 0.1 ppm Cu was added while O. pariflora showed increased tolerance of pH at 0.05 and 0.20 ppm Cu. Little or no changes in pH sensitivity were detected in the response of P. strobus, P. banksiana and T. canadensis pollen after copper addition. The pollen of D. lonicera demonstrated a small increase in sensitivity to acidity with increase in Cu from 0.05 to 0.2 ppm followed by a decrease in sensitivity on addition of 0.4 ppm Cu to the medium. These changes in LD_{50} pH indicate that the sensitivity of pollen of some species to pH is influenced by the presence of Cu at concentrations currently occurring at times in ambient precipitation and on the plant surface in urban and rural sites in northeastern North America. However, it is not known how these changes in pH sensitivity reflect the Cu nutrient status of the pollen or that of the individual that produced the pollen.

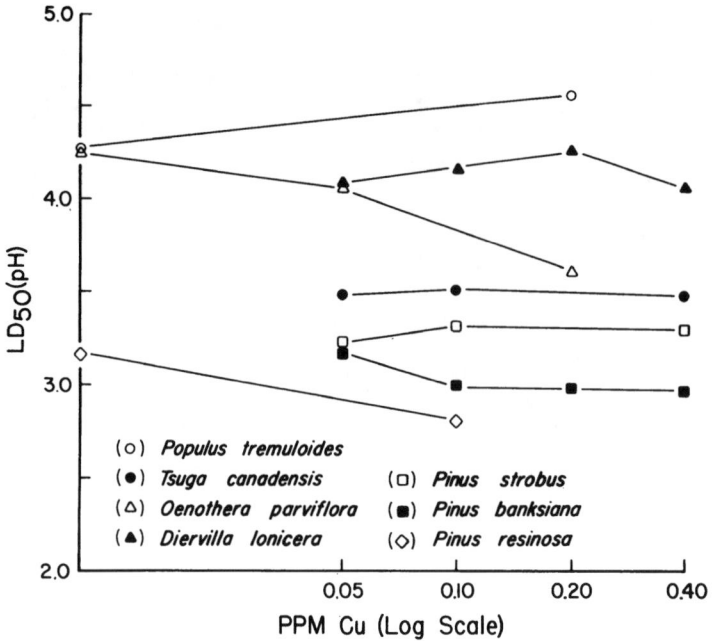

Fig. 2. The effect of copper concentration on the LD_{50} of the pollen germination response of seven forest plants in liquid culture.

Sensitivity to Combinations of pH and Aluminum

Plants are known to vary in their response to soil pH (Russell 1973) and may show differential adaptation to various soil types within their range (Snaydon 1962). These facts, together with the suggestion that pollen tolerances may reflect the tolerances of their sporophytes (Feder 1981; Murdy 1979) and the current interest in pH/Al interactions and toxicity, prompted an investigation into their combined effects on P. strobus pollen collected from adjacent sites with calcareous and acid soils. Pollen was sampled from sites no farther apart than 2.5 km in the Little Current area, Manitoulin Island, Ontario, and was air-dried and stored over silica gel at -15°C for no more than 2 weeks.

The factorial experiments were carried out using the same techniques as described in Cox (1983a,b). Two pairwise comparisons were made between pollen from sites of contrasting soil types using a factorial with five replicates, for each of the four pH treatments (pH 5.6, 4.6, 3.6 and 2.6). Five aluminum concentrations were used (0, 10, 50, 100, 300 µM Al). In the second comparison, pollen was assayed under the same conditions except no 300 µM Al treatments were used.

The pollen response surfaces for mean percent germination (arcsin) are shown in Fig. 3. Again, ANOVA indicated significant pH (p < 0.0001) and aluminum (p = 0.56 to 0.0001) effects and significant interaction terms (p = 0.05 to 0.0001). Significant synergism is seen at pH 4.6 and concentrations of aluminum greater than 50 µM.

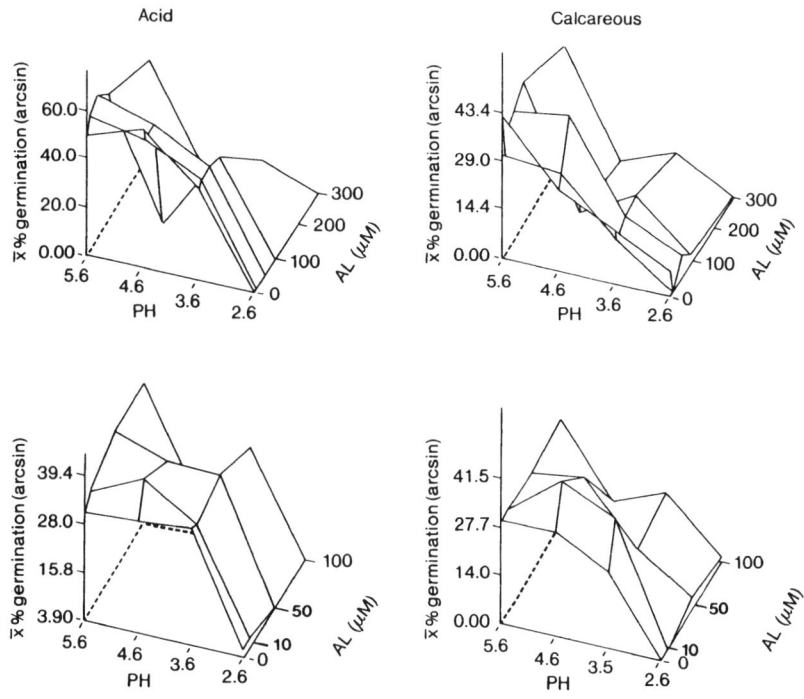

Fig. 3. Responses of pollen germination from two individuals of Pinus strobus from an acid and a calcareous soil population to combinations of pH and aluminum concentrations.

Apart from the synergistic interaction, the general slope of the response surfaces indicates the more susceptible nature of the pollen sampled from calcareous genotypes. These preliminary observations support the hypothesis that pollen tolerance may reflect edaphic tolerances of the pollen donors selected under calcareous and acid soil conditions.

Stigma Receptivity in Simulated Acid Rain

The effects of field simulations of 0.26 cm of precipitation on stigma receptivity of Oenothera parviflora were examined by Cox (1984). Virgin receptive stigmas of emasculated flowers of a Muskoka, Ontario, population of this species, whose petals had been removed to stop natural pollination, were treated with sprays of various pHs prior to controlled pollinations with known

loads of pollen. The percentage of potentially viable pollen
(darkly stained contents) that 1) germinated, and 2) produced
tubes long enough to penetrate the stigma (i.e., 3 x diameter of
grain) was recorded (Fig. 4). Acid rain simulants caused a
significant reduction in stigma receptivity using both criteria,
and pH 3.6 and 2.6 treatments significantly reduced receptivity
compared with pH 5.6 and unsprayed treatments. The LD_{50} dosages
of pH computed for viable pollen 1) that failed to germinate,
and 2) that failed to produce tubes long enough to penetrate the
style were pH 3.45 and pH 4.55, respectively.

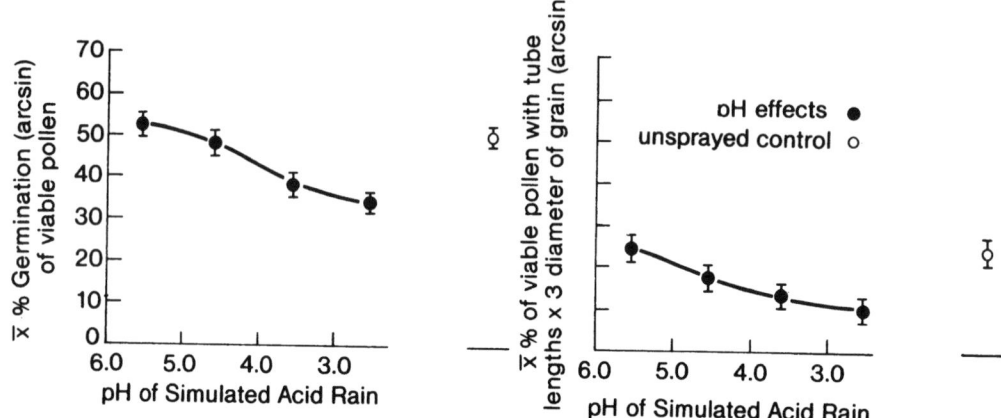

Fig. 4. The effects of simulated acid rain on stigma receptivity
of a Muskoka population of Oenothera parviflora in the field.
(Means ± SE derived from 8 replicate stigmas.)

DISCUSSION

The in vitro assays suggest variation between species in the pH
susceptibility of their pollen. In broad-leaved species, pollens
of canopy trees were more susceptible to pH than those of the
understory and ground flora perennials tested. As a group, conifers seem to have pollen that is more tolerant of acidity, with
the exception of P. strobus. The more generalized pollen sensitivities described in this study resemble the sensitivity of
similar groups of species described by Jacobson and van Leuken
(1977) and by Evans (1980) for foliar sensitivity to simulated
acid rain. It has been suggested (Cox 1983a,b) that a fuller
understanding of the relative sensitivities of reproductive processes of forest plants to acid precipitation can be obtained
only if species location in the canopy strata as well as composition of the canopy is considered because of demonstrated differences in pH of ambient rainfall and throughfall (Richter and
Granat 1978).

The combined in vitro effects of pH and cadmium on T.
grandiflorum pollen confirmed its sensitivity to pH and indicated
that Cd may be included with Pb and Mg among the cations found to
be stimulatory to pollen at concentrations found in acid

precipitation (Masaru et al. 1980). However, the synergistic interaction of pH and $\overline{Cd} > 50$ µM may be of concern if Cd pollution increases. This synergism indicates the importance of investigating combined effects.

Copper was also shown to influence the sensitivity of pollen to pH in some species at levels that are present, on occasion, in wet deposition. There is no reason to doubt that higher levels of copper may also accumulate on plant parts as a result of accumulated dry deposition and evaporite deposits left by rain and (or) fog events too light to wash the plant's surface as was suggested for Cd by Lindberg et al. (1981). It is likely that metal accumulation on stigmatic tissue or pollen is dependent on the length of time these structures are exposed to pollution and their position in the canopy. Variation between species in their in vitro pollen responses to copper was apparent, six species showed no overall response while four species showed highly significant ($p < 0.01$) inhibition. Trillium grandiflorum, however, showed significant ($p < 0.05$) stimulation of pollen germination. This variation in response may be due to differences in nutrient requirement for the metal and (or) toxicity. Furthermore, it must be noted that the effects of copper on pollen response are most obvious at pHs from 5.6-4.6.

Chaney and Strickland (1984) have reported significant inhibition of red pine pollen germination at copper concentrations in excess of 0.56 $\mu mol \cdot L^{-1}$ (0.36 $mg \cdot L^{-1}$) and an ED_{10} dosage of 0.68 $\mu mol \cdot L^{-1}$ (0.043 $mg \cdot L^{-1}$). In that study, it was indicated that germination was more sensitive than germ tube growth. Comparisons between these results and the ones for red pine in this investigation are difficult because of the lack of pH measurements, differences in pollen pretreatment, high incubation temperature, and the absence of organics in the incubation media. These studies do, however, underline the importance of further study not only of individual air pollutants on plant reproductive processes but also of combinations of air pollutants.

The combined action of aluminum and pH on germination of P. strobus pollen indicates both Al and pH effects together with a significant interaction. The synergistic interaction on pollen germination (Fig. 3) may result from a change in aluminum in the culture to a more toxic form from pH 5.6 to 4.6, i.e., $Al(OH)_2^+$. At pH 3.6 reduced toxicity may result from increased complexation of trivalent aluminum (Al^{3+}) which has high affinity to organics like those present in the 20% sucrose medium (Burrows 1977). Toxicity increases again as complexation is reduced at pH < 3.6. This indicates that in vitro assays of pollen germ tube may provide a useful way to examine aluminum toxicology of higher plants in soil as root hairs and pollen tubes are similar in many respects and may respond to acidification and Al in a similar fashion.

Of more relevance to direct effects of acid precipitation is the apparent greater pH sensitivity to the P. strobus pollens sampled from calcareous sites than those from acid sites. This supports the view that sporophyte tolerances may be reflected in their pollen and may indicate that reproduction in P. strobus adapted to calcareous soils may suffer more from pollen damage by acid rain than populations from acidic environments.

The direct exposure of the pine pollen to acidic rain, fog, or mist may occur while the pollen is adhering to the stigmatic tissue of the receptive cone scales. In addition, pine pollen is exposed to the accumulated acidity deposited on the cone scale by either wet or dry deposition during the exudation of the pollination droplet from the micropyle.

The reduced stigma receptivity of O. parviflora reported by Cox (1984) is the first such in vivo effect observed from simulated acid rain though similar effects have been observed by DuBay and Murdy (1983a,b) in response to SO_2 fumigations. This evidence supports the suggestion of Evans (1979) and Jacobson (1980) that reproductive structures exposed to acid precipitation may suffer direct damage.

Comparion of in vitro and in vivo pollen sensitivities in O. parviflora indicates that in both cases, pollen germination was inhibited by pHs equal to or lower than 3.6. LD_{50} dosages, however, indicate that in vivo germination may be somewhat less affected by initial pH than in vitro experimentation. This may reflect the extra buffering capacity of the stigmatic surface. This buffering explanation is also discussed by Karnosky and Stairs (1974) and DuBay and Murdy (1983a) in relation to SO_2 effects on pollen. The increased degree of in vitro inhibition of germination over the in vivo response, however, may be due to the heterogeneous distribution of acidity on the stigmatic surface. The LD_{50} dosages of the two in vivo pollen responses measured for O. parviflora were pH 4.66 and 3.45 for the mean percentage (arcsin) of grains that produced tubes three times greater than their diameter and the mean percentage (arcsin) that produced tubes greater than their diameter, respectively. This indicated that the critical stage of stigmatic penetration was more sensitive to simulated acid rain than germination of the pollen on the stigma.

The overall results then suggest that the threshold of response to pH of the various pollen tested in vitro and in vivo has been exceeded by the pH of ambient rainfall events in northeastern North America. Rain events approaching pH 3.0 already occur in southern Ontario (Chan 1982). Daily precipitation pH at Longwood near London, Ontario, had percentage frequencies of 0.98, 1.96, and 5.88 for pH 3.0 to 3.25, pH 3.25 to 3.50 and pH 3.5 to 3.75, respectively, for 1981 (Chan, pers. comm. 1982). The effects of these low pH events at the time of pollination are likely to reduce the seed set in at least the more sensitive species, as similar pollen inhibitions caused by SO_2 fumigation have been demonstrated (Murdy 1979; DuBay and Murdy 1983b). Simulated acid rain effects on seed set are currently under examination in a variety of species.

Reductions in pollen viability caused by air pollution be it SO_2, O_3, acid rain, or various combinations may be of special significance to reproduction in boreal forest herbs where fecundity may already be low due to pollination limitation. Barrett and Helenurm (1980) were able to increase fecundity of 11 out of 12 understory boreal forest herbs by controlled hand pollination as compared with open pollinations. Fecundity of only three of the herbs in this study were resource limited.

A further implication of air pollution effects on pollen on the stigmatic surface is the direct environmental selection of pollen thus reducing the genetic variability passing to the next generation. Such interference with natural gametic competition, should it occur, may affect fitness of future generations of plants (Mulcahy 1979), especially if genetic adaptation to base-rich calcareous environments is an important component of fitness.

Acknowledgements

The excellent technical assistance of D. Derbyshire and F. Vagnoni of the University of Toronto is gratefully acknowledged, as are the facilities of the Department of Botany and Institute for Environmental Studies at Toronto, together with the Maritimes Forest Research Centre at Fredericton. Financial support was provided by a contract through the Canadian Forestry Service, Environment Canada as part of the L.R.T.A.P. program.

REFERENCES

Antipov BG (1970) The effects of SO_2 on the reproductive organs of woody plants. Ohrana priody naturale, Sverdlovak 7: 31-35 (For Abstr, 32 (4), 752)

Arvind K, Malik CP (1976) Effects of metabolic inhibitors on pollen germination and pollen tube growth in Petunia alba. Plant Sci (Lucknow) 8: 26-27

Barrett SCH, Helenurm K (1980) Reproductive biology of some boreal forest herbs. In: Botany '80 Abstracts of papers presented at University of British Columbia, Vancouver, 12-16, July 1980. Botanical Society of America, Miscellaneous Series Publication, 158: 9-10

Benoit LF, Skelly JM, Moore LD, Dochinger LS (1982) The influence of ozone on Pinus strobus L. Can J For Res 13: 184-187

Brewbaker JL, Majumder SK (1961) Cultural studies of the pollen population effect and the self-incompatability inhibition. Amer J Bot 48: 457-462

Burrows WD (1977) Aquatic aluminum chemistry, toxicology and environmental prevalence. CRC Critical Reviews in Environmental Control 7: 167-216

Chan WH (1982) Acid precipitation in Ontario study. Daily precipitation chemistry listings and statistical summaries. July 15. 1980 - December 31, 1981. Air Resources Branch, Ontario Ministry of the Environment

Chancy WR, Strickland RC (1984) Relative toxicity of heavy metals to red pine pollen germination and germ tube elongation. J Environ Qual 13: 391-394

Cooper WC (1939) Vitamins and the germination of pollen grains and fungal spores. Bot Gaz 100: 844-852

Cox RM (1983a) Sensitivity of forest plant reproduction to long range transported air pollutants: In vitro sensitivity of pollen to simulated acid rain. New Phytol 95: 269-276

Cox RM (1983b) Sensitivity of forest plant reproduction to acid rain. In: Rennie PJ, Robitaille G (eds) Proc Int Conf Acid rain and forest resources. Quebec City 1983 (in press)

Cox RM (1984) Sensitivity of forest plant reproduction to long range transported air pollutants: In vitro and in vivo sensitivity of Oenathera parviflora pollen to simulated acid rain. New Phytol 97: 63-70

Datta PC, Naug A (1976) Staining pollen tubes in the style; cotton blue versus carmine for general use. Stain Technol 42: 81-85

Dopp W (1931) Uber die wirkung der schwefligen Saure auf Blutenorgane. Ber Dent Bot Ges 47: 173-221

DuBay DT, Murdy WH (1983a) The impact of sulphur dioxide on plant sexual reproduction: In vivo and in vitro effects compared. J Environ Qual 12: 147-149

DuBay DT, Murdy WH (1983b) Direct adverse effects of SO_2 on seed set in Geranium carolinianum L.: a consequence of reduced pollen germination on the stigma. Bot Gaz 144 (3): 376-381

Evans LS (1979) A plant developmental system to measure the impact of pollution in rainwater. J Air Pollut Control Assoc 29: 1145-1148

Evans LS (1980) Foliar responses that may determine plant injury by simulated acid rain. In: Toribara T, Miller M, Morrow P (eds) Polluted rain. Plenum Press, New York, p 239-257

Feder WA (1968) Reduction in tobacco pollen germination and tube elongation induced by low levels of ozone. Science 160: 1122

Feder WA (1981) Bioassaying for ozone with pollen systems. Environ Health Perspect 37: 117-123

Feder WA, Sullivan F (1969) Differential susceptibility of pollen grains to ozone injury. Phytopath 59: 399 (Abstr)

Galloway JN, Thornton JD, Norton SA, Volchok HL, McLean RAN (1982) Trace metals in atmospheric deposition: a review and assessment. Atmos Environ 16: 1677-1700

Houston DB, Dochinger LS (1977) Effects of ambient air pollution on cone, seed and pollen characteristics in eastern white and red pines. Environ Pollut 12: 1-5

Iwanami Z, Okada I, Iwamatsu M, Iwadore T (1978) Inhibitory effects of royal jelly acid and myrmicacim and their analogous compounds on pollen germination, pollen tube elongation and pollen tube mitosis. Cell Struct Funct 4 (2): 135-142

Jacobson JS (1980) The influence of rainfall composition on yield and quality of agricultural crops. In: Drablos D, Tollan A (eds) Ecological Impact of Acid Precipitation, Proc Intern Conf, Sandefjord, Norway, March 11-14, 1980, p 41-46

Jacobson JS, van Leuken P (1977) Effects of acidic precipitation on vegetation. In: Proc 4th Intern Clear Air Congress 1977, Tokyo, Japan, p 124-127

Karnosky DF, Stairs GR (1974) The effects of SO_2 on in vitro forest tree pollen germination and tube elongation. J Environ Qual 3: 406-409

Kratky BA, Fukunaga ET, Hylin JQ, Nakano RT (1974) Volcanic air pollution: deleterious effects on tomatoes. J Environ Qual 3: 138-140

Lindberg SE, Harriss RC (1981) The role of atmospheric deposition in an eastern U.S. deciduous forest. Water Air Soil Pollut 16: 13-31

Lindberg SE, Turner RR, Shriner DS, Huff DD (1981) Atmospheric deposition of heavy metals and their interaction with acid precipitation in a North American deciduous forest. Proc Intern Conf Heavy Metals in the Environment, Amsterdam, 1981, CEP Consultants, Edinburgh

Ma T-H, Khna H (1976) Pollen mitosis and pollen tube growth inhibition by SO_2 in cultured pollen tubes of Tradescantia. Environ Res 12: 144-149

Ma T-H, Isbandi D, Khan SH, Tseng YS (1973) Low level SO_2 enhanced chromatid aberration in Tradescantia pollen and seasonal variation in the aberration rates. Mutation Res 21: 93-100

Masaru N, Katsuhisa F, Sankichi T, Yukata W (1980) Effects of inorganic components in acid rain on tube elongation of Camellia pollen. Environ Pollut 21: 51-57

Masaru N, Syozo F, Suburo K (1976) Effects of exposure to various injurious gases on germination of lily pollen. Environ Pollut 11: 181-187

Mulcahy DL (1979) The rise of angiosperms: a genecological factor. Science 206: 20-23

Mumford RA, Lipke H, Loufer DA, Feder WA (1972) Ozone-induced changes in corn pollen. Environ Sci Tech 6: 427-430

Murdy WH (1979) Effect of SO_2 on sexual reproduction in Lipidium verginicum L. originating from regions with different SO_2 concentrations. Bot Gaz 140 (3): 299-303

Murdy WH, Ragsdale HL (1980) The influence of relative humidity on direct sulfur dioxide damage to plant sexual reproduction. J Environ Qual 9: 493-496

Mrkva R (1969) Einfluss der immission auf die saatgutgute der kiefer (Pinus sylvestris L.) im gebiet des forstbetriebes breclov (sudmahren). Acta Univ Agric Brno 38: 345-360

Pelz E (1963) Investigations of fruit development and smoke damaged spruce stands. Arch Forstwesen 18: 47-49

Podzoroz NV (1965) Effects of smoke in air and the quality of pine seeds. Lesn Chozjajstvo 18: 47-49

Powell P, Timms P (1974) The chemistry of non-metals. Chapman and Hall, London, p 103

Richter A, Granat L (1978) Pine forest throughfall measurements. Report AC-43 Dept Meterology, Univ Stockholm. International Meterological Institute, Stockholm

Russell EW (1973) Soil conditions and plant growth. Longmans, Green and Co, London

Sabachnikoff V (1912) Action de l'acide sulfureux sur le pollen. Comples rendus de la Societe de Biologie Paris. 72S: 191-193

SAS User's Guide (1979) SAS Institute, Gary, North Carolina, U.S.A.

Snaydon RW (1962) The growth and competitive ability of contrasting natural populations of Trifolium repens L. on calcareous and acid soils. J Ecol 50: 439-447

Stanley RG, Linskens HF (1974) Pollen biology biochemistry management. Springer-Verlag, New York, Heidelberg, Berlin, p 49

Strickland RC, Chaney WR (1979) Cadmium influences on respiratory gas exchange of Pinus resinosa pollen. Physiol Plant 47: 129-133

Taniyama T, Uman H (1973) The effects of exposure to sulphur dioxide on eggplant pollen. In: Studies on damage of agricultural vegetation. Ministry of Agriculture and Forestry, Japan, p 112-114

Varshney SRK, Varshney CI (1981) Effect of sulphur dioxide on pollen germination and pollen tube growth. Environ Pollut 24: 87-92

CONSEQUENCES OF CLOUD WATER DEPOSITION ON VEGETATION AT HIGH ELEVATION

M.H. Unsworth and A. Crossley

Institute of Terrestrial Ecology, Edinburgh Research Station,
Bush Estate, Penicuik, Midlothian EH26 0QB, Scotland, U.K.

ABSTRACT

The capture of wind-driven cloud by vegetation provides a pathway for pollutant deposition that has only recently been identified, and remains only poorly quantified. This paper reviews current knowledge of three aspects of the pathway: measurement and modelling the rates of deposition of cloud water to various vegetation types; techniques for monitoring the concentrations of soluble and solid material in cloudwater; and potential mechanisms for injury to vegetation by the deposited material.

Although there have been many measurements of amount of fog drip below trees, there are very few where there is sufficient detail of the environmental conditions and plant structure to allow generalizations to be drawn. Analysis of existing measurements supports the view that fog-water fluxes are essentially limited only by rates of turbulent transfer and so can be modelled realistically from a knowledge of momentum transfer. This can be shown to imply that fog-water fluxes are typically 1 $mg.m^{-2}.s^{-1}$ (4 $\mu m.h^{-1}$) over short grass and 10 $mg.m^{-2}.s^{-1}$ (40 $\mu m.h^{-1}$) over forests. Analysis of drag forces on leaves and shoots can be used to show that isolated trees and shrubs capture fog water at rates of up to 100 $mg.m^{-2}.s^{-1}$, consistent with the few useful observations.

These analyses suggest that there is substantial spatial variation in fog-water deposition, for example at the upwind edges of forests, and on dominant trees in canopies. There will also be large vertical variation in a dense canopy in the amount of water deposited per unit foliage density - current knowledge of in-canopy wind profiles can be used to estimate this. Chemical concentrations in cloud water can be much larger than in rain. Apart from data at Whiteface Mountain, there do not appear to be long series of records to establish the "climatology" of cloud-water chemistry, i.e., its dependence on synoptic meteorology and the seasonal variability at sites. The design of collectors probably leads to biases, because of differing variations of collection efficiencies with drop size. In Britain, cloud water with large concentrations of acidic substances often contains substantial amounts of solid particles, and so vegetation may have much larger inputs of particulate material in this occult deposition than we have previously estimated when considering only dry deposition.

The low pH of cloud water may, by itself, be sufficient to damage foliage, by eroding cuticular waxes and, for example, directly injuring epidermal cells, or allowing increased leaching, or altering gas or water vapour exchange. Evaporation from the water film on leaf surfaces, proceeding simultaneously with evaporation, may lead to a further concentration of solution on the surface. Evaporation from forests would be very sensitive to small alterations in humidity (in the range 97-100% rh). For extensive forest canopies, the concentration ratio (surface liquid/cloud drop) may exceed 10, leading to surface pH below 3. The spatial variation of the concentration phenomenon within canopies and on isolated trees is likely to be large, and could help to explain the positioning of observed visible injury.

Large inputs of sulphur and nitrogen to upland forests by occult deposition may also increase the probability of frost injury to trees, both directly by altering membrane strength and indirectly through changes in rates of cold hardening.

INTRODUCTION

At high elevations, plants experience many stresses that are not present, or are less severe, at lower locations. Examples are the stresses of wind, frost, winter dessication, ultra-violet radiation, and nutrient-impoverished soils. The observation that recent forest declines occurred initially at high elevations in Germany and the U.S.A. (Schutt and Cowling 1985; Johnson and Siccama 1983, 1984) has led to various theories of air pollution stresses that might be specific to such sites. Three pathways for deposition are likely to be more important at high elevation sites than at lower levels: dry deposition of photochemically generated pollutants, such as ozone or nitric acid vapour, which appear to be at higher concentrations and to show less diurnal variation than at low elevation (Ashmore et al. 1985; Smidt 1983); orographic enhancement of rainfall, in which concentrations of dissolved pollutants have been increased above those in low-level rain by the 'seeder-feeder' mechanism (Carruthers and Choularton 1984); and the capture by vegetation of wind-driven cloud drops containing high concentrations of pollutants (Dollard et al. 1983). Hydrologists coined the term 'occult precipitation' (occult = hidden) to describe the water input by this latter pathway because the precipitation was not detected in any standard rain gauges. By analogy we use the phrase 'occult deposition' to describe the previously hidden input of pollutant materials by this pathway.

The main section of this paper reviews current knowledge of three aspects of occult deposition: the rate of cloudwater capture by vegetation; the chemical content of cloudwater; and the physical processes which may alter the concentration of deposited cloudwater. We also discuss the mechanisms by which pollutant deposition may cause biological responses at high elevation, with particular reference to the possible alteration in frost hardiness.

RATES OF CLOUDWATER CAPTURE BY VEGETATION

The high efficiency with which tall vegetation captures wind-driven cloud drops has long been recognized by ecologists, and examples of areas where the vegetation species and diversity are uniquely related to the dominant role of occult precipitation have been thoroughly studied - for example the coastal forests of California (Cannon 1901; Byers 1953; Oberlander 1956), the mountains of Hawaii (Ekern 1964) and Table Mountain in South Africa (Marloth 1907; Nagel 1956). There is also a substantial German literature on occult precipitation (Rubner 1935; Baumgartner 1952; Grunow 1955), usually in relation to forest hydrology. Reviews of the subject by Penman (1963), Kerfoot (1968) and Rutter (1975) all emphasize the wealth of site-specific studies but there is a dearth of research from which generalizations can be drawn concerning rates of water capture of forests or isolated trees. Many of the problems arise because artificial collectors were used to capture the cloudwater and at present there is no simple method of relating such measurements to complicated structures such as trees.

In a long study of the effectiveness of trees in dispersing sea-fog, Hori and his colleagues (Hori 1953) developed several micrometerological techniques which allowed fog water fluxes to an extensive forest to be estimated; more recently Dollard used micrometeorological methods to estimate rates of occult precipitation to rough pasture (Dollard and Unsworth 1983) and moorland terrain (Dollard et al. 1983). These investigations provide only a very limited set of observations of deposition rates, insufficient to verify theoretical models. In the absence of any strong evidence otherwise, we will assume here that deposition proceeds at the maximum rate determined by momentum transport; the measurements of Dollard and Unsworth (1983) suggest that this assumption is unlikely to substantially overestimate true deposition. For extensive uniform surfaces, the rate of deposition by turbulent transfer can be described adequately with knowledge of windspeed, surface roughness and the liquid water content of the air. The rate of deposition D ($g.m^{-2}.s^{-1}$) is given by

$$D = W (v_t + v_s) \quad (1)$$

where W is the liquid water content in the cloud ($g.m^{-3}$) and v_t and v_s are deposition velocities ($m.s^{-1}$) by turbulent diffusion and sedimentation, respectively. Assuming that the factors limiting the deposition of momentum and droplets are identical, v_t can be derived in terms of the roughness characteristics of surfaces

$$v_t = \frac{k^2 u}{(\ln (z-d/z_0))^2} \quad (2)$$

where k is von Karman's constant (0.41), u is the windspeed at a reference height z, z_0 is the roughness length and d is the zero plane displacement (Monteith 1975). Table 1 shows calculations

Table 1. Calculated rates of cloudwater deposition to extensive uniform surfaces, based on typical roughness parameters. The calculations assume that the liquid water concentration in the cloud is 0.1 g.m^{-3} and the wind speed is 5 m.s^{-1} at a height 2 m above the top of the vegetation. Other details of the calculations are given in the text. For conversion, 1 mg.m^{-2}.s^{-1} = 3.6 μm.h^{-1}.

Surface	Vegetation height h(m)	Roughness length z_o(m)	Zero Plane displacement d(m)	Deposition velocity v_t(mm.s^{-1}) v_s(mm.s^{-1})		Water flux D(mg.m^{-2}.s^{-1})
Grass	0.15	0.03	0.10	48	20	6.8
Moorland	0.30	0.07	0.10	70	20	9.0
Forest	15.5	0.93	11.8	256	20	27.6

of v_t and D calculated from equations 1 and 2 for typical surfaces, using the value of v_s determined by Dollard and Unsworth (1983). The table indicates that forests are at least four times more efficient than grasslands at capturing cloudwater when both are exposed to the same wind speed.

More complex models of the capture of cloudwater by vegetation have been published by Shuttleworth (1977) and Lovett (1984), by adapting models originally developed for studying heat, momentum and gas transfer in canopies. Within the uncertainties and variabilities associated with specifying the aerodynamic characteristics of the canopy and the vertical distribution of foliage, the models agree with each other reasonably well, but there have been only very limited comparisons of models with direct observation.

Extending the assumption of similarity between droplet and momentum transfer to the interception by isolated trees and branches enables rough estimates to be made of the occult precipitation inputs to them. Appropriate terms to describe the interference to momentum transfer that is imposed by the various overlapping branches and leaves are the 'shelter factor' (Landsberg and Jarvis 1973) or the drag coefficient (e.g., Chamberlain 1975). Fraser (1962) measured the drag force on trees in a wind tunnel, and found that coefficients c_d at 15 $m.s^{-1}$ ranged from 0.57 for Norway spruce to 0.25 for Western hemlock. If the same coefficients are appropriate for the capture of cloudwater, the rate of capture F ($g.s^{-1}$) by a tree with a vertical cross-sectional crown area A_1 (m^2) in a cloud with liquid water content W ($g.m^{-3}$) and wind speed u ($m.s^{-1}$) is

$$F = 1/2 \, W \, u \, c_d \, A_1 \tag{3}$$

and if the ground area covered by the tree crown is A_2 (m^2), the maximum flux density D ($g.m^{-2}.s^{-1}$) of occult precipitation is

$$D = 1/2 \, W_A u \, c_d \, A_1/A_2 \tag{4}$$

Table 2 shows the variation of D with wind speed and aspect ratio A_1/A_2 assuming a liquid water content 0.1 $g.m^{-3}$ and c_d = 0.50.

The assumptions in Table 2 that c_d and A_1 are independent of wind speed are probably not good, but the limited experimental evidence does not merit more sophisticated models at present. Fraser (1962) found that when wind speed on full grown trees exposed in a wind tunnel increased from 9 to 26 $m.s^{-1}$, the foliage became increasingly streamlined, so that the drag force increased only linearly with wind speed rather than with the square of wind speed, as would be expected for rigid bodies. This relation implies that the product $c_d \, A_1$ in equation 3 was proportional to u^{-1}, in which case F would be independent of wind speed. On the other hand, the assumption that c_d is the same for momentum and drops is certainly poor at low wind speeds when drops are less efficiently captured (Chamberlain 1975); it may also overestimate capture at high wind speeds when drops shatter on impaction (Thorne et al. 1982). The estimates in Table 2 thus

Table 2. Calculated values of the maximum flux density ($mg.m^{-2}.s^{-1}$) of cloudwater below an isolated tree with crown area/ground area aspect ratio A_1/A_2, exposed at wind speed $u(m.s^{-1})$ in a cloud with liquid water content $0.1\ g.m^{-3}$. The capture coefficient for cloudwater is assumed to be 0.50 (see text). For conversion, $1\ mg.m^{-2}.s^{-1} = 3.6\ \mu m.h^{-1}$ precipitation.

Wind Speed $u(m.s^{-1})$	Flux density ($mg.m^{-2}.s^{-1}$) Aspect ratio A_1/A_2			
	0.5	1	2	5
2	25	50	100	250
5	63	125	250	625
10	125	250	500	1250
20	250	500	1000	1500

have considerable uncertainty but illustrate the potential that isolated and dominant trees have for capturing cloud water.

There have been very few experimental studies suitable for comparing with the estimates in Table 2. Yosida and Kuriowa (1953) found that an isolated conifer tree about 2 m high captured water at a rate of about $1.2\ L.h^{-1}$ when the wind speed was about $3.5\ m.s^{-1}$ and the liquid water content of the fog was $0.4\ g.m^{-3}$. If the ground area below the tree was about $1\ m^2$ and the aspect ratio of the tree was about 1, this value corresponds to $D = 330\ mg.m^{-2}.s^{-1}$, which is consistent with the values in Table 2, considering that the fog was four times as dense as that assumed in the table.

The differences between Tables 1 and 2 illustrate that isolated objects exposed to horizontally-driven drops are much more efficient collectors than objects within a uniform canopy where vertical transfer by sedimentation and turbulence is the only mechanism for droplet deposition. Consequently, dominant trees projecting above a canopy, and trees at the edges of forests where there is horizontal penetration of the wind will collect much more water than trees within a canopy. Similarly, because capture efficiency probably increases with wind speed, and wind speed increases with height, isolated tall trees are better collectors than lower shrubs with the same exposed surface area. These points emphasize that there is likely to be great heterogeneity in the rate of occult precipitation in natural ecosystems, depending on the heights of the various plant species and their distribution. The biological implications of this variability in input do not seem to have been studied in detail.

CHEMICAL CONTENT OF CLOUDWATER

There is currently considerable discussion between atmospheric chemists on the processes that lead to the incorporation of chemicals in cloudwater (Daum et al. 1984; Lazrus et al. 1983). There are two possible sources: soluble particles that act as condensation nuclei, and gases or particles that are scavenged by drops in the cloud. Recent evidence suggests that the cloudwater composition is not derived by dissolution of pre-cloud aerosol, but often involves in-cloud transformations of gases such as SO_2 and NO_x in the production of acidity.

Aircraft observations of specific cloud events have revealed that concentrations of major cloudwater ions vary by an order of magnitude between samples, as might be expected in view of the processes described above. Daum et al. (1984) reported that the highest concentrations, C, were associated with clouds with the lowest liquid water content, L, and that the product LC was relatively constant, consistent with the idea of cloud-drops growing around a soluble nucleus. However, they also noted that concentrations of sulphate and nitrate were highest when the NO_x and SO_2 concentrations in air were largest.

Although there have been several studies of the chemistry of individual cloud events at mountain sites (Houghton 1955; Mrose 1966; Okita 1968; Lovett et al. 1982; Falconer and Falconer 1980; Dollard et al. 1983; Waldman et al. 1982), there is at present no detailed chemical climatology of cloudwater to the same extent that there is similar information for rainfall. A major limitation on the routine sampling of cloudwater is the difficulty of excluding rainfall from cloudwater collectors. Rainwater usually has much lower chemical concentrations than cloudwater - a result of the microphysical processes leading to the growth of raindrops.

Devices for collecting cloudwater may be of the 'active' type, where relative motion between the drops and the collector is generated artificially (Okita 1968; Garland et al. 1973; Dollard and Unsworth 1983; Walters et al. 1983), or 'passive' type in which the natural wind speed causes the impaction of drops against a collector (Nagel 1956; Falconer and Falconer 1980; Dollard et al. 1983).

Active samplers can be arranged so that their sampling efficiency is independent of wind speed, whereas passive samplers become less efficient as wind speed decreases. The part of the passive sampler where drops impact is generally either a fine wire or gauze, to optimize the sampling efficiency for small drops (Chamberlain 1975). Nagel (1956) used a cylinder of gauze made with wires 0.25 mm diameter at a spacing of 1.6 mm, fixed above a rain-gauge to measure the volume captured, but a metal gauze collector would be unsuitable for chemical measurements. Falconer and Falconer (1980) described a cloudwater collector designed for use on Whiteface Mountain New York State, and Gervat (1985) used a similar design. The collector consists of two polypropylene discs, diameter 25 cm, separated and supported by polypropylene rods 1 m long. A continuous strand of Teflon fluorocarbon fibre (PTFE), diameter 0.4 mm, is strung back and forth between the discs at 3 mm spacing around their circumference. A plastic funnel at the base of the collector catches the intercepted water running down the fibres and channels it to collecting bottles and analysers.

Dollard et al. (1983) used a conical design of cloudwater collector consisting of nylon filament 0.16 mm in diameter wound continuously onto a P.V.C.-coated frame, and we have recently developed this design further; Figure 1a illustrates the main features. Two significant advantages of the conical gauge in high wind speeds are i) that a component of the horizontal wind acting on the strings accelerates the drops into the collecting funnel and ii) that the increase in droplet size by coalescence during movement down the strings results in capillary entrapment between the closer spaced strings, reducing the wind losses of large droplets that have been observed with the parallel filament design of Falconer and Falconer (1980). These benefits appear to outweigh the airflow interference problems resulting from the close spacing of the filaments near the base of the cone in our design. Nylon filament is available in a good range of diameters but may not be well-suited for use in fogwater collectors that are left exposed in dry conditions because of the affinity of nylon for nitric acid vapour (Appel et al. 1980). Dry deposition of NHO_3 and subsequent washoff by cloud drops would enhance the nitrate and pH in collected water. The problem would not arise if gauges were exposed only during periods of cloud because the concentrations of highly soluble HNO_3 would be negligible in the interstitial air of clouds.

Teflon filament is not readily available in fine diameters and lacks the tensile strength of nylon, but clean P.T.F.E. is a very poor absorber of dry deposition. To investigate whether the advantages of Teflon outweigh the disadvantages, we are comparing gauges with stringing in nylon and P.T.F.E. Table 3 shows preliminary results from gauges using either 0.16 mm diameter nylon or 0.60 mm P.T.F.E. The gauges were left in position on Dunslair Heights (602 m), about 40 km from Edinburgh; rainfall was recorded on an adjacent raingauge. Table 3 shows that over the whole period, the Teflon gauge collected about 3% fewer hydrogen ions than the nylon gauge, but there were several events when this general conclusion was reversed. These preliminary results are inconclusive because the two sizes of stringing have different efficiencies of collecting cloud drops, and it may be that the very fine drops that are collected less efficiently by the P.T.F.E. strings contained high concentrations of chemicals. However, irrespective of the initial relative merits of filament materials, we suspect that contamination of filaments with particles rapidly negates any potential advantages that P.T.F.E. may have, as the residues on the strings may allow ion exchange between dry and wet deposits.

The particles that we find in cloudwater typically have diameters in the range 0.5-5 um; dry particles of these sizes would be only inefficiently captured by the strings of the gauges, and so we

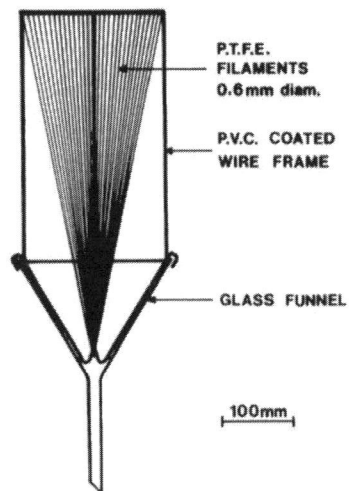

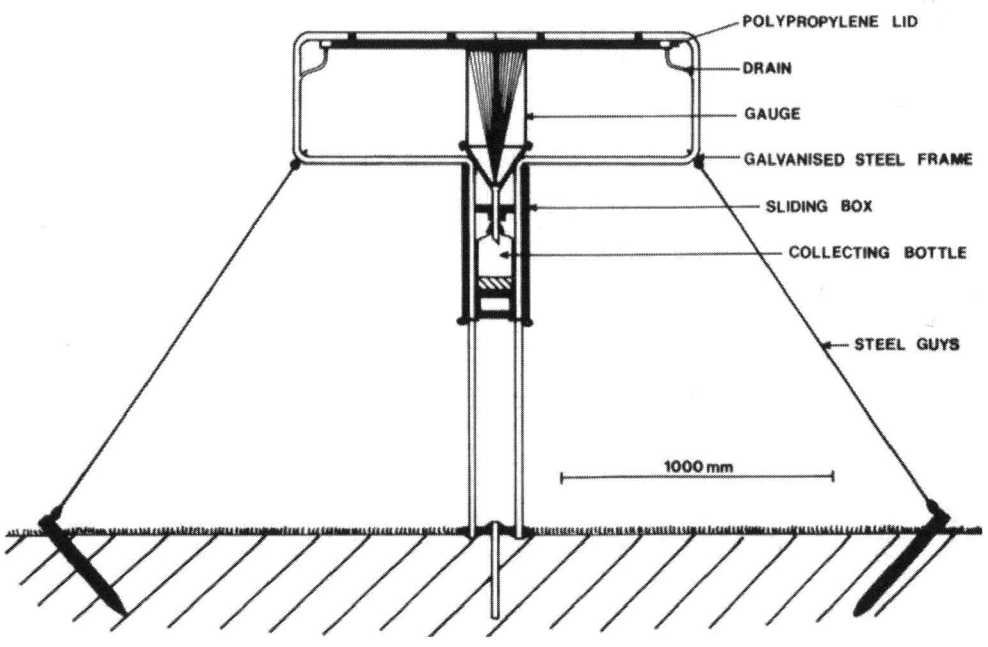

Fig. 1. Diagram of the gauge design used for cloudwater collection. (a) The framework was strung with either 0.16 mm diameter nylon filament or 0.6 mm diameter P.T.F.E. tubing. (b) The design incorporates a 1.2 m diameter lid to protect the strings from 'contamination' by rain drops larger than 0.5 mm at wind speeds up to 5 m sec^{-1}.

Table 3. Comparison of hydrogen ion concentrations (μeq.L^{-1} H$^+$) collected by 0.16 mm diameter nylon and 0.6 mm diameter P.T.F.E. filaments at Dunslair Heights.

Date	Wind direction	Rain	Cloudwater nylon open gauge	Cloudwater PTFE open gauge
26/2/85	WSW	0.03	10.5	1.0
4/3/85	NW	1.6	724.4	575.4
20/3/85	SE	0.01	41.7	41.7
25/3/85	SW	0.05	371.5	562.3
1/4/85	SW	13.2	30.2	19.9
11/4/85	N	0.6	128.8	134.9
17/4/85	-	1.1	3.0	39.8
23/4/85	NE	1.5	104.7	2.0
25/4/85	N	0.2	0.2	0.3
29/4/85	NW	6.6	7.1	2.4

conclude that they are probably collected during occult deposition, entrained within cloud droplets. Subsequent evaporation of liquid from the filaments would leave a residue of particles, held by electrostatic forces. Chemical washings in 5% HCl as suggested by Falconer and Falconer (1980) or ultrasonication in solvent fail to remove the deposits, although accessible filaments can be wiped clean with tissue; current design changes are tackling the problem of string contamination.

To illustrate the substantial amounts of particulate material collected in cloudwater, Table 4 gives details of events sampled at Castlelaw (408 m) near Edinburgh, including occasions when long range transport of pollutants from England and Europe was possible. The cloudwater was usually opaque, and electron microscope examination of the filtered material showed that the particles were dominated by carbon and pulverised fuel ash (P.F.A.), characteristic of domestic and industrial combustion. If the liquid water content of the cloud was 0.1 g.m^{-3}, then the observed particle concentration in the cloudwater would arise from the 1 μg particles from each 1 m^3 of air becoming incorporated into cloud drops. Since annual mean concentrations of particles in urban air are typically 17 μg.m^{-3}, it is not surprising that we and others measuring cloudwater frequently find that samples contain large amounts of particles (Gervat 1985; Harvey, pers. comm.).

Apart from the contamination that particles produce on the filaments of collecting gauges, they must also be collected very effectively via the occult deposition mechanism by vegetation with fine foliage, e.g., conifers. Although it does not appear to have been reported, we would expect there to be large gradients of particle deposition at the exposed edges of forests, etc., and there is a need to investigate the biological implications of this type of occult deposition of particles. It is well-known that dry particles of submicron size are only very

Table 4. Examples of hydrogen ion content and particulate loading for collections made at Castlelaw in rain, open cloudwater and lidded cloudwater gauges exposed to different wind directions (both cloud gauges strung with 0.6 mm diameter P.T.F.E.).

Collection Period	Wind Direction	Rain gauge		Cloudwater open gauge		Cloudwater lidded gauge	
		$\mu eq.L^{-1}$ H^+	particulates $mg.L^{-1}$	$\mu eq.L^{-1}$ H^+	Particulates $mg.L^{-1}$	$\mu eq.L^{-1}$ H^+	Particulates $mg.L^{-1}$
1/4/85-9/4/85	E	41.7	4.9	83.2	23.1	52.5	16.5
9/4/85-10/4/85	E	52.5	12.6	109.6	15.0	138.0	18.5
19/4/85-24/4/85	NW	39.8	0.2	81.3	78.8	109.6	71.4
24/4/85-30/4/85	W	11.5	6.8	131.8	35.2	144.5	67.9
30/4/85-6/5/85	SSE	125.9	2.5	363.1	81.0	436.5	94.8

inefficiently deposited on vegetation (Chamberlain 1975), and so the pathway of occult deposition makes particle inputs to vegetation growing at high elevation much larger than those to vegetation growing below cloud level.

RAIN AND CLOUDWATER

The capture of rain in cloudwater gauges 'dilutes' the cloudwater; correction using analysis of rain from an adjacent raingauge is not possible because the filaments of a cloudwater gauge have a capture efficiency for rain that depends on windspeed and raindrop size. For routine measurements, a lid over a cloudwater gauge would exclude large rain drops, but would need to be inconveniently large to eliminate rain contamination at high wind speeds. As a compromise, we have designed a lid 1.2 m diameter (Fig. 1b) which is sufficient to exclude drops larger than 0.5 mm at wind speeds of up to about 5 $m.s^{-1}$. Figure 2 illustrates some results of a comparison of a raingauge, an open cloudwater collector, and a lidded collector exposed for six weeks on Castlelaw. The measurements illustrate the large differences in concentration between rain and cloudwater, and show that lidded gauges consistently recorded the highest concentrations. Further analysis of volumes and concentrations from each event may allow better quantification of the benefits of the lid.

Analysis of other ions in the cloudwater samples has shown that on many occasions the dominant anions, sulphate and nitrate, are not balanced by hydrogen ion; full ionic balances are not yet available. Sulphate and nitrate concentrations for the cloudwater observations in Figure 3 were in the range 300-1200 $\mu g.L^{-1}$, comparable with values for northern England reported by Dollard et al. (1983) and Gervat (1985).

EVAPORATION AND THE CONCENTRATION OF DEPOSITED CLOUDWATER

In addition to the deposition of cloudwater by sedimentation and turbulent diffusion, water can be transferred in the vapour phase between the atmosphere and vegetation. For example, if the vegetation is below the dewpoint of the air, water will condense; similarly, if the air close to the surface is unsaturated, and/or if sufficient radiative energy is available at the surface, deposited water will evaporate. Shuttleworth (1977) described the physical principles involved in this process when vegetation is in cloud; Lovett (1984) reported a case where evaporation reduced the fog drip rate to zero in a balsam fir forest; and Unsworth (1984) calculated how the simultaneous evaporation of cloudwater would increase the concentration of non-volatile chemicals in water deposited on leaf surfaces.

Figure 3, based on the calculation methods described by Unsworth (1984), shows that the evaporation/condensation process depends strongly on available energy (predominantly solar radiation penetrating the cloud) and on relative humidity. For forests

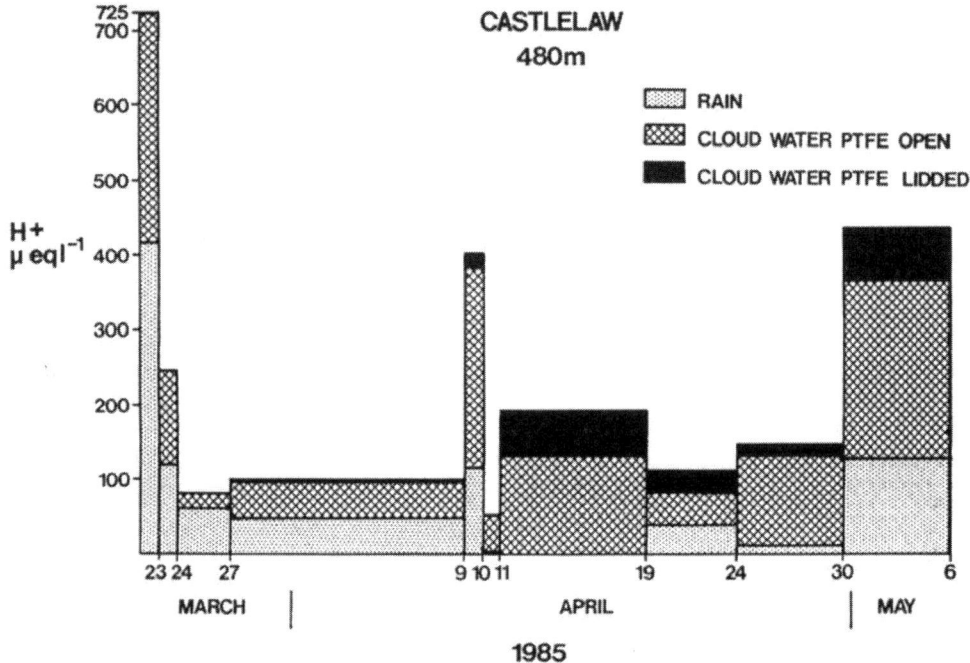

Fig. 2. Histogram of hydrogen ion concentration (μeq.L^{-1} H$^+$) collected by rain, cloudwater (open P.T.F.E.) and cloudwater (lidded P.T.F.E.) gauges at Castlelaw over six weeks. Each collection period spans between the sample dates indicated, thus producing time-averaged concentrations.

that are aerodynamically rough, the humidity dependence is very strong, whereas for smoother surfaces such as grass, the available energy dominates in driving the process. On mountains where cloud forms in layers, the radiative energy will be greatest near the top of the layer, so that the concentration mechanism would be pronounced where a cloud top intercepted a mountain side. The concentration mechanism may also be large close to cloud base where the entrainment of drier air from below the cloud lowers the humidity below saturation. It may also be important that thin cloud, in which available energy would be largest, is associated with the highest chemical concentrations in cloud drops (Daum et al. 1984).

BIOLOGICAL IMPLICATIONS

Direct effects of acid rain and mist on vegetation surfaces have seldom been observed unless the pH is below 3 (Jacobson 1980). It is becoming clear that cloudwater often has a lower pH than rain, but we do not yet have sufficient routine observations to be able to assess the likely frequency with which high elevation

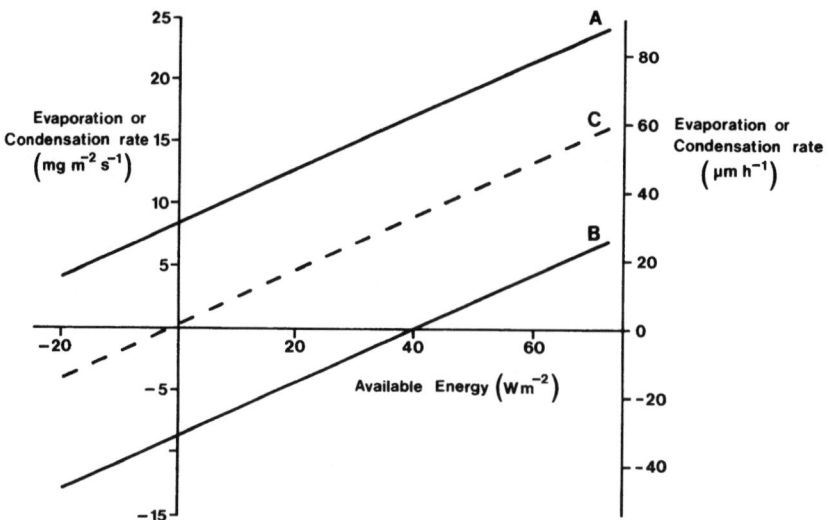

Fig. 3. Rates of evaporation and condensation onto forests (——, A,B) and grassland (---,C) for water vapour in wind-driven cloud. The calculations are based on the methods described by Unsworth (1984) and assume that the aerodynamic resistance to vapour transfer is 5 s.m^{-1} for forests and 100 s.m^{-1} for short grassland, corresponding to wind speeds of about 3-4 m.s^{-1}. The relative humidity is taken as 99% (line A), and 101% (line B). For grass, the evaporation rate is much less sensitive to humidity than for forests, and line C is appropriate for humidities between 99 and 101%. Temperature is assumed to be 10°C.

vegetation is exposed to low pH deposition. The significance of simultaneous evaporation in concentrating the solution on leaf surfaces has still to be demonstrated in the field.

Indirect effects of acidic deposition on plants would include disturbances of soil fertility (Matzner and Ulrich, this volume), leaching of essential nutrients from foliage (Prinz, this volume), and predisposition of vegetation to other stresses such as frost (Friedland et al. 1984; Nihlgard 1985). Because occult deposition occurs mainly at high elevation, where natural stresses are often also severe, it may be that ecosystems in these areas would be early indicators of regional pollution problems. Of the examples of indirect effects given above, least attention at present has been paid to the frost risk.

Johnson (1985, this volume), Friedland et al. (1984), and Nihlgard (1985) emphasize the hypothesis that dry and wet deposition of nitrogen may alter the sensitivity of foliage to frost. There do not appear to have been experiments yet to test this hypothesis. However, there is good evidence from laboratory and field experiments that exposure to atmospheric sulphur inputs increases frost sensitivity in grasses (Davison and Bailey 1982) wheat (Baker et al. 1982), and spruce (Keller 1981). The frost injury observed on wheat in the field occurred after only about 20 days of exposure to 100 ppb SO_2. Calculation suggests that the corresponding sulphur deposition over this time period to plant and soil was 10-20 kg $S.ha^{-1}$. Lovett (pers. comm.) estimated that annual sulphur deposition by occult deposition on Mount Moosilauke, New Hampshire, was about 46 $kg.ha^{-1}$, so that if total sulphur input rather than SO_2 concentration influences frost hardiness, there are strong reasons to suspect that some high elevation forests are at risk. Observations of the occurrence of frost damage support this view: Spaleny (1981) noted that frost injury to Norway spruce in Czechoslovakia and West Germany was worst close to areas where pollution concentrations were high, and at high elevation. Although Spaleny suggested that SO_2 appeared to be implicated, it is very likely that nitrogen deposition would also be high at these sites, and it is necessary to develop experimental work to clarify the mechanisms of action of sulphur and nitrogen deposition on frost hardiness.

CONCLUSION

Vegetation at high elevation is often subject to natural stresses (cold, wind, dessication) which can limit its distribution and productivity. Occult deposition is a pollution stress confined mainly to high elevations, and can result in large inputs of hydrogen, sulphate and nitrate ions. More observations are necessary of the chemistry of cloudwater so that frequencies of exposure to chemically concentrated material can be assessed. But this is only a small part of the research necessary to identify the risks to high elevation ecosystems. Methods of measuring and modelling occult deposition to various forms of vegetation must be developed; the importance of evaporation/ condensation must be investigated; and studies of direct and indirect responses of plants to occult deposition must be expanded.

Acknowledgements

This work is supported by the Natural Environment Research Council, the Commission of the European Communities, and the UK Department of the Environment. We thank Gary Lovett for several helpful comments in the preparation of this paper.

REFERENCES

Appel BR, Wall SM, Kothny EL, Tokiwa Y, Haik M (1980) Simultaneous nitric acid, particulate nitrate and acidity measurements in ambient air. Atmos Environ 14: 549-554

Ashmore M, Bell N, Rutter K (1985) The role of ozone in forest damage in West Germany. Ambio 14: 81-87

Baker CK, Unsworth MH, Greenwood P (1982) Leaf injury on wheat plants exposed in the field in winter to SO_2. Nature 299: 149-151

Baumgartner A (1952) Nebel und Nebelniederschlag als Standortfaktoren am Grossen Falkenstein (Bayerischer Wald). Forstw Cbl 77: 257-272

Byers HR (1953) Coast redwoods and fog drip. Ecology 34: 192-193

Cannon WA (1901) Relation of redwoods and fog. Torreya 1: 137-139

Carruthers DJ, Choularton TW (1984) Acid deposition in rain over hills. Atmos Environ 18: 1905-1908

Chamberlain AC (1975) The movement of particles in plant communities. In: Monteith JL (ed) Vegetation and the atmosphere, Vol 1. Academic Press, Lond, pp 155-203

Davidson AW, Bailey IF (1982) SO_2 pollution reduces the freezing resistance of ryegrass. Nature 297: 400-401

Daum PH, Schwartz SE, Newman L (1984) Acidic and related constituents in liquid water stratiform clouds. J Geophys Res 89: 1447-1458

Dollard GJ, Unsworth MH (1983) Field measurements of turbulent fluxes of wind-driven fog drops to a grass surface. Atmos Environ 17: 775-780

Dollard GJ, Unsworth MH, Harvey MJ (1983) Pollutant transfer in upland regions by occult precipitation. Nature 302: 241-242

Ekern PC (1964) Direct interception of cloudwater on Lanaihale, Hawaii. Proc Soil Sci Soc Am 28: 419-421

Falconer RE, Falconer PD (1980) Determination of cloud water acidity at a mountain observatory in the Adirondack Mountains of New York State. J Geophys Res 85: 7465-7470

Fraser AI (1962) Wind tunnel studies of the forces acting on the crowns of small trees. Report on Forest Research, HMSO, London, pp 178-183

Friedland AJ, Gregory RA, Karenlamp L, Johnson AH (1984) Winter damage to foliage as a factor in red spruce decline. Can J For Res 14: 963-965

Garland JA, Branson JR, Cox LC (1973) A study of the contribution of pollution to visibility in a radiation fog. Atmos Environ 7: 1079-1092

Gervat GP (1985) Clouds at ground level: samples from the southern Pennines. Central Electricity Research Laboratories Report No. TDRD/L/2700/N84

Grunow J (1955) Die Niederschlag in Bergwald. Forstwiss Zentbl 74: 21-36

Hori T (1953) Studies on fogs. Tanne Trading Co Ltd, Sapporo

Houghton HJ (1955) On the chemical composition of fog and cloudwater. J Meteorology 12: 355-357

Jacobson JS (1980) The influence of rainfall composition on the yield and quality of agricultural crops. In: Drablos D, Tollan A (eds) Ecological impact of acid precipitation, SNSF, As-NLH, Norway

Johnson AH (this volume) Deterioration of red spruce in the Northern Appalachian Mountains.

Johnson AH, Siccama TG (1983) Acid deposition and forest decline. Environ Sci and Tech 17: 294-305

Johnson AH, Siccama TG (1984) Decline of red spruce in the northern Appalachians: assessing the possible role of acid deposition. TAPPI: 68-72

Keller TH (1981) Folgen einer winterlichen SO_2 - Belastung fur die Fichte. Gartenbauwissenschaft 46: S170-178

Kerfoot O (1968) Mist precipitation on vegetation. Forestry Abstracts 29: 8-20

Landsberg JJ, Jarvis PG (1973) A numerical investigation of the momentum balance of a spruce forest. J Appl Ecol 10: 645-655

Lazrus AL, Haagenson PL, Kok GL, Huebert BJ, Kreitzberg CW, Liken GE, Mohnen VA, Wilson WE, Winchester JW (1983) Acidity in air and water in a case of warm frontal precipitation. Atmos Environ 17: 581-591

Lovett GM (1984) Rates and mechanisms of cloud water deposition to subalpine fir forest. Atmos Environ 18: 361-371

Lovett GM, Reiners WA, Olson RK (1982) Cloud droplet deposition in subalpine balsam fir forests: hydrological and chemical inputs. Science 218: 1303-1304

Marloth R (1907) Results of further experiments on Table Mountain. Trans S Afr Phil Soc Cape Town 16: 97-105

Matzner E, Ulrich B (this volume) Results of studies on forest decline in northwest Germany.

Monteith JL (1975) (ed) Vegetation and the atmosphere, Vol 2, Academic Press, London, p 439

Mrose H (1966) Measurements of pH, and chemical analyses of rain-, snow-, and fog-water. Tellus 18: 266-270

Nagel JF (1956) Fog precipitation on Table Mountain. Quart J R Met Soc 82: 452-460

Nihlgard B (1985) The ammonium hypothesis - an additional explanation to the forest dieback in Europe. Ambio 14: 2-8

Oberlander GT (1956) Summer fog precipitation of the San Francisco peninsula. Ecology 37: 851-852

Okita T (1968) Concentration of sulfate and other inorganic materials in fog and cloud water and in aerosol. J Met Soc Japan 46: 120-127

Penman HL (1963) Vegetation and hydrology. Commonwealth Agricultural Bureaux, London

Prinz B, Krause GHM, Jung K-D (this volume) Responses of German forests in recent years: cause for concern elsewhere?

Rubner K (1935) Nebelniederschlag in Wald und seine Messung. Tharandter Forstl Jahrb 86: 330-342

Rutter AJ (1975) The hydrological cycle in vegetation. In: Monteith JL (ed) Vegetation and the atmosphere, Vol 1. Academic Press, London, pp 111-154

Schutt P, Cowling EB (1985) Waldsterben - a general decline of forests in Central Europe: symptoms, development, an possible causes of a beginning breakdown of forest ecosystems. Plant Disease 69: 548-558

Shuttleworth WJ (1977) The exchange of wind-driven fog and mist between vegetation and the atmosphere. Boundary-Layer Meteorology 12: 463-489

Smidt S (1983) Uber das Auftreten von Ozon und Stickstoffoxiden in Waldgebieten Osterreichs. European J For Path 13: S 133-141

Spaleny J (1981) Forst hardiness of Norway spruce needles in the polluted area of the Ore Mountains. Archiwum Ochrony Srodowiska, Polish Academy of Sciences 2-4: 145-153

Thorne PG, Lovett GM, Reiners WA (1982) Experimental determination of droplet impaction on canopy components of Balsam fir. J Appl Meteor 21: 1413-1416

Unsworth MH (1984) Evaporation from forests in cloud enhances the effects of acid deposition. Nature 312: 262-264

Waldman JD, Munger JW, Daniel JJ, Flagan RC, Morgan JJ, Hoffman MR (1982) Chemical composition of acid fog. Science 218: 677-679

Walters PT, Moore MJ, Webb AH (1983) A separator for obtaining samples of cloud water in aircraft. Atmos Environ 17: 1083-1091

Yosida Z, Kuroiwa D (1953) Wind force on a conifer tree and the quantity of fog water thereby captured. In: Hori T (ed) Studies on fogs, Tanne Trading Co Ltd, Sapporo

ACID DEPOSITION, NURTRIENT IMBALANCE AND TREE DECLINE: A COMMENTARY

G.H. Tomlinson

Domtar Inc./Research Centre, Senneville, Quebec, Canada H9X 3L7

ABSTRACT

Syndromes involving yellowing and other colour changes in foliage, similar to those observed in German conifers, are now appearing in many locations in the North East U.S. and adjacent areas of Canada. In Germany, it has been shown in several instances by analysis of the affected needles that this colour change, which presages their early loss and the premature death of the tree, results from a nutrient deficiency, of which magnesium has been definitely implicated in some instances and in which calcium and potassium may play a role. However, this nutrient imbalance can be corrected by addition of the missing elements to the soil, thus revitalizing the tree and confirming the fact that an inadequate supply of nutrients is the predisposing factor in tree decline. Visible symptoms, coupled with foliage analysis of affected trees, indicate that similar deficiencies are occurring in North America. This effect may result from the continuous and cumulative loss of nutrients, as a consequence of acid deposition or of exposure of foliage to acidic fog at high altitudes.

INTRODUCTION

The macronutrients calcium, magnesium and potassium are leaching from the rooting zone of the trees in eastern North America on a continuing basis (Foster et al. 1983). Acidic deposition is accelerating this process in areas now receiving acidic atmospheric inputs. The cumulative loss of vital elements can result in a decreasing amount remaining in the soil in each successive growing season and this may be such that eventually one or more becomes inadequate to supply the trees' nutritional needs.[1] Although it has been assumed by some that compensation for such a critical loss may occur as a result of an increased rate of solution of "non-available" nutrients from rock, Abrahamsen (1983) on the basis of simulated soil acidification, has concluded that with acidic spodosols this is not the case. In southern Sweden, Tyler et al. (this volume) have compared analytical data on soil plots previously studied in 1949-53. They concluded that weathering processes have not been able to keep pace with the additional inputs of acid, thus accounting for the major changes which have occurred in the soil. Tamm and colleagues have reported similar soil acidification occurring in Central Sweden, which over a 50-year period has caused changes of over one pH unit extending to the B and C soil horizons (Tamm and Hallbäcken 1986).

---

[1] See also Zoettl (this volume) regarding the role of calcium.

Symptoms of deficiencies in trees are now being observed at many locations, particularly where, because of the initial nature of the soil, one or more nutrients are in short supply. Magnesium deficiency, characterized by the yellowing of older foliage followed by its premature loss, has been observed in conifers in North East Bavaria (Zech and Popp 1983) and in the Black Forest in Germany (Zottl and Mies 1983) and, more recently, in balsam fir on Whiteface Mt. in New York state (Tomlinson 1985a).

Zottl and Mies (1983) have found that trees growing on the granitic soils having low magnesium content in the Bärhalde Ecological Area in the Black Forest show strong magnesium deficiency symptoms, whereas these symptoms have not appeared in trees growing on immediately adjacent areas where the soil is dominated by shale with a higher magnesium content.

Zech and Popp (1983) have found that these magnesium deficient conifers can be revitalized by adding soluble magnesium salts to the soil or even by spraying the solution on the foliage. Huttl and Wisniewski (in press), Hüttl (1985) and Zöttl (1985 a,b), in an extensive series of diagnostic fertilization experiments, established that, unless the foliage has already become lethally damaged, addition of soluble salts of the deficient nutrient to the soil at the beginning of the growing season will increase the nutrient content of the needles to adequate levels. As a result, the needles become green again and improved growth is observed during the same growing season. In the adjacent untreated areas the trees continue to decline, still showing yellow needles and poor growth. Conifers of all ages and of several species in many areas of Germany responded to this treatment (Zöttl 1985; Hüttl 1985; Zech et al. 1983). This rapid revitalization at several locations resulting from increased nutrient availability, while trees in adjacent unamended soil remain in decline, suggests that other factors such as ozone, weather, insects and fungal pathogens are probably secondary to nutrient losses occurring both in the soil and in the foliage as a consequence of acid leaching.

Although the dying trees in Southern Germany often show magnesium deficiencies, potassium deficiency has been observed in the less acidic soils of the Schwabian Alps and foothills, with these trees also showing yellowed foliage, even though the magnesium level may be adequate. It was found that supplementation with potassium salts revitalized the trees in such areas (Zoettl this volume; Hüttl 1985).

These ameliorative experiments have diagnositic value in that they establish that the key factor in the decline is the inability of the root system to derive from the soil a supply of nutrients adequate for new growth and for compensation for foliar leaching. Thus, the acid-induced changes in the soil involving nutrient loss, which are to be expected from a theoretical standpoint and have been shown by lysimeter tests to occur (Molitor and Raynal 1983; Foster et al. 1983; Matzner et al. 1984), can result in nutrient deficiencies responsible for the decline. Ongoing acid deposition in the absence of adequate controls is likely to accentuate the problem and cause it to spread. This anthropogenic acidic stress is additional to the natural acid-generating processes in temperate, podsolic soils under conifers.

Tomlinson (1985 a,b) has discussed the reasons why inadequate calcium and magnesium have a deleterious effect on the trees. Because of the different structural, chemical and physiological roles of the different nutrients in the tree, the visible symptoms are characteristic, but quite different, depending on the specific nutrient in short supply.

The Different Roles of Individual Nutrients

It has been observed (Wallace 1961) that magnesium, a constituent of the chlorophyll molecule, is mobile in the plant system and when in insufficient supply will move from older foliage to supply the needs of the current year's foliage. Thus, the older needles near the trunk have the lowest magnesium content, turn yellow first and are prematurely lost.

Calcium is the key element reponsible for the production of calcium pectate, which forms the principle constituent of the cell walls of the cortex of the fine feeder roots and of the leaves (Wallace 1961). Calcium pectate also forms the primary cell wall of the xylem, on which the cellulose and lignin are subsequently deposited, and thus is involved at all points of new growth. In contrast to magnesium, calcium when once deposited in foliage, will not move out to supply the requirements of new growth, and when supplies in the soil are inadequate, the current year's needles become deficient and are thus the first to be lost.

Potassium does not enter into the actual composition of any major tree constituent, and most of it is contained in soluble form in the cell sap and cytoplasm where it is involved in the control of physiological processes and the turgor of the growing cells of roots, stem, twigs and foliage (Wallace 1961). An inadequate supply adversely affects the formation of chlorophyll, with the result that the needles become yellow. Because of the high mobility of potassium, the oldest needles are affected first by its deficiency as is the case with magnesium.

OCCURRENCE OF DEFICIENCY SYMPTOMS IN NORTH AMERICA

Since the appearance of discoloured foliage, when coupled with analytical data, has proved useful in tracing the cause of the tree decline syndrome in Germany, the author has collected foliage for analysis in some areas of North America where similar discolouration has been observed.

Trees Showing Low Magnesium Concentrations in the Foliage

Table 1 shows data obtained on foliage in which magnesium deficiency was suspected. Recently, and particularly in 1984, the foliage on balsam fir growing on the upper slopes of Whiteface Mt. in New York State developed a syndrome similar to that which had been attributed to magnesium deficiency by Zech and Popp (1983) and by Zöttl and Mies (1983). The current year's needles,

Table 1. Content of cationic macronutrients in needles and twigs of trees with discoloured foliage showing low magnesium concentration.

Date of sample	Needle growth year	Needle age	Colour of needles	Ca (mg.g^{-1}) twigs	Ca (mg.g^{-1}) needles	Mg (mg.g^{-1}) twigs	Mg (mg.g^{-1}) needles	K (mg.g^{-1}) twigs	K (mg.g^{-1}) needles
Balsam fir on upper slopes of Whiteface Mountain									
Oct. 12/84	1984	1st year	green	3.28	3.87	0.83	0.76	–	–
	1983	2nd year	yellow	3.58	5.26	0.48	0.33	–	–
	1982	3rd year	yellow	4.80	6.07	0.45	0.37	–	–
Engleman spruce on upper slopes of Mt. Germania, Alta, Utah									
Feb. 4/85	1984	1st year	green	3.19	2.73	1.03	0.78	6.23	5.85
	1983	2nd year	yellow	3.47	3.08	0.77	0.45	6.24	3.84
	1982	3rd year	yellow	4.81	3.17	0.69	0.43	5.52	3.50
	1981	4th year	yellow	3.87	5.97	0.56	0.35	3.50	3.36
For comparison, nutrient values for balsam fir needles, Maine (Young and Guinn 1966)									
1965	all years				7.7		0.86		

as observed in the autumn, were green, while the previous year's needles were bright yellow. As shown in Table 1, the magnesium content of the green current needles, pooled from 10 balsam firs of similar appearance, was found to be 0.76 mg.g^{-1} while that of a blend of the yellow needles which had been formed the previous year was 0.33 mg.g^{-1}. This suggests that the chlorotic appearance resulted from magnesium deficiency, probably due to an inadequate movement of magnesium from the soil and roots, and thus the alternative transfer from old to new needles. The twigs on which each year's needles had grown were also analyzed, and their magnesium content indicates a similar picture.

A similar syndrome was observed with Engleman spruce in February 1985 in Alta, Utah. As shown in Table 1, analyses of needles and twigs indicate a similar deficiency-induced transfer of magnesium from the yellow to the green foliage to provide for new growth. The development of this syndrome at Alta is of interest because of its location in relation to sources of potentially damaging pollutants. Unlike areas in the East, it is largely isolated from upwind areas where any substantial amounts of NO_x, or ozone derived from it, are likely to be generated. On the other hand, a large smelter with a 458 m stack is located at the south end of Great Salt Lake, approximately 50 km to the west. Considerable tree die-back has developed at Alta during recent years, a situation previously observed at locations down wind from other smelters. In the absence of additional data on SO_2 patterns of distribution and patterns of tree decline over the 50 km intervening area, this is merely a suggestion.

Trees Showing Low Calcium and Magnesium in the Foliage

Table 2 gives analytical data on the foliage of red spruce, a tree species which is now showing serious symptoms of decline in eastern North America (Johnston and Siccama 1983). The discolouration and loss of needles in this species starts with the youngest needles at the ends of the branches and later moves towards the trunk and from the top of the tree downwards, the opposite of the syndrome observed when a deficiency of magnesium only is involved. In early 1984, the most recently formed needles on many of the red spruce throughout its northern range had turned brown, while the previous year's needles remained green or yellowish green. This phenomenon, which had been previously observed, does not occur every year and it has been suggested that weather conditions involving either an early frost in the autumn or an early thaw in the late winter followed by very cold weather can be a contributing factor. Samples of brown 1983 and green 1982 needles from 10 red spruce growing on Camel's Hump Mt. in Vermont were sampled prior to bud opening in 1984. The brown needles showed a calcium content of only 1.05 mg.g^{-1} compared with 1.83 mg.g^{-1} in the previous year's needles. For comparison, data of Young and Guinn (1966) are shown in Table 1. Their values for calcium in needles with all years' growth combined for red spruce grown in Maine is 8.4 mg.g^{-1}. This large difference between present and the earlier values from Maine, together with the known inability of calcium to move to young foliage when in insufficient supply (Wallace 1961), indicates the possibility that calcium deficiency is now involved in the affected trees.

Table 2. Content of calcium and magnesium in needles and twigs of red spruce showing low concentrations in trees with discoloured foliage and higher values in trees with more normal foliage.

Date of sample	Needle growth year	Needle age	Colour of needles	Ca $(mg.g^{-1})$*		Mg $(mg.g^{-1})$	
				twigs	needles	twigs	needles
Red spruce on Camel's Hump Mt., Vt.							
June 25/84	1984	not formed	-	-	-	-	-
	1983	2nd year	brown	-	1.05	-	0.52
	1982	3rd year	green	-	1.83	-	0.43
Red spruce on Whiteface Mountain, N.Y. showing yellow foliage							
Oct. 12/84	1984	1st year	yellow	2.1	1.95	0.51	0.62
	1983	2nd year	green/yellow tip	3.0	2.70	0.43	0.47
	1982	3rd year	green/yellow tip	3.2	3.19	0.38	0.42
Red Spruce on Whiteface Mountain, N.Y. showing more normal foliage							
Oct. 12/84	1984	1st year	slightly yellow cast	2.3	2.91	1.14	0.84
	1983	2nd year	"	2.7	4.01	0.95	0.73
	1982	3rd year	"	3.3	5.15	0.72	0.63
	1981	4th year	"	3.5	5.10	0.57	0.43
For comparison, literature values for red spruce needles, Maine (Young and Guinn 1966)							
1965		all years			8.4		1.35

* Border range for calcium deficiencies, Norway spruce 1 to 2 $mg.g^{-1}$ dry solids in 1 year needles (Hüttl 1985).

Although concentrations at which deficiency symptoms appear may differ between different species of spruce it is of interest that Zoettl (this volume) considered that a Norway spruce growing in the Baden-Baden District containing 0.90 $mg.g^{-1}$ Ca and 0.41 $mg.g^{-1}$ of Mg in the current needles, values similar to those shown for red spruce in Table 2, was deficient in both of these nutrients. A diagnostic fertilization test involving the addition to the soil of calcium nitrate, together with a K-Mg-Ca compounded fertilizer, resulted in an increase of the needle calcium level to 1.9 $mg.g^{-1}$ and of the magnesium level to 0.61 $mg.g^{-1}$, bringing them above the deficiency concentrations, and revitalizing the tree.

During the autumn of 1984 it was observed that most of the mature red spruce on Whiteface Mt. were dead or dying but some of the smaller trees of about 1 m height on the lower slopes had survived. On some of these, 1984 needles showed a definite yellow colour and in many cases they were very short, 4 to 5 mm versus about 9 mm. The older needles on the same tree were green with yellow tips. Analysis of the 1984 needles on trees with discoloured foliage and trees with green foliage gave calcium values of 1.95 $mg.g^{-1}$ and 2.91 $mg.g^{-1}$, respectively. Magnesium values of the yellowed foliage were also low, the corresponding values being 0.62 $mg.g^{-1}$ and 0.84 $mg.g^{-1}$. All of these values are substantially less than those reported for red spruce needles harvested in Maine in 1965 (Young and Guinn 1966). Hüttl (1985), on the basis of diagnostic fertilization tests, placed the border range for calcium deficiency at 1-2 $mg.g^{-1}$ dry solids in 1 year needles of Norway spruce. The application of factorial diagnostic fertilization tests with red spruce could give important insights as to whether or not nutrient deficiencies are affecting the health status of this species.

Trees Showing Low Potassium Concentration in Foliage

During the late winter of 1985 yellowed foliage was observed on white spruce, balsam fir and pines at relatively low elevations in many locations in Quebec, Ontario, New York State, Vermont and New Hampshire. The yellowing of foliage was found to be site-specific. It may occur in a small group of trees surrounded by apparently normal trees, while in other locations relatively large areas are affected. The intensity of the yellowing was found to be variable. In areas in which the change in colour was most marked all sizes of trees were affected. As in Germany, there was substantial lichen growth on many trees, particularly those showing a major loss of older foliage and those already dead.

The syndrome was observed for several kilometers along the Blue Ridge to Newcombe road in New York State and also in a number of more isolated locations in the general area of St-Jovite, Quebec. It is not limited to locations near the highway. Analytical data for needles taken from trees in these areas are shown in Table 3. The calcium content of the 1984 foliage was relatively high, ranging from 3.84 to 6.79 $mg.g^{-1}$ while the potassium concentration was in the range of 0.90 to 1.65 $mg.g^{-1}$. Hüttl (1985) has found that the border range for potassium deficiency in Norway spruce is 3.5 - 4.5 $mg.g^{-1}$ in 1-year needles. Zoettl (this volume) showed that through the use of diagnostic fertilization with potassium the concentration of that element in current needles increased from 2.9 - 5.1 $mg.g^{-1}$ and that the deficiency symptoms disappeared during a single growing season. The pH of the soil in $CaCl_2$ was 5.2 and the base saturation was dominated by calcium ion. He related the potassium deficiency to the low concentration of available potassium and the high concentration of calcium in the soil which results in calcium-potassium "antagonism" in uptake.

The samples in the area of the Blue Ridge to Newcombe road were taken during the dormant period and the soil was covered with

Table 3. Content of cationic macronutrients in needles and twigs of trees with discoloured foliage showing low potassium concentration.

Date of sample	Needle growth year	Needle age	Colour of needles	Ca (mg.g^{-1})		Mg (mg.g^{-1})		K (mg.g^{-1})	
				twigs	needles	twigs	needles	twigs	needles
Balsam Fir - Road from Blue Ridge to Newcombe, N.Y.									
Mar. 24/85	1984	1st year	green	3.63	3.84	1.03	0.70	1.59	0.90
	1983	2nd year	yellow	4.93	6.72	1.13	0.79	1.34	0.74
	1982	3rd year	yellow	4.79	8.13	0.98	0.65	0.99	0.53
White spruce - road from Blue Ridge to Newcombe, N.Y.									
Mar. 11/85	1984	1st year	yellow	5.47	4.5	0.95	1.08	3.34	1.15
	1983	2nd year	yellow	5.86	5.4	0.66	0.93	3.43	1.29
White spruce - road from Lac Carré to Lac Supérieur, Quebec									
Mar. 22/85	1984	1st year	green	4.81	6.79	0.59	0.64	2.10	1.50
	1983	2nd year	yellow	5.17	11.28	0.50	0.73	1.49	1.48
	1982	3rd year	yellow	4.14	13.46	0.36	0.63	1.38	1.52
White spruce - road from Arundel to St-Jovite, Quebec									
Apr. 27/85	1984	1st year	yellow	5.12	5.62	0.67	0.75	2.47	1.65
	1983	2nd year	yellow	5.22	8.13	0.49	0.64	1.98	1.40
For comparison, white spruce showing potassium deficiency (Heiburg and Lowenstein 1958)									
		not stated							3.1

* Border ranges for potassium deficiency, Norway spruce, 3.5 to 4.5 mg.g^{-1} dry solids in first year needles (Hüttl 1985)

snow. However, a sample of soil taken on a subsequent visit gave a pH in $CaCl_2$ of 4.49 and in water of 5.23.

Historical data are available from studies carried out at the Pack Forest near Warrensburg, N.Y., where potassium deficiency and the beneficial effects obtained by the addition of potassium to the soil have been described. In 1928-1929 red and white pine and white spruce were planted on sandy glacial out-wash soils that had been previously farmed for over 100 years and abandoned (Heiberg and Lowenstein 1958). Some years later many of the white spruce and white pine died, while growth of those remaining, together with the red pine which appeared more resistant, became stagnant. It was found that this resulted from potassium deficiency which was attributed to its earlier removal from the soil in farm crops. Addition of potassium, either as the chloride, the sulphate or the nitrate resulted in increased foliar potassium content and in a marked and sustained increase in growth. The potassium content of a surviving 31 year-old white spruce was 3.1 $mg.g^{-1}$ while that of one which had been fertilized 5 years earlier was 4.5 $mg.g^{-1}$.

In a further paper dealing with the same site, Heiberg et al. (1964) compared the condition of red pine - unfertilized versus fertilized - using, in different experiments, 56-168 $kg.ha^{-1}$ of potassium (as K). Fertilization resulted in improved colour, growth and longevity of the needles. The trees increased in basal area and height and showed improved hardiness to drought and frost as well as resistance to physical damage and parasite attack. Twenty years after fertilization, the trees were 45% higher than controls.

According to White (1985) a low potassium content is characteristic of the out-wash soils which occur extensively in the Adirondacks and in adjacent areas of Quebec and Ontario. Most previous North American studies dealing with forest decline have focused on high elevation low pH soils. The present finding that the foliage of trees growing at lower elevations with less acidic soils have very low potassium content in their foliage brings a disturbing new factor into the situation.

CONCLUSION

A balanced uptake of calcium, magnesium and potassium from the soil is essential to the growth and well-being of a tree. As long as all of these nutrients are in ample supply the fine roots are capable of selectively taking up the tree's requirements in a ratio appropriate to its needs. The annual release in the soil of nutrients from the decay of litter and humus has traditionally maintained a balanced ratio which is stored by ion exchange on the clay and humus until required from recylcing in the tree.

The continued inputs to the soil of sulphuric acid and of nitric acid are resulting in changed equilibria in the soil. As a result, the quantity of one or more of the nutrient cations could be reduced to the point at which deficiency results in the tree.

Because of the separate roles of individual nutrients in the structure and metabolism of the tree, different symptoms become apparent, depending on the deficient nutrient and the tree species. Nevertheless, all of these symptoms have the same underlying cause - nutrient imbalance which develops in the soil as a result of the cumulative effects of acid deposition.

Studies in Germany have demonstrated that analysis of foliage and soil can be used to identify the cause of the specific decline syndrome in a tree and forest, and that this can be confirmed by diagnostic fertilization. The present study gives a signal that similar deficiencies may be now developing in North America.

REFERENCES

Abrahamsen G (1983) Effects of long range transported air on forests. In: Ecological effects of acid deposition, National Swedish Environmental Protection Board Report PM 1636, p 191-197

Foster NW, Nicholson JA, Morrison IK (1983) Acid deposition and plant cycling in Eastern North American forests. Paper presented at Effects of acid rain on forest ecosystems conference, Quebec City, June 1983

Heiberg SD, Lowenstein H (1958) Depletion and rehabilitation of a sandy outwash plain in Northern New York. In: Proc 1st N. Amer for soils conf., E. Lansing, MI, p 172-180

Heiberg SD, Madgwick HA, Leaf AL (1964) Some long-time effects of fertilization on red pine plantations. Forest Sci 10: 17-23

Hüttl R (1985) Jungste Waldschaden Ernahrungsstorung und Diagnostiche Dungung. Vortrag Anlasslich des VDI-Kolloquiums "Waldschäden", 18-21 June, 1985, Golsar, Harz, FRG

Hüttl R, Wisniewski J (in press) Fertilization as a tool to mitigate forest decline associated with nutrient deficiencies. Water Air Soil Pollut

Johnston AH, Siccama TG (1983) Acid deposition and forest decline. Environ Sci Technol 19: 294-305A

Matzner E, Ulrich B (1984) Raten der Deposition, der internen Production und des Umsatzes von Protonen, Z Pflanzernahr Bodenk 174: 290-308

Molitor AV, Raynal D (1983) Acid precipitation and ionic movement in Adirondack forest soil. Soil Sci Soc Am J: 137-141

Tamm CO, Hallbäcken L (1986) Changes in soil pH over a 50-year period under different foerst canopies in SW Sweden. Water Air Soil Pollut

Tomlinson GH (1985a) Calcium and magnesium deficiencies in conifers and their effect on foliage. Report 84-8031-01, Domtar Inc., Senneville, Quebec, 1-24

Tomlinson GH (1985b) Acid deposition and loss of nutrients from forest soils. TAPPI J 68: 54-58

Tyler G, Berggren D, Bergkvist B, Flakengren-Grerup U, Folkeson L, Ruhling A (this volume) Soil acidfication and metal solubility in forests of southern Sweden.

Wallace T (1961) The diagnosis of mineral deficiencies in plants by visual symptoms. Her Majesty's Stationary Office, London, UK, p 10-12

White E (1985) Private communication, Dept. of Soil Chemistry, New York State College of Environmental Science and Forestry, Syracuse, N.Y.

Young HE, Guinn VP (1966) Chemical elements in complete mature trees of seven species in Maine. TAPPI 49: 190-197

Zech W, Popp E (1983) Magnesiummangel einer Grunde fur das Fitchen-und Tannensterben in N.O. Bayern. Forstw Cbl 102: 50-55

Zoettl HW (this volume) Responses of forests in decline to experimental fertilization of forests in decline.

Zöttl HW (1985) Waldschäden und Nahrelementversorgung. Dusseldorfe Geobot Koll 2: 31-41

Zöttl HW, Mies E (1983) Nahrelementversorgung und Schadstoffbelastung von Fictenokosystemen im Sudschwarzwald unter Immissionseinfluss, Mitteilgn Dtsch Bodenkundl Gesellsch 38: 429-434

RESPONSE OF TREES TO DROUGHT

M.T. Tyree, L.B. Flanagan and N. Adamson

Department of Botany, University of Toronto, Toronto, Canada
M5S 1A1

ABSTRACT

The hydraulic architecture of trees is divided into segments of high and low hydraulic conductance. Segments of major stems are 50 to 1000 times more capable of supplying water to leaves downstream than are minor branches. The functional advantages of plant segmentation are: 1) to allow all leaves to compete on a more or less equal basis for water resources in the canopy of large trees and 2) to confine xylem cavitations and embolisms to the minor branches during times of water stress.

Growth of trees is always restricted by drought stress. The mediating factor most often influencing growth response to drought is turgor pressure, because turgor pressure is the force causing plastic enlargement of cell walls. Low turgor can prevent the enlargement of cells to their full potential. Low turgor can also prevent meristematic cells from reaching the minimum size for cell division. Reduced shoot and leaf growth in one dry season can reduce the vigour and growth potential of trees for several subsequent years. Net assimilation of carbohydrates is also reduced by drought. The mechanisms causing reduction of net assimilation by drought are: (1) stomatal closure reducing the net uptake of CO_2 into leaves, (2) reduction of electron flow in photosystems I and II and (3) disruption of enzyme activity thereby reducing the dark reactions of photosynthesis.

INTRODUCTION

Water deficits influence all phases of tree growth, and always in a restrictive way. Kramer (1980) estimated that water deficits probably limit tree growth more than the effects of all other causes combined. Zahner (1968) attributed up to 90% of the annual variation in xylem production of conifers in arid areas to water deficits, and up to 80% in humid areas. However, in temperate climates a broad mix of microenvironmental factors can determine the growth pattern of individual trees (Cook, this volume).

We have been asked to write a brief background paper regarding the response of forests to drought. Space does not permit a broad survey of the literature. Some subjects will have to be completely ignored, e.g., (a) the impact of various leaf, stem and root morphologies on the "drought tolerance" of trees, (b) the induction by drought of "plastic" morphological changes, such as changes in stomatal size and frequency, leaf size and drought deciduousness, and (c) the large literature on the mechanism of stomatal regulation of water loss and the role of plant hormones in drought tolerance. Drought stress alone or in combination with other environmental stresses can lead to forest decline. This paper is intended to provide a brief introduction to some of the relevant literature regarding drought stress for the benefit of scientists currently studying problems of acidic deposition and air pollutants on forests.

THE DYNAMICS OF WATER STRESS DEVELOPMENT IN THE CANOPY OF TREES

Water stress develops in trees because the leaves (the sites of evaporation) are at the end of a long pipeline for water conduction. To extract water from the soil, the leaf water potential must fall below that of the soil by an amount sufficient (a) to lift water against gravity and (b) to overcome frictional drag effects in the soil, xylem conduits and non-vascular tissues. This situation can be represented diagrammatically by the "unit pipe model" (Fig. 1) of a tree (Shinozaki et al. 1964).

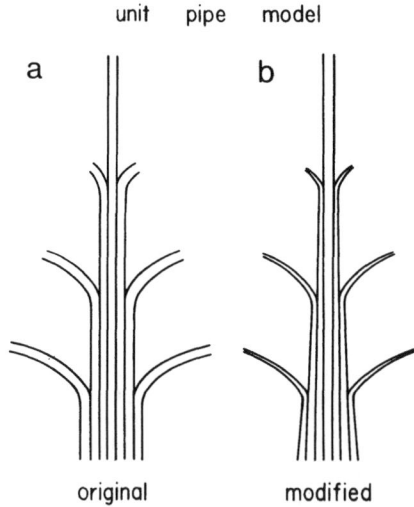

Fig. 1. On the left is a representation of the unit pipe model of Shinozaki et al. (1964) for a tree with apical dominance. On the right is a modified model in which the pipes taper towards the shoot apicies. The modified model better represents the anatomical facts and observations that the leaf specific conductance decreases acropetally.

A tree can be viewed as a bundle of unit pipes, each subtended by a unit amount of leaf area. Figure 1a shows a representation of a tree with apical dominance. In this model, it can be seen that the apical foliage is at the end of the longest unit pipe, whereas the basal foliage is at the end of the shortest unit pipe. In this model the apex is likely to be at a lower water potential than the rest of the canopy because (a) it is at the end of the longest conduit and (b) it is more exposed to sun, wind and dry air. If we assume that water stress limits growth more than any other factor, then it is difficult to understand how the unit pipe model of Fig. 1a could allow for apical dominance. The answer is that the original unit pipe model fails to conform to the experimental facts concerning the hydraulic architecture of trees as we currently understand them. Minor branches and nodes appear to offer a disproportionately large resistance to water flow within the canopy of trees (Zimmermann 1983). This situation is diagrammatically represented by Fig. 1b in which the unit pipes are shown to taper towards their ends. This is consistent with the observation that vessels or tracheids in minor branches tend to be smaller in diameter and shorter than recently formed conduits in older stems (Zimmermann 1983). Figure 1b would be even more realistic if we imagine that adjacent conduits are interconnected by pits through which water can flow.

The evidence in support of Fig. 1b is based on recent measurements of stem conductance or more specifically a quantity called leaf specific conductivity, LSC. The LSC is the hydraulic conductivity of a stem segment (e.g., kg water flow per sec per unit pressure gradient) expressed per unit of evaporative surface area (or leaf mass) supplied by the stem segment. The LSC is a direct measurement of the hydraulic capability of any given stem segment to supply water to the leaves downstream from the segment. To measure LSC a stem segment of length $\underline{l}$ is excised from a tree and water is pushed through the segment with pressure difference ΔP. The absolute conductance, K, is then calculated from $K = w / (\Delta P/l)$, where w is the measured water flow rate through the stem segment. If $\underline{A}$ is the area of all the leaves supplied downstream from the stem segment, then the LSC is given by

$$LSC = K/A = (w/A)/(\Delta P/l) = E/(\Delta P/l) \qquad (1)$$

where E is the average evaporation rate from the leaves supplied by the stem segment. From Eq. 1 it can be seen that if two different stem segments supply water to leaves at the same average evaporation rate, then there is an inverse relationship between the LSC values and the pressure gradient ($\Delta P/l$) needed to maintain water flow through the stem segment.

Reports of LSC are now available on a few diffuse porous trees (i.e., hardwoods with a uniform distribution of vessels throughout the growth ring, Zimmermann 1978a,b; Thompson et al. 1983) and on several conifers (Tyree et al. 1983; Ewers and Zimmermann 1984a,b). A consistent pattern has appeared in the range of values of LSC in the canopies of hardwoods and softwoods. The findings are that (1) there is a hydraulic constriction in nodes of trees, i.e., LSCs measured in stem segments including a node are usually 2 to 4 times less than in internodal stem segments of the same length, (2) the LSC values decrease from the trunk to major stems to minor stems, i.e., LSC values diminish by a factor of 50 to 1000 between the trunks of trees and the smallest branches. This is illustrated in Fig. 2 where LSC values for stems of Thuja occidentalis are plotted as a function of stem diameter. In Fig. 3 is a sketch of a small Abies balsamea tree showing measured values of LSC in some of the bigger stems. It is often found that trees showing apical dominance also show unusually large LSC values in stems leading to the dominant apex (see Fig. 3); this finding is represented diagramatically in Fig. 1 by unit pipes increasing in diameter towards the dominant apex. However, the anatomical basis for large LSC values near the dominant apex probably results from less leaf area per tracheid rather than the presence of larger tracheids.

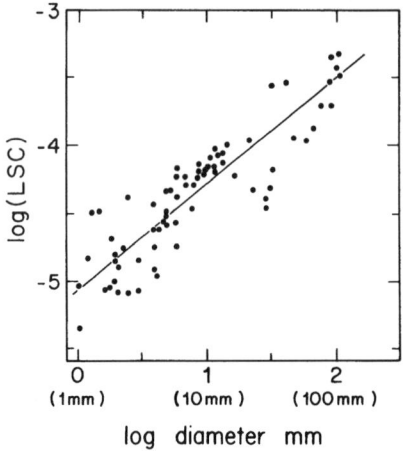

Fig. 2. Log base 10 of leaf specific conductance (LSC) versus log base 10 of stem diameter. On the ordinate the value of -5 corresponds to an LSC of 1×10^{-5} kg.s^{-1}.MPa^{-1}.m. Adapted from Tyree et al. (1983).

An important consequence of this hydraulic architecture is that the hydraulic resistance to water flow from the ground level to all minor branches is approximately the same for all twigs, whether the twig is located near the base of a canopy and at the end of a short hydraulic path or at the top of a canopy and at the end of a long hydraulic path. All shoots are approximately equally capable of competing for the water resources of the tree. One might view the minor shoots of a tree as a collection of approximately independent plants all rooted in the wood of larger stems rather than rooted in the soil.

Ever since the introduction of the cohesion theory of sap ascent in plants, it has been recognized that xylem conduits can be under tension (= $-\psi_{xylem}$ = minus the xylem water potential). Xylem has been aptly termed "the vulnerable pipeline," because as a water conducting system, it operates at the edge of physical possibility, i.e., it is in a non-equilibrium or metastable state. This instability of xylem water poses a direct threat to

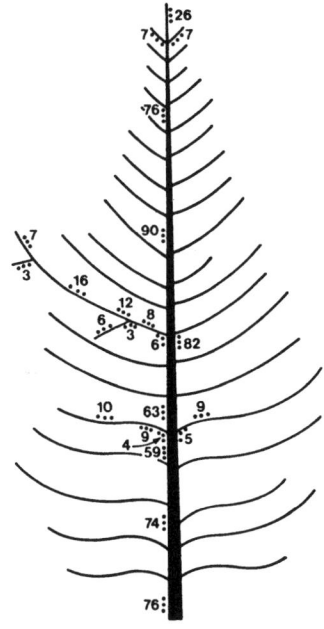

Fig. 3. Leaf specific conductivities of vigorously growing 20-year-old <u>Abies balsamea</u>. Dots mark locations where stem segments were excised and LSCs measured. The numbers are LSC values in multiples of 10^{-7} kg.s^{-1}.MPa^{-1}.m per kg dry weight of leaves.

the survival and photosynthetic productivity of plants. The degree of tension that can exist in the xylem before water makes a violent transition to the equilibrium vapour phase is limited. This explosive phase change is termed cavitation. Immediately after a cavitation, the xylem conduit begins to embolize, i.e., fill with air as gases come out of solution in surrounding tissue to fill the void left by the cavitation event. The danger of embolism for plants is that it impairs water conduction, the overall conducting capacity of the xylem is reduced and mid-day water stresses in leaves are increased. This has important consequences since leaf water stress is a major factor limiting terrestrial plant productivity on a worldwide basis (Fischer and Turner 1978; Turner and Kramer 1980; Boyer 1982; Hanson and Hitz 1982).

Observations of the hydraulic architecture of trees have led to the concept of "plant segmentation" and to hypotheses regarding its significance (Zimmerman 1983). Trees appear to be segmented into regions of high hydraulic conductance (the main stems) and regions of very low hydraulic conductance (minor branches). Since the main stems are hydraulically favoured over the minor branches and leaves, the minor branches and leaves will reach water potentials low enough to cause cavitations and embolisms well before the main stems. Consequently, at times of extreme drought stress, the minor branches and leaves (which are expendable) will die back leaving the conducting system of the main stem relatively intact. This strategy would not work if LSCs were the same for major and minor branches. If LSC values of all branches equaled that of the tree trunk, there would be much smaller differences in water potential between minor

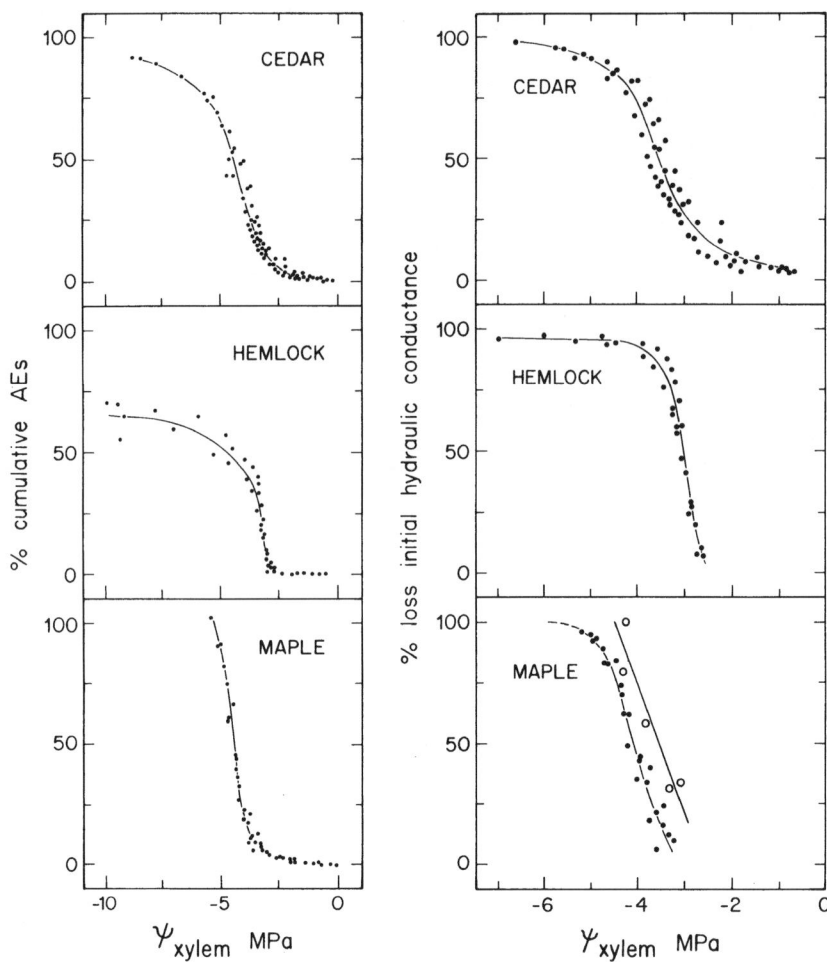

Fig. 4. Left: Percent cumulative acoustic emissions (AEs) <u>versus</u> xylem pressure potential during the dehydration of sapwood samples initially fully hydrated. Individual AEs are thought to correspond to individual cavitation events. In maple stems 100% AEs arbitrarily correspond to the number of AEs upon loss of 98% of the initial hydraulic conductance. About 5 times more AEs occur during subsequent air dehydration. These AEs are thought to correspond to cavitations in wood fiber lumina.

Right: Percent loss of initial hydraulic conductance <u>versus</u> xylem pressure potential during the dehydration of sapwood samples initially fully hydrated. The solid dots correspond to measurements made immediately upon attaining the plotted xylem pressure potential. The open circles correspond to samples held at the plotted water potential for about 12 h prior to measuring conductance. Water potentials were held constant by wrapping the stems in plastic sheets. During the hold time embolisms have more time to develop fully. From Tyree, MT and Dixon MA (1986) Water stress induced cavitation and embolism in some woody Plants. Physiol. Plant. 66: 397-405

branches and the trunk. At times of high soil water potentials
this would be a favourable circumstance. At low soil water
potentials this would tend to induce cavitation and embolism
uniformly throughout the stems and would lead to death of the
entire tree.

Recently studies have begun to determine the relative susceptibility of different species to xylem cavitation (Tyree and Dixon,
unpublished). Stem segments have been dehydrated over a period
of 24 to 48 hours while simultaneously measuring (a) water loss,
(b) stem water potential using an in situ stem hygrometer (Dixon
and Tyree 1984), (c) the number of cavitation events using
ultrasonic acoustic methods (Tyree et al. 1984) and (d) loss of
hydraulic conductance. Results comparing Thuja occidentalis,
Tsuga canadensis, and Acer saccharum are shown in Fig. 4. It can
be seen that the order of susceptibility to cavitation events and
loss of conductance versus water potential is Thuja > Tsuga >
Acer. It can be shown by perfusions of dye through partly
dehydrated stems that large earlywood tracheids of many conifers
embolize before the smaller latewood tracheids (Virginia Seymour
Berg, unpublished). This relationship does not appear to hold
among species. Large Acer vessels are more stable than the much
smaller tracheids of Tsuga and Thuja. It would be of interest to
know if species most affected by acidic precipitation or
pollution might be predisposed to dieback because of a greater
susceptibility to embolism. At the moment we can only speculate.

EFFECTS OF DROUGHT ON VEGETATIVE GROWTH OF TREES

Vegetative growth of roots and shoots can be divided into two
categories: (1) primary growth, which includes the elongation
and widening of the plant axis basipetal to the apical meristem,
and (2) secondary growth, the increase in thickness of the plant
axis as a result of new vascular tissue produced by the vascular
cambium.

Many of the processes related to primary and secondary growth are
affected by water deficits; of these, cell wall expansion is one
of the most sensitive (Hsiao et al. 1976; Bradford and Hsiao
1982; Tyree and Jarvis 1982). Cell expansion is, in part,
dependent on high turgor pressure. Turgor pressure is necessary
for the plastic deformation of cell walls. In the cambium and
apical meristems, high turgor pressure is needed to cause
sufficient enlargement for cells to reach the critical size for
division. Doley and Leyton (1968) found that cambial initial
cells of Fraxinus had to reach a diameter of 6 µm before they
could divide. It is presumed that a minimum turgor pressure is
required to permit such enlargement. Low turgor may also reduce
growth of developing tissue after cell division. A number of
physiological responses to water stress can be viewed as
mechanisms for maintaining sufficient cell turgor to permit the
maintenance of growth; among these responses are (a) increases in
ABA level induced by water stress, which cause stomatal closure
(e.g., Pierce and Raschke 1980; Henson 1983), (b) the direct

responses of stomates to humidity whereby low humidity levels cause stomatal closure in some species (Hall et al. 1976), and (c) instances of osmoregulation (Wyn Jones and Gorham 1983).

Primary Growth

During primary growth, cell division at the apical meristem and the subsequent elongation of newly formed cells result in the formation of leaf bud primordia. Water deficits may affect the formation of leaf buds and/or the subsequent elongation of preformed leaf buds.

In many tree species, leaf buds are formed the season before the primordia develop into leaves. Water deficits occurring during the formation of leaf primordia may affect the elongation of leaves in the subsequent season. Clements (1969) showed that the number of needle fasicles on the new shoot, shoot length and fasicle spacing were correlated to the size of the bud formed the previous season. The size of the bud formed was correlated to the level of moisture stress at the time it was formed (Kramer and Kozlowski 1960; Lotan and Zahner 1963). For species that undergo a single flush of growth only, there is evidence that the previous year's precipitation is closely correlated with extension growth, while the current year's precipitation has almost no influence on growth (Duff and Nolan 1958; Zahner and Donnelly 1967). In species that undergo multiple flushes of growth, the number of flushes and the amount of shoot growth are determined more by the availability of moisture during the current growing season (Zahner 1968) than by the previous year's precipitation.

Drought reduces both shoot growth and needle production in conifers. The reduction in leaf area of a tree may have significant effects on photosynthesis and growth for several years. Since needles persist for many years, reduction in shoot growth and needle production in any one year will affect the potential carbohydrate production for the entire life span of the needles (Lotan and Zahner 1963; Kozlowski 1964; Kozlowski and Keller 1966).

Secondary Growth

Three stages can be identified in the formation and development of secondary xylem and phloem tissue. Each stage is associated with a different growth zone. The first is the cambial zone in which the dividing mother cells and their most recent derivatives are located. As the cells are pushed out of this zone they enter the zone of enlargement in which they increase in diameter. The third zone is the maturation zone in which the cell walls increase in thickness (Wilson et al. 1966).

Five requirements for cambial activity and wood formation have been identified by Kramer (1964):
(1) a temperature resulting in a relatively high level of
    metabolic activity.

(2) a supply of growth regulators.
(3) a supply of carbohydrates and nitrogen containing substances.
(4) a supply of mineral nutrients.
(5) sufficient water to maintain high turgor in the cells.
Water availability and stress are in some way involved in all but the first of these requirements.

In conifers the transition from large, thin-walled earlywood tracheids to small, thick-walled latewood tracheids is mediated by a balance among water stress, nutrient levels and hormone levels. Large, thin-walled earlywood cells are formed in the spring when water stress is slight and competition for nutrients is highest because of rapid growth of cambium, roots, shoots and leaves. The high turgor pressure associated with high water potentials will cause the earlywood cells to have short lifespans; therefore they do not spend much time maturing or developing thick cell walls (Zahner 1963; Whitmore, Zahner 1966).

Kennedy (1961) related high production of earlywood in Douglas fir to years with high rainfall and low temperatures in the spring. In addition Zahner et al. (1964) showed that earlywood production in irrigated trees did not cease until late September. The development of the smaller, thicker latewood cells is related to moisture deficits. Zahner (1963; 1968) showed that there was a strong correlation between the transition from earlywood to latewood and the occurrence of the first severe water deficit. Under drought conditions cells spend more time in the zone of maturation and less time in the zone of enlargement than when exposed to high moisture levels (Whitmore and Zahner 1966).

The development of "false" growth rings (i.e., two or more alternating bands [earlywood - latewood] occurring within one growing season) appears to be caused by drought early in the growth season followed by adequate moisture. It is unclear whether water deficits have direct or indirect control over the transition from earlywood to latewood production (Larson 1963). The direct effect of water deficits would be the result of reduced expansion of differentiating xylem cells associated with low turgor pressure. The indirect effect would be the result of a decrease in auxin supply following the cessation of terminal shoot growth. Studies attempting to determine the role of auxins have not been totally conclusive (Zahner & Oliver 1962; Larson 1963; Zahner et al. 1964; Shepherd 1964; Doley and Leyton 1968; Evert and Kozlowski 1967; Whitmore and Zahner 1966 and 1967).

STOMATAL AND BIOCHEMICAL MECHANISMS OF PHOTOSYNTHETIC REDUCTION BY DROUGHT

Water availability may affect carbon fixation rates in several ways. The effects of drought on carbon fixation can be grouped into two broad categories: (1) stomatal limitation of photosynthesis and (2) nonstomatal limitation of photosynthesis. Commonly, stomatal closure in response to decreasing water availability has been thought to be the major contributor to reductions in photosynthesis. While it is clear that reductions in stomatal opening will restrict the flow of carbon dioxide into a leaf, photosynthesis may decline independently of stomatal closure in response to the same factors that cause stomatal closure (Farquhar and Sharkey 1982).

Stomatal limitation of photosynthesis is primarily mediated through changes in stomatal conductance. Stomatal conductance is affected by guard cell movement and stomata number. Nonstomatal limitation of photosynthesis can be of biochemical origin. Mesophyll assimilative capacity is affected by the concentrations and kinetic properties of photosynthetic enzymes and the activities of the light harvesting structures and electron transport components of the chloroplasts. Water stress may therefore act to reduce photosynthesis by: (1) causing stomatal closure and thereby reducing entry of carbon dioxide into the leaf, (2) reducing chloroplast activity and electron flow in the light reactions of photosynthesis or (3) disrupting enzyme activity and thereby reducing the dark reactions of photosynthesis.

Correlations among photosynthesis, transpiration and leaf conductance have often been interpreted to mean that the effect of water stress on photosynthesis is mainly controlled by stomatal closure (Brix 1962; Boyer 1976; Beadle and Jarvis 1977). Other studies using similar experimental approaches have shown increases in mesophyll resistance induced by water stress, i.e., a nonstomatal diffusional resistance to CO_2 uptake (Dykstra 1974; Bunce 1977; Osonubi and Davies 1980). However, the computation of mesophyll resistance is based on sometimes dubious computational estimates of CO_2 concentration in mesophyll air spaces and assumptions about the concentration at the reaction sites; therefore, "mesophyll resistances" may be erroneously estimated and may include biochemical rate limitations.

In the light reactions of photosynthesis, electron transport through photosystems I and II is responsible for the production of ATP, an energy source, and NADPH, a reducing agent, both of which are used in subsequent dark reactions. Therefore, reductions in electron transport may limit net assimilation. In a series of studies performed on Helianthus (Boyer 1971; Boyer and Bowen 1970; Boyer and Potter 1973; Keck and Boyer 1974) and on other species (Govindjee et al. 1981; Matorin et al. 1982), reductions in electron transport and quantum yield have been reported in vivo and in vitro in response to water stress. In Helianthus these reductions could not be accounted for by gross changes in chloroplast structure (Fellows and Boyer 1978).

Downton (1983) measured variable fluorescence (the relative fluorescence yield between I and P of the fluorescence induction transient; see Papageorgiou 1975) in a species capable of osmotic adjustment in response to water stress (grapevine) and one not capable of osmotic adjustment (oleander). Reduction of variable fluorescence is indicative of inhibition of electron transfer on the water side of photosystem II (Downton 1983). Rapid stressing of grapevine plants resulted in large reductions in variable fluorescence. Water stress imposed over several days did not disrupt variable fluorescence until much lower water potentials were attained. In the slow stress treatment, turgor was maintained in grapevine by osmotic adjustment. In oleander, a plant incapable of osmotic adjustment, the loss of variable fluorescence did not depend on whether the plants were slowly or rapidly stressed. This suggests that loss of turgor is an important mechanism affecting reductions in the light harvesting reactions of photosynthesis.

Sharkey and Badger (1982) studied the response of several component processes of photosynthesis in osmotically stressed mesophyll cells of Xanthium. They showed that photosynthetic electron transport was unaffected by water potentials as low as -4.0 MPa, but that photophosphorylation was very sensitive to water stress. However, the concentration of several intermediates of the photosynthetic carbon reduction cycle remained relatively constant, providing some evidence that ATP supply may not limit photosynthesis during water stress. Sharkey and Badger (1982) proposed that water stress may limit the utilization of ATP more than the supply of ATP.

In the dark reactions of photosynthesis, carbon dioxide is reduced to carbohydrate. This is accomplished by a series of reactions that are catalyzed by a number of enzymes. The activity of the various enzymes involved in photosynthesis is affected by several factors, particularly temperature, pH and the concentrations of ions in the cytoplasm. The leaf temperature increase caused by stomatal closure during water stress may not be large enough to cause changes in enzyme activity, but drought-induced changes in pH levels in the stroma of chloroplasts or drought-induced changes in ionic concentrations are more likely to be significant.

Kaiser (1982) has demonstrated that reductions in photosynthesis during water stress were correlated with changes in total protoplast volume of leaf tissue in a variety of species from different habitats. Species native to habitats with different moisture regimes (hygrophytes, mesophytes and xerophytes) had different patterns of reduction in photosynthesis associated with decreases in the osmotic potential of the culture media. The different pattern of response was a result of differences in the reduction of protoplast volume as water potential was reduced. Protoplast water and photosynthesis were reduced to a smaller degree in xerophytes than in mesophytes or hygrophytes. The smaller reduction in protoplast volume in xerophytes was a result of a lower osmotic potential in these species. The lower osmotic potential in xerophytes was mostly a result of higher concentrations of sugars (which are compatible with enzyme activity) in the protoplast. When changes in photosynthesis were related to the relative changes in protoplast volume, all species exhibited similar sensitivities to water stress. Photosynthesis declined when relative protoplast volume was reduced to 50%. The reduction in photosynthesis was probably a result of the increased concentration of ionic solutes, such as Na^+, K^+, and Cl^-, as protoplast water volume decreased. All species showed similar reductions in enzyme activity with increased ionic concentration. However, because of the relatively high elastic modulus of plant cells, only small (< 10-15%) volume changes normally occur with moderate water stress (Dainty 1979).

In a series of papers by Berkowitz and Gibbs (1982a,b, 1983a,b), the effect of water stress on the light and dark reactions of photosynthesis was reported in spinach. They showed that decreased production of ATP and reducing agent by the light reactions occurred in isolated chloroplasts in response to reduced water potential. Reductions in stromal pH had an even greater impact on photosynthesis in isolated chloroplasts. The decrease in stromal pH was responsible for reduced activity of two enzymes, fructose bisphosphatase (FBP) and RUBP carboxylase. The reduction in activity of FBP was the larger. Experimental additions of a strong alkalating agent NH_4Cl reversed the inhibition of FBP and RUBP carboxylase. The activity of another enzyme, phosphoribulokinase, was also reduced by water stress, but its activity was not increased by NH_4Cl additions (during water stress); it was therefore concluded that reduced stromal pH was not the mechanism limiting this enzyme. Phosphoribulokinase may have been affected by increased solute concentration in the cytoplasm.

Berkowitz et al. (1983) have shown that similar stromal acidification may be the mechanism by which photosynthesis is limited in vivo. These authors propose that the mechanism of stromal acidification may be related to potassium ion concentration in chloroplasts during dehydration. Increased K^+ in the chloroplasts could initiate an increased export of K^+ coupled with an increased import of hydrogen ions via the K^+/H^+ antiporter, resulting in lower pH. Berkowitz and Gibbs (1983a) also suggested that plants with higher concentrations of potassium in the cytoplasm would be less susceptible to increased influx of H^+ into the chloroplasts. In support of this suggestion, they presented data which indicates that KCl solutions of up to 50 mM reversed the effect of stromal acidification.

CONCLUSIONS

During drought, changes in plant turgor pressure may be the factor initiating reduction in growth and net assimilation of CO_2. Low turgor can reduce the plastic enlargement of cell walls and therefore prevent meristematic cells from reaching the minimum size for cell division. Changes in stomatal conductance and disruption of chloroplast biochemical activity are also linked with reductions in turgor pressure. The maintenance of high turgor pressure during water stress depends on efficient water conductance through the plant. Trees may be divided into segments of high and low water conductance. Major stems are 50 to 1000 times more capable of supplying water to leaves downstream than are minor branches. Reductions in stem conductance occur during water stress as a result of xylem cavitation and embolism. Different tree species have different susceptibility to cavitation. A greater susceptibility to xylem cavitation and embolism may predispose some species to the compounded effects of acid deposition, air pollution and drought.

REFERENCES

Beadle CL, Jarvis PG (1977) Effects of shoot water status on some photosynthetic partial processes in Sitka Spruce. Physiol Plant 41: 7-13
Berkowitz GA, Gibbs M (1982a) Effect of osmotic stress on photosynthesis studies with the isolated chloroplast. Generation and use of reducing power. Plant Physiol 70: 1143-1148
Berkowitz GA, Gibbs M (1982b) Effect of osmotic stress on photosynthesis studied with the isolated chloroplast. Site-specific inhibition of the photosynthetic carbon reduction cycle. Plant Physiol 70: 1535-1540
Berkowitz GA, Gibbs M (1983a) Reduced osmotic potential effects on photosynthesis. Identification of stromal acidification as a mediating factor. Plant Physiol 71: 905-911
Berkowitz GA, Gibbs M (1983b) Reduced osmotic potential inhibition of photosynthesis. Site-specific effects of osmotically induced stomatal acidification. Plant Physiol 72: 1100-1109
Berkowitz GA, Chen C, Gibbs M (1983) Stromal acidification mediates _in vivo_ water stress inhibition of nonstomatal-controlled photosynthesis. Plant Physiol 72: 1123-1126
Boyer JS (1971) Nonstomatal inhibition of photosynthesis in sunflower at low leaf water potentials and high light intensities. Plant Physiol 48: 532-536
Boyer JS (1976) Water deficits and photosynthesis. In: Kozlowski TT (ed) Water deficits and plant growth, vol 4. Academic Press, New York, p 154-191
Boyer JS (1982) Plant productivity and environment. Science 218: 443-448
Boyer JS, Bowen BL (1970) Inhibition of oxygen evolution in chloroplasts isolated from leaves with low water potentials. Plant Physiol 45: 612-615
Boyer JS, Potter JR (1973) Chloroplast response to low leaf water potentials. I. Role of turgor. Plant Physiol 51: 989-992
Bradford KJ, Hsiao TS (1982) Physiological responses to moderate water stress. In: Encyclopedia of Plant Physiology NS vol 12b, Springer-Verlag, Berlin Heidelberg New York, p 263
Brix H (1962) Effect of water stress on the rates of photosynthesis and respiration in tomato plants and loblolly pine seedlings. Physiol Plant 15: 10-20
Bunce JA (1977) Nonstomatal inhibition of photosynthesis at low water potentials in intact leaves of species from a variety of habitats. Plant Physiol 59: 348-350
Clements JR (1969) Shoot responses of young red pine to watering applied over two seasons. Can J Bot 48: 75-86
Cook ER (this volume) The use and limitations of dendrochronology in studying effects of air pollution on forests.
Dainty J (1979) The ionic and water relations of plants which adjust to a fluctuating saline environment. In: Jefferies RL Davy AJ (eds) Ecological processes in coastal environments. Blackwell Scientific Publications, Oxford, p 201-209

Dixon MA, Tyree MT (1984) A new stem hygrometer, corrected for temperature gradients and calibrated against the pressure bomb. Plant, Cell and Environ 7: 693-697

Doley D, Leyton L (1968) Effects of growth regulating substances and water potential on the development of secondary xylem in Fraxinus. New Phytologist 67: 579-594

Downton WJS (1983) Osmotic adjustment during water stress protects the photosynthetic apparatus against photoinhibition. Plant Science Letters 30: 137-143

Duff GH, Nolan NJ (1958) Growth and morphogenesis in the Canadian forest species. III. The time scale of morphogenesis at the stem apex of Pinus resinosa Ait. Can J Bot 36: 687-706

Dykstra GF (1974) Photosynthesis and carbon dioxide transfer resistance of Lodgepole Pine seedlings to irradiance, temperature and water potential. Can J For Res 4: 201-206

Evert RE, Kozlowski TT (1967) Effect of isolation of bark on cambial activity and development of xylem and phloem in trembling aspen. Am J Bot 54: 1045-1054

Ewers FW, Zimmerman MH (1984a) The hydraulic architecture of balsam fir (Abies balsamea). Physiol Plant 60: 453-458

Ewers FW, Zimmermann MH (1984b) The hydraulic architecture of eastern hemlock (Tsuga canadensis). Can J Bot 62: 940-946

Fellows RJ, Boyer JS (1978) Altered ultrastructure of cells of sunflower leaves having low water potential. Protoplasma 93: 381-395

Farquhar GD, Sharkey TD (1982) Stomatal conductance and photosynthesis. Ann Rev Plant Physiol 33: 317-345

Fischer RA, Turner NC (1978) Plant productivity in the arid and semiarid zones. Ann Rev Plant Physiol 29: 277-317

Govindjee, Downton WJS, Forks DC, Armond PA (1981) Chlorophyll A fluorescence transient as an indicator of water potential of leaves. Plant Science Letters 20: 191-194

Hanson AD, Hitz WD (1982) Metabolic responses of mesophytes to plant water deficits. Ann Rev Plant Physiol 33: 163-203

Hall AE, Schulze ED, Lange OL (1976) Current perspectives of steady-state stomatal responses to environment. In: Lange OL, Kappen L, Schulze ED (eds) Water and plant life. Springer-Verlag, Berlin Heidelberg New York, p 169

Henson I (1983) ABA and water relations of rice, Oryza sativa. Ann Bot 52: 247-256

Hsiao TC, Acevedo E, Fereres E, Henderson DW (1976) Water stress, growth, and osmotic adjustment. Phil Trans R Soc Lond B 273: 479-500

Kaiser, WM (1982) Correlation between changes in photosynthetic activity and changes in total protoplast volume in leaf tissue from hygro-, meso- and xerophytes under osmotic stress. Planta 154: 538-545

Keck RW, Boyer JS (1974) Chloroplast response to low leaf water potentials. III. Differing inhibition of electron transport and photophosphorylation. Plant Physiol 53: 474-479

Kennedy RW (1961) Variation and periodicity of summer wood in some second growth Douglas fir. TAPPI 44: 161-166

Kozlowski TT (1964) Shoot growth in woody plants. Bot Rev 30: 335-392

Kozlowski TT, Keller T (1966) Food relations of woody plants. Bot Rev 32: 293-382

Kramer PJ (1964) The role of water in wood formation. In: Zimmermann MH (ed) The formation of wood in forest trees. Academic Press, New York, p 519-532

Kramer PJ (1980) Drought, stress and the origin of adaptations. In: Turner NC, Kramer PJ (eds) Adaptations of plants to water and high temperature stress. Wiley & Sons, New York, p 7-20

Kramer PJ, Kozlowski TT (1960) Physiology of trees. McGraw-Hill, New York, p 642

Larson PR (1963) The indirect effect of drought on tracheid diameter in red pine. Forest Sci 9: 52-62

Lotan JE, Zahner R (1963) Shoot and needle responses of 20-year-old red pine to current soil moisture regimes. Forest Sci 9: 497-506

Matorin DN, Ortaoidze TV, Nikolaev GM, Venediktov PS, Rubin AB (1982) Effects of dehydration on electron transport activity in chloroplasts. Photosynthetica 16: 226-233

Osonubi O, Davies WJ (1980) The influence of water stress on the photosynthetic performance and stomatal behavior of tree seedlings subjected to variation in temperature and irradiance. Oecologia 45: 3-10

Papageorgiou G (1975) Chlorophyll fluorescence: an intrinsic probe of photosynthesis. In: Govindjee (ed) Bioenergetics of photosynthesis. Academic Press, New York, p 319-371

Pierce M, Raschke K (1980) Correlation between loss of turgor and accumulation of abscisic acid in detached leaves. Planta 148: 174-182

Sharkey TD, Badger MR (1982) Effects of water stress on photosynthetic electron transport, photophosphorylation, and metabolite levels of Xanthium strumarium mesophyll cells. Planta 156: 199-206

Shepherd KR (1964) Some observations on the effect of drought on the growth of Pinus radiata D Don. Aust Forestry 28: 7-22

Shinozaki K, Yoda K, Hozumi K, Kira T (1964) A quantitative analysis of plant form - the pipe model theory I. Basic analyses II. Further evidence of the theory and its application in forest ecology. Jpn J Ecol 14: 97-105, 133-139

Thompson RG, Tyree MT, LoGullo MA, Salleo S (1983) The water relations of young olive trees. Ann Bot 52: 399-406

Turner NC, Kramer PJ (1980) Adaptation of plants to water and high temperature stresses. John Wiley & Sons, New York, 482 pp

Tyree MT, Jarvis PG (1982) Water in tissues and cells. In: Encyclopedia of Plant Physiology NS vol 12b. Springer-Verlag, Berlin Heidelberg New York, p 35-77

Tyree MT, Graham MED, Cooper KE, Bazos LB (1983) The hydraulic architecture of Thuja occidentalis L. Can J Bot 61: 2105-2111

Tyree MT, Dixon MA, Tyree EL, Johnson R (1984) Ultrasonic acoustic emissions from the sapwood of cedar and hemlock: An examination of three hypotheses concerning cavitations. Plant Physiol 75: 988-992

Whitmore FW, Zahner R (1966) Development of the xylem ring in stems of young red pine trees. Forest Sci 12: 198-210

Whitmore FW, Zahner R (1967) Evidence for a direct effect of water stress on tracheid cell wall metabolism in pine. Forest Sci 13: 397-400

Wilson BF, Wodzicki TT, Zahner R (1966) Differentiation of cambial derivatives: Proposed terminology. Forest Sci 12: 438-440

Wyn Jones RG, Gorham J (1983) Osmoregulation. In: Encyclopedia of Plant Physiology NS vol 12a. Springer-Verlag, Berlin, Heidelberg, New York, p 35-58

Zahner R (1963) Internal moisture stress and wood formation in conifers. Forest Prod J 13: 240-247

Zahner R (1968) Water deficits and growth of trees. In: Kozlowski TT (ed) Water deficits and plant growth, vol 2. Academic press, New York, p 191-254

Zahner R, Donnelly JR (1967) Refining correlations of rainfall and radial growth in young red pine. Ecology 48: 525-530

Zahner R, Oliver WW (1962) The influence of thinning and pruning on the date of summer wood initiation in Red and Jack Pines. Forest Sci 8: 51-63

Zahner R, Lotan JE, Baughmann WD (1964) Earlywood - latewood features of red pine grown under simulated drought and irrigation. Forest Sci 10: 361-370

Zimmermann MH (1978a) Structural requirements for optimal water conduction in tree stems. In: Tomlinson PB, Zimmermann MH (eds) Tropical trees as living systems. Cambridge Univ Press, Cambridge, p 517-532

Zimmermann MH (1978b) Hydraulic architecture of some diffuse-porous trees. Can J Bot 56: 2286-2295

Zimmermann MH (1983) Xylem structure and the ascent of sap. Springer-Verlag, Berlin Heidelberg New York, p 143

RESPONSES OF AMERICAN FORESTS TO PHOTOCHEMICAL OXIDANTS

J.R. McBride and P.R. Miller*
Department of Forestry and Resource Management, University of
California, Berkeley CA

* Pacific Southwest Forest and Range Experiment Station,
USDA Forest Service, Riverside, CA

ABSTRACT

Photochemical oxidants produced over urban areas in North America contribute to the development of regional levels of air pollution which have been injurious to forests. Ozone data recorded since 1974 suggest that the Pacific Southwest and the Atlantic Northeast have the greatest potential for tree damage. Ozone persists longer in nonurban areas because of the absence of chemical scavengers. Factors influencing oxidant accumulation include terrain features which result in the formation of inversion layers, elevation, and seasonal patterns of air movement. Ozone concentrations typical of several mountain sites are presented. Information on the effects of ozone on North American forest species is limited. Caution is recommended in the use of sensitivity rankings from the limited available data because of variability in methods used to assess ozone injury. Tree and forest decline in response to ozone has been demonstrated in a limited number of studies in North America. Chronic injury to forests has been reported in the transverse mountain ranges of southern California, the Blue Ridge and southern Appalachian Mountains, and the Cumberland Plateau of Tennessee. Recent observations indicate that similar levels of chronic injury are occurring in the mountains southwest of Mexico City. Increased ability to characterize oxidant-caused foliar damage symptoms may also implicate ozone and associated oxidants as important elements of tree and forest decline observed in other regions of the United States and Europe.

INTRODUCTION

Studies of the origin and extent of ozone and associated photochemical oxidants in both urban and nonurban regions of North America since the 1960s have revealed the ubiquitous nature of these pollutants. During the same period, our knowledge of the effects of ozone on forest species has also increased. The purposes of this paper are to review the factors contributing to the accumulation of ozone in nonurban areas and to describe our current knowledge of the extent and nature of ozone injury to forests of North America. Effects of ozone on European forests is discussed elsewhere in this volume.

ORIGIN OF PHOTOCHEMICAL OXIDANTS

Ozone is the most important of the oxidants. It is unique because it can accumulate in the lower atmosphere from two sources, namely, through brief stratospheric intrusions (Viezee et al. 1982) and photochemical synthesis from the precursors, nitrogen dioxide and unsaturated hydrocarbons (Demerjian et al. 1974). The other oxidants: formic acid, peroxyacetyl nitrate (PAN), peroxypropionyl nitrate (PPN), and hydrogen peroxide, are formed by photochemical reactions in the troposphere and are not, with the exception of PAN, proven to be responsible for damage to forest vegetation at concentrations usually found in the ambient atmosphere. PAN is known to damage some sensitive herbaceous species (Temple and Taylor 1983), but not tree species (Kohut et al. 1976; Davis 1977). Formic acid, PAN, PPN, and hydrogen peroxide have the same diurnal concentration pattern as ozone with peak concentrations occurring at roughly the same time. However, Tuazon et al. (1981) have shown that ozone levels return to background much faster than formic acid, PAN, or hydrogen peroxide at night. It is important to acknowledge that these pollutants coexist with ozone because of the potential for joint action (Kohut et al. 1976; Temple and Taylor 1983).

Examples of Oxidant Pollution in North America

United States: SAROAD (Storage and Retrieval of Aerometric Data) files for ozone from 1974 to 1977 were used to delineate the size of geographical areas around individual SAROAD stations in which there is a 90% probability that the 0.12 ppm ozone standard will be exceeded (Ludwig and Shelar 1980). The resultant maps emphasize that the Pacific Southwest and Atlantic Northeast coastal regions have the greatest probability of exceeding the 0.12 ppm standard. More recent SAROAD data included the Gulf Coast with the northeast and west coasts as one of the regions with the highest 1-hour ozone concentration in the United States. The averages of the second highest 1-hour ozone concentration within each of the three areas averaged 0.15, 0.17, and 0.19 ppm, respectively (U.S.E.P.A. 1982). Evans et al. (1983) reported ozone data from eight relatively remote National Forest sites distributed throughout the United States. The concentrations during daylight hours were generally higher than 0.025 ppm during the April through October period but did not exceed an average of 0.054 ppm. In spite of the remote locations of these stations, several may have received ozone from urban areas.

The Los Angeles urban plume has been traced as far as 350 km east to the Colorado River Valley, near Needles, California, where daily maximum ozone concentrations ranged from 0.07 to 0.08 ppm during an episode of high pollution. The long-term record at this location showed a range of 0.019 to 0.044 ppm for the daily peak (Hoffer et al. 1982).

The concentration of ozone at 50 California and 15 Texas sites between 1973 and 1982 have been compared using the average daily maxima (Walker, 1985). These sites represent both urban and nonurban situations. There is no evidence for a decrease in ozone concentrations in California, and in Texas there has been a slight increase (2.3% per year) over the last 8 years.

Canada: In Canada ozone monitoring networks are in place for specific regions only. Episodes of phytotoxic ozone concentrations occur principally in southern Ontario and the greater Vancouver area (Rennie, 1985). As early as 1955 in the Point Dover region of southern Ontario, tobacco damage was associated with daily peak concentrations of 0.07 ppm ozone (Macdowall et al. 1963). Ontario has been partitioned into 4 regions from south to north, based on 7-hour daily ozone averages for the June through August period, of 0.05, 0.04, 0.03, and 0.02 ppm, respectively. The zone of highest concentration is located in the southern part of the Province. The annual averages for the southern region have been between 0.050 and 0.052 ppm during 1974 to 1978 and about 0.045 ppm during 1979 to 1981 (Linzon et al. 1984).

Some short-term measurements of ozone at both urban and remote sites in Alberta during spring and summer months showed concentrations in the range of 0.011 to 0.046 ppm. Concurrent measurements of PAN showed peaks as high as 0.023 ppm (Peake et al. 1983).

Mexico: The Valley of Mexico has all the ingredients for a photochemical oxidant problem. The surrounding mountains and frequent temperature inversions during the cooler months allow accumulation of precursor chemicals from 3 million cars and other sources. Data from two ozone monitoring stations located in the northern and southern sections of the city have shown consistently higher concentrations and later peak times at the southern station. Winds are generally from the northeast. During the 1983-84 season, the ozone concentration exceeded 0.12 ppm on 20 occasions (Bravo, pers. comm.).

FACTORS INFLUENCING OXIDANT ACCUMULATION AND TRANSPORT

Terrain and meteorological factors have important influences on ozone concentrations. One important terrain feature is the land-sea interface where low level inversions persist well into the day, thus the Pacific Coast, Northeast Coast and Great Lakes regions are more prone to the development of high ozone concentrations. In California, the inversions prevent vertical mixing of polluted air and onshore breezes transport pollutants inland across other metropolitan areas which add to the pollutant burden downwind (Neiburger et al. 1961). In such a situation, the maximum ozone concentrations are observed about 20 miles downwind from the major source areas. The Northeast states have episodes of high ozone during periods of air stagnation which may encompass a large region for several days (Lynn et al. 1964).

Evidence for Vertical Gradients of Ozone Concentration in Mountainous Terrain

Generally speaking, nocturnal ozone concentrations are higher at mountain than at basin sites (Berry 1964; Stasiuk and Coffey 1974). Under conditions common in the Northeastern United States, Samson (1978) explains that the reversed diurnal ozone fluctuation observed at mountain summits is related to vertical mixing of the layer of air formerly at the surface during the day; such mixing is forced by an increase in windspeed above the nocturnal surface inversion. In the coastal climate of

California, Edinger et al. (1972) observed that the afternoon, inland flow of cleaner marine air undercuts the ozone-polluted air at the same time the polluted air mass is transported eastward. Ozone remaining in the new marine layer is scavenged at night by fresh infusions of nitric oxide from basin sources, but ozone in layers aloft is partitioned from the continuing emissions of nitric oxide. Farber et al. (1982) described the nighttime inversion in the Los Angeles Basin as typically about 600 m thick; above this, the stable air layers containing ozone may extend up to 3000 m.

Even during daylight hours, the ozone concentration at Sky Forest in the San Bernardino Mountains east of Los Angeles is higher than the nearest basin station in the city of San Bernardino. The Sky Forest station is located at 1709 m elevation on a south-facing ridgecrest; it is 14.4 km north and 4.8 km east of the basin station (360 m elevation). The average hourly ozone concentrations for the daylight period (0600 to 2000 PST) for all available hours were compared by the month for the 1968 to 1977 period (Table 1). The mountain concentration was higher in all 12 months. Correlation coefficients describing the relationships of paired daily average concentrations from the two stations were generally very low.

Table 1. Comparison of daylight (0600 to 2000), monthly average ozone concentrations at a mountain (1709 m) and nearby basin station (360 m) for each month from 1968 to 1977. Sky Forest is located 15 km NNE and 1349 m higher than San Bernardino.

Month	Average Concentration (ppm)		R^2
	San Bernardino	Sky Forest	
January	0.003	0.005	0.59
February	0.011	0.020	0.17
March	0.014	0.026	0.30
April	0.019	0.044	0.46
May	0.031	0.071	0.37
June	0.055	0.091	0.40
July	0.064	0.105	0.19
August	0.062	0.100	0.23
September	0.048	0.064	0.25
October	0.026	0.038	0.26
November	0.011	0.017	0.23
December	0.005	0.008	0.37
Average	0.029	0.049	

Ozone Concentrations Typical of Several Mountain Sites Where Ozone Damage to Forest Species is Observed: Ozone was monitored from 1979 to 1982 at Big Meadows (1050 m elevation) in Shenandoah National Park and reported as monthly 8-hour ozone averages (1100-1800) EST) by Skelly et al. (1983). During this 4-year-period, the 8-hour averages for the growing season (May through September) ranged from 0.023 to 0.065 ppm with an average of 0.049. The peak 1-hour average for each month ranged from 0.06 to 0.13 ppm with an average of 0.09. Big Meadows (1050 m) was the highest of four stations where monthly mean ozone concentrations were compared for May through December, 1982. During each month, the mean ozone concentration was the largest at Big Meadows; thus the same relationship between mountain and valley sites pertains here as was observed in the San Bernardino Mountain area (Table 1). Foliar injury due to ozone was noted on native species at Big Meadows (Skelly et al. 1983).

In the southern portion of the Sierra Nevada, at locations influenced by oxidant-polluted air transported upslope from the San Jaoquin Valley, three monitoring stations were maintained from June through September from 1976 to 1981 (Vogler 1982). These stations were located in the mixed conifer forest type where ozone damage symptoms to foliage of ponderosa pine (Pinus ponderosa Laws.) and Jeffrey pine (P. Jeffreyi Grev. and Balf.) were considered to be mostly slight and occasionally moderate. The elevations of these stations were: Whitaker Forest, 1654 m, Mountain Home, 1805 m, and Greenhorn Summit, 1860 m. The daytime mean ozone concentrations are shown in Table 2.

Daily maximum hourly averages at the three mountain sites were often the same as the nearest urban sites in the San Jaoquin Valley, and as observed in the San Bernardino Mountains, the nocturnal concentrations ranged up to 0.08 ppm.

Table 2. Means of daytime hourly average ozone levels during the summer months at monitoring sites in the southern Sierra Nevada, 1976 to 1981 (from Vogler 1982).

Site Name	Mean daytime (0900-2000 PST) ozone value during June through September					
	1976	1977	1978	1979	1980	1981
Whitaker Forest	(0.070)[a]	0.084	0.078	0.079	0.075	0.082
Mountain Home	ND	[0.073]	0.068	0.071	0.057	0.073
Greenhorn Summit	ND	ND	0.085	0.084	(0.082)	(0.086)

[a] For the years shown, 85% or more of the hours were recorded, except as follows: ( ) = 60%; [ ] = 45%; ND = no data.

The daily maximum hourly average occurs between 1200 and 2000 hour PST at these stations and ranges from 0.10 to 0.16 ppm. The air quality data from the San Bernardino, Sierra Nevada, and Shenandoah National Park areas point out a problem that makes station-to-station comparisons difficult. The average times chosen for daylight periods were not the same. They were 0600-2000 in the San Bernardino, 0900-2000 in the Sierra Nevada and 1100-1800 in the Shenandoah National Park regions. In another analysis of the San Bernardino Mountain data base the 24-hour average ozone concentration ranged from a background of 0.03-0.04 ppm to a maximum of 0.10-0.12 ppm during the May through September period. Injury to needles of ponderosa pine was evident at a 24-hour average concentration of 0.05-0.06 ppm (Miller 1983). These different choices of average times make it difficult to compare the doses received at different sites.

More attention must be directed toward determination of the most meaningful ozone statistics for interpreting the ozone dose response of trees. Walker (1985) discussed 12 possible statistics for interpreting trends. The least robust was the annual maximum hour (1 value), and the most robust was the annual average (7000 values). The three statistics favoured for trend analysis were annual average, average daily maximum (May-October) and number of hours exceeding 0.12 ppm. Linzon et al. (1984) calculated the predicted seasonal (June through August) 7-hour mean using the annual 1 hour maximum value as the independent variable for regions in Ontario with three levels of crop loss. This approach has shown that if the seasonal mean is no greater than 0.03 ppm (with an annual maximum of 0.07 ppm) there will be little or no injury to crops.

STATUS OF INFORMATION ON EFFECTS OF OZONE TO FOREST SPECIES

Several areas of continuing interest are the knowledge of relative sensitivity derived from studies of container-grown seedling populations, the physiological responses of trees in forest stands to chronic ozone stress, and the evaluation of the current extent of ozone injury to forests in North and Central America.

Relative Sensitivity Determined by Seedling Fumigation

Listings of the relative sensitivity of tree species to ozone have been compiled (Jacobson and Hill 1970; Davis and Wilhour 1976). Caution is recommended in the use of these sensitivity of rankings because of the extreme variability of the methods used to assess ozone injury. The most common fault is that only one or two plants of a given species were exposed to any single treatment (Genys and Heggestad 1978). This prevents an estimate the population response, an important factor in forest species which have a large range of genetic variability.

Many of the published sensitivity rankings are based on foliar injury, which is not necessarily the best indicator of performance following chronic stress. For example, Jensen (1973) exposed nine deciduous tree species for 5 months to a low ozone dose and measured the differences in height growth. Davis and

Wilhour (1976) ranked the ozone sensitivity of the same species on the basis of foliar symptoms. Their ranking was similar to that of Jensen for only three of the nine species. Height growth is probably the most meaningful estimate.

In addition to the criteria for ranking sensitivity (foliar injury and growth) there are many other variables that should be acknowledged in a properly designed test of sensitivity. The size of the population treated, the age and source of plants, the environmental conditions prior to and during treatment (time of year), method of exposure, dose administration, moisture availability, and nutrition must all be considered in order to evaluate the representativeness of the results.

Tree Decline and the Physiological Response of Forest Species to Ozone

The deterioration in the health and vigor of a single forest species due to a number of abiotic and biotic factors is referred to as tree decline; when more than one species is involved in a synchronous deterioration, the term forest decline is applied (Manion 1981; Schutt and Cowling 1985). Further, Manion (1981) categorized factors involved in decline as predisposing (climate, air pollutants, site, tree age, and genetic potential), inciting (frost, drought, and air pollutants), and contributing (bark beetles, root, and stem diseases caused by fungi). Air pollutants can be important as both predisposing (long-term) and inciting (short-term) factors.

In the short-term, ozone can hasten the decline of photosynthesis and stomatal function normally related to aging of needles (Coyne and Bingham 1981). These workers found that the photosynthesis of needles from ponderosa pine, classed as having slight, moderate, and severe chronic injury, was reduced to about 10% of the maximum rate after experiencing 800, 700, and 450 pphm-hour ozone, respectively, at which point needle abscission occurred. In all needle ages of all three injury classes, losses in photosynthetic capacity exceeded reductions due to stomatal conductance, suggesting that injury to the mesophyll carboxylation, or excitation components of the carbon dioxide fixation pathway was greater than injury to the stomata. Measurements of the photosynthetic potential of the needles of plantation-grown eastern white pine (_Pinus strobus_ L.) showed no differences between trees with pollutant-caused chlorotic dwarf and adjacent healthy trees although total chlorophyll concentrations were different, especially in older needles. Growth decreases were attributed to lower foliar biomass (Mann _et al._ 1980).

Other physiological and cellular effects of ozone include deterioration of the cell membrane system, increased acid phosphatase activity, modification of amino acids, proteins, unsaturated fatty acids, and sulfhydryl residues, and localized accumulation of extracellular water (Heath 1975; Guderian _et al._ 1985). An increase in soluble sugars has been reported in leaves following ozone exposure (Tingey _et al._ 1976; Wilkinson and Barnes 1973). This increase may result from reduced translocation or the inhibition of starch synthesis.

Arndt et al. (1982) suggested that ozone was an important factor to consider in the evaluation of the possible causes of the Central European forest decline. Krause et al. (1983) found a relationship between intermittant treatments of 5-year-old spruce (Picea abies Karst) with ozone at 0.10 and 0.30 ppm, simulated acid rain at pH 3.5, and nutrient loss from needles, particularly magnesium. Possibilities for joint action between these and other pollutants must be actively investigated.

The severity of declines in vigor and growth can be estimated or judged by the reduction of foliar biomass as shown by studies of both eastern white pine (Mann et al. 1980) and ponderosa pine (Axelrod et al. 1980).

Evaluation of the Current Extent of Ozone Injury to North American Forests

United States: A review of the known examples of chronic injury by ozone to forest species in North America has been prepared by Smith (1984). Chronic injury has been reported in the San Gabriel, San Bernardino and the Sierra Nevada mountain ranges of California, the Blue Ridge and southern Appalachian Mountains, and the cumberland Plateau of Tennessee. Investigations in the San Bernardino Mountains in southern California during the 10 years from 1974 to 1983, showed that 33.2% of ponderosa pine in the young mature age class (50-99 years) died on plots located in the zone of high (0.08 to 0.12 ppm hourly average, May through September) ozone concentration. Only 6.9% of ponderosa pine trees in the same age class died on plots in the zone with lower (less than 0.08 ppm average) ozone concentration (McBride et al. in press). The less severe injury symptoms and the limited mortality observed on associated species (Abies concolor [Gord & Glend.] Lindl, ex Hildebr., P. lambertiana Dougl., Calocedrus decurrens (Torr.) Florin) and Quercus kelloggii Newb.) suggest a change in species composition as a result of prolonged exposure to ozone.

Research in the Blue Ridge and southern Appalachian Mountains has linked injury to eastern white pine with ozone (Skelly et al. 1983). Shifts in forest species composition may result here as well as in the Cumberland Plateau region (Mann et al. 1980). McClenahen (1978) reported that the importance of sugar maple (Acer saccharum Marsh) decreased along a gradient of increasing pollution in the upper Ohio River Valley. In each of these examples there is a proven relationship between ozone exposure, although other pollutants may have acted jointly with ozone to injure sugar maple. The crown deterioration of one or more tree species increases the possibility for forest community and ecosystem level effects.

Canada: In the southern Ontario region symptoms of ozone injury have been observed extensively on the foliage of white ash (Fraxinus americanus L.) and eastern white pine. The June through August, 7-hour daily ozone averages are 0.04-0.05 ppm (Linzon et al. 1984). The major areas occupied by conifer forest are further north where ozone is much lower and these forests may be protected under present circumstances.

The decline of sugar maple in Ontario has been related to many possible causes including insect defoliation, spring droughts, root rot, tree age and site management. Acidic precipitation is tentatively included in this list of causes (McLaughlin et al. 1985). A number of forest species show foliage symptoms of ozone damage in the southern region of Ontario. Since sugar maple is showing substantial dieback in southern Ontario and Quebec, it is possible that ozone is playing a role in this decline.

Mexico: Pine and fir forests occupying the mountains 20 to 30 km southwest of Mexico City may be severely damaged from present levels of ozone and associated pollutants. There are two National Parks in this region where investigations are underway to define the role of air pollution in the crown deterioration and eventual death of native tree species. In the Parque Nacional Cumbres De Ajusco, the cause of foliar injury to Pinus hartwegii Lindl. and P. montezumae Lamb., var. lindleyi was indicated to be ozone by using several lines of circumstantial evidence, including the decrease of symptom severity with distance southwest of the city and following enclosure of branches in chambers equipped with activated carbon filters (Hernandez 1984; de Bauer et al. 1985). A short distance away in the Parque Nacional Desierto de Los Leones there has been a large amount of mortality in stands of Mexican sacred fir, (Abies religiosa (H.B.K.) Schl. et Cham.) in the mid- to upper- elevations and also in Hartwegii pine at elevations up to 3000 m. The symptoms of needle injury, i.e. that older and lower crown needles are injured first (Miller per. observation), and the circumstantial evidence regarding urban ozone concentrations and direction of transport from the city strongly suggest that ozone is involved in this problem. A comprehensive research program is being coordinated by the Instituto Nacional De Investigaciones Forestales to determine the causes of the problem and furnish information for formulating management action (INIF 1984).

CONCLUSIONS

The current literature demonstrates that rural ozone concentrations, primarily at higher elevations, are responsible for chronic injury to forest species. The meteorological and topographic conditions which influence ozone distribution and persistence result in serious long-term threats to the productivity and composition of forests in specific regions. The current list of areas in North America of tree and forest decline related to ozone (San Bernardino and San Gabriel Mountains, the Sierra Nevada, the Blue Ridge and southern Appalachian Mountains, the Cumberland Plateau, and the Valley of Mexico) may be expanded as shifts in and growth of human populations occur. The possible interactive effects of both dry and wet acidic deposition with ozone and associated oxidants must be investigated thoroughly.

REFERENCES

Arndt U, Seufert G, Nobel W (1982) Die beteiligung von ozon an der komplexkrankheit der Tanne (Abies alba Mill.) -- eine prufenswerte hypothese. Staub-Reinhalt Luft 42: 243-247

Axelrod MC, Coyne PI, Bingham GE, Kercher JR, Miller PR, and Hung RC (1980) Canopy analysis of pollutant injured ponderosa pine in the San Bernardino National Forest. In: Miller PR (ed) Proc. Symposium effects of air pollutants on mediterranean and temperate forest ecosystems, USDA, Forest Service, PSW, Gen. Tech. Rpt. PSW 43, p 256

Berry CR (1964) Differences in concentrations of surface oxidant between valley and mountaintop conditions in the southern Appalachians. J Air Pollut Contr Assoc 14: 238-239

Bravo H (personal communication). Dep. Contaminacion Ambiental, Centro de Ciencias de la Atmosfera, Cd. Universitaria, Mexico, D.F., Mexico

Coyne PI, Bingham GE (1981) Comparative dose response of gas exchange in a ponderosa pine stand exposed to long-term fumigations. J Air Pollut Contr Assoc 31: 38-41

Davis DD, (1977) Response of ponderosa pine primary needles to separate and simultaneous ozone and PAN exposures. Plant Dis Reptr 61: 640-644

de Lourdes de Bauer M, Hernandez Tejeda T, Manning WJ (1985) Ozone causes needle injury and tree decline in Pinus hartweggii at high altitudes in the mountains arond Mecixo City. J Air Pollut Contr Assoc 35: 838

Davis DD, Wilhour RG (1976) Susceptibility of woody plants to sulfur dioxide and photochemical oxidants. USEPA Report EPA-600/3-76-102 p 72

Demerjian KL, Kerr JA, Calvert JG (1974) The mechanisms of photochemical smog formation. Adv Environ Sci Technol 4: 1-264

Edinger JG, McCutchan MH, Miller PR, Ryan BC, Schroeder MJ, Behar JV (1972) Penetration and duration of oxidant air pollution in the south coast air basin of California. J Air Pollut Contr Assoc 22: 882-886

Evans G, Finkelstein P, Martin B, Possiel N, Graves M (1983) Ozone measurements from a network of remote sites. J Air Pollut Contr Assoc 33: 291-296

Farber RJ, Huang AA, Bregman LD, Mahoney RL, Eatough DJ, Hansen LD, Blumenthal DL, Keifer WS, Allard DW (1982) The third dimension in the Los Angeles basin. Sci Total Environ 23: 345-360

Genys JB, Heggestad HE (1978) Susceptibility of different species, clones and strains of pines to acute injury caused by ozone and sulfur dioxide. Plant Dis Reptr 62: 687-691

Guderian R, Tingey DT, Rabe R (1985) Effects of photochemicals on plants. In: Guderian R (ed) Air pollution by photochemical oxidants. Springer-Verlag, New York, p 129-169

Heath RL (1975) Ozone. In: Mudd JB, Kozlowski TT (eds) Responses of plants to air pollution. Academic Press, New York, p 23-55

Hernandez Tejeda T (1984) Effecto de los gases oxidantes sobre algunas especies del genero Pinus nativas del Valle de Mexico. Tesis de Maestria en ciencias, Colegio de Postgraduados, Institucion de Ensenanza e Investigacion en Ciencis Agricolas, Chapingo, Mexico, p 109

Hoffer TE, Farber RJ, Ellis EC (1982) Background continental ozone levels in the rural U.S. Southwest desert. Sci Total Environment 23: 17-30

INIF (Instito Nacional Investigaciones Forestales) (1984) Plan integral de investigacion Parque Recreativo Y Cultural Desierto De Los Leones, Subsecretaria Forestal, Mexico D.F., p 54

Jacobson JS, Hill AC (eds) (1970) Recognition of air pollution injury to vegetation: a pictoral atlas. Air Pollut Contr Assoc, Pittsburgh, PA, p 41

Jensen, KF (1973) Response of nine forest tree species to chronic ozone fumigation. Plant Dis Reptr 57: 914-917

Kohut RJ, Davis DD, Merrill W (1976) Response of hybrid poplar to simultaneous exposure to ozone and PAN. Plant Dis Reptr 60: 777-780

Krause GHM, Prinz B, Jung KD (1983) Forest effects in West Germany. In: Davis DD, Millen AA, Dochinger LS (ed) Symposium on Air Pollution and the Productivity of the Forest. Izaak Walton League of America, Washington, DC, p 344

Linzon SN, Pearson RG, Donnan JA, Durham FN (1984) Ozone effects on crops in Ontario and related monetary values. ARB-13-84-Phyto, Ontario Ministry of the Environment, p 60

Ludwig FL, Shelar E. Jr (1980) Empirical relationships between observed ozone concentrations and geographical areas with concentrations likely to be above 120 ppb. J Air Pollut Control Assoc 30: 894-897

Lynn DA, Steigerwald BJ, Ludwig JH (1964) The November-December 1962 air pollution episode in the eastern United States. U.S. Public Health Service, National Air Pollution Control Administration, Cincinnati, OH, Pub No 999-AP-77

Macdowall FDH, Vickery LS, Runeckles VC, Patrick ZA (1963) Ozone damage to tobacco in Canada. Can Plant Dis Survey 43: 131-151

Manion PD (1981) Tree disease concepts. Prentice-Hall, Inc Engelwood Cliffs, NJ, p 399

Mann LK, McLaughlin SB, Shriner DS (1980) Seasonal physiological responses of white pine under chronic air pollution stress. Environ Exper Botany 20: 99-105

McBride JR, Miller PR, Laven R (in press) Effects of oxidant air pollutants on forest succession in the mixed conifer forest type of southern California. In: Proc. Efffects of air pollutants on forest ecosystems. May 8-9, 1985, Univ Minn, St Paul, MN

McClenahen JR (1978) Community changes in a deciduous forest exposed to air pollution. Can J For Res 8: 432-438

McLaughlin DL, Linzon SN, Dimma DE, McIlveen WD (1985) Sugar maple decline in Ontario. ARB-144-85-Phyto, Ontario Ministry of the Environment, p 18

Miller PR (1983) Ozone effects in the San Bernardino National Forest. In: David DD, Millan AA, Dochinger L (ed) Symposium on air pollution and the productivity of the forest. Izaak Walton League of America, Washington, DC, p 344

Neiburger M, Johnson DS, Chen CW (1961) Studies of the structure of the atmosphere over the Eastern Pacific Ocean in the summer. I. The inversion over the eastern Pacific Ocean. Univ Calif Pubs Meteor, 1: 1, Univ Calif Press, Berkeley, CA, p 94

Peake E, Maclean MA, Sandhu HS (1983) Surface ozone and peroxyacetyl nitrate observations at rural locations in Alberta, Canada. J Air Pollut Control Assoc 33: 881-883

Rennie PJ (1985) Statements on long-range transport of air pollutants and air pollution studies. In: 1985 Edition of the Canadian Research Catalogue, Canadian Forestry Service

Samson PJ (1978) Nocturnal ozone maxima. Atmos Environ 12: 951-955

Schutt P, Cowling EB (1985) Waldsterben - a general decline of forests in Central Europe: symptoms, development, and possible causes. Plant Dis 69: 548-558

Skelly JM, Yang Y-S, Chevone BI, Long SJ, Nellessen JE, Winner WE (1983) Ozone concentrations and their influence on forest species in the Blue Ridge Mountains of Virginia. In: Davis DD, Millen AA, Dochinger L (ed) Symposium on air pollution and the productivity of the forest. Izaac Walton League of America, Washington, DC p 344

Smith WH (1984) Ecosystem pathology: a new perspective for phytopathology. Forest Ecol Man 9: 193-219

Stasiuk WN, Coffey PE (1974) Rural and urban ozone relationships in New York State. J Air Pollut Control Assoc 24: 564-568

Temple PJ and Taylor OC (1983) World-wide ambient measurements of peroxyacetyl nitrate (PAN) and implications for plant injury. Atmos Environ 17: 1583-1587

Tingey DT, Wilhour RG, Standley E (1976) The effect of chronic ozone exposures on the metabolite content of ponderosa pine seedlings. For Sci 22: 234-241

Tuazon EC, Winer AM, Graham RA, Pitts JN, Jr (1981) Trace pollutant concentrations in a multiday smog episode in the California south coast air basin by long path length Fourier-transform infrared spectroscopy. Environ Sci Tech 15: 1232-1237

U.S. Environmental Protection Agency, SAROAD Data Files (1982) Office of Air Quality Planning and Standards, Research Triangle Park, NC

Viezee W, Singh HB, Shigeishi H (1982) The impact of stratospheric ozone on tropospheric air quality - implications from an analysis of existing field data. Final Report. Prepared for Coordinating Research Council, Atlanta GA, for Contract No. CAPA-15-76(1-80) by SRI International, Menlo Park, CA

Vogler D (1982) Ozone monitoring in the southern Sierra Nevada, 1976-1981. USDA, Forest Service Pacific Southwest Region, Forest Pest Management Report No. 82-17, p 43

Walker HM (1985) Ten-year ozone trends in California and Texas. J Air Pollut Control Assoc 35: 903-912

Wilkinson TC, Barnes RL (1973) Effects of ozone on $^{14}CO_2$ fixation patterns in pines. Can J Bot 9: 1573-1558

OZONE TOXICITY - IS THERE MORE THAN ONE MECHANISM OF ACTION?

Eva J. Pell
Department of Plant Pathology and Center for Air
Environment Studies, The Pennsylvania State University

ABSTRACT

The hypothesis that ozone injury to green leaves is determined by a single mode of action is examined. Chemical, genetic, histological and physiological lines of evidence are considered.

When ozone enters into solution it decomposes into a variety of chemical components which include hydroxyl and organic free radicals, superoxide anions, and hydrogen peroxide. Ozone and its decomposition products may attack one or several biochemical sites. Which biological molecule(s) is affected may be random, related to oxidizing potential of the attacking species and/or the biochemical vulnerability of the receptive site.

Evidence for single gene control of the ozone response, support for the hypothesis, is not strong. A number of researchers have suggested that heritability of the ozone response is a quantitative trait probably under multigenic control. Studies were conducted to determine the inheritance of the ozone response in potato. Since dominance does not become more important with increasing numbers of generations, it is unlikely that a single gene could control the inheritance of the ozone response. Selfed progeny of sensitive parents were often injured more severely than the respective parents, which provides further evidence of multigenic inheritance.

There is substantial evidence that ozone injures the plasma membrane but there is no proof that it is the only site of action. There are several ultrastructural studies in which "internal" cellular injury is observed in the absence of apparent effects on the plasma membrane or the related symptom of plasmolysis. In a study with soybean leaves some cells exhibited plasmolysis which probably reflected direct injury to the plasma membrane. Other cells exhibited no apparent effects on the limiting membrane. Internal membranes, including the endoplasmic reticulum, the limiting membrane of the chloroplast and both membranes of the mitochondria, all exhibited intense lead staining, indicative of lesions.

Additional evidence which supports the prospect that ozone toxicity results from injury to more than one site is supported by work with an _in vitro_ system. Isolated potato leaf mesophyll protoplasts were exposed to ozone. Protoplasts responded by rupturing or staining negatively with fluorescein diacetate (FDA). The two symptoms occurred simultaneously and in parallel which suggests that the responses were independent rather than causal. Protoplast lysis probably reflected a direct effect on the plasma membrane. The protoplasts which did not stain with FDA may well have sustained internal cellular injury in response to ozone rather than direct attack to the limiting cell membrane.

It is relevant to explore further the possibility that ozone injures foliage by more than one mechanism. It may expand our understanding of the basic way in which ozone injures plants. In addition, it may provide a different perspective as to how this information could be applied to develop ozone-tolerant plant material.

INTRODUCTION

The mechanism of action of ozone has been pondered since the gas was first described as a phytotoxic air pollutant in 1958 (Richards et al. 1958). Emphasis has been directed at identifying a single mechanism and/or the primary site of action of ozone. While several researchers have provided evidence that one or just a small number of genes control the ozone response (Engle and Gabelman 1966; Butler et al. 1979) results from other genetics studies do not support this conclusion (De Vos 1982; Dragoescu 1985; Dragoescu et al., in prep.; Sand 1960). Many researchers have conducted studies intended to characterize or define the primary site of action. Whereas these studies have described early events in the development of the ozone response, they have not eliminated the possibility of multiple sites of action of the air pollutant. In this essay, consideration will be given to the hypothesis that ozone injury to green leaves is determined by a single mode of action.

Relevance of Ozone Chemistry

An examination of the chemistry of ozone is critical when assessing the probability of the "single mechanism of action" hypothesis. In water, ozone decomposes into hydroxyl and organic free radicals, and superoxide anions (Tingey and Taylor 1982). Hydrogen peroxide has been suggested as an intermediate breakdown product of ozone in the presence of fatty acids (Tingey and Taylor 1982; Grimes et al. 1983). When ozone was introduced to an aqueous solution at pH 6.0, only hydroxyl free radicals and ozone could be detected (Grimes et al. 1983). The studies reflected by the citations above were conducted in water, buffer solutions or other simple mixtures. The potential exists for generation of any or all of these reactive species in the complex cytoplasm.

Ozone and its breakdown products all have high oxidizing potentials. There are a number of possible fates within the cell for these reactive species. All the ozone products could attack the same biochemical site within the cell. Alternatively, each chemical species could affect different but specific biochemicals within the cell. Given the high oxidizing potential of the ozone breakdown products, the oxidation process could be random. The cellular sites which ozone attacks could be determined by biochemical vulnerability, which is dependent on biological, chemical and physical constraints. Biochemical sensitivity could be the limiting factor determining the path of ozone attack. If the latter scenario were correct, cells within the leaf could react in a number of different ways.

In summary, ozone toxicity may result from one or several biochemical lesions within a single cell or multiple lesions occurring on different cells. The biochemical effects of ozone and/or its by-products may be site specific or random events. Due to the general oxidizing nature of ozone and its by-products it seems unlikely that one single site and/or a single biochemical account for the entire ozone response.

GENETIC EVIDENCE

Ozone response is, at least in part, under genetic control (De Vos et al. 1982). Examination of genetics studies may provide evidence germane to the hypothesis under consideration. However, it must be recognized that the objective of many genetics studies has been to provide useful information for plant breeders. Therefore, efforts have been directed toward identifying mechanisms of ozone tolerance rather than of plant sensitivity. Ozone response is, in most cases, measured by foliar injury which is a composite of primary and secondary biochemical responses to ozone. As such, data for single gene control of the ozone response would support the hypothesis. Evidence of multi-gene control would only partially refute the hypothesis since primary or secondary gene action could be involved.

Two studies support the hypothesis as presented earlier. Engle and Gabelman (1966) proposed that ozone tolerance in onion is determined by a single dominant gene which controls guard cell response to ozone. The proof of the one gene hypothesis is stymied by an intermediate field response which could not be accounted for by the authors, and by F_2 progeny of inadequate size. Stomatal closure will prevent ozone injury. However, it is not clear from the work of Engle and Gabelman (1966) that one gene controls leaf response when the stomates are open. Butler et al. (1979) also suggested that one or a few gene pairs are responsible for ozone-induced injury to the foliage of green bean. The small number of progeny utilized in the experiment and the inability to control environmental factors which could mimic genetic variation, prevent closure in the genetic interpretation presented by Butler et al. (1979).

Sand (1960) suggested that several genes were responsible for determining ozone injury on tobacco leaves. Sand's interpretation was based on the response of the F_1 population which exhibited average ozone tolerance greater than that of the mid parent, and the F_2 population which displayed transgressive segregation.

Since the early work of Sand (1960), many researchers have taken a quantitative approach to the study of inheritance of ozone resistance. The importance of additive genetic effects has been demonstrated for alfalfa, petunia, potato and tobacco (Aycock 1983; Campbell et al. 1977; Hanson 1973; Hanson et al. 1976; De Vos et al. 1982; Dragoescu 1985; Dragoescu, in prep.). Results of studies with alfalfa, petunia and potato also reflect importance of genes with dominance characteristics (Campbell et al. 1977; De Vos et al. 1982; Hanson 1973; Hanson et al. 1976). While these studies do not assure multiple gene action, they support the probability that more than a single gene is involved.

In our laboratory we have conducted two quantitative genetics studies with potato (De Vos et al. 1982; Dragoescu 1985; Dragoescu et al., in prep.). In the first study, 7 cultivars of potatoes were subjected to a diallel set of crosses (De Vos et al. 1982). General combining ability accounted for 70% of the variation in ozone response among hybrids. Specific combining ability, reflecting genes for dominance, accounted for 20% of the variation among hybrids. Dominance was in the direction of tolerance to ozone-induced foliar injury. In a sequential set of experiments, Dragoescu (1985; Dragoescu et al., in prep.) selected 2 parents and the respective F_1's, reciprocals and selfs from the study of De Vos et al. (1982); this material was used to generate the F_2 generation, backcrosses and second generation of selfs. The ozone sensitivity of the latter genetic material was tested in a controlled environment (De Vos et al. 1982) and foliar injury of young plants was quantified. Dragoescu (1985; Dragoescu et al., in prep.) developed a new autotetraploid genetic model which could be used to test more fully the inheritance of ozone sensitivity in potatoes. Using generation mean analysis, foliar injury ratings for the different generations were regressed on the parameters of the genetic model. Results of this experiment were similar to those of De Vos et al. (1982). The parameter which reflected genes expressing additive effects accounted for 65% of the total reduction in variation; dominance was not examined directly. If a single gene were responsible for inheritance of the ozone response, we would predict that with the study of increasing numbers of generations, dominance effects would become more important. Since the percentage of variation attributed to additive effects was sustained through studies with increasing numbers of generations, we concluded that it was unlikely that a single gene would have controlled inheritance of ozone injury in potato foliage.

The ozone response of potato plant foliage of selfed generations was compared with that of 14 clonal parents (De Vos et al. 1982; Dragoescu 1985) (Table 1). In almost every case, the selfs were more sensitive than the respective parents. Since tolerance appeared to be dominant, it was not surprising to find that selfed progeny were more responsive to ozone than the clonal parents (De Vos et al. 1982; Dragoescu 1985; Dragoescu et al., in prep.). If only one gene were involved, we would predict that the selfs of more sensitive material would behave the same as the respective parents; this was not always the case (Table 1).

The genetic literature provides indirect evidence which refutes the hypothesis that one gene controls ozone toxicity. It is significant that several studies with different species point in the same direction.

Table 1. Foliar reponse of parent (clonal) and selfed 3-week old potato plants to acute ozone exposure.

Cultivar	Parent	Self
Superior	8.2[a]	19.6
Chieftan	8.9 (10.6)[b]	15.0 (12.0)
Norchip	11.2 (15.3)	18.9 (19.7)
Monona	13.3	27.2
Norland	33.9	46.8
Haig	35.6	35.7
Cherokee	61.1 (48.7)	61.1 (57.7)
Pungo	5.5	17.0
Dazoc	7.5	12.8
Peconic	10.2	17.2
Blanca	13.3	13.7
Alamo	19.5	15.7
Seminole	24.8	35.3
Manota	47.0	48.5

[a] Area of foliar necrosis on most severely injured leaf of each plant visually estimated within 10% intervals (from De Vos et al. 1982).
[b] Numbers in parentheses represent data collected by Dragoescu (1985).

CELLULAR EVIDENCE

Since genes code for the biochemicals which perform physiological functions, it is important to explore the literature in search of molecular evidence to support or refute the hypothesis.

Cell membranes have long been thought of as likely targets for ozone action. The protein-lipid bilayer which comprises most membranes contains numerous sites of potential oxidation. Reviews by Tingey and Taylor (1982) and Heath (1984) include citations of research demonstrating the effects of ozone on cell lipids and proteins. While evidence of ozone-induced changes in foliar proteins and lipids abound, the biochemical extractions do not reveal specific site(s) of action. In an ultrastructural study of the effects of ozone on tobacco leaf mesophyll cells, Swanson et al. (1973) report that membrane proteins appear to be more sensitive than the respective lipids.

Giese and Christensen (1954) suggested that the plasma membrane, specifically, would be a logical site of action of ozone. Evidence of the impact of ozone on the physiological functioning of the plasma membrane has provided support for this hypothesis. Altered leaf conductance, ion exchange, and water relations have been reported for a variety of plant species in response to ozone (Heath 1980; Tingey and Taylor 1982). Recently Dominy and Heath (1985) reported that the activity of membrane bound ATPases was severely inhibited in plasma membranes isolated from leaves of a sensitive cultivar of green bean following an ozone exposure.

These changes could reflect direct effects on the plasma membrane. Alternatively, a change in behavior of the plasma membrane could result from damage to subcellular constituents.

It is probable that the plasma membrane is one site of ozone injury. Indirect evidence that proteins and lipids can be affected, that membrane function is altered and that the cell membrane is, in fact, one of the first cell structures to encounter the ozone molecule, all support this thesis. There is also evidence to suggest that the plasma membrane may not be the only susceptible site.

In 1973, Swanson et al. (1973) reported that ozone induced changes in the shape of mitochondria and chloroplasts without any visible impact on the ultrastructure of the plasma membrane. The authors interpreted the observed changes as a reflection of altered membrane function, i.e., permeability. This interpretation was consistent with Heath's (1975) discussion of "primary site of injury" in which he stated that "... a slight ozone-induced alteration in cell membrane function might eventually throw metabolism into complete disarray by a shift away from homeostasis. In this case, the primary site would be the cell membrane, even though this initial reaction alone would not irreversibly alter the cell." More recently Miyake et al. (1984) observed ozone-induced swelling of grana- and stroma-thylakoids of chloroplasts of spinach following ozone exposure. The latter researchers concluded that ozone did not injure the plasma membrane but rather penetrated the cell and damaged the chloroplasts.

Pell and Weissberger (1976) reported two kinds of ultrastructural symptoms resulting from ozone injury of soybean leaf mesophyll cells (Figs. 1-5). One type of symptom was described as cell plasmolysis, very possibly reflecting breakdown of the plasma membrane. The second type of symptom occurred in cells which appeared intact with no apparent effects on the limiting cell membrane. Internal membranes of these cells, including the endoplasmic reticulum, the limiting membrane of the chloroplast and both membranes of the mitochondria, all exhibited intense lead staining indicative of lesions. Both symptom types were observed 24 hours after an ozone exposure, at which time macroscopic symptoms were apparent. Therefore, it was unlikely that the two symptoms were sequential. We concluded that the plasma membrane may have been injured in the nonplasmolyzed cells but that conventional thin sectioning of leaf tissue presented such a limited membrane surface as to minimize the potential to observe injury. Alternatively, injury to internal membranes may have reflected independent action of ozone or its breakdown products, separate and distinct from injury leading to cell plasmolysis. Whether the nonplasmolyzed cells, described by their dense staining, were dead or alive could not be ascertained by ultrastructural observation. Consequently, the relationship between the ultrastructural symptoms and macroscopic foliar injury was unclear. Was the less severe symptom a reflection of metabolic malfunction in a cell which still contributed to the healthful appearance of the leaf, or was this symptomatic cell a contributor to foliar injury and, hence, a participant in the multi-gene response suggested to us by the genetics research?

Answers to the above question could not be addressed directly. In another, more dynamic experiment we have provided evidence which supports the interpretation that both forms of injury reflect dead cells. In order to examine the effects of ozone on cells unencumbered by stomata, leaf architecture etc., we exposed isolated potato leaf protoplasts to ozone in an in vitro setting (Illman 1983; Illman and Pell 1985). Two protoplast responses to ozone were observed and quantified: protoplasts ruptured and protoplasts intact but unable to retain and/or metabolize the vital stain, fluorescein diacetate (FDA). Living protoplasts fluoresce yellow green when stained with FDA while dead cells fluoresce red (Widholm 1972).

The impact of ozone on the protoplasts, as determined by the two criteria described above, was plotted against ozone dose (Fig. 6). Protoplast death as determined by FDA staining, and by rupturing were not sequential but rather parallel events (Fig. 6); this observation supports the independence of the two responses. Protoplast lysis probably resulted from direct attack by ozone to the plasma membrane. A similar phenomenon was reported for red blood cells exposed to ozone (Goldstein and Balchum 1967). Protoplast death, as defined by lack of a positive fluorescein diacetate stain reaction, appeared to be the result of a different mechanism of action of ozone. As already stated, it does not appear that cells died and then ruptured. Rather, these cells appeared to die with the plasma membrane intact. It was possible that the limiting membrane was impaired but to a degree which precluded lysis; altered permeability could still have doomed the cell.

Both the electron microscopy study with soybean leaves (Pell and Weissberger 1976) and the potato protoplast experiment (Illman 1983; Illman and Pell 1985) have revealed two apparently independent cell responses. How do we explain this phenomenon? If biochemical lesions are produced randomly, it is conceivable that sometimes the limiting membrane is injured by direct attack of a very reactive ozone by-product (e.g., hydroxyl radicals). In other cases, ozone and/or its reaction products could penetrate the plasma membrane and injure internal cellular components. It is also possible that all cells are not equally tolerant of ozone. The plasma membrane of some cells could be relatively tolerant of the air pollutant and those cells might, as a result, sustain injury to subcellular constituents. These subcellular components could be internal cell membranes e.g., endoplasmic reticulum, chloroplast or mitochondrial membranes. The vulnerability of these membranes has been suggested earlier (Pell and Weissberger 1976) (Fig. 3). It is also possible that some "internal" damage is sustained by injury to nonmembranous cellular components. It has been suggested that one ozone decomposition product may be hydrogen peroxide (Tingey and Taylor 1982). The latter compound is far less reactive than free radicals and, hence, may penetrate beyond the site of its formation (Tanaka et al. 1982). As such, hydrogen peroxide may react with important soluble cell proteins including ribulose bisphosphate carboxylase

Figs. 1-5. (1-3) Electron micrographs of primary leaf cells of Glycine max "Hark". (1) Spongy parenchyma cell of a nontreated primary leaf (x5,800). (2) Chloroplast of a spongy parenchyma cell form a nontreated plant. Note the limiting membrane (arrow) and grana (arrow) of the chloroplast, the plasma membrane (arrow) and mitochondrion (M) (x16,100). (3) A portion of a cell injured by ozone. Note the lead deposition on the endoplasmic reticulum (arrow), the limiting membrane of the chloroplast (arrow) and the inner and outer membranes of the mitochondrion (M) (x13,000). (4) A spongy parenchyma cell injured by ozone. The protoplast is collapsed but the chloroplasts retain their integrity (x9,000). (5) A palisade parenchyma cell totally disrupted by ozone. Note the concentration of lead at the junctions between disorganized chloroplasts (arrow) (x4,000).

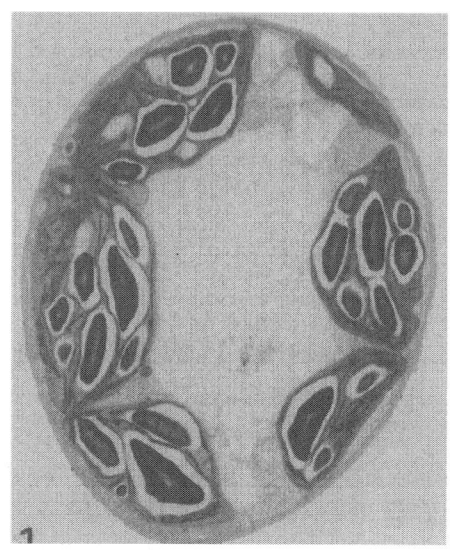
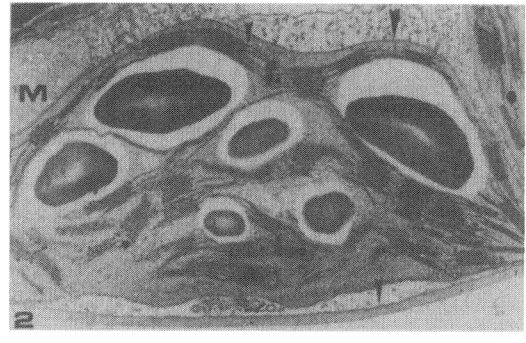
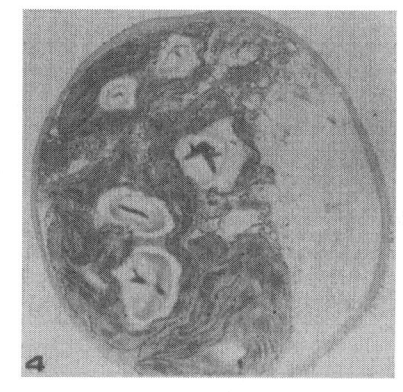
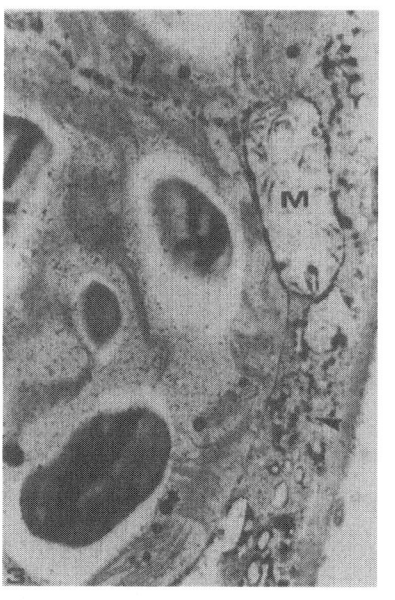
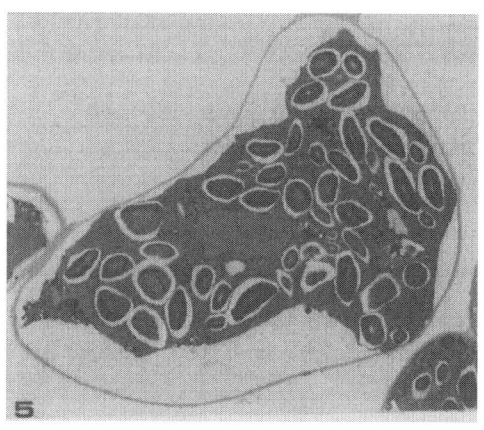

(Rubisco) (Tanaka et al. 1982). The sensitivity of Rubisco to ozone has been documented for rice and alfalfa (Nakamura and Saka 1978; Pell and Pearson 1983).

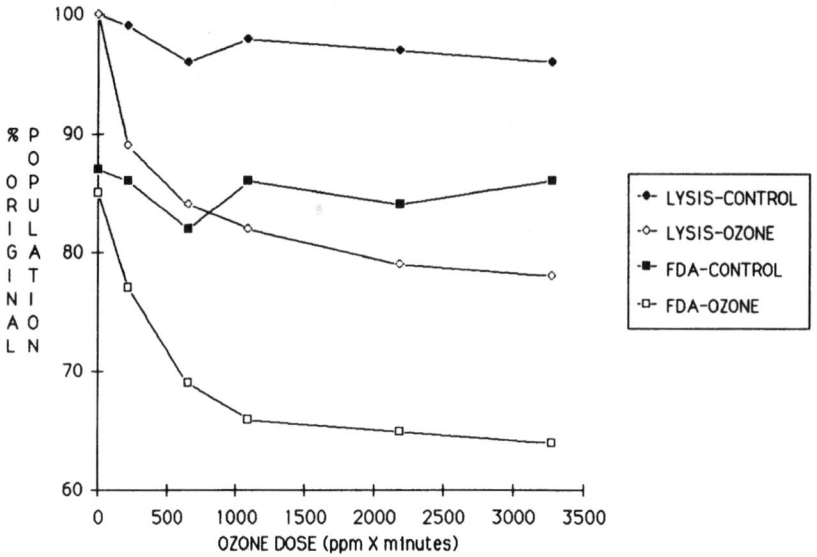

Fig. 6. Impact of ozone on viability of protoplasts of Solanum tuberosum L. cv. Cherokee following an exposure to 219 ± 31 ppm ozone for 0, 1, 3, 5, 10 or 15 minutes. Impact was characterized by % lysed protoplasts and % protoplasts which do not stain positively with fluorescein diacetate (FDA) (from Illman 1983).

CONCLUSION

In conclusion, how strong is the evidence that ozone is toxic by virtue of one mechanism of action? The most popular working hypothesis viz. that the plasma membrane is the primary site of action of ozone is not totally convincing. There is ample evidence that the plasma membrane is one site of action. The scientists whose research support this hypothesis have characterized the plasma membrane response to ozone and placed the changes early in the chain of cellular effects. The data do not exclude the possibility of other parallel responses. The genetic evidence certainly favors a multi-gene response. A multi-gene response could reflect several reactions on a single structure, e.g., plasma membrane, and in so doing would not confound the hypothesis that the plasma membrane is the primary site of action. Alternatively, ozone could be affecting several cell structures in unrelated ways. We have found, in studies with several species, that at least two different symptom types are apparent. One symptom reflects severe injury to the plasma membrane. The other symptom is expressed in cells with no direct evidence of effects on the plasma membrane; instead cells appear to have been affected "internally".

Evidence does not support the hypothesis that ozone injury is determined by a single mode of action. Instead I would propose an alternate hypothesis that ozone toxicity to foliage is the result of multiple distinct chemical reactions. The more reactive ozone by-products, including hydroxyl radicals, probably do damage the plasma membrane. Other reactive species penetrate the plasma membrane and injure subcellular constituents. Cells sustaining "internal" injury may have escaped plasma membrane damage by chance, and less toxic oxidizing species could penetrate the cell causing resultant internal injury. Alternatively, the plasma membrane of some cells could be less susceptible to injury and reactive oxidizing species could penetrate the cell causing internal injury and subsequent death.

Efforts to explain the ozone response in plants may have been hindered by the attempt to define one primary site of action followed by a series of sequential events. If the process is more complex, as has been proposed herein, apparent disparity in published results would be easier to comprehend. Information on mechanisms of action could be applied to strategies which would protect plants from ozone stress. In particular, such information would be useful to plant breeders involved in the development of ozone tolerant plants.

Acknowledgements

Contribution No. 1514, Department of Plant Pathology, the Pennsylvania Agricultural Experiment Station. Authorized for publication as Journal Series Paper No. 7206, and Contribution No. 747-85 from the Center for Air Environment Studies.

REFERENCES

Aycock Jr MK (1983) Genetics of weather fleck in Nicotiana tabacum. Can J Genet Cytol 25: 97-100

Butler LK, Tibbitts TW, Bliss FA (1979) Inheritance of resistance to ozone in Phaseolus vulgaris L. J Am Soc Hort Sci 104: 211-213

Campbell TA, Devine TE, Howell RK (1977) Diallel analysis of resistance to air pollutants in alfalfa. Crop Sci 17: 664-665

De Vos NE, Hill Jr RR, Pell EJ, Cole RH (1982) Quantitative inheritance of ozone resistance in potato. Crop Sci 22: 992-995

Dominy PJ, Heath RL (1985) Inhibition of the K^+-stimulated ATPase of the plasmalemma of pinto bean leaves by ozone. Plant Physiol 77: 43-45

Dragoescu N (1985) Development of an autotetraploid model for genetic analysis of ozone tolerance in potato, Solanum tuberosum L. Ph.D. Dissertation, The Pennsylvania State University, University Park p 42

Dragoescu N, Hill Jr RR, Pell EJ (in preparation) An autotetraploid model for genetic analysis of ozone tolerance in Solanum tuberosum L.

Engle RL, Gabelman WH (1966) Inheritance and mechanism for resistance of ozone damage in onion, Allium cepa L. J Am Soc Hort Sci 89: 423-430

Giese AC, Christensen E (1954) Effects of ozone on organisms. Physiol Zoo 27: 101-115

Goldstein BD, Balchum OJ (1967) Effect of ozone in the red blood cell. Proc Soc Exp Biol Med 126: 356-358

Grimes HD, Perkins KK, Boss WF (1983) Ozone degrades into hydroxyl radical under physiological conditions: a spin trapping study. Plant Physisol 72: 1016-1020

Hanson GP (1973) Inheritance of air-pollution tolerance in petunias. Genetics 74: 104

Hanson GP, Addis DH, Thorne L (1976) Inheritance of photochemical air pollution tolerance in petunias. Can J Genet Cytol 18: 579-592

Heath RL (1975) Ozone. In: Mudd JB, Kozlowski TT (eds) Responses of plants to air pollution. Academic Press Inc, New York, p 383

Heath RL (1980) Initial events in injury to plants by air pollutants. Ann Rev Plant Physiol 31: 395-431

Heath RL (1984) Air pollutant effects on biochemicals derived from metabolism: organic, fatty and amino acids. In: Koziol MJ, Whatley FR. (eds) Gaseous air pollutants and plant metabolism. Butterworths, London, p 466

Illman BL (1983) Ozone response of foliage and cells of sensitive and tolerant potato cultivars. Ph.D. Dissertation. The Pennsylvania State University, University Park, p 73

Illman BL, Pell EJ (1985) Characterization of the ozone response of potato leaf protoplasts. Can J Bot 63: 1936-1941

Miyake H, Furukawa A, Totsuka T, Maeda E (1984) Differential effects of ozone and sulphur dioxide on the fine structure of spinach leaf cells. New Phytol 96: 215-228

Nakamura H, Saka H (1978) Photochemical oxidants injury in rice plants: 3. effect of ozone on physiological activities in rice plants. Jpn J Crop Sci 47: 707-714

Pell EJ, Pearson NS (1983) Ozone-induced reduction in quantity of ribulose-1,5-bisphosphate carboxylase in alfalfa foliage. Plant Physiol 73: 185-187

Pell EJ, Weissberger WC (1976) Histopathological characterization of ozone injury to soybean foliage. Phytopathology 66: 856-861

Richards BL, Middleton JT, Hewitt WB (1958) Air pollution with relation to agronomic crops V. Oxidant stipple of grape. Agron J 50: 559-561

Sand SA (1960) Weather fleck in shade tobacco as a problem of interaction between genes and environment. Tobacco Sci 4: 137-146

Swanson ES, Thomson WW, Mudd JB (1973) The effect of ozone on leaf cell membranes. Can J Bot 51: 1213-1219

Tanaka K, Otsubo T, Kondo N (1982) Participation of hydrogen peroxide in the inactivation of calvin cycle SH enzymes in SO_2-fumigated spinach leaves. Plant and Cell Physiol 23: 1009-1018

Tingey DT, Taylor Jr GE (1982) Variation in plant response to ozone: a conceptual model of physiological events. In: Unsworth MH, Ormrod DP (eds) Effects of gaseous air pollution in agriculture and horticulture. Butterworths Scientific, London, p 532

Widholm JM (1972) The use of fluorescein diacetate and phenosafranine for determining viability of cultured plant cells. Stain Technol 47: 189-194

ETHYLENE - A POSSIBLE FACTOR IN THE RESPONSE OF PLANTS TO AIR POLLUTION AND ACID PRECIPITATION

D.M. Reid

Plant Physiology Research Group, Biology Department, University of Calgary, Calgary, Alberta, Canada T2N 1N4

ABSTRACT

Ethylene is a ubiquitous component of urban and rural air at concentrations varying from 0.6 µg.m^{-3} to over 8000 µg.m^{-3}. Rural levels are usually below 14 µg.m^{-3} but as one approaches burning vegetation, automobile exhaust, and city and industrial areas, levels rise rapidly. Levels of over 120 µg.m^{-3} have been often reported in such areas.

Since ethylene is also a naturally produced plant growth hormone, as well as an air pollutant of anthropogenic origin, there is a possibility that unnaturally high concentrations of gas will interfere with normal growth and development. Low levels of ethylene (well below 120 µg.m^{-3}) have been shown to cause stem dwarfing, slow leaf expansion, increased epinasty, leaf curling, promote premature chlorosis, senescence and abscission, and interfere with flowering and various processes concerned with seed and bud dormancy.

Many common environmental stresses such as SO_2, O_3, heavy metal toxicity, cold, drought, flooding, wind, and UV irradiation have been shown to stimulate endogenous ethylene biosynthesis. Perhaps such stress-induced increases in ethylene might act additively with ethylene from anthropogenic sources to produce some of the symptoms seen in forests subjected to acid precipitation.

INTRODUCTION

Before preparing this paper I picked up three volumes on air pollution and acidic deposition. Two of them had no mention of ethylene, while the other mentioned it only in passing. One could thus conclude that either there is no significant relationship between ethylene and the responses of plants to acidic deposition, or, if such a relationship does exist, it has been missed. I wish to present the argument that there may be connections between ethylene and acidic deposition. At the outset it is intriguing that some of the symptoms seen in trees, thought to be brought on by acid precipitation, can be promoted by ethylene alone. Examples of these symptoms are stunting of stem and leaf growth, leaf and needle chlorosis, and premature abscission.

HISTORICAL ASPECTS

Ethylene has many effects on plant growth and development (Abeles 1973). Since the gas is a natural byproduct of plant metabolism (Lieberman 1979), and at low concentration can markedly influence plant behaviour, it is considered to be a natural plant growth hormone. We thus have the odd situation of having a substance which is an ubiquitous component of air (Abeles 1973), yet is at the same time an endogenous controller of growth and development. A "plant hormone" is a substance which has its effects on metabolism at very low concentrations. To give one example, fumigation of tomato plants with 50 ppb (57 $\mu g.m^{-3}$) ethylene causes petiolar epinasty within 2 hours (Abeles 1973). Thus the possibility is raised that the small quantities of ethylene found in air might affect growth. It is interesting to note that the first suspicion that ethylene might play a role in plant growth emerged from air pollution problems. Observations on the harmful influence of illuminating gas on plants (first reported by Girardin 1864) led Neljubow (1901) to suggest that ethylene might affect plant growth. It was many years before ethylene was proven to be endogenously produced (Gane 1934) and through the pioneering work of workers at the Boyce Thompson Institute it was shown that the gas had a wide range of effects on many species (Crocker et al. 1932, 1935). A perusal of Abeles' (1973) monograph may leave the reader with the feeling that ethylene appears to be involved in a bewilderingly large array of different aspects of growth and development.

Before summarizing these data, it must be pointed out that for the most part they merely show that an exogenous supply of ethylene, often at unrealistically high concentrations, produces various reactions. The bulk of this research does not conclusively demonstrate that endogenous ethylene exerts any control over these processes. However, recent studies using chemicals that specifically block the action of ethylene or lower ethylene biosynthesis (Beyer 1979; Fabijan et al. 1981), do indeed show that ethylene produced by the plant influences many aspects of growth and development.

EFFECTS ON GROWTH AND DEVELOPMENT

At the molecular level, there is evidence that ethylene might bind to metaloprotein (Burg and Burg 1967; Sisler 1980) sites in membranes (Evans et al. 1982), and such receptor sites may be closely involved in the primary action of ethylene (Sisler and Yang 1983). There are few data on the precise nature or mode of operation of these sites, and little is known of the immediate effects of ethylene. Although the gas promotes and alters the synthesis of certain polypeptides (Zurfluh and Guilfoyle 1982) and new mRNA (Christoffersen and Laties 1982) and numerous enzymes (Liberman 1979; Eisinger 1983), it is far from clear that those represent the primary effects of ethylene.

The major effect of ethylene is to slow leaf expansion and stem growth (Abeles 1973), producing dwarfed plants. There are, however, a few reports showing that, especially in aquatic species, low levels of the gas can promote elongation (Ku et al. 1970). Ethylene might achieve these effects through its ability to alter microtubule orientation (Land et al. 1962; Steen and Chadwick 1981). Such microtubules may control orientation of the cellulose microfibrils in the growing cell wall (Pickett-Heaps 1967) which in turn influences the rates and directions of wall growth (Frey Wyssling 1976; MacLachlan 1977). This might in part explain how the gas can slow stem elongation, yet stimulate stem hypertrophy in many species including trees (Wallace 1928; Blake and Reid 1981), reduce leaf expansion and cause leaf rolling (Abeles 1973) and petiole epinasty (Crocker et al. 1932), and alter rates of root and fruit expansion (Abeles 1973; Jackson et al. 1981; Reid et al., in press). Exposure to the gas also slows cell division (Burg et al. 1971), decreases the number of stomata (Reid and Watson 1985), closes stomates in a few species (Browning 1974; Vitagliano 1975; Pallas and Kays 1982), promotes ageing processes such as fruit ripening (Hulme 1970), leaf and flower senescence (Abeles 1973, p. 174; Kende and Hanson 1976) and abscission in many species (Addicott 1982) including deciduous trees (Shonnard 1903), increases chlorophyll degradation (Nakagaki et al. 1970) and slows carotenoid synthesis (Kang and Burg 1972). Developmental processes are also affected. Ethylene alters the rates of shoot formation in callus (Huxter et al. 1981), rooting in annuals (Fabijan et al. 1981) and in willow (Zimmerman and Hitchcock 1933), promotes xylogenesis in lettuce (Miller and Roberts 1982), tillering in monocots (Harrison and Kaufman 1982), increased root hair production (Zimmerman and Hitchcock 1933), and causes male sterility in wheat (Rowell and Miller 1971). Ethylene can initiate flowering or promote femaleness (Abeles 1973) in Cucurbitaceae, Euphorbiaceae, pineapples and Canabinaceae. In some woody species ethylene causes hypertrophy of lenticels (Wallace 1928) and in pines stimulates light-wood production and resin formation (Wolter 1977). There is also good evidence that endogenous ethylene speeds up the darkening of Monterey pine heartwood (Shigo and Hillis 1973).

Application of ethylene can influence dormancy and in some species may break seed (Ketring and Morgan 1969) or bud dormancy (Hall et al. 1957), stimulate sprouting of tubers and corms (Vacha and Harvey 1927) and promote germination of peach pollen (Buchanan and Briggs 1969). However, the gas can also inhibit lateral bud growth in peas (Blake et al. 1983). A lowering of ethylene levels around dormant cold-stored loblolly pine seedlings increases their survival and root growth upon subsequent planting in the field (Barnett 1983).

DOSE-RESPONSE-TIME RELATIONSHIPS

Many factors affect the degree of response of an individual to ethylene, and different species show a wide range of sensitivity (Heck et al. 1970; Reid et al., in press). Furthermore the type and age of the tissue has a large influence and it has been suggested that the prior exposure to other environmental stresses might modify, perhaps reduce, the sensitivity of plants to ethylene (Abeles 1982; Reid et al., in press). The threshold doses are in the range of 50 ppb (58 µg.m^{-3}) to 200 ppb (230 µg.m^{-3}) for 1 h and perhaps as low as 10 ppb (12 µg.m^{-3}) to 50 ppb (58 µg.m^{-3}) for 24 h. Most of these data are for annual agricultural and horticultural species and unfortunately there is little information on forest tree species. That information which is available suggests that trees react in a similar manner. Angiosperms and gymnosperms both show chlorosis, dwarfing, changes in branch angle, premature senescence and abscission. The very high levels found adjacent to leaking gas lines can cause total defoliation and death of bark (Abeles 1973). In flower growing operations, economic loss is frequently reported, and in many cases the probable cause of this is ethylene (Abeles 1973).

SOME CONSEQUENCES OF THESE EFFECTS

It is clear that if plants are fumigated with sufficient ethylene at critical stages of development, survival will be reduced. For instance, in a northern area with a short growing season, any reduction in photosynthetic capacity caused by ethylene-induced abscission, premature senescence, alteration in leaf angle, leaf curling, reduction in numbers of stomata and/or leaf area, or decrease in pigmentation, could drastically affect long-term survival. Likewise, ethylene-induced premature bud break in spring, alterations in sex ratios or increases in male sterility, premature seed or pollen germination are all potentially harmful. Thus, although there is a paucity of data on trees, it is possible that even transient increases of ethylene in the environment might override the normal environmental cues and lower the chances of survival of a species.

SOURCES OF ETHYLENE IN THE ENVIRONMENT

There are many sources of ethylene; for example, most plants produce ethylene at rates of between 0.5 to 6 µL.kg^{-1}.h^{-1}. Ripening fruits, and tissues that have been wounded (Hanson and Kende 1976) or stressed (e.g., waterlogged; Reid and Bradford 1984) in various ways can often produce ethylene greatly in excess of these rates. Soils can act as sources and sinks for ethylene (Smith and Cook 1974) with soil fungi and bacteria both playing an important role. Overall, the soil probably acts as a major ethylene sink (Abeles et al. 1971). While ethylene production from biological sources (including animals and aquatic sources) must not be ignored, anthropogenic sources are very important. In Canada and the U.S., internal combustion engines account for well over half of the anthropogenically produced ethylene, with the other major sources being burning vegetation and petrochemical production (Reid et al., in press). Generally, rural environments, except nearby vegetation fires, have much lower concentrations of ethylene than urban and industrial zones or areas nearby roads.

LEVELS OF ETHYLENE IN THE ENVIRONMENT

In rural areas near-ground concentrations can vary between 0.5 ppb (0.6 µg.m^{-3}) to 14 ppb (16 µg.m^{-3}) (Cox et al. 1976, cited in Reid et al., in press) with higher levels being seen at noon (Reid and Watson, in press). This rural diurnal variation may reflect the daily rhythms in plant metabolism. As with any air pollutant, meteorological conditions have a large influence (Byrne 1960, cited in Hasek et al. 1969). During an inversion layer in the city of Calgary, ethylene levels were much higher than normal, and usually more ethylene was found downwind of the city than upwind (Reid and Watson, in press). Values for ethylene concentrations in industrial, urban zones, and near transport corridors are higher but vary enormously. Values from 5 ppb (6 µg.m^{-3}) to as high as 7100 ppb (8165 µg.m^{-3}) have been reported (Gordon and Meeks 1977). Near roads the daily variation in ethylene concentration correlates positively with density of road traffic (Scott et al. 1957; Altshuller and Bellar 1963; Menzies and Shumate 1976). Harbourn and McCambley (1973) measured between 280 and 410 ppb (320 to 470 µg.m^{-3}) near a main road (Swansea, UK). The highest level of ethylene that I have seen Calgary has been just below 100 ppb (115 µg.m^{-3}), measured 10 metres downwind of a busy road. (For details of environmental dynamics, see Reid et al., in press.)

EFFECTS OF AMBIENT ETHYLENE ON VEGETATION

We can now turn to the questions "might the quantities of ethylene often found in the air affect normal growth and development of plants, and might ethylene interact with other air pollutants?" We are now in a situation in which (a) there is a shortage of long-term studies of the influence of ethylene in the field; (b) laboratory studies with chronic ethylene fumigations are rare; (c) there are few studies with non-agricultural or forest species; (d) there are few studies of interactions of ethylene with the other air pollutants. (One exception is a study by Heck (1964) who found that ethylene did not seem to interfere with the injurious action of NO_2 on leaves.) Using the little information available, it is clear that almost any elevation of ethylene above 50 ppb (58 µg.m^{-3}) has the potential to influence growth. Abeles and Heggestad (1973) grew petunia, red kidney beans and cucumbers in chambers that were supplied with a range of ethylene levels below and above ambient. They found that all species grew to the greatest size in below ambient ethylene concentrations and size decreased steadily with an increase in the dose rate. This work suggested that almost any increase in ethylene levels caused dwarfing and reduction in fruit yield. More recently, Reid and Watson (1985) found similar results using Canola (rape seed) oats, barley, sunflower and brome grass. However, at certain stages of development the growth of some species was promoted by slightly above ambient ethylene concentrations. Other workers have shown that under certain conditions very small quantities of ethylene can promote root elongation of a few species (Konings and Jackson 1979) and stem growth of hydrophytes (Ku et al. 1970). Indeed it has been suggested that small quantities of endogenous ethylene might be necessary for normal growth and development (Lieberman and Kunishi 1972; Zobel and Roberts 1974; Huxter et al. 1979). However, concentrations above 50 ppb (58 µg.m^{-3}) over long periods are usually inhibitory and anything over 100 ppb (115 µg.m^{-3}) for more than a few days almost always retards growth. Thus, the work of Abeles and Heggestad (1973) and Reid and Watson (in press), and the numerous other short-term investigations using acute doses, support the possibility that ethylene, at the levels found near roads, cities and industrial areas, could have the following effects on growth:

Reductions in stem/leaf growth	...probable
Effects on leaf orientation, reduction in seed yield, slowing of developmental processes	...possible
Premature senescence	...possible at higher levels

Is there any field evidence for such predictions? Hall et al. (1957) observed that cotton plants grown near a polyethylene factory showed symptoms typical of ethylene fumigation (leaf abscission, stem dwarfing, chlorosis, prostrate habit), and premature fruit abscission of mandarin oranges near a gas works has been observed (Fukuchi and Yamamoto 1969). It has been known for many years that improper venting of oil or gas heaters in greenhouses can cause defoliation, epinasty, petal drop and other symptoms typical of exposure to ethylene (Hasek et al. 1969). The report of James (1963) which documents increasingly serious losses of flower growers in the San Francisco area (1952-1962) suggests that ethylene was in part responsible. Of course it is difficult to separate some of the effects of ethylene from the other gaseous pollutants, and in some of these situations it was not absolutely proven that ethylene was the only culprit. However, in many of these cases some of the symptoms are typical of and, sometimes even quite specific to, ethylene.

The growth chamber experiments purporting to show that small increases in ethylene harm plant growth (Abeles and Heggestad 1973; Reid and Watson, in press) are open to the criticism that growth chamber plants growing in optimized conditions may not respond as would field grown plants. Indeed Abeles (1982) has pointed out that the ethylene dose response curves for growth chamber-grown plants may be different from plants grown in the field. For example, ethylene is known to promote glucanase levels in bean leaves (Abeles and Forrence 1970). Later, Abeles et al. (1971) found that bean plants grown in an area shown to have high levels of ethylene in the air possessed more glucanase than did plants from an unpolluted district. Furthermore the former plants would not respond to yet more ethylene, suggesting an alteration in sensitivity to the gas. We are attempting to look at dose response curves of plants grown in optimized environment versus plants from more stressed environments.

Why might plants from the field respond to ethylene differently from the laboratory plants? Simple observation shows this often to be true with regard to overall appearance. Levitt (1980) has argued that mild stress may confer hardiness to other forms of stress. Certainly it is common for many plants in the field to be water stressed and to have different shoot/root ratios than their pampered pot-grown cousins in the greenhouse. Our own work has shown that unstressed sunflowers can survive a rather acidic nutrient solution but a combination of flooding stress plus low acidity was quickly fatal (Drakeford and Reid 1984). That experiment contrarily demonstrated that in at least one case, one stress did not "confer hardiness" to another, but rather, the opposite occurred. In any event, data obtained with laboratory experimental plants must be interpreted with care when one attempts to make extrapolations to those growing naturally in the outside environment. There are other criticisms of laboratory growth chamber experiments in that such conditions do not properly simulate the complexity of the natural environment (Kimball and Levin 1985). However, as these and many other authors point out, "laboratory bioassays" do provide a necessary first step.

INTERACTIONS WITH OTHER STRESSES

When trying to understand how plants respond to an external and elevated ethylene supply there is yet another complicating factor to be considered. This is the fact that many environmental stresses promote ethylene biosynthesis inside the stressed plant (including other air pollutants and heavy metals; Kimmerer and Kozlowski 1982).

All the following stresses promote ethylene biosynthesis, freezing (Elstner and Konze 1976), chilling (Wang 1983), pathogens (Pegg 1976), reorientation in respect to direction of gravitational force (as in lodging; Crocker et al. 1932; Hall et al. 1957; Clifford et al. 1983), drought (McMichael et al. 1972; Wright 1980), flooding and anoxia (Kawase 1974; Reid and Bradford 1984), herbicides (Abeles 1973), wounding (Hanson and Kende 1976), bending in wind and mechanical perturbation (Jaffee and Biro 1977), and pressure (Goeschl et al. 1966). Repetitive shaking (simulating wind) or mechanically stressing (by tying branches in knots) causes production of thicker but shorter tree stems and branches, probably due to elevated ethylene levels (Neel and Harris 1972; Leopold et al. 1972). There is a report showing a small stimulatory effect of simulated acid rain on ethylene production in potato, soybean and radish (Arny and Pell 1986). Ozone at 0.05 $\mu L^{-1} \cdot h^{-1}$ (Tingey et al. 1976), SO_2 at 0.3 $\mu L \cdot L^{-1}$ (Kimmerer and Kozlowski 1982), UV irradiation (Chalmers and Faragher 1977), Ca^{2+}, Fe^{3+}, and Hg^{2+}, all stimulate ethylene production.

Some of the symptoms of these environmental stresses are the direct result of the stress-induced increases in internal ethylene. A good example of this is to be found in many of the effects exerted by waterlogged soils on plants (including wetland species and trees), e.g., epinasty, hyponasty of stems, enlarged lenticels, chlorosis, leaf abscission, stem dwarfing, adventitious rooting. All of these are in part caused by high internal ethylene levels (Reid and Bradford 1984).

An example of the difficulty in deciding if it is external ethylene that is modifying the growth, or if it is the stress-induced endogenous ethylene that is active, is contained in the report of Fluckiger et al. (1979). They saw premature senescence in trees and shrubs adjacent to a road. As they moved nearer to the road, increased levels of ethylene in the plant tissues, higher levels of tissue peroxidase and more abscission were observed. What produced these effects? Was it the high levels of ethylene from the traffic (not measured), gaseous pollutants other than ethylene, or even constant traffic generated wind? Perhaps a combination of all these factors was the cause.

These reports therefore suggest that while many environmental stresses may have direct and primary damaging effects on growth, they secondarily cause elevations in endogenous ethylene production. This endogenous ethylene together with yet more from an anthropogenic source could act additively in further affecting growth. In Fig. 1 are shown three situations, all of which would increase, to varying degrees, internal levels of ethylene. While the quantity of anthropogenically derived ethylene in C might not in itself be sufficient to do much harm, in conjunction with the other stresses and elevated internal ethylene concentrations, it might be enough to be one of the factors inducing symptoms such as premature abscission and senescence in areas subjected to acid deposition. I am not aware of studies other than Heck's (1964) in which plants have been subjected to combinations of ethylene with other known harmful air pollutants, such as SO_2, O_3, NO_x, or PAN. Our preliminary work indicates that drought stress and low levels of ethylene act additively in reducing overall growth of sunflower seedlings.

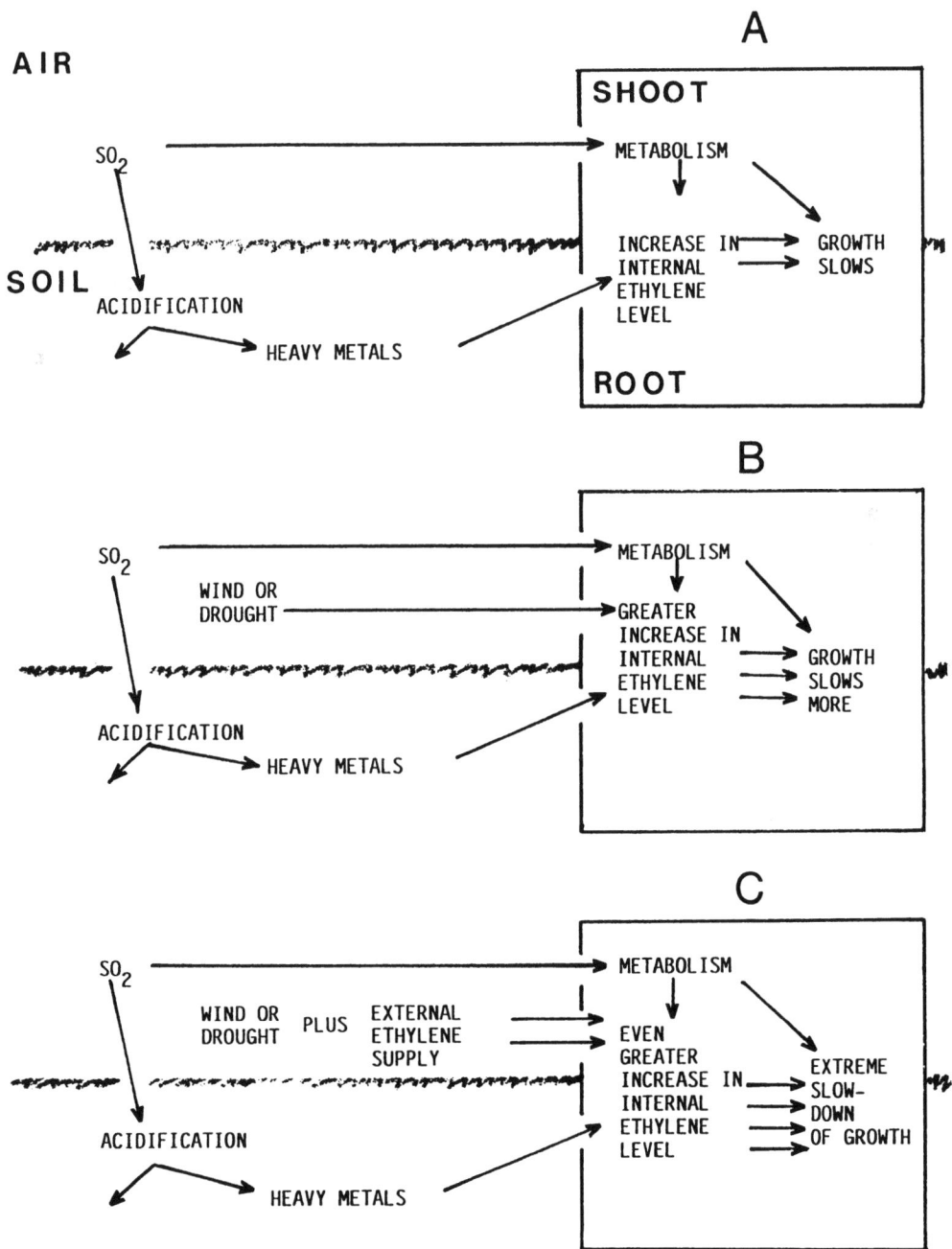

Fig. 1. Hypothetical additive effects of A, SO_2 alone; B, SO_2 + wind or drought; and C, SO_2 + wind or drought + an external supply of ethylene on plant growth.

REFERENCES

Abeles FB (1973) Ethylene in plant biology. Academic Press, New York

Abeles FB (1982) Ethylene as an air pollutant. Agric For Bull of University of Alberta 5(1): 4-12

Abeles FB, Forrence LE (1970) Temporal and hormonal control of β-1, 3-glucanase in Phaseolus vulgaris L. Plant Physiol 45: 395-400

Abeles FB, Heggestad ME (1973) Ethylene: an urban air pollutant. J Air Pollut Contr Assoc 23: 517-521

Abeles FB, Craker LE, Forrence LE, Leather GR (1971) Fate of air pollutants: removal of ethylene, sulphur dioxide and nitrogen dioxide by soil. Science 173: 914-916

Addicott FT (1982) Abscission. Univ of Calif Press, Berkeley

Altshuller AP, Bellar TA (1963) Gas chromatographic analysis of hydrocarbons in the Los Angeles Atmosphere. J Air Pollut Contr Assoc 13(2): 81

Arny CJ, Pell EJ (1985) Ethylene production by potato, radish and soybean leaf tissue treated with simulated acid rain. Enviro and Expt Bot 26: 9-15

Barnett JP (1983) Ethylene: a problem in seedling storage. Tree Planter's Notes. Winter: 28-29

Beyer EM Jr (1979) Effect of silver ion, carbon dioxide, and oxygen on ethylene action and metabolism. Plant Physiol 63: 169-173

Blake TJ, Reid DM (1981) Ethylene, water relations and tolerance to waterlogging of three eucalyptus species. Aust J Plant Physiol 8: 497-505

Blake TJ, Reid DM, Rood SB (1983) Ethylene, indole acetic acid and apical dominance in peas: a reappraisal. Physiol Plant 59: 481-487

Browning G (1974) 2-chloroethane phosphoric acid reduces transpiration and stomatal opening in Coffea arabica L. Planta (Berl) 121: 175-179

Buchanan DW, Briggs RH (1969) Peach fruit abscission and pollen germination as influenced by ethylene and 2-chlorethane phosphoric acid. J Amer Soc Hort Sci 94: 327-329

Burg SP, Apelbaum A, Eisinger WR, Kang BG (1971) Physiology and mode of action of ethylene. Hort Sci 6: 359-364

Burg SP, Burg EA (1967) Molecular requirements for the biological activity of ethylene. Plant Physiol 42: 144-152

Chalmers, DJ, Faragher JD (1977) Regulation of anthocyanin synthesis in apple skin. II Involvement of ethylene. Aust J Plant Physiol 4: 123-131

Christoffersen RE, Laties GG (1982) Ethylene regulation of gene expression in carrots. Proc Natl Acad Sci USA 79: 4060-4063

Clifford PE, Reid DM, Pharis RP (1983) Endogenous ethylene does not initiate but may modify geobending - a role for ethylene in autotropism. Plant Cell Environ 6: 433-436

Crocker W, Hitchcock AE, Zimmerman PW (1935) Similarities in the effects of ethylene and the plant auxins. Contrib Boyce Thomp Inst 7: 231-248

Crocker W, Zimmerman PW, Hitchcock AE (1932) Ethylene-induced epinasty of leaves and the relation of gravity to it. Contrib Boyce Thomp Inst 4: 177-218

Drakeford DR, Reid DM (1984) Changes in the ability of plants to alter the pH of the flooding medium as an early symptom of flooding stress in Helianthus annuus. Can J Bot 62: 2417-2422

Eisinger W (1983) Regulation of pea internode expansion by ethylene. Ann Rev Plant Physiol 34: 225-240

Elstner EF, Konze JR (1976) Effect of point freezing on ethylene and ethane production by sugar beet leaf discs. Nature 263: 351-352

Evans DL, Bengochea T, Cairns AJ, Dodds JH, Hall MA (1982) Studies on ethylene binding by cell-free preparations from cotyledons of Phaseolus vulgaris L. subcellular localization. Plant Cell Enviro 5: 101-107

Fabijan D, Taylor JS, Reid DM (1981) Adventitious rooting in hypocotyls of sunflower (Helianthus annuus) seedlings. II Action of gibberellins, cytokinins, auxins and ethylene. Physiol Plant 53: 589-597

Fluckiger W, Oretli JJ, Fluckiger-Keller H, Braun S (1979) Premature senescence in plants along a motorway. Environ Pollut 20(3): 171-176

Frey-Wyssling A (1976) The Plant Cell Wall. Borntraeger, Berlin

Fukuchi T, Yamamoto T (1969) J Pollut Control 5: 17 (cited in Abeles 1973)

Gane R (1934) Nature (Lond) 134: 1008

Girardin JPL (1964) Jahresber Agrikult-Chem Versuchssta Berlin 7: 199 (cited in Abeles 1973)

Goeschl JD, Rappaport L, Pratt HK (1966) Ethylene as a factor regulating the growth of pea epicotyls subjected to physical stress. Plant Physiol 41: 877-884

Gordon SJ, Meeks SA (1977) A study of gaseous pollutants in the Houston, Texas area. Am Inst Chem Symp Ser 73: 84-94

Hall WC, Truchelut GB, Leinweber CL, Herrero FA (1957) Ethylene production by the cotton plant and its effects under experimental and field conditions. Physiol Plant 10: 306-317

Hanson AD, Kende H (1976) Biosynthesis of wood ethylene in morning-glory flower tissue. Plant Physiol 57: 538-541

Harbourn CL, McCambley T (1973) Proc 3rd Int Clean Air Cong. Dusseldorf (FRG) S. C38-41

Harrison MA, Kaufman PB (1982) Does ethylene play a role in the release of lateral buds (tillers) from apical dominance in oats? Plant Physiol 70: 811-814

Hasek RF, James HA, Sciaroni RH (1969) Ethylene - its effect on flower crops. Florists Review 144 (3721): 21, 65-68 & 78-82

Heck WW (1964) Plant injury induced by photochemical reaction products of propylene-nitrogen dioxide mixtures. J Air Pollut Contr Assoc 14: 255-261

Heck WW, Daines RH, Hindawi IJ (1970) Other phyotoxic pollutants. In: Recognition of air pollution injury to vegetation. A pictorial atlas. Air Pollut Contr Assoc, Pittsburg, F1-F24

Hulme AC (1970) Ed. The biochemistry of fruits and their products. vols I & II. Academic Press, New York

Huxter TJ, Thorpe TA, Reid DM (1981) Shoot initiation in light- and dark-grown tobacco callus: the role of ethylene. Physiol Plant 53: 319-326

Huxter TJ, Reid DM, Thorpe TA (1979) Ethylene production by tobacco (Nicotiana tabacum) callus. Physiol Plant 46: 374-380

Jackson MB, Drew MC, Giffard SC (1981) Effects of applying ethylene to the root system of Zea mays on growth and nutrient concentration in relation to flooding tolerance. Physiol Plant 52: 23-28

Jaffe MJ, Biro R (1977) Thigmomorphogenesis: role of ethylene in wind induced growth retardation. Proc 4th Ann Meet Plant Growth Regul Work Group, p 118-124

James HA (1963) Flower damage - a case study. Bay Area Air Pollution Control District Information Bulletin, August (cited in Hasek et al. 1969)

Kang BG, Burg SP (1972) Involvement of ethylene in Phytochrome mediated carotenoid synthesis. Plant Physiol 49: 631-633

Kawase M (1974) Role of ethylene in induction of flooding damage in sunflower. Physiol Plant 31: 29-38

Kende H, Hanson AD (1976) Relationship between ethylene evolution and senescence in morning glory flower. Plant Physiol 57: 523-527

Ketring DL, Morgan PW (1969) Ethylene as a component of the emanations from germinating peanut seeds and its effect on dormant Virginia-type seeds. Plant Physiol 44: 326-330

Kimball KD, Levin S (1985) Limitations of laboratory bioassays: the need for ecosystem-level testing. BioSci 35: 165-171

Kimmerer TW, Kozlowski TT (1982) Ethylene, ethane, acetaldehyde, and ethanol production by plants under stress. Plant Physiol 69: 840-847

Konings H, Jackson MB (1979) A relationship between rates of ethylene production by roots and the promoting or inhibiting effects of exogenous ethylene and water on root elongation. Z Pflanzanphysiol 92: 385-397

Ku HS, Suge H, Rappaport L, Pratt HK (1970) Stimulation of rice coleoptile growth by ethylene. Planta 90: 333-339

Land JM, Eisinger WR, Green PB (1962) Effects of ethylene on the orientation of microtubules and cellulose microfibrils of pea epicotyl cell with polylamellate cell walls. Protoplasm 110: 5-14

Leopold AC, Brown KM, Emerson FH (1972) Ethylene in the wood of stressed trees. Hort Sci 7: 175

Levitt J (1980) Responses of plants to environmental stress. Vols I and II. Academic Press, New York, p 497 and 606

Lieberman M (1979) Biosynthesis and action of ethylene. Ann Rev Plant Physiol 30: 533-591

Lieberman M, Kunishi AT (1972) Thoughts on the role of ethylene in plant growth and development. In: Carr DJ (ed) Plant growth substances 1970. Springer-Verlag, New York, p 549-566

MacLachlan GA (1977) Cellulose metabolism and cell growth. In: Pilet PI, Plant growth regulation. Springer-Verlag, Heidelberg, p 12-20

McMichael BL, Jordan WR, Powell RD (1972) An effect of water stress on ethylene production by intact cotton petioles. Plant Physiol 49: 658-660

Menzies RT, Shumate MS (1976) Remote measurements of ambient air pollutants with a bistatic laser system. Appl Optics 15: 2080-2084

Miller AR, Roberts LW (1982) Regulation of tracheary element differentiation by exogenous L-methionine in callus of soya bean cultivars. Ann Bot 50: 111-116

Nakagaki Y, Hirai T, Stahmann MA (1970) Ethylene production by detached leaves infected with tobacco mosaic virus. Virology 40: 1-9

Neel PL, Harris RW (1972) Tree seedling growth: effects of shaking. Science 175: 918-919

Neljubow D (1901) Uber die horizontale nutation der stengel von Pisum Sativum und Einiger Ander Planzen. Bech Bot Centralbl 10: 128

Pallas JE Jr, Kays SJ (1982) Inhibition of photosynthesis by ethylene - a stomatal effect. Plant Physiol 70: 598-601

Pegg GF (1976) The involvement of ethylene in plant pathogenesis. In: Heitefuss R, Williams PH (eds) Physiological plant pathology. Springer-Verlag, Berlin, p 582-591

Pickett-Heaps JC (1967) The effect of colchine on the ultra structure of dividing cells, xylem wall differentiation, and distribution of cytoplasmic microtubules. Dev Biol 15: 206-236

Reid DM, Bradford KJ (1984) Effects of flooding on hormone relations. In: Kozlowski TT (ed) Flooding and plant growth. Academic Press, New York, p 195-219

Reid DM, Watson K (in press) Ethylene and Plant Development. In: Tucker G, Roberts J (eds) Ethylene as an air pollutant. Proc Of Nottingham Eastern School, Butterworths, London

Reid DM, Sheffer MG, Pierce RC, Bezdicek DF, Linzon SN, Reuvers T, Spencer MS, Vena F (in press) Ethylene in the environment. NRC Canada Publication No. NRCC 22496, Ottawa

Rowell PL, Miller DG (1971) Induction of male sterility in wheat with 2-chlorethyl-phosphoric acid (Ethrel). Crop Sci 11: 629-631

Scott WE, Stephens ER, Hanst PC, Doerr RC (1957) Further development in the chemistry of the atmosphere. Proc Amer Petrol Inst 37: 171-183

Shigo AL, Hillis WE (1973) Heartwood, discolored wood, and microorganisms in living trees. Ann Rev Phytopath 11: 197-222

Shonnard F (1903) Effect of illuminating gas on trees. Dept Public Works Yonders NY, p 48

Sisler EC (1980) Partial purification of an ethylene-binding component from plant tissue. Plant Physiol 66: 404-406

Sisler EC, Yang SF (1983) Effect of butanes and cyclic olefins on etiolated pea plants in relation to the ethylene response. Plant Physiol (Supp) 72: 40

Smith AM, Cook RJ (1974) Implications of ethylene production by bacteria for biological balance of soil. Nature 252: 703-705

Steen DA, Chadwick AV (1981) Ethylene effects in pea stem tissue, evidence of microtubule mediation. Plant Physiol 67: 460-466

Tingey DT, Standley C, Field RW (1976) Stress ethylene evolution: a measure of ozone effects on plants. Atmos Environ 10: 969-974

Vacha GA, Harvey RB (1927) The use of ethylene, propylene, and similar compounds in breaking the rest period of tubers, bulbs, cuttings, and seeds. Plant Physiol 2: 187-193

Vitagliano C (1975) Effects of ethephon stomata, ethylene evolution and abscission in olive (<u>Olea europala</u> L) cv coratina. J Am Soc Horti Sci 100: 482-485

Wallace RH (1928) Histogenesis of intumescence in the apple induced by ethylene gas. Amer J Bot 15: 509-524

Wang CY (1983) Use of 1-aminocyclopropane-1-carboxylic acid levels as an index for chilling exposure at various temperatures. Plant Physiol (supp) 72: 43

Wolter KE (1977) Ethylene-potential alternative to bipyridilium herbicides for inducing light wood in pine. Proc Lightwood Res Coord Council, p 90-99

Wright STC (1980) The effect of plant growth regulator treatments on the levels of ethylene emanating from excised turgid and wilted wheat leaves. Planta 148: 381-388

Zimmerman PW, Hitchcock AE (1933) Initiation and stimulation of adventitious roots caused by unsaturated hydrocarbon gases. Contrib Boyce Thompson Inst 5: 351-369

Zobel RW, Roberts LW (1974) Control of morphogenesis in the ethylene requiring mutant diageotropica. Can J Bot 52: 735-741

Zurfluh LL, Guilfoyle TJ (1982) Auxin- and ethylene-induced changes in the population of translatable messenger RNA in basal sections and intact soybean hypocotyl. Plant Physiol 69: 338-340

RESPONSES OF FORESTS IN DECLINE TO EXPERIMENTAL FERTILIZATION

H.W. Zoettl

Institute of Soil Science and Forest Nutrition, Albert-Ludwig University, D-7800 Freiburg i.Br., Federal Republic of Germany

ABSTRACT

For several characteristic types of forest damage in SW-Germany, nutritional disturbances of the stands are evident from chemical soil properties, needle analysis data, and visible symptoms. Under the relatively low impact of acidic deposition in the study area recent drastic changes in soil conditions are not probable, but the measured higher ozone concentrations and acid fog may cause increased leaching of ions from the foliage. These losses may not be compensated by the trees where available reserves of the elements concerned are low in the rooted substrate.

To test the role of the nutrient supply in these processes, a comprehensive programme of "diagnostic fertilization experiments" was started. Four experiments with Norway spruce are presented with the preliminary data. In two trials on very acid soils with extremely low base saturation, the gold-yellow-chlorosis of older needles was mainly due to magnesium deficiency. It disappeared after fertilizer application. One trial on soil with high calcium saturation but low potassium supply represents a different damage type with the symptoms of potassium deficiency. In the last trial on calcareous soil the trees were yellow-chlorotic in all needle age classes due to extreme deficiency of manganese. Also in these trials the response to the corresponding fertilization was clearly positive.

The results of these recent experiments and the chronological changes of the nutritional status of spruce stands within the last decades point out that nutrient deficiencies play a major role in the development of particular forest damage types.

INTRODUCTION

Studies of the recent forest damage ("neuartige Waldschäden") in the dominant tree species of Central Europe give evidence not only of some common disease phenomena but also of a pronounced variation of symptoms between species and regions (Schütt et al. 1983; Rehfuess 1985). This leads to a concept of different disease types caused by varying budgets of site and stress factors. The concept combines the knowledge of traditional forest site science and modern stress physiology (Manion 1981; Mohr 1984). A particular decline type can be characterized by careful observation of symptoms and analysis of dominant site (growth) factors or stressors (Zöttl 1985).

Based on our investigations of several characteristic damage types in SW-Germany, we consider soil nutrient deficiencies as an important predisposing stress factor (Zöttl and Mies 1983b; Zöttl and Hüttl 1985). Photooxidants and acid fog are believed to be inciting stress factors, causing increased foliar leaching of elements which are relatively mobile and related to the pre-damaged chloroplasts, i.e., Mg, K, Mn, and Zn. The trees can compensate for those increased nutrient losses through increased uptake from the soil only if the supply in the rooted substrate is sufficient and if the roots are undamaged. Weather conditions and pathogens (fungi) can also play an important role, but primarily as contributing stress factors.

DIAGNOSTIC FERTILIZATION TRIALS

To test this hypothesis, we have started a comprehensive program of "diagnostic fertilization experiments". The purpose of these experiments is to determine if particular nutrient deficiencies characterized by soil and needle analysis and visible symptoms can be corrected through application of corresponding fertilizers. In order to get a quick response, easily soluble chemical compounds are preferred. Response control is performed by needle analysis and observation of symptoms. Climatic field conditions (degree of air pollution) remain, of course, essentially unchanged.

Preliminary results are presented below from four trials (Fig. 1 and Tables 1 - 4). They demonstrate the different nutritional status of selected spruce stands in relation to the particular site/stress conditions and their reactions to the applied fertilizers. For chemical analysis needles were sampled in late autumn 1983 and 1984 from the first whorl of 5 - 8 trees per plot. Soil data are given for the main rooting zone. Fertilizers were applied in spring 1984 at all plots.

The type and level of air pollutants are rather uniform in our study area (southern part of Baden-Württemberg): low deposition of H^+ (bulk deposition 0.5 - 1 $kmol.ha^{-1}.y^{-1}$), S (10 - 20 $kg.ha^{-1}.y^{-1}$), and N (15 - 25 $kg.ha^{-1}.y^{-1}$), corresponding to low concentrations of SO_2 and NO_2, but a relatively high O_3 concentration in the ambient air (Fig. 1). The site conditions vary mainly with respect to parent rock and soil. There are also climatic differences.

Trial "Elzach 10a" represents the widely distributed "Norway spruce decline at higher altitudes on acidic soils" with gold-yellow-chlorosis of older needles under conditions of increased irradiation (Zöttl 1985). The site characteristics of "Elzach 10a" are typical for this kind of damage: very acid brown earth with extremely low base saturation (see Table 1), developed in periglacial solifluction layers derived from granite with very low Ca and Mg content, an altitude of 900 m a.s.l., and a mean annual precipitation of 1600 mm.

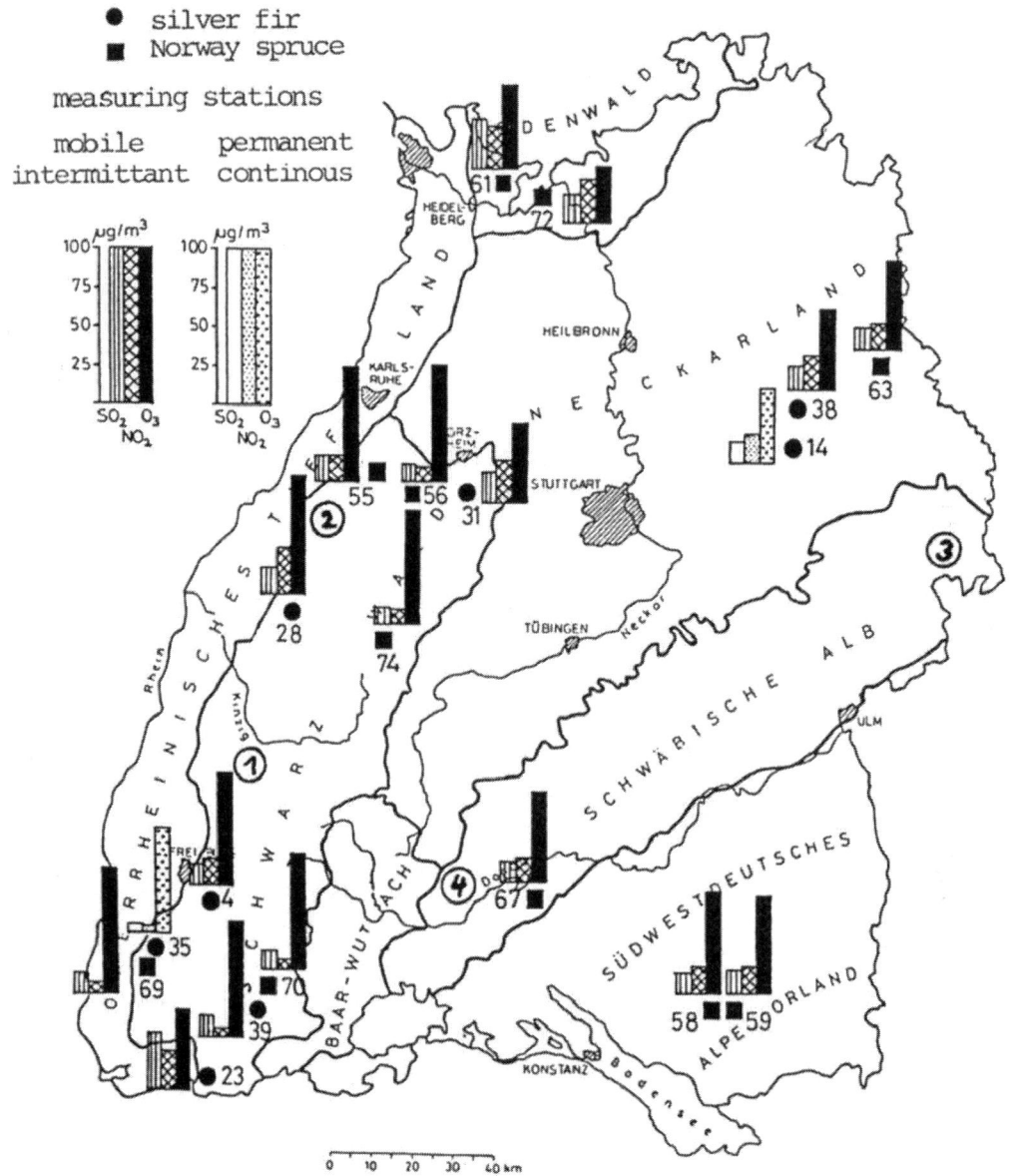

Fig. 1. Air pollution at Baden-Württemberg forest stations; 30 minute mean values during 1984 (data from Oblander and Hanss, 1985) Location of fertilization trials ①.

The nutritional status of the trees (Table 1) is characterized by optimum N, P and K values, low but sufficient Ca, regular levels of Mn, Fe and Al, low Zn, and strong Mg deficiency. The element contents in autumn 1983 show the nutritional status of all the plots before fertilization; the 1984 values show both the low yearly variation in the control plots and the differences following fertilization. Due to an adequate supply of all elements except Mg, only "Bittersalz", a quick release Mg fertilizer, was applied in this experiment. The Mg status of the fertilized trees improved considerably; almost no change occurred with the other elements. A great part of the former yellow needles greened up in the fertilized plot and needle loss was reduced. The level of good Mg supply is not yet fully reached with a magnesium content of 0.56 mg.g^{-1} needle dry weight but it appears to be a positive response.

Trial "Baden-Baden 2a" is another example of the above-mentioned type of damage. This site has podsolic brown earth with raw humus, also formed from granite, with an altitude of 600 m a.s.l., and mean annual precipitation of 1700 mm. The yellowing of older needles and a 30% "needle loss" was noticed in 1983. Little white-grey spots were detected on current needles which appeared as eroded craters under the electron microscope. The soil chemical data (Table 2) are similar to "Elzach 10a" but with an even lower Ca^{2+} saturation. This corresponds well with the low level of 0.9 mg Ca.g^{-1} needle dry weight. Again Mg is within the deficiency range and Zn is rather low, whereas N, P, and K are high or in the optimum range.

In spring 1984, a newly developed K-Mg-Ca-fertilizer was applied to improve the Mg and Ca supply. The K component would not have been necessary due to the good K supply at this site, but it was part of the fertilizer. Nitrate should help to establish a better anion:cation ratio in the rooting zone which is known to favour the Mg uptake from a substrate with a high concentration of exchangeable Al^{3+}. As a result of this fertilization, the nutritional status of Mg and Ca was raised above the deficiency level. The low Zn content also increased. Due to the high dosage of nitrate (corresponding to 220 kg.ha^{-1}) the N content of the needles also increased. The K content in the needles dropped a little, apparently due to Ca:K antagonism. These results show again that by fertilizer addition a higher uptake rate of basic cations is possible in spite of the extremely high Al saturation in the rooted substrate. The Al level in the needles remained unchanged. In conformity with the improved nutrient supply, the deficiency symptoms mostly disappeared with the exception of the already necrotic needles.

In "Ebnat 2" soil conditions are quite different. Pleistocene weathering of older sediments had formed a shallow layer of acid loam over the weathering zone of jurassic limestone. Base saturation, dominated by Ca^{2+}, is nearly complete. The site is 600 m a.s.l. and receives 800 mm mean annual precipitation. In autumn 1983, the trees were classified as having 30% "needle loss" and exhibiting a weak "silver tinsel symptom". Discolorations mainly of older needles into lemon-yellow or bright green

Table 1. Results of the Diagnostic Fertilization Trial* "Elzach 10a": (a) elemental analyses of the current needles of a 12-year old spruce stand; (b) chemistry of the control soil at 20 - 30 cm depth.

(a)

Treatment	N	P	K	Ca	Mg	Mn	Fe	Zn	Al
			$mg \cdot g^{-1}$				$\mu g \cdot g^{-1}$		
control 1983	16.0	2.7	7.5	2.4	0.20	640	54	18	105
control 1984	16.2	2.5	8.6	3.0	0.32	690	43	20	95
fertil. 1984	15.1	2.6	10.1	2.4	0.56	740	40	24	85

(b)

K^+	Ca^{2+}	Mg^{2+}	Mn^{2+}	Fe^{2+}	Al^{3+}	H^+	pH $CaCl_2$	base sat.%	K/Ca	Mg/Al (mol)
			$\mu eq \cdot g^{-1}$							
1.9	1.9	0.7	0.8	0.1	64.2	9.1	3.6	5.7	2.0	0.016

*Fertilizer used was 150 g $MgSO_4 \cdot 7H_2O$ per tree. Data from Hüttl (in press).

Table 2. Results of Diagnostic Fertilization Trial* "Baden-Baden 2a": (a) elemental analysis of the current needles of a 35-year old spruce stand, autumn 1984; (b) chemistry of the control soil at 20 - 30 cm depth.

(a)

Treatment	N	P	K	Ca	Mg	Mn	Fe	Zn	Al
			$mg \cdot g^{-1}$				$\mu g \cdot g^{-1}$		
control	16.0	2.0	10.0	0.9	0.41	450	58	16	115
fertilized	18.5	2.2	8.6	1.9	0.61	460	51	29	110

(b)

K^+	Ca^{2+}	Mg^{2+}	Mn^{2+}	Fe^{2+}	Al^{3+}	H^+	pH $CaCl_2$	base sat.%	K/Ca	Mg/Al (mol)
			$\mu eq \cdot g^{-1}$							
1.2	0.4	1.0	0.4	0.6	53.0	9.9	3.4	3.9	6.0	0.028

*Fertilizer used was 1700 kg$\cdot$ha^{-1} of [K_2O (12%) + MgO (18%) + CaO (7%)] and Kalksalpeter 1300 kg$\cdot$ha^{-1} of [17% N + CaO]. Data from Hüttl (in press).

were common. Some foliage was reddish-brown, due to infections of needle cast fungi (Butin, pers. comm.). The needle analysis (Table 3) revealed the deficient K supply; the level of all other elements was sufficient. Aluminum content of the needles was low because almost no Al^{3+} ions were available. The damage symptoms (cf. Baule and Fricker 1967) conform to the insufficient K supply, which is clearly demonstrated by the K content of the needles below 3.5 mg.g^{-1} and the extremely low ratio of exchangeable K:Ca in the soil. The needle cast fungi were epidemic in 1983-84, favoured by weather conditions and insufficient K nutrition (see Zöttl and Hüttl 1985). The fertilization produced a significant rise in K content within one vegetation period. The trees in the fertilized plot had dark green needles and vigorous shoots in autumn 1984.

"Immendingen 7" (750 m a.s.l., 750 mm mean annual precipitation) represents another type of damage. The trees have strongly yellow-chlorotic needles of all age classes, including the youngest. Needle analysis data (Table 4) show a weak N nutrition and extreme deficiency of Mn in the control plot. Symptoms and nutritional status coincide very well (cf. Kreutzer 1972). This is easily explained by the soil characteristics: high pH and redox potential of this clay-rich rendzina soil reduces the Mn availability to a minimum. Because of the high Mn fixation capacity of this soil, a foliar Mn chelate spray was applied in spring 1984 to improve the Mn nutrition. The effect was visible in summer 1984: green needles with significantly higher Mn contents (see Table 4).

From the results, it is very clear that damage symptoms which are related to nutritional disorders of the trees disappear following application of the deficient element(s). In our experiments, these positive reactions occurred without changing the impact of other stress factors, which, it seems, do not play a dominant role in our study area. The results confirm our hypothesis in so far as the contribution of a deficient nutrient supply to the development of forest damage is concerned.

DISCUSSION OF RESULTS

There are still many questions left unanswered. Is the lack of nutrients owing mainly to low soil supply or brought about by increased leaching from the needles? Are low soil reserves of available nutrients a result of the chemical composition of the parent rock and natural soil formation processes or caused by higher acid deposition?

No general answers are possible. We must again discuss every disease type or region separately. In the trials "Elzach" and "Baden-Baden" the soils are derived from granite with extremely low contents of Mg and Ca. For similar sites in southern Black Forest spruce stands, significant correlations among soil nutrient supply, tree nutritional status and damage symptoms were found by Zöttl and Mies (1983b). Atmospheric acid deposition, as measured by Mies (see Zöttl 1985), is low and may increase the

Table 3. Results of Diagnostic Fertilization Trial* "Ebnat 2": (a) elemental analyses of the current needles of a 76-year old spruce stand; (b) chemistry of the control soil at 20 - 30 cm depth.

(a)

Treatment	N	P	K	Ca	Mg	Mn	Fe	Zn	Al
			—mg.g^{-1}—			—µg.g^{-1}—			
control	16.3	1.9	2.9	4.9	1.30	715	29	27	<50
fertilized	15.5	1.9	5.1	5.9	1.09	480	46	25	<50

(b)

K^+	Ca^{2+}	Mg^{2+}	Mn^{2+}	Fe^{2+}	Al^{3+}	H^+	pH $CaCl_2$	base sat.%	K/Ca (mol)	Mg/Al
		—µeq.g^{-1}—								
1.4	124	5.0	0.9	0.01	0	5	5.1	95.7	0.022	∞

*Fertilizer used was 800 kg.ha^{-1} K_2O (28%) + MgO (7%). Data from Hüttl (in press).

Table 4. Results of Diagnostic Fertilization Trial* "Immendingen 7": (a) elemental analyses of the current needles of a 20-year old spruce stand; (b) chemistry of the control soil at 20 - 30 cm depth.

(a)

Treatment	N	P	K	Ca	Mg	Mn	Fe	Zn	Al
			—mg.g^{-1}—			—µg.g^{-1}—			
control	10.1	1.2	8.3	6.8	1.67	3	28	24	<50
fertilized	10.3	1.2	8.9	6.3	1.71	20	24	31	<50

(b)

K^+	Ca^{2+}	Mg^{2+}	Mn^{2+}	Fe^{2+}	Al^{3+}	H^+	pH $CaCl_2$	base sat.%	K/Ca (mol)	Mg/Al
		—µeq.g^{-1}—								
1.0	456	4.4	0.06	0.2	0	0	7.4	100	0.0009	∞

*Fertilizer used was 0.25 L/tree Mn-Chelate 7% spray to needles. Data from Hüttl (in press).

leaching of exchangeable Mg and Ca in the rooted soil only
slightly. The loamy soil in trial "Ebnat" has a very high Ca
saturation but low K reserves, probably due to relict components
in the upper horizons from intensively weathered tertiary
sediments. K:Ca antagonism should also reduce the K uptake.
Neither in "Ebnat" nor in the calcareous clay soil "Immendingen"
are negative influences of external acidity input to the ecosystem to be expected under the prevailing air pollution level.

In summary, there is no evidence that a recent drastic change in
soil conditions of the described sites had occurred. What may
have changed is the concentration of photooxidants (e.g., ozone)
due to the increased emission of precursors. In 1983-84 little
white spots and "flecking" of the needles, together with erosion
of the cuticles, were observed - symptoms which could be attributed to ozone (cf. Lang and Holdenrieder 1985). Unfortunately,
measuring of ozone started only in 1984 in our study area (see
Fig. 1) and no previous data are available. If we look for
comparable measuring stations with earlier ozone records there
are two stations in southern Bavaria (Reiter 1983, Attmannspacher
et al. 1984), but their sequences (since 1967 or 1977, respectively) of ozone (and nitrogen oxide) concentration in the
ambient air do not show any significant increase. Therefore, we
can only speculate that the impact of photooxidants has been
augmented in the last few years in our study area.

There is another weak point in the hypothesis concerning the
leaching of ions from the needles (Prinz et al. 1982). The
experiments of Krause et al. (1983) showed an increase of leached
Mg in Norway spruce after fumigation with high O_3 concentrations
(300 and 600 $\mu g.m^{-3}$), but no good dose-response relationship.
The results of ozone fumigation of Scots pine (Skeffington and
Roberts 1985) did not support the leaching hypothesis.
Therefore, the interaction of ozone and acid fog and their
effects on ion leaching still remain to be tested.

CHRONOLOGICAL CHANGES OF THE NUTRITIONAL STATUS

To find out "what has changed", the historic approach is useful,
i.e., comparisons of needle analysis data from recent years and
20 or more years ago. Zöttl and Hüttl (1985) pointed out that
the nutritional status of spruce stands on moraine soils in
southern Baden-Württemberg became drastically worse in terms of K
(current needles, mean value 1961: 9.8 $mg.g^{-1}$ dry weight; 1983:
3.7 $mg.g^{-1}$). In 1983-84 symptoms of K deficiency appeared and Zn
(not analyzed in 1961) had low values (12 - 25 $\mu g.g^{-1}$). Also in
the northern Black Forest recent studies showed decreased K
contents in spruce and silver fir needles (Aldinger, in prep.;
Evers, pers. comm.). The causes for these changes are still
unknown, especially as there are no earlier chemical soil data
available.

The decrease of the Mg status has been indicated by Zöttl and
Mies (1983a) for spruce stands in the southern Black Forest.
Zöttl (1985) described how the spread of the Mg deficiency

symptoms in spruce was evident only since the mid-1970s. Hüttl and Zöttl (1985) documented lower K and Mg contents in silver fir on silicate soils compared with data collected 20 years ago.

The above-cited publications indicate that the N status of Norway spruce and silver fir hardly changed in the last decades and remained near optimum. This corresponds with the weak increase of the N input at most of the measuring stations in our area. Our needle analysis data show no signs of overoptimum N nutrition, as is assumed in red spruce in the northern Appalachians by Johnson et al. (1984). But it must be mentioned that a higher N input is said to contribute to the better growth of pine stands in northern Bavaria (Pretzsch 1985) on soils with very low N reserves due to the frequent litter raking in earlier times.

An improvement of the phosphorous supply within the last 10 - 20 years seems evident from some needle analysis data (Evers, pers. comm.; Aldinger, in prep.). This is especially striking for the northern Black Forest where P fertilization has always produced a good response on acid sandstone soils. An explanation may be higher microbial P mineralization in these humus-rich soils caused by a priming effect of higher N input and warmer climate in some recent years.

Changes are also seen in the Pb contents. A comparison of needle analysis data over several years since 1975 from the same sites showed how Pb decreased after the last reduction of Pb in gasoline to 0.15 $g.L^{-1}$ in 1976 (Zöttl and Mies 1983a).

CONCLUSIONS

It can be seen from the results of the diagnostic fertilization experiments and sequences of needle analysis data that nutrient deficiencies play a major role in the development of particular forest damage types. This can be caused by low levels of available nutrient reserves in the rooting substrate and losses by leaching from the needles. Antagonisms between ions strongly influence nutrient uptake, but no direct effects of potentially toxic ions like Al^{3+} seem to be involved. Quick correction of the damage symptoms is possible by appropriate fertilizer application.

In forest practice, specific fertilization can be recommended at least as a tool to mitigate several decline phenomena. But fertilization cannot be seen as a general therapy for forest decline, especially not when there are conditions of high concentrations of pollutants or when the nutrient supply is not involved in the decline phenomena. On acid soils with raw humus, liming will help to buffer high acidity. This has been demonstrated largely in Germany. When liming is practised, a harmonic nutrient supply must be obtained; thus, additional rates of Mg and K fertilizer are recommended.

Acknowledgement

I would like to thank the Federal Ministry for Science and Technology (BMFT) for financial support.

REFERENCES

Attmannspacher W, Hartmannsgruber R, Lang P (1984) Langzeittendenzen des Ozons der Atmosphäre aufgrund der 1967 begonnen Ozonmessreihen am Meteorologischen Observatorium Hohenpeissenberg. Meterolog Rundschau 37: 193-199
Baule H, Fricker C (1967) Die Düngung von Waldbäumen. BLV Verlagsges, München
Hüttl R (in press) "Neuartige" Waldschäden und Nährelementversorgung von Fichtenbeständen (Picea abies Karst.) in Südwestdeutschland. Freiburger Bodenkundl Abh 16
Hüttl R, Zöttl HW (1985) Ernährungszustand von Tannenbeständen in Süddeutschland - ein historischer Vergleich. Allg Forstz 40: 1011-1013
Johnson AH, Friedland AJ, Gregory RA, Karenlampi L (1984) Winter damage to foliage as a factor in red spruce decline. Can J For Res 14: 963-965
Krause GHM, Jung KD, Prinz B (1983) Neuere Untersuchungen zur Aufklärung immissionsbedingter Waldschäden. VDI-Berichte Düsseldorf 500: 257-266
Kreutzer K (1972) Die Wirkung des Manganmangels auf die Farbe, die Pigmente und den Gaswechsel von Fichtennadeln (Picea abies Karst.). Forstwiss Centralbl 91: 80-98
Lang KJ, Holdenrieder O (1985) Nekrotische Flecken an Nadeln von Picea abies - ein Symptom des Fichtensterbens? Eur J For Pathol 15: 52-58
Manion PD (1981) Tree disease concepts. Prentice-Hall, Englewood Cliffs, p 399
Mohr H (1984) "Baumsterben" als pflanzenphysiologisches Problem. Biologie in unserer Zeit: 103-110
Obländer W, Hanss A (1985) Zwischenbericht über Schadstoffmessungen in Waldgebieten Baden-Württembergs. LFU Bericht 97/85
Pretzsch H (1985) Wachstumsmerkmale Oberpfälzer Kiefernbestände in den letzten 30 Jahren. Allg Forstz 40: 1122-1126
Prinz B, Krause GHM, Stratmann H (1982) Waldschäden in der Bundesrepublik Deutschland. LIS-Berichte, Essen 28: 1-154
Rehfuess KE (1985) On the causes of decline of Norway spruce (Picea abies Karst.) in Central Europe. Soil Use and Management 1: 30-32
Reiter R (1983) Basiserarbeitung zum Problem "Waldschäden im bayerischen Nordalpenraum". Bayer Staatsmin Landesentwicklung Umweltfragen Materialien 28
Schütt P, Koch W, Blaschke H, Lang KJ, Schluck HJ, Summerer H (1983) So stirbt der Wald. BLV Verlagsges, München
Skeffington RA, Roberts TM (1985) The effects of ozone and acid mist on Scots pine saplings. Oecologia 65: 201-206
Zöttl HW (1985) Waldschäden und Nährelementversorgung. Düsseldorfer Geobot Kolloq 2: 31-41

Zöttl HW, Hüttl R (1985) Schadsymptome und Ernährungszustand von Fichtenbeständen im südwestdeutschen Alpenvorland. Allg Forstz 40: 197-199

Zöttl HW, Mies E (1983a) Die Fichtenerkrankung in den Hochlagen des Südschwarzwaldes. Allg Forst-Jagdz 154: 110-114

Zöttl HW, Mies E (1983b) Nährelementversorgung und Schadstoffbelastung von Fichtenökosystemen im Südschwarzwald unter Immissionseinfluss. Mitt Dtsch Bodenkundl Ges 38: 429-434

DECLINE AS A PHENOMENON IN FORESTS: PATHOLOGICAL AND ECOLOGICAL CONSIDERATIONS

P.D. Manion

State University of New York, College of Environmental Science and Forestry, Syracuse, NY 13210, U.S.A.

ABSTRACT

Forest decline is a popular topic of the 1980s. Forest and tree decline was likewise a popular topic 20 to 30 years ago. What is forest decline and what makes it a popular topic at certain points in time and a neglected topic at others? This paper will attempt to provide some perspective on the decline topic by comparing and contrasting various concepts of decline. Forest decline and acidic deposition are often drawn together. An understanding of the possible role of acidic deposition as a causal factor in forest decline can only come from consideration of the interactions of acidic deposition with other biotic and environmental factors on tree health.

INTRODUCTION

In the 1980s we have a heightened awareness of industrial pollution in our environment. Acid rain is a topic at the forefront of public attention. Acid rain is only one segment of a larger air pollution topic involving heavy metals (primarily lead), gaseous air pollutants (ozone, sulphur dioxide, and others), nutrient deficiencies (primarily magnesium), and aluminum toxicity associated with soil acidification.

Awareness of pollution, in my opinion, has often generated a cause looking for an effect. Dying and dead forest trees provide an obvious place to begin. However, the complexity of the forest does not lend itself to simple cause-effect models, and this is indeed the case with all ecosystems.

Forest declines are not new or unusual. Over the past 100 years forest pathologists have been only partially successful in properly understanding the causes of the many tree decline and death situations that have occurred. In spite of these complexities and inadequacies, there has developed a body of tried and tested concepts to deal with these forest disease problems. This knowledge revolves around concepts of disease. These concepts are not universally agreed upon nor well understood by all forest pathologists. They likewise are not necessarily always accepted by other scientists. This situation should be recognized by anyone using or criticizing the concepts.

The foundations of the present day decline models were established in the 1960s by Sinclair (1964, 1966) and Houston (1967) although some of the ideas are traceable to their predecessors 20 years earlier. They and their associates were working on a region-wide flareup of decline diseases on northern hardwoods such as ash dieback, oak declines, birch dieback, maple declines, and beech bark disease. Similar problems were developing in other regions, such as littleleaf disease and sweetgum blight in the south and pole blight in the west. These major disease problems could not be explained on the basis of simple biotic disease concepts. No primary pathogens could be found despite detailed searches. These diseases had common symptoms, including stressed appearance, growth reductions, and gradual crown dieback eventually culminating in death. Predisposing stress agents, both biotic and abiotic, were also common features shared by these diseases.

Houston (1967) developed a concept of stress-triggered diseases in which healthy trees affected over time by stress become altered. Altered trees are colonized by organisms of secondary action which contribute to eventual death. This concept has been further developed and described over the years (Houston 1973, 1974, 1981a,b, 1984). The key features involve infection by weak pathogens (organisms of secondary action) that can lead to disease and contribute to mortality only after the host is altered (weakened) by some other stress. This interpretation, with its origins in basic plant pathology, treats tree decline very much like other diseases. Traditional pathological concepts involve an interaction of the plant, the pathogen, and the environment, often depicted as the disease triangle. The decline disease, using this model, has a large and specific requirement for an environmental stress.

Another interpretation of Houston's model suggests that declines are very different from standard plant diseases. If one assumes that the environmental components of a pathogen-based disease influences only the pathogen and the infection process, then Houston's decline model differs because the environmental component of a decline primarily affects the host thereby permitting weakly pathogenic organisms to attack.

The difference in the two interpretations is based on differences in interpretation of the disease triangle. I assume that the environment affects the plant and pathogen before, during and after initiation of disease. Houston assumes that host condition is not a requisite for disease.

Sinclair (1964, 1966) developed a concept that involved a three-step system of predisposing factors, inciting factors, and contributing factors. The stress triggering features and the organisms of secondary action features of the Houston concept are the inciting and contributing factors of the Sinclair concept. Predisposing factors account for the differences in response to inciting and contributing factors from one place to another. For example, oak decline triggered by drought in the northeast is most likely to occur on ridge tops where poor soil conditions and other long-term environmental stresses predispose the trees to effects of drought. Houston considers the localized occurrence of biotic and abiotic stress agents as the factor affecting differences from one place to another.

The Sinclair model describes a specifically ordered three factor system to produce death. No single agent could be found to alone account for the disease. A number of biotic and abiotic factors in combination produce the problem on sites where environmental factors have already predisposed the trees. Some factors like drought can play more than one role in declines.

These two models are quite similar. In both, the important role of adverse environmental factors and the contributions of secondary pathogens combine to produce deterioration and death of trees. Since disease concepts developed initially around the primary role of the pathogen as the causal agent, the absence of a single pathogen and the major role of environmental stress makes it difficult for some to consider these decline models as representing "disease".

Although these models provided a way of considering some very puzzling tree mortality problems, many people considered them simply as interim models for categorizing diseases that will eventually be better understood as simple pathogen cause-effect systems once we have done the necessary research to find and identify the actual cause.

I find these models very useful. They suggest that the factors involved in a tree decline do not have to be the same from one place to another (Manion 1981). The stress factors are to some degree interchangeable. The lack of consistency of causal factors from one place to another make tree declines quite different from other diseases. The initial foundations of disease etiology were based in Koch's Postulates for proof of pathogenicity. The first step in Koch's postulates states that one should establish a constant association of a suspected causal agent with the diseased host. Declines are based on interchangeable stress factors not pathogens.

Tree decline problems often have an apparent relationship to tree age. Therefore it seems appropriate to at least consider that declines may be a part of natural succession (Manion 1981).

The tree decline models enable us to better interpret a number of diseases that others would characterize in simple "cause-effect" systems. The use of decline concepts in considering the red spruce decline suggests that others also find them useful (Johnson and Siccama 1984).

It is appropriate to point out that other disciplines have developed their own models to describe tree and forest deterioration. Forest canopy dieback is described under a synchronous cohort senscence model by Muller-Dombois (1983). This model developed to interpret a tree deterioration problem of Hawaiian and other Pacific island forests that was difficult to interpret using a stress-triggered disease model.

The synchronous cohort senescence model involves the development of a population of trees as a cohort following some type of disturbance. The individuals in the population develop together, mature together, and likewise begin to senesce together. A sudden perturbation causes the population to deteriorate synchronously. This stressed population is attacked by a number biotic agents which contribute to and hasten the eventual death.

The model describes the Sinclair decline model in different terms. The cohort of plants that begins to senesce together develop to that point because of the predisposing factors of the site on which they are growing. The perturbation is the inciting factor, and the biotic agents that attack the stressed trees are the contributing factors.

The unique feature of the cohort senescence model is the idea that there is more than just common environment affecting the trees. Tree populations may originate from a common genotypic base following specific events. Lorimer (1980) suggests that what might look like an all-aged forest is really made up of a series of groups of trees with common origin following various kinds of disturbances. Strong local and regional winds, ice storms, drought, and fires are some of the natural means for synchronizing cohorts. Logging, fires and land abandonment are other means of synchronizing cohorts. The cohort population may be very small, as in the case of a population developing in the gap provided by the death of a single dominant tree, or it may be large as would occur after some major disturbance like spruce budworm that generates a synchrony to the balsam fir-spruce stands in about 40-year cycles.

Previous declines may also be a means of synchronizing the tree population. Red spruce in the northeast have been synchronized by declines in the past. Fox (1985) described a blight of the Adirondack spruce that killed one third to one half of the mature trees. It was so severe that needles fell from the trees while still green. Shedding of green foliage is a symptom that is sometimes considered unique to the current decline problems in Germany. Hopkins (1901) and Graves (1899) likewise described spruce blight episodes in the northeast during the nineteenth century. The accepted opinion was that spruce beetle was the cause, but drought and hard winters were also considered. Death was usually confined to the larger mature trees but Hopkins (1899) reported on a spruce mortality problem in West Virginia that involved both large and small trees. The spruce beetle is not mentioned in this mortality.

The most recent synchronizing event for red spruce, other than the current episode, was centered around the drought period of the 1950s which also spurred the development of the hardwood decline concepts described above (Oosting 1951; Mook and Eno 1956).

Drought periods are a common denominator of many declines of the past and present. Another feature that is prominent in declines of trees in northern climates is winter injury. The review of the literature on effects of acidic deposition on forest ecosystems in northeastern United States enumerates a number of recent winter injury episodes (Burgess 1984). Friedland et al. (1984) recognize winter damage as a component of the current decline problem. Curry and Church (1952) likewise describe winter drying of spruce in the Adirondacks during the winter of 1947-48. Winter injury has been proposed as a major component of decline of Alaska cedar in southern Alaska (Shaw et al. 1985).

Drought and winter injury are capable of causing problems that are easy to understand as simple cause-effect models. These same events also fit in as predisposing or inciting events in the decline or cohort senescence models. The same thing could be said for many of the other features of the decline model. There are no absolute lines separating declines from simple injury or standard disease models. It is largely a matter of judgement based on how intense or prolonged the environmental perturbation is judged to be or how effective the biotic agent is as a pathogen.

The lack of demarcation between a decline and simple biotic disease is illustrated by the oak decline system. Wargo et al. (1983) summarize the oak decline in a recent Forest Insect and Disease Leaflet. Oak wilt caused by Ceratocystis fagacearum (Bretz) Hunt. kills red oaks and is clearly differentiated from stress-triggered declines by leaf wilt over the entire crown, rapid progression of symptoms, and the presence of the primary pathogen. In red oak, decline and wilt are readily differentiated. In contrast, oak wilt of live oak in Texas has many of the characteristics of a decline disease including an apparent absence of the pathogen. Only by isolation from trees with incipient wilt can the pathogen be recovered (Lewis and Olivera 1979). This disorder initially was called live oak decline because the pathogen was not recognized at first and because of its similar appearance to other declines.

Armillaria mellea (Vahl ex Fr.) is a common organism of secondary action (contributing factor) in almost all declines. Recent examination of red spruce decline in the northeast revealed A. mellea, but the occurrence was inconsistent enough to suggest that it was not the principal agent involved (Carey et al. 1984). This root rot pathogen was less prevalent at higher altitudes than in the lower elevation forests. Other root rot organisms may be involved in the decline of spruce and other species but they are generally more difficult to identify than A. mellea. If one understands the role that A. mellea plays in a tree decline, then it does not have to be considered the sole root infecting agent of all dying trees. If one attempts to interpret A. mellea in a classic disease model, then one has to reject this organism as the pathogen responsible for the disease.

The same type of reasoning can be applied to the role of air pollutants. There are locations where these are primary factors producing injury and mortality. The Sudbury and Wawa regions of Ontario are excellent examples of forest destruction caused by air pollution. Trees nearest the pollutant source are killed, but on the edges of the zone of destruction, loss of tree health is associated with a subtle interaction of pollutants with other factors. In other regions, deterioration and mortality may be associated with air pollution and a complex of other factors such as bark beetles (Cobb et al. 1968).

An additional complicating factor in characterizing injury, typical disease, or decline-contributing roles for pollutants, is the varied syndrome of pollution effects on plants. Quite often the symptoms are not either particularly or absolutely diagnostic. Fluoride accumulates to toxic levels in plant tissues, causing local lesions such as tip burn in gladioli, so that chemical analysis of these tissues can determine excess levels of fluoride. Many other factors produce visual symptoms that are similar to those produced by air pollutants. Therefore pollution-related problems can be diagnosed indirectly by correlations with fumigation episodes, etc. Further complications arise since the criteria for indirect diagnosis of air pollution effects in the forest are not clearly defined.

Direct and indirect effects of acidic deposition are implicated in the decline syndrome model for spruce (Johnson and Siccama 1984). Recent reviews acknowledge possible roles for acidic precipitation in the forest but generally conclude that it is impossible to determine if forests are being affected (Burgess 1984; Morrison 1984; SAF 1984). Consideration of this factor like gaseous air pollutants is complicated by inconsistencies from one location to another. Acidic deposition should not be expected to explain all declines. Acid deposition effects on trees in the field are not clearly defined and therefore it is impossible to clearly distinguish such effects from among those due to other factors.

There are no absolutes for the recent forest decline problems of northeastern United States and Europe. One common feature of these current problems is the inadequate attention paid to simple biotic explanations. Schutt (1979) describes the importance of the scale insect and the fungus Nectria coccinea (Persoon ex Fries) Fries in beech mortality in Germany. Fir dieback associated with A. mellea and Heterobasidion annosum (Fr.) Bref. is still a problem. A poorly understood pathological wetwood condition of fir is also described in the European literature (Schutt 1979; Berchtold et al. 1981). These diseases could be a part of the present "Waldsterben". Schutt (1984) presents arguments for and against the possible role of infectious agents but there is no reason to assume that the forests are all affected by the same agent.

The role of pathogens in red spruce decline in high elevation forests in the northeastern United States is summarized by the statement, which regularly occurs in current literature, that no primary pathogens were present (Siccama et al. 1982). This is a misleading statement in that it assumes, incorrectly, that a "thorough" search has been made for primary agents capable of causing the problem. It is inappropriate to assume that no primary agents are involved (Burgess 1984).

A recent article from Germany identified rickettsia-like bacteria and mycoplasma-like structures in declining beech from the Black Forest and Saxony regions (Parmesauran and Liese 1984). These types of agents are capable of causing diseases alone or in combination with other factors. They are not readily diagnosed on the basis of external symptoms since these symptoms are similar to those produced by other disease agents. The consistent occurrence of mycoplasma-like structures in declining ash in New York suggests that this decline problem may have a significant biotic component (Matteoni and Sinclair 1985).

A final point that should be considered in interpreting the forest decline problems of today is to recognize that the agents that cause diseases, the factors that cause injury, and the stress and organisms that result in decline problems do not occur one at a time in the forest. They are all impinging on the association of tree species of different ages and in various successional stages. Therefore, it is inappropriate to assume that the deterioration of the different species in many different places is caused by some common denominator. It is also inappropriate to assume that all the dieback and mortality in one forest is caused by the same agent. The "fruit salad" of deteriorating, declining, and dying trees should not all be squeezed into a generic "fruit juice". To properly identify the unique disease-causing conditions, including atmospheric deposition effects, it is essential to properly characterize each of the specific conditions and events that interact to make up the total.

Acknowledgements

The author thanks Drs. W.A. Sinclair, D.R. Houston, D. Mueller-Dombois, D.J. Raynal, and R.A. Zabel and Mr. Hans Fuernkranz for their technical review and comments on drafts of this manuscript.

REFERENCES

Berchtold R, Alcubilla M, Foerst K, Rehfuess KE (1981) Standortskundliche studien zum tannesterben: Kronen-und stammerkmale von probebaumen aus funf bayerischen bestanden. Eur J For Pathol 11: 233-243

Burgess RL (ed) (1984) Effects of acidic deposition on forest ecosystems in the northeastern United States: An evaluation of current evidence. State University of New York, College of Environmental Science and Forestry, Syracuse, NY 13210, pub ESF 84-016, p 108

Carey AC, Miller EA, Geballe GT, Wargo PM, Smith WH, Siccama TG (1984) Armillaria mellea and decline of red spruce. Plant Dis 68: 794-795

Cobb FW, Wood AL, Stark RW, Parameter JR (1968) Photochemical oxidant injury and bark beetle (Coleroptera: Scolytidae) infestation of ponderosa pine. Hilgardia 39: 121-152

Curry JR, Church TW (1952) Observations on winter drying of conifers in the Adirondacks. J For 50: 114-116

Fox WF (1985) The Adirondack black spruce. Forest Commission, State of New York annual report for 1894, p 10-14

Friedland AJ, Gregory RA, Karenlampi L, Johnson AH (1984). Winter damage to foliage as a factor in red spruce decline. Can J For Res 14: 963-965

Graves HS (1899) Practical forestry in the Adirondacks. USDA Forest Service Bull No. 26, p 84

Hopkins AD (1899) Report of investigation to determine the cause of unhealthy condition of spruce and pine from 1880-1893. West Virginia Agr Exp Sta Bull 56: 197-254

Hopkins AD (1901) Insect enemies of spruce in the northeast. USDA Div Ento Bull No 28 (new series)

Houston DR (1967) Dieback and decline of northeastern hardwoods. Trees 28: 12-14

Houston DR (1973) Diebacks and declines: Diseases initiated by stress, including defoliation. Int Shade Tree Conf Proc 49: 73-76

Houston DR (1974) Diagnosing and preventing diebacks and declines of urban trees: Lessons from some forest counterparts. Morton Arbor Qtly 10: 55-59

Houston DR (1981a) Stress triggered tree diseases. The diebacks and declines. USDA Forest Service NE-INF-41-81, p 36

Houston DR (1981b) Some dieback and decline diseases of northeastern forest trees: Forest management considerations. In: Proc Nat Silviculture Workshop. Roanoke, VA, USDA Forest Service Division of Timber Management, Washington, DC, p 248-265

Houston DR (1984) Stress related to diseases. Arbor J 8: 137-149

Johnson AH, Siccama TG (1984) Decline of red spruce in the northern Appalachians: Assessing the possible role of acid deposition. Tappi J 67 (1): 68-72

Lewis R, Olivera FL (1979) Live oak decline in Texas. J Arboric 11: 241-244

Lorimer CE (1980) Age structure and disturbance history of a southern Appalachian virgin forest. Ecology 61: 1169-1184

Manion PD (1981) Tree disease concepts. Prentice-Hall, Inc. Inglewood Cliffs, NJ, p 399

Matteoni JA, Sinclair WA (1985) Role of the mycoplasmal disease, ash yellows, in decline of white ash in New York State. Phytopath 75: 355-360

Mook PV, Eno HG (1956) Relation of heart rots to mortality of red spruce in the Green Mountains National Forest. USDA Forest Service NE Forest Exp Sta For Res Note No 59, p 2

Morrison IK (1984) Acid rain; a review of literature on acid deposition effects in forest ecosystems. For Abstr 45: 483-506

Mueller-Dombois D (1983) Canopy dieback and successional processes in Pacific forests. Pacific Science 37: 317-325

Oosting HJ (1951) A comparison of virgin spruce-fir forests in the northern and southern Appalachian system. Ecology 32: 84-103

Parameswaran VN, Liese W (1984) Uber das vorkommen von rickettsien-ahnlichen bakterien und mykoplasmen-anlichen organismen in buchen aus waldschadens gebieten. Eur J For Pathol 14: 373-377

SAF (1984) Acidic deposition and forests. Soc American Task Force Report, p 48

Schutt P (1979) Buchenund tannensterben, zwei altbekannte waldkrankheiten von hochster aktualitat. Mitteilungen der Deutschen Dendrologischen Gesellshalt 71: 229-235 [Review of Plant Path 63: 80]

Schutt P (1984) Der wald stirbt and stress. C Bertelsmann Verlag GMbH, Munchen, p 264

Shaw III CG, Eglitis A, Laurent TH, Hennon PE (1985) Decline and mortality of Chamaecyparis nootkatensis in southeastern alaska, a problem of long duration but unknown cause. Plant Dis 69: 13-17

Siccama TM, Bliss M, Vogelmann HW (1982) Decline of red spruce in the Green Mountains of Vermont. Bull Torrey Bot Club 109: 163-168

Sinclair WA (1964) Comparison of recent declines of white ash, oaks and sugar maple in the northeastern woodlands. Cornell Plantations, Ithaca, NY 20: 62-67

Sinclair WA (1966) Decline of hardwoods: Possible causes. Proc Inter Shade Tree Conf 42: 17-32

Wargo PM, Houston DR, LaMadeline LA (1983) Oak decline. USDA Forest Service Forest Insect and Disease Leaflet 165, p 8

THE USE AND LIMITATIONS OF DENDROCHRONOLOGY IN STUDYING EFFECTS
OF AIR POLLUTION ON FORESTS

E.R. Cook

Tree-Ring Laboratory, Lamont-Doherty Geological Observatory
of Columbia University, Palisades, NY 10964, U.S.A.

ABSTRACT

The annual ringwidths of trees can be used to search for hypothesized air pollution effects on forests. This search is extremely complicated by the inherent statistical properties of ringwidth data and the high level of uncertainty regarding the sources of variance observed in the ringwidths. A linear aggregate model for ringwidths which highlights the general classes of variance that may be found in a tree-ring series is described. Dendrochronological principles and techniques that can be used to create a tree-ring chronology that is suitable for rigorous statistical analysis and hypothesis testing are described. An analysis of a red spruce tree-ring chronology indicates that a decline in ringwidths since 1968 cannot be explained by a linear temperature response model.

INTRODUCTION

The recent decline of red spruce (Picea rubens Sarg.) in northern New York and New England (Siccama et al. 1982; Scott et al. 1984) has prompted considerable speculation and research concerning the possible impact of acid deposition on tree growth. Aside from the obvious visual symptoms of decline such as foliar dieback, a marked reduction in radial increment since the mid-1960s is apparent in many trees (Johnson and Siccama 1983). Although this radial growth reduction appears to be anomalous compared to the radial growth of previous years, numerous uncertainties must be addressed before this phenomenon can be interpreted as a pollution-caused event. Growing in an uncontrolled environment, trees respond to numerous natural growth modifying influences that are both beneficial and detrimental to annual ring formation. These influences represent the background variability of the trees' operational environment into which pollutants have been introduced. The presence of this potentially confounding background variability makes the identification of pollution effects very difficult and uncertain.

The science of dendrochronology and the subdiscipline of dendroclimatology (Fritts 1976) have developed many principles and techniques for analyzing tree-ring data. The purpose of this paper is to describe a rational approach to tree ring/pollution, or dendropollution, research given the limitations of the data and prior knowledge, based on dendrochronological principles. This approach will assume from the start that the tree-ring series being analyzed have been precisely dated using accepted cross-dating techniques (Fritts 1976).

In order to clarify the problem of indentifying pollution-related changes in ringwidths from other factors, a linear aggregate model will be described for a hypothetical ringwidth series.

A LINEAR AGGREGATE MODEL FOR RINGWIDTH SERIES

Consider a hypothetical annual ringwidth series as an aggregation of several unobserved subseries representing the sources of variance that may be found in the observed process. Let this aggregated time series be expressed as

$$R_t = G_t + C_t + \delta D1_t + \delta D2_t + \delta P_t + E_t$$

where:
- R_t = the observed ringwidth series measured along a single radius
- G_t = the growth trend associated with increasing age and size of the tree
- C_t = the climatically-related growth variations common to a stand of trees
- $D1_t$ = the variance due to endogenous disturbances, which only affect a small subset of trees in a stand
- $D2_t$ = the variance due to natural exogenous disturbances, which have a standwide impact on radial growth
- P_t = the variance due to anthropogenic pollutants which have a standwide impact on radial growth
- E_t = the random variance unique to each tree or radius

The δ associated with $D1_t$, $D2_t$ and P is a binary indicator of the presence ($\delta = 1$) or absence ($\delta = 0$) of a subseries in R_t for some year or group of years. Thus $D1_t$, $D2_t$ and P_t are not time invariant features of R_t and need not be present at all.

The growth trend, G_t, is a non-stationary process that arises, in part, from the geometric constraint of adding a volume of wood each year to a stem of increasing radius. This constraint suggests that the trend in radial growth should possess a simple exponential decay as a function of time. Unfortunately, the growth trends of trees growing in closed-canopy forests are usually very complex and stochastic because of disturbances and competitive interactions within the forest. Therefore, G_t must be generalized to allow for a variety of linear and curvilinear growth trends of arbitrary slope and shape.

The climatically related subseries, C_t, reflects certain broad-scale meteorological variables that directly or indirectly limit the growth processes of trees in a stand. These variables are assumed to be uniformly important for all trees of a given species when the site characteristics of the stand, such as hydrology, elevation, exposure and soil, are more or less homogeneous. As a common signal in the ringwidths of all trees, C_t could be mistakenly identified as a pollution signal if the recent behaviour of C_t mimics the expected pollution effect on ringwidths. Thus, the effects of climate on ringwidth must be carefully modelled before a pollution effect can be inferred.

The variance accounted for by disturbances $D1_t$ and $D2_t$ can be split into two general classes of disturbance: endogenous and exogenous (White 1979). Endogenous disturbances are caused by factors related to characteristics of the vegetation that are independent of the environment (White 1979). Such disturbances occur when dominant overstory trees senesce, die and topple as a natural consequence of competition, aging and stand succession. In the context of searching for a pollution signal in tree rings, endogenous disturbances can be expected to occur randomly in space and time in forest communities. Thus, the loss of a dominant tree in one section of a stand is not likely to be related temporally or spatially to similar losses at other locations in the stand. This property suggests that endogenous disturbance pulses in tree rings will rarely be synchronous among separated trees in a stand except by chance alone. Thus, the lack of synchrony in ringwidth fluctuations between trees during a hypothesized pollution effect period may be used as evidence for rejecting the presence of a pollution signal.

Exogenous disturbances are caused by natural environmental forces that lie external to and are independent of the vegetation (White 1979). Unlike endogenous disturbances, these disturbances can affect large areas of forest. Some of the important causal agents are fire, windstorms, ice storms, frost damage, disease and insect infestation. Since the areal extent of an exogenous disturbance can be great, the resultant disturbance pulse, $D2_t$, may occur contemporaneously in virtually all trees in a stand. This property presents obvious difficulties for differentiating a pollution-caused ringwidth decline from that caused by a natural exogenous disturbance. Historical documentation of exogenous disturbances in forests may be needed to determine the presence or absence of this confounding source of variance.

The pollution signal, P_t, is assumed to be common to all ringwidth series in the sampled trees of a stand. This assumption could be criticized for being too restrictive in requiring a pollution effect on all samples trees. It could be argued, for example, that the crowns of canopy trees "scrub out" wet and dry atmospheric pollutants before they reach understory trees. If this were the case, then the understory trees might not show a pollution effect. This possibility indicates the need for the stratified sampling of trees based on a priori criteria such as crown class or canopy position.

SOME EXAMPLES AND PROPERTIES OF ACTUAL RINGWIDTH DATA

Figure 1 shows three examples of the kinds of non-stationarity frequently encountered in ringwidth series. The first series (Fig. 1a) is from an open-canopy stand of ponderosa pine (Pinus ponderosa Laws.) from a lower forest border site in New Mexico. It shows the classic negative exponential trend in ringwidth, which is principally a function of increasing tree age and size. The second series (Fig. 1b) is from a closed-canopy stand of red spruce in New Hampshire. It shows the suppression-release effects of gap-phase stand dynamics that are typical in stands of shade tolerant species such as red spruce. The third series (Fig. 1c) is from a closed-canopy stand of eastern hemlock (Tsuga canadensis Carr.) growing in Pennsylvania. The overall ringwidth

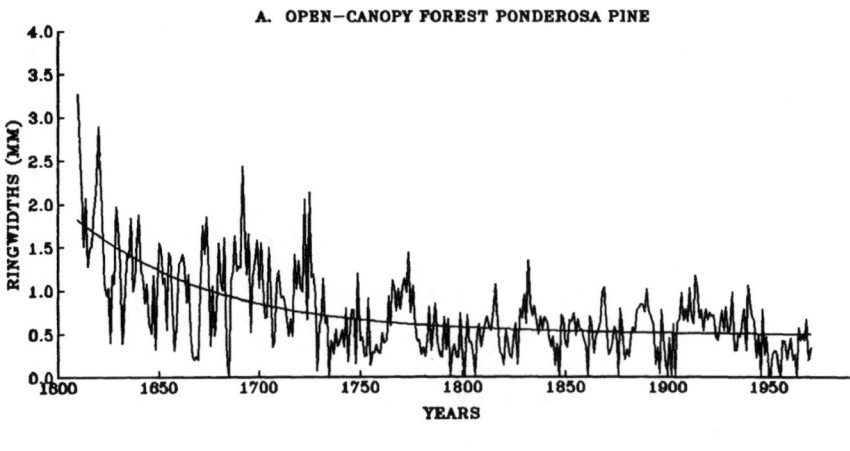
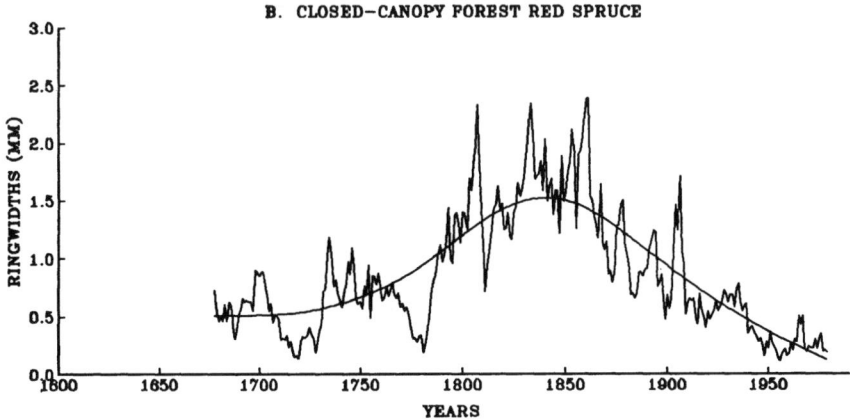
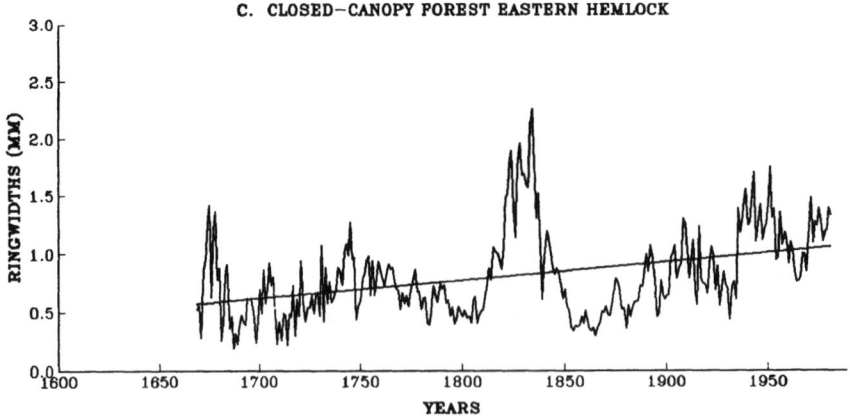

Fig. 1. Three examples of the kinds of non-stationarity frequently found in ringwidth series. The smooth curves are intended to highlight the general forms of the non-stationarity, not model them closely.

trend is positive for the 312 years of record, but erratic, shorter-term fluctuations are present.

These series illustrate that there is a high probability that ringwidth series will show a trend in radial growth rate due to age, size, and stand-related variables. Thus, a ringwidth decline cannot be used as prima facie evidence for abnormal decline in forest productivity or tree vigor due to pollution. In addition, it is not possible to differentiate endogenous disturbances from other sources of common (standwide) variance without additional ringwidth series collected from the stand. For example, the red spruce series in Fig. 1b shows two pronounced periods of suppressed radial growth beginning around 1720 and 1775. Are these periods unique to this tree or common to the stand? The answer is seen in Fig. 2 which shows three additional series from that stand. It is immediately apparent that neither of those suppression periods are common to the other series. Thus, they can be categorized as endogenous disturbance pulses.

When searching for hypothesized pollution effects in tree rings, the creation of a time series mean-value function is recommended, which concentrates the common signal among all series and simultaneously "averages out" the unique information or noise in the individual series. The creation of a statistically valid mean-value function requires that the individual ringwidth series are drawn from the same population of response variables. This criterion will usually be satisfied when the ringwidth series are from a stand of trees having reasonably homogeneous physiographic, edaphic, and hydrologic characteristics. In this case, the principal signal found in all trees may come from broad-scale environmental influences such as climate and, perhaps, natural exogenous disturbances and pollutants.

Returning now to Fig. 2, it is readily apparent that averaging ringwidths together to produce a mean-value function cannot be recommended. Series non-stationarity coupled with large differences in absolute ringwidth between series could produce a response variable mean-value function that is dominated and confounded by stand dynamics effects. The average correlation between these series for the common period 1716-1979 is -0.028, which is clearly not statistically significant. This result suggests that these series were not sampled from the same population of response variables. Yet, when the long-term trends are removed, the average correlation between series rises to 0.27, which is statistically significant at the 99% confidence level. Thus, non-stationarity due to age-size effects and stand dynamics must be removed before the tree-ring series are averaged into a mean-value function. This is the principal purpose of tree-ring standardization (Fritts 1976; Cook 1985).

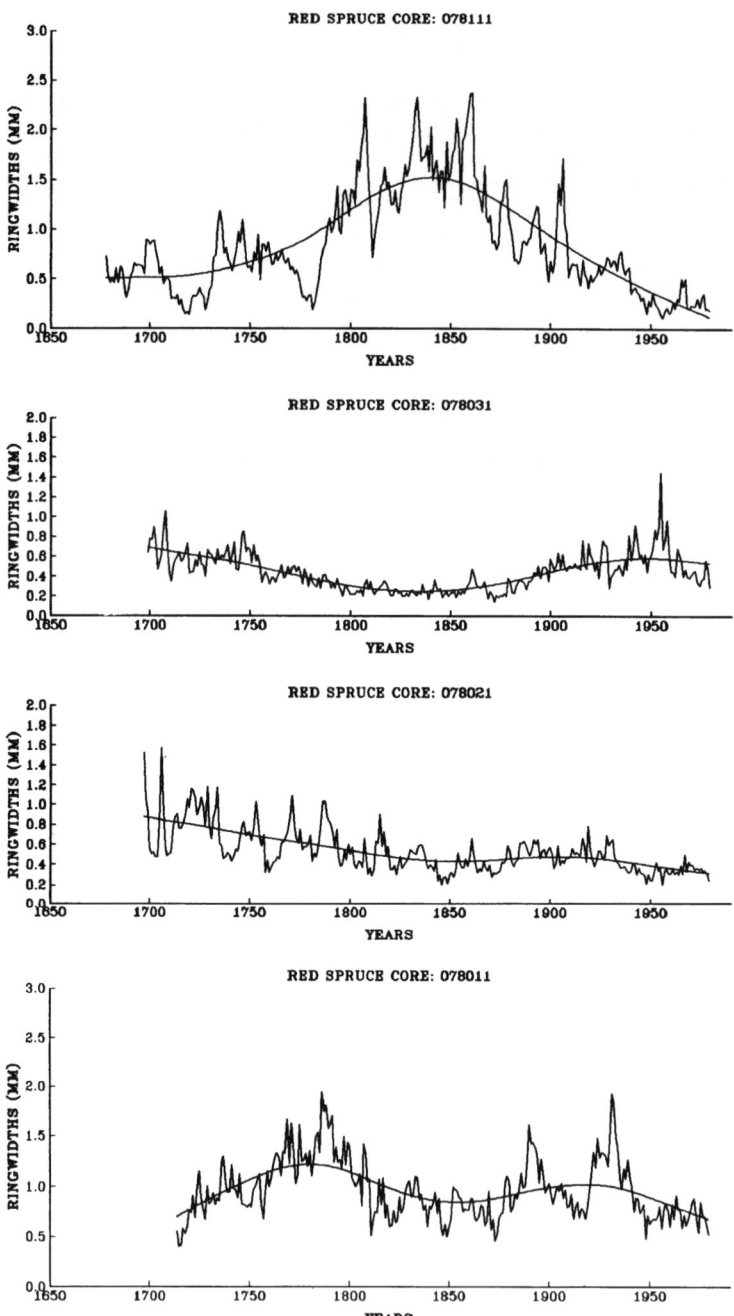

Fig. 2. Four red spruce ringwidth series from the same stand. These series show the low-frequency lack of agreement between series that are largely due to endogenous disturbances and tree histories.

TREE-RING STANDARDIZATION

Tree-ring standardization is designed to reduce the non-stationary ringwidths to a sequence of relative tree-ring indices that are stationary through time. This is accomplished by removing age-size and stand dynamics effects using smooth mathematical curves such as those seen in Figs. 1 and 2.

The relative tree-ring indices are computed as: $I = R_t/G_t$ where I_t is the tree-ring index, R_t is the measured ringwidth, and G_t is the fitted growth curve value, all for year t. Tree-ring indices produced by this procedure have a long-term mean of 1.0 and a variance that is largely stabilized by the division process. In the usual dendrochronological interpretation, the growth curve (G_t) is a sequence of expected ringwidth values that would be produced by the tree if shorter-term fluctuations such as those attributable to climate (I_t) were unchanging through time. Conversely, the I_t can be thought of as fractional departures from the expected growth curve, G_t, which are related to variations in climate and, possibly, pollution.

The principal difficulty in standardizing ringwidths is estimating a satisfactory growth curve. For trees growing in open-canopy environments, simple monotonic functions, such as the negative exponential curve (Fritts 1976), are often satisfactory (see Fig. 1a). Unfortunately, such models cannot be applied to closed-canopy forest trees, as Fig. 1 and 2 clearly show. Orthogonal polynomials (Fritts 1976) and cubic smoothing splines (Cook and Peters 1981; Cook 1985) can be used as data-adaptive methods for estimating the more complex and stochastic growth curves found in closed-canopy tree-ring series. These curves have little theoretical basis and implicity admit the ignorance of tree histories in most, if not all, unmanaged or natural forest stands. In addition, there is the very real danger that these more flexible curves will inadvertently remove some or all of the pollution signal being sought for study if they are allowed to be overly flexible. For this reason, the detrending curves should be quite "stiff" relative to the short-term variations possibly attributable to pollution.

MODELLING THE CLIMATIC SIGNAL IN TREE-RING SERIES

Once the mean-value function has been created, the modelling of the climatic signal within that series can proceed. This exercise is designed to create a prediction model for tree rings from climate, which can be used to test for the intervention of a non-climatic influence, such as pollution, on growth.

Climatic response models are usually based on multiple regression methods, such as stepwise regression analysis, which are empirical and, therefore, not theoretically based. The lack of _a priori_ theory means that a statistically significant regression model can be created by chance alone. Should a largely spurious climatic response model be developed for a tree-ring chronology, then incorrect conclusions may be drawn regarding the presence or absence of a pollution signal.

Spurious regression principally manifests itself through the lack of time stability of the regression model. In dendroclimatology, the time stability of regression equations is routinely tested. This procedure, known as verification (Fritts 1976) is based on withholding actual data from the estimation of the regression model. Once the parameters of the model have been estimated, the withheld data are predicted by the model. If the predictions are sufficiently similar to the actual data, then the time stability of the regression model has been verified, and the model can be used for prediction.

The choice of the verification period in dendropollution studies must be based on prior information regarding the known or suspected beginning of the pollution effect period. For the red spruce decline in New England, this hypothesized intervention date is probably sometime after 1960. Having specified an intervention date, the verification period should begin several years before that date. For example, if the pollution effect period is suspected to have begun in 1965, then the verification period should begin in, say, 1950. The reason for this extended verification period is twofold. First, the verification period is necessary to test the time stability of the regression model as described earlier. Second, the verification period can also be used to reveal significant changes in the relationships between the climate and tree-ring variables that may be reflecting the intervention of a non-climatic agent. However, it may be very difficult to separate the effects of spurious regression from non-climatic (e.g., pollution) effects if the verification period only includes the suspected pollution period. By including earlier tree-ring data in the verification period, which are believed to predate the onset of forest decline, the stability of the regression model can be tested up to the intervention date without the possibly confounding influence of the decline signal. If the regression model verifies up to the intervention date, then it is reasonable to expect the model to verify in the post-intervention period. Should this happen, then the pollution intervention hypothesis can be rejected. Conversely, a lack of model verification in the post-intervention period will provide evidence for a change in some aspect of the growth environment of the tree that may be related to pollution effects.

A lack of model verification in the pre-pollution verification period may doom the development of a meaningful climatic response model for predictive purposes. In such cases, an anomalous decline in ringwidth during the hypothesized pollution impact period cannot necessarily be interpreted as being non-climatic because the trees may still be responding to climate in a non-stationary or threshold sense. That is, some climatic variable such as drought may not have any significant impact on radial growth until it exceeds a critical threshold of severity. For example, the severe 1960s drought over the northeastern United States has been cited as a possible inducing stress mechanism of red spruce decline in New England (Johnson and Siccama 1983).

The possibility of threshold responses to climate suggests that the effects of climatic change on forests (Hepting 1963) could be responsible for red spruce decline. The 1960s and 1970s have been notable for anomalous climatic patterns throughout the northern hemisphere (Lamb 1977). In the United States, a series of extreme winters has occurred since 1975 that has a return time probability of more than 1000 years (Karl et al. 1984). Thus, a greater frequency of extremes in climate that cross some critical threshold for red spruce could explain the decline of that species. Wintertime damage to red spruce needles, which is associated with anomalously warm temperatures for several days in mid-winter, has been reported (Friedland et al. 1984) as part of the decline syndrome. This is a clear temperature threshold phenomenon that needs to be investigated in the context of climatic change.

AN EXAMPLE OF MODELLING THE DECLINE SIGNAL FOR RED SPRUCE

The principles and techniques just described for examining dendropollution hypotheses will now be utilized in an example. The tree-ring series in this example were obtained from a stand of red spruce growing at an elevation of 1150 metres in the Adirondack Mountains of northern New York. This stand is only 27.5 kilometres from Whiteface Mountain where symptoms of red spruce decline have been documented (Scott et al. 1984; Johnson and Siccama 1983). The trees, which were cored in July 1977, are from a highly stratified population of dominant and co-dominant trees. Two increment cores were extracted at DBH from each of 20 trees. The ringwidth series range from 142 years to 297 years long with the majority lying within the 200-250 year class. The ringwidth series were standardized using stiff spline detrending and autoregressive modelling (Cook 1985). The tree-ring chronology mean-value function is shown in Fig. 3 for the time period 1750-1976. The chronology reveals an abrupt and prolonged decrease in the tree-ring indices which appears to have begun in 1968. The pattern and timing of tree-ring index decline with no recovery is consistent with the ringwidth patterns seen in other red spruce stands exhibiting the decline syndrome (Johnson and Siccama 1983).

Monthly mean daily temperature data for the Northern Plateau Climatic Division of New York were used for modelling the climatic signal in the chronology. This record covers the period 1889-1976. The choice of monthly temperatures as candidate predictors of red spruce growth was based on previous dendroclimatic studies of the species in New England (Conkey 1979; Cook 1982). Due to well-documented lag responses of tree growth to climate through physiological preconditioning (Fritts 1976), the monthly temperatures were adjusted into a dendroclimatic year which begins in the previous March and ends in the current September of annual ring formation. This time window spans two complete growing seasons. This resulted in a pool of 19 candidate predictors of tree growth and the loss of one year of data after developing the lagged dendroclimatic year.

Two regression and verification analyses were run: one with existent autocorrelation left in the chronology and one with autocorrelation removed. These will be referred to as the unwhitened and prewhitened chronologies, respectively. The prewhitened chronology was developed by modelling the unwhitened chronology as an order-3 autoregressive (AR) process for the time period 1750-1960. The coefficients of the AR process were then used to prewhiten the entire series up through 1976. This method of prewhitening is based on the premise that the AR model for the 1750-1960 pre-decline period is a time invariant representation of the chronology autocorrelation structure in the absence of any anomalous decline signal. Any residual persistence in the post-1960 period may be due to the decline signal. The coefficients of the fitted AR model are: ϕ = .232, ϕ = .062 and ϕ = -.138. The prewhitened chronology is shown superimposed on the unwhitened chronology in Fig. 3. There is little difference between the two even in the anomalous post-1967 decline period. It is clear that the persistently below average 1968-1976 period is not consistent with the long-term persistence structure of the chronology.

Stepwise regression analysis was used to select a subset of predictors of tree growth. The time period used in the regression analysis was 1890-1950. This allowed data in the 1951-1976 period to be used for verification tests. The minimum F-level for entering variables into the regression equation was set at 2.0. The regression results are given in Table 1. Each model explains over 50% of the tree-ring chronology variance and is highly significant statistically. The prewhitened chronology model performs somewhat better in terms of R^2 and model parsimony. However, except for an extra variable in the unwhitened model, the models are extremely similar. Note the very important contribution of antecedent temperature to radial growth. These variables consistently show up in a posteriori dendroclimatic models of red spruce growth (e.g., Conkey 1979; Cook 1982).

Table 1. The stepwise regression results for the unwhitened and prewhitened chronologies. The standardized regression coefficients (b) and their standard error (se) are given for the temperature variables entered into the model. The fractional variance explained by the regression equation (R^2) and the residual degrees of freedom (DF) are also given below.

Variable	b	(se)	b	(se)
Previous July	-.261	(.106)	-.232	(.100)
Previous August	-.300	(.107)	-.271	(.099)
Previous October	.180	(.103)	-----	
Previous November	.202	(.107)	.293	(.101)
Previous December	.291	(.099)	.337	(.094)
Current May	.155	(.103)	.214	(.098)
Current September	.209	(.105)	.213	(.097)
R^2	.502		.538	
DF	53		54	

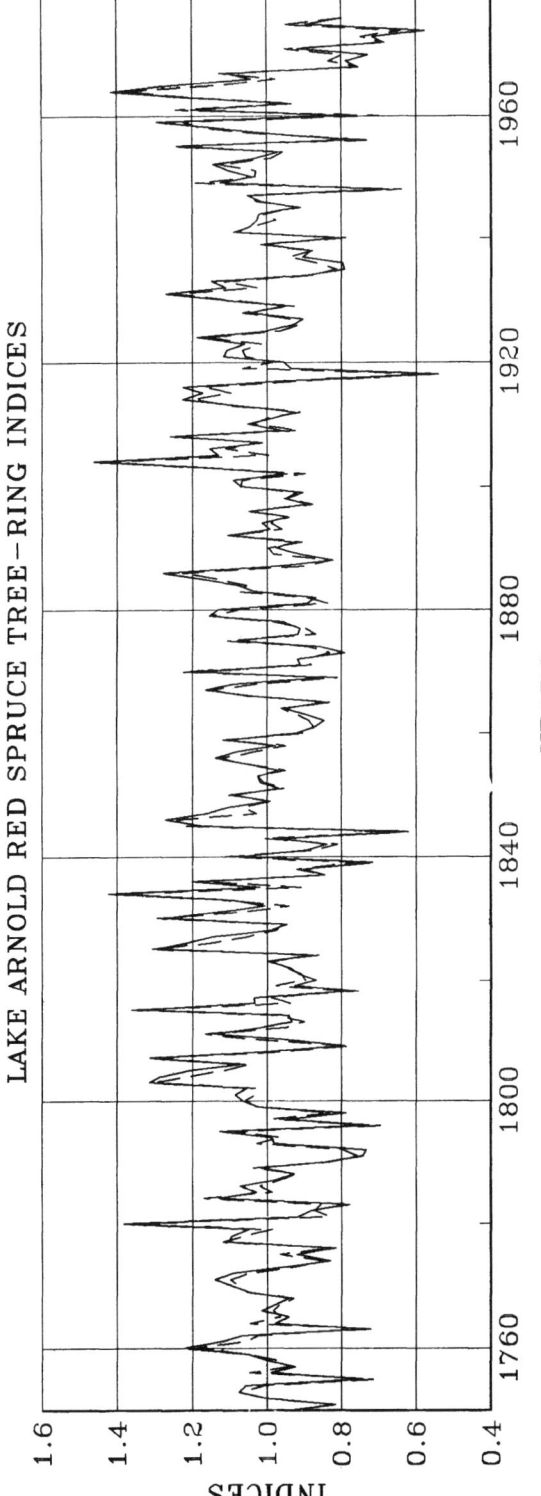

Fig. 3. The Lake Arnold red spruce chronology used in the climatic modelling example. The solid-line plot is the chronology with existent autocorrelation left in. The dash-line plot is the same chronology after the autocorrelation has been removed by autoregressive time series modelling.

The actual and estimated tree-ring indices are shown for each model in Fig. 4, including the model predictions in the 1951-1976 verification period. The models do a reasonably good job of predicting the indices up to 1967. After 1967, the predictions are noticeably inferior especially in estimating the mean level of the actual indices. These observations are confirmed by the verification test statistics in Table 2. In both the 1951-1960 and 1951-1967 periods, the difference between means ($\Delta \bar{x}$) is a very small and not statistically significant. In addition, the product-moment correlations (r) are statistically significant (p = 0.05) and the reduction of error statistics (RE) are positive indicating some predictive skill for each model. The 1968-1976 period shows a virtual reversal of verification results. The $\Delta \bar{x}$ for each model is significantly different from zero, indicating a substantial bias in the predictions of tree-ring index. The direction of bias is consistent with an abnormal decline in ringwidth as part of the red spruce decline syndrome. The r and RE also show degradation in both models. These results indicate that the observed ringwidth decline of red spruce in this stand cannot be explained by the verified climatic response models developed here. As a result, a change in the growth environment of these trees has probably occurred which had a standwide impact on the sampled trees by 1968.

This conclusion indicates that pollution is still a viable hypothesis for explaining red spruce decline at this site. However, climatic change and threshold response to climate cannot be ruled out at this point. In this regard, the wintertime foliar damage hypothesis (Friedland et al. 1984) needs to be tested within the context of climatic change.

Table 2. The verification test results of the two regression models. The difference between the means of the actual and predicted indices ($\Delta \bar{x}$), the product-moment correlation coefficient (r), and the reduction of error (RE) are estimated for three verification periods. The asterisks indicate those statistics which are significantly different from zero (p = 0.05) using a one-tailed test. The RE has no significance test, but an RE > 0 is an indication of some prediction skill when coupled with a significant r.

Verification period	N	Unwhitened			Prewhitened		
		$\Delta \bar{x}$	r	RE	$\Delta \bar{x}$	r	RE
1951-1960	10	.005	.585*	.322	-.011	.600*	.346
1951-1967	17	.046	.539*	.347	.032	.533*	.328
1968-1976	9	-.232*	.290	.148	-.184*	.490	.228

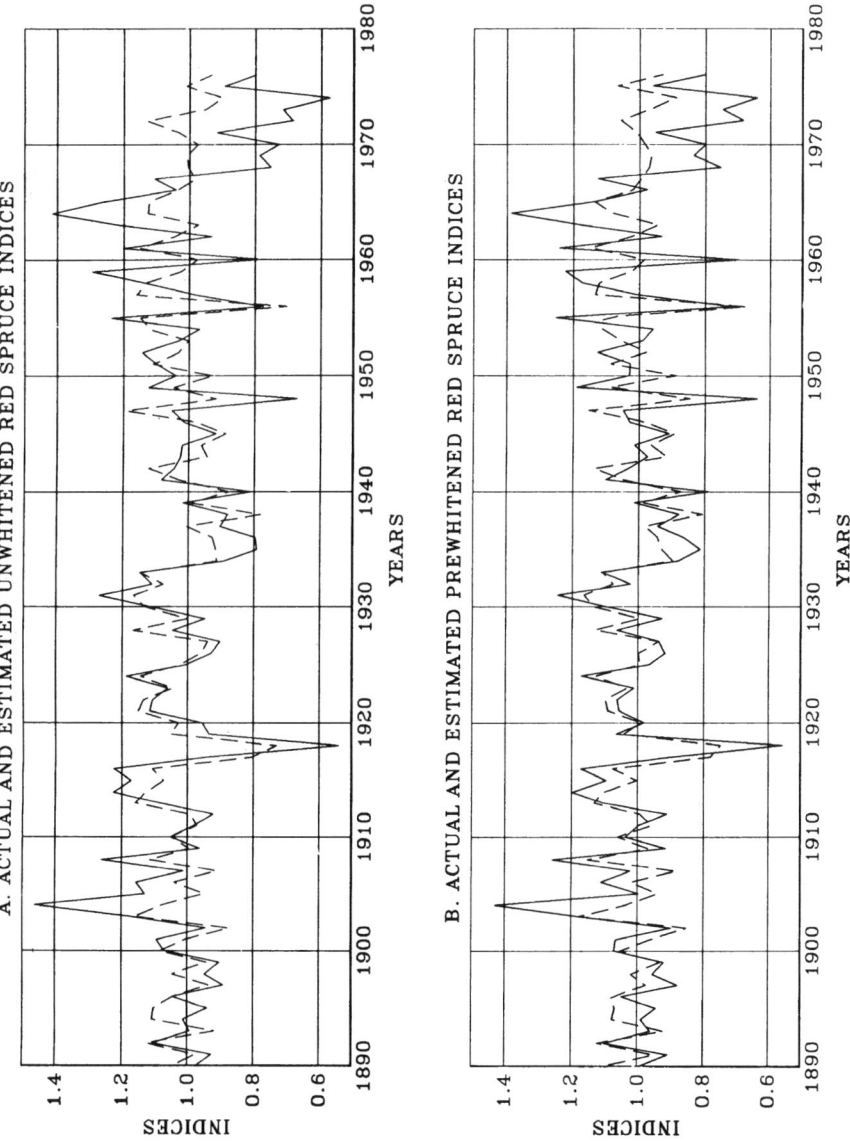

Fig. 4. The actual and estimated red spruce indices based on a temperature response model. The model was developed using the 1890-1950 data. Model predications run from 1951 through 1976. Solid lines are actual values; dashed lines are temperature estimates.

CONCLUSIONS

A general methodology has been presented for examining dendropollution hypotheses based on dendrochronological principles and techniques. In presenting this methodology, some inherent limitations of ringwidth data have been described which require careful treatment before the tree-ring data can be interpreted. The interpretation of tree-ring variations is complex and difficult because of the high level of ignorance regarding causal mechanisms. When a correlative approach, such as regression analysis can be used to model natural environmental influences on ringwidth, a much more rigorous assessment of the impact of pollution on tree growth is possible.

Acknowledgements

This research was supported by the National Science Foundation, Division of Climate Dynamics contract ATM83-09491. Dr. Gordon C. Jacoby of Lamont-Doherty Geological Obsevatory and Dr. T.J. Blasing of the Oak Ridge National Laboratory reviewed an earlier version of this manuscript. Lamont-Doherty Contribution No. 3929.

REFERENCES

Conkey LE (1979) Response of tree-ring density to climate in Maine, USA. Tree-Ring Bull 39: 29-38
Cook ER (1982) Eastern North America. In: Hughes MK, Kelly PM, Pilcher JR, LaMarche VC Jr (eds) Climate from tree rings. University Press, Cambridge, p 126
Cook ER (1985) A time series analysis approach to tree ring standardization. PhD thesis, Univ Arizona, Tucson
Cook ER, Peters K (1981) The smoothing spline: A new approach to standardizing forest interior tree-ring width series for dendroclimatic studies. Tree-Ring Bull 41: 45-54
Friedland AJ, Gregory RA, Karenlampi L, Johnson AH (1984) Winter damage to foliage as a factor in red spruce decline. Can J For Res 14: 963-965
Fritts HC (1976) Tree rings and climate. Academic Press, New York, p 567
Hepting GH (1963) Climate and forest decline. Ann Rev Phytopath 1: 31-50
Johnson AH, Siccama TG (1983) Acid deposition and forest decline. Environ Sci Tech 17: 294-306
Karl TR, Livezey RE, Epstein ES (1984) Recent unusual mean winter temperatures across the contiguous United States. Bull Am Met Soc 65: 1302
Lamb HH (1977) Climate: present, past and future, vol 2, Climatic history and the future. Methuen & Co, London, p 835
Scott JT, Siccama TG, Johnson AH, Briesch AR (1984) Decline of red spruce in the Adirondacks, New York. Bull Torrey Bot Club 111 (4): 438-444
Siccama TG, Bliss M, Voglmann HW (1982) Decline of red spruce in the Green Mountains of Vermont. Bull Torrey Bot Club 109: 163-168
White PS (1979) Pattern, process, and natural disturbance in vegetation. Bot Rev 45: 229-299

DENDROECOLOGICAL ANALYSIS OF ACIDIC DEPOSITION EFFECTS ON FOREST PRODUCTIVITY

D.C. LeBlanc, D.J. Raynal and E.H. White

SUNY College of Environmental Science and Forestry, Syracuse, New York, U.S.A. 13210

ABSTRACT

Dendroecological techniques have been employed in several studies to assess the effects of acidic deposition on tree growth. However, some of these studies have not evaluated the roles of other determinants of tree growth (ontogeny, stand development, competition, climate, soil-site variation) and therefore have not been able to adequately test hypotheses regarding the role of acidic deposition. In this study, detailed stem analysis was used to obtain complete historical records of terminal and radial growth patterns for trees of three species of plantation-grown conifers, representing a spectrum of potential susceptibility to acidic deposition effects. Analyses indicate that long-term variation in growth of the sample trees is related to stand dynamics and climatic factors, and there is little evidence of any long-term trend of diminished growth that might be related to acidic deposition effects. These results emphasize the importance of evaluating the roles of natural determinants of growth before inferences are made concerning the effects of acidic deposition, particularly in relatively young, second-growth forests.

INTRODUCTION

Over the last decade, it has become apparent that assessment of acidic deposition effects on forest tree growth is a very difficult and complex problem. Vast geographic areas receive acidic deposition and there are few, if any, unaffected areas with equivalent ecological and environmental characteristics that can be used as controls in comparative studies. The size and longevity of tree species make them very difficult to study in controlled experiments. Tree growth responses to environmental stimuli often vary with ontogeny, such that it is generally inappropriate to extrapolate experimental results obtained for one life-stage to predict growth responses of trees in other life-stages. Trees growing under natural conditions respond to an environmental complex; effects of acidic deposition on tree growth responses must be evaluated in the context of this environmental complex.

Several researchers have employed dendroecological techniques to address or circumvent difficulties encountered in assessing acidic deposition effects on tree growth (Baes and McLaughlin 1984; Cogbill 1977; Johnson et al. 1981; Jonsson 1977; Strand 1980). Radial annual ring width chronosequences, measured in increment cores taken at breast height (1.37 m), are employed as a relatively simple measure of historical tree growth responses over long periods of time. These dendroecological studies can be categorized into two groups based on whether tree growth during different time periods or in different geographical locations was compared. Jonsson (1977) and Strand (1980) compared historical growth trends of trees located in geographically separate areas of Scandinavia that receive different amounts of acidic deposition, to determine if increased pollutant loading is associated with decreased growth rates. Cogbill (1977) compared historical growth trends of several tree species growing in natural forests across the eastern United States to determine if a widespread, regional decline in forest tree growth has occurred over the past 20-30 years. Johnson et al. (1981) studied historical growth patterns of four coniferous tree species in the pinelands of New Jersey, comparing growth performance of trees during a pre-1950s "pre-acidic-deposition" period to growth of the same trees during a post-1950s "post-acidic-deposition" period. Baes and McLaughlin (1984) reported coincident variation in periodic radial growth of short-leaf pine (Pinus echinata Mill.), and pollutant emissions associated with past smelting activities (1863-1912) and recent increases in air pollution from coal-fired power plants. Metal content in growth rings also varied coincidentally with variation in air pollution emissions, but no significant correlations were found between metal concentrations in the phloem-cambium region and annual ring widths.

Some studies report evidence of decreased growth that is inferred to be associated with historical trends of increased acidic deposition (Baes and McLaughlin 1984; Jonsson 1977; Johnson et al. 1981), while others report no consistent evidence of diminished growth that might be associated with acidic deposition (Cogbill 1977; Strand 1980). Variations in historical growth patterns among the trees sampled were observed in all of these studies, both between species and within species, and were the major reason given for the inconclusive results reported by some. The nature and potential causes of these between-tree variations in historical growth patterns were left unexplained in these studies.

All dendroecological studies of acidic deposition effects on tree growth suffer from the confounding of variation associated with acidic deposition with that associated with other environmental factors. Studies involving different geographic locations (Cogbill 1977; Jonsson 1977; Strand 1980) have not accounted for differences in climate, soil, geology, hydrology, or land-use history. Comparisons of historical tree growth during different time periods have not accounted for environmental variation associated with climatic fluctuations, stand dynamics or stand history events (Johnson et al. 1981), nor has the effect of tree age been adequately evaluated (Baes and McLaughlin 1984; Johnson et al. 1981). While some past studies cited weather fluctuations as a major source of variation in the annual ring width data (Cogbill 1977; Johnson et al. 1981; Strand 1980), standard dendroclimatological techniques have not been consistently employed to evaluate such relationships quantitatively. The inability of these studies to distinguish acidic deposition effects on tree growth from growth fluctuations resulting from natural phenomena indicates that the roles of natural determinants of plant growth (climate, soils, biotic interactions) should be assessed before the effect of acidic deposition can be evaluated. Such an evaluation requires data pertaining to many aspects of tree growth. These data cannot be solely obtained from increment cores taken at breast height.

The choices of study sites and methodologies used in this study were made to address difficulties generally associated with dendroecological studies of acidic deposition effects on tree growth. These difficulties include: 1) the identification of site- and stand-related variation in annual ring width sequences, 2) selection of an appropriate standardization function to model "expected growth," and 3) an imperfect relationship between variation in relative growth measures (ring width indices) and absolute measures (wood volume or biomass), which are needed for economic and biological impact assessments. While stand dynamics may exert minimal influence on the growth of dominant, old-growth trees, the vast majority of forests in the northeastern United States are relatively young, second- and third-growth forests that developed after extensive logging in the 19th and early 20th centuries. Stand dynamics may be expected to exert influences on historical tree growth patterns in these forests. Standardization procedures used to remove long-term variation associated with tree age and stand dynamics (Cook and Peters 1981; Blasing et al. 1983) allow the investigator wide latitude in selecting the most appropriate model. The problem of standardization is addressed by Cook in this volume, and an improved, more objective standardization procedure is proposed for old-growth tree-ring chronologies. However, the standardization of tree-ring chronologies less than 200 years long remains problematic. While a sizeable body of dendroclimatology literature has demonstrated the efficacy of using increment core data to date and characterize historical climatic fluctuations, there are many reports in the forestry literature that indicate increment core data are not adequate to quantify growth responses to altered stand dynamics (thinning) or altered nutrient status (fertilization) (Mitchell and Kellogg 1972; Myers 1963). The results presented in these papers indicate that the magnitude, and sometimes the direction, of a growth response to silvicultural treatments may vary depending on the position at which the annual growth rings are measured on the stem. Hence, there is often a weak quantitative relationship between increment core data from one point on the stem (breast height) and volume increment data derived from measurements at multiple points on the stem.

To address the difficulties described above, we studied tree growth in simple plantation ecosystems where environmental variation is minimal, and employed detailed stem analysis to assess both relative and absolute growth responses quantitatively. A relative growth measure that does not require standardization to remove intrinsic growth patterns is presented.

METHODS AND MATERIALS

Study Site

A detailed study of historical tree growth and stand development was undertaken at the Charles Lathrop Pack Demonstration Forest, near Warrensburg, New York, located on the southeastern edge of the Adirondack Mountains. Plantations of various coniferous species, planted at Pack Forest between 1929-1936, on a 48.5 ha sandy outwash plain, have been the subject of a long-term research effort on soil fertility-growth relationships since 1937. The literature resulting from this 45-year research effort contains invaluable information on land-use history and soil-site relationships that is only rarely found in North America. The plantations are very simple systems: an even-aged overstory, no understory, an aggrading mor-type litter layer, and a deep, stratified, nutrient-deficient sandy outwash soil. Forest fertilization trials conducted in plantations of various species indicate that potassium is the major limiting factor for most species on this site (Leaf 1969). However, among the eight species of pine, spruce, and larch, some species do not appear to suffer any nutrient deficiency, some exhibit moderate nutrient deficiency symptoms, and others suffer both severe nutrient deficiency and are susceptible to drought on the well-drained outwash soil. We proposed that this spectrum of adaptation to the nutrient-deficient outwash soil exhibited by the various coniferous species may result in a corresponding spectrum of susceptibility to the potential effects of acidic deposition. Three species were selected for study: Scots pine (_Pinus sylvestris_ L.), red pine (_Pinus resinosa_ Ait.) and Norway spruce (_Picea abies_ Karst.). Scots pine did not respond to soil amendments and is considered well adapted to the nutrient deficient, well-drained soil. Red pine exhibits moderate potassium deficiency but is not particularly drought sensitive. Norway spruce suffers severe potassium deficiency and is drought sensitive on this site (Leaf 1969). We hypothesized that if acidic deposition is adversely affecting plant growth, either directly or via depletion of soil nutrient reserves, then those species that are poorly adapted to the site and already under stress should exhibit decline in growth before those that are well adapted to the site.

Growth Analysis

Trees selected for harvesting and detailed growth analysis were of dominant or codominant crown class, with no forks or evidence of damage on their stems, growing on soil profiles of more or less uniform characteristics, and located in unthinned, untreated plots used as controls in fertilization trials. All trees meeting the selection criteria were identified and a random sample of trees was subjected to detailed stem analysis. These sample trees were felled at ground level, all increments of annual terminal growth (hereafter referred to as internodes) were identified and measured, and cross-sectional disks were cut from the center of each internode. Disk surfaces were sanded smooth and ring widths were measured with an eyepiece micrometer at 15x to the nearest 0.1 mm along four radii on two randomly located perpendicular diameters which had been surfaced with a razor blade.

Internode length and ring width data were used to derive a complete record of annual growth increments and historical tree size for each sample tree, including height, diameter at breast height, annual stem wood volume increments, specific volume increments (volume of wood produced per unit area of cambium), and cumulative stem wood volume, as well as three different types of ring width sequences. Historical tree size data provide insight into the development of the even-aged stands and the position of the sample trees within the stand during this development. Gross annual increment of stem wood volume provides the best growth parameter for evaluating the impact of acidic deposition on forest production in terms that are easily related to economic impact. However, annual volume increment is highly dependent on tree size, is characterized by a high degree of auto-correlation, and has heteroscedastic variance over time (LeBlanc 1985). These statistical properties complicate the necessary time-series analyses required to evaluate relationships between growth performance and climatic fluctuations. The development of procedures to apply dendroclimatological statistical techniques to annual volume increment chronologies is an area of ongoing research.

A vertically-oriented ring width sequence, described by Duff and Nolan (1953), provides a measure of relative growth performance over time that is relatively unaffected by intrinsically-determined growth patterns within the tree, and is less affected by the undesirable statistical properties of autocorrelation and heteroscedasticity associated with annual volume increment sequences. Vertical ring width sequences are composed of a series of annual rings, one from each internode, which are an equal number of rings from the pith; for example, the first ring around the pith from each internode. There are many such sequences within a tree, each defined as being a certain number of rings from the pith, hence the alternate nomenclature "ring number sequence" proposed by Fayle (1973). Ring widths in a vertical sequence are produced by cambium of equal age and in equivalent position with respect to the photosynthetic centers of the crown. For example, the first ring around the pith in any internode was produced by new cambium when that internode was at the apex of the tree. Consequently, intrinsically-determined growth trends resulting from the effects of cambial age progression or a hormonal-nutritional gradient related to distance from the cambium to photosynthetic centers in the crown are greatly reduced in vertical ring width sequences (Duff and Nolan 1953). Intrinsically-determined allometric growth trends dominate radial ring width sequences obtained from increment cores, obscuring growth responses to environmental stimuli. The lack of adequate statistical procedures to remove this intrinsic pattern (standardization) without concurrently removing time trends associated with environmental factors, such as acidic deposition, has hindered the application of dendroecological techniques in assessing acidic deposition effects on tree growth. Vertical ring width sequences were selected as the indicator of historical growth performance because of their desirable statistical properties (homoscedastic, diminished autocorrelation) and because they do not require standardization to remove intrinsically-determined allometric growth trends.

Another vertically-oriented ring width sequence, called an oblique sequence (Mott et al. 1957), or a growth-layer profile (Fayle 1973), is particularly useful in assessing historical crown development and stand history events. Oblique ring width sequences are composed of multiple measurements of ring width made on the same annual sheath of wood, but at different heights on the stem (Duff and Nolan 1953). Thus, for each year of growth there is an oblique sequence composed of ring widths produced during that year, but at different positions along the stem.

Oblique sequences exhibit characteristic, intrinsically-determined patterns associated with the position of the cambium with respect to the photosynthetically active crown and the crown status of the tree. Larger ring widths are produced by cambium near the center of the photosynthetically-active crown, and ring width declines above and below this point on the stem. Variations in the characteristic pattern of oblique sequences have been described for trees in different crown classes (Duff and Nolan 1953). Suppressed trees produce small ring widths along the entire length of the stem and maximum ring width near the apex. Trees in dominant and codominant crown classes growing in forest stands produce oblique sequences that exhibit wide variation in ring width, with large ring width near the 6th - 10th internode and decreasing ring width to a more or less stable ring width along the branch-free bole. Open-grown trees produce oblique sequences which are characterized by large ring width over most of the length of the stem. Thus, a chronological series (chronoseries) of oblique sequences may be useful as an historical record of crown status, documenting variation in growth patterns associated with varying levels of competition for light. Some researchers have indicated that oblique sequences are also most useful in assessing the impact of stand history events which affect the crown, including infestation by defoliators and release from competition of competitor canopies by thinning (Mott et al. 1957; Mitchell and Kellogg 1972).

Stand Structure Analyses

Detailed data related to stand structure and stand history were taken within circular plots, with radii of 4 m and 8 m, circumscribed around each sample tree to evaluate the potential role of past and present competitive interactions as causes of variation in sample tree growth. These data include measurements of the diameter at breast height of all living and dead stems and stump diameters where the stems were missing. Competition intensity around the sample trees was estimated by a distance-independent diameter model described by Daniels (1976). These data were entered into correlation analyses with 5-year periodic growth variables (height, basal area, and volume increment) to infer the potential impact of current and past competitive interactions with neighbouring trees on the historical growth patterns of the sample trees. The potential impact of competitor mortality was assumed to be directly related to the number of competitors that had died and the relative size of the competitors compared to the size of the sample trees. Mortality among suppressed neighbours was presumed to have minimal impact on sample tree growth, while mortality among neighbours in size classes comparable to or greater than the size class of the sample trees was expected to result in alteration in growth patterns of the sample tree.

Dendroclimatological Analyses

Growth-climate relationships were assessed by dendroclimatological procedures described by Fritts (1976). A vertical ring width chronosequence was used as the growth response variable in these analyses. A simple normalization of individual tree chronologies to a common mean of 0 and a standard deviation of 1 was made by subtracting the 50-year mean from individual ring widths and dividing the difference by the standard deviation. An average chronology for each species was calculated as the average of all sample trees of that species. Regionally averaged climatic data (recommended by Blasing et al. 1981) were compiled from National Oceanic and Atmospheric Administration weather records from stations near Pack Forest. The weather variables, including monthly and seasonal temperature, precipitation, and degree days, as well as length of growing season, were subjected to principal components analysis (PCA) to reduce the dimensionality of the weather data set and to produce mutually orthogonal principal component scores. Weather variables included in PCA were selected from a large number of potential variables (>100) by correlation analysis between individual weather variables and individual and mean tree-ring sequences. Those variables having correlations (significance level of $p < 0.15$) with at least one-third of the trees were included in the PCA. Two sets of principal component scores were derived, one for Norway spruce and red pine (having 15 principal components) and one for Scots pine (with 12 principal components). Weather principal component scores and three years of lagged growth were entered into stepwise regression models (SAS Stepwise procedure) as predictor variables, with normalized mean ring widths from vertical sequences as the dependent variable representing annual growth performance. Residuals from these regression models represent variations in tree growth performance that are not explained by the growth-climate model, but may still contain some variation associated with climatic factors or events that were not included in the model. Such factors may be related to extraordinary weather conditions or events that cause non-linear responses in tree growth or are too infrequent to be represented adequately in the growth-climate model. Synchronous variation of residuals from growth-climate regression models for the three tree species studied, occurring in years of particularly poor or good growth, may indicate years when such extraordinary weather events occurred. Variation in growth of sample trees not related to these factors was evaluated to determine if unexplained decreases in growth have occurred, and if so, whether these decreases support the hypothesis that a spectrum of susceptibility to acidic deposition effects exists among the three coniferous species studied.

RESULTS AND DISCUSSION

Detailed stand structure analyses were undertaken to assess current competitive conditions around sample trees and to evaluate the possibility that competition levels have changed over the past two decades (a prime period of interest for acidic deposition effects). Correlations between competition levels in 1981 and periodic volume increment of individual trees (period 1976-1981) were -0.67, -0.53, and -0.78 for Scots pine, red pine, and Norway spruce, respectively (all significant at the $p < 0.05$ level). A new method for inferring historical changes in competition, based on stem analysis growth data and detailed stand structure analyses (LeBlanc 1985), indicates that competition levels experienced by Scots pine sample trees have been more or less constant from 1951-1981, those of red pine have been slowly increasing (probably due to low mortality and continued growth), and those of Norway spruce changed abruptly during the 1950s. Both Scots pine and Norway spruce stands have suffered 75% mortality since planting, but the timing and potential impact on sample tree growth differ between these species. Mortality in the Scots pine stand has been concentrated in suppressed size classes and occurred mainly during early stages of stand development; 59% of the positions in the planting grid did not even have a stump present. In contrast, mortality in the Norway spruce stand is evenly divided between early mortality (no stump present) and later mortality (stump or dead stem present); 33% of the stand has died since 1964. Furthermore, many dead Norway spruce stems are relatively large, compared to the size of living trees, and it is likely that surviving trees have responded in some manner to release from competition. There has been little mortality in the red pine stand (25%), most of which was concentrated in small size classes. Current density is 1613, 1588, and 2583 stems.ha^{-1} in Scots pine, Norway spruce and red pine, respectively.

Mean vertical ring width chronosequences, presented as the upper curve in each panel of Fig. 1, illustrate both differences and similarities between the three tree species. Scots pine has consistently produced larger ring widths over the entire 50-year period, compared to the other two species. Red pine exhibits better early growth than Norway spruce, but since the 1950s these two species have produced ring widths of comparable size. Both red pine and Norway spruce exhibit decreased growth during the mid-1960s, coincident with a very severe drought from 1964-1966 (Cook and Jacoby 1977), but growth of Scots pine does not appear to have been significantly affected (LeBlanc 1985). Subsequent to this drought period, red pine vertical sequence ring widths returned to pre-drought sizes within one year after the drought had ended. Recovery of Norway spruce appears to have been slower, occurring over several years, indicating that the trees

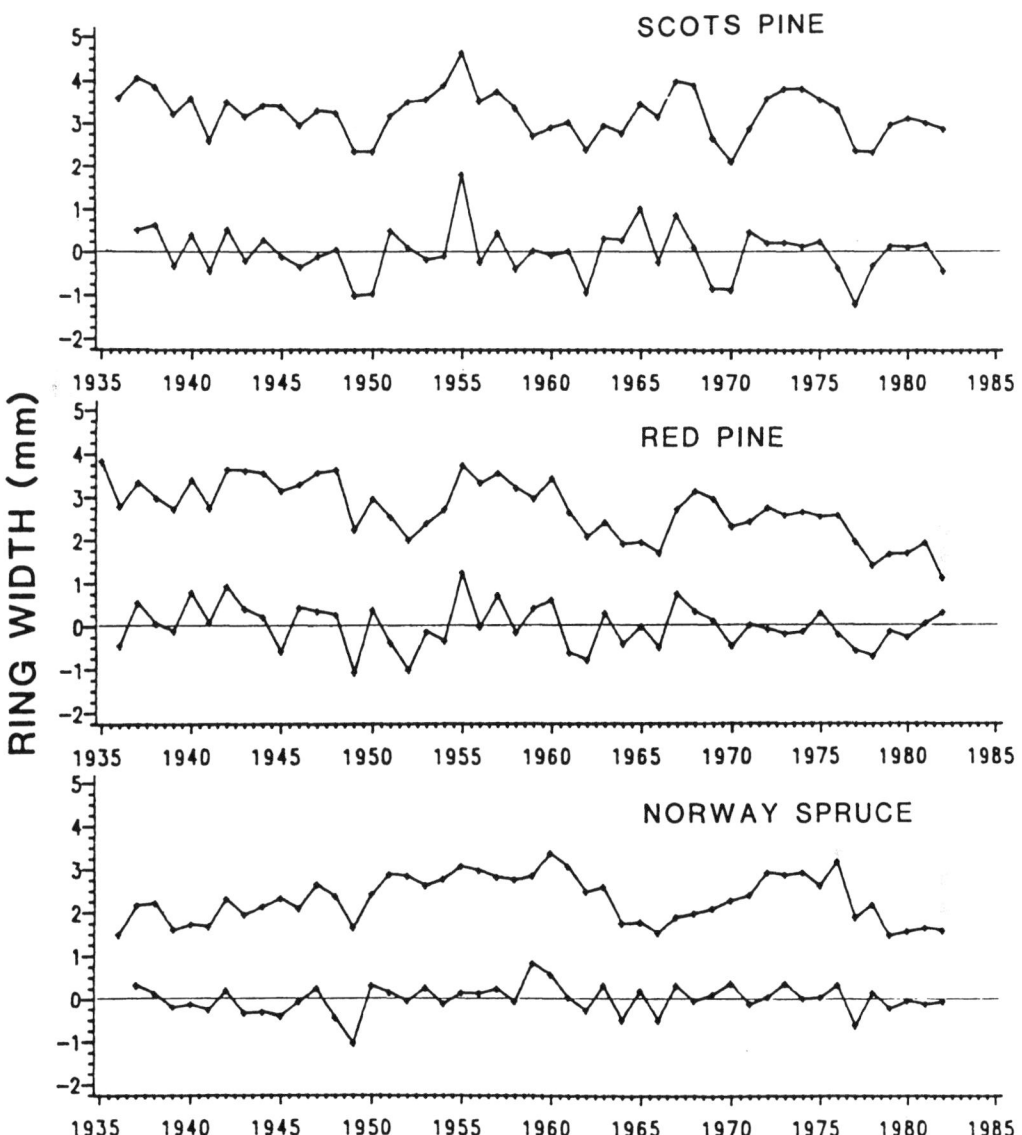

Fig. 1. Vertical sequence mean ring widths and residuals from growth-climate regression models for three coniferous species growing on the same site. The upper curve in each panel is a vertical ring width sequence and the lower curve in each panel is a plot of residuals.

may have suffered physical damage that required production of new tissues. Several instances of synchronous variation in growth of the three species, indicating the influence of the macro-scale environment, are apparent in these chronosequences, including the years of 1941, 1949, 1955, 1962-1966, 1970 and 1977. Both red pine and Norway spruce have exhibited decreasing growth from 1977 to present.

Residuals from growth-climate regression models for the three species, representing variation not explained by the models, are presented in the lower curve in each panel of Fig. 1. These residuals are the result of regressions on ring widths normalized to a mean of 0 and standard deviation of 1 and are hence dimensionless. Much of the variation in vertical ring width associated with such well-defined climatic events as the mid-1960s' drought has been accounted for by the regression models. However, synchronous variation of growth among the three species is still present in the residuals, indicating that the growth-climate model may not have accounted for all variation associated with weather factors. Years in which synchronous variation is particularly obvious include 1949, 1955, 1962, and 1977. The amount of variation accounted for by the growth-climate models varied between the three species, with coefficients of determination of 0.41, 0.51, and 0.69 for Scots pine, red pine, and Norway spruce, respectively. These differences in the amount of variation in growth accounted for by the growth-climate model parallel the spectrum of adaptation to the site and susceptibility to drought, with the species best adapted to the site exhibiting the weakest relationship with climatic variables. Nonetheless, all growth-climate regression models were significant at the $p < 0.0001$ level, with 45 degrees of freedom. See Table 1 for multiple regression coefficients of determination associated with lagged-growth variables and climatic principal component scores.

Statistical and graphical assessment of within-species variability indicates that Norway spruce sample trees included two subsets of trees with significantly different historical growth patterns, while Scots pine and red pine sample trees constituted essentially homogeneous groups. The difference in vertical ring width chronologies from trees in the two subsets of sample trees from the Norway spruce stand, illustrated in Fig. 2A, is representative of similar within-species differences in other growth parameters. The majority of Norway spruce sample trees (17 out of 20) fall into group 1, composed of trees that did not suffer from suppressed growth during the early phases of stand development. Group 2 is composed of 3 trees that appear to have suffered severe suppression during the first 20 years after planting, but since 1955 have exhibited growth performance equal to or exceeding that of the unsuppressed trees. Over the past 20 years the two groups of Norway spruce sample trees have exhibited divergent growth trends, with unsuppressed trees generally declining and the previously suppressed trees exhibiting generally increased growth. Trees in the suppressed group exhibit smaller decline in growth associated with the mid-1960s' drought, and subsequent to the drought these trees returned to pre-drought

Table 1. Dendroclimatic multiple regression coefficients of determination for STEPWISE regression models, including climate variables only, lagged-growth variables only, and both climate and lagged-growth variables. A significance level of $p < 0.05$ was required for entry in the STEPWISE models.

	Scots Pine	Red Pine	Norway Spruce
3-years of lagged-growth variables only	0.20	0.44	0.47
climate principal component scores only	0.26 (1)[a]	0.40 (2)	0.23 (1)
climate principal component scores and lagged-growth variables	0.41 (2)	0.51 (2)	0.69 (4)

[a]Numbers in brackets are the number of predictor variables in the models, including weather variable principal component scores and a 1-year lagged-growth variable in the third row (the only lagged-growth variable found to be significant at $p < 0.05$).

growth levels almost immediately. This indicates that the smaller trees were less affected by drought stress than their larger competitors, perhaps as a consequence of their smaller crowns and smaller respiratory biomass. This difference between the two groups is also evident in the growth-climate regression models. Fifty-two percent of the variation in growth of the unsuppressed group is accounted for by a variable representing growth during the previous year and 25% of the variation is accounted for by climatic principal component scores. In contrast, 82% of variation in growth of the suppressed Norway spruce is accounted for by growth during the previous year, and only 5% is accounted for by climatic principal component scores after this lagged-growth variable is entered in the stepwise regression procedure. Regression models for both groups of Norway spruce are significant at $p < 0.0001$. Residuals from separate growth-climate regression models applied to the two groups of Norway spruce are illustrated in Fig. 2B. Residuals for the suppressed groups are generally negative until 1950, and after the mid-1960s there is a slight tendency for the residuals of this group to be positive. There is no indication of any long-term trend in residuals from the growth-climate model for the unsuppressed group, although all but one residual for the past 6 years have been negative.

Chronoseries of oblique sequences for two representative trees from the two within-species groups identified in the Norway spruce sample trees are presented in Fig. 3A and 3B to illustrate the use of oblique sequences in assessing historical growth patterns. Each vertically-oriented curve in the chronoseries is a single oblique sequence, composed of multiple ring width measurements on the same annual sheath of wood. Variation along

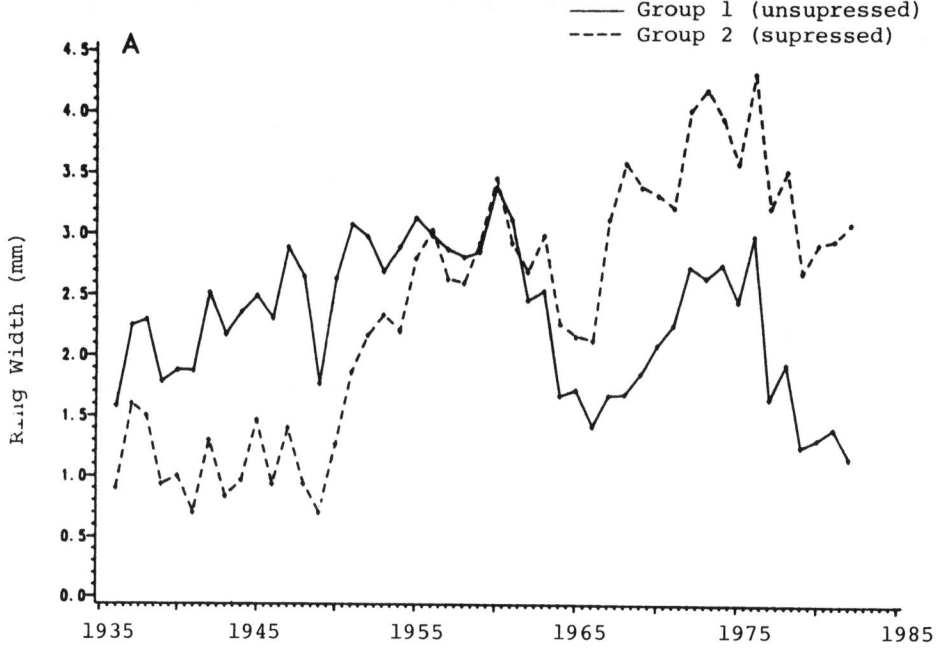
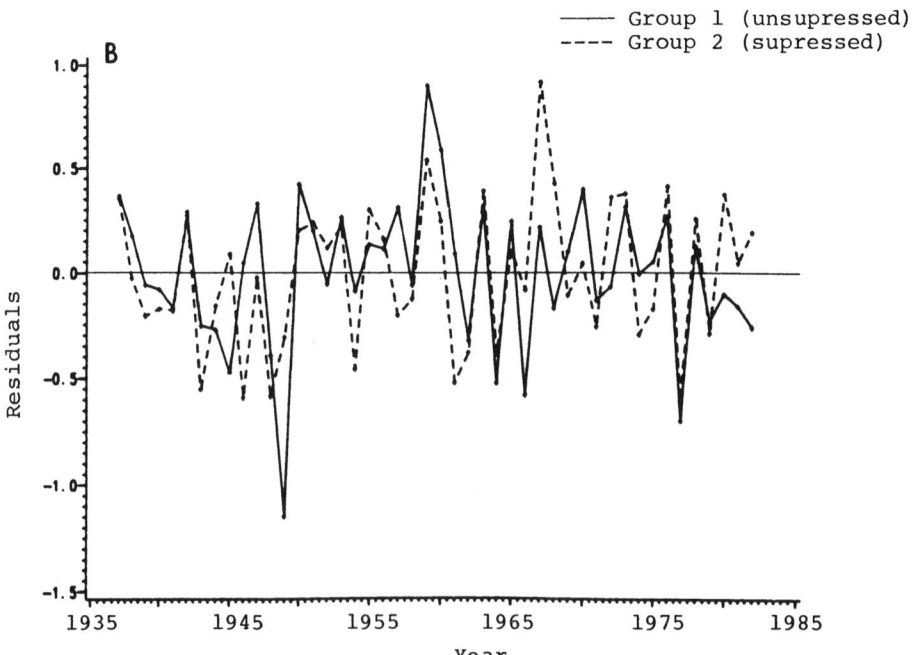

Fig. 2. Vertical sequence mean ring widths and residuals from growth-climate regression models for suppressed and unsuppressed Norway spruce sample trees growing in the same stand.

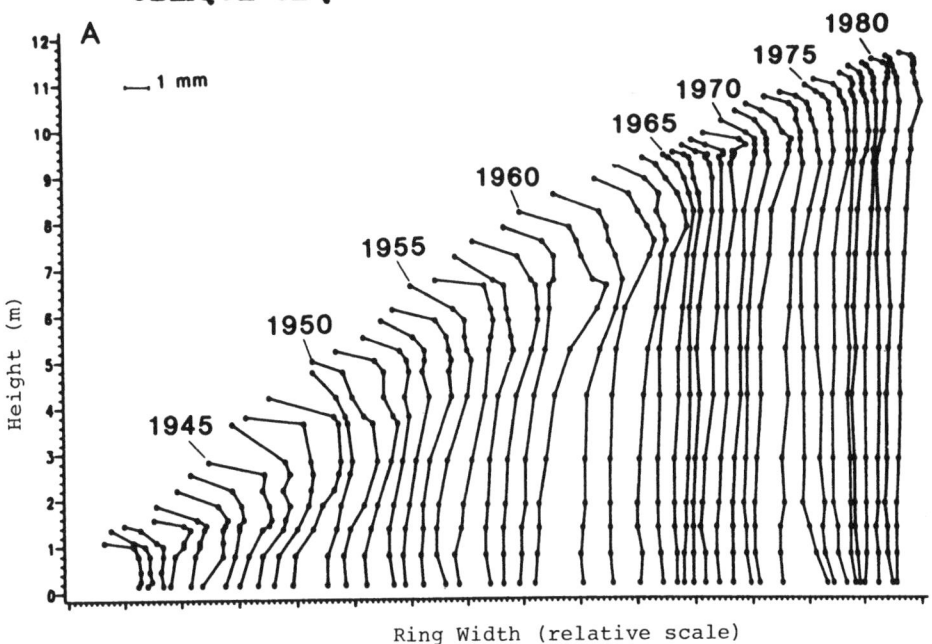

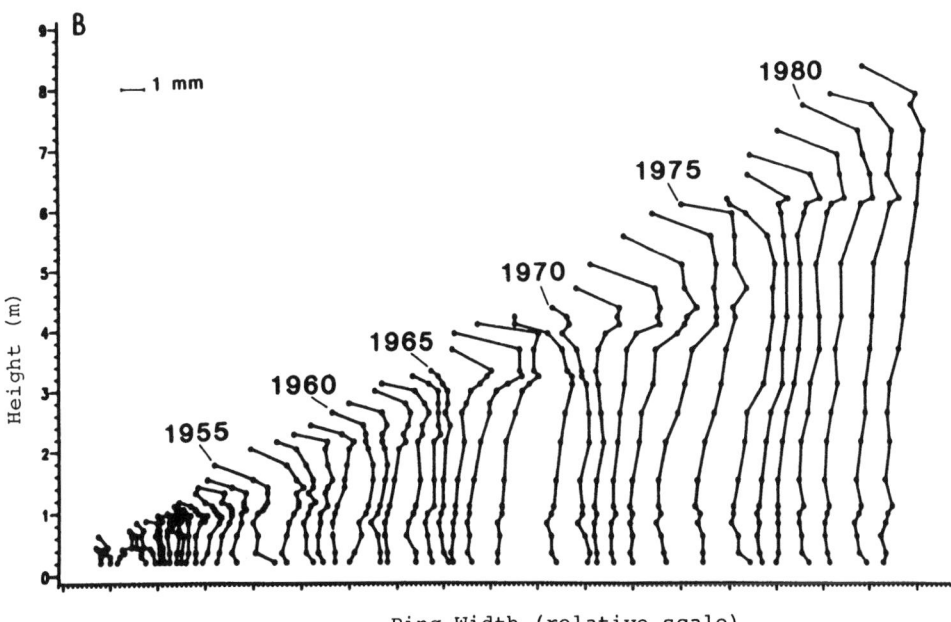

Fig. 3. Oblique sequence chronoseries of representative trees from suppressed (Norway spruce 19) and unsuppressed (Norway spruce 3) Norway spruce sample trees. See text for details of chronoseries construction and interpretation.

the Y-axis represents variation in the height of measurement, and variation along the X-axis is proportional to variation in ring width, with the 1 mm scale mark provided for reference. Each curve depicts the radial growth allocation pattern along the stem during that year. The most relevant characteristics of these oblique sequence curves are the amount of variation in each curve along the X-axis, the consistency of the spacing between curves of consecutive years, and the height of the curves, an indicator of tree height during that year. Large variations in ring width along the length of the stem are indicative of good growth in stand-grown trees, while consistently small ring widths indicate severe growth reduction. Consistent spacing between consecutive curves indicates that no alteration of growth allocation patterns has occurred from one year to the next. Norway spruce 3 represents the majority of Norway spruce sample trees. It exhibits consistently good growth during the first 35 years of life, but a decline for most of the years after the mid-1960s' drought. Norway spruce 19 was severely suppressed until 1955, exhibited improved growth from 1955-1965, and after the mid-1960s' drought, exhibited dramatically improved growth. These results are consistent with stand history data which indicate that mortality among competitors of relatively large size classes occurred subsequent to the mid-1960s' drought and has resulted in improved growth among the survivors, particularly those which were suffering from severe competition. Even Norway spruce 3 exhibits increased ring widths and increased longitudinal variation in ring width during the post-drought period from 1970-1976, indicative of improved growing conditions. It is interesting to note that data from increment cores taken from red spruce on Whiteface Mt., in the Adirondack Mountains, also exhibit a sequence of decline associated with the mid-1960s' drought, with some subsequent recovery, followed by continued decline since the late 1970s (E.R. Cook, pers. comm.). The striking similarity between these observations and the growth patterns of Norway spruce presented in this study indicate that both climate and stand history may need to be evaluated as possible causes of fluctuations in recent growth performance of these species.

In conclusion, there is little evidence of any long-term trend of diminished growth rates in the data from the three tree species studied that might be related to effects of acidic deposition. This apparent lack of long-term adverse effects on such a potentially susceptible site indicates that it is unlikely that such effects will be found in other low altitude forests growing on more fertile soils. However, the correlative analyses used here do not preclude the possibility that air pollutants may adversely affect tree growth during short-term episodic events or in conjunction with climatic factors included in the growth-climate model.

SUMMARY

In this paper, we have stressed the importance of studying historical growth patterns within the context of the total environment complex as a means for assessing the impact of acidic deposition on forest productivity. Studies in relatively simple plantation ecosystems diminish the number of natural environmental factors that must be accounted for and, in the case of our study of Pack Forest, management records for plantations may serve as a valuable source of historical data. However, even in natural, uneven-aged forests, valuable historical information on stand dynamics processes may be obtained if detailed analysis of stand structure, including living and dead stems and stumps, is done in the area around the sample trees. Most dendroecological studies could benefit from inclusion of such analyses.

Vertical ring width sequences for the three species defining a spectrum of adaptation to the nutrient-deficient soil at Pack Forest exhibit interspecific variation in historical growth patterns, with Norway spruce exhibiting generally declining growth trends over the past 20 years, while Scots pine and red pine data show little or no decline. Regression analyses of growth-climate relationships indicate that the recent decline in growth performance of Norway spruce is closely correlated with climatic factors, particularly a severe drought during the mid-1960s. Further synchronous variation in the residuals from growth-climate regression models for the three species indicate that these models have not accounted for all variation in growth that might be related to climatic events. Analyses of within-species variation indicate that improved growth performance of some Norway spruce during the past 25 years is probably related to competitive release associated with significant mortality in this stand. This assessment of growth fluctuations unassociated with climatic factors was made possible by detailed growth analyses and stand descriptions. These analyses require large expenditures of time and energy, and further research is needed to develop ways to integrate these intensive procedures with the less demanding, more extensive standard dendroecological methods.

REFERENCES

Baes CF III, McLaughlin SB (1984) Trace elements in tree rings: evidence of recent and historical air pollution. Science 224: 494-497
Blasing TJ, Duvick DN, West DC (1981) Dendroclimatic calibration and verification using regionally averaged and single station precipitation data. Tree-Ring Bull 41: 45-53
Blasing TJ, Duvick DN, Cook ER (1983) Filtering the effects of competition from ring-width series. Tree-Ring Bull 43: 19-30
Cogbill CV (1977) The effect of acid precipitation on tree growth in eastern North America. Water, Air and Soil Pollut 8: 89-93
Cook ER, Jacoby GC (1977) Tree ring-drought relationships in the Hudson Valley, New York. Science 198: 399-401
Cook ER, Peters K (1981) The smoothing spline: a new approach to standarizing forest interior tree ring width series for dendroclimatic studies. Tree-Ring Bull 41: 45-53
Daniels RF (1976) Simple competition indices and their correlation with annual loblolly pine tree growth. Forest Sci 22: 454-456

Duff GH, Nolan NJ (1953) Growth and morphogenesis in the Canadian forest species. I. The controls of cambial and apical activity in Pinus resinosa Ait. Can J Bot 31: 471-513

Fayle DCF (1973) Patterns of annual xylem increment integrated by contour presentation. Can J For Res 3: 105-111

Fritts HC (1976) Tree rings and climate. Academic Press, New York, p 567

Johnson AH, Siccama TG, Wang D, Turner RS, Barringer TH (1981) Recent changes in patterns of tree growth rate in the New Jersey Pinelands: a possible effect of acid rain. J Environ Qual 10(4): 427-430

Jonsson B (1977) Soil acidification by atmospheric pollution and forest growth. Water, Air and Soil Pollut 7: 497-501

Leaf AL (1969) America's oldest intensive fertilization experiments: some lessons learned. Advanced Frontiers of Plant Sci (India) 24: 1-37

LeBlanc DC (1985) Dendroecological assessment of acidic deposition effects on tree growth using detailed stem analysis. Ph.D. dissertation, State University of New York College of Environmental Science and Forestry, Syracuse, New York, p 400

Mitchell KJ, Kellogg RM (1972) Distribution of area increment over the bole of fertilized Douglas-fir. Can J For Res 2: 95-97

Mott DG, Nairn LD, Cook JA (1957) Radial growth in forest trees and effects of insect defoliation. Forest Sci 3(3): 286-304

Myers CA (1963) Vertical distribution of annual increment in thinned Ponderosa pine. Forest Sci 9(4): 394-404

Strand L (1980) The effect of acid precipitation on tree growth. In: Drablos D, Tollan A (eds) Ecological impact of acid precipitation. Proc Int Conf, Sandefjord, Norway. SNSF Project, PO Box 61, 1432 As-NLH, Norway

TRACE METAL UPTAKE AND ACCUMULATION IN TREES AS AFFECTED BY ENVIRONMENTAL POLLUTION

C.F. Baes III and S.B. McLaughlin

Environmental Sciences Division, Oak Ridge National Laboratory, Oak Ridge, Tennessee, U.S.A. 37831

ABSTRACT

Regional-scale increases in combustion of fossil fuels, trace metal deposition profiles in lake sediments, and metal accumulation patterns in forest soils suggest that forests have been exposed to increasing levels of trace metals in recent decades. Such increases are of concern because metals such as Al, Cd, Cu, Fe, Mn, Ni and Zn typically have long residence times in soil-plant systems and can adversely affect plant physiological and litter decomposition processes at sufficiently high concentrations. Chronic effects of these metals at present regional concentrations are largely unknown, and their increased availability to trees can occur through multiple pathways, most of which are poorly characterized. Our surveys of changes in tree-ring chemistry of conifers show that increased tissue concentrations of some metals are quite similar to temporal increases in local and regional fossil fuel combustion, suggesting a relationship between the two. Highest metal concentrations in tree rings were found near a power plant and at high-elevation sites remote from point sources. Levels of many trace metals, particularly Al, Fe, and Mn, in tree phloem tissues are at or exceed levels reported as toxic thresholds in other plant species. In reviewing the literature, it is clear that toxic thresholds are strongly influenced by soil characteristics, plant species, and plant part. Thus, until more information is obtained on toxic thresholds for individual and combined levels of trace metals on plant physiological and forest ecosystem-level processes, their role, if any, in forest declines must be considered conjectural.

INTRODUCTION

Increased combustion of fossil fuels and emissions from smelting activities during the past century have substantially augmented natural inputs of trace metals to forest ecosystems over extensive geographic areas, particularly in the northeastern United States. Analysis of sediment layers from lakes in Maine and Vermont (Hanson et al. 1982) revealed that deposition of metals started to increase on a regional scale in the United States as early as 1880. Norton et al. (1980) report 200 to 800% increases of zinc and lead in sediments from New England during the past 100 years.

Atmospheric deposition of pollutants increased most rapidly in many areas of the eastern United States in the last two to three decades. Friedland et al. (1984) found increases of Pb, Cu, and Zn of up to 148% in organic soils at Camels Hump Mountain in Vermont during the past 14 years. Chemical analysis of short-leaf pine tree cores from east Tennessee (Baes and McLaughlin 1984) showed both local and regional coincidence of increased trace metal accumulations in tree rings with local and regional increases in combustion of fossil fuels.

While the possibility that forest trees may be important biological indicators of trends in atmospheric deposition offers a potentially useful tool for providing biological characterization of historical trends in atmospheric deposition, there are still substantial gaps in our understanding of the linkages between metal inputs to forests and their uptake by trees. This paper will address some aspects of the uncertainties regarding trace metal effects, with particular emphasis on (1) the scientific basis of concern about trace metal accumulation in forests, (2) interpretation of observed patterns of accumulation of metals in tree tissues relative to temporal changes in atmospheric deposition of fossil-fuel-derived pollutants, and (3) examination of the likelihood of adverse effects of specific heavy metals on forests at current ambient soil and tissue levels.

BASIS OF CONCERN

Heavy metals, which may be deposited or increased in availability by dry and wet deposition, have the potential to significantly affect forest ecosystems through multiple pathways (McLaughlin 1985). Their potential significance is based on regional increases in precipitation acidity and its effects on metal cycling (Norton et al. 1980), recent accumulations in litter and soil (Johnson et al. 1982; Friedland et al. 1984), long residence times in soil systems (Hutchinson 1980), and possible toxicity to plant growth processes (Bowen 1979; Whitby and Hutchinson 1974; Foy et al. 1978). The mobilization of aluminum and iron (Ulrich et al. 1980) and other elements (Norton et al. 1980) in forest soils by acid deposition focused considerable attention on possible adverse effects of metals to forest trees. Aluminum has been of major concern because of its well-documented toxic effects on root growth of crops and its role in affecting availability of essential plant nutrients, most notably phosphorus and calcium (Foy et al. 1978).

Several mechanisms for increased trace metal availability to forest ecosystems exist (Fig. 1). These include increased wet and dry deposition to vegetative surfaces and soil, increased availability of foliar-deposited materials through either altered cuticular surface characteristics or increased (acid-induced) chemical availability on the leaf surface, and increased chemical mobilization in soils as a function of soil-acidifying reactions. In reviewing the potential phytotoxicity of trace elements released by coal combustion, Vaughan et al. (1975) list the following principal considerations: (1) emission rate, (2) dilution level in endogenous soil pools, (3) bioaccumulation

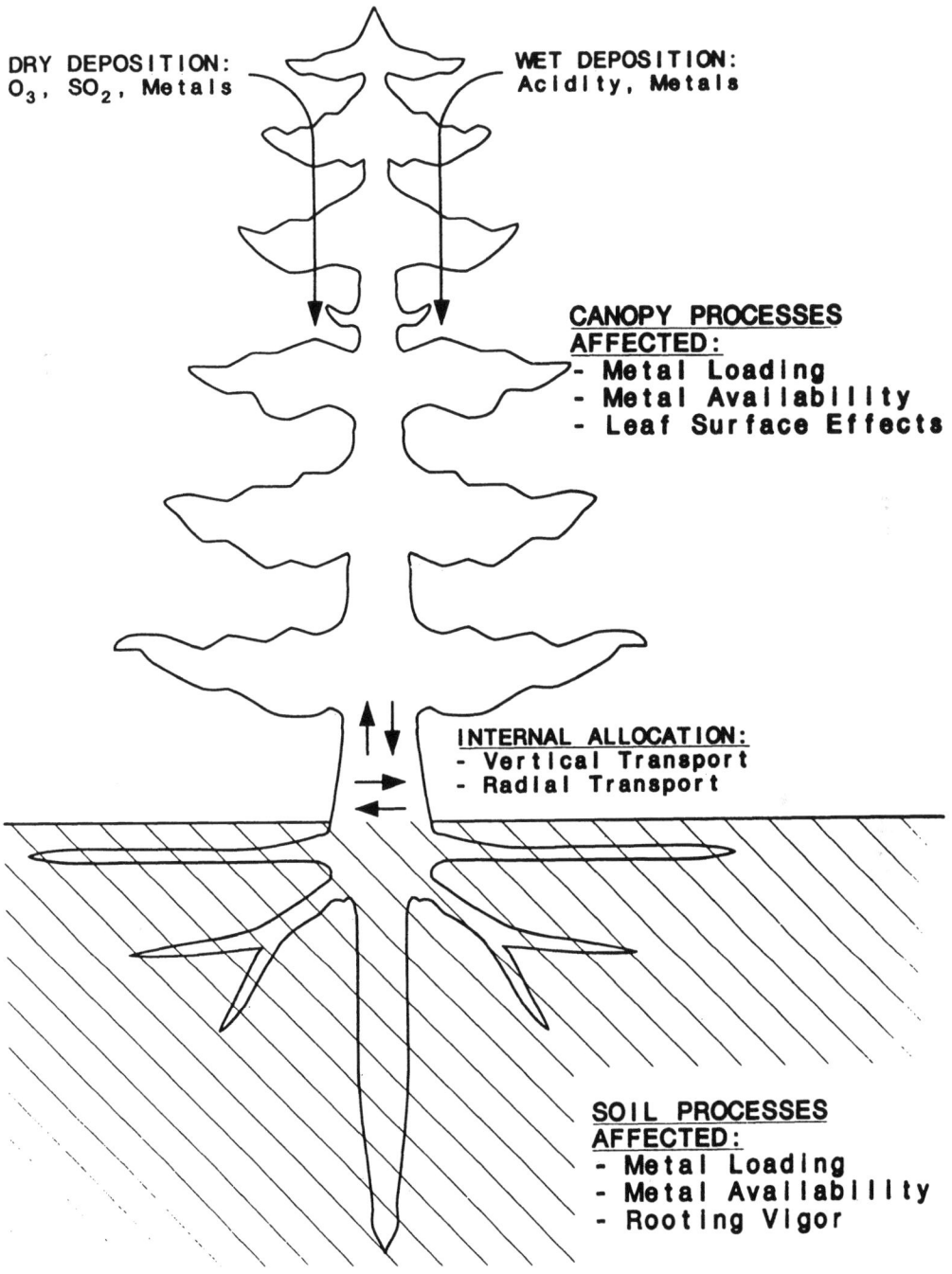

Fig. 1. Conceptual model of interactions between trees and depositing materials.

rate, and (4) toxicity. The interplay of factors controlling chemical and biological availability, and ultimately biological effects, is, however, still poorly understood.

The potential for trace elements to accumulate in forest systems is high, due to both long residence time and multiple pathways of entry into forest trees. Hutchinson (1980) lists residence times in soils ranging from 280 years for cadmium to 2100 years for zinc and 6300 years for chromium in the Upper Thames Basin in England. Friedland et al. (1984) recently reported increases in lead and zinc by 124 and 148%, respectively, in the organic horizons of forest soils in the Green Mountains of Vermont during the past 14 years. Copper and zinc also increased in approximate proportion to deposition rates during that time span.

Indirect effects of trace metals on forest nutrient cycles through disruption of litter decomposition processes were documented in the vicinity of smelting operations (Jackson and Watson 1977), but the significance of regional-scale inputs to mineral cycling is largely unknown. Chaney et al. (1978) found that addition of cadmium and zinc to soil microcosms at levels below those found in industrial areas reduced soil respiration by 30%. Nutrient availability may thus be influenced by trace element deposition to forests as a function of altered root vigor, altered nutrient uptake, or altered mineralization of nutrients by litter decomposers. The significance of these processes at regional-scale levels of trace metal input is, however, currently unquantified.

Studies of distribution patterns of trace elements in spruce in the German Black Forest indicate that Pb, Cd, and Cu accumulate mainly from atmospheric inputs, Mn from the soil, and Zn by both sources (Raisch and Zöttl 1983). The relative contributions of wet:dry deposition of trace inputs to a deciduous forest canopy may vary widely for different elements, ranging from a minimum of 0.1 for manganese and 0.8 for lead to 3 to 4 for cadmium and zinc (Lindberg et al. 1982). Trace metal mass balance measurements in several deciduous forest watersheds indicate a significant net accumulation of many trace metals, with estimates of retention ranging from 20 to 70% for Mn to 88 to 99% for Cd, Pb, and Zn (Turner and Lindberg 1985).

Accumulation of trace metals may vary widely among tree species and among tissue types within species. For example, in a forested watershed in east Tennessee (Van Hook et al. 1977) the distribution of lead and zinc varied widely among tissue types (15 to 45-fold) as well as among tissues of the same type for different species (three- to six-fold). In general, elemental contents decreased in the following order: O_1 litter > O_2 litter > fine roots > large roots > leaves > branches > bole.

The distribution of trace metals within tree boles varies by tissue age and type (Baes et al. 1984). These variations are of particular interest from two perspectives. First, concentration variations in the annual rings may provide an historical record of metal availability to forest trees as influenced by depositional patterns or chemical reactions in soil or on leaf surfaces. Second, concentrations in metabolically active tissues may provide some indication of the potential for adverse physiological effects resulting from metal toxicity. The remainder of this paper details some of our recent investigations of trace metal concentrations in bole tissue and evaluates the results in light of the above perspectives.

TRACE ELEMENT ACCUMULATION PATTERNS

We examined the tree-ring chemistry of short-leaf pine (Pinus echinata Miller) at various distances from a coal-fired power plant and downwind of a copper smelter in east Tennessee, and red spruce (Picea rubens Sarg.) in both the Great Smoky Mountains National Park (GSMNP) and at Camels Hump Mountain in Vermont. Also sampled in the Smokies were Fraser fir (Abies fraseri (Pursh) Poir.) at high-elevation sites, eastern hemlock (Tsuga canadensis (L.) Carr.) at mid-elevation sites, and pitch pine (Pinus rigida Mill.) and eastern white pine (Pinus strobus L.) at low-elevation sites. The trees were sampled with a 40.6 cm in length, 12 mm in diameter, nickel-plated, steel increment corer. In the laboratory the cores were cleaned and sectioned into phloem and 5- and 10-year xylem sections, dry ashed at 400°C, taken up in 10% HNO_3, and analysed for 31 elements by inductively coupled plasma optical emission spectroscopy. Details of the sampling, analysis, and quality control procedures are available elsewhere (Baes et al. 1984; Baes and McLaughlin 1986).

Our surveys revealed increased concentrations of Al, Cu, Fe, and Ni in rings formed since the 1950s in almost all trees, regardless of age, in a wide variety of sites. At some sites, zinc and manganese also showed increases since the 1950s. The greatest post-1950s increases in tissue concentrations of these metals were found in trees near a point source of trace metal emissions and at high-elevation sites. Also, high concentrations of potentially toxic metals were observed in stem phloem tissues at a number of sites.

Concentration patterns of trace metals in xylem may reflect temporal changes in direct deposition of trace metals on leaf surfaces or on soil. For example, in short-leaf pine growing 1.6 km downwind of the 1700-MWe coal-fired Kingston Steam Plant, which began operations in 1956, xylem that formed after 1956 shows elevated levels of Cu (threefold), Fe (fourfold), and Zn (sixfold, Fig. 2a), compared to xylem formed before this date. Trees growing >16 km downwind of the plant show only slight increases of copper and iron and no increase in zinc after 1956. Figure 2a compares the patterns for zinc only, but the patterns for copper and iron are very similar.

The concentration pattern for zinc near Kingston Steam Plant is of particular interest. Zinc has been shown to translocate across xylem rings from younger to older tissue (Symeonides 1979). The typical pattern of zinc concentrations in xylem is

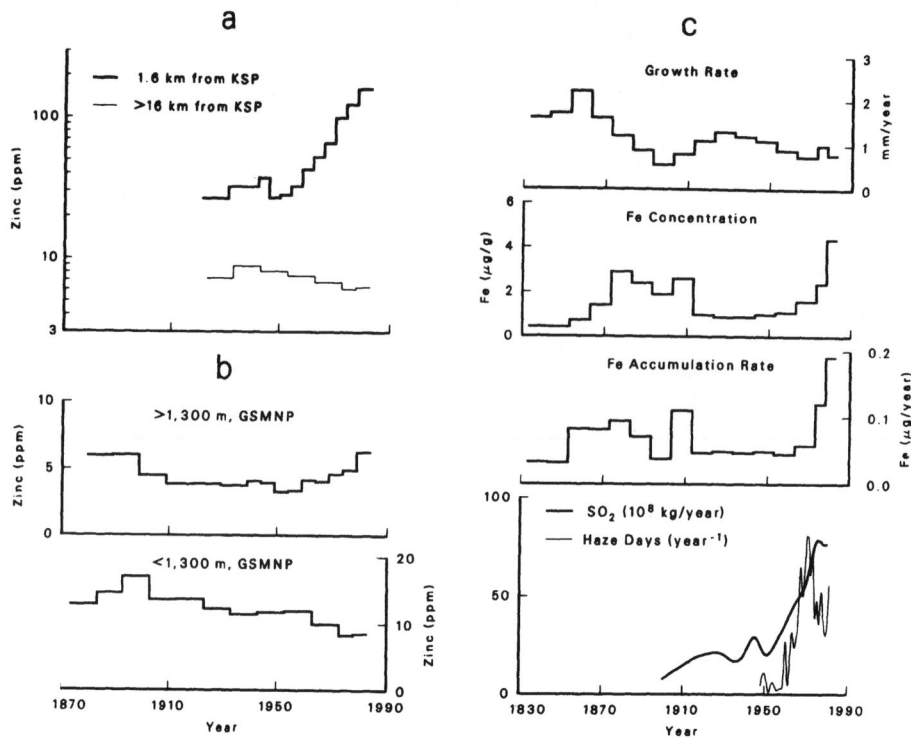

Fig. 2. (a) Comparison of temporal concentration patterns for zinc in short-leaf pine 1.6 and > 16 km downwind of Kingston Steam Plant; (b) comparison of temporal concentration patterns for zinc in Fraser fir at high-elevation and in hemlock, short-leaf pine, pitch pine, and white pine at low-elevation sites in the GSMNP; and (c) comparison of growth rate and iron in short-leaf pine downwind of the Copper Hill smelters with temporal patterns of increase in sulphur dioxide emissions within 900 km southeast-southwest of the GSMNP and number of haze days (U.S. Department of Commerce 1982) near the site.

rather flat, with highest concentrations in older tissue and lowest concentrations in the most recent tissue, which is consistent with across-ring translocation from younger to older tissues (Fig. 2b). In our studies, exceptions to this typical pattern were observed only twice: in short-leaf pine near Kingston Steam Plant (Fig. 2a) and in Fraser fir at high-elevation sites in the GSMNP (Fig. 2b). In both instances the deviation from the norm suggests that the xylem concentrations show a short-term response to increased zinc deposition, despite the likely slower rate of zinc translocation.

Concentration patterns of trace metals in xylem may also reflect temporal changes in soil availability of trace metals in response to acid deposition. For example, suppressed growth and increased iron concentration between 1863 and 1912 were observed in short-leaf pine growing 88 km downwind of a copper smelter which released large amounts of SO_2 during this period of time (Fig. 2c). The iron concentrations cannot be explained solely as a function of reduced growth. When variation in growth rate is factored out of the concentration pattern, the remaining temporal "xylem accumulation rate" (μg Fe.g^{-1} wood x g wood.year^{-1} or μg Fe.year^{-1}) pattern (Baes and McLaughlin 1984) still shows the same temporal response shown by the iron concentration pattern (Fig. 2c).

We believe that the historical changes in iron concentration reflected changes in iron availability, either through soil or foliar pathways, associated with inputs of sulphur and associated acidity (Baes and McLaughlin 1984). Similar increases of iron (and other metals) were found in rings formed after the 1950s, a period when regional fossil-fuel combustion emissions increased roughly 200% (Gschwandtner et al. 1984, Fig. 2c).

At all sites, highest trace metal concentrations in bole tissues were found in living phloem (plus cambium) tissues at levels one to two orders of magnitude higher than the highest concentrations in xylem tissue. Levels of copper in the phloem were generally less than levels considered toxic in aboveground plant tissues (Table 1), but levels of Al, Cd, Fe, Mn, Ni, and Zn, particularly at Camels Hump Mountain, Vermont, often exceeded levels indicated as toxic in herbaceous vegetation. It is not known whether such concentrations are toxic in these tree tissues (iron, manganese, and zinc are essential nutrients and these levels may be normal in phloem); however, we believe that metal concentrations of such magnitude in living tree tissues are of potential concern, and further research is warranted.

In the GSMNP, statistically significant differences in phloem concentrations of Al, Mn and Zn ($p < 0.01$) and Cd and Ni ($p < 0.05$) were found among tree species. This is particularly evident in comparisons of red spruce and southern pines (Table 1). Statistically significant site differences (within a single species) were noted for Cd, Fe, and Mn in the GSMNP and for Al, Fe, Mn and Zn in red spruce at Camels Hump. At the latter site, higher concentrations of all metals but zinc were found at the higher elevation. Our studies in the GSMNP indicate that phloem

Table 1. Comparison of trace metal concentrations (ppm, dry weight) in tree bole tissues based on recent surveys of conifers in Tennessee and Vermont.

Site[a]	Species[b]	Number	Al	Cd	Cu	Fe	Mn	Ni	Zn
Phloem (including cambium)									
KSP, far	SLP	19	400±40	2.3±0.7	2.5±0.6	34±13	290±20		46±4
GSMNP, high	RS,FF,H	42	99±32[c]	1.2±0.2	2.6±0.3	17±4	960±160	24±5	61±9
GSMNP, low	H,SLP,WP,PP	29	500±80	0.9±0.1	2.7±0.3	10±2	210±30	41±17	39±7
CHM, high	RS	5	34±9	3.2±0.5	3.5±0.4	250±55	1600±170		110±11
CHM, low	RS	5	11±2	3.8±0.7	3.2±0.7	140±39	1100±120		200±24
Toxicity thresholds in various plant tissues[d]			>62	2-80	>22	>100	>400	12-250	>150
Xylem (most recently formed)									
KSP, near	SLP	3	24±8	0.3±0.1	1.8±0.3	66±9	73±14		160±50
KSP, far	SLP	19	14±1	0.6±0.1	1.3±0.2	12±6	62±5		66±38
GSMNP, high	RS,FF,H	42	5.8±0.7		1.8±0.1	2.6±0.7	120±17	8.9±1.1	6.7±0.8
GSMNP, low	H,SLP,WP,PP	29	7.9±1.1		1.4±0.1	1.7±0.7	30±3	17±4	4.4±1.1
CHM, high	RS	5	22		1.0	37±11	170±15	2.5	8.6±0.8
CHM, low	RS	5				23±5	120±20		14±2

[a] KSP = Kingston Steam Plant, GSMNP = Great Smoky Mountains National Park, CHM = Camels Hump Mountain; high = high elevation (≥ 1,300 m), low = low elevation (< 1,300 m), near = 1.6 km downwind, and far ≥ 16 km downwind.

[b] RS = red spruce, FF = Fraser fir, H = eastern hemlock, SLP = short-leaf pine, WP = white pine, and PP = pitch pine.

[c] Mean ± standard error of the mean of all trees sampled.

[d] Dry weight tissue concentrations which have been reported as being toxic, based on Pavan et al. (1982) for Al, Sommers (1980) for Cd, Embleton et al. (1976) for Cu, Foy et al. (1978) for Fe, Gough et al. (1979) for Mn, Soane and Saunder (1959) for Ni, and National Research Council (1979) for Zn.

tissue concentrations of Cu, Ni, and Zn are linearly dependent on the metals' availability in soil as measured by dilute acid (0.1 N HCl) extractability.

POTENTIAL TRACE METAL EFFECTS

The most critical problem in determining the impacts of increasing trace metals in forest ecosystems is quantifying their toxicity thresholds, both individually and in combination. Determination of toxicity thresholds for tissue levels of trace metals is further complicated by their direct and indirect influence on nutrient uptake by plants. Aluminum, for example, may reduce phosphorus or calcium availability to plants (Foy 1971). Aluminum may also replace calcium and magnesium in cell walls because of its much greater affinity for pectins (Hütterman, in press).

A comparison of the relative toxicity (concentrations that induce deficient growth or toxic reactions in seed plants) of trace metals, based on nutrient solution culture tests (Bowen 1979), indicates that several elements, including Cd, Cu, Cr, Fe, and Pb, have toxicity ranges that overlap that of aluminum (Table 2). Of more interest from a field assessment standpoint are dry weight tissue concentrations at which toxic reactions occur. Unfortunately, there are very limited data available on soil and tissue concentrations at which metals become toxic to trees in forest soils. It is known, however, that there are substantial differences among plant species and soil types (Table 2).

Experiments with hardwood and conifer tree seedlings cultured in nutrient solutions indicate that toxicity thresholds of aluminum may be rather high (10 to 160 ppm, McCormick and Steiner 1978; Schier 1985). By contrast, toxicity thresholds for Norway spruce may be on the order of 1.0 to 2.0 ppm (Rost-Siebert 1984), and maintenance of a Ca:Al ratio $\geq$ 1.0 may be essential to minimize root damage. Considerably lower toxicity thresholds have been noted for agricultural crops for which the toxicity of aluminum in acid soils is a well-recognized problem. A comparison of aluminum toxicity to cotton in soil solutions from various southeastern soils in the United States (Adams and Moore 1983) indicates that toxicity is highest in some B horizons, with soil solution concentration toxicity thresholds as low as 0.01 ppm. Aluminum in upper soil horizons is 20 to 330 times less toxic, presumably because of the formation of organic complexes.

Table 2. Toxicity thresholds and typical soil solution concentrations for aluminum, based on a review of the literature.

Concentration (ppm)	Media	Specific for	Reference
Reported toxicity thresholds:			
0.27	Nutrient culture	Coffee seedlings	Pavan et al. 1982
1-2	Nutrient culture	Norway spruce	Rost-Siebert 1984
3	Nutrient culture	Poplar clones	Steiner et al. 1984
3	Nutrient culture	Peach seedlings	Kirkpatrick et al. 1975
10-160	Nutrient culture	Six tree genera	McCormick and Steiner 1978
62-100	Leaf tissue	Coffee seedlings	Pavan et al. 1982
113	Leaf tissue	Peach seedlings	Kirkpatrick et al. 1975
130	Needles	Red spruce	Schier, in press
250	Needles	Balsam fir	Schier, in press
Typical soil solution concentrations:			
0.1-6.8	Humus	Beech forest	Rost-Siebert 1984
0.2-17	Humus	Spruce forest	Rost-Siebert 1984
0.3-5.7	Groundwater	Beech forest	Rost-Siebert 1984
0.38	A_2 horizon	Fir-zone forest	Cronan and Schofield 1979
0.55	E horizon	Hardwood forest	David and Driscoll 1984
0.72	B horizon	Conifer forest	David and Driscoll 1984
1.0-49.5	50 cm depth	Spruce forest	Rost-Siebert 1984
2-7	A horizon	Swedish podzols	Tyler 1981

Because of the uncertainties of the influences of elemental ratios and rhizosphere reactions in nutrient solutions, the levels of trace metals in plant tissues may be a more reliable and realistic indicator of toxic thresholds for comparison between laboratory and field studies. Aluminum concentrations in peach seedlings growing in a 3-ppm aluminum solution, for example, were at levels of 1700, 48, and 113 ppm (dry weight) in roots, stems, and leaves, respectively, and were associated with reduced growth of both shoots and roots (Kirkpatrick et al. 1975). Leaf concentrations at which coffee seedlings grown in acid soils showed toxic effects were 62 and 100 ppm for inhibition of root and shoot growth, respectively (Pavan et al. 1982).

Soil solution aluminum levels are generally below those used in nutrient culture studies. Aluminum concentrations in soil leachates were found to be 0.32 and 0.69 ppm for B horizons from hardwood and conifer forests, respectively, in the Adirondack Mountains of New York (David and Driscoll 1984), and 0.38 ppm for an A_2-horizon percolate from a fir-zone forest in New Hampshire (Cronan and Schofield 1979). These values are somewhat lower than the 1- to 2-ppm aluminum soil solution concentrations reported by Ulrich et al. (1980) as occurring on a continuous basis in the Solling Forest in Germany. Tyler (1981) reported substantial seasonal fluctuations of aluminum in soil solutions from Sweden, ranging from a minimum of 2 ppm in the spring to a maximum of 7 ppm in late summer. Rost-Seibert (1984) reported levels as high as 49.5 ppm at a depth of 50 cm in a spruce forest in Germany. In general, however, aluminum concentrations from these forest soil solutions are much lower than reported toxicity thresholds for nutrient culture experiments with tree seedlings, but much higher than injury thresholds determined for agricultural species (Table 2). The importance of such differences in species tolerances and the applicability of nutrient culture tests to field situations have yet to be evaluated.

SUMMARY AND CONCLUSIONS

There is evidence of recently increasing deposition of trace metals to forest ecosystems based on historical changes in fossil fuel consumption and increasing trace metal concentrations in lake sediments and soils. Our studies of tree-ring chemistry chronologies in conifers show a general trend of trace metal concentration increases since the 1950s that follow very closely regional historical increases in emissions of fossil-fuel combustion products. These increases in tree-ring trace metals may reflect increases in metal deposition or availability or both.

Of particular interest are recent increases in concentrations of Al, Cd, Cu, Fe, Mn, Ni, and Zn. Copper, iron, and zinc were significantly increased in modern rings of short-leaf pine near a coal-fired power plant as compared to earlier formed rings. Because these recent trace metal concentration increases are seen in conifers of various ages, occur at a variety of sites, and show sensitivity to local emissions, we believe that conifers may serve as bioindicators of trace metal inputs to forest ecosystems.

Also found were high concentrations of trace metals in tree phloem at a number of sites in east Tennessee and at Camels Hump Mountain in Vermont. Of particular significance are aluminum that often exceeds 200 ppm, iron that often exceeds 100 ppm, and manganese that often exceeds 400 ppm, levels known to be toxicity thresholds in foliage of agricultural plants. It is not possible to say whether levels found in phloem are normal or typical for these tissues, but the potential for direct or indirect toxicity to reduce tree vigor should be investigated.

The mechanisms of increased availability of trace metals to forests are many, but our understanding of their uptake, cycling, and effects are poorly understood. It is known that excess trace metals can reduce soil respiration, alter root and shoot vigor, and affect nutrient uptake. However, it is unclear what the impacts of sublethal concentrations of trace metals are and whether they might contribute directly or indirectly to recent forest declines. Part of the problem is assembling a coherent picture of exactly what toxicity thresholds are. The numerous studies of aluminum toxicity, for example, involve different plant species, tissue culturing techniques and chemical forms of aluminum, and produce varying indications of aluminum toxicity thresholds to plants. It is quite difficult to integrate such results to assess the impacts of current levels of trace metals in forest ecosystems. By increasing our understanding of trace metal cycling, uptake and effects in forests and of influences of soil type, plant species, and abiotic factors, however, a clearer understanding of the role, if any , that trace metals play in forest declines will emerge.

Acknowledgements

We wish to thank A.J. Freidland who provided us with the red spruce cores from Camels Hump. This research was sponsored by the National Park Service, U.S. Department of the Interior, under Interagency Agreement NPS 0492-082-2 (DOE 40-1249-82) with Martin Marietta Energy Systems, Inc., under Contract No. AE-AC05-84OR21400 with the U.S. Department of Energy.

Publication 2571, Environmental Sciences Division, Oak Ridge National Laboratory.

REFERENCES

Adams F, Moore BL (1983) Chemical factors affecting root growth in subsoil horizons of coastal plain soils. Soil Sci Soc Am J 47: 99-102
Baes CF III, McLaughlin SB (1984) Trace elements in tree rings: evidence of recent and historical air pollution. Science 224: 494-497
Baes CF III, McLaughlin SB (1986) Multielemental analysis of tree rings: a survey of coniferous trees in the Great Smoky Mountains National Park. ORNL-6155, Oak Ridge National Laboratory, Oak Ridge, TN
Baes CF III, McLaughlin SB, Hagan TA (1984) Multielemental analysis of tree rings: temporal accumulation patterns and relationships with air pollution. In: Davis DD, Millen AA, Dochinger L (eds) Air pollution and the productivity of the forest, Proceedings of the Symposium, October 4-5, 1983, Washington, DC, p 273

Bowen HJM (1979) Environmental chemistry of the elements. Academic Press, New York, p 333

Chaney WR, Kelly JM, Strickland RC (1978) Influence of cadmium and zinc on carbon dioxide evolution from litter and soil from a black oak forest. J Environ Qual 7: 115-119

Cronan CS, Schofield CL (1979) Aluminum leaching response to acid precipitation effects on high elevation watersheds in the Northeast. Science 204: 304-305

David MB, Driscoll CT (1984) Aluminum speciation and equilibria in soil solutions of a haplorthod in the Adirondack Mountains (New York, U.S.A.). Geoderma 33: 297-318

Embleton TW, Jones WW, Platt RG (1976) Leaf analysis as a guide to citrus fertilization. In: Reisenauer HM (ed) Soil and plant-tissue testing in California. University of California, Division of Agricultural Science Bulletin 1879, p 4

Foy CD (1971) Effects of aluminum on plant growth. In: Carson EW (ed) The Plant Root and its Environment. Proceedings of an institute, sponsored by the Southern Regional Education Board, held at Virginia Polytechnic Institute and State University, July 5-16, 1971, University Press of Virginia, Charlottsville, chapter 20

Foy CD, Chaney RL, White MC (1978) The physiology of metal toxicity in plants. Ann Rev Plant Physiol 46: 511-566

Friedland AJ, Johnson AH, Siccama TG (1984) Trace metal content of the forest floor in the Green Mountains of Vermont: spatial and temporal patterns. Water Air Soil Pollut 21: 161-170

Gough LP, Shacklette HT, Case AA (1979) Element concentrations toxic to plants, animals, and man. USGS Bulletin 1466, U.S. Government Printing Office, Washington, DC

Gschwandtner G, Gschwandtner KC, Eldridge K (1984) Historic emissions of sulfur and nitrogen oxides in the United States from 1900 to 1980. Pacific Environmental Services, Inc, Durham, NC

Hanson DW, Norton SA, Williams JS (1982) Modern and paleolimnological evidence for accelerated leaching and metal accumulation in soils in New England, caused by atmospheric deposition. Water Air Soil Pollut 18: 227-239

Hütterman A (in press) The effects of acid deposition on the physiology of the forest ecosystem. Experientia

Hutchinson TC (1980) Impact of heavy metals on terrestrial and aquatic ecosystems. In: Miller PR (ed) Effects of air pollutants on mediterranean and temperate forest ecosystems. USDA Forest Service Report PSW-43, p 158

Jackson DR, Watson AP (1977) Disruption of nutrient pools and transport of heavy metals in a forested watershed near a lead smelter. J Environ Qual 6: 331-338

Johnson AH, Siccama TG, Friedland AJ (1982) Spatial and temporal patterns of lead accumulation in the forest floor in the northeastern United States. J Environ Qual 11: 557-580

Kirkpatrick HC, Thompson JM, Edwards JH (1975) Effects of aluminum concentration on growth and chemical composition of peach seedlings. Hortscience 10: 132-134

Lindberg SE, Harriss RC, Turner RR (1982) Atmospheric deposition of metals to forest vegetation. Science 215: 1609-1611

McCormick LH, Steiner KC (1978) Variation in aluminum tolerance among six genera of trees. For Sci 24: 565-568

McLaughlin SB (1985) Air pollution effects on forests: a critical review. J Air Pollut Control Assoc 35: 512-534

National Research Council (1979) Zinc. University Park Press, Baltimore, MD

Norton SA, Hanson DW, Campana RJ (1980) The impact of acidic precipitation and heavy metals on soils in relation to forest ecosystems. In: Miller PR (ed) Effects of air pollutants on mediterranean and temperate forest ecosystems. USDA Forest Service Report PSW-43, p 152

Pavan MA, Bingham FT, Pratt PF (1982) Toxicity of aluminum to coffee in Utisols and Oxisols amended with $CaCO_3$, $MgCO_3$ and $CaSO_4 \cdot H_2O$. Soil Sci Soc Am J 46: 1201-1207

Raisch W, Zöttl HW (1983) Heavy metal distribution in spruce stands in the Barholde Watershed (Southern Black Forest). Mitt Dtsch Bodenkundl Ges 38: 399-406

Rost-Siebert K (1984) Aluminum toxicity in seedlings of Norway spruce (Picea abies Karst.) and beech (Fagus silvatica L.). In: Proceedings of the Workshop on Aluminum Toxicity to Trees. Uppsala, Sweden, May 14-16, 1984

Schier GA (1985) Response of red spruce and balsam fir seedlings to aluminum toxicity in nutrient solutions. Can J For Res 15: 29-33

Soane BD, Saunder DH (1959) Nickel and chromium toxicity of serpentine soils in southern Rhodesia. Soil Sci 88: 322-330

Sommers LE (1980) Toxic metals in agricultural crops. In: Sludge - Health Risks of Land Application. Ann Arbor Science, Ann Arbor, MI, p 105

Steiner KC, Barbour JR, McCormick LH (1984) Response of Populus hybrids to aluminum toxicity. For Sci 30: 404-410

Symeonides C (1979) Tree-ring analysis for tracing the history of pollution: application to a study in northern Sweden. J Environ Qual 8: 482-486

Turner RR, Lindberg SE, Coe JM (1985) Comparative analysis of trace metal accumulation in forest ecosystems. Heavy metals in the environment, Vol 1, CEP Consultants, Edinburgh, p 356-358

Tyler G (1981) Leaching of metals from the A-horizon of a spruce forest. Water Air Soil Pollut 15: 353-369

Ulrich B, Mayer R, Khanna TK (1980) Chemical changes due to acid precipitation in a loess-derived soil in central Europe. Soil Sci 130: 193-199

U.S. Department of Commerce (1982) Tape Deck 1440, Airways surface observations. National Oceanic and Atmospheric Administration, Ashville, NC

Van Hook RI, Harris WF, Henderson GS (1977) Cadmium, lead, and zinc distributions and cycling in a mixed deciduous forest. Ambio 6: 281-286

Vaughan BE, Abel KH, Cataldo DA, Hales IM, Hane CE, Rancitelli LA, Routson RC, Wildung RE, Wolf EG (1975) Review of potential impact on health and environmental quality from metals entering the environment as a result of coal utilization. Battelle-Pacific Northwest Laboratories, Richland, WA

Whitby LM, Hutchinson TC (1974) Heavy metal pollution in the Sudbury mining and smelting region of Canada. II. Soil toxicity tests. Environ Conserv 1: 191-200

AIR POLLUTION AND SOIL ACIDIFICATION

G. Abrahamsen

Norwegian Forest Research Institute, 1432 AAS-NLH, Norway

ABSTRACT

Evaluation of effects of acid deposition on soil acidity can be based on general considerations of acid-producing and acid-consuming processes in the soil, theoretical calculations, reanalyses of soils previously analysed for soil acidity and experiments with artificial acidification.

Use of these different approaches indicates that many soils exposed to atmospheric acid deposition are likely to become more acidic. However, theoretical considerations, as well as experimental results, show that a fast-growing forest stand has a very significant natural acidifying effect on the soil. In areas with moderate acidic deposition, this effect is likely to override the effect of atmospheric deposition. Interpretation of changes in soil acidity with time is, for this and other reasons, sometimes very difficult.

INTRODUCTION

Soil acidity is the primary factor to be considered when evaluating the effects of acid deposition on soil. Soil acidity influences a large number of biological and chemical processes which are of utmost importance for plant productivity.

The acidity of natural soils is a function of long-term leaching combined with processes which produce and consume acids. Net leaching from soils is restricted to areas with a surplus of precipitation. The acid-producing processes include decomposition of organic material through production of organic and carbonic acid, mineralization of plant nutrients especially of N and S, and accretion of base cations in the forest biomass. The acid-consuming processes are mainly weathering of soils minerals, and under anaerobic conditions, reduction of N and S compounds. In most soils in areas with a surplus of precipitation, the processes producing acids have been greater than those consuming acids. During the period of soil formation, these soils have therefore gradually become acidic.

Natural soil acidification is commonly thought to be a slow process. The acidity we observe today, such as in many soils of the temperate region, is generally a result of soil-forming processes during the last 10,000 years. There are, however, indications that soil acidity is increasing faster at present than what is considered to be natural. The cause is often ascribed to atmospheric acid deposition.

The intention of this paper is to discuss effects of acid deposition on soil acidity. This is done by discussing: 1) the theoretical effects of acid deposition on soil acidity; 2) what is known about changes in soil acidity in recent times; and 3) the experimental evidence for relations between acid deposition and soil acidity. Also, alternative reasons for increased soil acidity are briefly discussed.

THEORETICAL EFFECTS OF ACID DEPOSITION ON SOIL ACIDITY

The theoretical effects of acid deposition on soil acidity are discussed in a number of papers during the last decade. In the present paper, therefore, only the general principles will be briefly discussed. There is general agreement that the main effect of acid deposition on soil acidity is due to the increased input of SO_4^{2-} and H^+ and, in some cases, NO_3^- and NH_4^+. The mobility of anions is the driving force in the acidification process. It has been shown that in most temperate soil/plant systems SO_4^{2-} is much more mobile than NO_3^- (e.g., Abrahamsen 1980). This is not due to sorption in the soil, but to the higher biological need for N than for S. It should not be forgotten, however, that there are also soils in the temperate region that, similar to most tropical soils, have the ability to sorb large amounts of SO_4^{2-} (Johnson et al. 1982). Nor should it be forgotten that there are also temperate forest ecosystems that are saturated with N and in which significant amounts of NO_3^- are leached from the soil (van Breemen and Jordens 1983; van Miegroet and Cole 1984).

Anions cannot be leached from the soil without being accompanied by equivalent amounts of cations. The effect of acid deposition on soil acidity is due to the deposition of H^+ which exchanges with other cations in the soil. With a surplus of water they will be leached from the soil together with the anions. In cases where the leached cations are mainly base cations, the soil is likely to become more acidic. In other cases where the leached cations are mainly H^+ and Al ions, the soil will not become more acidic, but the surface and ground water will be influenced. The acidifying effect of acid deposition, therefore, depends on the efficiency of H^+ to exchange with base cations in the soil. The replacing efficiency of H^+ depends on the mineral composition and base saturation of the soil (Wiklander and Andersson 1972). Low base saturation corresponds to low replacing efficiency. Although we know the general relationship between the replacing efficiency and the acid/base status of the soil, this is not enough to tell us whether a given soil will become more acidic due to the atmospheric deposition. The reason is that we do not know to what extent the leached base cations are replaced by cations released by weathering processes. There are studies indicating that weathering of feldspar is not much influenced by acidity as long as the pH of the soil solution varies between 3 and 8 (Helgeson et al. 1984). This suggests that acid precipitation, which often has a pH of about pH 4.2, is not sufficiently acidic to affect the weathering rate.

This problem can also be evaluated from a slightly different point of view. Most natural soils in the temperate forest area are much more acidic than the parent material. This means that during soil formation, acid-producing processes have dominated over acid-consuming processes. Putting more acid into such a system would most likely increase the acidity.

In order to predict the long-term effect of acid deposition on soil acidity, calculations have been carried out on the theoretical reductions in soil pH and base saturation due to atmospheric H^+ inputs (McFee et al. 1977; Altshuller and Linthurst 1984). These calculations also indicate a reduction in base saturation of the upper part of the soil.

When evaluating short-term changes in soil acidity it is also important to realize that changes can take place without the influence of atmospheric acid deposition. For example, the production of acid due to the build-up of plant biomass and humus material may amount to annual average values of 50-350 mmol $H^+.m^{-2}.y^{-1}$ for a rotation period (Andersson et al. 1980; Nilsson et al. 1982; Driscoll and Likens 1982; Matzner and Ulrich 1981; van Breemen et al. 1984). If the biomass returns to the soil, base cations will be released and the net H^+ increase will be minimal in the long run. However, if the biomass is removed there will be a net production of acid which increases with the amount of biomass exported. If only stems are harvested, the net H^+ production probably will be less than 10-50 mmol.$m^{-2}.y^{-1}$. However, if whole trees are removed in clear cuttings, the H^+ production may be up to 120 mmol.$m^{-2}.y^{-1}$ (Nilsson 1983).

Significant soil acidification may also take place in alder forest ecosystems where large amounts of N are fixed biologically and where nitrification produces both acidity and a mobile anion (van Miegroet and Cole 1984). Also, variations in the amounts of precipitation from one year to another affect the acidity of soils. This effect is probably most significant in the part of the soil influenced by fluctuating groundwater.

CHANGES IN SOIL ACIDITY IN RECENT TIME

During the last 10-12 years, an increasing number of studies have been carried out to estimate changes in soil acidity over time. In the first studies, no change in soil pH was observed (Wiklander 1973/74; Linzon and Temple 1980). Later, however, quite a few studies from Central Europe have indicated significant reduction in soil pH over the last decades (Blume 1981; Butzke 1981; von Zezschwitz 1982; Evers 1983; Reichmann and Streitz 1983; Grenzius 1984; Reibeling and Schaefer 1984; Glatzel et al. 1985). These studies have shown that decreases in pH are larger with high original pH's than with low ones. pH depressions during a 10-30 year period of the order of 0.5 to >1 pH unit appear to be common.

In Scandinavia as well, resampling of soils has given some disquieting results. In a National Forest Survey in Sweden, a comparison was made between soil chemical properties found in 1961/63 with those found in 1971/73 (Troedsson 1980). During this period, a slight increase in exchangeable H^+ and Al was found together with a more significant decrease in exchangeable Ca, Mg and K. The author was very cautious in explaining the reason for the changes and emphasized the effect of forest age on soil acidity. The change observed was an average change for a large number of soil samples distributed randomly over a large territory. In another study in southwestern Sweden an area was sampled in 1927 and resampled in 1982/83 (Tamm and Hallbacken, in press). In this area the deposition of S and N is of the order of 25 and 20 $kg.ha^{-1}.y^{-1}$, respectively. Soil pH (H_2O) was measured exactly the same way on both occasions. The results showed that changes in pH over time was as large or sometimes even larger in the C-layer than in the organic layer. In the O-horizon reduction in pH from about 4.2 to about 3.7 was found in different stands of Norway spruce and beech. In the C-horizon, soil pH was reduced from about 5.2 to about 4.6. Due to the lower pH in the upper horizon the change in H^+ concentration in the C-layer is about 10-20% of that in the organic layer. Increasing age involved reduced soil pH. However, the correlation between age and soil pH decreased with increasing soil depth. Thus, in the C-layer, no correlation was found. The regressions between age and pH of the 1927 material had the same slope as that of 1982/83, but the entire 1982/83 material showed significantly lower pH values than the older material. The author's interpretation of this was that the difference between 1927 and 1982/83 especially for B- and C-horizons, must be explained by something else than just natural biological effects.

In Norway, podzolic soil from a mountain area (Rondane) above the tree line has also been resampled (Dahl, in preparation). The geology of the area is dominated by very nutrient-poor gray feldspatic metasandstone with beds of quartz conglomerates. Also in this area significant changes in soil pH (H_2O) were found over the period 1942-49 to 1984. Soil pH was measured the same way on both occasions. The change in pH increases with increasing soil depth. In the top soil (O-layer) the pH was reduced from 4.3 to 4.0 and in the C-layer from 5.7 to 4.9. The observed changes in soil pH in this study are surprising in view of the relatively small deposition of pollutants in the area (3-4 $kg\ S.ha^{-1}.y^{-1}$). Therefore, one may wonder if other factors such as long-term climatic fluctuations have been of importance.

EXPERIMENTAL RESULTS OF SOIL ACIDIFICATION

As discussed previously, it is extremely difficult to give quantitative estimates of the impact of acid deposition on soil acidity. Therefore, it is reasonable to use experimental methods. Long-term effects of acid deposition may be simulated by applying water with higher concentrations of acids than in normal acid precipitation. Laboratory and greenhouse studies of this kind have been carried out with the use of simulated acid rain (Wood and Bormann 1977; Ogner ad Teigen 1980; Bjor and Teigen 1980; Lee and Weber 1982). All studies have shown increased leaching of base cations with increasing acidity of the "rain". Compared to the effect of "control rain", effects of "pH 4 rain" on the exchangeable base cation content of the soil were only slight. With "rain" of pH 3 and lower, however, all studies gave similar results: significant reduction in the base saturation of the soil. However, at such low pH-values plants are damaged, and part of the reduction in base saturation is likely to be indirectly influenced by this.

Field experiments with acid application have results similar to those of laboratory experiments. In Swedish experiments the acid applied was in the form of a fairly strong solution of sulphuric acid, in amounts corresponding to a total of about 200 kg $S.ha^{-1}$ over a six-year period (Farrell et al. 1980). This resulted in a statistically significant decrease in the content of exchangeable "base cations" in all soil layers down to the C-horizon, except in the eluviated layer.

Figures 1 and 2 give some new results from two similar Norwegian field experiments where artificial rain of different acidity has been applied. The artificial rain was made from groundwater with added H_2SO_4 to obtain the pH levels of the different treatments. Some characteristics of these experiments are shown in Table 1.

Table 1. Description of two Norwegian field experiments with application of artificial rain of different acidity. Further description of the experiments is given by Abrahamsen et al. (1976) and Stuanes and Sveistrup (1979).

	Experiment	
	A-2	B-1
Location	Nordmoen, 11°7'E, 60°15'N	Aamli, 8°30'E, 58°48'N
Forest	Norway spruce	Scots pine
Year of planting	1956	1968-70
Height Autumn 1984	7.3 m	5.2 m
Soil Type		
U.S.D.A.	Typic Udipsamment	Typic Udorthent
Canada	Eluviated dystric brunisol	Orthic Humo-ferric podzol
FAO-UNESCO	Cambic arenosol	Cambic arenosol
Size of expt. plots	150 m^2	75 m^2
Treatments	Not watered, pH 6, pH 4, pH 3, pH 2.5	Not watered, pH 6, pH 4, pH 3, pH 2.5, pH 2.0
No. of replicates	3	3
No. of irrigations	27, 50 mm each time	38, 50 mm each time
Period of irrig.	July 1973 - Sept. 1978	August 1974 - Sept. 1981
Atmospheric Depos.		
S (kg.ha^{-1}.y^{-1})	10 kg	15 kg
N (kg.ha^{-1}.y^{-1})	8 kg	13 kg

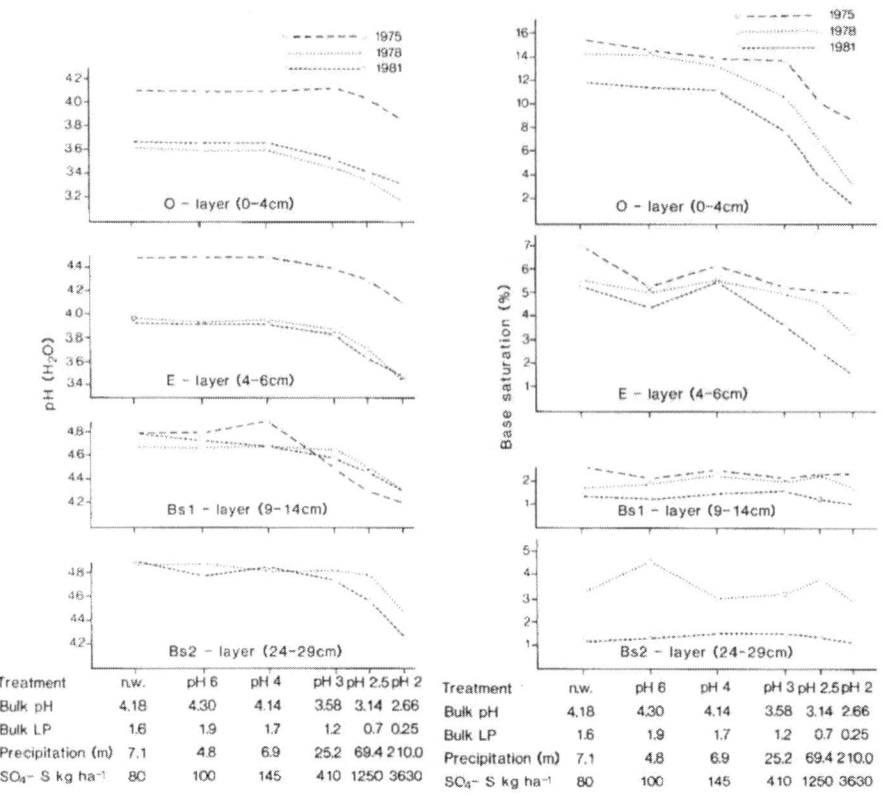

Fig. 1. Effects of different pH in the artificial rain on soil pH and base saturation in various soil layers in experiment B-1. pH and lime potential (LP) of bulk precipitation (artificial plus natural) are shown together with the equivalent amount of precipitation with pH 4.2 and [Ca + Mg] = 7 µM. Also, the total amount of SO_4^{2-}-S supplied in the different treatment is given.

The canopy in experiment A-2 is almost closed, but this has not yet occurred in experiment B-1. The atmospheric deposition values given in the table are for wet plus dry deposition. Dry deposition is estimated to be 25% of the total deposition (Overrein et al. 1980). The watering was carried out five times annually in the frost-free period. The groundwater used had a higher content of K^+, Na^+, Ca^{2+}, Mg^{2+} and Cl^- than normal precipitation and a lower content of NO_3^- and NH_4^+.

Soils from these experiments were sampled in 1975, 1978, 1981 and 1984. From each plot a composite sample of 20 subsamples has been analysed. At present only soil pH (H_2O) in experiment A-2 has been measured in the 1984 sample. pH was measured on air-dried, sieved soil with water added in a soil/water ratio of 1/2.5. The exchangeable "base cations" were determined after extracting the soil with 1 M NH_4Ac (pH 7). Total acidity was determined by titrating the extract back to pH 7. Further descriptions of the chemical procedures used are given by Ogner et al. (1984).

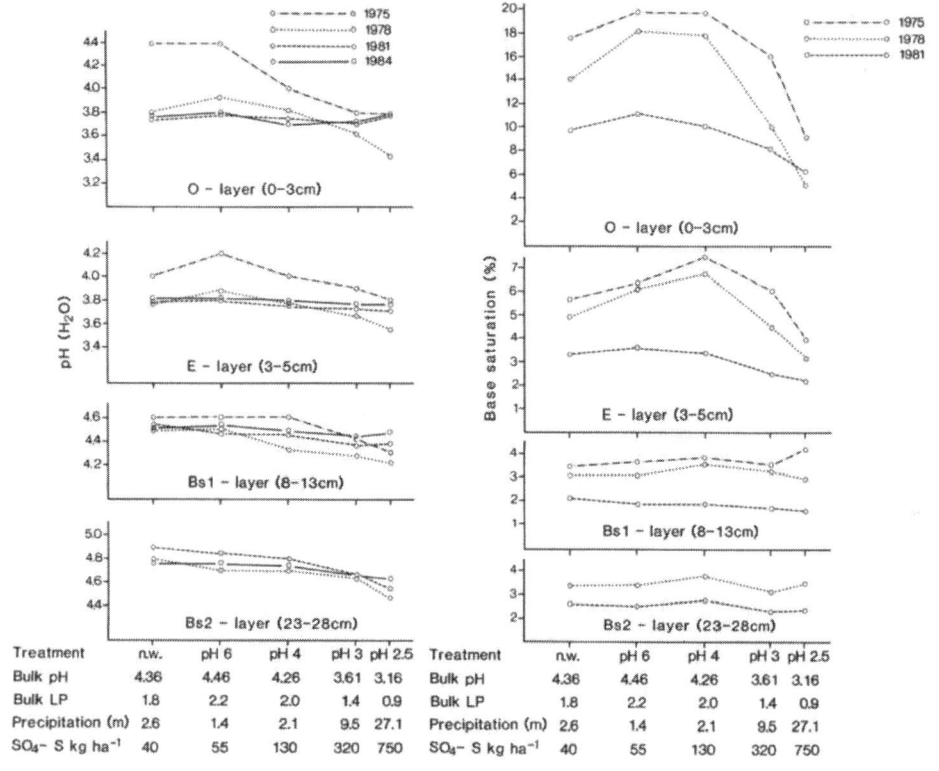

Fig. 2. Effects of different pH in the artificial rain on soil pH and base saturation in various soil layers in experiment A-2. Further information is given in Fig. 1.

The cation exchange capacity (at pH 7) has not been changed by the treatments. For experiment A-2 it varies from 560 mmol $(+).kg^{-1}$ in the O-layer to about 40 mmol $(+).kg^{-1}$ in the lower B-layer. In experiment B-1 the variation in CEC is from 1350 mmol $(+).kg^{-1}$ in the O-layer to about 20 mmol $(+).kg^{-1}$ in the lower B-layer.

Figure 1 shows results from experiment B-1. In 1975, one year after the experiment was started, soil pH had dropped in the two or three most acidic treatments. The reduction in pH was observed down to 9-14 cm soil depth. Deeper soil samples were not collected that year. The sampling in 1978 and 1981 confirmed the results observed in 1975. Effects of the two most acidic treatments could now also be observed in the 24-29 cm soil depth. However, from 1975 to 1978 a significant reduction in soil pH had taken place in the O- and E-layers in all plots irrespective of the different treatments. This effect is probably due to increased build-up of forest biomass. From 1978 to 1981 the change in soil pH was insignificant compared to the former three-year period. The base saturation shows more or less the same reaction as discussed for soil pH. In the B-horizon, however, little effect due to the treatments can been seen.

Figure 2 shows the variation in soil pH and base saturation in the A-2 experiment. In general, the same pattern appears as for experiment B-1. In A-2, application of artificial rain was terminated in Sept. 1978. After 1978 many of the differences between the treatments have disappeared. Base saturation, in contrast to soil pH, was also significantly lower in the most acidic treatments in 1981, but the dramatic differences found in 1975 and 1978 have greatly diminished.

The greater effects on soil chemistry observed in experiment B-1 compared to A-2 can partly be attributed to the large amounts of acid applied.

The experiments indicate that large changes in acidity created by application of sulphuric acid to the soil may be moderated during a relatively short period after the application. The most likely explanation for this is that the chemical imbalance created in the soil by applying sulphuric acid may have accelerated weathering processes.

CONCLUSION

The discussion in this paper shows some of the difficulties involved in determining the effect of acid deposition on soil acidity. Soil acidity is significantly influenced by a growing forest stand. The effect of the stand is largest during the first part of the rotation period, before closure of the canopy. Later, soil acidification should be reduced because of a more balanced uptake and recycling of nutrients. This effect complicates interpretation of changes in soil acidity over time. However, the study of Tamm and Hällbacken (in press) are very interesting as they have been able to sort out the effect of the growing forest.

In most of the resampling studies, significant soil acidification appears to have also occurred in the deeper soil layers. One should expect that when most of the H^+ is exchanged in the upper horizon, little acidification would take place deeper in the soil. The explanation may be that some H^+ is transported, together with anions, to deeper soil layers. In poorly buffered sands this may be enough to increase the acidity. This explanation is supported to some extent by experiments with artificial acidification. However, it is also important to realize that long-term climatic changes may involve changes in the redox potential of the soil. This effect is of special importance in deeper soil layers.

A number of arguments can be made against all methods used for studying the effect of atmospheric acid deposition on soil acidity. However, when considering the results of these various studies, I think the conclusion must be that the soil acidity of large areas exposed to acid deposition is increasing faster than natural rates of acidification. In the long-run, this is likely to have a number of negative effects on both soil productivity and freshwater quality.

REFERENCES

Abrahamsen G (1980) Acid precipitation, plant nutrients and forest growth. In: Drabløs D, Tollan A (eds) Proc int conf ecol impact acid precip. SNSF-project, Norway, p 58

Abrahamsen G, Bjor K, Teigen O (1976) Field experiments with simulated acid rain in forest ecosystems. I. Soil and vegetation characteristics, experimental design and equipment. SNSF-project, FR 4/76, Norway

Altshuller AP, Linthurst RA (eds) (1984) The acidic deposition phenomenon and its effects: Critical assessment of review papers. vol II Effects Sciences EPA-600/8-83-016BF

Andersson F, Fagerstrøm T, Nilsson I (1980) Forest ecosystem responses to acid deposition - hydrogen ion budget and nitrogen/tree growth model approaches. In: Hutchinson TC, Havas M (eds) Effects of acid precipitation on terrestrial ecosystems. NATO Conference Series Plenum Press, New York, p 319

Bjor K, Teigen O (1980). Effects of acid precipitation on soil and forest. 6. Lysimeter experiment in greenhouse. In: Drablos D, Tollan A (eds) Proc int conf ecol impact acid precip. SNSF-project, Norway, p 200

Blume HP (1981) Alarmierende versauerung Berliner Forsten. Berliner Naturschutzblatt 1981: 713-715

Butzke H (1981) Versauern unsere Wälder? Erste Ergebnisse der überprüfung 20 Jahre alter pH-Wert-Messungen in Waldböden Nordrhein-Westfalens. Der Forst- und Holzwirt 36: 542-548

Driscoll CT, Likens GE (1982) Hydrogen ion budget of an aggrading forested ecosystem. Tellus 34: 283-292

Evers FH (1983) Orientierende Untersuchungen langfristiger Bodenreaktionsänderungen in südwestdeutschen Düngungs-Versuchsflächen. Der Forst- und Holzwirt 38: 317-320

Farrell EP, Nilsson J, Tamm CO, Wiklander G (1980) Effects of artificial acidification with sulphuric acid on soil chemistry in a Scots pine forest. In: Drabløs D, Tollan A (eds) Proc int conf ecol impact acid precip. SNSF-project, Norway, p 186

Glatzel G, Kilian W, Sterba H, Stöhr D (1985) Waldbödenversauerung in Österreich: Ursachen - Auswirkingen. Allgemeine Forstzeitung 96: 35-36

Grenzius R (1984) Starke Versauerung der Waldböden Berlins. Forstw Cbl 103: 131-139

Helgeson HC, Murphy WM, Aagaard P (1984) Thermodynamic and kinetic constraints on reaction rates among minerals and aqueous solutions. II Rate constants, effective surface area, and the hydrolysis of feldspar. Geochim Cosmochim Acta 48: 2405-2432

Johnson DW, Turner J, Kelly JM (1982) The effects of acid rain on forest nutrient status. Water Resour Res 18: 449-461

Lee JJ, Weber DE (1982) Effects of sulphuric acid rain on major cation and sulphate concentrations of water percolating through two model hardwood forests. J Environ Qual 11: 57-64

Linzon SN, Temple PJ (1980) Soil resampling and pH measurements after an 18-year period in Ontario. In: Drabløs D, Tollan A (eds) Proc int conf ecol impact acid precip. SNSF-project, Norway, p 176

Matzner E, Ulrich B (1981) Effects of acid precipitation on soil. In Fazzolare RA, Smith CB (eds) Beyond the energy crisis. Opportunity and challenge. vol II, Pergamon Press, Oxford, New York, p 555

McFee WW, Kelley JM, Beck RH (1977) Acid precipitation effects on soil pH and base saturation of exchange sites. Water Air Soil Pollut 7: 401-408

Nilsson SI (1983) Effects on soil chemistry as a consequence of proton input. In: Ulrich B, Pankrath J (eds) Effects of accumulation of air pollutants in forest ecosystems. Reidel, Dorecht, p 150

Nilsson SI, Miller HG, Miller JD (1982) Forest growth as a possible cause of soil and water acidification: An examination of the concepts. Oikos 39: 40-49

Ogner G, Teigen O (1980) Effects of acid irrigation and liming on two clones of Norway spruce. Expanded version with basic data included. Nor inst skogforsk, 1432 Ås-NLH, Norway

Ogner G, Haugen A, Opem M, Sjøtveit G, Sørlie B (1984) The chemical analysis program at The Norwegian Forest Research Institute, 1984. Nor inst skogforsk, 1432 Ås-NLH, Norway

Overrein LN, Seip HM, Tollan A (1980) Acid precipitation - effects on forest and fish. Final report of the SNSF-project 1972-1980. SNSF-project, Norway

Reichmann H, Streitz H (1983) Fortschreitende Bodenversauerung und Waldschäden im industrienahen Stadwald Wiesbaden. Forst- und Holzwirt 38: 322-328

Riebeling R, Schaefer C (1984) Jahres- und Langzeitentwicklung der pH-Werte von Waldböden in hessischen Fichtenbeständen. Forst- und Holzwirt 39: 177-182

Stuanes A, Sveistrup TE (1979) Field experiments with simulated acid rain in forest ecosystems. 2. Description and classification of the soils used in field, lysimeter and laboratory experiments. SNSF-project, FR 15/79, Norway

Tamm CO, Hallenbäcken L (in press) Changes in soil pH over a 50-year period under different forest canopies in SW Sweden. Water Air Soil Pollut

Troedsson T (1980) Long-term changes of forest soils. Annls Agric Fenn 19: 81-84

van Breemen N, Jordens ER (1983) Effects of atmospheric ammonium sulphate on calcareous and noncalcareous soils of woodlands in the Netherland. In: Ulrich B, Pankrath J (eds) Effects of accumulation of air pollutants in forest ecosystems. Reidel, Dordrecht, p 171

van Breemen N, Driscoll CT, Mulder J (1984) Acidic deposition and internal proton sources in acidification of soils and waters. Nature 307: 599-604

van Miegroet H, Cole DW (1984) The impact of nitrification and cation leaching in a red alder ecosystem. J Environ Qual 13: 586-590

von Zezschwitz E (1982) Akute Bodenversauerung in den Kammlagen des Rothaargebirges. Forst- und Holzwirt 37: 275-276

Wilklander L (1973/74) The acidification of soil by acid precipitation. Grundförbättring 26: 155-164

Wilklander L, Andersson A (1972) The replacing efficiency of hydrogen ion in relation to base saturation and pH. Geoderma 7: 159-165

Wood T, Bormann FH (1977) Short-term effects of a simulated acid rain upon the growth and nutrient relations of _Pinus strobus_ L. Water Air Soil Pollut 7: 479-488

A DISCUSSION OF THE CHANGES IN SOIL ACIDITY DUE TO NATURAL PROCESSES AND ACID DEPOSITION*

D.W. Johnson

Environmental Sciences Division, Oak Ridge National Laboratory, Oak Ridge, TN 37831, U.S.A.

ABSTRACT

It is generally accepted that in extremely acid soils, soil solution acidity can change rapidly in response to inputs of sulphate due to the "salt effect". However, there is little evidence of rapid change in bulk soil acidity due to acid deposition, except where acid inputs are extremely high. Over many decades, soil acidity can be increased due to uptake of base cations by the vegetation, by humus formation, and by natural as well as acid-deposition-enhanced soil leaching wherever annual mean precipitation exceeds mean annual evaporation. The role of acid deposition in accelerating soil acidification rates can be estimated from elemental budget approaches, but the actual changes in soil acidity due to acid deposition are difficult to document because so few studies combine the measurement of element and proton fluxes with long-term periodic remeasurements of soil acidity. Most empirical observations to date show little change in forest soil acidity from any cause in less than 20 to 30 years (cf. Tyler et al., this volume). However, some increases in exchangeable or soluble Al^{3+} without concomitant reductions in exchangeable base cations have been noted at the Solling site in West Germany. These Al^{3+} releases may be due to dissolution of interlayer Al^{3+} in 2:1 clays. All attempts to monitor long-term changes in soil acidity need to take into account the possibility of seasonal variations due to annual cycling of base-cations by vegetation. The effects of seasonal uptake and return of base cations by vegetation upon the base cation status of surface soils can be considerable and might be confused with long-term trends if not properly accounted for.

INTRODUCTION

The question of whether acid deposition has caused or can cause changes in soil acidity has been the subject of considerable debate (e.g., Wiklander 1974; Malmer 1976, Krug and Frink 1983). It has been argued, quite correctly, that most soils have such high exchangeable acidity compared with acid deposition inputs that little change can be expected to occur (Krug and Frink 1983). On the other hand, it is argued that soil solutions and, consequently, surface waters can be acidified by the leaching of a mobile anion (sulphate) through an acid soil, even if the acidity of the soil itself is not measurably affected (Seip 1980; Reuss and Johnson 1985). While the mechanism involved in the latter phenomenon is generally well accepted and has been modelled (Christopherson and Wright 1981; Goldstein et al. 1984; Cosby et al. 1985; Reuss and Johnson 1985), the actual magnitude of acidification that can be produced by this mechanism under field conditions will remain open to debate until further field

---

* Copyright of this paper: U. S. Government

testing is accomplished.

Much of the confusion and controversy surrounding the soil and water acidification issue can be dispelled by considering the capacity and intensity factors of acidity as described by van Breemen et al. (1983) and Reuss (1985). According to van Breemen et al. (1983), "Intensity factors are determined by chemical properties and are independent of the quantity or size of the system considered, while, in contrast, capacity factors are a function of the quantity or size of the system" (p 284). We can think of soil solution acidity as an intensity factor that can change readily when acid deposition is added, whereas soil acidity is a capacity factor that will change much more slowly. Changes in solution acidity (and associated changes in aluminum concentration) resulting from acid deposition are dicussed in a number of other papers (e.g., Reuss 1983; Seip 1980; Galloway et al. 1983; Krug and Frink 1983). This paper will focus on the capacity factor, soil acidity, as it may be affected by acid deposition as well as a number of other processes.

COMPONENTS OF SOIL ACIDITY

Soil acidity is generally thought to consist of two major components: (1) acidity due to the presence of exchangeable Al^{3+} on clays (exchangeable acidity) and (2) acidity due to the dissociation of H^+ from amorphous iron and aluminum oxides and organic matter (titratable acidity) (Coleman and Thomas 1967). The latter may be thought of as a weak acid component of soil acidity and is responsible for pH-dependent cation exchange capacity (CEC) of soils (Colemen and Thomas 1967). Cation exchange capacity due to permanent charge of clay minerals (i.e., the charge produced by isomorphous substitution of Al for Si in clay lattice structures) is independent of pH (Coleman and Thomas 1967). A third component of acidity in some soils (e.g., highly weathered soils of the southeastern United States) is that due to the presence of adsorbed anions (Coleman and Thomas 1967; Mehlich 1964). The "acidity due to anions" was noted by Mehlich (1964) in liming experiments on soils pretreated with SO_4^{2-}, OH^-, and H^+. Soils containing adsorbed SO_4^{2-} had higher lime requirements and lower pHs than those containing adsorbed OH^-. This was a logical result (at least in retrospect), and one that has implications for soils that adsorb atmospherically-deposited SO_4^{2-}.

Sulphate adsorption and precipitation reactions cause a reduction in acidification due to base cation leaching, but soil "acidity due to anions" is increased. The way in which this anion acidity will ultimately manifest itself depends on a number of factors. If input concentrations of SO_4^{2-} are reduced, and adsorbed SO_4^{2-} desorbs and leaches, the anion acidity will ultimately have the same (but delayed) effects on soil exchangeable acidity as if SO_4^{2-} had leached freely through the soil. In effect, the desorption of sulphate can be thought of as a net production of anions in the soil, analogous to internal HCO_3^- or NO_3^- formation. On the other hand, if adsorbed SO_4^{2-} does not desorb when input SO_4^{2-} concentrations are lowered, anion acidity remains elevated and will be manifested only if the soil is limed.

PROCESSES CONTRIBUTING TO SOIL ACIDITY

Several investigators have proposed schemes for the construction of hydrogen ion budgets to assess the relative contribution of acid deposition to natural acidification processes (Reuss 1976; Ulrich 1980; van Breemen et al. 1984; Nilsson et al. 1982; Driscoll and Likens 1982). There are some variations in how these budgets are constructed, but basically these schemes incorporate production or consumption of anions within the soil, plant uptake, humus formation, and atmospheric deposition. The papers cited above provide detailed descriptions of these processes, and only a few points will be commented on here.

Production and Consumption of Anions

In natural systems, the net production of H^+ associated with bicarbonate, organic, and sometimes nitrate anions within the soil is a major contributor to soil leaching and acidification (Johnson et al. 1977; van Miegroet and Cole 1984). Acid deposition will increase the leaching rate in soils if (1) the associated anion (SO_4^{2-} and/or NO_3^-) is mobile in the soil and (2) atmospheric anion inputs do not cause equivalent reductions in the rate of natural anion leaching.

Krug and Frink (1983) argued that, in acid soils, the introduction of sulphate (and presumably also nitrate) causes a reduction in the rate of natural leaching by carbonic and organic acids. This is logical, because the introduction of a mobile anion to a soil must result in an increase in the concentration of cations, including Al^{3+} and H^+, to maintain charge balance. The increase in H^+ and Al^{3+} concentration will be most pronounced in extremely acid soils where it will cause the protonation of some proportion of the weak acid (e.g., carbonic and organic acids) which reduces their concentrations in soil solution. However, it is not reasonable to assume that this process quantitatively negates the effect of sulphate or nitrate inputs on soil base cation leaching rates, as Krug and Frink (1983) suggest. This would imply a one-for-one equivalent increase in soil solution H^+ concentration for each increase in sulphate or nitrate concentration (i.e., absolutely no base cation displacement by H^+). Indeed, in moderately acid or circumneutral soils, the anion shift is probably of negligible importance because most H^+ rapidly exchanges for base cations that enter solution to balance the additional anions. Thus, while anion shift must occur to varying degrees in all soils regardless of their acidity (depending on base saturation), in no case will it completely offset the effects of acidic deposition on soil leaching rates. Thus, acidic deposition will very likely cause an increase in the rate of soil base cation leaching.

Plant Uptake

Plant uptake is acidifying to the rhizosphere if the sum of base cation uptake exceeds the sum of non-bicarbonate anion uptake (Ulrich 1980; Nye 1981). In this case, plants release H^+ or take up HCO_3^- (which results in dissociation of H_2CO_3 to $H^+ + HCO_3^-$ and is equally acidifying). The form in which N is taken up

(NO_4^+ or NO_3^-) is crucial to the charge balance of roots (and, consequently, the acidification of the rhizosphere) because uptake of the other major nutrient anions (PO_3^{4-} and SO_4^{2-}) normally cannot balance base cation uptake. Unfortunately, the form of N uptake under field conditions is seldom known (although frequently guessed at), making estimates of rhizosphere acidification due to plant uptake extremely uncertain.

On a whole-soil basis, it can be shown that N mineralization followed by uptake has no net effect on H^+ production. That is, the H^+ consumed during mineralization of organic N to NH_4^+ is offset by H^+ produced during NH_4^+ uptake. If nitrification occurs, there is still an H^+ balance if NO_3^- is taken up [ie., 2 H^+ produced during nitrification are balanced by 1 H^+ consumed during N mineralization and 1 H^+ consumed during NO_3^+ uptake (Reuss 1976; Ulrich 1980)]. Another way of looking at it is that the transformation from organic N in soil to organic N in plants has no net effect on H^+ production, and the same principle applies to transformations of organic P, S, Ca, K and Mg (Ulrich 1980). Thus, the net acidifying or alkalizing effects of plant uptake on whole-soil (i.e., not rhizosphere) acidity are due to the uptake of nutrient ions from inorganic sources (i.e., atmospheric deposition and soil exchange sites).

One way to avoid the uncertainties associated with the form of N uptake and sources of nutrients for uptake is to consider only the accumulation of base cations in vegetation and the potential depletion of exchangeable base cations as a result. Alban (1982) found an apparent effect of forest Ca^{2+} cycling patterns on soil exchangeable Ca^{2+} and pH. High rates of Ca^{2+} uptake and cycling by aspen and white spruce stands apparently caused a depletion of exchangeable Ca^{2+} (and a lowering of pH) in mineral soils and an enrichment of exchangeable Ca^{2+} (and an increase in pH) in surface soils over a 40-year period in glacial outwash soils in Minnesota. In contrast, red and jack pine forests, which took up and cycled much less Ca^{2+}, have a more acidic litter but less acid subsoil than the aspen or spruce stands.

Ulrich (1983) argues that the rhizosphere is the zone at which much of the acidifying effects of acid deposition are ultimately manifested. This argument is based on the assumption that acid deposition causes accelerated foliar cation leaching, which in turn causes accelerated cation uptake, which in turn causes accelerated H^+ release from roots.

It should be noted, however, that root turnover, and consequently, rhizosphere turnover are very rapid (Harris et al. 1977) and thus one could argue that whole-soil acidification calculations are most meaningful for long-term considerations of soil acidification.

Humus Formation
Humus formation adds to the titratable acidity and pH-dependent CEC of soil but it does not, in isolation from the nutrient cycle, cause a reduction in exchangeable base cations as leaching and plant uptake do. (Base cations accumulated in humus were, of

course, at one time taken up by vegetation.)

The rate of humus formation in coniferous forests with a mor-type forest floor is relatively easy to quantify: the weight of humus is determined by destructive sampling and divided by the length of time it has accumulated (usually the stand age). The acidity due to forest floor humus formation can be determined empirically as the pH-dependent CEC (Ulrich 1980) or the accumulation of base cations as in vegetation (Nilsson et al. 1982). Values for humus formation in northern temperate forest calculated by Ulrich (1980) and Nilsson et al. (1982) range from 0.8 to 4 $keq.ha^{-1}.y^{-1}$.

Forest floor humus accumulation is much less in warmer climates and in deciduous forests with mull-type forest floors than in northern temperate forests, but accumulation of humus in the soil itself may increase titratable acidity in these cases. Unfortunately, estimation of acidity due to humus accumulation within mineral horizons is difficult, due both the sampling problems (i.e., what is the rate of change in soil humus?) and to uncertainties in the method of calculating acidity (how does one separate the base cation content of humus from that in mineral phase of the soil?).

PROCESSES MITIGATING AGAINST SOIL ACIDIFICATION

Until now, the discussion has centered on processes (both natural and anthropogenic) that tend to acidify soils. The actual extent to which these processes will acidify soils depends on the rate at which compensating mechanisms like deep rooting and recycling by vegetation, atmospheric cation inputs, and soil weathering offset the potentially acidifying effects of the processes described previously.

Deep Rooting and Recycling by Vegetation

Stone (1975) argued that rooting beyond depths of 1 m is common in forest stands. Thus, if we draw the system boundary at 1 m we must view uptake by deep rooting and subsequent recycling as an input to the system. It is commonly observed that most root biomass is in surface soils, but it is probable that these roots are primarily involved in uptake of recycled nutrients. Long-term tendencies for surface soil acidification may be compensated by base cation uptake via deep rooting followed by recycling.

Atmospheric Base Cation Inputs

Wiklander (1974) and Bache (1980) point out that the ability of precipitation to acidify or alkalize soils depends on the pH and base cation concentrations in both precipitation and soil. The lime potential, $K_L = pH - 1/2p(Ca^{2+} + Mg^{2+})$, provides an easily measured index of the ability of precipitation to acidify or alkalize soils. If the K_L of the incoming solution exceeds that of the soil, the incoming solution will increase the base saturation of the soil (assuming equilibrium), whereas the opposite is true if the K_L of the incoming solution is less than that of the soil. It is important to note that the ability of a solution to acidify or basify a soil is <u>not</u> solely a function of either solution or soil pH.

Soil Weathering

Soil weathering is one of the most important, yet least understood, processes offsetting soil acidification. The pathways by which primary minerals (those formed beneath the earth's crust or beneath the ocean) are "weathered" or transformed into secondary minerals (those formed in the soil, such as clay minerals) are fairly well known (e.g., Jackson 1963). However, this knowledge has not as yet been successfully translated and applied to the specific problem with which we are faced in the acidic deposition issue; namely, to what extent and at what rate are exchangeable base cations in the soil replenished from sources <u>within</u> the soil, be they primary minerals, secondary minerals (clays), or amorphous materials. The most rigorous and accurate way to measure the latter is to monitor inputs, outputs, and changes in the soil of H^+, aluminum, and base cations over a period of time. Stuanes (1980) used this approach on the Norwegian acid irrigation study plots and found that weathering rate (defined here as replenishment of exchangeable base cations from within the soil) increased with increasing inputs of sulphuric acid. Primary mineral weathering rates over geologic time have been estimated through a chronosequence approach, wherein the current mineral (or element) content of a soil is simply subtracted from what is assumed to have been present initially (based on soil parent material) and divided by estimated soil age (Bockheim 1980). Mazzarino <u>et al.</u> (1983) used this approach to estimate historical rates of H^+ consumption by weathering at the Solling site in West Germany and obtained values of 0.6 to 1.9 $kmol.ha^{-1}.y^{-1}$, which equalled 30 to 50% of current atmospheric H^+ inputs to the ecosystem. However, it is not known if this overall average value is representative of current weathering rates.

MEASUREMENTS OF CHANGES IN SOIL ACIDITY

Due primarily to the uncertainties associated with soil weathering rates, projections of changes in soil acidity from element budget data are very tentative at best. Unfortunately, there are very few studies in which changes in soil acidity have actually been measured by periodic sampling through time, let alone in conjunction with element budget measurements. One notable exception is the work in a beech forest at the Solling site in West Germany by Ulrich <u>et al.</u> (1980). Other examples are studies from Sweden by Tyler <u>et al.</u> (this volume) and by C.O. Tamm. Although Ulrich's sampling intensity was low (n = 2-3), the authors noted increases in Al^{3+} in "equilibrium soil solution (ESS)" (a water extract of field-moist soil) and, to a much lesser degree, increases in exchangeable Al^{3+} (1 $\underline{M}$ NH_4Cl extraction) between 1966 and 1979 (with one intermediate sampling in 1973). There were slight decreases in pH, larger decreases in exchangeable Mn^{2+}, increases in exchangeable Mg^{2+}, and initial decreases (1969-1973) then increases (1973-1979) in exchangeable Ca^{2+}. The authors interpret these seemingly contradictory results as a result of dissolution of interlayer polyhydroxyl Al (with associated occluded Ca and Mg) by protons. Accompanying these results were detailed element budgets showing high rates of proton inputs from both atmospheric and internal sources (see also Matzner and Ulrich, this volume).

Wiklander (1974) and Linzon and Temple (1980) re-sampled soils at time intervals of 39 and 18 years in Sweden and Canada, respectively, and found either no changes or inconsistent changes in soil pH. Troedesson (1980) noted a slight trend toward increasing acidity in Swedish forest soils over a period of 10 years, but added that the time of forest occupancy is the most important variable affecting soil acidity. Jenkinson (1970) reported a pH decline from 7.1 to 4.5 over a period of 82 years of natural reforestation of abandoned agricultural land at the famous Rothamsted Experimental Station in England. Associated with this pH decline was an increase in soil S content, which was attributed to accumulation of atmospherically deposited S by SO_4^{2-} adsorption and which, in turn, was enhanced by increasing soil acidity. There was no implication that the pH decline was due to atmospheric deposition, only that it enhanced the adsorption of atmospherically deposited S. The pH decline in itself was simply attributed to natural acidification processes due to reforestation. The Tyler et al. study (this volume) reports pH decreases of up to 0.7 pH units in southern Sweden and suggests that acid deposition may well be a factor in this. Hallbäcken and Tamm (submitted) have similar data for samples 50 years apart.

When measuring long-term changes in soil acidity, base cation status, or other indices of soil fertility, it is important to account for seasonal variations, especially in surface soil. Haines and Cleveland (1981) noted significant seasonal variations in pH, cation exchange capacity, and exchangeable Ca^{2+} and Mg^{2+} in 0 - 20 cm depths of forest soils in Georgia. While there will always be questions as to confounding effects due to sampling inconsistencies in either seasonal or long-term soil changes, the results of Haines and Cleveland (1981) are worthy of serious consideration. We noted apparent seasonal variations in nonchangeable Al^{3+} and extractable P (NH_4F/HCl) in a deciduous forest soil (Typic Fragiudult) on Walker Branch Watershed, Tennessee, (Johnson and Todd 1984). We also noted large seasonal variations in exchangeable Ca^{2+}, K^+, and pH in the 0 - 15 cm level of a Fullerton series soil (Typic Paleudult) on Walker Branch Watershed (Figs. 1-3). Seasonal variations were minimal deeper in the soil (45-60 cm; Figs. 1-3), giving us confidence that surface variations were not due merely to sampling or analytical inconsistencies.

The seasonal variations in surface-horizon exchangeable Ca^{2+} are thought to be due to annual cycling of Ca^{2+} via uptake and return (litterfall and root turnover) by trees. The difference in exchangeable Ca^{2+} from April to December in the 0 to 5 cm depth in Fig. 1 amounts to approximately 70 to 130 $kg.ha^{-1}.y^{-1}$, values which are in the same ranges of estimates of Ca uptake and return for this forest type (60-100 $kg.ha^{-1}.y^{-1}$; Henderson et al. 1978; Cole and Rapp 1981). Similarly, the differences in exchangeable

K^+ (approximately 15 kg.ha^{-1}.y^{-1}) are thought to be due to K^+ uptake and return (34-44 kg.ha^{-1}.y^{-1}).

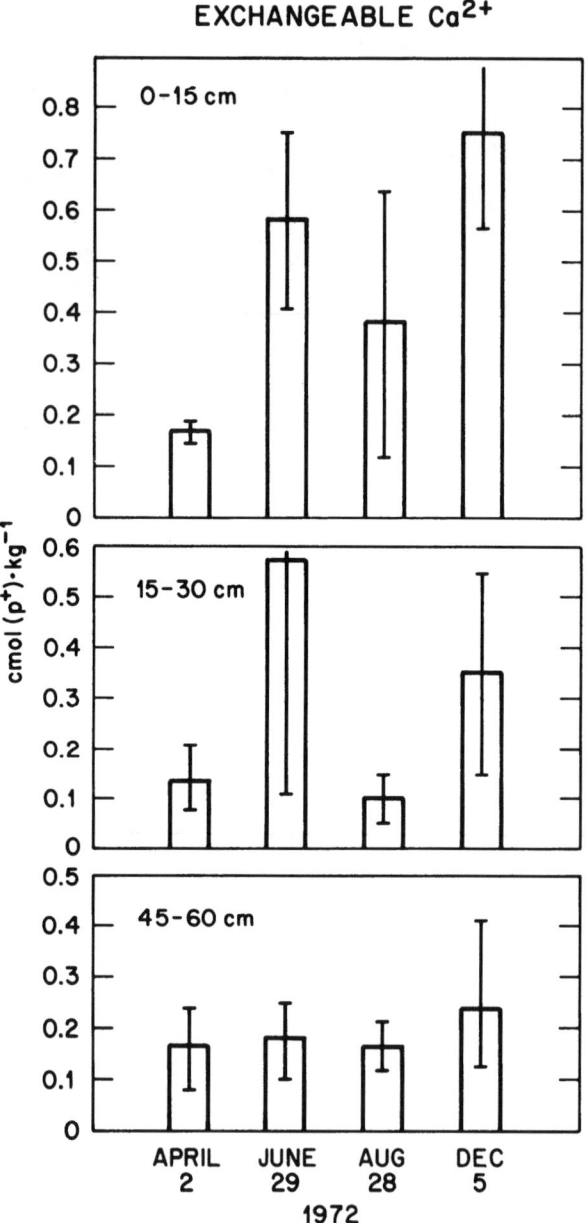

Fig. 1. Seasonal variations in exchangeable Ca^{2+} (1 $\underline{M}$ NH$_4$Cl extraction) in a Fullerton soil (Typic Paleudult) on Walker Branch Watershed, Tennessee. (D.W. Johnson and D.E. Todd, unpublished data).

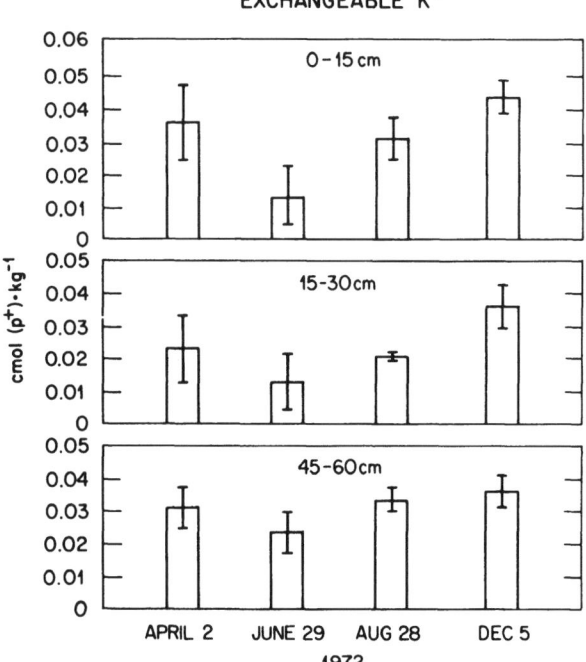

Fig. 2. Seasonal variations in exchangeable K^+ (1 $\underline{M}$ NH_4Cl extraction) in a Fullerton soil (Typic Paleudult) on Walker Branch Watershed, Tennessee (D.W. Johnson and D.E. Todd, unpublished data).

These large seasonal variations greatly exceeded differences in exchangeable Ca^{2+} in surface soils sampled in March-April 1971 versus March 1982 in a nearby long-term monitoring plot (D.W. Johnson, unpublished data). Further details on long-term changes in soils of Walker Branch Watershed will be reported in a later paper.

CONCLUSIONS

There is little doubt that acid deposition causes increases in soil leaching. However, any changes in soil acidity caused by acidic deposition must be much smaller than that caused by the rate of soil leaching for three reasons: (1) soil leaching is only one of three major soil acidification processes (the other two being humus formation and nutrient uptake by trees), (2) soil weathering (which may be stimulated by increased leaching) affects the potential acidification of soils by leaching (or any other acidifying process); and most important, (3) the acidity of even the "youngest" soils (i.e., glacial soils) is the cumulative result of thousands of years of acidification by natural processes, to which acidic deposition has probably added very little as yet.

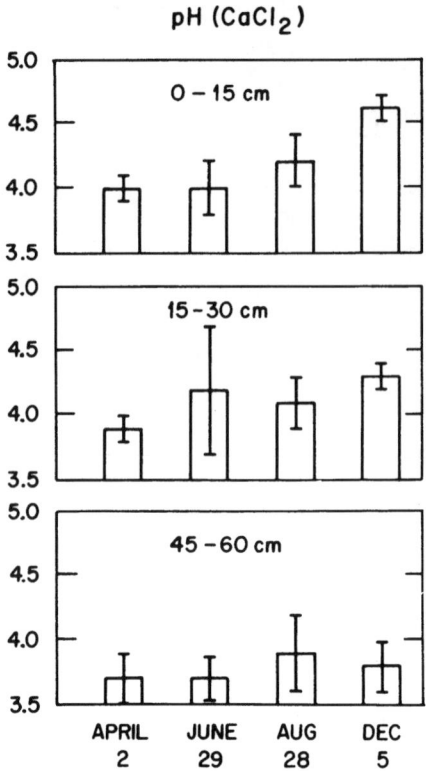

Fig. 3. Seasonal variations in pH (0.01 M $CaCl_2$) in a Fullerton soil (Typic Paleudult) on Walker Branch Watershed, Tennessee (D.W. Johnson and D.E. Todd, unpublished data).

The question of whether acidic deposition has caused or will cause changes in soil leaching rates (intensity) or in soil acidity itself (capacity) becomes a matter of quantifying H^+ inputs, leaching rates, and soil reserves.

That acidic deposition has caused increased soil acidity near point sources like smelters is beyond doubt (e.g., Wolt and Lietzke 1982; Hutchinson and Whitby 1974). Ulrich et al. (1980) and Ulrich (1983) contend that the high rates of acidic deposition in central Europe have caused soil acidification since the beginning of industrialization. However, care must be taken to determine seasonal variations versus long-term changes in soil chemical properties. The annual cycling of base cations (especially Ca^{2+}) by trees is often considerably greater than annual base cation losses by leaching, making it difficult to ascertain long-term trends from short-term fluctuations in soil base cation status.

In the final analysis, the degree to which acid deposition has caused or will cause soil acidification depends on site-specific conditions (i.e., soil properties and amount of input). Thus, broad generalizations as to the amount of soil acidification caused by acid deposition have little meaning.

Acknowledgements

Research supported by the Electric Power Research Institute (RP 1813-1) and the Office of Health and Environmental Research, U.S. Department of Energy, under contract DE-AC05-840R21400 with Martin Marietta Energy Systems, Inc. Publication No. 2723, Environmental Sciences Division, Oak Ridge National Laboratory.

REFERENCES

Alban DH (1982) Effects of nutrient accumulation by aspen, spruce, and pine on soil properties. Soil Sci Soc Am J 46: 853-861
Bache B (1980) The acidification of soils. In: Hutchinson TC and Havas M (eds) Effects of acid precipitation on terrestrial ecosystems. Plenum Press, New York and London, p 183-202
Bockheim JG (1980) Solution and use of chronofunctions in studying soil development. Geoderma 24: 71-85
Christopherson N, Wright RF (1981) Sulfate budget model for sulfate concentrations in streamwater at Birkenes, a small forested catchment in southernmost Norway. Water Resour Res 17: 377-389
Cole DW and Rapp M (1981) Elemental cycling. In: Reichle DE (ed) Dynamic properties of forest ecosystems. Cambridge University Press, London, p 341-409
Coleman NT, Thomas GW (1967) The basic chemistry of soil acidity. In: Pearson RW, Adams F (eds) Soil acidity and liming. American Soc of Agronomy, Madison, Wisconsin, p 1-41
Cosby BJ, Hornberger, GM, Galloway JN, Wright RF (1985) Modeling the effects of acid deposition: Assessment of a lumped parameter model of soil water and streamwater chemistry. Water Resour Res 21: 51-63
Driscoll CT, Likens GE (1982). Hydrogen ion budget of an aggrading forest ecosystem. Tellus 34: 283-292
Galloway JN, Norton SA, Church MR (1983) Freshwater acidification from atmospheric deposition of sulphuric acid: A conceptual model. Environ Sci Technol 17: 541-545
Goldstein RA, Gherini SA, Chen CW, Mok L, Hudson RJM (1984) Integrated acidification study (ILWAS): A mechanistic ecosystem analysis. Phil Trans R Soc Lond B 305: 259-279
Haines SG, Cleveland G (1981) Seasonal variation in properties of five forest soils in southwest Georgia. Soil Sci Soc Amer J 45: 139-143
Hallbäcken L, Tamm CO (submitted) Changes in soil acidity from 1927 to 1982-84 in a forest area of southwest Sweden. Water Air Soil Pollut
Harris WF, Kinerson RS, Edwards NT (1977) Comparison of belowground biomass of natural deciduous forest and loblolly pine plantations. Pedobiologia 17: 369-381

Henderson GS, Swank WT, Waide JB, Grier CC (1978) Nutrient budgets of Appalachian and Cascade region Watersheds: A comparison. Forest Sci 24: 385-397

Hutchinson TC, Whitby LM (1974) A study of heavy metal pollution in the Sudbury mining and smelting region of Canada. I. Contamination of air, soil and vegetation. Environ Conser 1: 123-132

Jackson ML (1963) Aluminum bonding in soils: A unifying principle in soil science. Soil Sci Soc Am Proc 27: 1-10

Jenkinson DS (1970) The accumulation of organic matter in soil left uncultivated. Commonwealth Bureau of Soil Science, Rothamsted Experiment Station for 1970, Rothamsted, England, p 113-137

Johnson DW, Todd DE (1984) Effects of acid irrigation on carbon dioxide evolution, extractable nitrogen, phosphorus, and aluminum in a deciduous forest soil. Soil Sci Soc Am J 48: 664-666

Johnson DW, Cole DW, Gessel SP, Singer MJ, Minden RV (1977) Carbonic acid leaching in a tropical, temperate, subalpine and northern forest soil. Arct Alp Res 9: 329-343

Krug EC, Frink CR (1983) Acid rain on acid soil: A new perspective. Science 221: 520-525

Linzon SN, Temple PJ (1980) Soil resampling and pH measurements after an 18-year period in Ontario. In: Drabløs D, Tollan A (eds) Ecological impact of acid precipitation. Johs Grefslie Trykkeri, Mysen, Norway, p 176-177

Malmer N (1976) Acid precipitation: Chemical changes in soil. Ambio 5: 231-233

Matzner E, Ulrich B (this volume) Results of studies on forest decline in northwest Germany.

Mazzarino MJ, Heinrichs H, Folster H (1983) Holocene versus accelerated actual proton consumption in German forest soils. In: Ulrich B, Pankrath J (eds) Effects of accumulation of air pollutants in forest ecosystems. Reidel, Dorecht, p 113-132

Mehlich A (1964) Influence of sorbed hydroxyl and sulphate on liming efficiency, pH, and conductivity. Soil Sci Soc Am Proc 27: 496-499

Nilsson SI, Miller HG, Miller JD (1982) Forest growth as a possible cause of soil and water acidification: An examination of the concepts. Oikos 39: 40-49

Nye PH (1981) Changes of pH across the rhizosphere induced by roots. Plant Soil 61: 7-26

Reuss JO (1976) Chemical/biological relationships relevant to ecological effects of acid rainfall. Water Air Soil Pollut 7: 461-478

Reuss JO (1983) Implications of the calcium-aluminum exchange system for the effect of acid precipitation on soils. J Environ Qual 12: 591-595

Reuss JO (1985) Modeling the effects of acid deposition on soil and water acidification. In: Johnson DW (ed) Predicting soil and water acidification: Proceedings of a Workshop, ORNL/TM-9258. Oak Ridge National Laboratory, Oak Ridge Tennessee

Reuss JO, Johnson DW (1985) Effect of soil processes on the acidification of water by acid deposition. J Environ Qual 14: 26-31

Seip HM (1980) Acidification of freshwaters: Sources and mechanisms. In: Drabløs D, Tollan A (eds) Ecological impact of acid precipitation. Johs Grefslie Trykkeri A/S, Mysen, Norway, p 358-366

Stone EL (1975) Effects of species on nutrient cycles and soil change. Phil Trans R Soc Lond Ser B 271: 149-162

Stuanes A (1980) Effects of acid precipitation on soil and forest S release and loss of nutrients from a Norwegian forest soil due to artificial rain of varying acidity. In: Drabløs D, Tollan A (eds) Ecological impact of acid precipitation. SNSF Project, Sandefjord, Norway, p 198-199

Troedesson T (1980) Ten years acidification of Swedish forest soil. In: Drabløs D, Tollan A (eds) Ecological impact of acid precipitation. Johs Gerfslie Trykkeri, Mysen, Norway, p 184

Tyler G, Berggren D, Bergkvist B, Falkengren-Grerup, Folkeson L, Ruhling A (this volume) Soil acidification and metal solubility in forests of southern Sweden.

Ulrich B (1980) Production and consumption of hydrogen ions in the ecosphere. In: Hutchinson TC, Havas M (eds) Effects of acid precipitation on terrestrial ecosystems. Plenum, New York, p 255-282

Ulrich B (1983) Soil acidity and its relation to acid deposition. In: Ulrich B, Pankrath J (eds) Effects of accumulation of air pollutants in ecosystems. Reidel, Dodrecht, p 127-146

Ulrich B, Mayer R, Khann RK (1980) Chemical changes due to acid precipitation in a loess-derived soil in central Europe. Soil Sci 130: 193-199

van Breemen N, Driscoll CT, Mulder J (1984) Acidic deposition and internal proton sources in acidification of soils and waters. Nature 307: 599-604

van Breemen N, Mulder J, Driscoll CT (1983) Acidification and alkalization of soils. Plant Soil 75: 283-308

van Miegroet H, Cole DW (1984) The impact of nitrification on soil acidification and cation leaching in red alder ecosystem. J Environ Qual 13: 586-590

Wiklander L (1974) The acidification of soil by acid precipitation. Grundforbaettring 26: 155-164

Wolt JD, Lietzke DE (1982) The influence of anthropogenic sulfur unputs upon soil properties in the Copper Basin Region of Tennessee. Soil Sci Soc Am J 46: 651-656

SOIL ACIDIFICATION AND METAL SOLUBILITY IN FORESTS OF SOUTHERN SWEDEN

G. Tyler, D. Berggren, B. Bergkvist, U. Falkengren-Grerup, L. Folkeson, Å. Rühling

University of Lund, Metal ecology group, Östra Vallgatan 14, S-22361 LUND, Sweden

ABSTRACT

Far-reaching acidification of forest soils has occurred in southern Sweden during the last decades. The decreasing pH has, directly or indirectly, increased the solubility of several elements in the soil, including magnesium, aluminum, cadmium and zinc. This has resulted in high concentrations of these elements in the soil-water and a considerably greater output than input from the forest ecosystems. Both deciduous and coniferous forest soils have become acidified but the solubilization and flow of metals is greater in spruce stands than in beech and birch stands on originally similar soil.

INTRODUCTION

Until recently, soils were widely thought to be quite resistant to and well buffered against changes caused by acid rain. Acidification of freshwater aquatic systems was considered the main ecological consequence of the excessive atmospheric input of acidic and acidifying substances. Therefore, the attention was invariably directed towards chemical and biological changes in aquatic ecosystems. The terrestrial environment was largely ignored.

During the last decade, however, the rapidly expanding damage to forests in Central Europe and, recently, in Scandinavia, and the activity of several scientific groups, e.g., B. Ulrich and co-workers in western Germany, have raised a justified, sometimes almost overheated interest by both the general public and the regulatory authorities of many countries in damages to forested terrestrial systems. Part of this increasing interest is directed towards possible changes in the soil.

SOIL ACIDIFICATION

In a recent study by Hallbäcken and Tamm (in press) of numerous forest soils in a limited area (Tönnersjöheden) in southwestern Sweden a considerable decrease in soil pH during the last 50 years was demonstrated. The study aroused considerable attention in Scandinavia, since a pH decrease extending down as far as the B and upper C horizons had probably not been demonstrated before, except in sites close to great sources of acidic pollutants. In western Germany, consistent pH decreases over the last 20 years were also recently reported for 35 forest topsoils (Butzke 1981).

However, the general validity of these findings in areas seemingly less exposed to high acidic inputs was not quite obvious. In 1984 the pH of sixteen deciduous and coniferous forest sites in the province of Skåne, southern Sweden, originally determined in 1949 by Linnermark (1960) and in 1953 by O. Andersson (unpublished), was redetermined using identical methods (Falkengren-Grerup, in press). Considerable pH-H_2O and pH-KCl decreases were recorded in the topsoils (0-5 cm), on average amounting to ca. 0.7 pH units (Fig. 1). In the edaphically intermediate mull soils (pH 5.0 to 6.5) the range of the pH-H_2O decrease was 0.7 - 1.2 units, while in the mor soils (originally much more acidic - pH 4.0 to 4.3) it was 0.3 - 0.5 units.

A pH-KCl decrease of 0.7 units in the topsoil is accompanied by a decrease of the base saturation by about 50% (Fig. 2). Half of the sum of base cations (Ca + Mg + K) in 1949/53 should therefore have been replaced by hydrogen and aluminum in 1984.

In nine of the forest soils studied in 1949, pH was measured throughout the soil profile down to a depth of 80-100 cm (the upper part of the morphological C horizon). Again, without exception, considerable pH decreases were observed in 1984 (Fig. 3). In most horizons of the gray-brown forest soils and in the B and upper C horizons of the podsols studied, the pH-KCl decrease during the last 35 years was 0.7-0.9 units. In the mull layer of the gray-brown soils the pH-KCl has dropped about 1 unit and is now approaching the pH of the mor soils in 1949.

It might be asked to what extent these considerable pH changes are due to acid deposition since 1949 or to forest successional processes. However, most sites of the 1949/53 studies were unchanged with respect to dominating tree species in 1984. Several sites were mature beech (<u>Fagus sylvatica</u>) forest. A few were dominated by hornbeam (<u>Carpinus betulus</u>), oak (<u>Quercus robur</u>) or spruce (<u>Picea abies</u>). One heathland soil, wholly dominated on both sampling dates by heather (<u>Calluna vulgaris</u>), was also studied with the same general result: a considerable pH decrease in all horizons, particularly in B and upper C.

There is also indirect evidence in support of the conclusion that a substantial pH decline has occurred in the forest soils of southern and southwestern Sweden. A comparison was made of oak forest topsoils in Skåne and in an area of the province of Småland, which is about 250 km NE, being less exposed to deposition of acidic pollutants. In spite of the fact that the same criteria were used for site selection in the two areas, the frequency curves of soil pH differ considerably between the areas. The topsoils of Skåne are, on average, 0.7 pH-KCl units lower than those of eastern Småland (Fig. 4). This could be due to primary differences in the acid-base conditions of the soils. However, the Småland area is developed on siliceous bedrock, whereas part of the Skåne sites are in areas with bedrock and tills of calcareous origin. Thus, the opposite pH difference was to be expected, other things being equal.

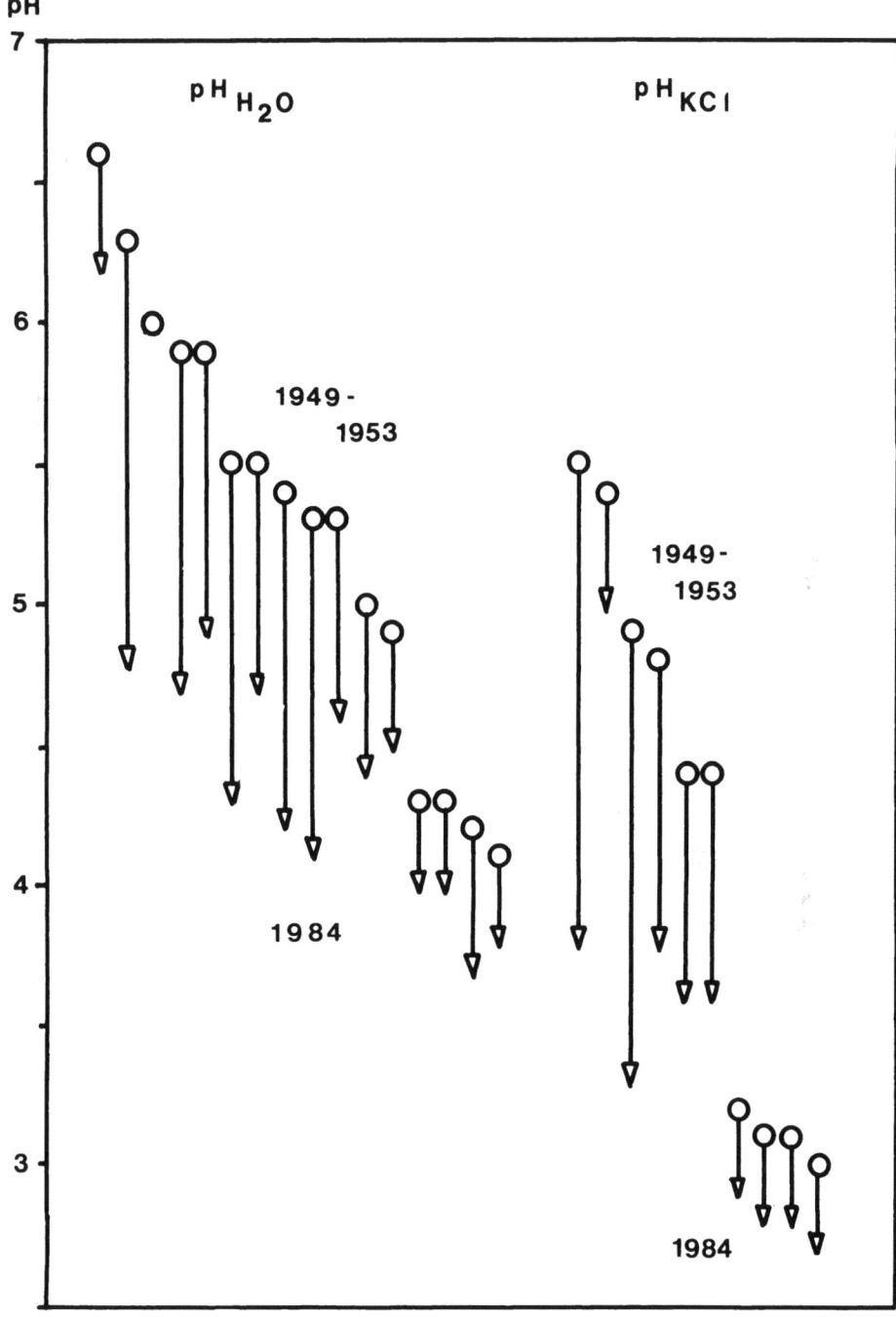

Fig. 1. Changes in soil pH (topsoil, 0-5 cm) of sixteen forest sites in southern Sweden between 1949/53 and 1984. The pH-KCl was only determined in ten of the sites. With one exception there was a decrease, on average of about 0.7 units, with both extractants used. Both deciduous and coniferous forest soils are included. (Redrawn from Falkengren-Grerup, in press).

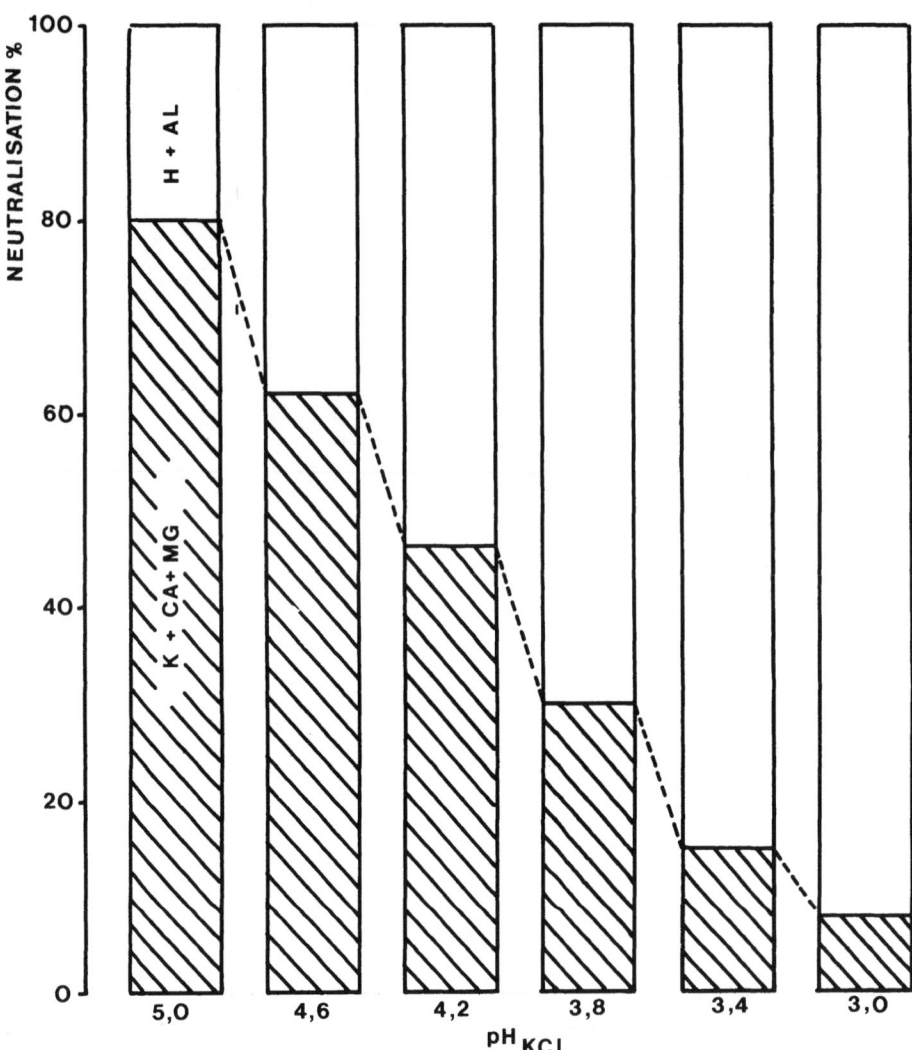

Fig. 2. There is a close relationship between the degree of neutralization (base saturation in M ammonium acetate) and the pH-KCl of top-soils. A decrease of 0.7 units means a decrease of the base saturation by about 50%. The graph is based on data from 230 deciduous forest soils (Tyler, unpublished data).

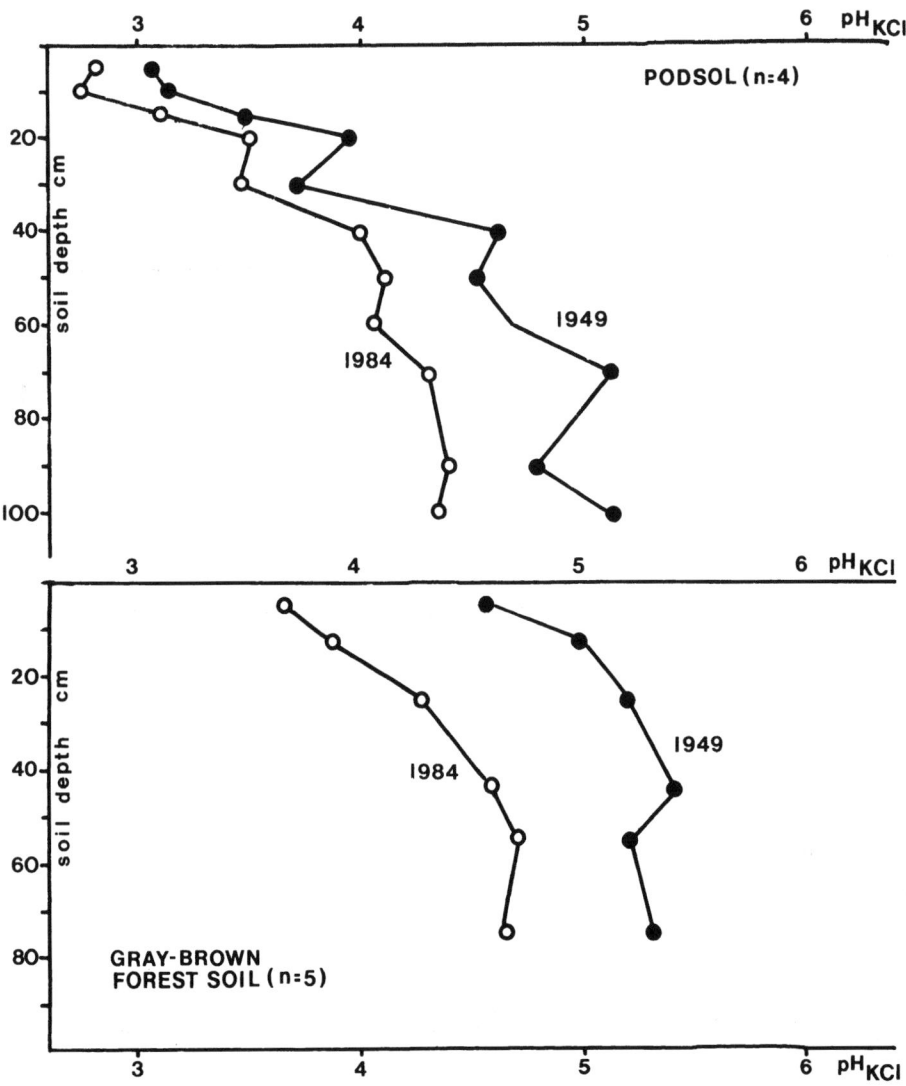

Fig. 3. The pH changes at different depths between 1949 and 1984 in forest soils of southern Sweden. Means of four podsols (above) and five gray-brown forest soils (below). The average pH decrease is 0.7-0.9 units in most horizons of the gray-brown soils and the B horizon of the podsols. (Redrawn from Falkengren-Grerup, in press).

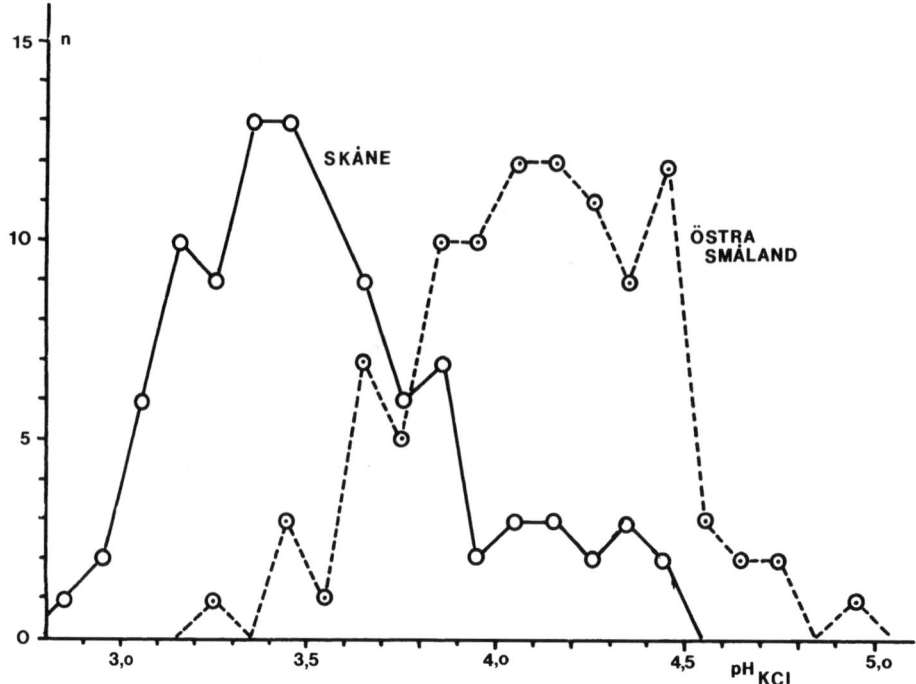

Fig. 4. Differences in present-day (1983) pH-KCl of oak forest soils (0-5 cm) between two areas in Sweden, Skåne (91 sites) and eastern Småland (101 sites). The latter area is less exposed to deposition of long-range transported pollutants, which may be a reason for the difference. N (y-axis) is the number of sites per each 0.1 pH-unit class (Rühling and Tyler, unpublished)

It seems increasingly difficult to explain all these consistent pH decreases and differences as solely or even depending mainly on ecosystem successional processes in and primary differences between the soils. The changes in total biomass and proportion of the quantitatively important species have not been large since the middle of this century at most of the sites considered. It seems reasonable to conclude that acid deposition is, to a large extent, accountable for the observed pH changes in forest soils of Skåne. Weathering processes have not been able to keep pace with the additional input of acids and acidifying constituents during the last decades.

METAL SOLUBILITY

Soil acidification increases the solubility of many metal elements, including both macronutrients and potentially toxic heavy metals. Cationic species may be released to the soil solution by the action of hydrogen ions and the presence of easily soluble anions. Simple ion exchange is one of the mechanisms involved; cations adsorbed by polyanionic colloids are replaced by hydrogen. Other mechanisms include an increased disintegration rate of minerals (weathering) and of complexes or chelates becoming less stable at a low soil pH. Also the anion introduced with the acid is of importance to ionic release.

The complexity of soil systems reduces the possibility of applying pure equilibrium chemistry in predicting release rates. Empirical studies of soil solutions from forests, however, have produced many results of interest in this context. As illustrated in Figs. 5-7 there is a close, though usually non-linear relationship between the pH and the total concentrations of several metals in acid soil solution. There is usually a more or less distinct 'bend' in the regression curves, indicating a rapidly increasing metal solubility below a particular solution pH. The position of this 'bend' along the pH axis depends not only on metal species but also, and in particular, on the soil horizon from which the solution originates. A much greater acidity is necessary in the A horizon (15 cm) of podsolic soils than in the B and upper C horizons to 'produce' a given concentration of, for example, magnesium or cadmium. The illustrations (Figs. 5-7), show the relationship between pH and percolating water content of elements for a mature spruce forest ecosystem in south Sweden, though similar trends have been obtained for other soils and for a few other elements (manganese, calcium and zinc).

There are at least two reasons for these conditions. The throughfall water from a spruce canopy is very acid. The average pH of rainwater in south Sweden is ca. 4.2. After passage through the spruce canopy the pH is normally lowered to 3.5-3.7. In percolating the soil the solution is gradually neutralized and it becomes increasingly saturated with metal cations until an equilibrium might be obtained in the C horizon. This process, however, gradually diminishes the exchangeable and easily weatherable pools of these metals in the biosphere (upper) part of the soil, in particular the A horizon, where most roots are located. A higher hydrogen ion concentration is, therefore, a prerequisite for the release of one unit of magnesium (or cadmium, zinc, etc.) from the A horizon than from the deeper soil horizons. In the case of magnesium this might gradually result in a shortage of plant-available forms in the upper soil layers.

The sharp 'bend' in the curves of potentially toxic elements like cadmium and aluminum (Figs. 6-7) is also of interest. In the pH range of 4.0-4.5 there is a drastic change of the solubility of these elements in the lower soil horizons. A drop in soil solution pH by 0.2 units results in at least a two- to three-fold increase in the aluminum and cadmium concentrations in the solution. Many, perhaps most, forest soils of southern Scandinavia presently have a solution pH within this range in the B horizon, at least during part of the year. This was evidently not the case earlier this century considering the data presented in Figs. 1 and 3. The changes of soil pH offer an explanation for the considerable net output from the soil of aluminum, cadmium, zinc and magnesium, which we have recorded in mineral budget studies of several forest ecosystems in recent years. Acidification processes have lowered the pH of soils to a level where every additional pH decrease results in a considerable increase of the metal solubility.

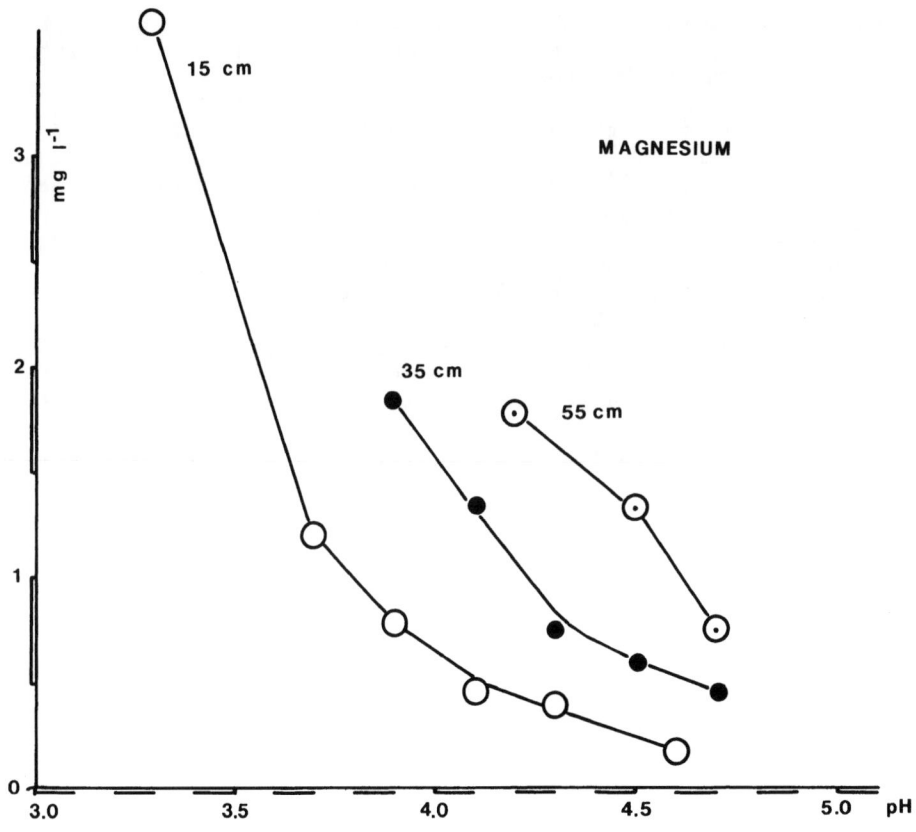

Fig. 5. The relationship between the pH and the magnesium concentration of percolation water at different soil depths. A considerably lower pH is needed for a given magnesium concentration in the A horizon (15 cm) than deeper in the soil. Lysimeter data are from a spruce forest (Bergkvist, unpublished).

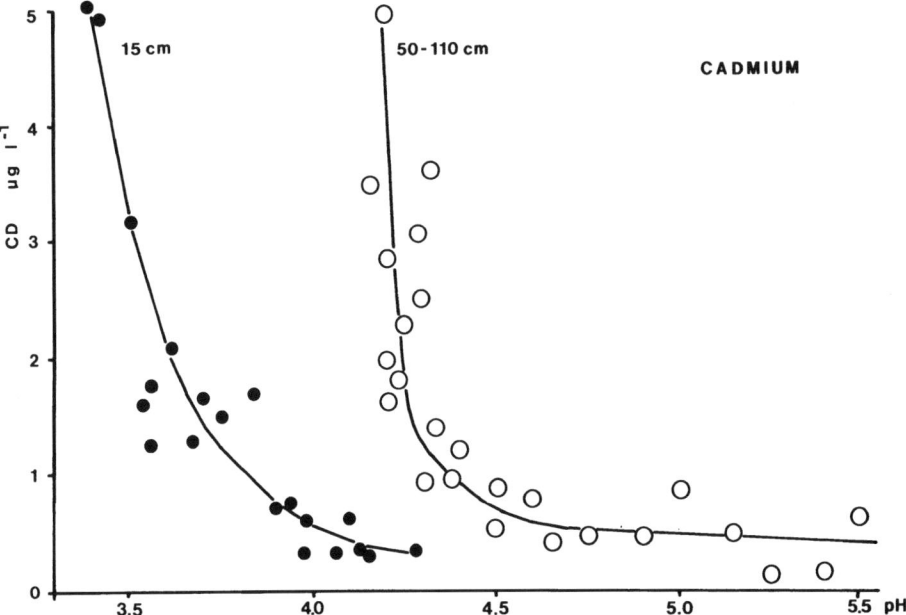

Fig. 6. The relationship between pH and cadmium concentrations of percolating water at different soil depths. A considerably lower pH is needed for a given cadmium concentration in the A horizon (15 cm) than at greater depths (lower B and upper C horizons). Lysimeter data are from a spruce forest. (From Tyler 1981; Bergkvist, unpublished).

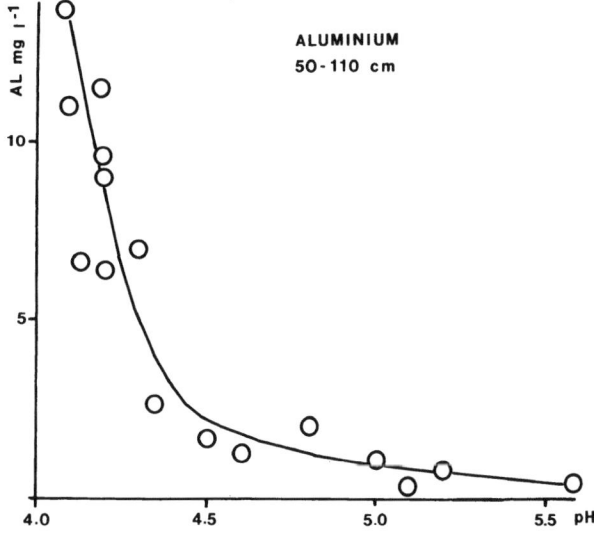

Fig. 7. The relationship between the pH and the aluminum concentration of percolating water in the lower horizons of a spruce forest soil (Bergkvist, unpublished).

METAL SPECIATION

For elements forming stable complexes with organic constituents ('fulvic acid'), the relationship between pH and metal concentration in the soil solution is less obvious in the A horizon than in other horizons. This is true of lead and copper for example, and for which complexation with soluble organic compounds seems to be the main mechanism of vertical transport through the soil. Only 8 - 10 percent of total copper in a solution from the A horizon of acid forest soils was present as an active cationic form extractable with a duolite gel (Table 1). The corresponding percentage of cationic zinc was 93 - 95 percent.

Table 1. Metal speciation of lysimeter water from the A horizon (15 cm) of acid forest soils (composite sample from several sites in southern Sweden). The pH was 4.6 and dissolved organic carbon 80 mg.L^{-1}. The gel type used for species separation was Duolite 225 SRC 9. (From Berggren, unpublished).

Metal	Active cationic species, %	Total concentration µg.L^{-1}
Copper	8 - 10	5.0
Lead	29 - 36	20
Zinc	93 - 95	24

An interesting situation is evidenced by aluminum. There is a close relationship between the content of dissolved organic matter (DOM) and total aluminum in the soil solution from the A horizon. As the seasonal differences in DOM are considerable, there is a corresponding seasonal fluctuation in total aluminum (Fig. 8). The direct relationship with pH is very weak or non-existent (Tyler 1981), as opposed to the situation encountered at greater depths in the soil (Fig. 7). This difference in behaviour has been proven to be due to the occurrence of at least two different aluminum species (Nilsson and Bergkvist 1983). The limited amount of aluminum which leaves the A horizon of a podsol is not an active cationic species but most probably a soluble organic complex. This species disappears gradually from the solution downwards in the soil as the content of DOM diminishes (Fig. 9). Instead, an inorganic aluminum species with a great cation activity appears and it increases in concentration through the B horizon. The occurrence of this species is evidently quite pH dependent (Fig. 7) and it may reach concentrations as high as 15-20 mg.L^{-1} in forest soils with a solution pH of 4.0-4.2 in the B horizon. This high level of soluble inorganic aluminum seems to be an abnormal state, as it causes a considerable net release of aluminum from the B horizon, where an accumulation of sesquioxides should occur according to pedological theory.

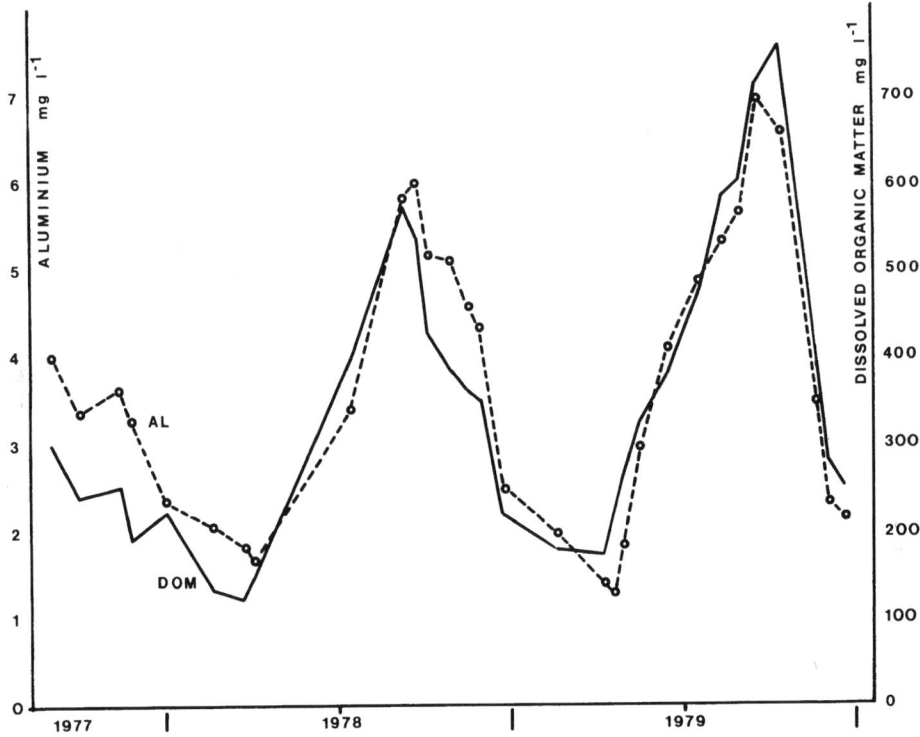

Fig. 8. Seasonal variability of the aluminum and the dissolved organic matter (DOM) concentrations in percolating water at 15 cm soil depth (the A horizon) in a spruce forest site. (Redrawn from Tyler 1981).

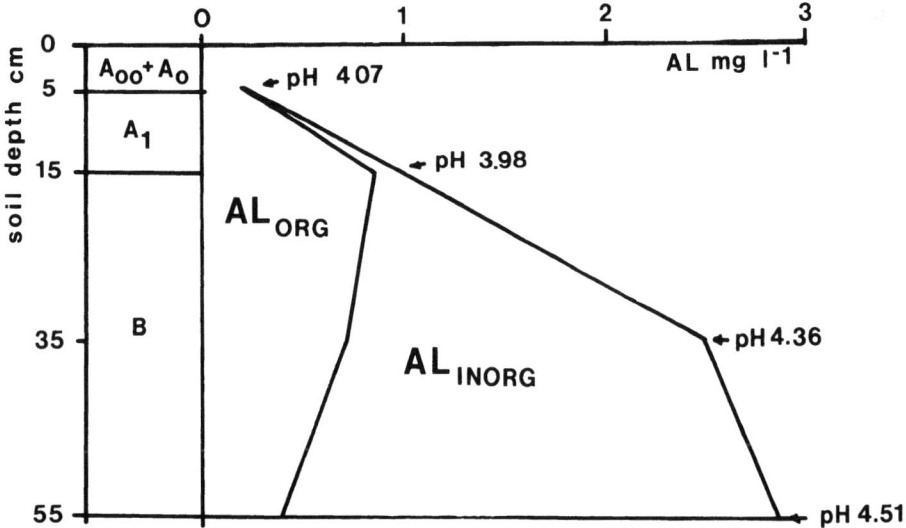

Fig. 9. Aluminum speciation at various depths in a shallow podsolic coniferous forest soil in southwestern Sweden. (Redrawn from Nilsson and Bergkvist 1983).

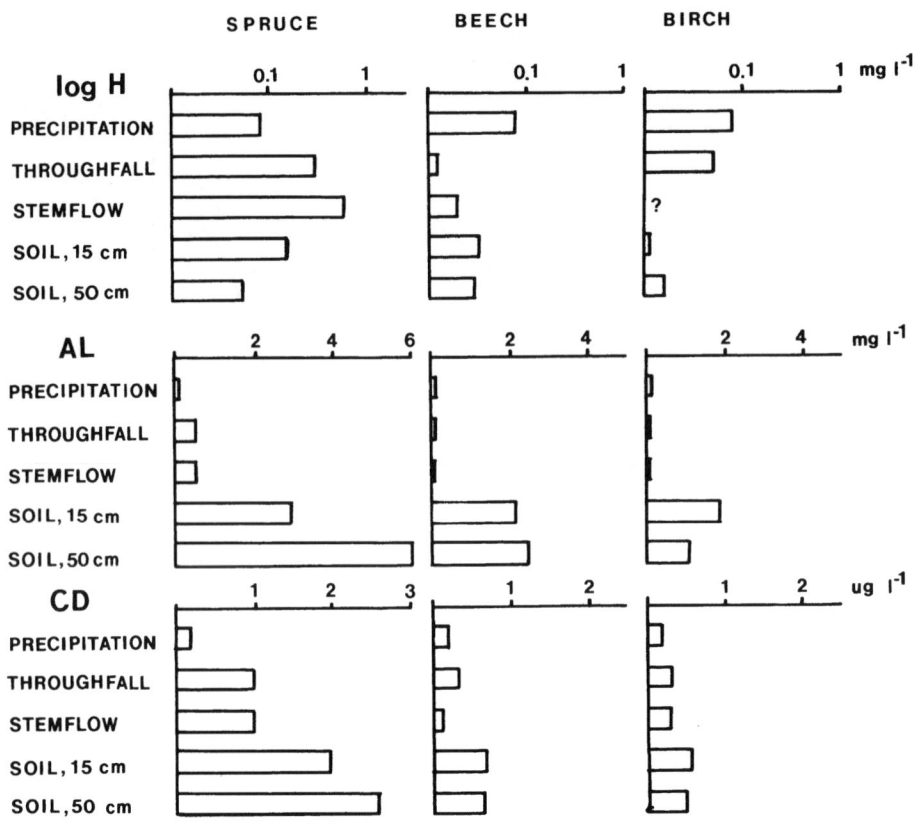

Fig. 10. Changes of the hydrogen, aluminum and cadmium concentrations from rainwater (open-field precipitation) over throughfall and stemflow water to percolating soil water at 15 and 50 cm depth. A comparison between adjacent stands of spruce, beech and birch developed on an originally similar soil. Preliminary means for a six-month period in 1984. The ecosystem flow of these and other elements differs considerably, being invariably highest in the spruce stand and usually lowest in the birch stand. (From Folkeson, Bergkvist, unpublished).

TREE SPECIES AND METAL SOLUBILITY

Though substantial decreases of soil pH seem to have occurred in all types of forest in southern Sweden, the species of tree which is dominant at each site is by no means unimportant to soil acidification and metal solubility. The acidifying power of spruce (Picea abies) is definitely greater than that of beech (Fagus sylvatica) and birch (Betula spp.), partly due to the greater aerosol-capturing capacity of conifers and partly to a greater internal production of acidic constituents in the spruce forest ecosystem. Since autumn 1983 we have been comparing elemental budgets of adjacent stands of spruce, beech and birch, developed on an originally similar soil in five localities in southern Sweden. Preliminary concentration data for hydrogen, aluminum and cadmium in water from various ecosystem levels are shown in Fig. 10 for one of these sites. It is situated in an area of archaean siliceous till close to the southwestern fringe of the South Swedish Uplands, where the deposition of long-range transported acidic pollutants is comparatively great. It is evident from the graph that the rainwater becomes enriched in both hydrogen, aluminum and cadmium on its passage through the spruce canopy, whereas its hydrogen is partly neutralized by the foliage of the deciduous trees. Moreover, the solubilization of cadmium and aluminum in the soil is much greater in the spruce stand, as was to be expected from the higher hydrogen ion concentration of the soil water.

CONCLUSIONS

Far-reaching acidification of forest soils has occurred in southern Sweden during the last decades. The decreasing pH has, directly or indirectly, increased the solubility of several elements, including magnesium, aluminum, cadmium and zinc. This has resulted in high concentrations of these elements in the soil water and a considerably greater output than input from the forest ecosystems. Both deciduous and coniferous forest soils have become acidified but the solubilization and flow of metals is greater in spruce stands than in beech and birch stands developed on an originally similar soil. Organic complexes of aluminum, copper and lead predominate in the A horizon of acidic forest soils, while in the B and C horizons cationic forms of aluminum predominate at low pH.

REFERENCES

Butzke H (1981) Versauern unsere Wälder? Erste Ergebnisse der Überprüfung 20 Jahre alter pH-Wert-Messungen in Waldböden Nordrhein-Westfalens. Der Forst- und Holzwirt 21: 542-548
Falkengren-Grerup U (in press) Long-term changes in pH of forest soils in southern Sweden. Environ Pollut
Hallbäcken L, Tamm CO (in press) Resampling of soil profiles in south-west Sweden for pH-measurements. Geologiska Föreningens i Stockholm Förhandlingar
Linnermark N (1960) Podsol och brunjord I-II. (With a summary in English). Publications from the Institute of Mineralogy, Paleontology and Quarternary Geology, University of Lund, Sweden 75: 1-233
Nilsson SI, Bergkvist B (1983) Aluminium chemistry and acidification processes in a shallow podzol on the Swedish westcoast. Water Air Soil Pollut 20: 311-329
Tyler G (1981) Leaching of metals from the A-horizon of a spruce forest soil. Water Air Soil Pollut 15: 353-369

DIFFERENCES IN ALUMINUM MOBILIZATION IN SPODOSOLS IN NEW HAMPSHIRE (USA) AND IN THE NETHERLANDS AS A RESULT OF ACID DEPOSITION

J. Mulder and N. van Breemen

Department of Soil Science and Geology, Agricultural University, P.O. Box 37, 6700AA Wageningen, The Netherlands

ABSTRACT

We studied aluminum mobilization in two Spodosols from the Netherlands (one under forest, and one under heather), and compared the results with aluminum mobilization data for North American Spodosols at the Hubbard Brook Experimental Forest (HBEF), New Hampshire. Annual budgets for aluminum, silica and sulphate were calculated from measured chemical fluxes in precipitation and throughfall, soil solution chemistry and estimated soil water fluxes. In the Netherlands, due to canopy entrapment, the forested Spodosol received higher inputs of atmospheric $(NH_4)_2SO_4$ (1.3 $kmol.ha^{-1}.y^{-1}$) than the heathland Spodosol (0.5 $kmol.ha^{-1}.y^{-1}$). Assimilation of NH_4^+, the main source of acidity, caused a strong mobilization of aluminum, which resulted in a net removal of Al (as sulphate) from the illuvial B horizon. Soil solutions in the B horizons at both Dutch sites were slightly undersaturated with natural gibbsite, but reached saturation with jurbanite $(Al(SO_4)(OH).5H_2O)$. Spodosols from New Hampshire (USA) have atmospheric acid inputs similar to the Dutch Spodosol under heather, but neutralization mainly takes place through basic cation solubilization, and only to a small extent via aluminum solubilization. In the North American podzol B horizons organic aluminum transport to and inorganic aluminum export from the B horizon are about equal, so there is no net breakdown of this horizon. Soil solutions from the B horizons at HBEF are slightly undersaturated with natural gibbsite and highly undersaturated with jurbanite. A lower content of weatherable silicate minerals is probably the main cause for the higher levels of dissolved aluminum in the Dutch Spodosols compared to those at HBEF, U.S.A. If this is true, high inorganic aluminum fluxes are expected in North American Spodosols with similar acid inputs but with much lower contents of bases, than at HBEF.

INTRODUCTION

In areas where precipitation exceeds evapotranspiration, soil acidification is a natural phenomenon. Acid inputs from natural and anthropogenic sources can be neutralized fully by basic cation dissolution (exchange, weathering) in base-rich soils. In the course of soil development basic cation solubilization decreases, due to depletion of easily weatherable minerals and to an associated decrease in weathering rates. Incomplete neutralization results in a lower soil pH, which in turn causes an increase in the solubility of aluminum. Organic acids effectively chelate aluminum and iron at low pH, resulting in increased solubility of these elements (Martin and Reeve 1958). Aluminum chelates migrate downward with percolating soil water until the organic acids become saturated with iron and/or aluminum and precipitate (Petersen 1976; Mokma and Buurman 1982; Buurman 1984). This theory of transport and subsequent precipitation of aluminum and iron in Spodosols is primarily based on chemical studies of soil samples and on laboratory experiments on the interaction between dissolved sesquioxides, organic acids and the soil solid phase (Schnitzer and Skinner 1963; Petersen 1976). The podzolization theory is supported by data on total dissolved aluminum in Washington Spodosols, which peaked in the E (eluvial) and decreased in the B (illuvial) horizon (Ugolini et al. 1977). The export of aluminum from the soil with drainage water is reported to be small (Ugolini et al. 1977; David and Driscoll 1984). Using a newly developed aluminum fractionation technique (Driscoll 1984), David and Driscoll (1984) and Driscoll et al. (1985) demonstrated the importance of Al-chelates in the transport of aluminum from the E to the B horizon in New York and New Hampshire Spodosols.

Observations on strongly weathered, acid soils (Spodosols, acid brown forest soils), developed in glacial till or in aeolian deposits in northeastern North America and in central and northwestern Europe, indicate that both total dissolved aluminum and dissolved inorganic monomeric aluminum concentrations are higher in soils affected by acid deposition, than in natural Spodosols, where polymeric and organic monomeric aluminum predominate (Cronan and Schofield 1979; Johnson et al. 1981; David and Driscoll 1984; Driscoll et al. 1985; Ulrich et al. 1979; Ulrich and Matzner 1983; Nilsson and Bergkvist 1983; Mulder et al., submitted). The major anions accompanying dissolved inorganic Al at sites with high acid deposition are SO_4^{2-} and NO_3^-. Apparently, aluminum chemistry changes in response to acid deposition. Such changes may be of great importance from a pedological and an ecological viewpoint.

In this paper we present data on the soil solution chemistry of two Spodosols from the Netherlands, developed in coversands, and affected by different regimes of acid deposition. Dutch Spodosols are intrazonal: the extremely poor soil material (further impoverished by centuries of biomass export by man), rather than climatic factors (as in most North American and Scandinavian Spodosols), is the primary factor predisposing the soil for podzolization. To demonstrate the role of both parent material and the regime of acid atmospheric deposition on the recent Spodosol solution chemistry, the data presented will be compared with literature values for zonal Spodosols in North America developed in parent material somewhat richer in bases. The three sites studied differ in vegetation cover, but the associated difference in biocycling will only be of minor importance for the net element budgets of the soil (Ulrich et al. 1979).

MATERIALS AND METHODS

Two Spodosols (Hasselsven and Tongbersven), developed in coversands in the eastern part of Brabant, the Netherlands, were investigated. Literature values were taken for three Spodosols developed in glacial till at the Hubbard Brook Experimental Forest (HBEF), New Hampshire, USA. All soils are well drained and can be classified according to Soil Survey Staff (1975) as Typic Haplorthods (Hasselsven and Tongbersven) and Typic Fragiorthods (HBEF). The vegetation cover consists of heather (Calluna vulgaris) at Hasselsven, pine forest (Pinus sylvestris) at Tongbersven and a deciduous forest (Fagus grandifolia, Acer saccharum and Betula alleghaniensis) at HBEF.

Soil solutions were collected monthly from October 1982 to October 1983 at Hasselsven and Tongbersven, and weekly from March 1982 to July 1982 at HBEF. At all sites sampling was done by the same type of tension lysimeter made of synthetic material. Details on lysimeters, installation and operation in the field are given by Driscoll et al. (1985). Analytical procedures for the soil samples are given elsewhere (Driscoll et al. 1985; Mulder et al., submitted). Procedures for the analyses of soil solutions of the HBEF sites are given by Driscoll et al. (1985) and for solutions of Hasselsven and Tongbersven by van Breemen et al. (submitted) and Mulder et al. (submitted). Fractionation of dissolved monomeric aluminum into an organic and an inorganic fraction was done by cation exchange according to Driscoll (1984), slightly modified by Mulder et al., submitted).

Disequilibrium indices (I_p), indicating the degree of saturation of a solution with mineral phases (Johnson et al. 1981), were calculated, using the program MINEQL (Westall et al. 1976).

$I_p = \log(Q_p/K_p)$,

where: Q_p = ionic activity product for solid phase p;

K_p = solubility product for the solid phase p; and

$I_p = 0$ indicates equilibrium with the phase considered, $I_p > 0$ supersaturation and $I_p < 0$ undersaturation.

Total wet deposition at Hasselsven and Tongbersven was estimated from open bucket precipitation data at a nearby station (Boxtel). Dry deposition on NH_3 and SO_2 at the forested site Tongbersven was estimated from $(NH_4)_2SO_4$ present in throughfall (van Breemen et al. 1982), measured from October 1982 to October 1983. Dry deposition is only of minor importance at the heather site (Hasselven). Moreover, since (1) equivalent amounts of NH_4^+ and SO_4^{2-} occur in throughfall, (2) net nitrification in the soil is negligible and (3) essentially all NH_4^+ is transformed in the soil into organic N, the total acid atmospheric deposition at Tongbersven and Hasselsven could be estimated as:

$H^+_{precip} + NH_4^+_{throughfall}$ (Tongbersven), or

$H^+_{precip} + NH_4^+_{precip}$ (Hasselsven) (van Breemen et al. 1983).

Annual acid inputs at HBEF, including dry deposition of SO_2, were taken from van Breemen et al. (1984).

Total atmospheric acid inputs at HBEF and Hasselsven are similar (1.4 and 1.2 $kmol.ha^{-1}.y^{-1}$), while at Tongbersven the annual atmospheric acid input is estimated as 2.9 $kmol.ha^{-1}.y^{-1}$.

Chemical fluxes in the soil at Hasselsven and Tongbersven were calculated by multiplying monthly solute concentrations and estimated monthly soil water fluxes. At HBEF chemical fluxes were estimated in the same way, using mean concentrations and water fluxes for the dormant and for the growing season. Because detailed physical soil data for the HBEF and the Dutch sites, necessary for a deterministic soil water transport model (van Grinsven et al., submitted) are still lacking, a first estimate of the water fluxes was made. At the Dutch sites a cascade model was used. For the HBEF soils, water fluxes were estimated after Wood (1980) by partitioning total transpiration (Likens et al. 1977) to each horizon on the basis of the distribution of fine root biomass (Driscoll et al. 1985). In this way, the estimated drainage water fluxes from the Bhs were 59, 47 and 25% of precipitation input for HBEF, Hasselsven and Tongbersven, respectively. The ratio for the forested Tongbersven site, lies well within the range of the 22 to 39% reported for Dutch woodland soils under a mixed oak and birch stand (van Grinsven et al., submitted). The higher drainage at Hasselsven is due to the lower transpirational demand of the heather vegetation.

RESULTS AND DISCUSSION

Table 1 shows selected characteristics of the soil solid phase for three HBEF profiles and the Hasselsven and Tongbersven soils. Soil pH profiles are similar for all sites, except that the eluvial E horizon is slightly more acidic at Hasselsven and Tongbersven. Free aluminum is mainly in a $Na_4P_2O_7$ - extractable form, believed to represent the organic aluminum fraction in the soil solid phase, and peaks in the Bhs horizon at all sites. At Hasselsven and Tongbersven, Spodosol development, as indicated by the levels of organic aluminum and organic carbon in the Bhs, is less pronounced than at HBEF. A relatively high contribution of aluminum, iron and basic cations to the soil mineral phase (around 15%) at HBEF, compared to Hasselsven and Tongbersven (around 3%), indicates that silicate mineral contents are much higher in the HBEF than in the Dutch soils, which essentially consist of quartz sand.

The cold and wet climate at HBEF climatic condition is favourable for Spodosol formation: mean annual precipitation is 1.32 m, mean annual runoff is 0.84 m (Likens et al. 1977) and mean monthly temperature at 50 cm soil depth is 7°C (Federer, 1973). Under the milder climate at Hasselsven and Tongbersen (October 1982/October 1983: precipitation of 0.925m; drainage to 0.44 and 0.23 m, respectively; mean monthly temperature at 50 cm soil depth of 9°C) podzolization could take place mainly because the parent material has a high quartz content and a very low content of basic cations, aluminum and iron.

The composition of drainage water leaving the soil systems is indicative of the net buffering mechanisms operating (Table 2). At HBEF only 23% of the cationic charge in Bhs2/3 drainage water is due to Al^{3+}, while in the drainage water of the Bs horizons at Hasselsvens and Tongbersven 43 and 72% of the cationic charge is due to Al^{3+}. The rest of the cationic charge is due to basic cations and to H^+, the latter accounting for less than 10% of the total charge. At HBEF more than half of the aluminum in the drainage water is organically complexed (Johnson et al. 1981), whereas at Hasselsven and Tongbersven this is less than 10%. So neutralization of atmospheric acid inputs at HBEF is accomplished primarily by solubilization and leaching of basic cations, while the role of aluminum is limited. At Hasselsven, with a similar total acid input, neutralization of acid inputs is primarily through solubilization and leaching of monomeric inorganic aluminum. At Tongbersven, with double the atmospheric acid input of the other sites, aluminum solubilization and leaching are even more pronounced.

Table 2. Total input of atmospherically derived strong mineral acids ($kmol.ha^{-1}.y^{-1}$) and the ionic equivalent contribution, in % of basic cations (K^+, Na^+, Ca^+ and Mg^{2+}), Al^{3+} and H^+ to the total cationic charge in the drainage water from the Bhs2/3 horizons (HBEF), from May 1979 to September 1979 (Mulder 1980), and from the Bs horizon (Hasselsven and Tongbersven) from October 1982 - October 1983.

Location	H^+ input ($kmol.ha^{-1}.y^{-1}$)	% cationic charge in drainage water		
		Na + K + Ca + Mg	Al	H
HBEF	1.4	72	23	5
Hasselsven	1.2	45	43	7
Tongbersven	2.9	23	72	3

Soil solution pH values in the B horizons are highest at HBEF (around 5) and lowest at Hasselsven and Tongbersven (around 4). They reach a minimum in the eluvial E horizon (around 4.6 at HBEF and around 3.4 at the Dutch sites), while pH values in the leachates of the O horizons range from 3.6 to 5.0 (Table 3). Soil solution pH values may differ by +1.1 to -0.5 units from the pH of an aqueous soil extract (pH(H_2O)). The difference may be due to (1) a higher degree of acid neutralization in air-dried, stored soil samples, (2) soil N transformations during air drying and storage, and (3) dilution and ion exchange effects during extraction with H_2O. We believe that soil solution pH values are most relevant for in situ ecological conditions.

Monomeric aluminum levels in the O-horizon leachates are of the same order of magnitude at all sites (7 to 37 $mmol.m^{-3}$). Water percolating from the O- and E horizons in HBEF soils has similar Al contents. At Hasselsven and Tongbersven, however, dissolved aluminum is distinctly higher in the E (108 and 54 $mmol.m^{-3}$, respectively) than in the O horizon. So, while the forest floor appears to be an important source of dissolved aluminum in HBEF soil solutions, it is not at Hasselsven and Tongbersven. The

Table 1. Selected chemical characteristics of the soil solid phase for the profiles HBEF I, HBEF II, HBEF III, Hasselsven and Tongbersven. Values are in mass fraction (%) of the fine earth.

Location	hor.	depth (cm)	pH(H$_2$O)	C	KCl	Extractable "free" Al$_2$O$_3$ (successive extraction)			
						Na$_4$P$_2$O$_7$	am.ox	DCB	total
HBEF I	O	+10-0	3.58	38	0	0.1	0.1	0	0.2
	E	0-7	4.00	5	0.1	0.5	0.1	0.1	0.8
	Bhs1	7-13	4.14	4	0.1	1.6	0.1	0.1	1.9
	Bhs2	13-40	4.39	3	0	1.4	0.1	0.1	1.6
	Bhs3	40-60	4.18	1	0	0.4	0.1	0	0.5
HBEF II	O+Al	+8-3	4.31	14	0.1	0.2	0	0	0.3
	E	3-16	4.21	1	0	0	0	0	0
	Bhs1	16-22	4.20	4	0.1	1.3	0	0.1	1.5
	Bhs2	22-48	4.39	3	0	1.6	0.2	0.1	1.9
HBEF III	O	+8-0	4.21	38	0	0.3	0	0.1	0.3
	E	0-4	3.90	1	0.1	0	0	0.1	0.1
	Bhs1	4-9	4.05	2	0.1	0.7	0.1	0.2	1.1
	Bhs2	9-24	4.24	2	0.1	1.2	0.2	0.2	1.7
	Bhs3	24-33	4.29	2	0	1.0	0	0.1	1.1
Hasselsven	O	+3-0	—	—	—	—	—	—	—
	E	0-9	3.90	2	0	0.1	0	0	0.1
	Bhs	9-15	4.17	2	0	0.3	0.1	0	0.4
	Bs1	15-33	4.29	1	0	0.5	0.1	0	0.6
	Bs2	33-70	4.56	0	0	0.3	0.2	0.1	0.6
Tongbersven	O	+5-0	—	—	—	—	—	—	—
	E	0-20	3.82	2	0	0	0	0	0
	Bhs	20-40	3.93	5	0	1.0	0.1	0	1.1
	Bs	40-60	4.20	1	0	0.6	0.2	0.1	0.9

Table 1 continued.

Location	horizon	Elemental concentration, mass fraction (%)					
		SiO_2	Al_2O_3	Fe_2O_3	MgO	CaO	K_2O
HBEF I	O	19	2.4	0.6	0.1	0.4	0.6
	E	77	7.7	0.7	0.1	0.4	2.6
	Bhs1	72	9.0	2.4	0.2	0.6	2.6
	Bhs2	73	11.0	3.1	0.3	0.8	2.7
	Bhs3	78	12.0	2.2	0.4	0.9	3.0
HBEF II	O+Al	72	2.9	1.5	0.1	0.7	0.6
	E	87	7.9	0.9	0.1	0.5	2.5
	Bhs1	71	9.9	2.8	0.3	0.7	2.3
	Bhs2	74	9.6	3.6	0.2	0.6	2.3
HBEF III	O	22	2.9	0.6	0.1	0.7	0.6
	E	87	10.3	1.4	0.1	0.7	0.6
	Bhs1	69	10.5	5.1	0.5	0.6	2.3
	Bhs2	67	12.2	4.8	0.8	0.6	2.3
	Bhs3	73	12.0	2.9	0.8	0.6	2.6
Hasselsven	O	--	--	--	--	--	--
	E	94	1.2	0.3	0.1	0.1	0.4
	Bhs	93	1.4	0.0	0.0	0.0	0.5
	Bs1	95	1.8	0.4	0.0	0.1	0.5
	Bs2	94	2.4	0.5	0.1	0.1	0.8
Tongbersven	O	--	--	--	--	--	--
	E	98	0.5	0.1	0.0	0.0	0.3
	Bhs	87	2.3	0.5	0.0	0.1	0.6
	Bs	96	2.4	0.4	0.1	0.1	0.7

importance of the forest floor as a source of dissolved aluminum in Spodosols from New York and New Hampshire was noted earlier by David and Driscoll (1984) and Cronan and Schofield (1979). Total monomeric aluminum levels peak in the E (HBEF II and III) and Bhs1 (HBEF I, Hasselsven and Tongbersven) horizons and decrease again at greater depth.

Monomeric aluminum in all forest floor leachates is primarily (60 to 100%) in organic form. In HBEF soils, dissolved aluminum at greater depth is also mainly organically complexed, but in the two Dutch soils, it is largely (90 to 95%) in an inorganic form. Although the concentrations of DOC in the Dutch soils are higher than those at HBEF (Table 3), these concentrations are probably not high enough to complex a considerable part of the dissolved aluminum. Furthermore, the lower pH of the Dutch soil solutions depresses the dissociation of organic acids and the complexation of aluminum. Dissolved inorganic aluminum at Hasselsven and Tongbersven is mainly aquo-aluminum (Al^{3+}), primarily balanced by dissolved sulphate (mean concentrations of 307 and 672 $mmol.m^{-3}$, respectively). These sulphate concentrations are much higher than at HBEF (mean 57 $mmol.m^{-3}$).

Dissolved SiO_2 concentrations are similar at HBEF and Hasselsven (means between 79 and 112 $mmol.m^{-3}$) and higher at Tongbersven (mean 169 $mmol.m^{-3}$).

Chemical fluxes can give valuable information on the systems' response to acid deposition (van Breemen et al., submitted; Mulder et al., submitted). Estimated soil water fluxes and fluxes of dissolved organic and inorganic monomeric Al, SO_4 and Si are given in Table 4. At all sites, inorganic monomeric aluminum fluxes reach a maximum in the Bhs horizon, and decrease again at greater depth. However, fluxes of total monomeric Al in the mineral soil are 5 to 20 times higher in the Dutch than in the HBEF Spodosols. For the solubilization of 1.34 $kmol\ Al^{3+}.ha^{-1}.y^{-1}$ at Tongbersven, 4.02 $kmol\ H^+.ha^{-1}.y^{-1}$ are needed. Atmospheric acid inputs amount to only 2.9 $kmol\ H^+.ha^{-1}.y^{-1}$. The discrepancy can be explained by an additional H_2SO_4 input (1.1 $kmol\ H^+.ha^{-1}.y^{-1}$) from the O horizon. The higher SO_4^{2-} fluxes from the O horizon than from the atmosphere (Table 4) may be due to net mineralization of organic sulphur. Matzner et al. (1984) reported that input (throughfall)/output(drainage at 75 cm depth) ratios for sulphate in pine stands varied from 50% in one year to 200% in another year. These differences must be due to annual variations in sorption or assimilation/mineralization of SO_4^{2-}. An underestimation of dry-deposited S, measured via throughfall, due to direct assimilation of S on the leaf surface, may also contribute to the relatively high dissolved sulphate fluxes in the soil. At HBEF, only small amounts of aluminum are mobilized by mineral acids and transported in inorganic form: maximally 0.03 $kmol.ha^{-1}.y^{-1}$.

Fluxes of inorganic aluminum (Table 4) decrease markedly in the subsoil (Bhs2/3 or Bs) at all sites. At HBEF this was also noted by Johnson et al. (1981), who proposed that neutralization of acid rain in the HBEF soils is a two-step process. The first

Table 3. Means and standard deviations of pH and concentrations of monomeric Al, SiO_2, and SO_4, and dissolved organic carbon in precipitation and soil solutions, collected from March through July, 1982 (HBEF profiles) and from October 1982 to October 1983 (Hasselsven and Tongbersven). Concentrations in $mmol.m^{-3}$.

Location	horizon	pH		Al_{mono}		SiO_2		SO_4		DOC	
HBEF	Rain[a]	4.13	(0.02)	--		--		30	(1)	--	
HBEF 1	O	3.62	(0.45)	37	(17)	112	(52)	85	(15)	3800	
	E	4.36	(0.95)	24	(21)	70	(50)	72	(9)	1200	
	Bhs1	4.67	(0.16)	32	(9)	103	(50)	50	(10)	840	(410)
	Bhs3	4.77	(0.79)	24	(9)	80	(15)	52	(6)	1033	
HBEF II	O	4.95	(0.50)	15	(7)	133	(72)	48	(11)	450	(450)
	E	4.63	(0.20)	19	(14)	107	(27)	49	(5)	330	(130)
	Bhs1	5.05	(0.09)	12	(5)	105	(14)	52	(4)	190	(120)
	Bhs2	5.12	(0.20)	9	(3)	123	(20)	52	(14)	220	(70)
HBEF III	O	4.49	(0.56)	7	(4)	50	(37)	71	(35)	2400	(1100)
	E	4.93	(0.43)	13	(9)	55	(48)	56	(24)	1000	(670)
	Bhs1	5.19	(0.16)	10	(3)	77	(27)	47	(16)	500	(120)
	Bhs3	5.22	(0.26)	10	(4)	112	(18)	48	(9)	520	(180)
Hass'ven	Rain[b]	4.44		--		--		58		--	
Hass'ven	O	3.81	(0.13)	22	(19)	150	(146)	201	(105)	3460	(1440)
	E	3.52	(0.14)	108	(87)	98	(54)	404	(112)	1670	(450)
	Bhs	4.37	(0.05)	170	(28)	62	(26)	336	(43)	750	(330)
	Bs2	4.33	(0.22)	92	(29)	76	(7)	286	(53)	1580	(590)
Tong'ven	O	4.33	(0.49)	13	(8)	33	(29)	514	(252)	5600	(1630)
	E	3.34	(0.14)	54	(29)	137	(47)	600	(326)	7320	(1230)
	Bhs	4.07	(0.08)	580	(120)	204	(67)	800	(180)	2110	(1020)
	Bs	4.12	(0.18)	552	(101)	167	(34)	773	(167)	2120	(1050)

[a] From Likens et al. (1977).
[b] Flux weighted mean for Hasselsven and Tongbersven.

Table 4. Estimated water fluxes (in cm.y^{-1}) and fluxes of organic and inorganic monomeric aluminum, silica and sulphate (in kmol.ha^{-1}.y^{-1}), as well as the Al/Si flux ratios for HBEF-, Hasselsven- and Tongbersven soils. Annual precipitation was 132 cm at HBEF (Likens et al. 1977) and 92.5 cm at Hasselsven and Tongbersven.

Location	hor.	water(cm)	Al$_{inorg}$	Al$_{org}$	Si	SO$_4$	Al/Si
HBEF	O	96.1	0.007	0.078	1.06	--	0.08
	E	92.3	0.011	0.074	0.98	--	0.09
	Bhs1	89.7	0.026	0.085	0.97	--	0.11
	Bhs2/3	78.1	0.018	0.067	0.96	--	0.09
Hass'ven-precipitation		92.5	--	--	--	0.52	--
Hasselven	O	77.4	0.03	0.10	0.94	1.47	0.15
	E	69.3	0.41	0.09	0.64	2.85	0.78
	Bhs	55.0	1.00	--	0.38	1.90	2.62
	Bs	43.5	0.42	--	0.34	1.30	1.24
Tong'ven-throughfall		63.3	--	--	--	1.33	--
Tongbersven	O	63.3	0.03	0.06	0.20	2.86	0.45
	E	50.9	0.24	0.05	0.75	3.43	0.39
	Bhs	34.5	2.12	--	0.73	2.93	2.89
	Bs	22.9	1.34	--	0.39	1.93	3.48

step is the transformation of incoming H^+ into a mixture of H^+ and Al^{3+} acidity in the surface soil. The second step is the final neutralization of virtually all acidity (H^+ and Al^{3+}) by basic cation solubilization, due to silicate weathering or cation exchange. At Hasselsven and Tongbersven the decreased aluminum fluxes in the subsoil are associated primarily with a loss of SO_4^{2-} from the percolating water, without an appreciable increase in the flux of basic cations.

Molar Al/Si flux ratios are around 0.1 in the HBEF soil solutions at all depths, and higher at Hasselsven and Tongbersven, reaching a maximum of 2.6 and 2.9 in the Bhs horizon. Weathering of feldspars or micas, the dominant aluminum silicate minerals present in these soils, could cause Al/Si flux ratios not exceeding 0.33 (albite, microcline) to 1 (anorthite, muscovite). The higher Al/Si ratios at Hasselsven and Tongbersven indicate that solubilized aluminum originates primarily from free (non-silicated) Al forms, which occur in high concentrations in the Bhs horizon.

Additional information on the interaction between solutions and mineral phases can be obtained from disequilibrium indices (I_p) in soil solutions (Table 5). At all sites, the trend in solution saturation with natural gibbsite was similar: highly undersaturated in the surface horizons (O and E) and approaching saturation in the B horizon. At all sites, solutions in the surface layers were undersaturated with kaolinite, whereas at greater depth supersaturation with kaolinite occurred. The differences in I_p values for the HBEF and the Dutch sites were most pronounced for the basic aluminum sulphates jurbanite ($Al(SO_4)(OH).5H_2O$) and basaluminite ($Al_4(SO_4)(OH)_{10}.5H_2O$). Driscoll et al. (1984, 1985) reported HBEF and Adirondack Spodosol solutions to be undersaturated with jurbanite and basaluminite. At Hasselsven and Tongbersven, soil solutions approach saturation (Hasselsven) or even reach slight supersaturation with jurbanite, with increasing depth, and change from highly undersaturated with respect to basaluminite in the surface horizons, to highly supersaturated in the subsoil. The importance of basic aluminum sulphates in regulating the soil solution aluminum activity, suggested by the I_p values, is supported by the removal of both Al and SO_4 from the soil solution in the lower B horizons of Hasselsven and Tongbersven, as discussed above (Table 4). For the lower B horizon of a shallow Swedish Spodosol, receiving acid inputs similar to those at HBEF and Hasselsven, Nilsson and Bergkvist (1983) also reported solutions to be supersaturated with basaluminite and nearly in equilibrium with jurbanite. These Spodosols have SO_4^{2-} and inorganic monomeric Al concentrations around 170 $mmol.m^{-3}$ and 80 $mmol.m^{-3}$, levels intermediate between those at HBEF I and at Hasselsven. Solutions of acid brown forest soils in the Netherlands also reach saturation with jurbanite in the subsoil, but remain undersaturated with natural gibbsite at all depths. The saturation of soil solutions with jurbanite should be interpreted with caution, however, since high inputs of sulphate and acidity in base-poor soils will automatically lead to ionic activity products close to the equilibrium constant of jurbanite (Mulder et al., submitted).

Table 5. Mean saturation indices (I_p) and standard deviations for natural gibbsite (May et al. 1979), kaolinite (Stumm and Morgan 1981) and jurbanite (van Breemen 1973) of soil solutions from various horizons at HBEF, Hasselsven and Tongbersven. HBEF solutions used in the calculations were from March 1982 through July 1982; for Hasselsven and Tongversven data from November 1982 and May 1983 were used.

Location	hor.	nat. gibbsite		kaolinite		jurbanite	
HBEF I	O	-5.33	(3.18)	-3.07	(0.43)	-2.83	(0.49)
	E	--	--	--	--	--	--
	Bhs1	-0.54	(0.50)	3.21	(1.28)	-1.88	(0.43)
	Bhs3	-0.03	(0.21)	3.93	(1.15)	-1.72	(0.36)
HBEF II	O+A1	-1.50	(1.45)	1.50	(3.24)	-3.19	(1.49)
	E	-1.64	(0.20)	1.56	(0.34)	-2.76	(0.16)
	Bhs1	-0.03	(0.25)	3.95	(1.93)	-2.29	(0.93)
	Bhs2	-0.38	(0.49)	4.01	(0.93)	-2.50	(0.28)
HBEF III	O	-2.57	(1.21)	-2.47	(1.79)	-3.38	(0.85)
	E	-3.11	(1.73)	-3.03	(4.59)	-4.92	(1.98)
	Bhs1	-1.48	(1.45)	1.23	(2.89)	-3.82	(1.53)
	Bhs3	-0.87	(0.64)	2.86	(3.58)	-2.98	(0.72)
Hasselsven	O	-3.27	--	-1.47	--	-2.61	--
	E	-3.47	--	-1.51	--	-1.10	--
	Bhs	-0.49	--	3.78	--	-0.10	--
	Bs	-0.56	--	4.16	--	-0.58	--
Tongbersven	O	-4.27	--	-3.11	--	-1.95	--
	E	-3.88	--	-2.11	--	-1.35	--
	Bhs	-0.95	--	3.86	--	0.37	--
	Bs	-0.12	--	4.93	--	0.64	--

To evaluate the present rate of Spodosol formation we can compare present-day rates of immobilization of organic dissolved Al in Bhs horizons at HBEF with the mean rate of podzolization estimated from the pool of pyrophosphate-extractable aluminum and the maximum time available for soil formation (13,000 yrs). The present-day rates (0.007 $kmol.ha^{-1}.y^{-1}$) are much smaller than the mean rates over the past 13,000 years (0.04-0.09 $kmol.ha^{-1}.y^{-1}$). From these data Driscoll et al. (1985) concluded that either the Spodosol development had decreased in recent times, or that other processes (e.g., root turnover) contributed to the formation of pyrophosphate extractable aluminum. For Hasselsven and Tongbersven soils we estimated the mean rate of accumulation of organic Al in the Bhs horizons over 6,000 years to be 0.11 and 0.16 $kmol\ Al.ha^{-1}.y^{-1}$. We have too few fractionation data on dissolved Al to compare mean historic and present-day rates of podzolization.

CONCLUSIONS

The way in which strong atmospheric inputs are buffered in Spodosols strongly depends on the base status of the soil. Apparently, Spodosols from the Netherlands that are rich in silica and poor in basic cations buffer acidity primarily through dissolution of aluminum. In the HBEF Spodosols, which are richer in bases than those from the Netherlands, acid neutralization is mainly accomplished by the solubilization of basic cations.

Aluminum solubilization in both soil types occurs to a large extent in the Spodosol B horizons. The base-poor Dutch Spodosol at the site with the highest acid input (2.9 $kmol.ha^{-1}.y^{-1}$) shows a much higher Al output (1.34 $kmol.ha^{-1}.y^{-1}$) than at the site with the lower acid input (input 1.4 $kmol.ha^{-1}.y^{-1}$ of H^+, output 0.4 $kmol.ha^{-1}.y^{-1}$ of Al^{3+}). Moreover, there is a net removal of Al from Bhs horizons amounting to 0.6 and 1.9 $kmol\ Al.ha^{-1}.y^{-1}$, respectively, which is 5 to 12 times the mean rate of organic aluminum deposition during podzolization in these Spodosols. This Al, mobilized in the Bhs, is transported in an inorganic form and partially immobilized in the Bs horizons (98% at Hasselsven and 41% at Tongbersven), possibly by precipitation of basic aluminum sulphates.

In the HBEF soils Driscoll et al. (1985) found no evidence for a breakdown of the Spodosol B horizon: net organic aluminum transport to the B horizon equals net export of inorganic Al from this horizon.

In both the HBEF and the Dutch Spodosols, soil solutions in the B horizon are slightly undersaturated with gibbsite. At the Dutch sites, but not at HBEF, soil solutions in the B horizon reach saturation with jurbanite. This difference is related to the higher levels of sulphate and aluminum in the Dutch Spodosols.

If the difference between Al mobilization at Hasselsven and HBEF is indeed mainly due to the availability of basic cations for neutralization of atmospheric acid inputs, North American Spodosols with parent material poorer in bases than HBEF soils, but with similar acid inputs, should have dissolved Al levels of the same order of magnitude as the Hasselsven soils. However, base-poor Spodosols with high acid inputs may be hard to find in North America where areas with a high acid load coincide with glaciated terrain and soils with a rich mineral assemblage.

Acknowledgements

Thanks are due to E.J. Velthorst, N. Nakken, H. Sliepenbeek and A.J. Kuyper for help during sampling and chemical analyses, to Th. Pape for help with data management and to J.J. M. van Grinsven for estimating soil water fluxes. RIN-Leersum and RIVM-Bilthoven collected and analysed throughfall at Tongbersven. Staats Bosbeheer and Mr. Ryken are gratefully acknowledged for making the Hasselsven and Tongbersven sites available for our research. We thank Dr. P. Buurman for his comments on the first draft of this paper.

This research was supported in part by the EEC (ENV 650 NL) and by the Netherlands Directorate General for Science. This is contribution 10 of the Hackfort project on Effects of Acid Atmospheric Deposition on Soils and Waters, Department of Soil Science and Geology, Agricultural University, Wageningen, The Netherlands.

REFERENCES

Buurman P (1984) Podzols. Van Nostrand Reinhold Comp. Inc., New York, p 450
Cronan CS, Schofield CL (1979) Aluminum leaching response to acid precipitation: effect on high elevation watersheds in the Northeast. Science 204: 305-306
David MB, Driscoll CT (1984) Aluminum speciation and equilibria in soil solutions of a Haplorthod in the Adirondack Mountains (New York, U.S.A.). Geoderma 33: 297-318
Driscoll CT (1984) A procedure for the fractionation of aqueous aluminum in dilute acidic waters. Int J Environ Anal Chem 16: 267-284
Driscoll CT, van Breemen N, Mulder J (1985) Aluminum chemistry in a forested Spodosol. Soil Sci Soc Am J 49: 437-444
Federer CA (1973) Annual cycles of soil and water temperatures at Hubbard Brook. USDA Forest Service Res. Note NE-167
Johnson NM, Driscoll CT, Eaton JS, Likens GE, McDowell WH (1981) "Acid Rain", dissolved aluminum and chemical weathering at the Hubbard Brook Experimental Forest, New Hampshire. Geochim Cosmochim Acta 45: 1421-1437
Likens GE, Bormann FH, Pierce RS, Eaton JS, Johnson NM (1977) Biogeochemistry of a forested ecosystem. Springer, New York, p 146
Martin AE, Reeve R (1958) Chemical studies of podzolic illuvial horizons 3. Titration curves of organic matter suspensions. J Soil Sci 9: 89-100
Matzner E, Khanna PK, Meiwes KJ, Cassens-Sasse E, Bredemeier M, Ulrich B (1984) Ergebnisse der Fluessemessungen in Waldoekosystemen. In: Berichte des Forschungszentrums Waldoekosysteme/Waldsterben, Bd. 2: 29-49

May HM, Helmke PA, Jackson ML (1979) Gibbsite solubility and thermodynamic properties of hydroxy- aluminum ions in aqueous solution at 25 C. Geochim Cosmochim Acta 43: 861-868

Mulder, J (1980) Neutralization of acid rain in the Hubbard Brook Exp. Forest. MSc. Thesis, Agricultural Univ. Wageningen, the Netherlands

Mulder J, van Grinsven JJM, van Breemen N (submitted) Hydrochemical budgets of woodland soils affected by atmospheric acid deposition. III: Aluminum chemistry. Soil Sci Soc Am J

Mokma DL, Buurman P (1982) Podzols and podzolization in temperate regions. ISM monograph 1, Int. Soil Museum, Wageningen, p 126

Nilsson SI, Bergkvist B (1983) Aluminum chemistry and acidification processes in a shallow podzol on the Swedish westcoast. Water Air Soil Pollut 20: 311-329

Petersen L (1976) Podzols and podzolization. DSR Forlag, Copenhagen

Schnitzer M, Skinner SIM (1963) Organo-metallic interactions in soils: 2. Reactions between different forms of iron and aluminum and the organic matter of a podzol Bh horizon. Soil Sci 96: 181-186

Soil Survey Staff (1975) Soil taxonomy. USDA Handbook 436. p 754

Stumm W, Morgan JJ (1981) Aquatic Chemistry: an introduction emphasizing chemical equilibria in natural waters, 2 edn. Wiley, New York, p 780.

Ugolini FC, Minden R, Dawson H, Zachara J (1977) An example of soil processes in the _Abies amabilis_ zone of Central Cascades, Washington. Soil Sci 124: 291-302

Ulrich B, Matzner E (1983). Abiotische Folgewirkungen der weitraumigen Ausbreitung von Luftverunreinigungen. Luftreinhaltung Forschungsbericht 104 02 615

Ulrich B, Mayer R, Khanna PK (1979) Die Deposition von Luftverunreinigungen und ihre Auswirkungen in Waldoekosystemen im Solling. Schriften Forstl Fak Univ Gottingen u Nieders Forstl Vers Anst, Band 58: 1-291

van Breemen N (1973) Dissolved aluminium in acid sulphate soils and acid mine waters. Soil Sci Soc Am Proc 37: 694-697

van Breemen N, Burrough PA, Velthorst EJ, van Dobben HF, de Wit T, Ridder TB, Reynders HFR (1982) Soil acidification from atmospheric ammonium sulphate in forest canopy throughfall. Nature 299: 548-550

van Breemen N, Mulder J, Driscoll CT (1983) Acidification and alkalinisation of soils. Plant & Soil 75: 283-308

van Breemen N, Driscoll CT, Mulder J (1984) Acidic deposition and internal proton sources in acidification of soils and waters. Nature 307: 599-604

van Breemen N, Mulder J, van Grinsven JJM (submitted) Hydrochemical budgets of woodland soils affected by atmospheric acid deposition. II: N-transformations. Soil Sci Soc Am J

van Grinsven JJM, van Breemen N, Mulder J (submitted) Hydrochemical budgets of woodland soils affected by atmospheric acid deposition. I: Simulation of unsaturated water transport. Soil Sci Soc Am J

Westall JC, Zachary JL, Morel FMM (1976) MINEQL: a computer program for the calculation of chemical equilibrium of aqueous systems. Ralph M. Parsons Laboratory for water resources and environmental engineering, Civil Engineering Dept., M.I.T. Technical Note 18

Wood TE (1980) Biological and chemical control of phosphorus cycling in a northern hardwood forest. PhD Thesis. Yale Univ., New Haven, CT

LIMITS ON CATION LEACHING OF WEAKLY PODZOLIZED FOREST SOILS: AN EMPIRICAL EVALUATION

I.K. Morrison and N.W. Foster

Government of Canada, Canadian Forestry Service, Great Lakes Forestry Research Centre, Sault Ste. Marie, Ontario, P6A 5M7

ABSTRACT

Chemical properties of two weakly podzolized sandy soils, one a Humo-Ferric Podzol, the other a less well-developed Dystric Brunisol, both from beneath mid-aged jack pine (Pinus banksiana Lamb.) stands in northern Ontario, Canada, are given. In a 7 1/2-year-old column-lysimeter experiment (reported elsewhere), it had been noted that both soils, but the Podzol in particular, initially exhibited strong resistance to SO_4^{2-} leaching. A hypothesis of anion immobilization by SO_4^{2-} adsorption was advanced. Data to support the hypothesis are presented in the present paper on SO_4^{2-} adsorption characteristics of the two soils. During the initial phase of SO_4^{2-} loading to these soils, cation leaching is effectively blocked by selective removal of SO_4^{2-} ions from the leaching solution and their adsorption chiefly into the Bf_1 or Bm_1 horizons, and the leaching solution is thereby robbed of counterions. Once SO_4^{2-} adsorption capacity is reached, bases move freely with surplus SO_4^{2-} ions, with the chief limitation to removal being the upper limit imposed by the supply of exchangeable ions themselves. Evidence suggests that prolonged exposure to acid solutions may result in increased weathering of silicate minerals of sufficient magnitude to compensate eventually for losses associated with the stripping of exchangeable reserves.

INTRODUCTION

There exists a substantial body of well-founded theory, with origins dating back over a century, that views cation retention and leaching of soils largely in terms of physical chemistry. In addition, considerable empirical data on cation leaching are accumulating at field sites (e.g., Johnson and Henderson 1979, Mollitor and Raynal 1982, Bockheim et al. 1984, Foster et al., in press). Data on soil chemical properties exist in varying quantities (depending on locale), providing the opportunity, once relationships between properties and processes have been determined, to apply information gained from a few sites to many sites.

Forest and other wildland soils in the Boreal Region of eastern Canada are, in comparison with many other soils, neither well documented with respect to properties nor well understood with respect to processes. The current heavy loadings of acid-forming air pollutants such as sulphate (SO_4^{2-}) ions, estimated in the Algoma District of northern Ontario, Canada, at 30 kg.ha^{-1}.y^{-1} (Foster et al., in press) and even higher elsewhere, have prompted various attempts to predict ecosystem response to acid rain or to estimate soil 'sensitivity' (Anon. 1980, 1981, 1983; Shilts 1981). Wide variation in soil types within the region makes interpretation of laboratory experiments difficult and extensions from theory to practice imprecise. The extent of the geographic area involved and the importance of making accurate predictions has caused us to focus on soil processes in general.

Results from a column-lysimeter experiment, now in its eighth year of operation and involving two acid sands, were reported earlier (Morrison 1981, 1983, in press). This work suggested a stage-by-stage process of element loss from these soils: first, both soils exhibited considerable initial resistance to base leaching, with SO_4^{2-} movement hampered by strong SO_4^{2-} retention (a hypothesis proposed by McColl (1969), Johnson and Cole (1976, 1977), Singh et al. (1980) and others); second, when SO_4^{2-} adsorption capacity reached saturation, bases moved freely with excess SO_4^{2-} ions; third, as exchangeable bases were depleted, H^+ ions increasingly dominated the charge composition of the leaching solution and pH declined; fourth, with pH of the percolating solution reduced, there was substantial mobilization of trace metals and indications of more rapid weathering of primary minerals. The purpose of the present paper is to evaluate the principal processes controlling the leaching of cations from these weakly-podzolized forest soils, in light of measurements of soil properties.

METHODS

The study soils, both acid sands supporting well stocked, mid-aged, Site Class II (Plonski 1974) jack pine (Pinus banksiana Lamb.) stands of fire origin, are (1) an Orthic Humo-Ferric Podzol (Canada Soil Survey Committee 1978) of the Wendigo Series, developed in gravelly, loamy sand of glacio-fluvial origin in Wells Township, Algoma District, and (2) an Orthic Dystric Brunisol, unnamed as to Series, developed in silt loam over loamy sand in Dupuis Township, Sudbury District, 135 km to the north. For convenience, the soils are hereinafter referred to as WELLS and DUPUIS.

Details of the lysimeter experiment are given in earlier reports (Morrison 1981, 1983, in press). In brief, the study soils were reconstructed into 1-m-deep column-lysimeters, then leached with artificial 'acid rain' consisting of sulphuric acid (H_2SO_4) diluted to pH 2, pH 3, and pH 4, as well as a distilled water control calibrated to approximate ambient precipitation levels. Soils were leached and percolates were collected weekly and analyzed monthly over 7 1/2 years.

For purposes of the present analysis, cation exchange capacity (CEC), by horizon for each soil, was determined by ammonium saturation, with 1 M NH_4Cl, unbuffered (Nõmmik 1974). Exchangeable potassium (K^+), calcium (Ca^{++}), magnesium (Mg^{++}) and sodium (Na^+) concentrations were determined in the same extracts by flame emission/atomic absorption spectrophotometry.

Pyrophosphate-extractable iron (Fe) and aluminum (Al) were determined according to McKeague (1978). Total K, Ca, Mg and Na in soil, including soil minerals, were determined following digestion in a sulphuric-nitric-perchloric acid mixture. Sulphate adsorption was by sequential leaching with Na_2SO_4, according to the method of Johnson and Henderson (1979). After elution of soluble SO_4^{2-}, adsorbed SO_4^{2-} was then extracted with NaH_2PO_4 containing 500 ppm phosphorus (P). Sulphate was determined by ion chromatography with a Dionex Model 2110i Ion Chromatograph.

In input solutions and in percolates, element concentrations were expressed in conventional terms. Total amounts were calculated by multiplying concentration by input (or percolate) volume. For soils, exchangeable and total concentrations were calculated on a horizon and on a whole-soil basis (to the same depth as the lysimeter columns) by multiplying 'concentration' by horizon weight, the latter determined from measures of horizon thickness and bulk density.

RESULTS

In general, in a horizon-by-horizon comparison, the WELLS soil is higher in CEC than the DUPUIS soil but lower in exchangeable K^+, Ca^{++} and Mg^{++} concentrations (Table 1). Concentrations of Na^+ were about equal. Pyrophosphate-extractable Fe and Al concentrations, with the exception of the Ae horizon, were consistently higher throughout the WELLS profile than they were in the DUPUIS profile. This result is also reflected in the profile analysis to 100 cm, with the DUPUIS soil being higher in exchangeable and total bases than the WELLS soil (Table 2). Quantities of SO_4^{2-} added and SO_4^{2-}, K^+, Ca^{++}, Mg^{++} and Na^+ recovered over 388 weeks of leaching were calculated from solution concentrations and input and percolate volumes (Table 3). Over a 7 1/2-year period, substantial quantities of SO_4^{2-} were retained by both soils, with significant loss occurring only in relation to the pH 2 treatment. Significant loss of bases likewise occurred only in relation to this treatment. Figure 1 illustrates that, for any given level, the Bf_1 horizon of the WELLS soil adsorbed more SO_4^{2-} than the equivalent Bm_1 of the DUPUIS soil. The Bf_1 and Bm_1 horizons both contained and adsorbed more SO_4^{2-} than did other horizons (Table 4). Retention decreased with depth in both soils. No adsorption maximum, within the range of 18 to 600 $mg.L^{-1}$, was observed in any horizon examined (Fig. 1). Below the Bf_1 and Bm_1 horizons, however, there was little difference between soils in their capacity to adsorb additional SO_4^{2-}. The WELLS soil retained all of the additional SO_4^{2-} in the pH 4 and 3 treatments, the DUPUIS soil all of the SO_4^{2-} in the pH 4 treatment. Base cations were leached from both soils in association with SO_4^{2-} in the pH 2 treatment.

Table 1. Selected chemical properties, by horizon, of Wells Township and Dupuis Township soils.

Horizon	Depth (cm)	CEC[a]	Exchangeable cations[a] (c mol (+).kg^{-1})				Pyrophosphate-extractable (%)	
			K	Ca	Mg	Na	Fe	Al
WELLS TOWNSHIP								
H	-	12.92	0.26	2.60	0.83	0.08	-	-
Ae	0-2	4.47	0.06	0.30	0.08	0.02	0.10	0.09
Bf$_1$	2-17	3.58	0.04	0.37	0.02	0.03	0.27	0.59
Bf$_2$	17-30	2.59	0.04	0.33	0.02	0.01	0.11	0.45
IIBC	30-45	1.10	0.04	0.06	0.02	<0.01	0.02	0.10
C	>45	0.80	0.02	0.02	<0.01	0.01	0.02	0.11
DUPUIS TOWNSHIP								
H	-	26.19	0.36	9.20	1.44	0.17	-	-
Ae	0-2	4.79	0.09	1.06	0.25	0.02	0.17	0.13
Bm$_1$	2-10	2.28	0.08	0.48	0.25	0.03	0.10	0.36
Bm$_2$	10-30	1.90	0.07	0.15	0.04	0.02	0.04	0.21
IIBC	30-44	1.00	0.02	0.17	0.06	0.01	0.01	0.08
C	>44	0.50	0.01	0.18	0.08	<0.01	0.01	0.05

[a]NH$_4$Cl, unbuffered

Table 2. Potassium, calcium, magnesium and sodium content, exchangeable and total, to a depth of 100 cm (kg.ha^{-1}) in Wells Township and Dupuis Township soils.

Soil	Base content to 100 cm				Total
	K	Ca	Mg	Na	
Exchangeable[a]					
Wells	39	259	12	24	334
Dupuis	142	511	120	27	800
Total					
Wells	112780	109258	92151	196654	510843
Dupuis	123528	99613	114689	254889	592719

[a]NH$_4$Cl, unbuffered

Table 3. Input of sulphate and output of sulphate and cations over 388 weeks in relation to treatment.

Treatment	Input SO_4^{2-}	Output over 388 weeks (kg.ha^{-1})				
		SO_4^{2-}	K^+	Ca^{++}	Mg^{++}	Na^+
WELLS						
DW	0	172.0	93.7	882.3	64.2	281.5
pH4	358	151.4	118.5	860.6	67.5	276.8
pH3	3581	171.8	109.9	926.8	83.3	337.1
pH2	35838	9628.6	218.2	1151.5	229.2	349.5
DUPUIS						
DW	0	199.7	35.0	581.4	141.1	203.3
pH4	358	188.8	36.6	630.0	155.3	191.7
pH3	3581	510.6	31.0	501.3	122.9	218.6
pH2	35838	17241.4	266.2	2475.5	539.1	389.4

Table 4. Fractions of SO_4^{2-}, by horizon, in Wells Township and Dupuis Township soils.

Horizon	SO_4^{2-} (µg.g^{-1} soil) extracted by				SO_4^{2-} absorbed[a] (µg.g^{-1} soil)	
	H_2O		NaH_2PO_4			
	Wells	Dupuis	Wells	Dupuis	Wells	Dupuis
Ae	34	37	4	6	36	0
$Bf_1(Bm_1)$	14	15	216	118	271	133
$Bf_2(Bm_2)$	16	8	208	103	170	163
IIBC	6	3	36	11	111	85
C	1	3	26	6	83	76

[a]Equilibrated with 225 mg.L^{-1} SO_4^{2-}

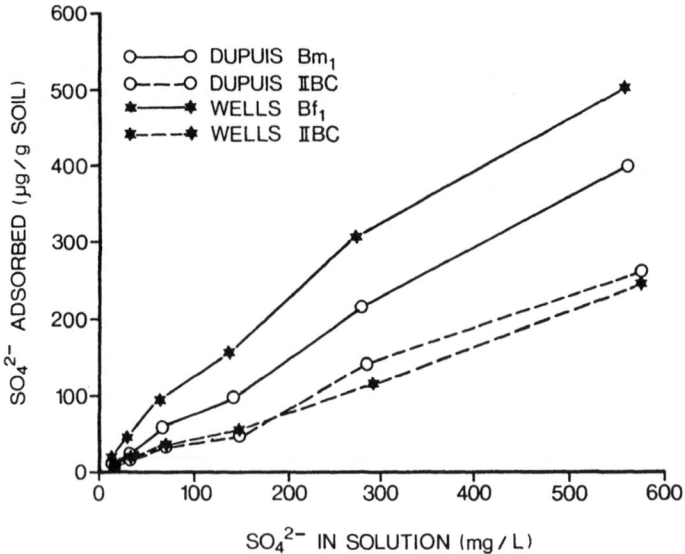

Fig. 1. Adsorption of SO_4^{2-} by B and II BC horizons of Wells Township and Dupuis Township soils.

DISCUSSION

Conventionally, cation exchange in soils is portrayed in terms of equilibria between large pools of slowly weathering mineral (or slowly decomposing organic) compounds, and smaller pools containing the same elements in ionic form held in negatively charged regions on the surfaces of soil particles or, in varying degrees of dissociation, more or less free in the soil solution. Displacement, diffusion and mass flow are seen as the main processes whereby elements move from the solid phase into the drainage. Our previous work with the two study soils generally conforms to this view (Morrison 1981, etc.).

In nature, limits are placed on rate of loss of cations by an array of external and internal factors. Two factors remaining largely outside the scope of the present discussion are: (1) mass removal of reaction products associated with gross differences in water flow, and (2) removals associated with higher plant uptake or with chemical or microbial immobilization. In the first instance, to simulate some degree of environmental realism, volume input in the experiment referred to herein had been adjusted to an equivalent of 1000 mm per annum (approximately equal to ambient precipitation in the area) applied in equal weekly 'events'. In nature, however, input is usually sporadic, with large portions of the annual runoff penetrating the profile within restricted time periods, giving limited opportunity for equilibration. Hence, the 8-year frame of the present experiment should not be likened to 8 years in nature, but, rather, a substantially longer period. With respect to plant uptake, this is better estimated by means of a balance-sheet approach.

Earlier, we had estimated gross uptake of K, Ca, Mg by the WELLS stand at 17.8, 20.7 and 3.2 $kg.ha^{-1}.y^{-1}$, respectively (Foster and Morrison 1976).

Internally, limits are imposed by a number of soil-related factors including rate of reaction, both exchange and chemical, rate of removal of reaction products, and the magnitudes of reserves. During the first phase, wherein both the WELLS and DUPUIS soils exhibited considerable initial resistance to SO_4^{2-} leaching, we borrowed a hypothesis from McColl (1969), Johnson and Cole (1976) and others with respect to SO_4^{2-} mobility. It was tentatively concluded (Morrison 1981) that the leaching of bases from the two study soils was effectively blocked during the initial period by the selective removal of SO_4^{2-} ions from the leaching solution and their adsorption onto exchange sites associated with the presence of Fe and Al sesquioxides in the B-horizons of the soils. At the time, the results were noted as consistent with the general perception (incorporated into the Canadian System of Soil Classification) that Podzols possess more highly developed B-horizons (as determined by the pyrophosphate-extractable Al and Fe test) than Brunisols. In the present paper, this is further substantiated by the higher estimates of adsorption of SO_4^{2-} in the Bf_1 horizon of the WELLS soil than in the corresponding Bm_1 horizon of the DUPUIS soil. Therefore, it is concluded that differences in loss-rate of SO_4^{2-} during the initial phase of loading are explained by differences in soil SO_4^{2-} adsorption capacity.

Sulphate loss from soil treated with distilled water was continuous over the 7 1/2 years of observations on the lysimeter experiment. Soluble SO_4^{2-} values from the soil analyses were highest in the surface horizon and decreased with depth. These observations suggest that the source of SO_4^{2-} leached from the soil is SO_4^{2-} mineralized in the surface mineral horizons. Further, desorption of SO_4^{2-} with water and phosphate solutions did not remove all SO_4^{2-} adsorbed from solutions onto the soils. Further, in the lysimeter experiment, substantial quantities of SO_4^{2-} remain unaccounted for. Precipitation of $AlOHSO_4$ (van Breemen 1973) or $Al_4OH_{10}SO_4$ (Singh and Brydon 1969) in addition to anion exchange may contribute to SO_4^{2-} retention in these soils.

With respect to ion mobility and mass transfer, rate of reaction and factors pertaining thereto were less dominant. The chief limit placed on cation loss during the phase wherein bases moved freely with surplus anions was the upper limit imposed by the exchangeable supply itself. The greater yield of exchangeable bases from the DUPUIS soil than from the WELLS soil is consistent with the determinations of exchangeable base reserves in both. It is interesting that decreasing the pH of the leaching solution to 3.0 and raising SO_4^{2-} additions to 480 $kg.ha^{-1}.y^{-1}$ failed to increase cation leaching significantly over 7 1/2 years. At pH 2.0 and 4800 $kg.ha^{-1}.y^{-1}$ of SO_4^{2-}, leaching was substantial. The higher CEC of the WELLS soil, however, and the observed tendency of that soil to resist pH change, is consistent with a hypothesis of colloidal buffering, although this did not come into effect until the soil was well depleted of nutrients.

After prolonged acid treatment, the composition of the drainage water appeared to be dominated by the weathering products of silicate minerals rather than by products associated with the decomposition of organic matter. The relative leaching of base cations was similar in both soils: $Ca^{++} > Mg^{++} > Na^+ > K^+$. On the subject of base depletion in general, it should be noted that, while acid treatment eventually stripped both soils of their exchangeable base content, 7 1/2 years of intensive treatment (i.e., pH 2.0) were sufficient to remove less than 1% of the total base content of either soil.

Acknowledgments

The authors wish to acknowledge the following members of the Canadian Forestry Service for their assistance during this project: Messrs. D. Ropke, J.R. Ramakers, and D.J.S. Kurylo.

REFERENCES

Anon. (1980) Second report of the United States-Canada research consultation group on the long-range transport of air pollutants. US-Canada Research Consultation Group on the Long-Range Transport of Air Pollutants. p 40
Anon. (1981) United States-Canada memorandum of intent on transboundary air pollution. Impact assessment work group 1. Phase II. Interim working paper, October.
Anon. (1983) United States-Canada memorandum of intent on transboundary air pollution. Impact assessment work group 1. Final report, January.
Bockheim JG, Leide JE and Esser JM (1984) Acidic deposition and ion movements in forest soils of northwestern Wisconsin. In: Stone EL (ed) Forest soils and treatment impacts. Proc. Sixth North Am For Soils Conf, Univ Tennessee, Knoxville, June 1983, p 291-312
Canada Soil Survey Committee (1978) The Canadian system of soil classification. Can Dep Agric, Ottawa, ON. Publ 1646, p 164
Foster NW and Morrison IK (1976) Distribution and cycling of nutrients in a natural *Pinus banksiana* ecosystem. Ecology 57: 110-120
Foster NW, Nicolson JA, Morrison IK (in press) Acid deposition and element cycling in eastern North American forests. In: Proc Acid Rain and Forest Resources Conf, Quebec, Canada, 14-17 June 1983
Johnson DW, Cole DW (1976) Sulfate mobility in an outwash soil in western Washington. USDA For Serv Northeastern For Exp Stn, Gen Tech Rep No NE-23, p 827-835
Johnson DW, Cole DW (1977) Sulfate mobility in an outwash soil in western Washington. Water, Air, Soil Pollut 7: 489-495
Johnson DW, Henderson GS (1979) Sulfate adsorption and sulfur fractions in a highly weathered soil under a mixed deciduous forest. Soil Sci 128: 34-40

McColl JG (1969) Ion transport in a forest soil: models and mechanisms. Ph.D. Dissert, Univ Washington, Seattle, p 214

McKeague JA (ed) (1978) Manual on soil sampling and methods of analysis. Can Soc Soil Sci Subcommittee on Methods of Analysis, p 212

Mollitor AV, Raynal DJ (1982) Acid precipitation and ionic movements in Adirondack forest soils. Soil Sci Soc Am J 46: 137-141

Morrison IK (1981) Effect of simulated acid precipitation on composition of percolate from reconstructed profiles of two northern Ontario forest soils. Dep Environ, Can For Ser, Ottawa, ON. CFS Res Notes 1: 6-8

Morrison IK (1983) Composition of percolate from reconstructed profiles of two jack pine forest soils as influenced by acid input. In: Ulrich B, Pankrath J (eds) Effects of accumulation of air pollutants in forest ecosystems, D Reidel Publ Co, p 195-206

Morrison IK (in press) Effects of artificial acid rain on percolate chemistry of two jack pine forest soils. In: Proc Acid Rain and Forest Resources Conf, Quebec, Canada, 14-17 June 1983

Nõmmik H (1974) Ammonium chloride-imidazole extraction procedure for determining titratable acidity, exchangeable base cations, and cation exchange capacity in soils. Soil Sci 118: 254-262

Plonski WL (1974) Normal yield tables (metric) for major forest species of Ontario. Ont Min Nat Resour, Div For, Toronto, ON, p 40

Shilts WW (1981) Sensitivity of bedrock to acid precipitation: modification by glacial processes. Geol Surv Can, Pap 81-14. 7 p + maps 1549A, 1550A, 1551A

Singh BR, Abrahamsen G, Stuanes A (1980) Effect of simulated acid rain on sulfate movement in acid forest soils. Soil Sci Soc Am J 44: 75-80

Singh SS and Brydon JE (1969) Solubility of basic aluminum sulfates at equilibrium in solution and in the presence of montmorillonite. Soil Sci 107: 12-16

van Breemen N (1973) Dissolved aluminum in acid sulfate soils and in acid mine waters. Soil Sci Soc Am Proc 37: 694-697

EFFECTS OF HEAVY METALS AND ALUMINUM ON THE ROOT PHYSIOLOGY OF SPRUCE (PICEA ABIES KARST.) SEEDLINGS

D.L. Godbold, R. Tischner* and A. Hüttermann
Forstbotanisches Institut
*Botanishes Institut
Universität Göttingen, 3400 Göttingen, F.R.G.

ABSTRACT

The toxicity of the heavy metals Zn, Cd, Hg and Pb and organic complexes of Hg and Pb to four-week old Picea abies seedlings was assessed by estimating the effects of these metals on root elongation. The toxicity of these metals when supplied as inorganic salts was found to differ considerably, the order of toxicity being Hg > Pb > Cd > Zn, with Hg being over a 100 times more toxic than Zn. Methyl-Hg was considerably more toxic than $HgCl_2$. This was shown to be due to its chemical form and not to a greater uptake. Lead supplied as triethyl, diethyl and dimethyl lead complexes was also more toxic than $PbCl_2$. However, there was no significant difference between the toxicity of the organic lead complexes. In short-term uptake experiments (5 h), the uptake of Cd into roots of Picea abies was found to be influenced by the ionic composition and pH of the uptake solution. Cd uptake decreased with decreasing pH, and increased with decreasing Ca supply. Al and Mn inhibited Cd uptake, whereas Hg and Zn produced a slight stimulation in uptake.

With the use of X-ray microprobe techniques, the ion contents of vacuoles in root tissue were found to be influenced by both pH and Al. At 2 mM Al (Ca/Al ratio 0.75) the vacuolar contents of Mg, K and P declined significantly compared to untreated plants. Toxic levels of Cd were found not to influence the K or Mg contents of the roots of Picea abies. Comparison of the metal concentrations shown to inhibit root elongation in nutrient solution to those found in the soil solution from the humus layer taken from under trees showing the symptoms of decline, suggests that the levels of Hg, Pb and Zn are sufficiently high to influence root growth.

INTRODUCTION

Investigations of roots at different soil levels in declining forests have shown a dramatic change in the fine roots with increasing soil depth (Hüttermann 1985). A good fine root development could only be found in the humus and Ah soil layers. In contrast, in undamaged areas of conifer forest intensive fine root development was found throughout the mineral soil layers (Vogt et al. 1980, 1981). Investigations of forest soils in areas of decline have shown that these soils often contain high levels of aluminum and heavy metals. In the soil profile these potentially toxic elements tend to have a different spatial distribution. The highest concentrations of aluminum in the soil solution are found in the mineral soil, whereas heavy metals accumulate principally in the humus layer. Hence, a potentially inhibitory chemical environment for root growth is present in all layers of the soil profile.

The work of Rost-Siebert (1985) and Murach (1984) have shown that the levels of aluminum found in forest sites in West Germany are sufficient to be a primary factor in forest decline, and the previously published studies of Tischner et al. (1983) have shown that aluminum affects both the morphology of, and enzymatic processes in roots of Picea abies. In particular, damage to the endodermis has been found both in roots of Picea abies taken from acidified soils and from roots grown in nutrient solutions containing Al (Fig. 1). At present it is not known whether the levels of heavy metals in the humus layer are sufficiently high to affect the physiology of forest trees, and hence be a contributing factor together with aluminum to forest decline.

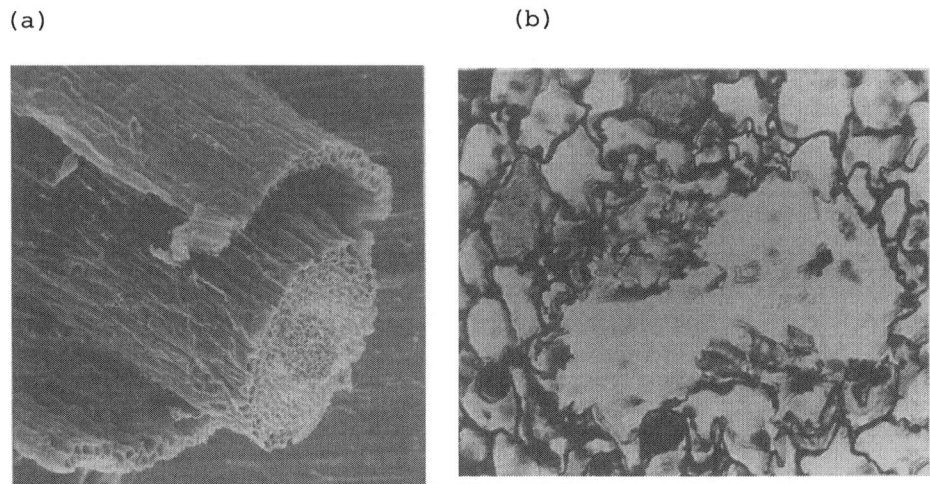

Fig. 1. (a) S.E.M. picture of a fine root of spruce collected from a stand growing on an acidified soil. (b) Transverse section of a root of a spruce seedling grown for 6 weeks in nutrient solution containing 2.5 mM Al (Ca/Al ratio 0.5) at pH 4.

APPROACH

All work was carried out on seedlings of Picea abies grown under sterile conditions (Godbold and Hüttermann 1985). Inhibition of root elongation was used to estimate the toxicity of heavy metals to spruce seedlings grown in nutrient solution. Nutrient solutions were adjusted to pH 4.5 and changed daily or twice daily to prevent depletion of added metals. Details for each experiment are included in figure captions.

EFFECT ON HEAVY METALS ON ROOT ELONGATION

Inorganic Complexes

Root elongation was greatly inhibited by 30 and 60 µM Zn (Fig. 2); inhibition was evident after only 24 h of Zn treatment. Cadmium was found to inhibit root elongation when supplied at levels between 5 - 60 µM (Fig. 3); the degree of inhibition being dependent on the concentration of Cd in the nutrient solution. Similar to Zn, Cd inhibited root elongation within 24 h. At 30 and 60 µM Cd, root elongation almost completely ceased after 5 days. At 0.5 and 0.05 µM Cd a significant inhibition of root elongation could not be found (data not presented).

Lead supplied as $PbCl_2$ (Fig. 4) inhibited root elongation within 24 h at a concentration of 0.5 µM Pb. Thereafter, the degree of inhibition compared to the control remained constant over the 7-day treatment period.

Mercury was found to be considerably more toxic than Zn, Cd or Pb. Root elongation was severely inhibited by 0.1 and 0.5 µM Hg after 24 h (Fig. 5). At 5 and 15 µM Hg, root elongation ceased completely within 24 h, and a radial shrinkage was observed 1-5 mm behind the root tip.

These data show that the elements Zn, Cd, Hg and Pb differ in toxicity to _Picea abies_ seedlings. The order of toxicity is Hg > Pb > Cd > Zn, with Hg being over 100 times more toxic than Zn.

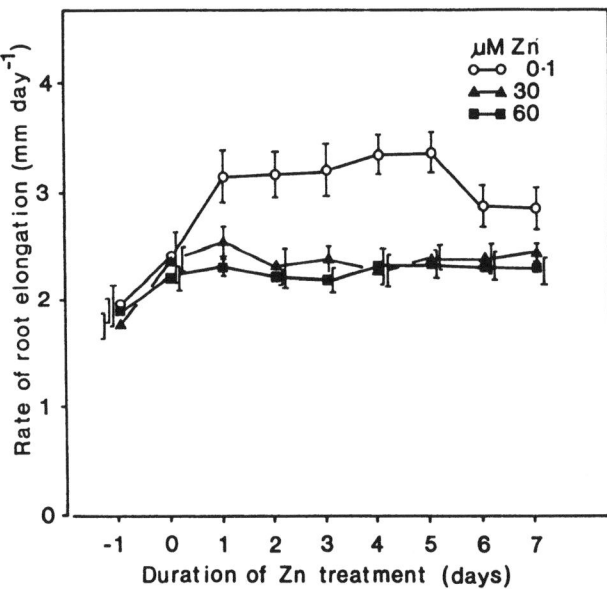

Fig. 2. Root elongation rate of _Picea abies_ seedlings grown for 7 days in nutrient solutions containing a range of $ZnSO_4$ concentrations (0.1 - 60 µM Zn). Bars indicate standard error.

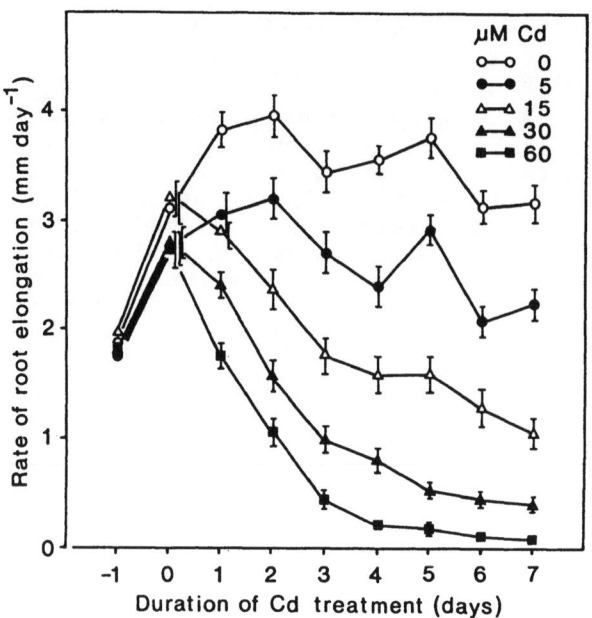

Fig. 3. Root elongation rate of <u>Picea abies</u> seedlings grown for 7 days in nutrient solutions containing a range of $CdSO_4$ concentrations (0 - 60 µM Cd). Bars indicate standard error.

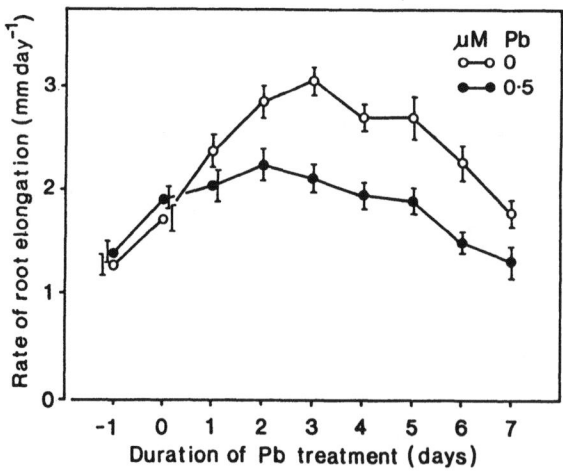

Fig. 4. Root elongation rate of <u>Picea abies</u> seedlings grown for 7 days in nutrient solutions containing 0 or 0.5 µM Pb as $PbCl_2$. Bars indicate standard error.

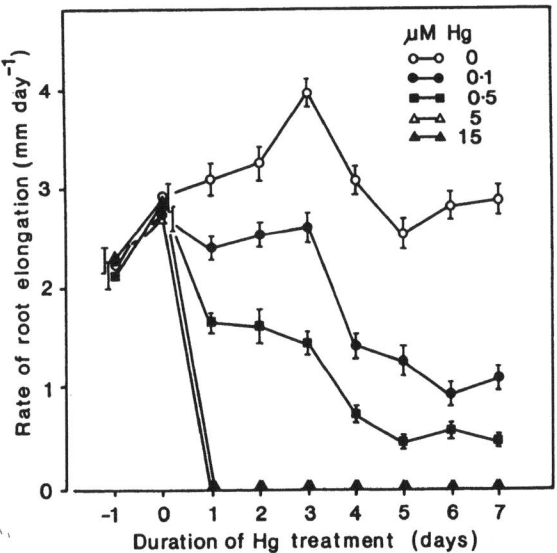

Fig. 5. Root elongation rate of *Picea abies* seedlings grown for 7 days in nutrient solution containing a range of $HgCl_2$ concentrations (0 - 15 µM Hg). Bars indicate standard error.

Organo-metallic Complexes

Organic complexes of heavy metals have been found in many heavy metal contaminated areas and are often considered to be more toxic than inorganic complexes. When lead was supplied as the organic complexes triethyl, diethyl, and trimethyl lead chloride, all were found to inhibit root elongation at concentrations of 2 µM Pb (Fig. 6). No significant differences were found between the various organic forms after 4 days of exposure, although a more severe inhibition was induced by the methyl form than by the ethyl forms in the first 3 days of treatment. However, all organic forms induced a greater inhibition of root elongation than did $PbCl_2$.

Elongation of *Picea abies* seedling roots exposed to 0.05 µM $HgCl_2$ or methyl-HgCl (Fig. 7) was significantly inhibited within 1-2 days. At 0.05 µM $HgCl_2$, the degree of inhibition remained relatively constant, whereas at 0.05 µM methyl-HgCl root elongation completely ceased after 3 days of exposure. At 0.01 and 0.001 µM methyl-HgCl, the rate of root elongation was inhibited after 2 and 4 days of treatment, respectively (Fig. 8). After 7 days of 0.01 µM methyl-HgCl treatment, root elongation almost completely ceased. In short-term experiments, the greater toxicity of methyl-HgCl compared to $HgCl_2$ was shown to be due to its chemical form and not due to a greater uptake. At 0.1 µM methyl-HgCl the rate of root elongation was inhibited by 4 h of exposure, whereas 0.1 µM $HgCl_2$ did not inhibit root elongation until 8 h (Table 1). A comparison of the amount of Hg taken up into the roots shows that after 4 and 8 h of exposure, similar amounts of Hg as $HgCl_2$ and methyl-HgCl had been accumulated in the root tissue. A greater accumulation of methyl-HgCl than $HgCl_2$ was found only after 8 h.

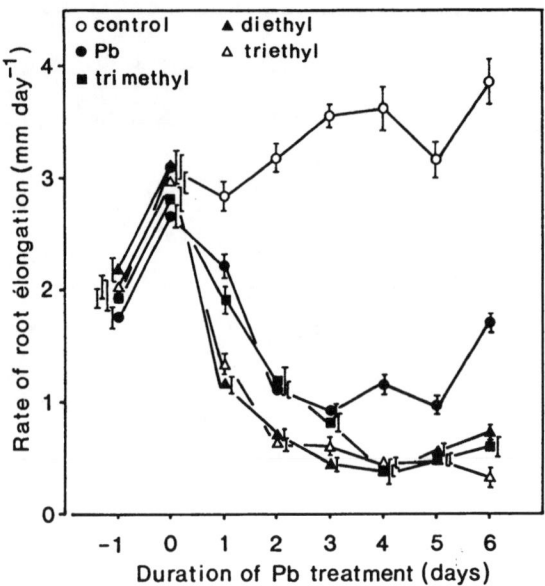

Fig. 6. Root elongation rate of Picea abies seedlings grown for 6 days in nutrient solutions containing 2 µM Pb supplied as either $PbCl_2$, triethyl, diethyl, or trimethyl lead chlorides. Bars indicate standard error.

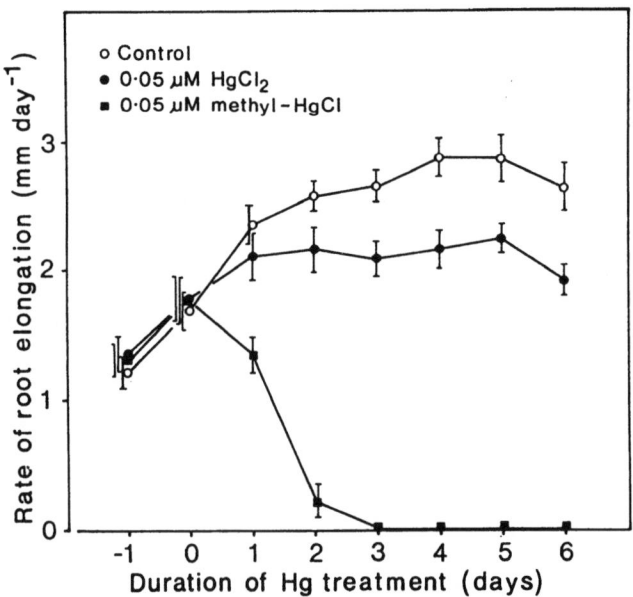

Fig. 7. Root elongation rate of Picea abies seedlings grown for 6 days in nutrient solutions containing 0 (control) or 0.05 µM Hg as $HgCl_2$ or methyl-HgCl. Bars indicate standard error.

Table 1. Mercury contents and percentage root growth (control 100%) of the roots of Picea abies seedlings maintained for 16 h in nutrient solutions containing 0 (control) or 0.1 µM Hg as $HgCl_2$ or methyl-HgCl. Standard error in parentheses.

Treatment duration	Hg concentration ($\mu mol.g^{-1}$ dwt)			% root growth	
	Control	$HgCl_2$	methyl-HgCl	$HgCl_2$	methyl-HgCl
4 hours	0	0.45 (0.01)	0.46 (0.07)	116.0	69.8
8 hours	0	0.76 (0.05)	0.73 (0.06)	86.4	18.9
16 hours	0	0.83 (0.07)	1.03 (0.02)	63.2	13.8

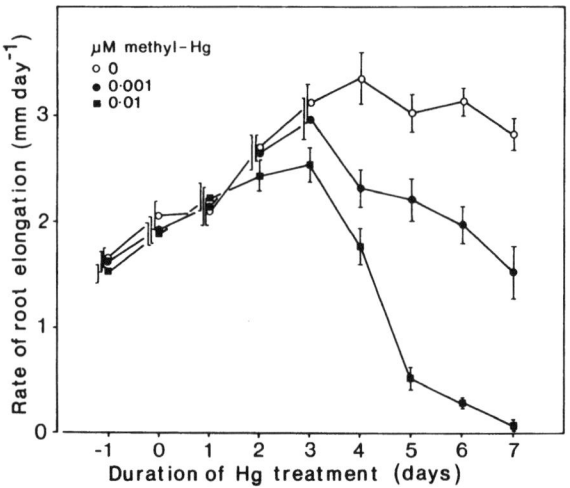

Fig. 8. Root elongation rate of Picea abies seedlings grown for 7 days in nutrient solutions containing a range of methyl-HgCl concentrations (0-0.01 µM). Bars indicate standard error.

Factors Affecting the Uptake of Heavy Metals

Many factors in soils and nutrient solutions have been shown to influence the uptake of heavy metals into plants. Cadmium uptake has been found to be influenced by soil pH, soil cation exchange capacity and competition from other ions. These effects tend to vary with plant species and conditions of growth (Jastrow and Koeppe 1980).

The effect of pH on Cd uptake into roots of Picea abies seedlings grown in nutrient solution is shown in Table 2. A strong relationship between decreasing pH and decreasing Cd uptake was found. The depression of Cd uptake is a result of both a reduced uptake into the apoplast, due to a reduced availability of COO^- groups in the cell wall, and a reduced symplastic uptake (unpublished data).

An increase in Al availability and leaching of Ca are two of the major changes in the chemical mileau of acidified soils. Cadmium uptake was found to be decreased by the presence of Al in the uptake medium (Fig. 9a). At 0.1 and 0.5 mM Al, decreases in Cd uptake of 40 and 86% of the control, respectively, were found. Similarly, the concentration of Ca in the nutrient solution was also found to influence Cd uptake (Fig. 9a). At 225 µM Ca, an increase in Cd uptake of 114% of the control value (450 µM Ca) was found. Conversely, at 900 µM Ca, Cd uptake decreased to 81.5% of the control. The action of Ca on Cd uptake was principally due to Ca influencing Cd uptake in the apoplast (unpublished data).

Table 2. Effect of pH on Cd uptake into roots of <u>Picea abies</u> seedlings from a nutrient solution containing 0.05 µM Cd (^{109}Cd), 5 h uptake at 25°C. Standard error in parentheses.

pH	Cd contents of roots (nmol.g^{-1} dwt)
2.5	1.1 (0.02)
3.5	7.0 (0.1)
4.5	16.8 (0.2)
5.5	32.8 (1.8)
6.5	85.2 (2.8)

The influence of other heavy metal cations on Cd uptake can be seen in Fig. 9b. Zn and Hg did not significantly influence Cd uptake compared to the control value, whereas Mn strongly inhibited Cd uptake. Mn inhibition of Cd uptake has been suggested to be due to competition for carrier sites at the plasmalemma (Cataldo <u>et al</u>. 1983).

An in-depth discussion of the kinetics of ion uptake is not relevant at this point. The data presented here merely emphasize that the uptake of heavy metals into <u>Picea abies</u> roots, for which Cd was used as a model, is strongly influenced by the ionic composition of the surrounding medium.

EFFECTS OF ALUMINUM AND CADMIUM ON NUTRIENT UPTAKE

Analysis of the ion contents of needles of trees showing symptoms of decline has shown that these trees often have lower concentrations of K and Mg than "healthy" trees (Zöttl and Mies 1983; Schulte-Bisping and Murach 1984). At present little is known about the effects of aluminum and heavy metals on the uptake and transport of nutrients in forest trees. Evers (1983) has shown that roots of Al-treated, one-year old Picea abies seedlings had lower concentrations of Mg, Ca and Fe. Examination of the ion contents of vacuoles of root tissue of P. abies seedlings using X-ray microprobe techniques (reported in full by Tischner et al., in press), have shown them to be sensitive to both pH and Al treatments. The K contents of vacuoles were found to decrease dramatically in the presence of 2 mM Al (Table 3). This was greatest in the vacuoles of the rhizodermis, and was found in all tissues except in vacuoles of the xylem parenchyme.

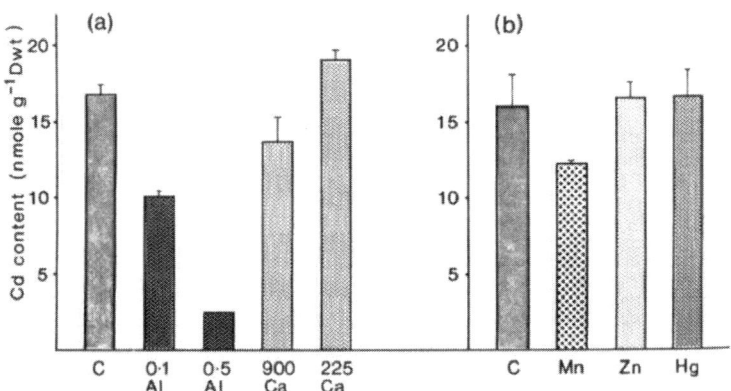

Fig. 9. (a) Effect of 0.1 or 0.5 mM Al (as $AlSO_4$) and 225 or 900 µM Ca (as $CaCl_2$, $Ca(NO_3)_2$) on Cd uptake into roots of Picea abies seedlings. (b) Effect of 5 uM Mn (as $MnSO_4$) or 5 µM Zn (as $ZnSO_4$), or 0.1 µM Hg (as $HgCl_2$) on Cd uptake into roots of Picea abies seedlings. Uptake solutions contained 0.05 uM Cd (^{109}Cd) at pH 4.5, 25°C, 5 h duration of uptake. "C" is the control treatment.

Table 3. Peak/background ratio of K contents in vacuoles of roots of Picea abies seedlings grown in the presence and absence of 2 mM Al (Ca/Al ratio 0.75) at pH 4 for 6 weeks.

	-Al	+Al
Rhizodermis	1.98	0.58
Cortex (middle)	1.70	1.14
Cortex (inner)	1.87	1.49
Endodermis	1.18	1.00
Pericycle	2.16	1.54
Xylem parenchyma	1.14	1.43

The ratio of P/S in the vacuole was also found to be sensitive to Al (Table 4). This was most extreme in the vacuoles of the endodermis and middle cortex, and was more pronounced at pH 3 than at pH 4.

Table 4. P/S ratio in different root tissues from roots of Picea abies seedlings grown in the presence and absence of 2 mM Al (Ca/Al ratio 0.75) at pH 3 or 4 for 6 weeks.

	pH 4, -Al	pH 4, +Al	pH 3, +Al
Cortex (middle)	11.8	0.14	0.30
Endodermis	10.2	3.3	0.21

The ratio shifts in favour of the S content and drops to less than one-tenth of untreated plants. The decrease in the P content of the cortex cell vacuoles may be due to an inhibition of P metabolism, or to a reduced storage of P. The latter is supported by the results of Ferguson and Clarkson (1975) who found the cortex cells to be the major site of P accumulation in roots. However, low pH and Al may reduce P metabolism directly through inhibition of hexokinase (Clarkson 1966) and acid phosphatase or ATPase (Woolhouse 1969).

The Ulrich hypothesis on the effects of Al on plant growth includes an uptake of Al ions into the roots (Ulrich 1981). Evidence is given that Al exchanges with Ca in the cell wall, altering their microstructure. The Ca/Al ratio in vacuoles was found to be sensitive to both Al and pH. The most obvious effects were detected in the rhizodermis and endodermis (Table 5). In both tissues, the ratio shifted in favour of Al, resulting in a more than ten-fold decrease in Al-treated plants at pH 3. The same Al treatment at pH 4 reduced the ratio to only one third and one tenth in the rhizodermis and endodermis, respectively. The sensitivity of the endodermis to Al is in accordance with the disturbance of the endodermis cell walls already reported (Tischner et al. 1983). Ca/Al ratios of a similar magnitude to those shown in Table 5 have been reported in roots of dying spruce trees (Gehrmann et al. 1984).

Table 5. Ca/Al ratio in different root tissues from roots of Picea abies seedlings grown in the presence and absence of 2 mM Al (Ca/Al ratio 0.75) at pH 3 or 4 for 6 weeks.

	pH 4, -Al	pH 4, +Al	pH 3, +Al
Rhizodermis	0.67	0.29	0.063
Endodermis	1.67	0.143	0.045

In contrast to the results presented for Al, no significant inhibition of Mg and K uptake in the short-term (24 h) could be demonstrated for Cd when supplied at 0.05 and 5 µM Cd (data not presented).

However, longer-term studies are clearly necessary to fully assess the influence of heavy metals on nutrient uptake.

INVOLVEMENT OF HEAVY METALS IN FOREST DECLINE

Rate of root elongation was used to estimate the toxicity of heavy metals to Picea abies seedlings. Direct comparison of concentrations of metals in nutrient and soil solutions is difficult as the speciation and activity of the metals in the soil solution are not known, although on some heavily-polluted soils, Pb and Cd in the soil solution were found to be mainly present as chlorides (Tills and Alloway 1983a, 1983b). It is probable that Zn and Hg also exist as inorganic complexes in the soil solution. When the levels of metals shown to inhibit root elongation in nutrient solution are compared to those in the humus layer under stands of Picea abies showing decline symptoms (Table 6), it can be seen that the Zn, Pb, and Hg contents of the soil and soil solutions are sufficiently high to influence root growth.

Table 6. Zn, Pb and Cd concentration in soil solution (µM) and total soil Hg contents (µmol.kg^{-1}) of the humus layer of a forest soil at Solling[a], and Pb in the soil solution (1 m) of a forest soil at Hamburg[b], F.R.G.

	Soil solution				Total soil
	Zn	Cd	Pb[c]	Pb[d]	Hg
Mean	5.1	0.01	0.14	1.05	4.6
Minimum	2.2	0.01	0.07	0.36	4.1
Maximum	11.2	0.02	0.30	3.52	4.8

[a] E. Matzner (pers. comm.)
[b] N. Rastin (pers. comm.)
[c] Solling
[d] Hamburg

The concentration of mercury in the soil solution is not known. However, if only 1% of the total mercury in the humus layer was available to plants as Hg^{2+} this would be sufficient to influence root growth. Mercury methylation in soils has been shown to be principally an abiological process (Rogers 1977) mediated by humus and fulvic acids (Nagase et al. 1982). The rate of methylation is increased by low pH and an increase in Hg concentration. Conversely, loss of methyl-Hg via demethylation and conversion to more volatile dimethyl-Hg is a biological process (Rogers 1976).

Hence, factors associated with soil acidification, decrease in soil pH, an increase in the depth of the humus layer and a decrease in microbial activity (Chet et al. 1984) would favour the production of methyl-Hg. In estuarine sediments methyl-Hg constitutes about 0.01% of the total Hg (Andren and Harriss 1973). If 0.01% of the total Hg in the humus layer was in the form of methyl-Hg in the soil solution, a level would be found similar to that of the lowest methyl-Hg level (0.001 µM) found to inhibit root elongation in nutrient solution. The level of lead shown to inhibit root elongation is similar to that in the soil solution at Solling, and below that at forest sites near Hamburg, indicating this to be a factor which could inhibit root growth of forest trees. The speciation of lead in the soil solution is unknown. Lead may be present, in part, as the more toxic alkylated forms, which are derived in the environment from leaded petrol.

All data presented here were obtained from plants grown in solutions with the addition of one metal and good availability of nutrients. Combinations of heavy metals have been shown to be more toxic than metals singly in maize (Hassett et al. 1976) and poplar (Carlson and Bazzaz 1977). Thus, under conditions of lower nutrient availability, low pH and in combination with other heavy metals, heavy metals in the soil solution may be more toxic than in nutrient solution. The accumulation of heavy metals in the humus layer is due principally to the high affinity of humic and fulvic acids for heavy metals. The majority of the total metals in the humus layer will be bound to organic material, and hence present in a less toxic form than the free ions. However, Kerndorff (1980) has shown that even when humic and fulvic acids are in great excess, all the metals available are not complexed. The percentage of metals bound was found to be pH-dependent; a decrease in pH decreased the percentage of bound metals. Hence, any sudden decrease in the pH of the soil, due to an acidification push (Hüttermann and Ulrich 1984), for example, is likely to release bound heavy metals. The rapidity of metal uptake by Picea abies and of growth inhibition, suggests that such occurrences are likely to be significant for root growth and turnover.

Due to problems of comparing data obtained from nutrient solution with data obtained in the field, it cannot be conclusively shown that heavy metal toxicity is a causal factor in forest decline. However, the data show that the levels of metals are sufficiently high in the humus layer that they can no longer be excluded as a possible factor contributing to forest decline.

Acknowledgements

We thank E. Matzner and N. Rastin for free access to the soils data. This work was supported by the Royal Society (London) and the European Community (Grant ENV-726-D(B)).

REFERENCES

Andren AW, Harriss RC (1973) Methyl mercury in estuarine sediments. Nature, London 235: 256-257

Carlson RW, Bazzaz FA (1977) Growth reduction in American Sycamore (Platanus occidentalis L.) caused by Pb-Cd interaction. Environ Pollut 12: 243-253

Cataldo DA, Garland TR, Wilding RE (1983) Cadmium uptake kinetic in soybean. Plant Physiol 54: 844-848

Chet I, Merg G, Hüttermann A (1984) The effect of acid rain on microbial population, biomass and activity. In: Ulrich (ed) Berichte des Forschungszentrums Waldökosysteme/Waldsterben, vol 3. Gottingen, p 135

Clarkson DT (1966) Effect of aluminum on the uptake and metabolism of phosphorous by barley seedlings. Plant Physiol 41: 165-172

Evers FH (1983) Ein Versuch zur Aluminium-Toxizität bei Fichte. Der Forst-und Holzwirt 12: 305-307

Ferguson IB, Clarkson DT (1975) Ion transport and endodermal suberization in the roots of Zea mays. New Phytol 75: 69-79

Gehrmann J, Gerriets M, Puhe J, Ulrich B (1984) Untersuchungen an Boden, Wurzeln, Nadeln und erste Ergebnisse von Depositionsmessungen im Hils. In: Ulrich B (ed) Berichte des Forschungszentrum Waldökosysteme/Waldsterben, vol 2, p 169

Godbold DL, Hüttermann A (1985) Effect of zinc, cadmium and mercury on root elongation of Picea abies (Karst.) seedlings, and the significance of these metals to forest dieback. Environ Pollut A 38: 375-381

Hassett JJ, Miller JE, Koeppe DE (1976) Interaction of lead and cadmium by roots. Environ Pollut 11: 297-302

Hüttermann A (1985) Symptome des Waldsterbens. Allgem Forstz 4: 67-70

Hüttermann A, Ulrich B (1984). Solid-phase- solution- root interaction in soils subject to acid deposition. Phil Trans Roy Soc Lond B305: 353-368

Jastrow JD, Koeppe DE (1980) Uptake and effects of cadmium in higher plants. In: Nriagu JO (ed) Cadmium in the environment, John Wiley and Sons, Toronto, p 607

Kerndorff H (1980) Analytische und experimentelle Untersuchungen zur Bedeutung der Humin- und Fulvosäuren als Reaktionspartner für Schwermetalle in anthropogen belasteten und unbelasteten Regionen. PhD Thesis, Universität Mainz, F.R.G.

Murach D (1984) Die Reaktion der Feinwurzeln von Fichten (Picea abies Karst.) auf zunehmende Bodenversauerung. PhD Thesis, Universität Göttingen, F.R.G.

Nagase H, Ose Y, Sato T, Ishikawa T (1982) Methylation of mercury by humic substances in the aquatic environment. Sci Total Environ 24: 133-142

Rogers RD (1976) Methylation of mercury in agricultural soils. J Environ Qual 5: 454-458

Rogers RD (1977) Abiological methylation of mercury in soil. J Environ Qual 6: 463-467

Rost-Siebert K (1985) Al-Toxizität bei Fichten- und Buchenkeimlingen, PhD Thesis, Universität Göttingen, F.R.G.

Schulte-Bisping H, Murach D (1984) Inventur der Biomass und ausgewählter chemischer Elemente in zwei unterschiedlich stark versauerten Fichtenjungbestanden im Hils. In: Ulrich B (ed) Berichte des Forschungszentrum Waldökosystem/Waldsterben, vol 2, p 207

Tills AR, Alloway BJ (1983a). The use of liquid chromatography in the study of cadmium speciation in soil solution from polluted soils. J Soil Sci 34: 769-781

Tills AR, Alloway BJ (1983b). The speciation of lead in soil solution from very polluted soils. Environ Tech Letts 4: 529-534

Tischner R, Kaiser U, Hüttermann A (1983) Untersuchungen zum Einfluss von Aluminium-Ionen auf das Wachstum von Fichtenkeimlingen in Abhangigkeit vom pH-Wert. Forstw Cbl 102: 329-336

Tischner R, Stelzer R, Hüttermann A (in press). Distribution pattern of different elements in root vacuoles of spruce seedlings dependent on combined Al and pH treatments. J Plant Physiol

Ulrich B (1981) Eine ökosystemare Hypothese uber die Ursachen des Tannensterbens (Abies alba Mill.). Forstw Cbl 11: 228-236

Vogt KA, Edmonds RL, Grier CC, Piper SR (1980) Seasonal changes in mycorrhizal and fibrous-textured root biomass in 23 and 180 year old Pacific silver fir stands in western Washington. Can J For Res. 10: 523-529

Vogt KA, Edmonds RL, Grier CC (1981) Seasonal changes in biomass and vertical distribution of mycorrhizal and fiberous-textured conifer fine roots in 23 and 180 year old subalpine Abies amabilis stands. Can J For Res 11: 223-229

Woolhouse HW (1969) Differences in the properties of the acid phosphatases of plant roots and their significance in the evolution of edaphic ecotypes. In Rorison IH (ed) Ecological aspects of the mineral nutrition of plants, Blackwell, Oxford p 357

Zöttl HW, Mies E (1983) Die Fichtenerkrankungen in Lagen des Südschwarzwaldes. Allgem Forst- und Jagdzeitg 154: 6-7

ALUMINUM TOXICITY IN FOREST TREE SEEDLINGS

T. Eldhuset*, A. Göransson, and T. Ingestad

Section of Forest Ecophysiology, Swedish University of Agricultural Sciences, P.O. Box 7072, S-750 07 UPPSALA, Sweden
* Present address: Norwegian Forest Research Institute, Section of Forest Ecology, P.O. Box 61, N-1432 Ås/NLH, Norway

ABSTRACT

The response of forest tree seedlings to aluminum concentrations was investigated. A growth technique was used in which nutrients were present in low concentrations in a circulating solution. Plant nutrition and relative growth rate were maintained in steady-state by adding the nutrients in optimum proportions at a constant relative addition rate. After a period of steady-state growth, aluminum was added to the nutrient solution in different concentrations.

Permanent decrease in growth rate occurred if aluminum concentration exceeded 0.5 to 1 mmol.dm^{-3} (Norway spruce), 1 to 3 mmol.dm^{-3} (European birch), and 3 to 5 mmol.dm^{-3} (Scots pine). Lethal concentrations were 6 to 10 mmol.dm^{-3} (Norway spruce), 10 to 12 mmol.dm^{-3} (European birch) and 25 to 30 mmol.dm^{-3} (Scots pine). The response was the same for birch whether grown under optimum or nutrient-stress conditions and for Scots pine grown with or without the mycorrhiza Suillus bovinus.

Thus, Norway spruce appears to be the species most sensitive to aluminum and Scots pine the most tolerant of the studied species. However, the results indicate that aluminum is probably not a serious hazard for Norway spruce, European birch or Scots pine since reported aluminum concentrations in soil solution are seldom higher than 0.5 mmol.dm^{-3}.

INTRODUCTION

For the last 60 years aluminum, especially Al^{3+}, has been regarded as an important growth reducing factor in acid soils, as reviewed by Pratt (1973) and Foy (1974), for example. Recently, in several investigations, aluminum was found to be toxic to different tree species at comparatively low concentrations (e.g., Pavan and Bingham 1982; Rost-Siebert 1983; Steiner et al. 1984).

Ulrich (1981) suggested that increased Al^{3+} concentrations in soil solution, due to acid precipitation, is one of the main causes of forest dieback in central Europe. Results from long-term lysimeter studies in the Solling area in West Germany showed increased aluminum concentrations in the soil solution with aluminum concentration increasing rather rapidly from about 0.2 mmol.dm^{-3} to about 0.6 mmol.dm^{-3} during the mid-seventies (Ulrich 1980). Also the Ca/Al ratio in soil solution decreased during the same period (Matzner and Ulrich 1981). Rost-Siebert (1983) found that spruce seedlings suffered severe root growth reductions when the molar Ca/Al ratio was less than 1.

Nilsson and Bergkvist (1983) reported high aluminum concentrations in a shallow Swedish podzol. The aluminum concentrations varied considerably during the year but occasionally reached levels of about 0.3 to 0.4 mmol.dm^{-3} in lysimeter measurements. They also found a large variation in the aluminum concentrations at different soil depths. There was a 300-fold difference in inorganic aluminum between the uppermost 5 cm and at a depth of 55 cm. Almost all inorganic aluminum was found beneath 15 cm.

These findings motivate a thorough study to establish at which concentrations Al has specific effects on forest trees. The aim of this paper is to give a preliminary account of experiments with elevated Al^{3+} concentrations on European birch, Norway spruce, and Scots pine. These experiments are published in detail elsewhere (Eldhuset, in press; Göransson and Eldhuset, in press). A growth technique was used in which the nutrients were present in low concentrations and in which all nutrients were frequently added at a specified relative rate. Thus, plant nutrition and relative growth rate were maintained in steady-state (Ingestad 1981, 1982; Ågren 1985; Kanazawa 1985). Under such conditions, secondary effects of Al, such as effects on phosphorus availability, are avoided as well as interactions with uncontrolled plant nutrition.

MATERIALS AND METHODS

Plant Culture

For the experiments, seeds of Betula pendula Roth., Picea abies (L.) Karst. and Pinus sylvestris L. were collected from specified sources (Ingestad and Lund 1979; Ingestad 1979), thus minimizing the genetic variability. The seeds were germinated in continuous light on wet filter paper for 10-14 days (European birch) or in a sand/vermiculite mixture for 12-17 days (conifers). After 10 to 17 days the plants were placed in optimum conditions (Ingestad and Lund 1979; Ingestad and Kähr 1985).

Pregrowth

Pregrowth and experiments were carried out in growth chambers with artificial light (Osram HQIR 250 W/D), irradiance ≈300 µmol.m^{-2}.s^{-1}, 24h.day^{-1}; air and root temperature 20°C; relative air humidity 75%. Most experiments were carried out at the Section of Forest Ecophysiology in Uppsala. The experiments with Scots pine and mycorrhiza were carried out in the phytotron in Stockholm (irradiance ≈250 µmol.m^{-2}.s^{-1}, 24 h.day^{-1}; air temperature 20°C; relative air humidity 75%).

The pregrowth period was about 12 days for birch and about 21 days for spruce and pine. After pregrowth, 60 seedlings of equal size were selected and divided into 12 groups for each experiment.

Experiments

The growth technique was essentially the same as described by Ingestad and Lund (1979), and Ingestad and Kähr (1985). Nutrient additions were made with a relative addition rate (R_A) of a complete fertilizer with optimal nutrient proportions for each species (Ingestad 1979, 1981). This allowed the seedlings to grow with stable internal nutrient concentrations and relative growth rates during the whole experimental period before Al additions, whether the R_A was optimal or growth rate limiting. The nutrient solution was not changed during the experimental period but was continuously circulated and sprayed on the roots. The nutrient additions were frequent (generally once per hour) and in small amounts. The conductivity of the nutrient solutions was usually <30 (European birch) or <50 $\mu S.cm^{-1}$ (Norway spruce and Scots pine) and never higher than 150 $\mu S.cm^{-1}$ before aluminum additions.

Three series of experiments were carried out: 1) optimum growth conditions where nutrients were in free access (Ingestad and Kahr 1985) but in low concentrations (Norway spruce, Scots pine and European birch); 2) optimum growth conditions with Scots pine inoculated with the mycorrhizal fungus <u>Suillus</u> <u>bovinus</u>; 3) nutrient-stressed European birch (R_A = 10%.day^{-1}). Aluminum was added (as one dosage) to different concentrations in the range 0.2-15 mmol.dm^{-3} (Norway spruce), 1-30 mmol.dm^{-3} (Scots pine, with and without mycorrhiza), and 1-15 mmol.dm^{-3} (European birch, optimum and stress). The additions were made as 50/50 mixtures of $AlCl_3$ and $Al(NO_3)_3$ or as $AlCl_3$ alone. After the aluminum additions conductivity increased to about 300-3000 $\mu S.cm^{-1}$.

To keep aluminum in the Al^{3+} form, pH was adjusted to 3.8±0.2 with HCl or NaOH. At the end of each experiment, samples of nutrient solutions were analyzed for total Al by plasma atomic emission spectrometry (ICP-AES, Instrumentation Laboratory IL P-200) and in some cases for monomeric Al by a pyrocatechol violet method (Driscoll 1984). Figure 1 demonstrates that most of the aluminum added was found in the solutions. At high concentrations a small loss of aluminum from the solution was recorded. It may be concluded that aluminum generally remained in solution during the whole experimental period and that the aluminum uptake by the seedlings was small. Figure 1 shows total aluminum in the solutions but measurements on samples of nutrient solution taken at the end of the experiments showed that total Al was equal to Al^{3+}.

Measurements

From the beginning of the experiments (D_0, day zero) to the day of Al addition (D_A, day of addition), three harvests, of 2 seedling groups each, were carried out to determine the relative growth rate (R_G) before Al addition. After D_A three harvests of 2 groups were carried out until the end of the experiment (D_E, day end of experiment) to determine R_G after Al addition. The time between each harvest varied from 3 days to 3 weeks depending on species and nutrient stress (see Ågren 1985). Growth was determined by curve fitting as R_G, %.day^{-1} (Jia and Ingestad 1984).

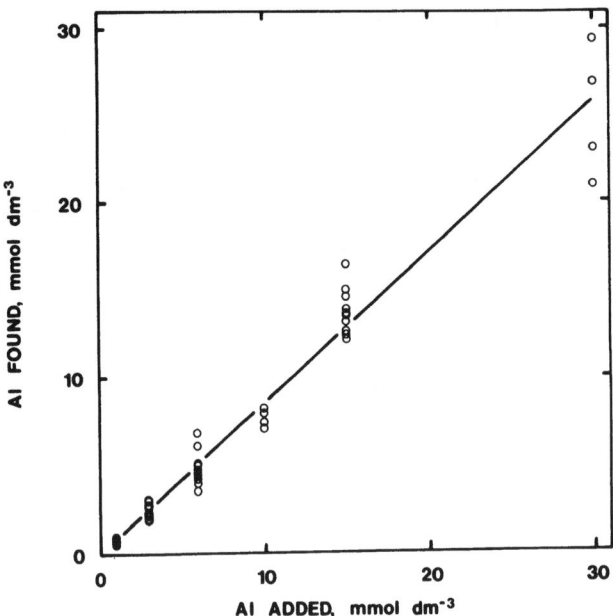

Fig. 1. Aluminum concentration in nutrient solution as added and as found by analyses at the end of the experiment.

At each harvest, fresh weights and dry weights of root, stem and leaf (European birch) or root and shoot (Norway spruce and Scots pine) were measured. Detailed results, including chemical analyses, will follow in later papers.

RESULTS

Visual Observations

Visual symptoms of disturbance first appeared on the roots with increases in aluminum concentration. In the conifers, thickening of the roots was a characteristic symptom. Grey to brown zones appeared behind the apical meristems 1 to 3 days after the aluminum addition even at low Al concentrations. In birch these effects disappeared after a few days. No transient effects were recorded on formation and growth of new fine roots within this range of the dose/response relationship (0.2 to 0.5 $mmol.dm^{-3}$ in spruce; 1 to 3 $mmol.dm^{-3}$ in birch; 6 to 10 $mmol.dm^{-3}$ in pine).

At higher aluminum concentrations, the symptoms became more pronounced. The injury to the apical root meristems became more severe and occasionally the roots died from the tip inwards at concentrations >3 $mmol.dm^{-3}$ (Norway spruce), >6 $mmol.dm^{-3}$ (European birch), or >15 $mmol.dm^{-3}$ (Scots pine). However, new roots were formed behind the tips at very high concentrations. Many of these roots seemed, at least for some time, to be normal in appearance.

The leaves also showed symptoms of disturbances in connection with dying root tips. In birch, the leaves turned yellow to red in colour, and older leaves wilted and fell off. Necrotic spots appeared, especially in the leaf margins. Needles of Norway spruce and Scots pine also yellowed. The tips, especially on older needles, became brown and the needles died from the tips inwards. At very high aluminum concentrations, new shoots grew out from the lower part of the birch stems, indicating that the apical shoot meristems were injured.

There was no difference in response of birch seedlings to high aluminum concentrations whether the plants were growing under optimum conditions or under nutrient stress. In mycorrhizal Scots pine, symptoms on the roots appeared at higher concentrations of Al than in seedlings without mycorrhizae.

Growth

The relative growth rate (R_G) was very stable during the period before Al additions. The experiments with Norway spruce were less accurate and the coefficient of determination, r^2, was >0.99 in European birch and Scots pine and >0.70 in Norway spruce.

Figure 2 shows the changes in fresh weight of birch under nutrient stress over time. Similar results were obtained in other experiments in which seedlings were growing under optimal conditions. Following the Al additions, growth was less stable and there was a tendency to decreasing R_G at higher concentrations, indicating a remaining or increasing injury with time (Fig. 2). Low Al concentrations caused relative growth rate to decline for some time but growth rates recovered again in such cases. Permanent decrease of growth rate was recorded at Al^{3+} concentrations of >0.5, >1, and >3 mmol.dm^{-3}, for Norway spruce, European birch, and Scots pine, respectively.

The effect on growth of increased aluminum concentration followed the same pattern in the three species. The lethal concentration was defined as the point of intersection between the curve and the abscissa. It was 6 to 10 mmol.dm^{-3} for Norway spruce, 10 to 12 mmol.dm^{-3} for European birch, and 25 to 30 mmol.dm^{-3} for Scots pine (Fig. 3).

DISCUSSION

About 85% of the added aluminum was found in the solutions at the end of the experiments (Fig. 1). Preliminary results of chemical analyses indicate that about 10% of the added aluminum was found in the plants. It is therefore justifiable to say that the plants were under a continuous Al^{3+} influence according to treatments. The growth technique used implied that the roots had immediate access to the nutrients added. Thus, there was no sign of any interaction between Al and P, because of precipitation, for example. This problem has often been mentioned by authors who used traditional batch solutions.

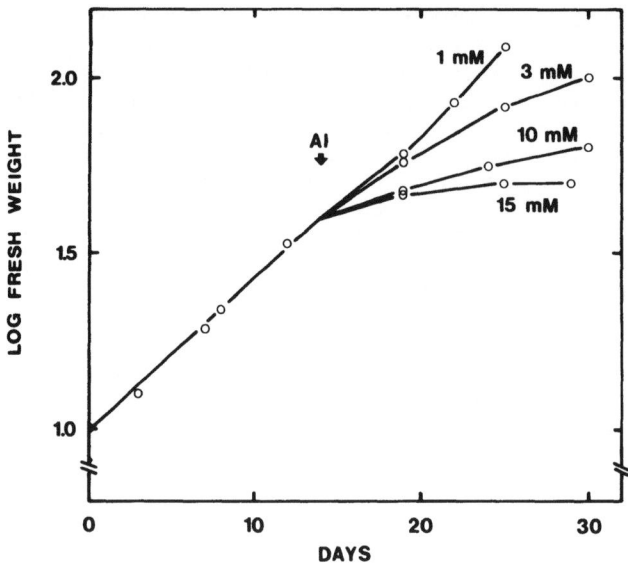

Fig. 2. Growth of birch seedlings under nutrient stress (relative nutrient addition rate 10%.day^{-1}) before and after addition of Al in different concentrations.

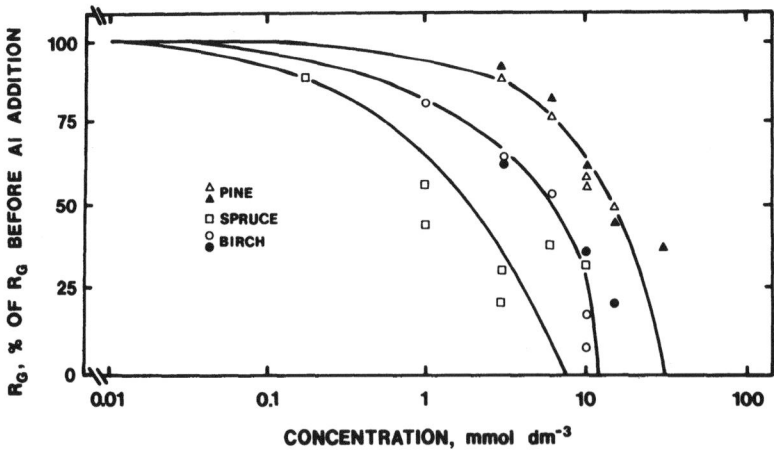

Fig. 3. Growth effects of different Al concentrations. The open symbols represent seedlings at maximum growth before Al additions. The filled symbols represent pine seedlings inoculated with mycorrhizal fungus (<u>Suillus bovinus</u>) or birch seedlings under nutrient stress (relative nutrient addition rate 10%.day^{-1}.

It has been claimed that the Ca/Al ratio in the root medium is important. For example, Rost-Siebert (1983) found that the molar Ca/Al ratio had to be higher than 1.0 for normal growth of Norway spruce. In all our experiments, the molar Ca/Al ratio was <0.025 even at the lowest aluminum concentration. Thus, it is probably not the Ca/Al ratio in itself which is important, but rather that a sufficient amount of Ca in relation to plant requirement must be available to the roots, as in the growth technique used.

The visual symptoms of the roots and shoots which we observed are very similar to those described by others (e.g., McCormick and Steiner 1978; Steiner et al. 1980, 1984; Rost-Siebert 1983; Schier 1985; van Praag et al. 1985).

Growth reductions due to aluminum have been measured in various ways (Foy 1974). Beyer and Hutnik (1969) measured the increase in dry weight of two birch and two pine species including Pinus sylvestris. They found that growth was affected at 3 and 6 mmol $Al^{3+}.dm^{-3}$ in pine and birch, respectively. Hoyle (1971) recorded dry weight reduction in Betula alleghaniensis at 3 mmol $Al^{3+}.dm^{-3}$. McCormick and Steiner (1978) compared root elongation of several tree species, with Al^{3+} concentration ranging from 0 to 10.4 $mmol.dm^{-3}$. In Quercus, Pinus and Betula species they found that there was no significant decrease in root length at 3 $mmol.dm^{-3}$ or lower. At 7.4 $mmol.dm^{-3}$ and higher, root elongation was approaching zero. However, the Pinus species had a root length reduction of only 50% even at 10.4 $mmol.dm^{-3}$. At 4.4 and 5.9 $mmol.dm^{-3}$, the Quercus species were more tolerant than Pinus species, which, in turn, were more tolerant than Betula species. Steiner et al. (1980) found that the tolerance to 4.4 mmol $Al^{3+}.dm^{-3}$, measured as root elongation differed significantly among provenances of Betula papyrifera. Rost-Siebert (1983) found reduction in elongation of the primary root of Picea abies at 0.4 $mmol.dm^{-3}$ (provided the Ca concentration was low). Schier (1985) found decreases in root length in Picea rubens at Al^{3+} concentrations of 2 $mmol.dm^{-3}$.

Despite differences in growth technique and response measurements, there is a general agreement between the present results and those cited above. Considering that aluminum concentrations in soil solution have seldom been reported to be higher than 0.5 $mmol.dm^{-3}$, there is little support for the idea that aluminum is an important factor in connection with forest die-back in Europe.

Acknowledgements

The authors are very grateful to Miss Ann-Britt Lund for her work with data revisions and to Mrs. Annette Hovberg for help with the technical work. The investigation was supported by grants from the Swedish Environmental Protection Board, Swedish Council for Forestry and Agricultural Research and the Agricultural Research Council of Norway.

REFERENCES

Ågren GI (1985) Theory for growth of plants derived from the nitrogen productivity concept. Physiol Plant 64: 17-28.
Beyer LE, Hutnik RJ (1969) Acid and aluminum toxicity as related to strip-mine spoil banks in Western Pennsylvania. Penn State Univ Res Briefs 3: 69-72.
Driscoll CT (1984) A procedure for the fractionation of aqueous aluminum in dilute acidic waters. Int J Environ Anal Chem 16: 267-283.
Eldhuset T (in press) Effects of aluminium on nutrition and growth of coniferous seedlings. Physiol Plant.
Göransson A, Elduset T (in press) Effects of aluminium on nutrition and growth of Betula pendula (Roth) seedlings. Physiol Plant.
Foy CD (1974) Effects of aluminum on plant growth. In: Carson EW (ed) The plant root and its environment. University Press of Virginia, Charlottesville. p 601.
Hoyle MC (1971) Effects of the chemical environment on yellow-birch root development and top growth. Plant Soil 35: 623-633.
Ingestad T (1979) Mineral nutrient requirements of Pinus silvestris and Picea abies seedlings. Physiol Plant 45: 373-380.
Ingestad T (1981) Nutrition and growth of birch and grey alder seedlings in low conductivity solutions and at varied relative rates of nutrient addition. Physiol Plant 52: 454-466.
Ingestad T (1982) Relative addition rate and external concentration; Driving variables used in plant nutrition research. Plant Cell Environ 5: 443-453.
Ingestad T, Kahr M (1985) Nutrition and growth of coniferous seedlings at varied relative nitrogen addition rate. Physiol Plant 65: 109-116.
Ingestad T, Lund AB (1979) Nitrogen stress in birch seedlings. I. Growth technique and growth. Physiol Plant 45: 137-148.
Jia H, Ingestad T (1984) Nutrient requirements and stress response of Populus simonii and Paulownia tomentosa. Physiol Plant 62: 117-124.
Kanazawa Y (1985) Theoretical consideration of the concepts proposed by Ingestad in his nutrition experiments. J Jap For Soc 67: 105-107.
Matzner E, Ulrich B (1981) Bilanzierung jährlicher Elementflüsse in Waldökosystemen im Solling. Z Pflanzenernähr Bodenkunde 144: 660-681.
McCormick LH, Steiner KC (1978) Variation in aluminum tolerance among six genera of trees. For Sci 24: 565-568.
Nilsson SI, Bergkvist B (1983) Aluminium chemistry and acidification processes in a shallow podzol on the Swedish westcoast. Water Air Soil Pollut 20: 311-329.
Pavan MA, Bingham FT (1982) Toxicity of aluminum to coffee seedlings grown in nutrient solution. Soil Sci Soc Am J 46: 993-997.
Praag HT van, Weissen F, Sougnez-Remy S, Carletti G (1985) Aluminium effects on spruce and beech seedlings. II. Statistical analysis of sand culture experiments. Plant Soil 83: 339-356.
Pratt PF (1973) Aluminum. In: Chapman HD (ed) Diagnostic criteria for plants and soils, 2nd edn. Abilene, Texas, p 3.

Rost-Siebert K (1983) Aluminium-Toxizität und -Toleranz an Keimpflanzen von Fichte (Picea abies Karst.) und Buche (Fagus silvatica L.) Allg Forstz 38: 686-689.

Schier GA (1985) Response of red spruce and balsam fir seedlings to aluminium toxicity in nutrient solutions. Can J For Res 15: 29-33.

Steiner KC, McCormick LH, Canavera DS (1980) Differential response of paper birch provenances to aluminum in solution culture. Can J For Res 10: 25-29.

Steiner KC, Barbour JR, McCormick LH (1984) Response of Populus hybrids to aluminum toxicity. For Sci 30: 404-410.

Ulrich B (1980) Die Wälder in Mitteleuropa: Messergebnisse ihrer Umweltbelastung, Theorie ihrer Gefährdung, Prognose ihrer Entwicklung. Allg Forstz 35: 1198-1202.

Ulrich B (1981) Eine Ökosystemare Hypothese über die Ursachen des Tannensterbens (Abies alba Mill.) Forstwiss Centralbl 100: 228-236.

THE EFFECT OF SIMULATED ACID RAIN ON BOREAL FOREST FLOOR FEATHER MOSS AND LICHEN SPECIES

T.C. Hutchinson, M. Scott, C. Soto and M. Dixon

Dept. of Botany and Institute for Environmental Studies, University of Toronto, Toronto, Ontario, Canada M5S 1A1

ABSTRACT

The effect of five years of treatment with simulated acid rains ranging in pH from pH 2.5 to 5.6 on Canadian boreal forest floor feather mosses and lichens was studied. The species examined were Pleurozium schreberi, Cladina stellaris, C. rangiferina, and C. mitis. The feather moss P. schreberi appears especially sensitive to simulated acidic rains in the boreal forest, while the lichens are somewhat less sensitive.

Severe effects on P. schreberi were noted at pH <3.5, with pH 2.5 causing an almost total elimination of the feather moss mat within three years. At pH 5.6, the mosses performed better than in unsprayed plots. In a second experiment, the growth and chlorophyll content of P. schreberi was measured after two years of pH 3.2 sprays. The mosses responded positively when the spray solution was acidified with nitric acid; inhibition of growth was noted in the sulphuric acid-only treatment compared with a nitric acid or a pH 5.6 spray. All acidic sprays caused substantial depletion of Ca and Mg in shoot tips.

Photosynthesis was inhibited in the surviving podetia of the three lichen species following the five-year spray programme at treatments of less than pH 3.5. Substantial declines in podetial height and dry weight were also observed in all three lichen species exposed to repeated sprays of pH 3.0 or 2.5. Morphological and anatomical abnormalities were found in C. rangiferina sprayed at pH 3.0 and 2.5. These included dwarf branching and proliferation of algal cell clusters. A high percentage of the algal cells sprayed at these low pHs had algal clusters in which complete loss of membrane integrity had occurred. The algal cells often contained numerous enlarged peripheral lipid bodies and starch bodies.

INTRODUCTION

In both eastern North America and in northern Europe, the boreal forest regions are subjected to acidic deposition. While wet and dry deposition of sulphates are the prime factors in this, the contribution of nitrates to the overall acidity of the rain, snow and dry deposition is substantial (Rodhe 1983; Husar and Holloway 1983). Since there are numerous natural sources of acidity in these forested ecosystems, it would seem, on theoretical grounds, that dominant circumboreal genera such as spruce (Picea), pine (Pinus), fir (Abies) and birch (Betula) should be well-adapted to tolerate additional inputs of acidic radicals due to acid precipitation. Much of the boreal forest occurs on acidic, podsolic soils developed since the last ice-age over an underlying bedrock composed of acidic Pre-Cambrian granitic or gneissic rocks with a low base status. The natural downward leaching of the soil solution is accentuated by the acidic litter of the coniferous boreal forest, which decomposes slowly at these cool, northern latitudes and yields a percolate rich in humic, fulvic and tannic acids. Therefore, the pH values of the humus and organic mat are normally in the range 3.5 to 5.2.

A general lack of concern about the sensitivity of northern forested ecosystems, based on this known tolerance to acidic soils, may be a factor in the very small number of experimental studies which have been carried out to determine reponses and sensitivities to acidic stress. However, such assumptions and theoretical considerations always require testing and validation.

An additional argument in support of experimental studies in boreal ecosystems arises from a consideration of the ground flora beneath the predominantly coniferous tree canopy. Huge areas of the boreal forest understory are dominated by either feather mosses, such as Pleurozium schreberi or Hylocomium splendens, or in open more xeric sites, by lichens of the genus Cladina, particularly C. mitis, C. stellaris and C. rangiferina. These understory mosses and lichens are characterized by their lack of a protective cuticle, thereby exposing photosynthetic cells to direct interaction with incoming precipitation and with the throughfall and stemflow initially intercepted by the tree canopy. The Cladina spp. are also ecorticate (lacking an upper cortex of fungal tissue), so that atmospheric moisture quickly hydrates the exposed clusters of algal cells. Both the feather mosses and the caribou forage lichens (Cladina) are strongly dependent upon rainfall for their bursts of photosynthetic activity, and thus for their growth opportunities. It is also believed that these groups of cryptogamic boreal forest floor species are largely independent of their substrata for nutrient uptake. They are, in fact, dependent upon both wet and dry precipitation for nutrients and are clearly capable of growing on strongly acidic soils. Their dependence, however, on atmospheric inputs for nutrients suggests that they may be especially vulnerable to changes in the acidity of these atmospheric inputs and that they may react very directly to increased rain acidity and to acidic dry deposition. Therefore, these cryptogams may be early warning, indicator species of stresses induced by acid deposition in boreal systems.

To assess the response of mosses and lichens to prolonged acid rain stresses, we commenced in 1981 a five-year programme of simulated acid rain sprays to the floor of mature jack pine (Pinus banksiana) forests. These stands were selected for their ground flora dominance by either the feather moss Pleurozium schreberi or by the caribou-forage lichens referred to earlier. The long-term responses of these plants and their associates to a series of deliberate acid rain stresses were determined. A range of sulphate to nitrate ratios in shorter-term sprays was also used to determine the influence of these anions on plant response.

METHODS

Mosses and Lichens - Five-Year Sprays

Sites were chosen in mature jack pine forest approximately 700 km north of Toronto (Lat. 48°09'N, Long. 80°01'W). At these sites, the trees were approximately 20-28 m tall, growing on deep sand deposits. The uniform age classes at the sites are the result of post-fire invasion following major forest fires in 1916. A rather xeric, open site was chosen in which the ground flora was dominated by Cladina species and a second more mesic site was selected a few kms away with an 80-90% ground cover of Pleurozium. Triplicate plots, each 5x2 m were marked out in the spring of 1981 for each of 5 pH spray treatments. In addition, triplicate unsprayed plots were set out as controls. All spray treatments were then randomly assigned to the plots.

Simulated rain sprays of pHs of 2.5, 3.0, 3.5, 4.0, and 5.6 were then sprayed on to the plots at a height of less than 2 m. In 1981, the sprays were given once monthly from June to October (a time period which approximates the growing season). Each spray was given to a depth of 5.0 cm, delivered over approximately 1 hour. From 1982-1985 the spray was given twice a month to a depth of 2.5 cm each time. The sprays consisted of double-deionized water from a nearby lake adjusted to treatment pH with a 2:1 molar solution of $H_2SO_4:HNO_3$. Details are given in Hutchinson et al. (1986).

The plots were also exposed to ambient rainfall and to dry deposition. The rainfall over the five-year period had an average pH of 4.2. The experimental spray represented an addition in water supply of approximately 30% for the growing season.

The percentage cover for each species present in the plots was estimated visually by TCH in late August of each year. Details are given in Hutchinson et al. (1986).

Feather Moss Experiment Using pH 3.0 and 5.6 Sprays with Different $H_2SO_4:HNO_3$ Ratios

In the feather moss site, triplicate 2x2 m plots were marked out in the spring of 1984, and treated with one of four spray treatments, randomly determined. These consisted of a pH 5.6 treatment using the 2:1 $H_2SO_4:HNO_3$ molar addition, or a pH 3.2 treatment composed of H_2SO_4 alone, or pH 3.2 composed of HNO_3 alone or pH 3.2 composed of a 2:1 ratio of $H_2SO_4:HNO_3$. These sprays were applied twice monthly from June to September in 1984 and 1985. Applications were to a depth of 2.5 cm per event.

Thirty Pleurozium schreberi shoots per replicate plot were selected randomly and a cotton thread was tied 5 mm from the growing tip of each of these shoots in June 1984 before the first spray. These shoots were used for measurements of subsequent increment growth and for elemental analysis. The analytical techniques used were inductively coupled plasma emission spectrophotometry, atomic absorption spectrophotometry and ion chromatography.

Finally, the chlorophyll content of the terminal 2 cm of the moss fronds was determined on triplicate samples for each plot. Since the feather moss carpet within the plots differed noticeably in greenness, the samples were taken to represent the range of colour present. Chlorophyll extractions were made on fresh shoots using an 80% acetone extraction.

Five-Year Lichen Spray Plots

At the more xeric jack pine site, with a dominant ground flora of the caribou-forage lichens, an experimental design similar to that used for the moss plots was implemented. Percent cover estimates were also recorded annually by TCH. In 1985, at the conclusion of the five-year spray programme, surviving lichens of each of Cladina stellaris, C. mitis and C. rangiferina were taken for more detailed assessment and for growth, morphological and anatomical studies. The methodologies for these are described in Scott and Hutchinson (in prep). Maximum net photosynthesis at 20°C and 250 ueinsteins.$m^{-2}.s^{-1}$ of the three lichen species from each of the spray treatments was determined using a Binos portable IRGA system.

Anatomy and Morphology of Field-Sprayed Lichen Podetia

Some of the material harvested for the growth and physiology study was used for the anatomical investigation. The top 2.5 cm of each podetium was removed, misted with double distilled water and placed in a growth chamber maintained at 10°C and 400 ueinsteins.$m^{-2}.s^{-1}$ for a period of 12 hours prior to fixation. Pieces (1 mm^3) of the lichen thallus were fixed for transmission electron microscopy using a previously published protocol (Scott and Larson 1986). Thin sections were examined with a Philips EM 300 transmission electron microscope operated at 60 kV. For each species, a total of 12 blocks of tissue (n=3 x 4 blocks per replicate) were examined per treatment. Micrographs showing gross morphological features were taken using a dissecting microscope.

RESULTS AND DISCUSSION

Response of Pleurozium schreberi to Five Years of Acidic Spray

The effect of the spray treatments on cover are presented in Table 1. The initial heterogeneity between the spray sites (1981) and unsprayed (1982) resulted in cover values of P. schreberi from 51 to 89%. It is necessary, therefore, to examine the responses over time within each spray treatment to determine comparative effects. A clear decline in cover values from the pre-treatment values is seen for the three treatments pH 2.5, 3.0 and 3.5. Under the most acidic spray of pH 2.5, a dramatic decline occurred by the August sampling date of the second year. Indeed, effects occurred during the first year (1981), with browning of the moss shoots towards the end of the summer in this

Table 1. Vegetative cover of _Pleurozium schreberi_ and bare ground (mean ± standard deviation) of Site 1 - recorded annually as % 1m² plots (n = 9 per pH treatment).

Year	Treatment (pH spray)					
	5.6	4.0	3.5	3.0	2.5	Unsprayed
	% Cover					
1981	83.9 ± 13.2	51.1 ± 20.1	83.9 ± 16.7	70.6 ± 12.9	66.7 ± 28.7	—
1982	86.7 ± 12.7	61.1 ± 23.3	68.0 ± 23.8	47.2 ± 20.0	1.6 ± 2.5	84.4 ± 11.3
1983	87.2 ± 9.7	52.8 ± 28.1	66.1 ± 29.6	33.1 ± 20.7	0.7 ± 1.6	83.3 ± 5.0
1984	90.0 ± 7.3	74.4 ± 12.0	66.9 ± 24.7	50.7 ± 27.6	0.2 ± 0.4	76.1 ± 16.4
1985	91.0 ± 6.7	75.6 ± 13.6	63.6 ± 25.5	40.0 ± 21.0	0.0	81.1 ± 9.6
1986	86.2 ± 7.8	66.7 ± 12.5	66.1 ± 23.0	41.7 ± 23.8	0.0	77.2 ± 12.0
	% Bare Ground					
1981	3.9 ± 5.5	21.7 ± 19.0	4.1 ± 7.2	9.2 ± 17.3	7.9 ± 9.5	—
1982	0.9 ± 0.9	5.8 ± 5.8	2.2 ± 3.4	10.3 ± 9.6	46.1 ± 11.4	1.8 ± 1.5
1983	2.9 ± 3.2	9.8 ± 12.1	5.4 ± 4.5	27.9 ± 16.1	64.4 ± 8.5	4.8 ± 2.7
1984	0.8 ± 0.8	6.9 ± 7.6	2.2 ± 1.6	24.2 ± 17.8	54.4 ± 33.8	4.7 ± 4.9
1985	0.9 ± 1.1	4.6 ± 3.4	8.0 ± 10.9	25.1 ± 16.2	88.0 ± 14.4	3.6 ± 3.8
1986	3.9 ± 3.4	11.7 ± 6.1	5.7 ± 4.1	38.8 ± 23.4	87.8 ± 7.1	10.4 ± 7.8

treatment. The cover values declined from 66.7% in August 1981 to 1.6% in 1982 and continued to decline to zero by 1985 (Fig. 1). Interestingly, in the 5 x 2 m plots treated with pH 2.5 sprays, it was always possible to find a few living shoots of the Pleurozium. Initially, such survival seemed possibly to be due to shading effects caused by shrub and tree trunks, despite the sprayers' attempts to avoid such refuges. By 1984, however, it seemed much more likely that the very acidic spray was acting as a powerful selective factor for feather moss survival, and that a few genotypes were much more tolerant than the rest. These may be the future propagules for a spread of a more acid tolerant form of the moss should the ambient rainfall pH decline further. It could even be argued that the ambient pH of rain of 4.2 may already have acted as a selective factor on the feather moss population structure.

Fig. 1. The effects of four years of simulated acid rain on P. schreberi. The mean cover value of the moss in the pH 2.5 plots was 0.2%. (a) pH 5.6 treatment (b) pH 2.5 treatment

At pH 3.0, the decline from 1981 to 1983 was severe, with cover reduced from 70.9% to 33.1%. Thereafter, cover values seemed to increase somewhat, but with a good deal more variation between samples than previously. This was due to some plots being especially d maged by pH 3.0 sprays compared with others. Microhabitat differences may be a factor in this, or possibly the more tolerant 1 x 1m plots represent clonal material from one or more tolerant genotypes.

At pH 3.5, the cover declined initially from 83.9% to 68%. Thereafter it was maintained at around 66%. This suggests an initial elimination of the least tolerant genotypes. The data for pH 4.0 (which is quite variable from year to year), and for pH 5.6 both suggest a positive response by the feather moss to the sprays. Such a positive response may be due to the additional water supplied by the spray above that of the ambient rains. The larger response at pH 4.0 could be due to a "water plus nitrogen" effect due to the nitric component of the spray. At the lower pHs (i.e., <3.5), these benefits seem to be more than overcome by the H^+ ion concentrations of the sprays.

The unsprayed plots did not show the positive response noted above (Table 1), perhaps due to their receiving only ambient rainfalls. The data for bare ground clearly show that the acidic sprays cause not only death of the feather moss, but produced extensive bare patches in the plots, changing the microclimate including insolation, water relations, etc.

In the only other field experiment which has involved acid sprays onto a forest floor, Abrahamsen et al. (1976) report the loss of cover of Pleurozium schreberi at pH 3.0 and pH 2.0 in a Norwegian Scots pine forest. They also report foliar lesions at pH 2.0.

Growth of Pleurozium schreberi Shoots Sprayed with Different Sulphuric to Nitric Ratios

In this experiment, the 5 mm shoot tip of 30 shoots per plot were tied with cotton thread in June 1984 and their growth measured in September 1984 and late August 1985. The plots were sprayed either at pH 5.6 or with an acidic pH 3.2 spray composed of one of three differing nitric:sulphuric ratios. The data are shown in Table 2. In 1984, the only difference between treatments was the enhanced growth of Pleurozium in plots receiving pH 3.2 as nitric acid only. The smallest growth, 20.6 mm, was in the pH 3.2 with sulphuric acid only. By August 1985, these relative differences had increased so that the pH 3.2 nitric acid treatment gave a mean shoot increase of 35.8 mm, while the equivalent sulphuric response was only 28.4 mm. The sulphuric:nitric mixture was intermediate in effect, and the pH 5.6 control with its trace quantities of sulphuric and nitrogen was significantly less than the nitric alone.

Table 2. Mean and standard deviation of moss frond length after two years of acidic spray of 5 mm frond tips initially marked in June 1984 (n=30).

Treatment	Moss frond length (mm)		
	Sept. 1984	Aug. 1985	Annual Increase
Nitric (pH 3.2)	24.72 ± 2.57	35.8 ± 7.0	+ 11.08
Sulphuric (pH 3.2)	20.56 ± 3.46	28.4 ± 3.3	+ 7.84
Sulphuric/ Nitric (pH 3.2)	21.56 ± 4.07	31.8 ± 7.2	+ 10.24
Control (pH 5.6)	21.37 ± 0.04	32.9 ± 1.9	+ 11.53

The positive response to the nitrogen addition of pH 3.2 sprays was emphasized by the chlorophyll determinations. pH 3.2 nitric acid sprays gave moss shoots with a mean chlorophyll content of 262 µg.g^{-1} while the control and the other two pH 3.2 treatments gave values of 112.6, 116.6 and 134.5 µg.g^{-1} chlorophyll, respectively.

Effects of Acid Sprays on Elemental Composition of Shoots

The top 6 cm segments of Pleurozium shoots from the pH 3.2 and 5.6 ratio experiment were analysed at the end of the experiment (August 1985) for their elemental composition, i.e. Ca, Mg, Al, Mn, Zn and Fe. Two cm segments back from the tip were analysed separately. Data are shown in Table 3. For the essential bases, calcium and magnesium, a consistent pattern is seen of highest concentrations in the apical 2 cm, and lowest in the next 2 cm. All pH 3.2 treatments yielded substantially lower levels of calcium and magnesium than the pH 5.6 treatment. The smallest reduction in relation to the pH 5.6 treatment was found in the nitric only pH 3.2 spray, in which the top 2 cm nevertheless showed a decrease of 29.1%. The greatest decrease in Ca was in the sulphuric/nitric mixture, with a 41.6% decrease compared with the pH 5.6 shoots. It is assumed that these deviations from the pH 5.6 'control' represent a substantial loss of structural calcium due to acid leaching.

Table 3. Elemental concentration, in ppm dry weight, of the top 6 cm segments of shoots of Pleurozium schreberi sprayed at either pH 5.6 or at pH 3.2 with sulphuric acid-only, nitric acid-only, or with a 2:1 mixture of the two (n=9).

Spray treatment	Region from top of shoot(cm)	Concentration (ppm)					
		Ca	Mg	Al	Mn	Zn	Fe
pH 5.6	0 - 2	3118	1178	802	195	68	368
	2 - 4	2528	758	924	145	44	1093
	4 - 6	2810	840	1207	144	40	1473
pH 3.2 suphuric only	0 - 2	2211	807	488	151	91	453
	2 - 4	1952	604	934	124	45	1131
	4 - 6	2258	694	1331	137	41	1469
pH 3.2 nitric only	0 - 2	2544	978	385	192	70	326
	2 - 4	2079	687	845	149	46	974
	4 - 6	2548	785	1119	174	49	1398
pH 3.2 2:1 sulphuric + nitric	0 - 2	1822	789	452	148	39	370
	2 - 4	1631	579	798	134	35	1012
	4 - 6	2096	683	1034	166	37	1343

The same pattern is seen for the magnesium contents of shoots, with a 17% difference in the nitric acid-only spray, cf. pH 5.6, and a 33.1% decline in the S:N pH 3.2 mixture. The Al concentration in the top 2 cm was also reduced in the pH 3.2 treatments, while for Mn-only the sulphuric acid only and the S:N pH 3.2 spray showed significant decreases compared with the pH 5.6 treatment. Iron and Zn showed little effect of the acid sprays, except that in the S:N mixture at pH 3.2 a significant decline in Zn occurred compared with all other treatments.

The general conclusion is that the pH 3.2 treatments caused a substantial and significant loss of Ca and Mg, which may be of physiological significance, while Mn was only lost by the apparently harshest leaching solutions, i.e., those containing sulphuric acid. The lower Ca and Mg levels in the slower growing sulphuric acid-only treatment (Table 2), suggests that, relatively, the sulphuric acid sprayed plants suffered an even greater loss of these bases when growth is taken into account. It would be invaluable to know how much Ca and Mg can be leached from healthy P. schreberi shoots before chronic damage becomes manifest. It would also be valuable to know the threshold rain pH below which cumulative losses of bases occur, resulting in damage, and above which losses can be adequately replaced. Leaching of bases from the foliage of higher plants when exposed to acid rain or acid mists has been widely reported (Eaton et al. 1973; Hornvedt and Joranger 1974; Abrahamsen et al. 1976). On certain soils in Germany this leaching of bases from foliage, coupled with accelerated soil drainage of bases, has led to acute Mg and Ca deficiencies in forest trees, as well as losses of K, Mn and Zn (Zoettl and Huettl 1986).

Net Photosynthesis and Growth of Acid-Sprayed Lichens

Substantial declines in podetial height, dry weight and net photosynthesis were observed for all three Cladina species exposed to repeated acid sprays of pH 3.0 or 2.5. It should be noted that these parameters were measured on the surviving population in the experimental plots, and that over the duration of the five-year spray programme substantial losses in percentage cover were recorded for Cladina species (Scott and Hutchinson, 1987).

C. stellaris underwent the largest decline in cover in pH 2.5-sprayed plots, compared with unsprayed control plots. By 1985, pH 2.5-treated podetia of this species showed a 33% decline in height and a 55% decline in dry weight compared with the pH 3.5-sprayed group which contained the largest individuals of any treatment (Table 4). However, there were no significant differences in growth or photosynthesis amongst the pH 5.6, 4.0 and 3.5 treatments. Simulated rain of pH 4.0 sprayed over the short-term has been reported by Lechowicz (1982) to cause a 72% decline in net photosynthesis of C. stellaris. Our long-term, field studies indicate a 60% decline in photosynthesis in pH 2.5-treated plots and a 30% reduction for pH 3.0-sprayed podetia of this species.

Table 4. Net photosynthesis and growth characteristics of Cladina species exposed to annual simulated rain events in the field (1981-1985), where $\bar{x}$ = mean, SE = standard error and SD = standard deviation).

Parameter		Unsprayed	pH Treatment 5.6	4.0	3.5	3.0	2.5
C. stellaris							
NPR ($mgCO_2 \cdot g^{-1} \cdot h^{-1}$)	$\bar{x}$ SE	1.69 0.12	1.92 0.28	1.68 0.17	1.66 0.24	1.34 0.32	0.76 0.62
Total podetial height ($p<.001$)*	$\bar{x}$ SE	88.4 4.2 a*	92.4 4.0 ac	92.8 4.2 ac	96.5 5.6 bc	88.6 4.4 ac	64.2 2.5 d
Total podetial dry weight (mg) ($p<.01$)	$\bar{x}$ SE	1370 144 ac	1124 158 a	1280 120 a	1630 172 bc	1412 195 ac	730 96 d
C. rangiferina							
NPR ($mgCO_2 \cdot g^{-1} \cdot h^{-1}$)	$\bar{x}$ SD	1.44 0.20	1.34 0.37	1.31 0.22	1.23 0.19	1.02 0.33	0.65 0.52
Total podetial height (mm) ($p<.001$)	$\bar{x}$ SE	64.3 2.6 ab	71.2 3.9 b	61.1 2.8 abc	60.2 2.6 abc	59.8 2.5 ac	52.5 2.8 c
Total podetial dry weight (mm) ($p<.001$)	$\bar{x}$ SE	129 18 ab	139 20 a	81 11 bc	74 9 c	99 10 ac	63 7 c
C. mitis							
NPR ($mgCO_2 \cdot g^{-1} \cdot h^{-1}$)	$\bar{x}$ SD	1.27 0.16	1.70 0.57	1.31 0.01	no sample		0.62 0.19
Total podetial height (mm) ($p<.001$)	$\bar{x}$ SE	40.9 1.8 ac	55.8 1.9 b	52.1 1.8 b	49.8 2.4 b	48.2 2.5 bc	38.6 1.9 a
Total podetial dry weight (mg) ($p<0.01$)	$\bar{x}$ SE	46 6 abc	66 6 ab	68 8 b	62 9 ab	56 5 ab	38 3 a

* Significantly different pairs of treatments ($p < 0.05$) are indicated by different lower case letters. Significance levels for each ANOVA are included in parentheses for each growth parameter.

Individuals of C. rangiferina, sprayed with pH 5.6 rain, were the largest and heaviest of any pH treatment of this species. Compared with mean values for the pH 5.6 group, a 26.3% reduction in height and a 55% reduction in dry weight were recorded for pH 2.5-sprayed podetia. A slight stimulation in mean height and dry weight was recorded in the pH 3.0-sprayed plots. This accords with previously published photosynthetic data for this species which showed that a slightly elevated NPR was obtained one week following a single spray with pH 3.0 simulated acid rain. While a less complete data set is presented for C. mitis, the severe depression of NPR with the pH 2.5 sprays is akin to that shown in C. rangiferina. NPR was decreased by 64%, podetial height by 31% and dry weight by 42% at pH 2.5 as compared with pH 5.6 sprays (Table 4).

The gross morphological and anatomical features of Cladina exposed for five years to simulated acid rain were investigated in the belief that changes at the cellular and structural levels would represent an integration of the organism's long-term response to low pH stress.

Morphology

The typical morphology of podetial tips of Cladina rangiferina is illustrated in Figure 2. This micrograph is representative of individuals from unsprayed, pH 5.6 or pH 4.0 treatments. Note that the recurved branch tips all point in the same direction and that reproductive structures (apothecia) occur at the ends of the branches. Middle and lower internodes of these podetia (not shown) are sparsely branched because the podetium increases in length by an annual dichotomy of the branch tips. Figures 3 and 4 are representative of C. rangiferina podetia which received either pH 2.5 or 3.0 acid sprays. These podetia developed morphological abnormalities not previously reported for environmentally-stressed lichens. 'Middle' internodes, defined as 5 to 7 years growth back from the tip based on the formation of one internode per year, were stimulated to produce numerous, stunted lateral branches which each contained 2 to 3 dichotomies (Figure 3). These newly-initiated 'dwarf' branches ranged in length from 0.25 mm to 15 mm. The central axis appeared deteriorated and had enlarged, bright green algal colonies interspersed with necrotic black or yellow clusters. Figure 4 shows a higher magnification of a 'dwarf' branch from a podetium of C. rangiferina sprayed at pH 2.5. In the majority of cases, short, newly-formed branches of 1 - 2 mm in length were already terminated by apothecia (Ap).

This rapid formation of a reproductive structure is extremely unusual for such a slow-growing, cryptogamic species which generally invests most of its fixed carbon in the formation of vegetative structures. In addition, a proliferation of algal cell clusters (dark masses in Figure 4) as compared to fungal hyphae is observed. Algal cells have also spread to the ends of the branches, in contrast to the situation for higher pH and unsprayed treatments in which podetial branch tips are composed exclusively of the mycobiont. It is possible that the low pH rain sprays caused nutrient leaching within the horizontally-

Figure 2. Bright field microscopy of an unsprayed podetium (Po) of C. rangiferina showing branching pattern of hyphal tips. Dark-coloured apothecia (AP) are terminal at branch ends.

Figure 3. Middle internode, representing five years' growth back from the tip, of a podetium of C. rangiferina which had been sprayed for five years with artificial rain of pH 2.5. Note the proliferation of numerous 'dwarf' branches (Db) arising from the main axis (Ax). Newly-initiated branched have already dichotomized (Di). Algal cell cluster (Ac).

Figure 4. Higher magnification of a dwarf lateral branch (Db). An apothecium (Ap) has formed at the branch tip. Dark colouration of the branch has occurred because of enlarged algal cell clusters.

Figure 5. Transmission electron microscopy (TEM) of an algal cell (A), from Cladina sprayed with rain of pH 4.0, containing the normal complement of cell organelles. Chloroplast (Chl). Pyrenoid body (Pb). Nucleus (Nu). Peripheral lipid (Li). Mitochondrion (Mi). Vacuole (Va). A fungal haustorium is present (Hy).

Figure 6. Higher magnification TEM of a pyrenoid body (Pb) in an unsprayed podetium with peripheral pyrenoglobuli (Py) and a granular stroma (Sr). Intrusive thylakoids (Th) and small starch grains (St) are evident.

Figure 7. TEM of an algal cell from Cladina stellaris sprayed for five years with rain of pH 2.5. Membrane integrity is preserved. However, extremely large lipid deposits (Li) are present in the peripheral cytoplasm.

Figure 8. TEM of algal cell (A) from the same plane of section as the cell in Figure 7. Cellular contents are completely degenerated. Persistent starch grains (St) are still recognizable.

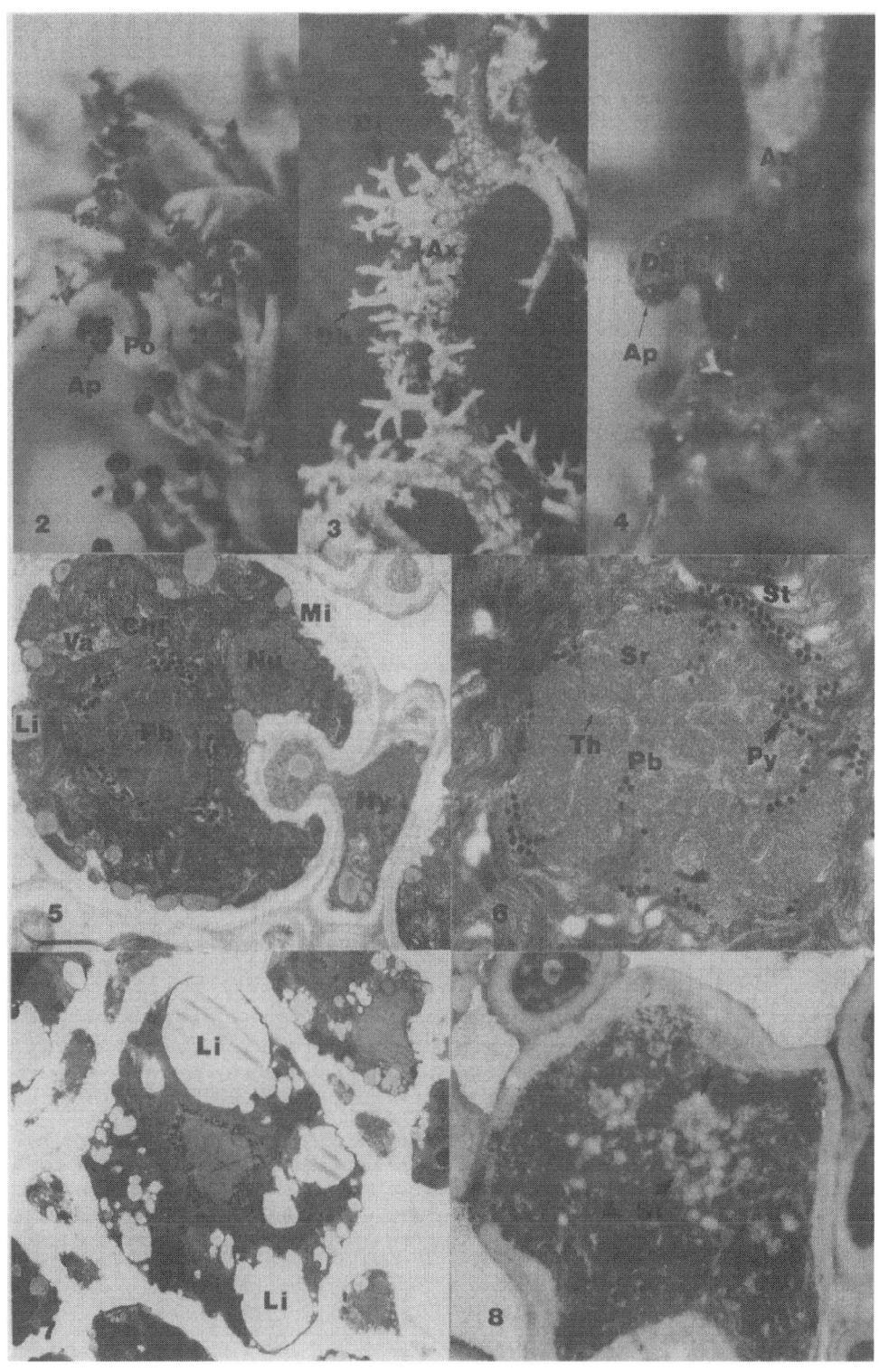

oriented podetia, thereby unbalancing the symbiosis in the lower internodes and resulting in algal outgrowths. _C. stellaris_ sprayed with acidic solutions did not show branch stimulation of the middle internodes. However, conspicuous stunting and thickening of the tips containing the past 2 - 3 years' growth were observed in some of the surviving population. This stunting may be related to the lower synthetic rates measured for pH 2.5- and 3.0-treated tips of this species noted earlier. Although mechanisms which control branching patterns in fruticose lichens have not been studied, the phenomenon of lateral branch initiation described here for stressed lichens suggests that future research on dichotomization should involve studies of nutritional factors and possibly growth hormones, such as auxins and cytokinins.

Transmission Electron Microscopy

Transmission electron micrographs (TEM) of algal cells in the genus _Trebouxia_ are shown in Figures 5 - 8 for _C. stellaris_ and _C. rangiferina_. Figures 5 and 6 are representative of algal cells from either unsprayed or high pH (pH 4.0 or 5.6) treatments. A single, stellate chloroplast containing a central pyrenoid body occupies most of the cell volume (Figure 5). In both species, lipidic pyrenoglobuli are peripherally distributed around the granular stroma of the pyrenoid body (Figure 5 and 6). Intrusive thylakoids (Figure 6) are common. A normal complement of cell organelles including small lipid bodies are commonly present in the peripheral cytoplasm of algal cells. Figure 5 shows a haustorial penetration of the algal cell. However, direct wall-to-wall contact between the two symbionts is the more common form of association in this genus. In both fall- and spring-harvested material, small starch grains are normally distributed between thylakoid membranes both in the region of the pyrenoid body and throughout the entire chloroplast. Starch is observed in healthy cells of both fall and spring-harvested material.

Several types of cellular abnormalities are observed in algal cell clusters of low-pH treated _C. stellaris_ and _C. rangiferina_. _C. stellaris_ appears to be more sensitive to repeated sprays of pH 3.0 or 2.5 rain than is _C. rangiferina_. Cells may have normal chloroplast morphology but contain numerous enlarged peripheral lipid bodies (Fig. 7). This phenomenon was also reported for the _Trebouxia_ cells of _Umbilicaria vellea_ exposed to moist, winter field conditions under the snowmelt, which were atypical of conditions encountered in its natural, desiccated microhabitat (Scott and Larson 1986). Loss of lipid from damaged plasma membranes may contribute to these enlarged peripheral deposits. A high percentage of the algal cells in pH 2.5- and 3.0-sprayed _C. stellaris_ had algal cell clusters in which complete loss of membrane integrity had occurred (Figure 8). These senescent cells were characterized by the accumulation of numerous starch bodies which often persisted in dead cells after the remaining cytoplasm had degenerated. Fungal hyphae were also observed in the vicinity of these persistent starch deposits, possibly indicating that the symbiotic balance had been upset in favour of fungal parasitism.

Cellular plasmolysis, coagulation of pyrenoglobuli, dilation of thylakoid membranes and wall dissolution were other characteristics of cells sprayed at low pH. Some of these symptoms may be the result of improper penetration of the gluteraldehyde into physiologically altered cells. Within any one plane of section, healthy-looking and senescent algal cell clusters regularly co-occurred, suggesting that differential sensitivity of algal cell phenotypes or genotypes to the low pH treatments may have occurred. Unsprayed and high pH-treated podetia did not contain senescent algal clusters. Accumulation of starch in stressed lichens has previously been reported by Scott and Larson (1986) and by Silva-Pando and Ascaso (1982) who studied lichens transplanted to heavily polluted urban environments.

Accumulation of photosynthate in Trebouxia cells suggests that the mechanism for translocation of ribose to the fungal symbiont may have been affected by acid treatment. Since both individual algal cells and algal cell clusters of low-pH treated podetia are often of larger diameter than those of pH 4.0 or 5.6-sprayed individuals, it is possible that algal cell growth is temporarily stimulated by the acidity. Since the mycobiont generally comprises about 95% of the total lichen thallus, algal sequestering of photosynthate would also result in internode stunting because of subsequent fungal starvation. Autoradiographic studies are needed to test these translocation hypotheses. Changes in fungal wall chemistry may also have altered the ability of the fungus to receive photosynthetic products.

It is apparent from consideration of these data, that long-term exposure to acid rain in the field may seriously restrict growth and photosynthetic potential of caribou-forage lichens. In addition, the symbiotic balance that normally exists between alga and fungus has been disrupted; potentially leading to collapse of the podetial structure. In comparison with the highly sensitive feather moss, Pleurozium schreberi, however, the slower-growing lichens are considerably more tolerant.

Since cryptogams form an important moisture-retaining gound cover in boreal forest ecosystems, the loss of the integrity of this cover will cause a sequence of events including increased downward leaching, microclimate changes including higher soil surface temperatures, greater drying out of the soils, changes in the microbial flora and fauna, loss of seed bed material, etc. Nutrient losses would be accelerated. We have found changes in the nutrient status of the soil solutions beneath the moss mat as well as increased levels of such potentially toxic elements as Al (Hutchinson, unpublished).

Acknowledgements

The invaluable help of a group of undergraduate summer assistants, 'the spray team', is gratefully acknowledged for performing the twice monthly sprays of the plots from 1981 to 1985. Kersti Meema patiently helped in the field measurements and in tying threads on innumerable Pleurozium shoots. Financial assistance for the study was provided by a grant from the Canadian

Sportsman's Fund 1982-1984, and by an NSERC Operating grant to
TCH. An NSERC Post-Doctoral Fellowship supported MGS. The
Ontario Ministry of Natural Resources made the field sites available, as well as accommodation at the Burt Lake Junior Ranger
Camp.

REFERENCES

Abrahamsen G, Bjor K, Horntvedt R, Tveite B (1976) Effects of acid precipitation on coniferous forest. In: Braekke FH (ed) Impact of acid precipitation on forest and freshwater ecosystems in Norway. SNSF Project, Oslo-As, p 37-63

Eaton JS, Likens, GE, Bormann, FH (1973) Throughfall and stemflow chemistry in a northern hardwood forest. J Ecol 61: 495-508

Horntvedt R, Joranger E (1974) Nedborens fordeling og kjemiske innhold under traer: juli-november 1973. SNSF Project TN/74, p 29

Husar R, Holloway JM (1983) Sulphur and nitrogen over North America. In: Ecological effects of acid deposition. National Swedish E.P.A. - Report PM 1636, p 95-115

Hutchinson TC, Dixon M, Scott M (1986) The effect of simulated acid rain on feather mosses and lichens of the boreal forest. Water Air Soil Poll 31: 409-416

Lechowicz MJ (1982) The effects of simulated acid precipitation on photosynthesis in the caribou lichen Cladina stellaris (Opiz) Brodo. Water Air Soil Poll 18: 421-430

Rodhe H (1983) Emission, transport and deposition of acidifying air pollutants. In: Ecological effects of acid deposition. National Swedish E.P.A. - Report PM 1636, p 77-88

Scott M, Hutchinson TC (in prep) The effects of long-term exposure to simulated acid precipitation on photosynthesis, growth and morphology of three caribou-forage lichens.

Scott MG, Larson DW (1986) Comparative morphology and fine structure of a group of Umbilicaria lichens. Can J Bot 62: 1947-1964

Silva-Pando FJ, Ascaso C (1982) Modificaciones ultraestructurales de liquenes epifitos transplantados a zonas urbanas de Madrid. Collectanea Botanica 13: 351-374

Zoettl HW, Huettl RF (1986) Nutrient supply and forest decline in southwest Germany. Water Air Soil Pollut 31: 449-462

ANNUAL ABSORPTION OF GASEOUS AIR POLLUTANTS BY MOSSES AND VASCULAR PLANTS IN DIVERSE HABITATS

W. E. Winner and C. J. Atkinson

Department of Plant Pathology, Physiology and Weed Science, Laboratory for Air Pollution Impact to Agriculture and Forestry, Virginia Polytechnic Institute and State University, Glade Road, Blacksburg, Virginia 24061, U.S.A.

ABSTRACT

Mosses are used as bioindicators of air pollution stress for plant communities because field studies of moss distribution around SO_2 sources show that they are more sensitive to gaseous pollution than vascular plants. Differences between moss and vascular plant sensitivity to SO_2 is thought to reflect greater SO_2 absorption by mosses; however, direct evidence supporting this hypothesis is limited.

This paper uses a general model of gaseous pollutant uptake to make comparisons between absorption of SO_2 for mosses and vascular plants. Annual SO_2 absorption is characterized for both types of plants for a number of diverse habitats in which the water status of plants would be expected to change with time. The estimation of SO_2 absorption showed how habitat or weather patterns may change absorption capacity of these two groups of plants.

We conclude that mosses are likely to absorb more SO_2 and other gaseous pollutants during the course of a year than are vascular plants. This was apparent with all habitats examined, from the tropics to the arctic tundra. The difference in annual SO_2 absorption between these types of plants varied with habitat and ranged from 400 to 4000-fold. This greater absorption potential for SO_2 occurs despite variations in annual moss water content being taken into account. SO_2 absorption calculated for mosses can help account for total sulphur content in mosses growing near SO_2 sources.

INTRODUCTION

Mosses and lichens are commonly thought to be more sensitive to gaseous air pollutants than are vascular plants. This idea has led to the use of mosses as bioindicators of air pollution, since their distribution is thought to be affected by air pollution dispersal patterns. Field studies documenting these observations have been conducted in urban areas of Europe and near air pollution point sources associated with ore smelters and natural gas refineries in North America (see LeBlanc and Rao 1974; Rao 1982).

The concept that mosses and lichens are more sensitive than vascular plants to air pollutants is based on the premise that cryptogams have higher rates of air pollution absorption because they lack the extensive cuticles which restrict gas exchange of true leaves. There is little direct evidence available to test this hypothesis. Very few measurements are available for describing gaseous air pollution absorption either into moss cells or into true leaves by absorption through the stomata. Recent studies suggest that gaseous pollution absorption by mosses is strongly influenced by normal changes in water content which occur during the desiccation-rehydration cycle (Winner and Koch 1982; Winner and Bewley 1983). In addition, studies with vascular plants indicate that the cuticle is an effective barrier to the absorption of gaseous pollutants into the leaf. However, absorption occurs through stomata, and stomatal conductance values can be used to estimate pollution-flux rates through stomata and into leaves (Winner and Mooney 1980).

The recognition of the role of moss water content and vascular plant stomata in regulating the absorption of gaseous pollutants make comparisons possible of the capacity of these two types of plants to absorb phytotoxic gases. This paper will estimate SO_2 absorption using methodology applicable to air pollutants in general, and in so doing characterize how annual, seasonal air pollution absorption by mosses and vascular plants may differ in habitats which range widely in precipitation and growing season. These estimates describe the difference in gaseous pollutant absorption capacity between vascular and non-vascular plants and may provide a foundation for predicting habitats or weather patterns that may change absorption capacity for both types of plants.

PRINCIPLES GOVERNING SO_2 ABSORPTION BY MOSSES AND VASCULAR PLANTS

Mosses, unlike vascular plants, lack the ability to restrict water loss through a cuticle or stomata, and water is lost to the atmosphere whenever mosses are hydrated and the relative humidity of the air is less than 100%. Loss of water induces a number of morphological and physiological changes (Bewley and Pacey 1978; Winner and Koch 1982; Winner and Bewley 1983; Fig. 1). The rate of desiccation and extent to which water loss occurs depends on the vapour pressure deficit between the air and moss and the duration of the desiccation period. Rehydrated moss cells can rapidly (30 sec to 24 hrs) return to their fully hydrated state, both morphologically and physiologically.

Whereas stomata limit gas exchange between vascular plant leaves and the atmosphere, only changes in moss water content influence gas exchange in cryptograms. The potential for mosses to absorb atmospheric gases, including CO_2 and air pollutants, declines as the plants desiccate (Fig. 2). In addition, the physiological effects of any absorbed gaseous pollutant are likely to be related to the moss hydration state; SO_2 toxicity may increase as mosses desiccate because decreases in cytoplasmic volume that occur during desiccation may act to concentrate absorbed pollutants or alter intrinsic stress resistance of cells (Winner and Koch 1982). Although the effects of SO_2 absorbed by mosses during desiccation are retained in plants rehydrated in SO_2-free air (Winner and Bewley 1983), it is not known whether this phenomenon is related to retained SO_2 or retention of SO_2-caused changes in moss physiology.

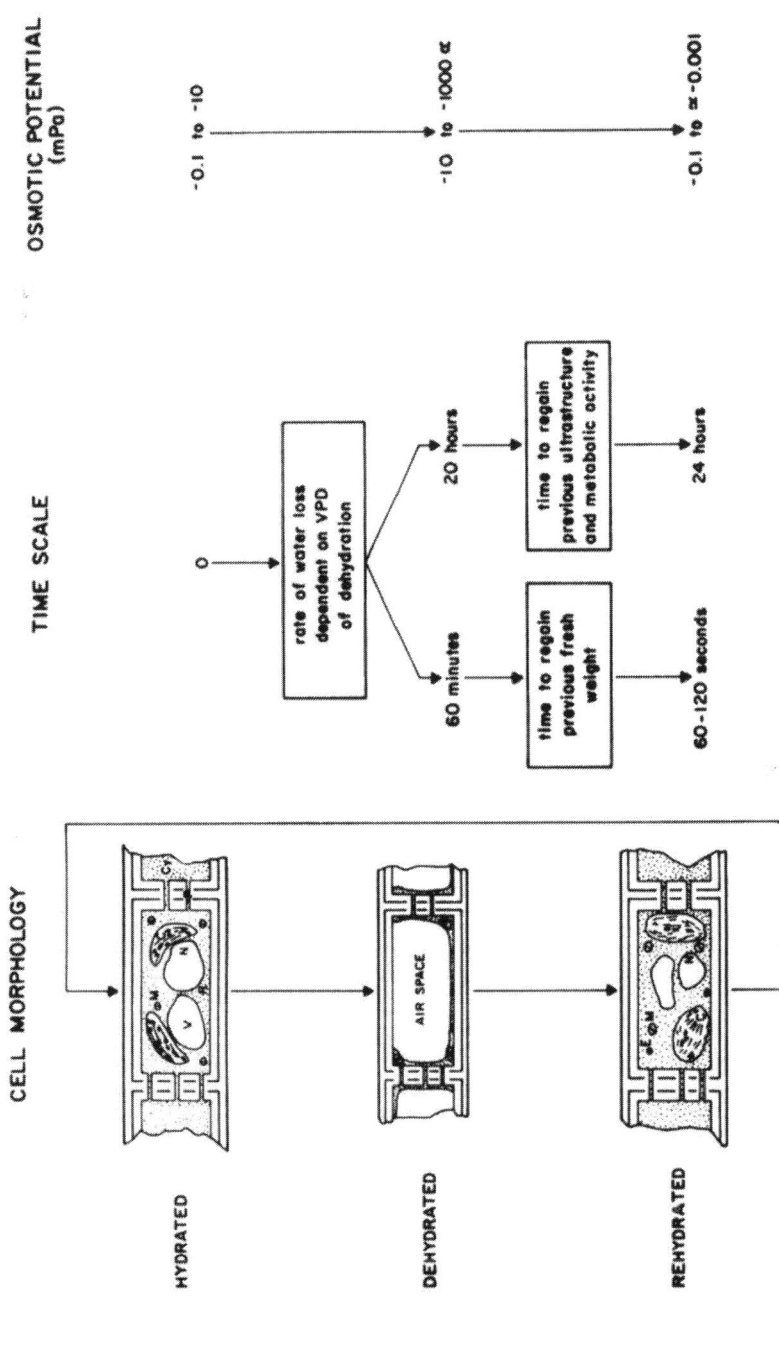

Fig. 1. Changes in cell morphology and osmotic potential as terrestrial moss cells procede through the desiccation-rehydration cycle. These processes are referenced to a time-scale. Symbols refer to electron dense bodies (E), chloroplasts (C), mitochondria (M), nucleus (N), ribosomes (R) and vacuoles (V).

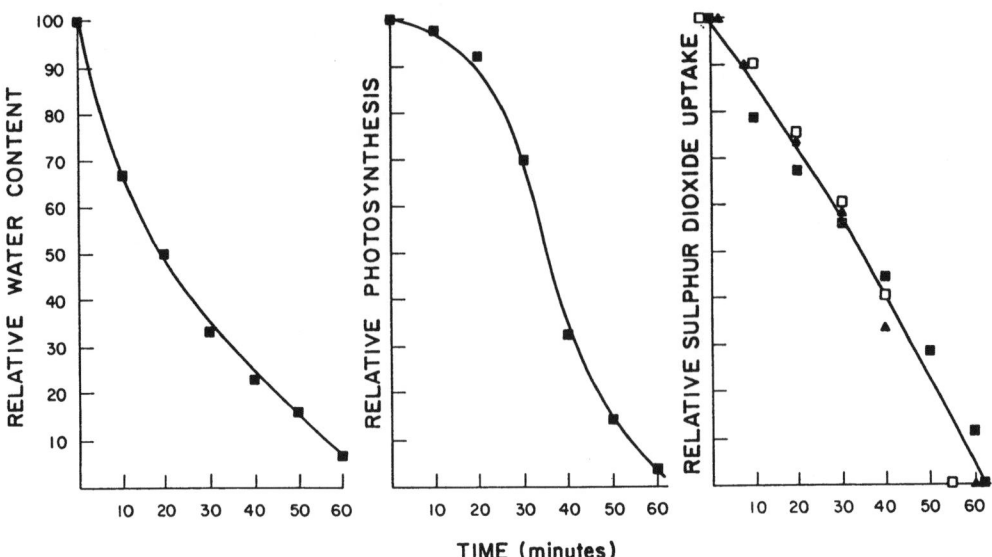

Fig. 2. Changes in relative water content (100% = fully hydrated, i.e., 1 g fwt of moss = 0.2 g dwt moss + 0.8 g water), relative photosynthesis (100% = 2.5 mg $CO_2 \cdot g^{-1}$ $dwt \cdot h^{-1}$), and relative SO_2 absorption (100% = 0.02 µg $SO_2 \cdot g^{-1}$ $dwt \cdot min^{-1}$ when SO_2 is 0.7 ppm [■]; 0.08 µg $SO_2 \cdot g^{-1}$ $dwt \cdot min^{-1}$ when SO_2 is 2.7 ppm [▲]; and 0.2 µg $SO_2 \cdot g^{-1}$ $dwt \cdot min^{-1}$ when SO_2 is 7.0 ppm [□]). Desiccation proceeds over a 1 hour period. All data are from Winner and Koch (1982) and Winner and Bewley (1983).

The term "SO_2 absorption" refers to total SO_2 uptake by moss cells, and we presume that since mosses lack extensive cuticles, all SO_2 deposited on them can potentially influence their metabolism. Thus for mosses, the terms absorption and uptake are synonymous. In the analysis of gaseous pollution fluxes between vascular plants and air, total SO_2 uptake is partitioned between SO_2 absorbed through stomata and SO_2 adsorbed on to exterior surfaces of leaves. The SO_2 which enters through the stomata is more directly damaging than that interacting at the leaf surface. Although we recognize the differences between these terms, both absorption and uptake are used to describe SO_2 flux through stomata.

CALCULATION OF SO_2 ABSORPTION

Mosses

SO_2 absorption curves for <u>Climacium dendroides</u> (Hedw.) Web & Mohr. (Winner and Koch 1982) and for feather mosses (Winner and Bewley 1983) were used to calculate a rate of SO_2 absorption at an ambient SO_2 concentration of 0.7 ppm. Total areas under these

SO_2 absorption curves were integrated. Absorption during the linear-phase (first 10 mins.) was expressed relative to the total absorption. From this, a rate of SO_2 absorption (26.1 mg.g^{-1} dwt.day^{-1}) was determined, and annual absorption was estimated from the number of potential growing days for a specific habitat (Table 1). Initially it was assumed that mosses were fully hydrated on every growing day, and therefore the absorption remained unchanged throughout the growing season. Subsequent calculations took into account seasonal changes in moss water content.

Our first set of calculations were made using 0.7 ppm SO_2 as the hypothetical annual mean concentration for calculations because SO_2 absorption rates were available at this concentration. Evaluating SO_2 absorption rates at higher concentrations (2.7 ppm and 7.0 ppm) suggests that absorption rate is proportional to concentration (Winner and Bewley 1983). This seems to be true even as the moss hydration state changes (Fig. 2). We assume that absorption rate and concentration are proportional at concentrations lower than 0.7 ppm and perhaps as low as 0.03 ppm, which is similar to the annual average concentration of some sites (U.S. E.P.A. 1984). Calculations of annual SO_2 absorption were made at 0.03 ppm for both mosses and vascular plants to take into account realistic ambient SO_2 concentrations. Thus, absorption for mosses and vascular plants at 0.03 ppm was determined by taking 4.2% of the value calculated for SO_2 at 0.7 ppm.

Vascular Plants

Stomatal conductances measured for water vapour for a number of vascular plants from diverse habitats were used to determine potential SO_2 absorption (Winner and Mooney 1980; see Table 1). Water vapour conductance was modified by multiplying by 0.5 to account for the differing diffusivity of gaseous SO_2 as compared to water vapour. SO_2 absorption on a leaf area basis was calculated by multiplying the SO_2 conductance value by the hypothetical exposure concentration of SO_2 (0.7 ppm), and annual absorption was calculated by multiplying the absorption on a leaf area basis by the potential number of growing days at a specific habitat. To compare the annual absorption of SO_2 for vascular plants with that of mosses, absorption on a leaf area basis was multiplied by specific leaf area to yield absorption on a leaf fresh weight basis. Values selected for stomatal conductance and specific leaf area were obtained from reported measured values; an effort was made to choose values which were most representative of the habitats considered in this article (Table 1).

Water vapour conductance values for plants increase with increasing capacity for leaves to exchange gases with the air and typically range from high values of 7.6 mm.s^{-1} in tropical regions to 3.5 mm.s^{-1} in deserts. Leaf specific areas were typically highest in arctic tundra (2.0 dm^2.g^{-1}) and lowest in deserts (0.4 dm^2.g^{-1}), indicating transition from thin, light leaves to thick, heavy leaves. These physiological and morphological features of plants, along with growing season, are important components in the estimation of seasonal absorption of gaseous pollutants.

Table 1. Habitats found world-wide characterized by climactic factors, phenological factors and calculation of annual SO_2 uptake for mosses and vascular plants.

Habitat	Average total annual precipitation[a]	Growing days[a]	Moss annual SO_2 uptake[b]	Water vapour conductance[a]	Vascular plant SO_2 absorption[c]	Vascular plant specific leaf area[a]	Vascular plant annual SO_2 uptake
	(mm)		($g.g^{-1}$ dwt)	($mm.s^{-1}$)	($\mu g.cm^{-2}.y^{-1}$)	($dm^2.g^{-1}$)	($mg.g^{-1}$ dwt)
Tropics	1900 – >2000	365	9.5	7.6	204	1.0	20.4
Evergreen forest	730 – 800	180	4.7	2.7	38	0.8	3.0
Deciduous forest	600 – 700	155	4.0	2.8	33	1.1	3.6
Arctic tundra	180 – 1000	63	1.6	5.8	16	2.0	3.2
Desert	50	33	0.9	3.5	5	0.4	0.2

[a] Data from: Larcher 1983; Holdridge et al. 1971; Busby et al. 1978; Rastofer 1978; MacMahon 1979; Korner et al. 1979; Miller et al. 1976; Fitter and Hay 1981; Cooper 1975; Lawrence et al. 1978.

[b] 26.1 $mg.g^{-1}$ $dwt.day^{-1}$; 0.7 ppm SO_2

[c] SO_2 concentration 0.7 ppm

CLIMATIC CHARACTERISTICS OF DIVERSE HABITATS

The relationships that exist between climatic factors, such as precipitation and growing season, are likely to determine plant water status by controlling the potential water availability within a specific habitat. The water status of plants define their capacity to absorb gaseous pollutants. Differences in water availability between habitats allowed us to explore the potential range of gaseous pollutant absorption by mosses and vascular plants. We identified five habitats that could be characterized with respect to precipitation and growing season (Table 1). This facilitated the systematic comparison of climate with growing season and, subsequently, plant water status. The habitats that we evaluated ranged in annual precipitation from 50 mm per year in deserts to more than 2,000 mm per year in the tropics.

ANNUAL SO_2 ABSORPTION BY VASCULAR PLANTS

Total annual SO_2 absorption by vascular plants growing in 0.7 ppm SO_2 ranged from a low of 0.2 mg $SO_2 \cdot g^{-1}$ dwt in desert habitats to more than 20 mg $SO_2 \cdot g^{-1}$ dwt in tropical rain forests (Table 1). SO_2 absorption was highest for tropical habitats where growing seasons are longest and stomatal conductance highest. With the exception of the desert habitat where SO_2 absorption was 0.2 mg $SO_2 \cdot g^{-1}$ dwt, absorption fell into the narrow range of 3.0 to 3.6 mg $SO_2 \cdot g^{-1}$ dwt. Thus stomata appear to regulate total gas exchange over a growing season so that total SO_2 absorption is similar for plants found in many temperate and arctic habitats.

ANNUAL SO_2 ABSORPTION BY MOSSES

Maximum Estimate

Annual estimates of SO_2 absorption for mosses, assuming a hypothetical mean SO_2 concentration of 0.7 ppm, ranged from values less than 1 g $SO_2 \cdot g^{-1}$ moss dry weight in desert habitats to almost 10 g $SO_2 \cdot g^{-1}$ moss dry weight in tropical habitats (Table 1). This ten-fold difference amongst mosses is predicated on their being fully hydrated during all growing days and that they are infinite SO_2 sinks.

These calculations illustrate the extent to which regional climatic factors such as precipitation and growing season can hypothetically influence absorption of gaseous pollutants. Moss SO_2 absorption is greatest in habitats where precipitation is high and the growing season is long. Thus mosses in tropical habitats are most likely to be hydrated for the longest periods during the year and will absorb the greatest amount of SO_2.

Effect of Desiccation-rehydration Cycles on SO_2 Absorption

The annual SO_2 absorption values for mosses shown in Table 1 were calculated with the assumption that mosses are fully hydrated during the total growing season. These estimates are, therefore, the maximum potential absorption that would occur. We would expect, however, that mosses would go through the desiccation-rehydration cycle repeatedly during the growing season and that as they did so, their SO_2 absorption capacity would reflect changes in their water content. We have tried to estimate the seasonal mean water content of mosses in several habitats in order to estimate the potential for the dehydration-rehydration cycle to reduce seasonal SO_2 absorption (Table 2). The calculated maximum SO_2 absorption values for mosses (Table 2) were reduced by amounts reflecting either the likely number of hydrated growing days or the seasonal mean water content. This process results in an estimate of the minimum SO_2 absorption (Table 2).

Maximum and minimum annual SO_2 absorption values do not differ widely in the tropics because mosses are likely to be hydrated for 300 days of the year (Table 2). Estimates of moss hydration states in deciduous and evergreen forests were about 50% of the maximum annual absorption value. Absorption is likely to be slightly greater in evergreen forests (2.8 $g.g^{-1}$ dwt) as compared to deciduous forests (2.0 $g.g^{-1}$ dwt) because the growing season is slightly shorter in the latter case. Maximum absorption in the arctic is reduced from 1.6 g to 0.6 g based upon the estimate that mean seasonal water content is about 40%. Thus in these forests, the arctic tundra and, more importantly, the desert where minimum absorption was only 10% of the maximum estimate, seasonal changes in moss water content play important roles in defining their capacity to absorb gaseous pollutants. Exceptionally wet growing seasons could increase annual SO_2 absorption as could episodic air pollution concentrations which coincided with times when mosses were most likely to be hydrated.

COMPARISONS OF CALCULATED AND MEASURED SULPHUR CONCENTRATIONS IN PLANT TISSUES

We calculated SO_2 absorption for mosses and vascular plants assuming the annual mean SO_2 concentration was 0.03 ppm rather than the value of 0.7 ppm used in initial calculations. This concentration (0.03 ppm) is a realistic approximation of annual mean SO_2 levels in habitats near SO_2 sources. Our calculations were made assuming rates of SO_2 absorption are proportional to concentration.

Annual SO_2 absorption by mosses exposed to SO_2 at 0.03 ppm ranged from 334 $mg.g^{-1}$ dwt in the tropics to 4 $mg.g^{-1}$ dwt in the desert (Table 2). Values for SO_2 absorption in forests were about 100 $mg.g^{-1}$ dwt of SO_2 or 50 $mg.g^{-1}$ dwt of S, which is lower than values reported for mosses growing near SO_2 sources. The sulphur content of mosses near SO_2 sources range from 15 $mg.g^{-1}$ dwt (Nash 1972), to 5.6 $mg.g^{-1}$ dwt (Freedman and Hutchinson 1980), to 1.5 $mg.g^{-1}$ dwt (LeBlanc et al. 1974). Calculated annual SO_2 uptake values may be greater than the measured S-content values due to a number of factors. At present we have little knowledge of the fate of S during desiccation and rehydration. Moss S-content may be reduced by leaching during rehydration. Additionally, gaseous S may be reemitted in various forms (e.g., Winner et al. 1981).

Table 2. Annual SO$_2$ uptake for mosses and vascular plants calculated for different habitats and with mean SO$_2$ concentrations of either 0.7 ppm and 0.03 ppm.

Habitat	Annual maximum SO$_2$ uptake by mosses	Hydrated days[a,b]	Annual minimum SO$_2$ uptake by mosses	Annual SO$_2$ uptake 0.03 ppm	
	0.7 ppm SO$_2$		0.7 ppm SO$_2$	Vascular plants	Mosses
	(g.g^{-1} dwt)		(g.g^{-1} dwt)	(μg.g^{-1} dwt)	(mg.g^{-1} dwt)
Tropics	9.5	300	7.8	870	334
Evergreen forest	4.7	107	2.8	12.8	120
Deciduous forest	4.0	77	2.0	15.4	86
Arctic tundra	1.6	25	0.6	13.7	26
Desert	0.9	5	0.1	0.9	4

[a] Calculated as a fraction of total growing season shown in Table 1.
[b] Data from: Pocs 1982; Busby et al. 1978; Tallis 1959; Miller et al. 1978; Alpert 1979.

The calculated SO_2 absorption by vascular plants growing over the period of one year with an SO_2 concentration of 0.03 ppm ranges from 870 $\mu g.g^{-1}$ dwt (435 μg $S.g^{-1}$ dwt) in the tropics to 0.9 $\mu g.g^{-1}$ dwt (0.45 μg $S.g^{-1}$ dwt) in the desert (Table 2). These values are 30- to 200-fold lower than those observed for mosses and are also considerably lower than S concentrations reported for vascular plants growing near SO_2 sources. For example, Picea mariana growing near an SO_2 point source had a sulphur concentration of 1.3 $mg.g^{-1}$ dwt (LeBlanc et al. 1974). Vascular plants would then appear to be acquiring most of their S from soils. As with mosses, our knowledge of gaseous S remission and S redistribution within the plant is limited. These calculations and measurements suggest that vascular plants have lower sulphur concentrations than mosses. The potential for leaves to translocate S and the perennial nature of evergreen foliage make it difficult to determine origins of S extracted from vascular plant leaves.

CONCLUSIONS

The main conclusion reached from this analysis is that mosses are likely to absorb more SO_2 during the course of a year than are vascular plants. Differences in absorption between these two types of plants could be as much as 4000-fold in any habitat. These conclusions take into account the change in SO_2 absorption capacity of mosses with changes in their water content, analysis of SO_2 absorption using principles of gas exchange for vascular plants, and a survey of SO_2 absorption by mosses and vascular plants in diverse habitats ranging from the tropics to the deserts. More work is needed to further explain why field studies suggest that mosses are more sensitive to SO_2 than are vascular plants. Such studies should be physiological in nature and probe the capacity of mosses and vascular plants to assimilate SO_2.

Acknowledgements

We acknowledge Dr. Thomas Nash, III for stimulating thinking about comparisons of SO_2 absorption rates of mosses and vascular plants. This work was supported by a research contract from the Electrical Power Research Institute to W.E.W.

REFERENCES

Alpert P (1979) Desiccation of desert mosses following a summer rainstorm. Bryologist 82: 65-71

Bewley JD, Pacey J (1978) Desiccation induced ultrastructural changes in drought-sensitive and drought-insensitive plants. In: Crowe JH, Clegg JS (eds) Dry biological systems, Academic Press, New York, p 51-72

Busby, JR, Bliss LC, Hamilton CD (1978) Microclimate control of growth rates and habitats of the boreal forest mosses, Tomenthypnum nitens and Hylocomium splendens. Ecol Monogr 48: 95-110

Cooper JP (1975) Control of photosynthetic production in terrestrial systems. In: Cooper JP (ed) Photosynthesis and productivity in different environments, Cambridge University Press, London, p 715

Fitter AH, Hay RKM (1981) Environmental physiology of plants. Academic Press, London, p 355

Freedman B, Hutchinson TC (1980) Pollutant inputs from the atmosphere and accumulation in soils and vegetation near a nickel-copper smelter at Sudbury, Ontario, Canada. Can J Bot 58: 108-132

Holdridge LR, Grenke WC, Hatheway WH, Liang T, Tosi Jr JA (1971) Forest environments in tropical life zones: A pilot study. Pergamon Press, New York, p 747

Korner CH, Scheel JH, Bauer H (1979) Maximum leaf diffusive conductance in vascular plants. Photosynthetica 13: 15-82

Larcher W (1983) Physiological plant ecology. Springer-Verlag, New York, p 303

Lawrence BA, Lewis MC, Miller PC (1978) A simulation model of populations of arctic tundra graminoids. In: Tieszen LL (ed) Vegetation and production ecology of an Alaskan arctic tundra. Springer-Verlag, New York, Ecological Study #29, p 686

LeBlanc F, Rao DN (1974) A review of the literature on bryophytes with respect to air pollution. Soc Bot Fr Coll Bryol 121: 237-255

LeBlanc F, Robitaille G, Rao DN (1974) Biological responses of lichens and bryophytes to environmental pollution in the Murdochville copper mine area, Quebec. J Hattori Bot Lab 38: 405-433

MacMahon JA (1979) North American deserts: their floral and faunal components. In: Goodall DW, Perry RA (eds) Arid land ecosystems, vol 1. Cambridge University Press, New York, p 881

Miller PC, Stoner WA, Tieszen LL (1976) A model of stand photosynthesis for the west meadow tundra at Barrow, Alaska. Ecology 57: 411-430

Miller PC, Oechel WC, Stoner WA, Sveinbjornsson B (1978) Simulation of CO_2 uptake and water relations of four arctic bryophytes at Point Barrow, Alaska. Photosynthetica 12: 7-20

Nash E (1972) Effects of effluents from a zinc smelter on mosses, PhD Thesis, Rutgers University, p 196

Pocs T (1982) Tropical forest bryophytes. In: Smith AJE (ed) Bryophyte ecology. Chapman and Hall, London, p 551

Rao DN (1982) Response of bryophytes to air pollution. In: Smith AJE (ed) Bryophyte ecology. Chapman and Hall, London, p 551

Rastorfer JR (1978) Composition and bryomass of the moss layer of two wet-tundra-meadow communities near Barrow, Alaska. In: Tieszen L L (ed) Vegetation and production ecology of an Alaskan arctic tundra. Springer-Verlag, New York, p 686

Tallis JH (1959) Studies in the biology and ecology of **Rhacomitrium lanuginosum** Brid. II. Growth and physiology. J Ecol 47: 325-350

US EPA (1984) National air quality and emission trends report, 1982, US EPA, Research Triangle Park, North Carolina, Publication #450/5-84-002, p 128

Winner WE, Bewley JD (1983) Photosynthesis and respiration of feather mosses fumigated at different hydration levels with SO_2. Can J Bot 61: 1456-1461

Winner WE, Koch GW (1982) Water relations and SO_2 resistance of mosses. J Hattori Bot Lab 52: 431-440

Winner WE, Mooney HA (1980) Ecology of SO_2 resistance I: effects of fumigations on gas exchange of deciduous and evergreen shrubs. Oecologia 44: 290-295

Winner WE, Smith CL, Koch GW, Mooney HA, Bewley JD, Krouse HR (1981) Rates of emission of H_2S from plants and patterns of stable isotope fractionation. Nature 289: 672-673.

EFFECTS OF QUANTITATIVE AND QUALITATIVE CHANGES IN AIR POLLUTION ON THE ECOLOGICAL AND GEOGRAPHICAL PERFORMANCE OF LICHENS

M.R.D. Seaward

School of Environmental Science, University of Bradford,
Bradford, BD7 1DP, Yorkshire, UK

ABSTRACT

Considerable data exist linking the decline of the lichen flora throughout Europe and elsewhere over the past two centuries to an increase in air pollution. Quantitative and qualitative changes in air pollution have differed spatially and temporally during this period, but never more so than over the past two decades, when changes in national energy policies, economic factors and implementation of clean air legislation have resulted in atmospheric regimes having markedly different effects on the lichen flora. On the one hand, the opportunity has existed for relatively small areas experiencing improved air quality to be re-invaded by a limited number of lichen species which exploit selective habitats; for example, the colonization by saxicolous lichens of high pH substrata which buffer potentially harmful atmospherically-derived pollutants. On the other hand, dilution of emissions, adopted by some authorities as a solution to local air pollution, has resulted in blanket pollution over major geographical areas, with profound effects on the lichen flora. Even a small rise in sulphur dioxide levels can cause a decline in diversity, species responding according to their sensitivity to this pollutant. More recently, the differing effects on lichens of both wet and dry acidic deposition have been detected in the field, but as yet little experimental work has been carried out to substantiate these observations. However, a comprehensive lichen mapping programme in the British Isles has shown that some species have extended their ecological and geographical range by exploiting acidified substrata.

INTRODUCTION

Lichens have been employed almost exclusively to monitor the extent or spread of air pollution, particularly sulphur dioxide. Bioindicational scales, based on species diversity and/or simple phytosociological analyses, have been developed for this purpose (e.g., Hawksworth and Rose 1970; Gilbert 1970b). Although the lichen flora of areas suffering air pollution is poor, the few species to be found can usually be related to a mean annual or winter sulphur dioxide level. There is a clearly defined negative relationship between species diversity and the concentration of the pollutant (Seaward 1976b). It has also been shown that it is possible to monitor air pollution concentrations by critical appraisal of the ecological performance of a single species (Seaward 1976a).

URBAN/INDUSTRIAL STUDIES

Zonal Mapping

The major approach to the study of lichens in areas affected by air pollution has been based on distributional studies. These have resulted in the construction of zonal maps, in which the distribution of one or more species correlates well with prevailing levels of pollution. Such studies have mainly resulted in the formulation of bioindicational scales for evaluating air pollution levels (e.g., Hawksworth and Rose 1970; Gilbert 1970b; Seaward 1972). In most polluted areas, the inner limit for the distribution of many lichen species is quite clearly defined; the ecological factors operating in such areas at, or immediately preceding, a particular date must be critical for the lichen's performance or even for its existence. However, when investigating incipient changes in the inner distributional limits of such plants, care should be taken in interpreting which of those species survive in microclimatic niches by chance establishment, which are relicts, and which have responded to a change in atmospheric pollution.

The decline in the lichen flora along transects running into urban and industrial areas has been investigated extensively throughout the world (Hawksworth 1971; Ferry et al. 1973; Hawksworth and Rose 1976; see also "Literature on air pollution and lichens" in The Lichenologist, 1974 onwards). These studies have been mainly directed towards epiphytic lichens (e.g., Rao and Blanc 1967; Gilbert 1970a,b; Hawksworth and Rose 1970), but some attention has been given to saxicolous species (e.g., Gilbert 1970a; Seaward 1972, 1975). By far the most important urban habitats for lichens are those involving a wide variety of saxicolous substrata, the majority being dressed stone and man-made materials, which possess specific and important chemical and physical properties (e.g., asbestos-cement roofs). Details of these properties and the factors operating on the flora of such habitats are provided in Seaward (1972, 1979a).

Substrate Ecology

Calcareous substrates are more favourable than non-calcareous ones for lichen colonization, and the high pH of such substrata as mortar and asbestos-cement provides a buffering effect on the toxicity of the 'urban climate'. It would appear that acidic deposition has to be counteracted by a substrate with an artificially high pH (about pH 9-12 in the case of asbestos-cement) if lichens are to succeed in a polluted environment. This is borne out by field observations. Distinctive distributions of lichen species can be correlated with the pH of the saxicolous substratum colonized and the ambient air pollution level (Seaward 1976a).

Not only is it necessary for lichens to colonize man-influenced and man-made substrata, but they must also tolerate constantly changing environmental factors, some species being more sensitive to short-term extremes and others to more long-term enhanced pollution levels. The strategy of toxitolerant species in adapting to the multiplicity of problematic factors operating in polluted environments varies considerably; their reproductive spores and propagules readily colonize certain substrata, particularly those which can buffer or neutralize potentially harmful acidic deposition. However, from an ever-present rich air spora, only a few species are capable of surviving the crucial stages of establishment, germination and development. Once established, such species are often aggressive, and, under conditions of amelioration, have generally proved to be highly competitive, to the exclusion of those species which would otherwise normally be growing under such regimes.

A further strategy adopted by the limited number of successful species is to exploit niches where the micro-environment differs, albeit marginally, from conditions prevailing elsewhere. Existing data on environmental factors operating in polluted areas are only available at a macro-level. Little is known of the conditions at the micro-level where significant differences in climate, pollution levels and substrata are known to occur over short distances.

Despite these micro-environmental differences, patterns created by lichen assemblages which would relieve the architectural severity of many building materials and vary the monotonous green 'film' generally covering tree trunks and branches in air-polluted landscapes do not attain their potential richness. Substrata are dominated by a very few lichen and algal species, and often a single species prevails. The mellowing effect produced by a mosaic of a variety of different species on substrata in unpolluted areas cannot be achieved. Notwithstanding considerable amelioration of air pollution in recent years, these species-poor communities, with few exceptions, continue to dominate such areas.

Ameliorating Conditions

It has been shown that not only can the ecological performance of a single species be effectively employed as a bioindicator of several air pollution regimes (Seaward 1976a), but it can also be used to monitor improvements in an environment (Brightman and Seaward 1977; Henderson-Sellers and Seaward 1979; Seaward 1979b). The epiphytic lichen Lecanora muralis, which has a distinctive distribution in urban/industrial areas, has inner limits, defined according to its substrate preference, which can be periodically mapped to determine its rate of spread (or otherwise) and hence the ambient air pollution level (Seaward 1976a). This species has recolonized the West Yorkshire conurbation at a rate of about 9 $km^2.y^{-1}$ over the past 13 years (with SO_2 levels declining by about 10 $\mu g.m^{-3}.y^{-1}$, mainly through its ability to exploit man-made calcareous substrata and thus extend its geographical and ecological ranges in polluted areas. A biological scale to monitor sulphur dioxide pollution and rainfall pH, based entirely on the substrate preference limits of L. muralis, has been developed (Seaward 1976a). This scale has proved most effective in determining rising, stable and falling levels of sulphur dioxide.

Depending on the species, lichens take several to many years to respond to ameliorating conditions, whereas a sudden increase in air pollution produces a much more rapid effect. The time-lag between pollution levels dropping below an identifiable threshold and the successful colonization of a lichen has therefore to be credibly established before any effective use can be made of lichen maps for monitoring a fall in sulphur dioxide levels. Computer simulations, based on relatively long-range field data, have shown that a time-lag of about 5 years seems to be most realistic for L. muralis (Henderson-Sellers and Seaward 1979), but other species (particularly epiphytes) will take much longer to respond, and many may never recolonize.

Further simulations to consider variation in sulphur dioxide threshold and colonization criteria, employing the 5-year time-lag, have been undertaken with L. muralis. In areas implementing clean air legislation, it is difficult to identify a direct relationship between ambient pollution levels and the success of lichens, and recently-collected data may be ineffective in demonstrating any relationship between species diversity and air pollution level (Seaward 1976b, 1979b). However, the undoubted improvement in some lichen floras during recent years, as illustrated by species diversity relative to the distance from the centre of urban/industrial complexes, has been clearly demonstrated (Henderson-Sellers and Seaward 1979; Seaward 1981; Showman 1981). The spread of several lichen species into the suburbs of such cities as Leeds, Bradford and Salford has been spectacular over the past 15 years. Data in respect of the performance of other plant groups under ameliorating conditions are lacking.

The rate and mode of spread of L. muralis at sites experiencing decreasing air pollution are being carefully monitored by means of time and space analyses of propagule influx and thallial growth measurements (Seaward 1980, 1982). Some tentative conjectures may reasonably be made from these observations. Although L. muralis is a successful lichen in urban/industrial environments, with a relatively rapid growth rate, it seems likely that in the long term it would be unsuccessful in competition with other slower-growing species. In these studies the intraspecific growth differences of the earlier cohorts (cf. urban ecad described in Seaward 1976a), and the later cohorts may be considered in the same context. Further factors which may determine the rate and mode of spread are discussed in Seaward (1982).

Several other lichen species are known to be highly successful in urban and industrial areas subject to relatively high levels of air pollution. Although most of these are equally successful elsewhere, some species (see below) which formerly had narrower ecological requirements and/or more restricted distributions appear to thrive on substrata subjected to new atmospheric regimes. They are probably favoured to some extent by reduced competition.

REGIONAL STUDIES

Lichen Decline

The above considerations refer mainly to urban/industrial complexes, where in the past pollution has undoubtedly devastated the lichen flora over relatively small areas. There is also a wealth of evidence to demonstrate the selective decline of lichen species over much wider areas during the past two centuries, mainly attributable to the rise in sulphur dioxide levels, although other pollutants and changing forestry and agricultural practices are implicated. More recently, implementation of clean air legislation and changes in energy policies and industrial practices have resulted in qualitative changes in, and dramatic reductions in levels of air pollution with consequent reinvasion by lichens (albeit slowly and as yet confined to a limited number of species). At the same time a less dramatic but nevertheless significant rise in pollution levels over wider areas due to changes in techniques for pollutant dispersion has resulted in the loss of a number of lichen species, although others have been able to exploit/tolerate the new atmospheric regimes.

The distribution of Usnea spp. in the British Isles (cf. Fig. 1) illustrates the type of change which has occurred since about 1800. The genus, at one time widespread and luxuriant, had over the course of the next 160+ years almost entirely disappeared from a major area covering at least 68,000 km^2, mainly as a result of the increase in atmospheric pollution. However, since the implementation of the Clean Air Acts of 1956 and 1968 in the UK, sulphur dioxide concentrations have declined dramatically in urban/industrial areas (>60%), with less dramatic reductions, or indeed occasional rises, in rural areas, with the result that much of the British Isles is experiencing a more homogeneous distribution of this pollutant.

Blanket Pollution

Regional pollution over wide areas has reduced lichen diversity, but has favoured expansion of a relatively small number of species formerly having narrower ecological requirements and/or more restricted distributions, e.g., Buellia punctata, Lecanora conizaeoides (Fig. 2), L. muralis, Parmelia incurva (Fig. 3), Parmeliopsis ambigua (Fig. 4), Phaeophyscia orbicularis, Scolisciosporum chlorococcum, Xanthoria candelaria and X. elegans (Fig. 5).

The re-establishment of Usnea spp. (mainly U. subfloridana), usually on Fraxinus and Salix, at numerous sites throughout England during recent years (Fig. 1) may largely reflect decreases in pollution levels in rural areas. There is some evidence that Ramalina farinacea, Bryoria fuscescens and Pseudevernia furfuracea are exhibiting a similar response. It would appear, however, that the stability of these taxa at some of these sites is tenuous, since small thalli may succeed in establishing themselves for only one or two years (Seaward 1980). On the other hand, more stable situations have been reported (e.g., Gilbert 1977), and it may well prove feasible in the near

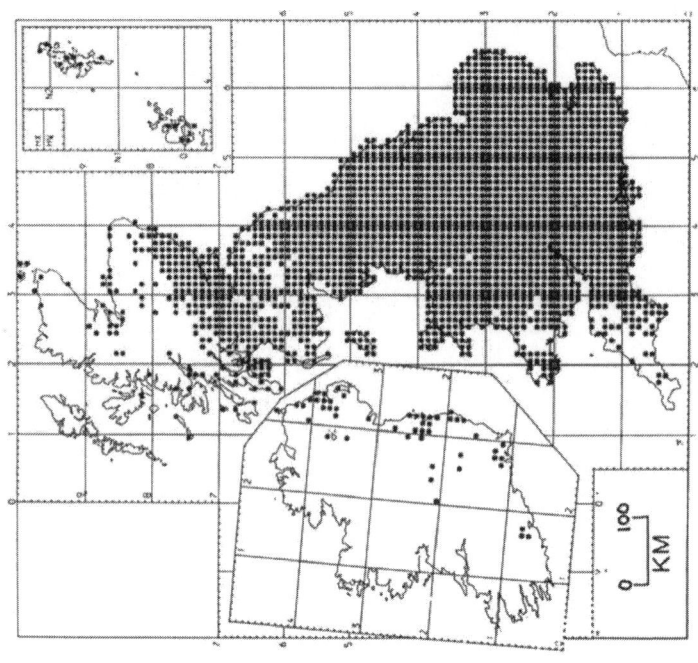

Fig. 1. Distribution of Usnea subfloridana Stirton in the British Isles, showing its disappearance from c. 68,000 km² during the period c. 1800-1970, and its subsequent re-establishment in more than 40 recording units (10 km x 10 km) within the area delineated. (O = pre-1960, usually 19th century, records; ● = 1960 onwards records.)

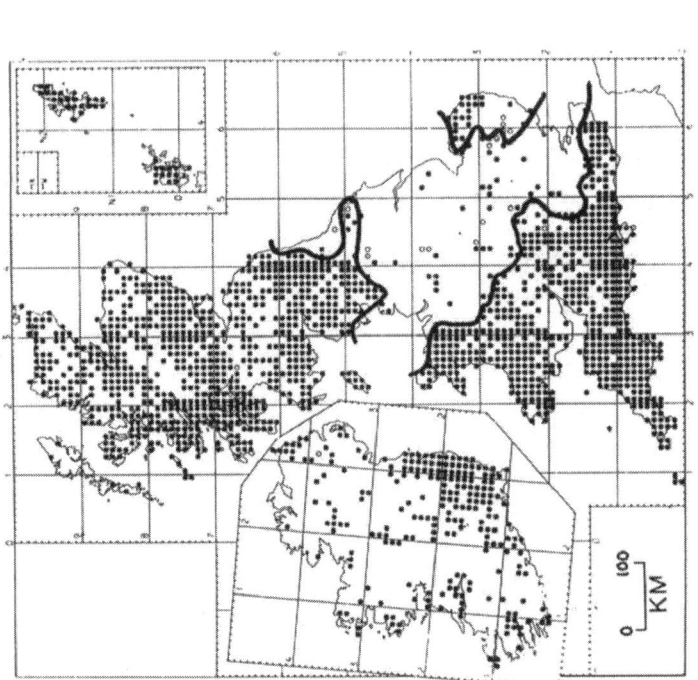

Fig. 2. The spread of Lecanora conizaeoides Nyl. ex Crombie since 1860 has been dramatic: today it is ubiquitous in England and spreading rapidly into Scotland, Wales and Ireland in those areas experiencing winter sulphur dioxide levels of 50-150 µg.m^{-3}; it is often the dominant, or indeed exclusive, member of many urban lichen communities.

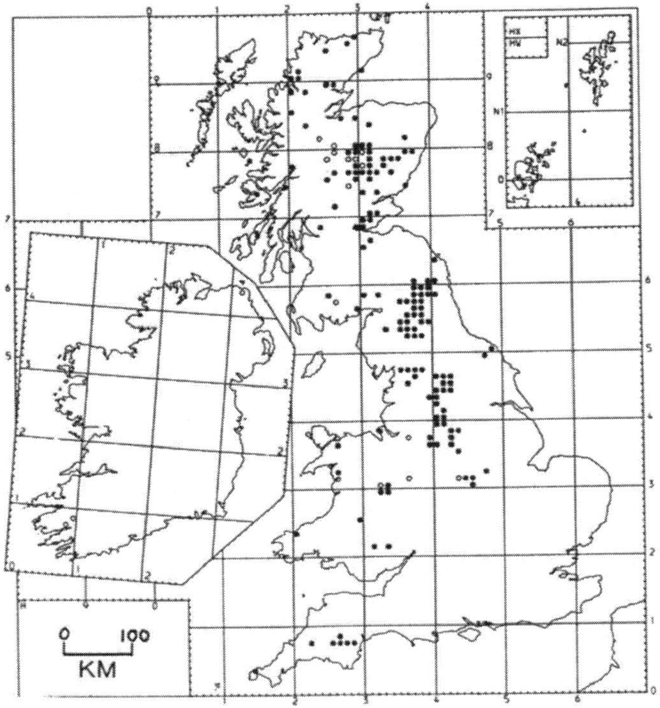

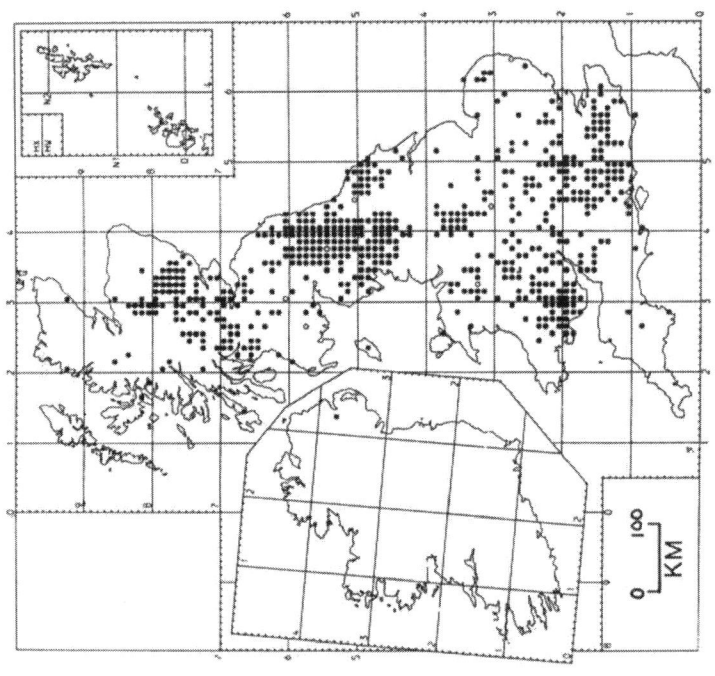

Fig. 3. Parmeliа incurva (Pers.) Fr. is extending its distribution in the British Isles, apparently due to its tolerance to mean sulphur dioxide levels up to about 70 µg.m^{-3} and aided by the elimination of less tolerant lichen competitors at such sites.

Fig. 4. Formerly a species of decorticate conifer wood in central and eastern Scotland, Parmeliopsis ambigua (Wulfen) Nyl. has conspicuously extended its range during the present century; an accelerated expansion (calculated at a rate of c. 15% over the past decade) demonstrates its adaptation to deciduous trees in moderately air-polluted areas, mainly in response to increased bark acidification.

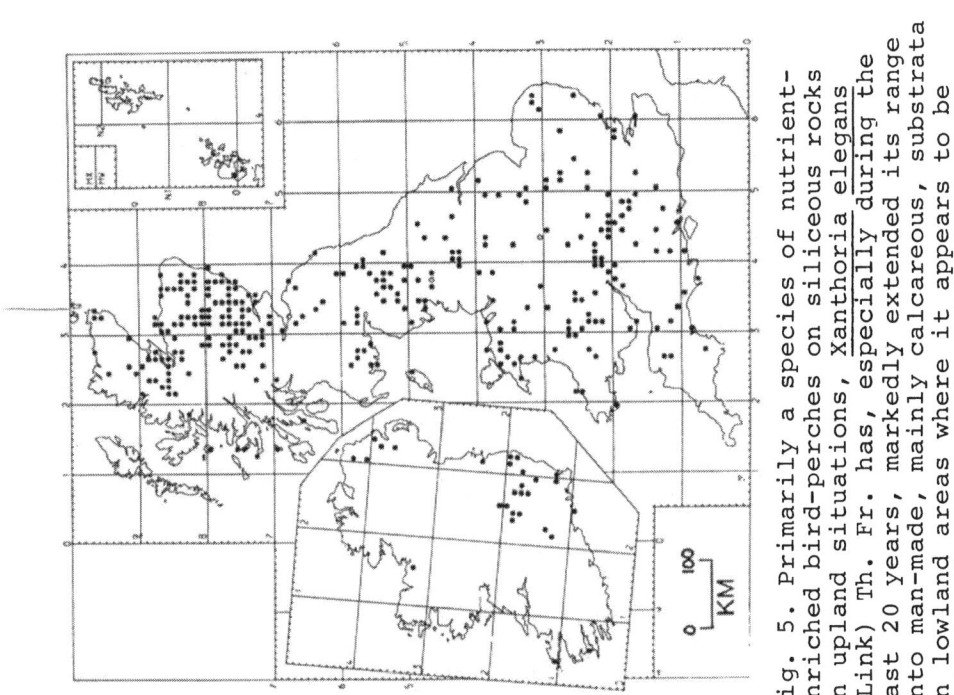

Fig. 5. Primarily a species of nutrient-enriched bird-perches on siliceous rocks in upland situations, Xanthoria elegans (Link) Th. Fr. has, especially during the past 20 years, markedly extended its range onto man-made, mainly calcareous, substrata in lowland areas where it appears to be fairly tolerant of air pollution.

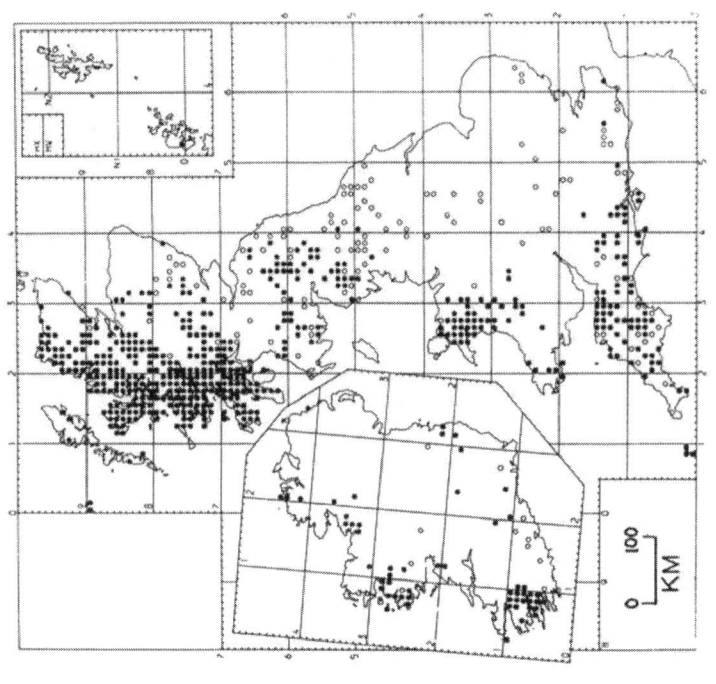

Fig. 6. More widely distributed in the 19th century, Lobaria pulmonaria (L.) Hoffm., is as a result of air pollution, tree-felling and other changes in forest management, now mainly restricted to higher rainfall areas in the west; in moderately polluted areas it is more or less confined to more basic bark (pH around 5.5) of old Fraxinus, Ulmus and Acer.

future to monitor ameliorating environments on a relatively large scale through the use of lichen mapping.

More recently, it has been observed that fruticose lichens such as Usnea (and possibly Ramalina) spp. survive at the expense of certain foliose and crustose species which would normally occur as epiphytic associates at the outer limits of areas experiencing 'blanket pollution'. From our knowledge of the behaviour of lichens subjected to sulphur dioxide air pollution, it has been possible to rank species according to their sensitivity to this pollutant, and it is apparent that the fruticose species are more susceptible to this gas than many foliose and crustose species. A situation is developing whereby foliose lichens, such as Parmelia sulcata, a species tolerant of a moderate level of sulphur dioxide, have succumbed to the new atmospheric regime. The original assemblage has often been replaced by one which includes a few thalli of fruticose lichens known to be less tolerant of sulphur dioxide. In the light of this, standard bioindicational scales will require certain modifications in the future to account for these hierarchical changes.

Acid Rain

There is strong reason to believe that the local reducing smogs, dominated by high SO_2, sulphate and sulphuric acid aerosols, have now given way to different mixtures, with regional acid rain caused by SO_2 and NO_x emissions. Lichens react differently to gaseous sulphur dioxide compared with the acidic substances in precipitation; furthermore, a synergistic effect may develop from the interaction of these two factors (Türk and Wirth 1975), since sulphur dioxide is much more toxic to lichens at pH values below 4.

'Acid rain' can affect lichens both directly, as described above, or indirectly by the acidification of substrata (Barkman 1958; Skye 1968; Grodzińska 1971; Robitaille et al. 1977). The exploitation of calcareous substrata by saxicolous species in polluted environments has already been referred to, but similar phenomena have been observed with respect to epiphytes. Parmeliopsis ambigua, for example, which was formerly to be found on decorticate conifer wood in Scotland, has undergone over the past few decades a remarkable expansion onto deciduous trees in moderately polluted areas throughout Britain (Fig. 4), presumably in response to increased bark acidification (to pH 3.0 - 4.0) (Seaward and Hitch 1982). On the other hand, some pollution-sensitive species, such as Lobaria pulmonaria (Fig. 6) have often disappeared from tree barks which have a poor buffering capacity (e.g., Quercus spp.) in areas subjected to subtle changes in the atmospheric regime, but nevertheless appear to be healthy on tree barks with a higher pH (e.g., Fraxinus excelsior) at the same site. Other demonstrable effects, such as the development of more oligotrophic epiphytic lichen floras, have been reported throughout Europe as a result of changes in the nature of pollutant emissions.

FUTURE ENQUIRY

A firm distinction must be drawn between factual information and speculative inferences: interpretation of field observations have yet to be substantiated by adequate field and laboratory experiments. However, preliminary laboratory experiments involving subjecting lichens to 'simulated acid rains' have demonstrated that physiological activities can be impaired (e.g., Denison et al. 1976; Lechowicz 1982; Fritz-Sheridan 1985), possibly in some cases through the mobilization of ions which under a higher pH regime would otherwise have remained inert and therefore harmless within the thallus (e.g., Sigal, in Lawrey 1984). Like so many in vitro experiments concerned with the response of lichens to air pollution (cf. Sigal 1984), the simulations are artificially severe, of short duration, and unlikely to reproduce the complexity of both the pollutant and the factors operating in the field.

Profitable lines of enquiry by means of which the effects on lichens of the new air pollution regimes may be determined could be achieved through long-term field techniques involving stringent ecological and phytogeographical criteria. To a large extent, such studies have been made possible in the British Isles by a comprehensive on-going programme of detailed lichen mapping carried out by members of the British Lichen Society since 1963. The information is stored on a major data-base at Bradford University, backed up by sophisticated facilities for cartographic reproduction and interpretation. However, even such complex systems must be supported by extensive field studies in a wide range of selected habitats in critical areas where more immediate changes can be scientifically diagnosed. Unhappily, the current situation regarding atmospheric pollution may be too urgent to allow sufficient time for relatively long-term investigations, and reliance will have to be placed on short-term evaluations. Whereas earlier air pollution episodes principally affected the urban and urban-rural regions, the newer regional air pollution-acid rain occurrences threaten hitherto relatively unpolluted remote areas.

REFERENCES

Barkman JJ (1958) Phytosociology and ecology of cryptogamic epiphytes. Van Gorcum, Assen, p 628
Brightman FH, Seaward MRD (1977) Lichens of man-made substrates. In: Seaward MRD (ed) Lichen ecology, Academic Press, London, p 253-293
Denison W, Caldwell B, Bormann B, Elched L, Swanberg C, Anderson S (1976) The effects of acid rain on nitrogen fixation in western Washington coniferous forest. Proc 1st Int Symp on Acid Precipitation in Forest Ecosystems, USDA Technical Report NE-23
Ferry BW, Baddeley MS, Hawksworth DL (eds) (1973) Air pollution and lichens. Athlone Press, London, p 389
Fritz-Sheridan RP (1985) Impact of simulated acid rain on nitrogenase activity in Peltigera aphthosa and P. polydactyla. Lichenologist 17: 27-31

Gilbert OL (1970a) Further studies on the effect of sulphur dioxide on lichens and bryophytes. New Phytol 69: 605-627

Gilbert OL (1970b) A biological scale for the estimation of sulphur dioxide pollution. New Phytol 69: 629-634

Gilbert OL (1977) Lichen conservation in Britain. In: Seaward MRD (ed) Lichen ecology, Academic Press, London, p 415-436

Grodzińska K (1971) Acidification of tree bark as a measure of air pollution in southern Poland. Bull Acad Polon Sci, ser sci bio II 19: 189-195

Hawksworth DL (1971) Lichens as litmus for air pollution: a historical review. Int J Environ Studies 1: 281-296

Hawksworth DL, Rose F (1970) Qualitative scale for estimating sulphur dioxide air pollution in England and Wales using epiphytic lichens. Nature, Lond 227: 145-148

Hawksworth DL, Rose F (1976) Lichens as pollution monitors. Edward Arnold, London, p 60

Henderson-Sellers A, Seaward MRD (1979) Monitoring lichen reinvasion of ameliorating environments. Environ Pollut 19: 207-213

Lawrey JD (1984) Biology of lichenized fungi. Praeger, New York, p 408

Lechowicz MJ (1982) The effects of simulated acid precipitation on photosynthesis in the caribou lichen *Cladina stellaris* (Opiz) Brodo. Water Air Soil Pollut 18: 421-430

Rao DN, LeBlanc F (1967) Influence of an iron-sintering plant on corticolous epiphytes in Wawa, Ontario. Bryologist 70: 141-157

Robitaille G, LeBlanc F, Rao DN (1977) Acid rain: a factor contributing to the paucity of epiphytic cryptogams in the vicinity of a copper smelter. Rev Bryol Lichénol 43: 53-66

Seaward MRD (1972) Aspects of urban lichen ecology. PhD thesis, University of Bradford, p 353

Seaward MRD (1975) Lichen flora of the West Yorkshire conurbation. Proc Leeds Phil Lit Soc, sci sec 10: 141-208

Seaward MRD (1976a) Performance of *Lecanora muralis* in an urban environment. In: Brown DH, Hawksworth DL, Bailey RH (eds) Lichenology: progress and problems, Academic Press, London, p 323-357

Seaward MRD (1976b) Lichens in air polluted environments: multivariate analysis of the factors involved. In: Kärenlampi L (ed) Proc Kuopio meeting on plant damages caused by air pollution, Univ Kuopio, Kuopio, p 57-63

Seaward MRD (1979a) Lower plants and the urban landscape. Urban Ecol 4: 217-225

Seaward MRD (1979b) Lichens as monitors of environments with decreasing sulphur dioxide levels. In: Int symp on sulphur emissions and the environment. Soc Chemical Industry, London, p 255-258

Seaward MRD (1980) The use of lichens as bioindicators of ameliorating environments., In: Schubert R, Schuch J (eds) Bioindikation auf der Ebene der Individuen, Martin-Luther-Universität, Halle-Wittenberg, p 17-23

Seaward MRD (1981) Lichen flora of the West Yorkshire conurbation - supplement II (1978-80). Naturalist 106: 89-92

Seaward MRD (1982) Lichen ecology of changing urban environments. In: Bornkamm R, Lee JA, Seaward MRD (eds) Urban ecology, Blackwell Scientific Publications, Oxford, p 181-189

Seaward MRD, Hitch CJB (eds) (1982) Atlas of the lichens of the British Isles, vol 1, Natural Environment Research Council, Cambridge, p 196

Showman RE (1981) Lichen recolonization following air quality improvement. Bryologist 84: 492-497

Sigal LL (1984) Lichen research and regulatory decisions. Bryologist 87: 185-192

Skye E (1968) Lichens and air pollution. Acta phytogeogr suec 52: 1-123

Türk R, Wirth V (1975) The pH dependence of SO_2 damage to lichens. Oecologia (Berl) 19: 285-291

Gilbert OL (1970a) Further studies on the effect of sulphur dioxide on lichens and bryophytes. New Phytol 69: 605-627
Gilbert OL (1970b) A biological scale for the estimation of sulphur dioxide pollution. New Phytol 69: 629-634
Gilbert OL (1977) Lichen conservation in Britain. In: Seaward MRD (ed) Lichen ecology, Academic Press, London, p 415-436
Grodzińska K (1971) Acidification of tree bark as a measure of air pollution in southern Poland. Bull Acad Polon Sci, ser sci bio II 19: 189-195
Hawksworth DL (1971) Lichens as litmus for air pollution: a historical review. Int J Environ Studies 1: 281-296
Hawksworth DL, Rose F (1970) Qualitative scale for estimating sulphur dioxide air pollution in England and Wales using epiphytic lichens. Nature, Lond 227: 145-148
Hawksworth DL, Rose F (1976) Lichens as pollution monitors. Edward Arnold, London, p 60
Henderson-Sellers A, Seaward MRD (1979) Monitoring lichen reinvasion of ameliorating environments. Environ Pollut 19: 207-213
Lawrey JD (1984) Biology of lichenized fungi. Praeger, New York, p 408
Lechowicz MJ (1982) The effects of simulated acid precipitation on photosynthesis in the caribou lichen Cladina stellaris (Opiz) Brodo. Water Air Soil Pollut 18: 421-430
Rao DN, LeBlanc F (1967) Influence of an iron-sintering plant on corticolous epiphytes in Wawa, Ontario. Bryologist 70: 141-157
Robitaille G, LeBlanc F, Rao DN (1977) Acid rain: a factor contributing to the paucity of epiphytic cryptogams in the vicinity of a copper smelter. Rev Bryol Lichénol 43: 53-66
Seaward MRD (1972) Aspects of urban lichen ecology. PhD thesis, University of Bradford, p 353
Seaward MRD (1975) Lichen flora of the West Yorkshire conurbation. Proc Leeds Phil Lit Soc, sci sec 10: 141-208
Seaward MRD (1976a) Performance of Lecanora muralis in an urban environment. In: Brown DH, Hawksworth DL, Bailey RH (eds) Lichenology: progress and problems, Academic Press, London, p 323-357
Seaward MRD (1976b) Lichens in air polluted environments: multivariate analysis of the factors involved. In: Kärenlampi L (ed) Proc Kuopio meeting on plant damages caused by air pollution, Univ Kuopio, Kuopio, p 57-63
Seaward MRD (1979a) Lower plants and the urban landscape. Urban Ecol 4: 217-225
Seaward MRD (1979b) Lichens as monitors of environments with decreasing sulphur dioxide levels. In: Int symp on sulphur emissions and the environment. Soc Chemical Industry, London, p 255-258
Seaward MRD (1980) The use of lichens as bioindicators of ameliorating environments., In: Schubert R, Schuch J (eds) Bioindikation auf der Ebene der Individuen, Martin-Luther-Universität, Halle-Wittenberg, p 17-23
Seaward MRD (1981) Lichen flora of the West Yorkshire conurbation - supplement II (1978-80). Naturalist 106: 89-92
Seaward MRD (1982) Lichen ecology of changing urban environments. In: Bornkamm R, Lee JA, Seaward MRD (eds) Urban ecology, Blackwell Scientific Publications, Oxford, p 181-189

Seaward MRD, Hitch CJB (eds) (1982) Atlas of the lichens of the British Isles, vol 1, Natural Environment Research Council, Cambridge, p 196
Showman RE (1981) Lichen recolonization following air quality improvement. Bryologist 84: 492-497
Sigal LL (1984) Lichen research and regulatory decisions. Bryologist 87: 185-192
Skye E (1968) Lichens and air pollution. Acta phytogeogr suec 52: 1-123
Türk R, Wirth V (1975) The pH dependence of SO_2 damage to lichens. Oecologia (Berl) 19: 285-291

CROP RESPONSES TO OZONE - SULPHUR DIOXIDE MIXTURES

D.P. Ormrod*, J.L. Deveau*, O.B. Allen† and D.W. Beckerson*

*Department of Horticultural Science, †Departments of Animal and Poultry Science and Mathematics and Statistics, University of Guelph, Guelph, Ontario, Canada N1G 2W1

ABSTRACT

A factorial experiment in controlled environment chambers with three levels of ozone and four levels of sulphur dioxide was used to obtain response surfaces describing the effects of these pollutants on foliar injury and growth of 'New Yorker' tomato, 'Maple Arrow' soybean and 'Golden Jubilee' maize. Non-destructive covariate measurements were used to increase precision. The two pollutants usually did not act independently in their effects on the plants. All three species demonstrated an increase in foliar injury as the concentration of the two gases increased. Leaf area and leaf dry weight for both soybean and maize were increased at low to medium levels of pollutants and decreased at higher levels. Visible injury was greater than growth retardation on a percentage basis. Research of this kind is essential to the development of more efficient designs to evaluate interaction and dose-response patterns. Research projects are underway to determine the effects of sequential treatments with ozone and sulphur dioxide which may be more realistic than simultaneous treatments with these two gases in terms of ambient atmospheric chemical composition.

INTRODUCTION

Ozone (O_3) is a photochemically produced oxidant gas while sulphur dioxide (SO_2) is a product of the burning of sulphur-containing fossil fuels and smelting of ores. Both gases occur widely in many regions and both are toxic to plants. Reviews of literature indicate that much information is available on plant responses to each (Treshow 1984). As well, the combined effects of these two gases have been researched more widely than those of any other combination of pollutant gases (Ormrod 1982).

Both O_3 and SO_2 have dynamic roles in the chemistry of polluted air masses (Hesketh 1972). Ozone, while involved in numerous atmospheric reactions, plays a major role in contributing oxygen atoms for the oxidation of nitric oxide to nitrogen dioxide and the oxidation of SO_2 to sulphate in aerosols. The presence of O_3 greatly accelerates both these reactions. At any one time, O_3 is being generated photochemically while being used in oxidation processes. The SO_2, while also involved in many atmospheric reactions, is dissolving in water and being oxidized to sulphite and then to sulphate, an ingredient of acidic precipitation. On this basis, the combination of O_3 and SO_2 in the atmosphere would not provide a stable gas mixture, so there is some doubt as to the validity of concurrent O_3 and SO_2 mixture studies in relation to the ambient environment. Few examples of co-occurrence of O_3 and SO_2 have been found within monitoring data (Lefohn and Tingey 1984). Sequential treatments with O_3 and SO_2 may be much more realistic in terms of the relationship of atmospheric chemical composition to meterological factors.

Nevertheless, studies of concurrent exposures to O_3 and SO_2 mixtures have been pursued in many laboratories. This mixture may provide a model system for development of research techniques for studies of various other mixtures. For example, we have used O_3 and SO_2 mixtures to test the use of the central composite rotatable design as a more efficient alternative to a full factorial design (Ormrod et al. 1984) and to test the use of non-destructive covariate measurements for increasing the precision of experiments with mixtures (Ormrod et al. 1983). This mixture has also been used widely for physiological mode-of-action studies as well as for the illustration and development of terminology for pollutant interactions. The synergistic effects that can be demonstrated experimentally with O_3 and SO_2 are of concern to regulatory agencies when they are asked to decide whether or not air quality criteria levels for single gases are applicable to mixtures.

The objective of our current research is to determine the nature of crop response to long-term exposures to combinations of O_3 and SO_2 and to extend the research to include effects of sequential treatments with these two gases. We report herein the results of long-term exposures of tomato, soybean and maize to a full-factorial arrangement of O_3-SO_2 mixtures. We also describe current studies of sequential treatments and provide, for discussion, our future research plans.

MATERIALS AND METHODS

Seeds of 'New Yorker' tomato, 'Maple Arrow' soybean and 'Golden Jubilee' maize were sown in a peat-vermiculite-perlite mixture in 10-cm diameter plastic pots. Pots were placed in 12 controlled-environment chambers. Environmental conditions for growing these plants were similar to those used for the base line growth studies of lettuce (Hammer et al. 1978), i.e., $25/20 \pm 1°C$ day/night temperature, $70 \pm 5\%$ relative humidity, and 325 ± 10 $\mu mol \cdot m^{-2} \cdot s^{-1}$ photosynthetically active radiation.

After emergence, seedlings were thinned to one plant per pot. Tomato and soybean were irrigated with 1/2 strength Hoagland's complete nutrient solution. Corn was irrigated with double strength Hoagland's complete nutrient solution with 1/4 strength potassium. Prior to pollutant treatments, covariate parameters were measured non-destructively on all plants to adjust each plant to a common initial size for the analysis of variance at harvest (Ormrod et al. 1983). Covariates included planar leaf area for tomato; height for tomato, soybean and corn; and plastochron index for tomato and soybean. Pollutant treatments were administered using, as a guide, the range of actual concentrations monitored in the Canadian environment. Ozone was supplied via a Grace high voltage generator and was continuously monitored with a Dasibi ozone analyzer. Sulphur dioxide was supplied from a cylinder (1500 ppm SO_2 in N_2) and was continuously monitored with a Thermo Electron pulsed fluorescent SO_2 analyzer. The pollutant treatments began 16 days, 10 days, and 6 days after planting tomato, soybean and maize, respectively. Each of the 12 plexiglass chambers was assigned a pollutant treatment. Plants were exposed to the pollutants for 7 consecutive days from 9 a.m. to 3 p.m. daily. Prior to harvest, all plants were assessed for the amount of visible injury. For tomato and corn the first four leaves only were assessed; for soybean two unifoliate leaves were assessed using a Horsfall-Barratt rating scale (Hofstra and Ormrod 1977).

At harvest, which occurred on the day following the 7-day exposure period, measurements included plant height, plastochron index, fresh weight (leaves and stems), dry weight (leaves and stems), and leaf area (Table 1). Plant harvest data were used to calculate dry matter content of leaves (leaf dry weight per leaf fresh weight) and specific leaf weight (leaf dry weight per leaf area). The symbols and units used for each response variable discussed in the text are listed in Table 1.

Table 1. Symbols and units for variables.

Symbols	Variables	Units
Response variables		
VI	Visible injury	%
LA	Leaf area	$cm^2 \cdot plant^{-1}$
LDW	Leaf dry weight	$mg \cdot plant^{-1}$
PI	Plastochron index	
DMC	Dry matter content	%
SLW	Specific leaf weight	$mg \cdot cm^{-2}$
Covariates		
HT	Plant height	cm
PI	Plastochron index	
PLA	Planar leaf area	cm^2

To normalize the responses and stabilize their variances, data for all growth variables were transformed to natural logarithms. Foliar injury did not require transformation. The experimental design chosen was an incomplete block with all combinations of three levels of O_3 and four levels of SO_2, three blocks of four chambers each, and three replications. There were 8 plots of each species per replicate for a total of 24 subsamples per species per treatment. The data were first analyzed as an incomplete block design to test for possible block and replication effects. A second order multiple regression response surface was then fitted to each variable. Finally, a reduced model was fitted by dropping nonsignificant terms out of the model and comparing the reduced model to the full model. Response surfaces for all variables were then computer drawn.

RESULTS AND DISCUSSION

There was joint action of the two pollutants, implying that both O_3 and SO_2 had some effect on plant response. The simplest type of joint action was a linear additive response, illustrated by visible injury for soybean (VI-S) (Fig. 1). The relative importance of the two pollutants can be estimated from the regression equation by using the linear regression coefficients. In this example, O_3 was the most injurious since it had the largest coefficient (Table 2), that is, increasing O_3 by 0.1 $\mu L \cdot L^{-1}$ resulted in a greater increase in injury than increasing SO_2 by 0.1 $\mu L \cdot L^{-1}$.

Table 2. Effects of O_3 and SO_2 mixtures on visible injury on leaves (% of leaf area injured).

O_3 ($\mu L \cdot L^{-1}$)	SO_2 ($\mu L \cdot L^{-1}$)				Mean
	0.00	0.03	0.12	0.30	
'Maple Arrow' Soybean					
0.00	4.5	3.7	5.4	10.7	6.1
0.08	9.4	7.5	9.6	10.5	9.3
0.15	7.8	9.4	12.3	20.6	12.5
Mean	7.2	6.9	9.1	13.9	
'New Yorker' Tomato					
0.00	5.8	13.2	9.9	11.8	10.2
0.08	13.4	21.5	18.1	29.3	20.6
0.15	27.4	34.5	31.8	40.0	33.4
Mean	15.5	23.1	19.9	27.0	
'Golden Jubilee' Maize					
0.00	3.4	6.3	3.9	5.1	4.7
0.08	10.3	13.2	12.1	18.1	13.4
0.15	13.4	21.3	24.6	31.1	22.6
Mean	9.0	13.6	13.5	18.1	

Quadratic polynomial equations for fitted response surfaces and R^2 values:

Soybean: $VI = 4.44 + 34.19\ O_3 + 19.98\ SO_2$
$R^2 = 0.727$

Tomato: $VI = -75.78 + 20.48\ \ln PI^a - 29.16\ O_3 + 1016\ O_3^2 + 9.02\ SO_2 - 17.29\ SO_2^2 + 258.8\ O_3 \times SO_2$
$R^2 = 0.885$

Maize: $VI = -11.39 + 5.47\ \ln HT^a - 9.59\ O_3 + 490.97\ O_3^2 + 11.81\ SO_2 - 86.63\ SO_2^2 + 428.95\ O_3 \times SO_2$
$R_2 = 0.895$

[a] PI and HT as covariates measured prior to exposure.

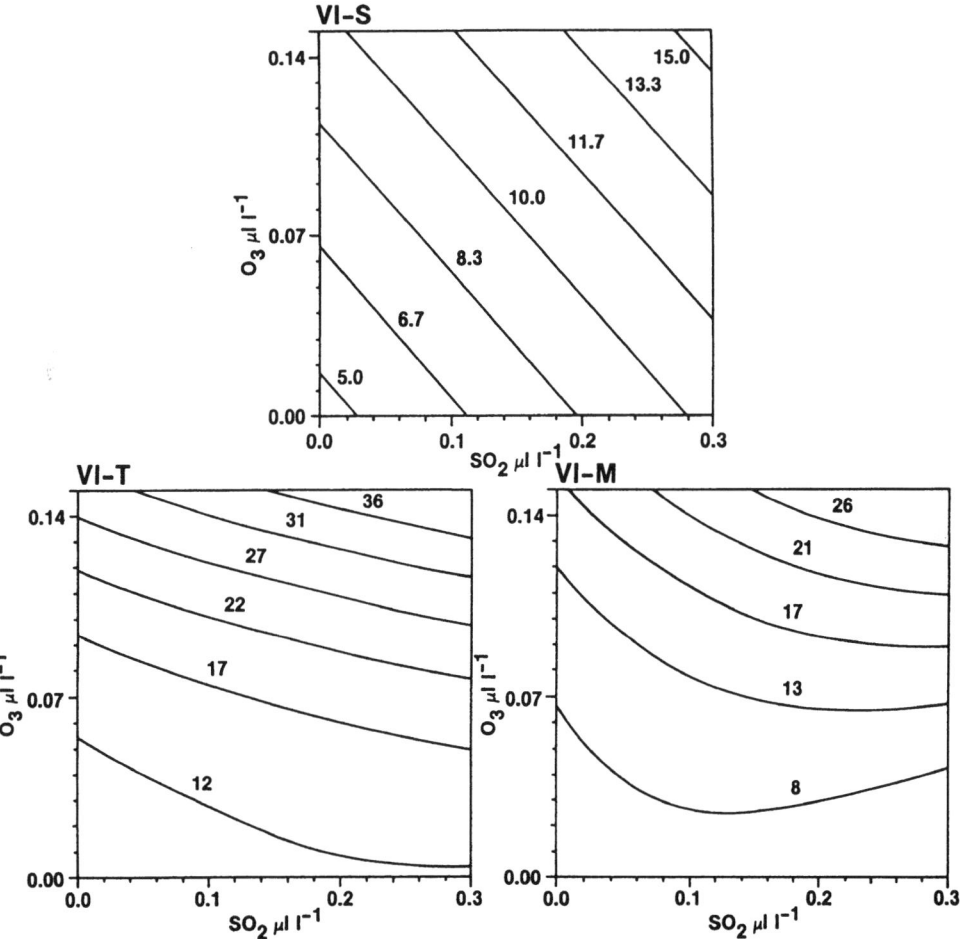

Fig. 1. Contour plots illustrating the visible injury (% of leaf area injured) responses of 'Maple Arrow' soybean (VI-S), 'New Yorker' tomato (VI-T), and 'Golden Jubilee' maize (VI-M) to combinations of O_3 and SO_2. The numbers on the isoeffect lines indicate the fitted value for the line.

A more complex form of joint action was the interaction demonstrated by leaf dry weight (LDW-T) of tomato (Fig. 2). At low O_3 and SO_2 levels, the response was due primarily to SO_2. Once SO_2 reached 0.14 ppm, most of the effect was caused by O_3. Then at high concentrations of both pollutants, there was an approximate linear additive response.

Response patterns were dependent on species and measured response variable as noted in an earlier study to determine the effects of O_3 and SO_2 on injury to crops (Ormrod et al. 1984). Soybean demonstrated an increase in visible injury as the concentrations of the two gases increased (Fig. 1; Table 2). Leaf area was increased at low to medium levels of pollutants and decreased at higher levels (Fig. 3; Table 3). Leaf dry weight increased at medium levels of O_3 and decreased at higher levels of this pollutant (Fig. 2; Table 4). The apparent growth stimulations by low levels of pollutants has had limited attention (Bennett et al. 1974) and will need further study.

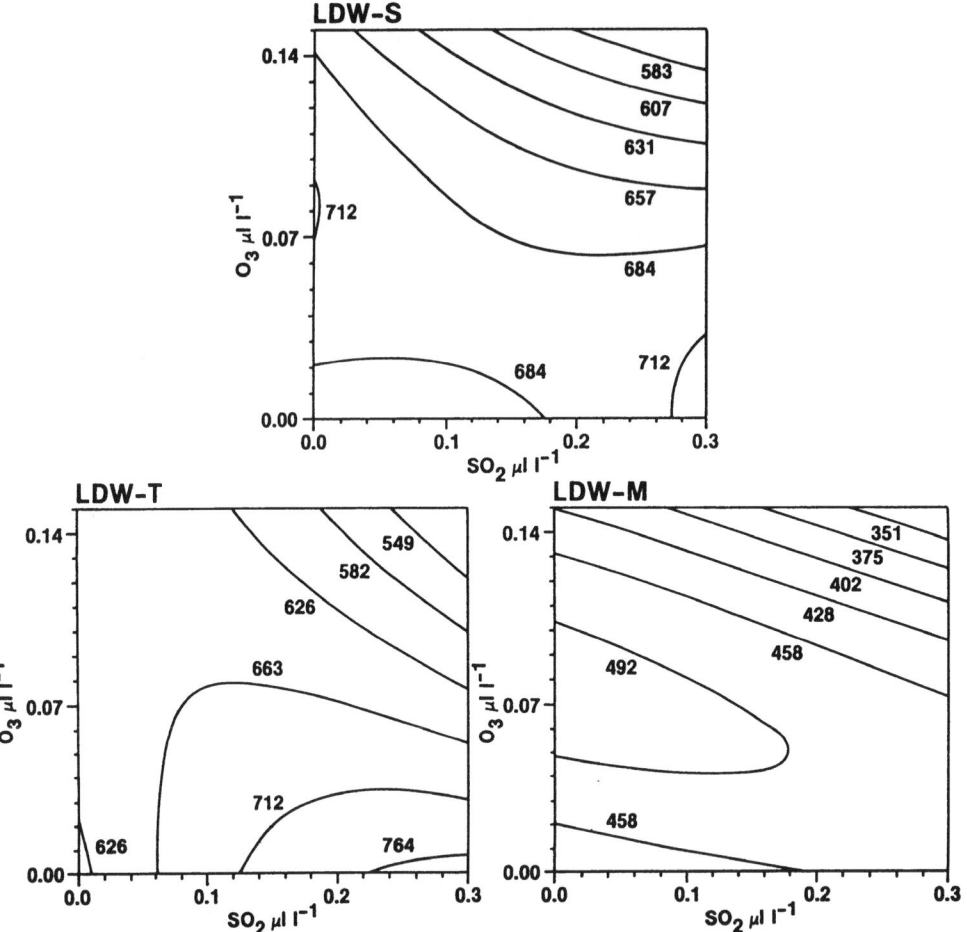

Fig. 2. Leaf dry weight (mg.plant^{-1}) responses of soybean (LDW-S), tomato (LDW-T), and maize (LDW-M) to combinations of O_3 and SO_2.

For tomato there was an increase in visible injury as the levels of both gases increased (Fig. 1; Table 2). Leaf area showed no joint action; however, as O_3 increased, leaf area decreased (Fig. 3; Table 3). As described earlier, leaf dry weight demonstrated a complex interacton.

Visible injury on maize leaves was enhanced by both pollutants (Fig. 1; Table 2). Both leaf area and leaf dry weight were increased at low to medium O_3 concentration and then decreased as the pollutants increased (Fig. 3; Table 3 and Fig. 2; Table 4, respectively). The peak in the response surface for maize leaf dry weight at the 0.08 $\mu l \cdot L^{-1}$ O_3 level was not significant and probably resulted from fitting a quadratic equation to the data.

Table 3. Effects of O_3 and SO_2 mixtures on leaf area ($cm^2 \cdot plant^{-1}$).

O_3 ($\mu L \cdot L^{-1}$)	SO_2 ($\mu L \cdot L^{-1}$)				
	0.00	0.03	0.12	0.30	Mean
'Maple Arrow' Soybean					
0.00	258	268	261	256	261
0.08	279	287	279	280	281
0.15	260	257	279	223	254
Mean	266	270	273	252	
'New Yorker' Tomato					
0.00	251	299	258	335	284
0.08	243	250	293	300	267
0.15	273	205	233	222	232
Mean	255	248	260	282	
'Golden Jubilee' Maize					
0.00	141	136	144	156	144
0.08	154	159	147	155	154
0.15	151	146	141	95	131
Mean	148	147	144	132	

Quadratic polynomial equations for fitted response surfaces and R^2 values:
Soybean: $\ln LA = 5.56 + 2.18\ O_3 - 15.62\ O_3^2 + 0.62\ SO_2 - 2.68\ SO_2^2$
$R^2 = 0.773$
Tomato: $\ln LA = 2.44 + 0.82 \ln PLA^a - 0.025 \ln PI^a + 0.031\ O_3 - 10.44\ O_3^2$
$R^2 = 0.783$
Maize: $\ln LA = 2.87 + 0.61 \ln HT^a + 4.36\ O_3 - 23.66\ O_3^2 + 0.95\ SO_2 - 1.32\ SO_2^2 - 12.32\ O_3 \times SO_2$
$R^2 = 0.860$

[a] PLA, PI, and HT as covariates measured prior to exposure.

As O_3 concentration increased, specific leaf weight of soybean and maize decreased and specific leaf weight of tomato increased (Table 5). The decreases in specific leaf weight were generally associated with a decrease in leaf dry matter accumulation (Table 6) and reduced leaf area (Table 3). Although both leaf dry weight and leaf area decreased in tomato, there was a greater decrease in leaf area than leaf dry weight resulting in an increased specific leaf weight for this species.

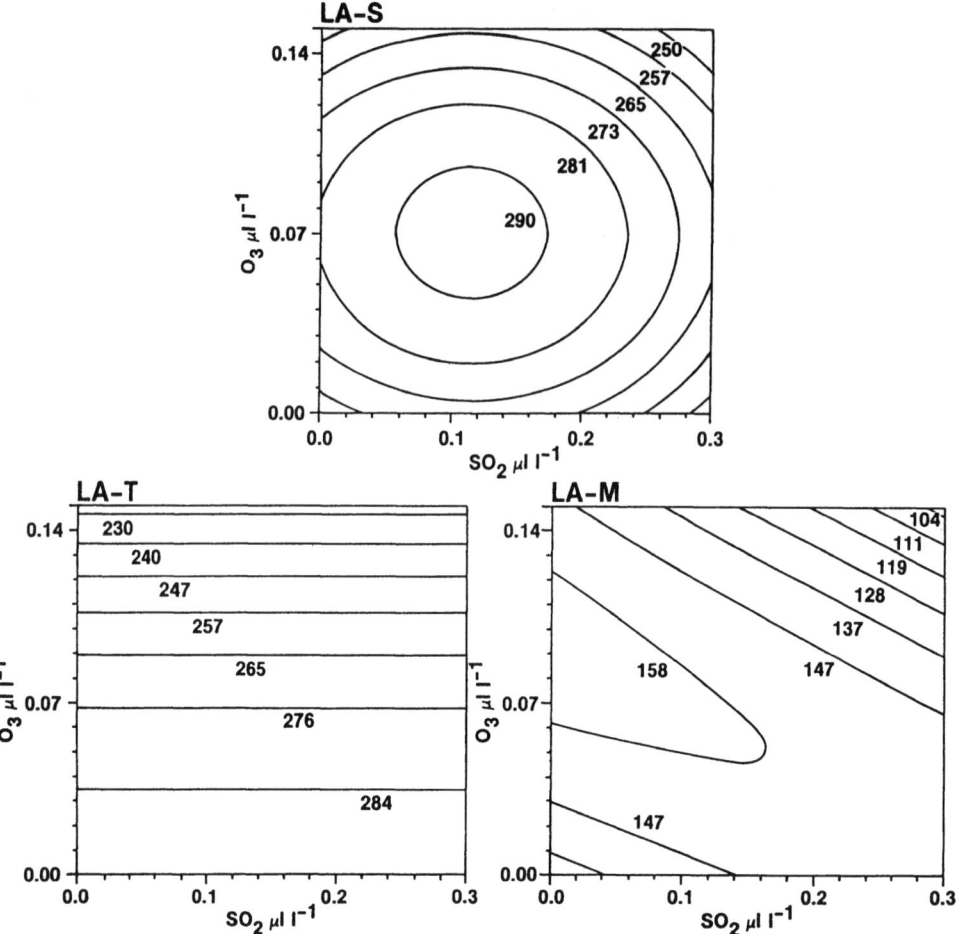

Fig. 3. Leaf area ($cm^2 \cdot plant^{-1}$) responses of soybean (LA-S), tomato (LA-T), and maize (LA-M) to combinations of O_3 and SO_2.

The pollutants had a greater effect on growth than on leaf development as indicated by changes in plastochron indices of only 0.12% to 2.5% (Table 7).

There are limitations to the use of response surface analysis for the interpretation of pollutant effects on plant growth variables. One is that there is a limited number of patterns that this model can produce. Polynomials also do not model thresholds or asymptotes well. For example, plant growth may not be affected when O_3 is below a threshold level but may respond strongly above this level. As a second example, injury cannot exceed 100%. Thus, at high O_3 levels, there must be an asymptote or plateau at 100%. Another limitation is that the model sometimes shows false peak values as a result of fitting a curve to the data set. The latter limitation may explain, in part, the

Table 4. Effects of O_3 and SO_2 mixtures on leaf dry weight (mg.plant^{-1}).

O_3 (µL.L^{-1})	SO_2 (µL.L^{-1})				
	0.00	0.03	0.12	0.30	Mean
'Maple Arrow' Soybean					
0.00	608	711	683	708	678
0.08	719	712	650	673	689
0.15	667	680	594	552	623
Mean	665	701	642	644	
'New Yorker' Tomato					
0.00	580	655	698	791	681
0.08	607	730	625	622	646
0.15	625	727	614	506	618
Mean	604	704	646	640	
'Golden Jubilee' Maize					
0.00	457	390	431	476	439
0.08	553	521	461	474	502
0.15	346	488	387	301	381
Mean	452	466	426	417	

Quadratic polynomial equations for fitted response surfaces and R^2 values:

Soybean: $\ln LDW = -0.42 + 1.87\, O_3 - 11.53\, O_3^2 + 0.055\, SO_2 + 0.79\, SO_2^2 - 6.36\, O_3 \times SO_2$
$R_2 = 0.407$

Tomato: $\ln LDW = -0.72 + 0.16 \ln PLA^a - 0.095 \ln PI^a + 0.67\, O_3 - 0.93\, O_3^2 + 1.47\, SO_2 - 2.28\, SO_2^2 - 11.53\, O_3 \times SO_2$
$R_2 = 0.761$

Maize: $\ln LDW = -1.64 + 0.23 \ln HT^a + 4.44\, O_3 - 29.18\, O_3^2 + 0.57\, SO_2 - 0.81\, SO_2^2 - 8.67\, O_3 \times SO_2$
$R_2 = .0700$

[a] PLA, PI, and HT as covariates measured prior to exposure.

apparent stimulation of growth by O_3 as noted for leaf dry weight of maize. An alternative model, the Weibull dose response model, which is more flexible and biologically realistic, has been proposed for the analyses of dose response data (Rawlings and Cure 1985).

CURRENT AND FUTURE RESEARCH

A further aim of this research project was to develop more efficient experimental designs to permit the full evaluation of interactions and dose-response patterns. The data set reported herein will be used to test various alternative designs in terms of suitability as alternatives to the full factorial design.

Table 5. Effects of O_3 and SO_2 mixtures on specific leaf weight ($mg.cm^{-2}$).

O_3 ($\mu L.L^{-1}$)	SO_2 ($\mu L.L^{-1}$)				
	0.00	0.03	0.12	0.30	Mean
'Maple Arrow' Soybean					
0.00	2.35	2.65	2.61	2.77	2.60
0.08	2.58	2.48	2.33	2.40	2.45
0.15	2.56	2.65	2.13	2.47	2.45
Mean	2.50	2.59	2.36	2.55	
'New Yorker' Tomato					
0.00	2.33	2.23	2.70	2.38	2.41
0.08	2.45	2.89	2.10	2.08	2.38
0.15	2.28	3.52	2.68	2.26	2.69
Mean	2.35	2.88	2.49	2.24	
'Golden Jubilee' Maize					
0.00	3.66	2.66	2.93	3.48	3.18
0.08	3.32	3.00	2.78	3.02	3.03
0.15	2.29	3.27	3.09	3.31	2.99
Mean	3.09	2.98	2.93	3.27	

Table 6. Effects of O_3 and SO_2 mixtures on dry matter content of leaves (%).

O_3 ($\mu L.L^{-1}$)	SO_2 ($\mu L.L^{-1}$)				
	0.00	0.03	0.12	0.30	Mean
'Maple Arrow' Soybean					
0.00	13.6	15.1	15.9	15.0	14.9
0.08	15.5	15.2	14.8	13.8	14.8
0.15	15.9	16.1	12.5	14.5	14.8
Mean	15.0	15.5	14.4	14.4	
'New Yorker' Tomato					
0.00	7.96	7.85	9.52	8.60	8.48
0.08	8.75	9.99	8.34	7.55	8.66
0.15	7.82	11.50	9.35	8.37	9.26
Mean	8.18	9.78	9.07	8.17	
'Golden Jubilee' Maize					
0.00	8.34	6.03	6.94	7.84	7.29
0.08	7.86	7.08	6.89	7.01	7.21
0.15	4.88	9.08	8.51	8.79	7.82
Mean	7.03	7.40	7.45	7.88	

An experiment was conducted recently to determine how growth of leaves and stems responded to the stress of O_3 and/or SO_2 and if any components could adapt or "compensate" (Bennett and Runeckles 1977) for the stress during the vegetative growth period and whether the adaptation was dependent on the species, length of exposure, or the existence of pollutants as single gases or as a mixture of the two gases. Obvious evidence of adaptation or compensatory growth was not found. This indicates that either this phenomenon does not exist in a major way in these cultivars or that the growth parameters used were inappropriate for the detection of compensatory growth.

The objective of a subsequent experiment is to determine the effects of sequential exposures to gaseous pollutants on the growth of crops. Plants are exposed to O_3 for seven days. During that period, SO_2 is introduced every second day in one-day doses. Following the completion of this experiment, the procedure will be repeated except that the order of presentation of the two pollutants will be reversed. This approach is designed to determine how the order of presentation of the gases affects response. These studies represent a partial response to the recent compilation of research recommendations for gaseous air pollutant mixtures (Lefohn and Ormrod, 1984).

Table 7. Effects of O_3 and SO_2 mixtures on the plastochron index.

O_3 ($\mu L \cdot L^{-1}$)	SO_2 ($\mu L \cdot L^{-1}$)				
	0.00	0.03	0.12	0.30	Mean
'Maple Arrow' Soybean					
0.00	6.30	6.37	6.49	6.26	6.36
0.08	6.49	6.38	6.46	6.37	6.43
0.15	6.25	6.41	6.56	6.30	6.38
Mean	6.35	6.39	6.50	6.31	
'New Yorker' Tomato					
0.00	8.42	8.64	8.69	8.46	8.55
0.08	8.56	8.52	8.59	8.19	8.47
0.15	8.84	8.70	8.44	8.43	8.60
Mean	8.61	8.62	8.57	8.36	

Quadratic polynomial equations for fitted response surfaces and R^2 values:

Soybean: $\ln PI = 4.18 + 0.63 \ln PI^a + 2.53 SO_2 - 8.73 SO_2^2$
$R^2 = 0.874$

Tomato: $\ln PI = 4.54 + 0.12 \ln PLA^a + 0.85 \ln PI^a - 0.22 SO_2 - 2.02 SO_2^2$
$R^2 = 0.613$

[a] PI and PLA as covariates measured prior to exposure.

REFERENCES

Bennett JP, Runeckles VC (1977) Effect of low levels of ozone on growth of crimson clover and annual ryegrass. Crop Sci 17: 443-445

Bennett JP, Rush HM, Runeckles VC (1974) Apparent stimulations of plant growth by air pollutants. Can J Bot 52: 35-41

Hammer PA, Tibbitts TW, Langhans RW, McFarlane JC (1978) Baseline growth studies of 'Grand Rapids' lettuce in controlled environments. J Am Soc Hort Sci 103: 649-655

Hesketh HE (1972) Understanding and controlling air pollution. Ann Arbor Science, Ann Arbor, Michigan

Hofstra G, Ormrod DP (1977) Ozone and sulphur dioxide interaction in white bean and soybean. Can J Plant Sci 57: 1193-1198

Lefohn AS, Ormrod DP (1984) A review and assessment of the effects of pollutant mixtures on vegetation - Research recommendations. US Environ Prot Agency EPA-600/3-84-037, p 104

Lefohn AS, Tingey DT (1984) The co-occurrence of potentially phytotoxic concentrations of various gaseous air pollutants. Atmos Environ 18: 2521-2526

Ormrod DP (1982) Air pollutant interactions in mixtures. In: Unsworth MH, Ormrod DP (eds) Effects of gaseous air pollution in agriculture and horticulture. Butterworths, London, p 532

Ormrod DP, Tingey DT, Gumpertz M (1983) Covariate measurements for increasing the precision of plant response to O_3 and SO_2. HortScience 18: 896-898

Ormrod DP, Tingey DT, Gumpertz ML, Olszyk DM (1984) Utilization of a response-surface technique in the study of plant responses to ozone and sulfur dioxide mixtures. Plant Physiol 75: 43-48

Rawlings JO, Cure WW (1985) The Weibull function as a dose-response model to describe ozone effects on crop yield. Crop Sci 25: 807-814.

Treshow M (ed) (1984) Air pollution and plant life. Wiley, London, p 486

STOMATE-DEPENDENT AND STOMATE-INDEPENDENT UPTAKE OF NO_x, AND EFFECTS ON PHOTOSYNTHESIS, RESPIRATION AND TRANSPIRATION OF POTTED PLANTS

H. Saxe

Institute of Plant Physiology and Anatomy, Royal Veterinary and Agricultural University, Thorvaldsensvej 40,
DK-1871 Frederiksberg C, Denmark

ABSTRACT

Cultivars of the most commonly grown potted plants in Danish commercial greenhouses were exposed to CO_2, NO, NO_2 gases, alone or in combination. Gas uptake and effects on net photosynthesis, dark respiration and transpiration were measured before, during and after the exposures. Carbon dioxide stimulated net photosynthesis (41%) and subsequent dark respiration (24%), while transpiration was reduced (10%). The NO exposure of 1 $\mu L.L^{-1}$ NO reduced photosynthesis (20%) and transpiration (18% at high CO_2), but did not affect respiration. NO_2 rarely had significant effects. All the effects of the gases on photosynthesis and transpiration were reversible and had independent mechanisms, while effects on respiration were non-reversible in clean air.

NO_2 uptake by the eight cultivars was related to transpiration, i.e., open stomates, while NO uptake was not dependent on stomatal opening. NO was taken up at a constant rate throughout the light period, while NO_2 uptake decreased towards the end of the day, as did transpiration. Uptake of NO_2 in the dark was reduced as much as transpiration.

At the same concentration, NO was four times more inhibitory to photosynthesis than NO_2. Relative to uptake, NO was 22 times more toxic than NO_2. NO must therefore have a potent mechanism of toxicity, different from that of NO_2.

The higher NO-dose, which reduced transpiration because the stomata closed, reduced photosynthesis less than a lower NO-dose. Since total NO uptake was not dependent on stomatal aperture, but stomatal closure reduced the NO-effect on photosynthesis, it could be suggested that stomata regulate the effective NO uptake (i.e. the small fraction affecting photosynthesis).

The difference in the sensitivity of the cultivars to NO and NO_2 did not correlate with the uptake of either pollutant gas.

Short-term effects (four days) of CO_2 + NO on photosynthesis were significantly correlated ($p < 0.05$) with previously observed long-term effects (four to five months) on the dry weight of the same cultivars in similar treatments.

INTRODUCTION

There is relatively little information in recent literature on the effects of NO_x (NO + NO_2) on plants. This is probably because NO_2 is less toxic than SO_2, and because the effects of NO_2 most often occur in combination with those of SO_2 (Carlson 1983: Freer-Smith 1985: Whitmore 1985) or at high concentrations of NO_2 alone (Reinert et al. 1982). Due to circumstantial evidence, NO has been considered even less harmful than NO_2, partly because of its very low water solubility, which results in low uptake (Mansfield and Freer-Smith 1981) and partly because ozone easily converts NO to NO_2 locally around plants.

The NO_x emissions in industrialized countries are steadily increasing, while SO_2 emissions have been reduced due to legislative regulation over the past decade. Several countries now experience levels of NO_x emission equal to or greater than those of SO_2 (Lane and Bell 1984a; Martin and Barber 1984). In the rural environment, medium distance transport of NO from motor vehicles (Martin and Barber 1984) and local emissions from fertilized soils (Johansson 1984) are the primary contributors to emission levels.

In northern latitudes such as Scandinavia, England and Canada, where ozone concentrations are relatively low, NO to NO_2 conversion is slow. Rural concentrations of NO may be nearly as high as the NO_2 concentrations (Martin and Barber 1984), and NO is the dominant species in urban air (Lane and Bell 1984b). The possible toxicity of NO should therefore be considered, in spite of its low water solubility. Effects of NO have, with few exceptions, (Lane and Bell 1984b) been studied only at high concentrations, where it significantly inhibits plant growth (Law and Mansfield 1982; Mortensen 1984; Saxe and Christensen 1985). Only a part of any NO_x emission penetrates the plant leaf cells and affects the photosynthetic process. The relative toxicity of NO and NO_2 must therefore be based on the proportion of the pollutant gas taken up, which affects physiological and biochemical processes, i.e., effective uptake.

This paper reports on differential uptake mechanisms of NO and NO_2, and the effects of NO (with or without CO_2-enrichment) and NO_2 relative to atmospheric concentration and to total uptake. The high concentrations applied are similar to those found in commercial greenhouses with direct hydrocarbon (kerosene, propane, methane) combustion for CO_2 enrichment. In this environment, where ozone is neglible, the NO produced by the burner makes up 95-100% of the NO_x (Saxe and Christensen 1985). Since potted plants are the most important greenhouse crop in Denmark, eight of its most important species and cultivars were included in the study.

MATERIALS AND METHODS

Plant Growth

Plants were obtained from commercial growers and kept at least 3 days prior to the experiments in a controlled environment similar to that to be used during the experiments. The plants were Ficus elastica 'Robusta', F. benjamina, Hedera helix 'Anne Marie', H. canariensis 'Montgomery', Hibiscus rosa-sinensis 'Red', H.

rosasinensis 'Moesiana', Dieffenbachia maculata 'Compacta' and Nephrolepis exaltata 'Bostoniensis'.

All plants were grown from cuttings, and used in experiments when actively growing at 8-12 weeks of age. Six to 40 plants in six to eight 11-cm pots with soil made up the canopy in each treatment, and the average canopy area was 35 dm^2.

In the controlled environment, plants were grown under light-dark cycles (12:12, 250 $\mu mol.m^{-2}.s^{-1}$ PAR, saturating photosynthesis) provided by Philips HPI lamps with incandescent supplement. During the light period the average temperature was 26.6°C with 70% RH, and during the dark period it was 23.0°C with 60% RH.

Ecosystem Exposure and Gas Measuement

A diagram of the exposure system is given in Fig. 1. Compressed dry air enters the system through two filters (F_o and F_c) that remove liquid and vaporous oil. The flow is split into separate supplies to roots and aerial parts, through independent reduction valves, giving 20 $L.min^{-1}$ air to roots, and approximately 200 $L.min^{-1}$ air to aerial parts. Air to roots is evenly distributed to all pot-soil-systems through a single purgemeter with a needle valve. Air to aerial parts is humidified by passing it over lukewarm water in three successive 40 L tanks encased in a thermostated (30°C) polystyrene box. The resultant relative humidity depends on temperature and air flow rate. The air then passes a water drain valve (F_w), a dust filter, a charcoal filter (F_c) and a Purafil filter (F_p) for removal of background air pollution.

The clean air is diverted into 3 $L.min^{-1}$ reference air, which is fed directly to the gas selector through a precision needle valve and 40 $L.min^{-1}$ to each of five 100 L glass exposure chambers through five with separate needle valves. The flow to the two chambers receiving additional CO_2 is combined in a 120 L mixing chamber, and a sample of the air going to each of the high CO_2 assimilation chambers can be selected as reference air for these chambers. There is a mixing fan (f) at the air inlet. Further distribution of air from chambers and reference air is controlled by an 8-channel gas selector.

Any selected gas channel is diverted in two directions. One goes directly to a TECO chemiluminescent $NO-NO_2-NO_x$ analyser. The other goes to a sealed Vaisala humidity probe in a thermostat waterbath at 30.0°C. Linear temperature transducers register waterbath and growth chamber temperature, permitting precise calculations of the water content of the chambers relative to the reference air (i.e., evapo-transpiration), and the RH% at given temperatures. Prior to IRGA CO_2-measurement the water vapour of the sample air is reduced to a constant, low level on a peltier-cooled (4°C) plate (P).

The Real Time Clock of a Gemini Multiboard microcomputer selects the open/closed state of all solenoid valves indicated by an asterisk and the light period in the air-conditioned room where the assimilation chambers are placed. The microcomputer also registers 8 electrical signals every 5 minutes (CO_2, RH%, NO,

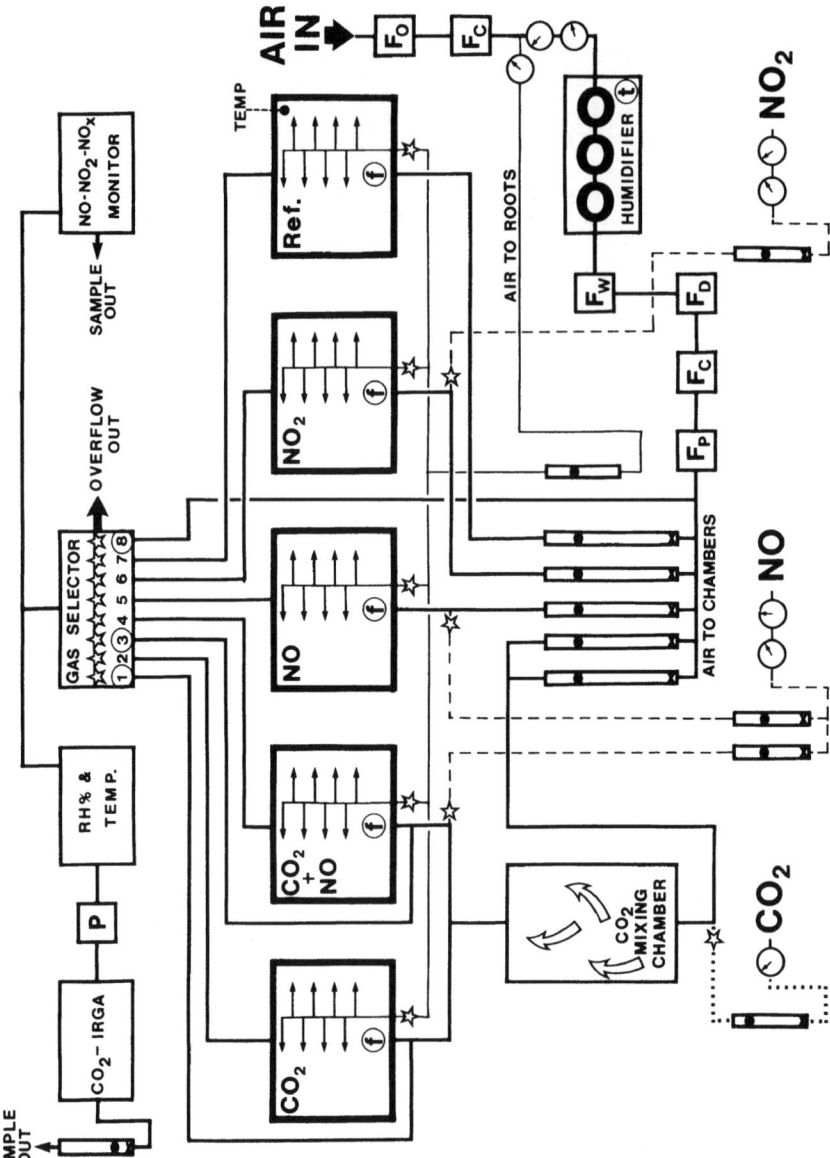

Fig. 1. Computerized gas exposure system with continuous measurements of net photosynthesis, dark respiration, transpiration and gas uptake. Details are given in text.

NO_2, NO_x, waterbath temperature, air temperature near plants and light on/off) through a high-speed 16-bit A/D-converter. Air to roots in a given chamber is turned off by a solenoid valve 10 minutes before air is selected from the chamber. After running the air through the monitoring system for 5 minutes, measurement is made immediately before selecting the next chamber, and an average of 10 measurements from each of the eight electrical signals are stored on a magnetic disc for further calculations. Root air is then turned on for the following 25 minutes. The 40 min. cycle (with its eight different air samples) is then repeated.

NO and NO_2 are dispensed from the pressurized cylinders with 0.1% NO or NO_2 in N_2 and CO_2 from a pressurized cylinder with liquid CO_2. Gases are led to the microcomputer-regulated solenoid valves placed on the air line just before the inlet to the growth or mixing chambers. The NO, NO_2 and CO_2 flow to individual chambers is controlled by purgemeters with precision needle valves. A reference chamber receives only clean air.

Exposure

Exposure was either given as a 1 $\mu L.L^{-1}$ NO (with or without 1 $mL.L^{-1}$ CO_2) or NO_2 throughout the 12 h light period of day 3-6 in 7-day experiments, or as approximately 4 $\mu L.L^{-1}$ for 5 h in the middle of a 12 h light period and in the middle of the following 12 h dark period. The days or hours of clean air before exposure indicate reference activity of individual canopies, and the days or hours after exposure reveal the extent of recovery in clean air. Experiments with each of the eight species and cultivars were repeated 2-4 times.

Calculations

Gas uptake, net photosynthesis, dark respiration and transpiration were registered with one value per cycle (30 or 40 min) by the differential gas-signals of chamber air and the relevant reference air. Values were transformed to the center-time of each cycle by linear regression.

For some purposes, the values were further transformed in a double-relative manner: first, the values of each chamber in each cycle during the four exposure days and the recovery before gas treatments (this reference is called zero); then, each of these relative values in the polluted chambers was calculated relative to values in the clean air chamber. All values were corrected for individual canopy leaf areas and air flow rates of individual chambers, with a final correction for effects due to the living soil-root-system by subtracting blind-experiments run with aerial plant parts cut away, or the pots removed. The double-relative values in themselves indicate the direct gas effects on canopies, excluding effects of natural variation between individual canopies in physiological and morphological characteristics, continuous natural senescence and possible chamber effects.

Statistical Analysis

In the evaluation of the significance of gas effects on photosynthesis, respiration and transpiration, both F- and t-tests for any plant and treatment were based on the degree of freedom and mean square for error between replicates in an analysis of variance with crossed sampling (Snedecor and Cochran 1976), i.e., the tests were based not only on the variance between individual measurements, but also on the variance between replicate experiments. For F-tests 680 values and for t-tests 272 values were used in each test.

All correlations between gas uptake, transpiration and photosynthesis, and between effects of short-term and long-term exposures, were evaluated by the significance of a Spearman rank correlation coefficient (Siegel 1956), based on either 17 sets of values each day in the 4-day exposure and on 8 sets of values in the 5-hour exposures, or on the average values for each of the 8 species and cultivars.

RESULTS AND DISCUSSION

Net Photosynthesis

NO and CO_2 affected photosynthesis significantly (Fig. 2, Table 1). On average, in all the species and cultivars studied, a concentration of 1 $mL.L^{-1}$ clean CO_2 increased photosynthesis by 40.9% relative to plants in untreated air. CO_2 + NO also increased photosynthesis but less than the clean CO_2 treatment. The reduction in photosynthesis due to NO, both in combination with CO_2 and alone, was about 20%, though the reaction of individual species and cultivars was very variable. Clearly some plants were resistant to NO pollution, while others were not. In accordance with earlier findings (Carlson 1983), NO_2 did not significantly inhibit photosynthesis. It only had significant effects on one of the eight cultivars studied (Table 1), and on the average reduced photosynthesis 3.7 times that of the treatment with NO, at comparable emission concentrations.

The stimulation of photosynthesis by CO_2 (Fig. 2) was seen as a stable plateau, except in one species (both cultivars of Ficus), which "adapted" to the high CO_2, and gradually lost the advantage of high CO_2 during the experiment. The daily pattern of CO_2+ NO closely followed that of clean CO_2, indicating that NO inhibited what CO_2 stimulated. The effect of CO_2 with or without NO was quite reversible, with a significant tendency however, to end up with a lower (average 9%) photosynthetic rate after CO_2 treatment than at the beginning of the treatment.

The effect of either NO or NO_2 alone was consistant (except for Ficus benjamina) with only a slight tendency to increase or decrease during the experiment (Fig. 2). The effects of both gases were completely reversible.

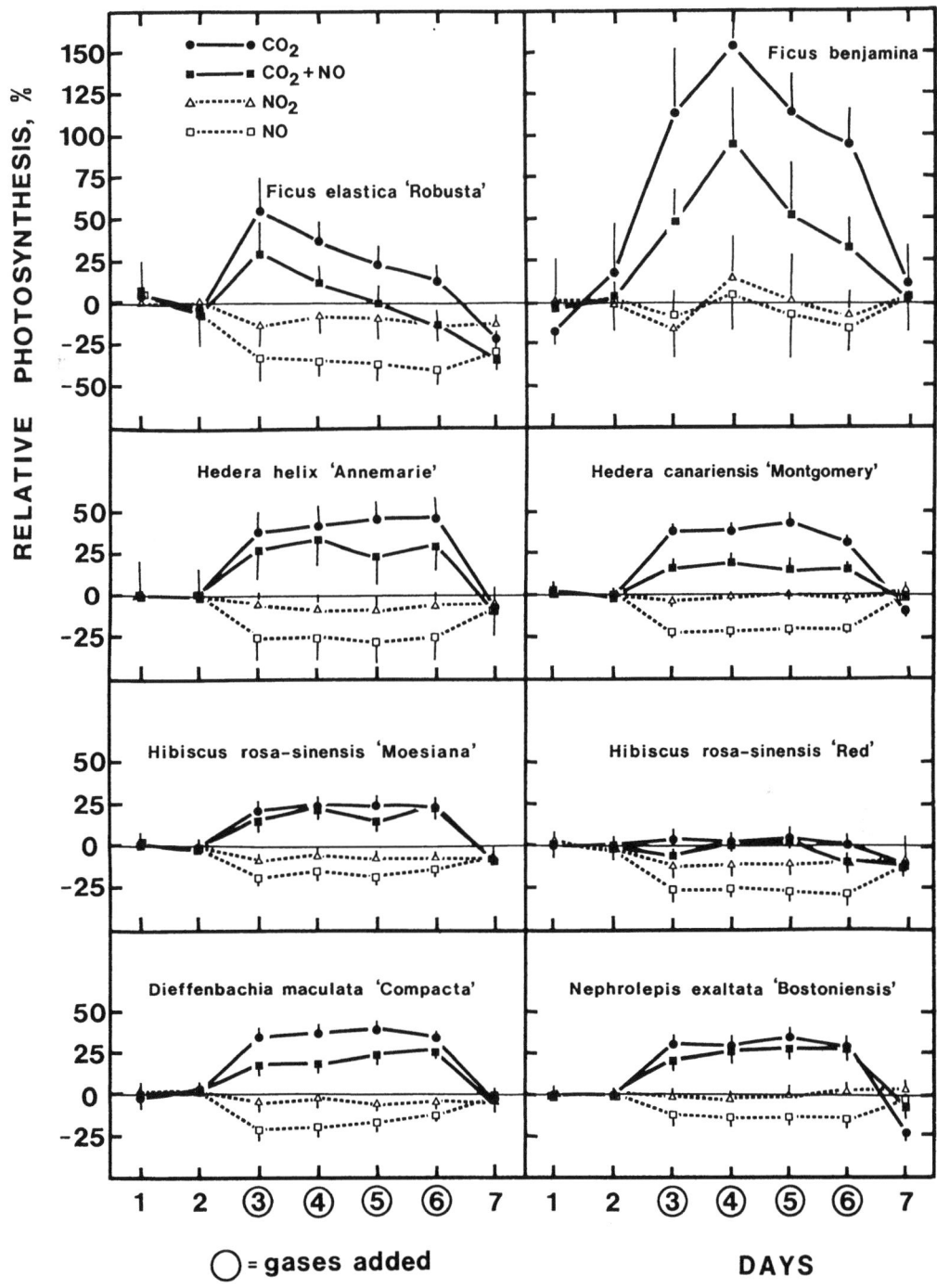

Fig. 2. Average effect during 12-h light periods on photosynthesis of 1 mL.L^{-1} CO_2, 1 mL.L^{-1} CO_2 + 1 µL.L^{-1} NO and 1 µL.L^{-1} NO_2 relative to clean air in a double relative plot. Vertical bars indicate average SD of 2-4 experiments.

Table 1. Effects of 4 days of 12 h or 1 $mL.L^{-1}$ CO_2 + 1 $\mu L.L^{-1}$ NO, 1 $\mu L.L^{-1}$ NO and 1 $\mu L.L^{-1}$ NO_2 relative to a clean air reference (REF), and CO_2 relative to CO_2 + NO, and NO relative to NO_2. Effects are given on net photosynthesis (P), transpiration (T) and dark respiration (R). F- and t- tests are based on crossed sampling. NS: Not significant, * p < 0.05, ** p < 0.01, *** p < 0.001.

Species		F-test	CO_2-REF	CO_2/NO-REF	NO-REF	NO_2-REF	CO_2-CO_2/NO	NO-NO_2
Ficus elastica	P:	NS	+31.8%*	+ 6.8%NS	-36.8%***	-11.5%NS	+25.0%NS	-25.3%*
	T:	NS	- 6.0%NS	-12.7%*	+ 0.5%NS	+ 1.7%NS	+ 6.7%NS	- 1.2%NS
	R:	NS	+11.9%NS	-12.3%NS	- 0.4%NS	+13.7%NS	+24.2%*	-13.3%NS
Ficus benjamina	P:	**	+119.5%***	+57.9%***	- 6.1%NS	- 1.8%NS	+61.6%**	+ 4.3%NS
	T:	NS	+ 5.2%NS	- 5.4%**	+ 6.9%NS	+ 3.2%NS	+10.6%**	+ 3.7%NS
	R:	***	+35.3%***	+32.7%***	+ 2.9%NS	+ 8.4%**	+ 2.8%NS	- 5.5%NS
Hedera helix	P:	**	+43.4%***	+28.6%**	-26.1%***	- 7.7%NS	+14.8%**	-18.4%**
	T:	NS	- 3.9%NS	-19.6%*	+ 1.3%NS	+ 8.3%NS	+15.7%*	- 7.0%NS
	R:	*	+39.7%***	+27.5%**	- 3.8%NS	+ 8.2%NS	+12.2%*	-12.0%NS
Hedera canariensis	P:	***	+37.5%***	+16.0%***	-21.9%***	- 1.7%NS	+21.5%***	-20.2%***
	T:	*	- 4.6%**	-10.8%***	- 3.5%NS	+ 1.4%NS	+ 6.2%NS	- 4.9%*
	R:	*	+18.5%***	+11.4%***	- 2.9%NS	- 1.0%NS	+ 7.1%*	- 1.9%NS
Hibiscus "M"	P:	***	+23.4%***	+19.1%***	-16.5%***	- 7.0%*	+ 4.3%**	- 9.5%*
	T:	**	-16.6%***	-20.7%***	-14.3%***	- 4.7%NS	+ 4.1%NS	- 9.6%**
	R:	NS	+11.1%NS	+ 2.7%NS	- 8.8%NS	- 9.4%NS	+ 8.4%NS	+ 0.6%NS

Table 1 continued.

Species		F-test	CO_2-REF	CO_2/NO-REF	NO-REF	NO_2-REF	CO_2-CO_2/NO	NO-NO_2
Hibiscus "R"	P:	NS	+ 3.5%*	- 2.9%NS	-26.8%**	-10.9%NS	+ 6.4%NS	-15.9%*
	T:	**	- 5.7%****	-17.8%****	- 4.0%***	- 7.8%***	+12.1%****	+ 3.8%*
	R:	NS	+29.5%*	+18.9%*	- 5.1%*	- 4.8%NS	+10.6%NS	+ 0.3%NS
Dieffenbachia	P:	***	+36.5%****	+21.9%****	-17.3%**	- 3.9%NS	+14.6%**	-13.4%**
	T:	***	-40.3%****	-48.7%****	- 3.7%**	- 2.0%NS	+ 8.4%***	+ 1.7%NS
	R:	NS	+12.7%**	+13.9%**	- 1.8%NS	+ 2.5%NS	- 1.2%NS	- 4.3%NS
Nephrolepis	P:	**	+31.5%****	+26.0%****	-13.2%**	- 0.2%NS	+ 5.5%NS	-13.1%**
	T:	NS	-11.6%**	- 8.5%**	- 1.6%NS	+ 2.2%ns	- 3.1%NS	- 2.8%NS
	R:	**	+29.4%****	+22.5%****	- 0.6%NS	+ 8.0%*	+6.9%NS	- 8.6%NS
AVERAGE	P:	**	+40.9%****	+21.7%****	-20.6%****	- 5.6%NS	+19.2%**	-15.0%**
	T:	**	-10.8%***	-18.0%****	- 2.3%NS	+ 0.3%NS	+ 7.2%***	- 2.0%NS
	R:	**	+23.5%****	+14.7%**	- 2.6%NS	+ 7.8%*	+ 8.8%NS	-10.4%NS

Dark Respiration

Elevated levels of clean CO_2 increased subsequent dark respiration significantly (23.5%), even though plants were only exposed to the gas during the day (Table 1). Including NO with CO_2 reduced this effect. The effect of CO_2 with or without NO was to gradually increase dark respiration. The increased respiration was hardly reversed after day-time CO_2 exposure was discontinued, which may explain why net photosynthesis, as mentioned above, at the same time finally ends below zero. The effect on respiration is thought to be a matter of available respiratory substrate.

Neither NO nor NO_2 alone affected dark respiration significantly, and the effects were much smaller than those of CO_2. The only significant NO_2-effect was stimulation, which is not in accordance with effects on soybean (Carlson 1983). There was no evidence of recovery from the slight NO_x effects on respiration.

Transpiration

Clean CO_2 significantly reduced transpiration by, on average, 10.8%. The addition of NO increased this effect to 18.0%, which may explain partly the NO inhibition of photosynthesis at elevated CO_2 levels. The effect of NO and NO_2 alone was negligible (Table 1), in accordance with some earlier finding (Amundson and Weinstein 1981; Natori and Totsuka 1984), but not with all (Carlson 1983).

The daily effects of CO_2 with or without NO typically assumed a somewhat unstable plateau, until a reasonably good recovery was reached on the last day.

Correlation Between Photosynthesis and Transpiration

The correlation between transpiration (stomatal opening) and photosynthesis was significant in all canopies before gas treatment. However, the 4-day exposures to CO_2, CO_2 + NO and NO alone (but not NO_2) reduced the correlation coefficients significantly to a level of no significant correlation. This is explained by the following observations: 1 $\mu L.L^{-1}$ NO decreased photosynthesis without significant effects on transpiration. Reduction of the correlation coefficients during CO_2 treatment was due to increased daily variance of photosynthesis with decreased variance of transpiration.

Gas Uptake

The correlation between average daily patterns of absolute transpiration and gas uptake is illustrated for all eight species and cultivars in Fig. 3. The correlation between transpiration (stomatal opening) and NO_2 uptake was significant ($p < 0.05$) for six plants and highly significant ($p < 0.01$) for two (Hedera canariensis and Nephrolepsis exaltata). The correlation between transpiration and NO uptake was significant ($p < 0.01$) for only one plant (Ficus benjamina) and not significant for the other seven. For all species and cultivars, NO_2 uptake correlated significantly ($p < 0.01$) with transpiration, while the NO uptake did not correlate with transpiration. Uptake of NO_2 was therefore stomate-dependent, while uptake of NO was stomate-independent.

The adsorption of NO_2 to pots and soil was greatest in the morning, but reached a constant, low level within 4 hours, while that of NO was consistently insignificant. Adsorption to dead material therefore does not interfere with the above conclusions.

Five-hour exposures to NO and NO_2 during both the day and the night confirmed that only NO_2 uptake is stomatal-dependent (Table 2). The average of all species and cultivars showed that NO_2 uptake was reduced by 86% during the night. This correlates significantly with an 85% reduction of transpiration. This does not preclude that light/dark conditions affect NO_2 flux in other ways than by stomatal regulation (Murray 1984). The NO uptake during the day was not significantly different from the NO uptake during the night, i.e., NO uptake was not dependent on stomatal aperture. Exposure concentration was the same during the day as it was during the night.

Table 2. Average total canopy area (dm^2), and average gas uptake ($\mu L.dm^{-2}.h^{-1}$) of 5-hour exposures and replicate experiments, with the same emission level day and night.

	dm^2	NO_2		dm^2	NO	
		day	night		day	night
Ficus elastica 'Robusta'	9.1	99.0	11.8	9.9	49.0	58.8
Ficus benjamina	45.1	27.6	2.8	44.1	14.6	14.8
Hedera helix 'Anne Marie'	38.6	41.4	3.0	38.5	15.0	16.4
Hedera canariensis 'Montgomery'	29.5	50.6	8.8	30.1	19.4	22.8
Hibiscus rosa-sinensis 'Moesiana'	27.2	78.4	8.4	26.5	32.6	25.6
Hibiscus rosa-sinensis 'Red'	28.2	94.0	17.4	28.6	30.8	24.2
Dieffenbachia maculata	23.7	44.8	5.8	25.7	22.2	22.8
Nephrolepis exaltata	27.0	32.0	8.2	27.8	19.0	19.0
pots and soil	-	169	47	-	392	378
exposure concentration, $uL.h-1/1800\ L.h^{-1}$	-	7794	7855	-	7276	7488

The average uptake of each cultivar during the 4-day exposures is given in Table 3. On average, the uptake of NO and NO_2 was 3.8 $\mu L.dm^{-2}.h^{-1}$ and 22.3 $\mu L.dm^{-2}.h^{-1}$, respectively. Therefore, the uptake of NO_2 was approximately 6 times that of NO. In the same exposures, photosynthesis was reduced almost 4 times more by NO than by NO_2. NO was therefore (5.9 x 3.7) 22 times more toxic than NO_2 relative to total uptake for all cultivars.

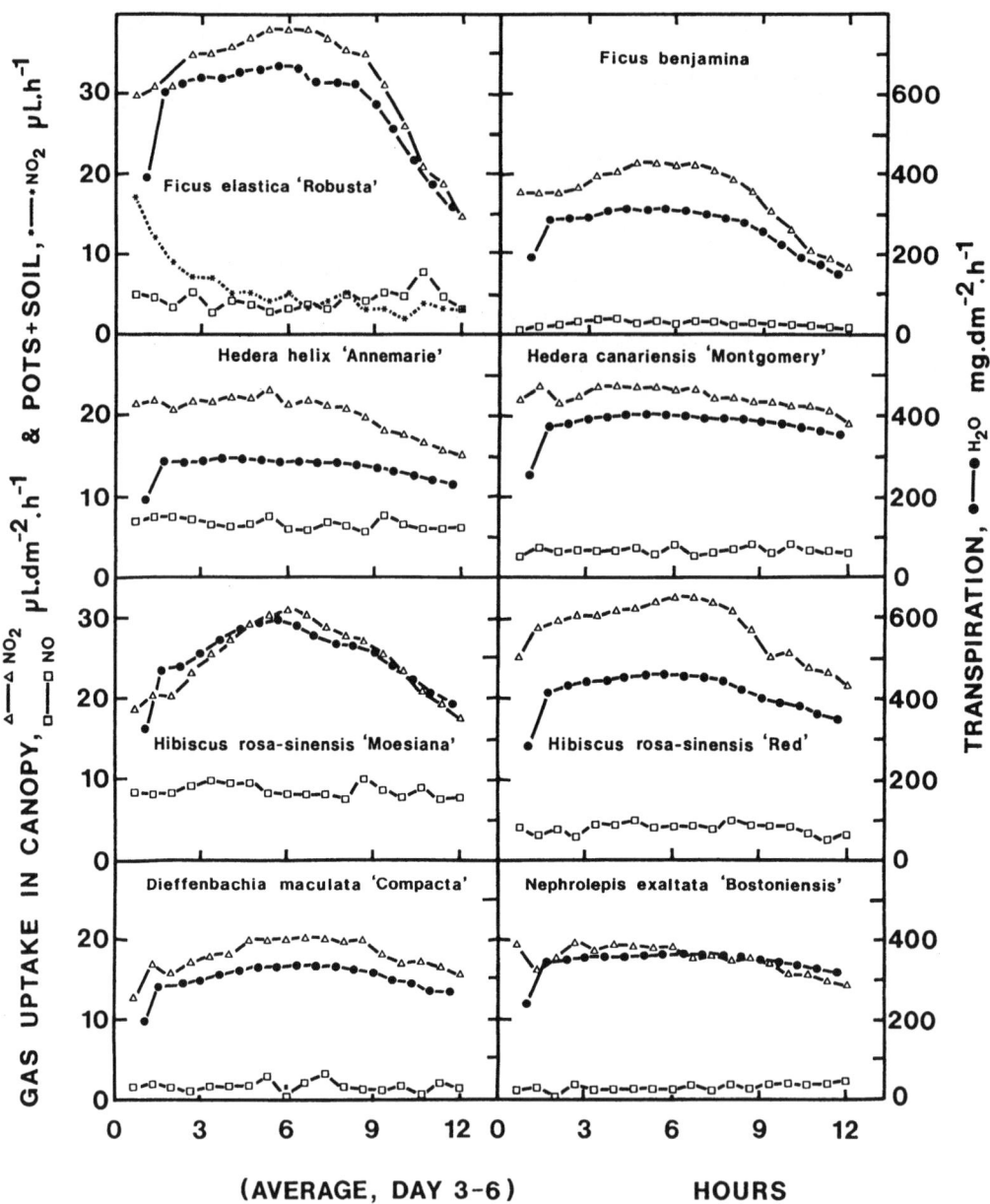

Fig. 3. Daily pattern of gas uptake and transpiration, average of four exposure days and replicate experiments.

Table 3. Average total canopy area (dm^2), and average gas uptake ($\mu L.dm^{-2}.h^{-1}$) of four 12-hour exposure days and replicate experiments, and average total transpiration (mg $H_2O.dm^{-1}.h^{-1}$) for each cultivar.

Species	dm^2	NO_2	dm^2	NO	dm^2	H_2O
Ficus elastica 'Robusta'	19.6	30.9	21.8	4.23	19.6	568
Ficus benjamina	52.2	17.1	56.9	1.37	52.2	265
Hedera helix 'Anne Marie'	46.7	20.2	50.5	6.60	46.7	273
Hedera canariensis 'Montgomery'	33.3	22.2	34.5	3.46	33.3	382
Hibiscus rosa-sinensis 'Moesiana'	24.6	24.8	27.5	8.59	24.6	503
Hibiscus rosa-sinensis 'Red'	33.1	28.6	35.0	3.99	33.1	414
Dieffenbachia maculata	37.9	17.5	39.7	1.33	37.9	296
Nephrolepis exaltata	50.5	17.3	48.4	1.27	50.5	338

In contrast to 4-day exposures, the NO_2 uptake in 5-h exposures was only twice that of the NO uptake, while the inhibition of photosynthesis, which was constant and completely reversible, was nearly the same for both gases (Table 4). Thus, NO was only twice as toxic to photosynthesis as NO_2 at the high exposure concentration relative to total uptake. The main reason for this is that NO_2 tripled its negative effect on photosynthesis primarily because NO_2 caused the stomates to close. Both NO and NO_2 affected stomatal aperture at the high emission level (only), and there is a very significant ($p < 0.01$) correlation between both NO and NO_2 effects on photosynthesis and on transpiration. Whether the primary effect of the gases was on photosynthesis or on transpiration is not clear. The fact remains, however, that the NO effect on photosynthesis was smaller by a third at the high dose (5 h, 3.5 $\mu L.L^{-1}$ NO) than at the low dose (12 h, 1 $\mu L.L^{-1}$, i.e., first day of the 4-day exposures). Since NO closed stomates rapidly within 30 minutes, as did NO_2, I propose, that the effective NO uptake is partly stomate-dependent. Since the total NO uptake is not stomate-dependent, I assume that the effective uptake makes up a very small proportion of the total uptake. If this is true, the specific toxicity of NO relative to NO_2 may be far more than the factor of 22 calculated above. The larger, ineffective portion of the NO uptake is thought to be adsorbed to the outer leaf surface.

The fact that there was no significant correlation between either total NO or total NO_2 uptake and photosynthesis or transpiration (in a cultivar-to-cultivar comparison) contradicts earlier findings by Elkiey and Ormrod (1981 a,b) that uptake correlates with sensitivity. The observed individual sensitivity may instead depend upon cultivar-specific biochemical properties (Wellburn 1982).

Table 4. Effect of 5-hour exposures of approximately 4 $\mu L.L^{-1}$ NO_2 or NO on net photosynthesis (PS) and transpiration (TR) relative to activity of plants in clean air, calculated in a double relative manner.

	NO_2		NO	
	PS	TR	PS	TR
Ficus elastica 'Robusta'	-30%	-37%	-21%	-30%
Ficus benjamina	-36%	-40%	-16%	-16%
Hedera helix 'Anne Marie'	-7%	-6%	-12%	-10%
Hedera canariensis 'Montgomery'	-18%	-11%	-19%	-9%
Hibiscus rosa-sinensis 'Moesiana'	-7%	-4%	-6%	-2%
Hibiscus rosa-sinensis 'Red'	-10%	-5%	-5%	0%
Dieffenbachia maculata	-11%	-11%	-17%	-17%
Nephrolepis exalta	-8%	-11%	-9%	-7%
TOTAL (average)	-16%	-16%	-13%	-11%

Predicting Long-term Sensitivity

The interest in comparing effects of clean CO_2 with those of CO_2 + NO is that these two treatments reflect existing conditions in commercial greenhouses. Clean CO_2 from a pressurized tank is more expensive, but the grower avoids the risk of by-products like NO, ethylene and propylene, which result from CO_2 produced in the greenhouse by hydrocarbon combustion. Ethylene and propylene levels can be minimized by proper servicing of burners. NO is produced due to the high combustion temperature, where atmospheric N_2 and O_2 combine to make NO, which cannot efficiently be reduced below 0.3-1.0 $\mu L.L^{-1}$ in the greenhouse atmosphere.

Since it became clear that NO affects some, but not all, plants (Saxe and Christensen 1985), it was desirable to be able to predict which species and cultivars are sensitive over the long-term to NO-pollution with the CO_2. The long-term effects of NO with CO_2 only rarely gave visible damage, such as leaf scorching. A major purpose of the present study was to determine whether short-term effects (4 days) of CO_2 + NO relative to clean CO_2 correlates with effects of long-term (4-5 months) exposure. From Table 5, it is clear that the long-term/short-term connection is not completely reliable, but the statistical significance ($p < 0.05$) of the correlation coefficient ($R = 0.66$) is surprisingly good.

Effects of 5-hour exposures with 4 $\mu L.L^{-1}$ NO on photosynthesis also correlate ($p < 0.05$) with effects of including 1 $\mu L.L^{-1}$ NO in 4-5 months exposures of the 8 potted plant cultivars grown at 1 $mL.L^{-1}$ CO_2 ($R = 0.69$). However, there is no correlation between long-term effects of CO_2 + NO relative to clean CO_2 and short-term effects of 1 $\mu L.L^{-1}$ NO alone ($R = 0.14$).

Table 5. Comparison of short-term with long-term exposure effects of $1 mL.L^{-1}$ CO_2 and $1 mL.L^{-1}$ CO_2 + $1 \mu L.L^{-1}$ NO relative to a clean air reference (REF) and CO_2 relative to CO_2 + NO. NS: not significant, * p < 0.05, ** p < 0.01, *** p < 0.001.

	CO_2-REF		CO_2/NO-REF		CO_2-CO_2/NO	
	4 days	4-5 months	4 days	4-5 months	4 days	4-5 months
Ficus elastica	+32*	+34*	+ 7NS	- 3NS	+19	+28
Ficus benjamina	+120***	+25*	+58***	- 6NS	+27	+25
Hedera helix	+43***	+28**	+29**	+ 6NS	+10	+17
Hedera canariensis	+38***	+ 9NS	+16***	+ 8NS	+16	+ 1
Hibiscus rosa-sinensis 'M'	+23***	+33*	+19***	+31*	+ 3	+ 2
Hibiscus rosa-sinensis 'R'	+ 3*	+30*	- 3NS	+35*	+ 6	-4
Dieffenbachia maculata	+37***	+16*	+22***	-15*	+11	+27
Nephrolepis exaltata	+32***	+15*	+26***	+19**	+ 5	- 3

CONCLUSIONS

Short-term effects on photosynthesis of NO with CO_2 predict the long-term effect on dry weight under the same conditions, i.e., the "commercial sensitivity" of potted plants to combustion (NO)-CO_2 versus clean CO_2. Based on the present results and those of Saxe and Christensen (1985), I recommend that growers use CO_2-enrichment whenever possible, but avoid combustion-CO_2 without at least a short-term trial of the CO_2/NO-effects on the particular crop.

Total NO_2 uptake depends on stomatal opening, while total NO uptake is independent of stomates. However, there is reason to believe that effective NO uptake (NO uptake that affects photosynthesis) is indeed stomate-dependent. In that case, the effective NO-uptake constitutes only a very small fraction of the total NO-uptake. Total NO or NO_2 uptake does not correlate with species/cultivar sensitivity.

NO easily converts to NO_2, but the fact that NO is more toxic than NO_2 indicates that NO, contrary to past and current belief (Lane and Bell 1984b), does not affect the plant after being transformed to NO_2 or any of its reaction products.

NO on the average is 4 times more toxic relative to emission and 22 times more toxic relative to NO_2 at comparable emissions. Also in many areas of plant growth, the occurrence of NO is as frequent as NO_2 and NO_x emission is generally increasing. As a result, I recommend that more attention be paid to NO_x-effects, not only in the greenhouse environment, in the future.

Acknowledgements

The author wishes to thank Kristian Kristensen, A. Madsen, R. Rajagopal and A.S. Andersen for helpful discussions on this paper, and the staff of our physics department for constructing the computer system. The work was supported by the Danish Agricultural and Veterinary Research Council J. No. 13-3031/3252/3180/3345/3504, Fa. Union, Thomas B. Thriges Fond, Tuborg Fonden, Esso Chemicals A/S. Kai Hansens Fond and Svejsecentralen.

REFERENCES

Admuson RG, Weinstein LH (1981) Joint action of sulfur dioxide and nitrogen dioxide on foliar injury and stomatal behavior in soybean. J Environ Qual 10 (2): 204-206

Carlson RW (1983) Interaction between SO_2 and NO_2 and their effects on photosynthetic properties of soybean Glycine max. Environ Pollut (A) 32: 11-38

Elkiey T, Ormrod DP (1981a) Sorption of O_3, SO_2, NO_2 or their mixture by nine Poa pratensis cultivars of differing pollutant sensitivities. Atmos Environ 15 (9): 1739-1743

Elkiey T, Ormrod DP (1981b) Absorption of ozone, sulphur dioxide, and nitrogen dioxide by Petunia plants. Environ Exper Bot 21: 63-70

Freer-Smith PH (1985) The influence of SO_2 and NO_2 on the growth, development and gas exchange of Betula pendula Roth. New Phytol 99: 417-430

Johansson C (1984) Field measurements of emission of nitric oxide from fertilized an unfertilized forest soils in Sweden. J Atmos Chem 1: 429-442

Lane PI, Bell JNB (1984a) The effects of simulated urban air pollution on grass yield: Part I - Description and simulation of ambient pollution. Environ Pollut (B) 8: 245-263

Lane PI, Bell JNB (1984b) The effects of simulated urban air pollution on grass yield: Part 2 - Performance of Lolium perenne, Phleum pratense and Dactylis glomerate fumigated with SO_2, NO_2 and/or NO. Environ Pollut (A) 35: 97-124

Law RM, Mansfield TA (1982) Oxides of nitrogen and the greenhouse atmosphere. In: Unsworth MH, Ormrod DP (eds) Effects of gaseous air pollution in agriculture and horticulture. Butterworth, London, p 93-112

Mansfield TA. Freer-Smith PH (1981) Effects of urban air pollution on plant growth. Biol Rev 56: 343-368

Martin A, Barber FR (1984) Acid gases and acid in rain monitored for over 5 years in rural east-central England. Atmos Environ 18(9): 1715-1724

Mortensen LM (1984) The effect of nitrogen oxides (NO_x during CO_2 enrichment on some greenhouse plants. Acta Horti 162: 285-287

Murray AJS (1984) Light affects the deposition of NO_2 to the Flacca mutant of tomato without affecting the rate of transpiration. New Phyto 98: 447-450

Natori T, Totsuka T (1984) Effects of mixed gas on transpiration rate of several woody plants. 2. Synergistic effects of mixed gas on transpiration rate of Euonymus japonica. Research Reports from the National Institute for Environmental Studies, Japan, 65

Reinert RA, Shriner DS and Rawlings JO (1982) Responses of radish to all combinations of three concentrations of nitrogen dioxide, sulphur dioxide, and ozone. J Environ Qual 11(1): 52-57

Saxe H, Christensen OV (1985) Effects of carbon dioxide with and without nitric oxide pollution on growth, morphogenesis and production time of pot plants. Environ Pollut (A) 38: 159-169

Siegel S (1956) Nonparametric statistics for the behavioural sciences, McGraw-Hill, London, p 202-213

Snedecor GW, Cochran WG (1976) Statistical methods, 6th edn, New York, Iowa State University Press

Wellburn AR (1982) Effects of SO_2 and NO_2 on metabolic function. In: Unsworth MH, Ormrod DP (eds) Effects of gaseous air pollution in agriculture and horticulture. Butterworth, London, p 169-187

Whitmore ME (1985) Relationship between dose of SO_2 and NO_2 mixtures and growth of Poa pratensis. New Phytol 99: 545-553

THE EFFECTS OF ACID RAIN, ALONE AND IN COMBINATION WITH GASEOUS POLLUTANTS, ON GROWTH AND YIELD OF CROP PLANTS

D.S. Shriner and J.W. Johnson, Jr.

Environmental Sciences Division, Oak Ridge National Laboratory, Oak Ridge, TN 37831, U.S.A.

ABSTRACT

Greenhouse, growth chamber, and field experiments were conducted to determine the response of crop plants to levels of acidity in simulated rain. The major objectives of this research effort were: (1) to determine the levels of acidity in rain that alter crop productivity, (2) to evaluate varietal differences in crop response, and (3) to determine the response of crop plants to the combined stress of acid rain and gaseous pollutants [primarily ozone (O_3)]. In the greenhouse, plants were exposed to either four or six levels of acidity twice per week with a rain simulator. In the growth chamber experiments, gaseous pollutant fumigations were conducted either once or three times weekly with a continuously stirred tank reactor exposure system. In the field, open-top field chambers were used in conjunction with an automatic rain exclusion-simulant distribution system to expose crops to controlled levels of gaseous pollutants and rain acidity. In an experiment with tomato cultivars, pH 2.3 rain applied to both foliage and soil reduced growth, while the same solution applied to the soil without foliar contact increased growth. The growth response of radish and alfalfa plants to the combined stress of acid rain and ozone was additive. The visible injury response of younger radish leaves to combinations that included pH 3.3 rain and 0.2 and 0.4 ppm of O_3 was less than additive. In field studies, Davis soybeans showed no significant growth or yield responses to ambient O_3 levels, or simulated rain at pH 5.2, 4.2 or 3.2. There was a small (-3%) rain treatment (pH) effect on weight per seed in the pH 4.2 simulated rain and the ambient rain treatments (also ~pH 4.2).

Several conclusions may be drawn from this research. Differences in sensitivity among crop cultivars may be an important factor in efforts to assess the impact of acid rain on crop productivity. Acid rain may affect plant growth by a direct effect on the foliage, or indirectly through an effect on soil. The net effect of acid rain on plant growth may be the result of offsetting hydrogen ion-induced negative effects and nitrate- and sulphate-induced nutritional effects. If so, the nutrient status of vegetation is an important determinant of the direction and magnitude of response to acid rain. While questions remain unanswered regarding the range of potential interactions between acid rain and gaseous pollutants, our research to date suggests that plant response to combinations of gaseous pollutants and acid rain is more often additive than interactive.

INTRODUCTION

Researchers assessing the short- and long-term consequences of acidic precipitation on terrestrial and aquatic ecosystems now recognize that vegetation is seldom exposed to phytotoxic doses of a single pollutant in isolation from other pollutants. As a result, assessment of the importance or significance of an individual pollutant should likewise take into consideration the multiplicity of pollutant combinations potentially involved in plant response.

Research is in progress at several locations across the United States to develop data for assessing the impact of either gaseous pollutants or acidified rain on crop productivity. In general, ongoing field research involves controlling or altering the levels of the pollutant of greatest interest (e.g., O_3 or H^+) and allowing the others to vary according to ambient conditions.

Open-top chambers have been used at several sites to expose crops to several levels of O_3 (Heagle et al. 1979). The application of these chambers to the development of regional estimates of crop loss due to O_3 has been described by Heck et al. (1982). An essential feature of the open-top chamber methodology is the inflow of ambient precipitation. By design, therefore, plants exposed to gaseous pollutants in open-top chambers have also been exposed to ambient acid rain. Several studies have been conducted utilizing rain simulation techniques in experiments designed to minimize environmental modification of the experimental plots. Ambient air pollution was not excluded, allowing for multiple pollutant exposures in those experiments.

Early greenhouse experiments suggested that there was potential for a greater-than-additive effect from combined exposure to O_3 and acid rain on the growth of bush bean (Phaseolus vulgaris) compared to effects from single exposures (Shriner 1978). Based on those experiments and recognition that significant portions of the agricultural mid-section of the United States are regularly exposed to combinations of O_3 and acidic precipitation, we initiated greenhouse, growth chamber, and field studies to evaluate the combined effects of these pollutants on crops. Along somewhat parallel paths, techniques for studying the effects of acid rain on crop productivity have evolved rapidly over the past decade (Lee et al. 1981; Irving and Miller 1981; Heagle et al. 1983; Jacobson et al. 1980; Evans et al. 1983). However, only in the case of the work by Jacobson et al. (1980) was an attempt made to avoid the uncontrolled exposure of the test crop to ambient gaseous pollutants. Proof of an additive response of crops to the combined stress of O_3 and acid rain would mean that knowledge gained from previous single pollutant studies would be valid for interpretation of the combined effect. If, however, the crop response to O_3 and acid rain is not additive, more complex experiments will be required to model the effects of both forms of pollution on crop productivity. This paper summarizes the results of four years of experimentation in both greenhouse and field facilities aimed at determining whether acid rain alone or in combination with O_3 affects crop yield.

MATERIALS AND METHODS

Seven common crops [bush bean, alfalfa, radish, tomato, wheat, soybean, and corn (see Table 1)] were used for the greenhouse experiments. Plants were grown from seed in a commercial potting medium (ProMixBX®) in a charcoal-filtered-air greenhouse or in a walk-in growth chamber. After one to five weeks of growth, a uniform group of plants was chosen and randomly assigned to treatment groups. Factorial statistical experiments were used to test the specific hypothesis that each experiment addressed. In the greenhouse experiments, rain exposures were conducted twice weekly for three or more weeks with a rain simulator (Johnston et al. 1981). The minor ionic constituents of the simulated rain duplicated the characteristic chemistry of rainfall for the growing season in the eastern United States (Johnston and Shriner 1985). Nitric and sulphuric acids were used to adjust the pH to desired levels. In each experiment, plant responses to the following pH levels were tested: (1) a pH 5.2 - pH 5.6 treatment (control); (2) pH at levels characteristic of average ambient rain in the eastern United States (pH 3.8-4.6); and (3) pH levels characteristic of very acidic rain (pH 3.0-3.6). In addition, rain acidity in some experiments was adjusted to pH 2.3-2.6 to simulate highly polluted rain. Each rain simulant application was equivalent to 1.1 cm of deposition and was 35-40 min in duration. The O_3 exposures were conducted in a continuously stirred tank reactor fumigation system (Taylor et al. 1983). Exposures lasting three or five hours were conducted either one or three time(s) weekly. Measured parameters included visible injury, foliar chlorophyll concentration, dry weight, marketable yield, photosynthetic rate, and gaseous pollutant uptake. Data was analysed with analysis of variance (SAS®) and significant differences among treatment means were determined with least significant difference tests.

In one tomato cultivar experiment, the significance of the mode of rainfall application was also evaluated. One set of plants received five weeks of treatment with simulated rain with deposition to both the foliage and soil, while an otherwise identical set of plants received the rain simulant solutions solely as a soil drench.

Field Experiments

The field study site for the open-top chambers was located on the Oak Ridge National Environment Research Park (Department of Energy) Roane County, Tennessee. Soils in the study area are moderately well-drained loams, silty loams, and clay loams of the Elk and Lindside Series.

The experiment, which was conducted first in 1983 and repeated in 1984, was a randomized complete block 3 x 4 factorial. All combinations of three gaseous pollutant treatments and four rain treatments (pH 3.2, 4.2, 5.2 and ambient) were represented in each of three blocks.

The gaseous pollutant treatments, which were administered in open-top field chambers (Heagle et al. 1973), included charcoal-filtered (CF) and nonfiltered (NF) air chambers and ambient air (AA) plots without chambers (Fig. 1). The fans supplying air to the chambers were operated continuously during 1983 and from 9000 - 2100 h during 1984.

In three of the rain treatments, ambient rainfall was excluded. In these, fiberglass panels automatically covered the test plot at the onset of a rain event (Fig. 1). A rain detector activated industrial door openers that moved the rain exclusion covers into position over the appropriate test plots (Fig. 2). A tipping bucket rain gage located adjacent to the plot area activated the rain simulant distribution system after 0.25 mm of deposition (one trip of the tipping bucket) was measured. Pumps delivered rain simulants at one of three pH levels (5.2, 4.2 and 3.2) to the plots from which the natural rain was excluded. Rain simulants were dispensed by the system until a rain gage located inside a rain exclusion plot registered as many counts as the outside rain gage, at which time the simulant dispensing system automatically stopped. A gentle shower would intermittently activate the simulant dispensing system throughout the rain event until its completion, with the final amount of simulated rain in the rain exclusion plots equalling that deposited by the natural rain in the ambient plots. Rain simulants were similar in chemistry to those described by Irving (1985). When the rain event ceased, the drying of the rain sensor re-activated the system and returned the exclusion covers to the open position. The fourth was an ambient rain (AR) treatment, in which the natural rain was not excluded from the experimental plots and no simulated rain was applied.

The soybean, Glycine max (L.) Merr. cv Davis, crop was planted in the late spring of 1983 (6/17) and 1985 (5/24). Fertilizer, lime and herbicides were applied according to soil test and Tennessee Agricultural Extension recommendations. Insecticides (Seven and Kelthane) were applied as required to control Japanese beetles and spidermites.

Soybeans were harvested as they neared physiological maturity. Measured parameters included stem weight, pod weight, pod number, and weight per seed. Statistical analyses included F-tests to determine the significance of effects and least significant difference (LSD) tests for mean separation where warranted by significant F-tests.

RESULTS

Greenhouse and Growth Chamber Experiments

Our results indicated that the threshold for a significant effect of rain acidity on crop growth (plant dry weight) occurred at pH levels more acidic than those usually found in ambient rain (pH $\leq$ 3.3) (Table 1). This generality held true for Blue Lake (BBL) 274 bush bean (Johnston et al. 1981), Buffalo alfalfa, Scarlet Globe radish (Johnston et al. 1986), Cherry Belle radish, two of three tomato cultivars (Better Boy and Roma), and Oasis wheat

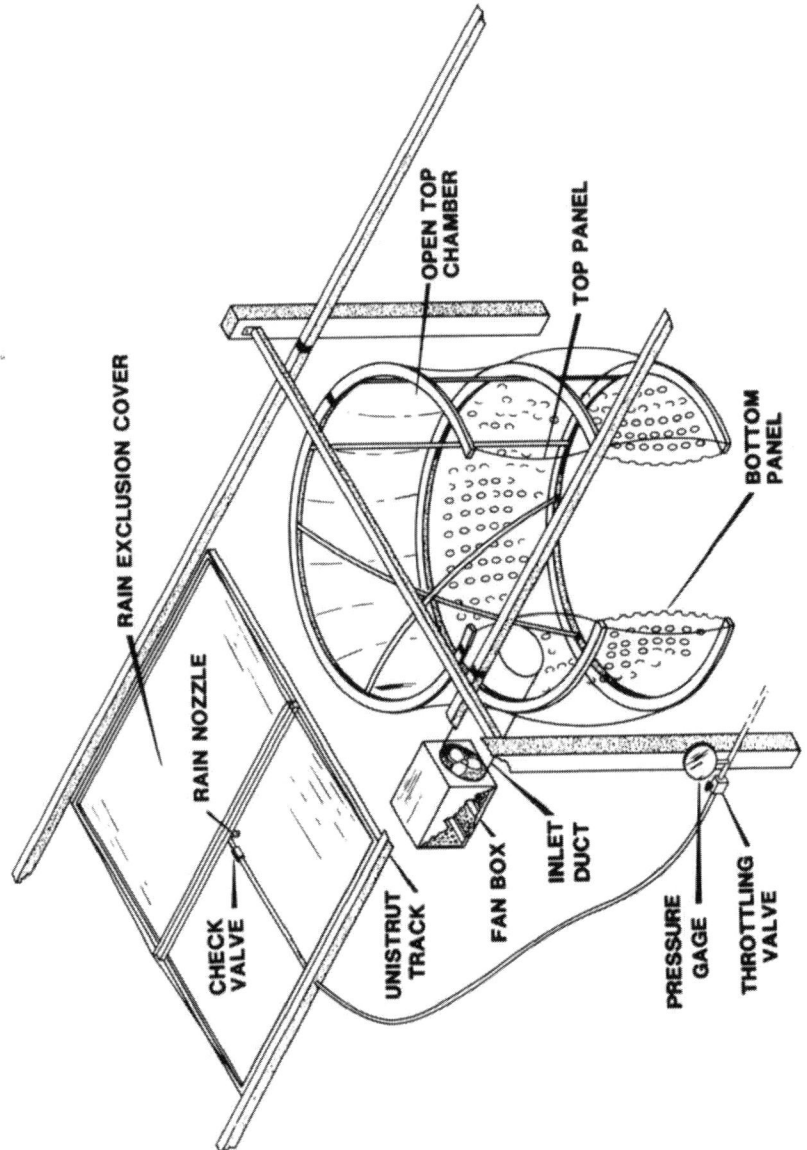

Fig. 1. Detail of an open-top chamber, rainfall exclusion cover, and rain simulant dispensing apparatus.

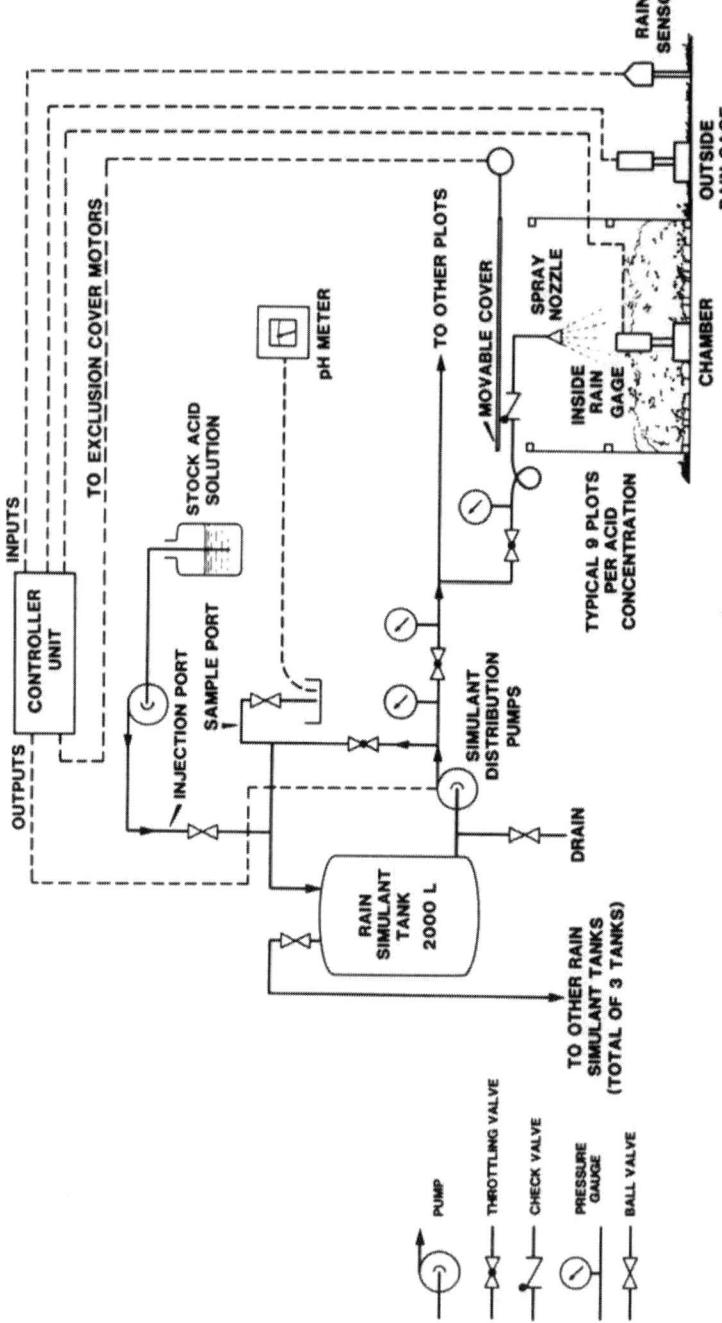

Fig. 2. Schematic illustrating control circuitry and water flow paths for the automatic operation of the rain exclusion and rain simulant dispensing systems.

(Johnston and Shriner 1985). The threshold for a significant growth reduction of Tiny Tim tomato was between pH 3.3 and 4.0. Growth of Davis soybean was not affected by pH levels as low as 2.6. Growth reduction was also observed for Aurthur 71 (pH 3.3), and Abe (pH 4.0 and 3.3) wheat, and Bantam Cross corn (pH 3.8) accompanied by growth stimulation at pH levels $\leq$ 3.0.

The experiment with wheat (Johnston and Shriner 1985) demonstrated that different cultivars may not exhibit the same pattern of response to acid rain (Fig. 3). The three tomato cultivars exhibited a similar pattern of response to rain acidity, but the threshold for a significant growth reduction for Tiny Tim was at a lower level of acidity than for Better Boy or Roma (Table 1).

In the experiment with tomato cultivars receiving simulated rain as either a soil drench or as a foliage/soil drench, plants receiving simulated rain via foliar and soil deposition exhibited growth reduction at pH 2.3, while those receiving the soil drench treatments exhibited increased growth (plant dry weight).

The effects of acid rain and O_3 combinations on growth of radish c.v. Scarlet Globe and alfalfa c.v. Buffalo was additive. Only visible injury of younger leaves (positions 4 and 5) of radish was less-than-additive for combinations that included pH 3.3 rain and 0.4 ppm of O_3. The injury and foliar chlorophyll response of Scarlet Globe radish to acid rain and O_3 was also additive.

Table 1. Levels of acidity in simulated rain that caused statistically* significant growth responses.

Species	Cultivar	pH level Growth reduction	pH level Growth stimulation
Wheat	Arthur	–	–
	Abe	4.0	2.3
	Oasis	–	3.3, 2.3
Tomato	Better Boy	$\leq$ 3.0	–
	Roma	$\leq$ 3.0	–
	Tiny Tim	$\leq$ 3.3	–
Radish	Scarlet Globe	$\leq$ 3.0	–
	Cherry Belle	$\leq$ 3.3	–
Bean	BBL 274	$\leq$ 3.2	–
Soybean	Davis	–	–
Alfalfa	Buffalo	$\leq$ 3.0	–
Corn	Bantam Cross	3.8	$\leq$ 3.0

* Analysis of variance and least significant difference tests ($p \leq 0.05$).

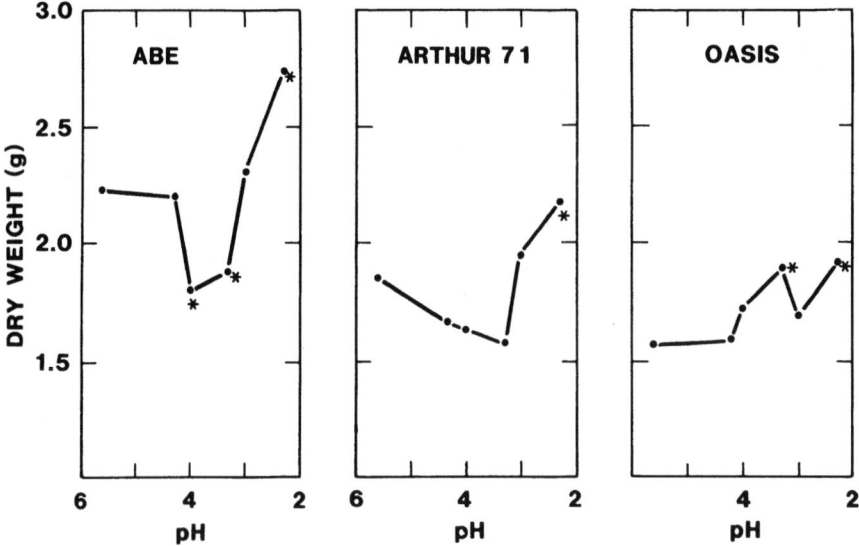

Fig. 3. Foliar dry weight response of wheat cultivars to simulated acid rain. Asterisks denote significant difference from the pH 5.6 control treatment based on LSD = 0.24 (p = 0.05; df = 17).

Field Experiments

Results of the field experiments with soybean c.v. Davis showed no evidence of an ozone main effect on yield parameters in an analysis of variance. Examination of treatment means did reveal a significant chamber effect on yield when ambient air plots were compared to open-top chambered plots, but revealed no apparent differences between charcoal-filtered and nonfiltered chamber plots.

Rain treatment effects were also small, and statistically insignificant with the exception of seed weight (Table 2). Weight per seed was lower in the pH 4.2 and ambient rain (also ~pH 4.2) plots than in the pH 5.2 or 3.2 plots. Analysis of variance revealed a statistically significant interaction between weight per seed and year (reflecting meteorological and cultural variables). This suggests that the weight per seed may be a more sensitive indicator of environmental stress than is yield. As a result, experiments which use predetermined and invarient simulated rain application schedules may avoid environmental stress that could affect sensitivity to either acid rain or gaseous pollutants, and in turn provide an unrealistic estimate of the actual pollutant stress.

There were no significant chamber by rain treatment interactions for any parameter measured, suggesting that at least for the one parameter where a significant effect was noted (weight per seed), the interaction was additive.

Table 2. Mean weight (g) per 100 seeds of Davis soybeans exposed to ambient or simulated acid rain in ambient, nonfiltered or charcoal-filtered air open-top field chambers.

Chamber treatment	Ambient	pH of simulated rain treatment		
		5.2	4.2	3.2
Ambient air	14.53abcd*	14.79cd	14.93d	15.06d
Nonfiltered air	14.30abc	14.72bcd	14.07ab	14.60abcd
Charcoal-filtered air	14.13ab	14.63abcd	14.05a	15.09d

*Means followed by the same letter are not significantly different (LSD = 0.65 g; p = 0.05).

For yield, since there were no significant effects of either pollutant (rain acidity or gaseous pollution), conclusions cannot be drawn concerning their interaction.

DISCUSSION

Efforts to assess the impact of air pollutants on crop productivity typically attempt to regulate the levels of a given pollutant with minimal disruption of other environmental factors capable of altering plant response. Open-top chambers have been used for this purpose in the assessment of ozone effects on crop yield (Heck et al. 1982). A variety of approaches have been used by researchers evaluating the effect of rainfall chemistry on crop growth and yield. The current state of the art methods combine rain exclusion techniques with methods for applying rain simulants to the crops in the field. To date, assessment of the potential effects of combined gaseous and wet deposition stress to vegetation has had to assume an additive response with little research to support that assumption (Lefohn and Brockson 1984). The methodology described herein can be used to test the hypothesis that the relationship of ozone and acid rain stress to crop growth is additive. This paper has presented preliminary confirmatory evidence to support this hypothesis.

Examination of the results of four years of research on the response of crop plants to rain acidity indicates that effects of acid rain may be positive or negative, or have no effect, at pH levels spanning the range found in ambient rain. The net effect of acid rain on plant growth may be a result of the negative influence of hydrogen-ion induced foliar effects and the positive influence of sulphate- and/or nitrate-induced nutrient effects, which may be foliar or soil mediated. The nutrient status of a crop may have a great influence on the magnitude and direction of its response to rain acidity. Further studies on these subjects are warranted.

Because ozone and acid rain may interact to affect forest vegetation as well, methodology of this type should prove useful in testing hypotheses of forest damage.

Acknowledgements

This research was sponsored by the U.S. Environmental Protection Agency (EPA) under Interagency Agreement EPA No. DW89930152-01-2 and the Electric Power Research Institute under Contract No. RP-1908-2 with the U.S. Department of Energy under Contract No. DE-AC05-840R21400 with Martin Marietta Energy Systems, Inc. Publication No. 2754, Environmental Sciences Division, ORNL.

Funding was obtained in part as part of the National Acid Precipitation Assessment Program by the U.S. Environmental Protection Agency, through Interagency Agreement DOE No. 40-740-78, EPA No. 79-X0533 with the U.s. Department of Energy. It has not been subjected to EPA review and therefore does not necessarily reflect the views of EPA, and no official endorsement should be inferred.

REFERENCES

Evans LS, Lewin KF, Patti MJ, Cunningham EA (1983) Productivity of field-grown soybeans exposed to simulated acidic rain. New Phytol 93: 377-388

Heagle AS, Body DE, and Heck WW (1973) An open-top field chamber to assess the impact of air pollution on plants. J Environ Qual 2: 365-368

Heagle AS, Philbeck RB, Rogers HH, Letchworth MB (1979) Dispensing and monitoring ozone in an open-top field chambers for plant-effect studies. Phytopath 69: 15-20

Heagle AS, Philbeck RB, Brewer PF, Ferrell RE (1983) Response of soybeans to simulated acid rain in the field. J Environ Qual 12: 538-543

Heck WW, Taylor OC, Adams R, Bingham G, Miller J, Preston E, Weinstein P (1982) Assessment of crop loss from ozone. J Air Pollut Control Assoc 32: 353-361

Irving PM (1985) Modeling the response of greenhouse-grown radish plants to acidic rain. Environ Exp Bot 25: 327-338

Irving PM, Miller JE (1981) Productivity of field-grown soybeans exposed to acid rain and sulfur dioxide alone and in combination. J Environ Qual 10: 473-478

Jacobson JS, Troiano J, Colavito LJ, Heller LI, McCune DC (1980) Polluted rain and plant growth. In: Toribara TY, Miller MW, Morrow PE (eds) Polluted rain. Plenum, New York

Johnston JW, Shriner DS (1985) Response of three wheat cultivars to simulated acidic rain. Environ Exp Bot 25: 349-353

Johnston JW, Shriner DS, Klarer CI, Lodge DM (1981) Effect of rain pH on senescence, growth and yield of bush bean. Environ Exp Bot 22: 329-337

Johnston JW, Shriner DS, Kinnerly CK (1986) The combined effects of simulated acid rain and ozone on injury, chlorophyll, and growth of radish. Environ Exp Bot 26: 107-113

Lee JJ, Neely GE, Perrigan SC, Grothaus LC (1981) Effect of simulated sulfuric acid rain on yield, growth and foliar injury of several crops. Environ Exp Bot 21: 171-185

Lefohn AS, Brockson RW (1984) Acid rain effects research - a status report. J Air Polut Control Assoc 34: 1005-1013

Shriner DS (1978) Interactions between acidic precipitation and SO_2 or O_3: Effects on plant response. Phytopathol News 12: 153

Taylor GE Jr, McLaughlin SB Jr, Shriner DS, Selvidge WJ (1983) The flux of sulfur-containing gases to vegetation. Atmos Environ 17: 789-796

THE NATURAL AND ANTHROPOGENIC ACIDIFICATION OF PEATLANDS

E. Gorham, J.A. Janssens, G.A. Wheeler, and P.H. Glaser

Department of Ecology and Behavioral Biology, University of Minnesota, Minneapolis, MN, 55455, U.S.A.

ABSTRACT

Peatlands are ubiquitous in northern landscapes, and decomposition of their plant remains produces complex, coloured organic acids that acidify their waters and those of the streams and lakes into which they drain. Fens with weakly acid surface waters (pH about 6) and low alkalinity (about 40 μeq.L^{-1}) are vulnerable to rapid change, and may be acidified by invasion of carpet-forming <u>Sphagnum</u> mosses that bring about major alterations in their biotic communities. The plant communities of such fens include a mixture of species characteristic of both minerotrophic and ombrotrophic peatlands.

Because mosses exhibit widely differing pH tolerances, stratigraphic examination of their remains in peat profiles (coupled with dating by various techniques) can reveal anthropogenic and natural acidification of peatlands. Decreasing concentrations of metals upward in peat profiles indicate concurrent impoverishment of lithophile elements (calcium, iron, etc.).

Acid deposition falling upon peatlands is largely neutralized (except where unusually heavy) by plant uptake and - beneath the water-table - by microbial reduction of associated nitrate and sulphate. Whether fen peats above the water-table can be leached sufficiently by acid deposition to initiate or accelerate invasion by <u>Sphagnum</u> and consequent acidification remains to be seen, but is to be expected at least under exceptionally severe conditions of acid loading.

INTRODUCTION

Peatlands are deposits of partially decomposed organic detritus in waterlogged environments that are anoxic a short distance beneath the surface (Clymo 1984a). They accumulate by terrestrialization - the filling in of shallow lakes - and in the case of extensive peatlands by paludification - the swamping of moist deglaciated soils, riverine flood-plains, and even truly upland soils (Gore 1984). In very wet climates paludification can take place on distinctly sloping terrain, forming the blanket bogs of western Europe and southeastern Newfoundland. Peatlands are an important component of landscapes in North America north of about 45° latitude, and in Europe north of about 50° (Gore 1984). They overlap areas of significant acid deposition in northeastern North America (Maine, southern Quebec, New Brunswick, Nova Scotia), and also northern Europe (Britain, southern Scandinavia,

the northern fringe of Germany and Poland, the westernmost part
of the Soviet Union).

In humid climates with an excess of precipitation over actual
evaporation, processes of peat formation may lead to development
of raised bogs dominated by the peat-forming Sphagnum mosses and
receiving their mineral supply solely from atmospheric deposition. These bogs overlie reed, sedge and shrub fens that once
received inputs of bases (chiefly calcium bicarbonate) leached
from adjacent mineral soils (Gore 1984). The remaining fens
still receive such bases and are therefore circumneutral to
weakly acidic; they are said to be minerotrophic. The raised
Sphagnum bogs, which retard water flow sufficiently to develop
water-table mounds above normal ground-water level, lack inputs
of bases from mineral soils and become strongly acid (pH normally
less than 4.5). They are described as ombrotrophic.

Such striking differences in mineral supply and acidity are
reflected in strongly contrasting assemblages of plant species
(Table 1). Whereas some species are found only in fens, no
species is confined strictly to bogs, although a few (e.g.,
Sphagnum fuscum) may be restricted to ombrotrophic hummock
microsites in fens (Sjörs 1963).

THE NATURAL ACIDIFICATION PROCESS

Acid/base balance in peatlands is determined by complex biogeochemical interactions: nutrient uptake by plants, decomposition,
redox reactions, cation and anion exchange, and rock and soil
weathering (Hemond 1980; Stumm et al. 1983; Gorham et al. 1984;
Schnoor and Stumm 1985; Urban et al., this volume). However, the
acidity of bog waters is associated chiefly with coloured organic
acids (Fig. 1), among which fulvic acids predominate (Oliver et
al. 1983; McKnight et al. 1985).

As fens are converted into bogs pH declines sharply because of
alterations in local hydrology. Removal of hydrologic inputs to
a peatland from adjacent or underlying mineral soils cuts off the
supply of bases that neutralize polyuronic acids produced by
Sphagnum (Clymo, this volume) and the complex, yellow-brown
organic acids produced by decomposition of plant remains (Urban
et al., this volume). Moreover, retardation of flow consequent
upon removal of such hydrologic inputs allows organic acids to
reach higher concentrations before being flushed from the
peatland; surface waters of acid bogs are often distinctly more
tea-coloured and higher in dissolved organic carbon (DOC) than
waters of circumneutral fens (Glaser et al. 1981).

Although there are no records of actual change in pH as a fen is
transformed into bog, it may be inferred from the relationship of
pH to the concentration of calcium and to the alkalinity (largely
bicarbonate) of surface waters in small kettle-hole peatlands in
northwestern Minnesota (Fig. 2; cf. Gorham et al. 1984, 1985).
These lie in hilly, calcareous terrain and many have distinctly
alkaline surface waters. However, gradual isolation of these

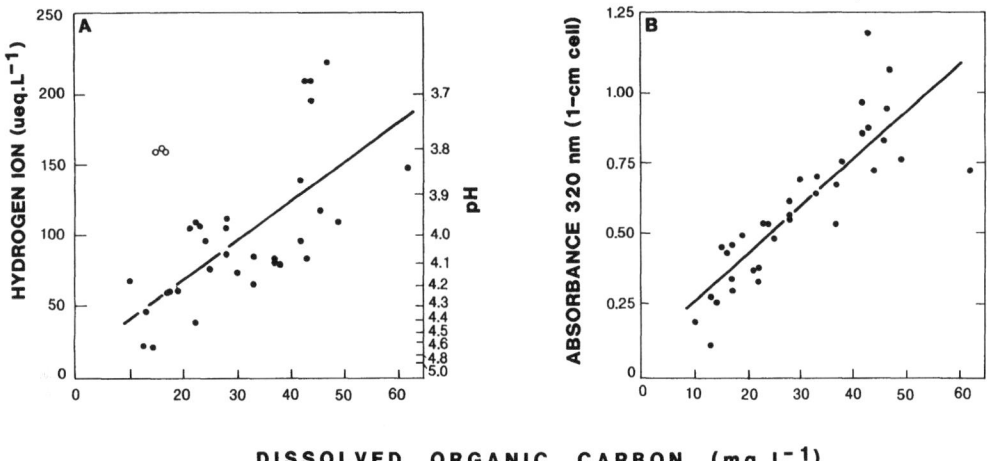

Fig. 1. The relationship of (a) hydrogen-ion concentration, and (b) colour (absorbance) to dissolved organic carbon in bog waters. In (a) the open circles represent British bogs subject to severe acid deposition; excluding those sites $H^+ = 2.74$ DOC + 13.9; n = 32, r^2 = 0.48. In (b), ABS = 0.0166 DOC + 0.0932; n = 35, r^2 = 0.72.

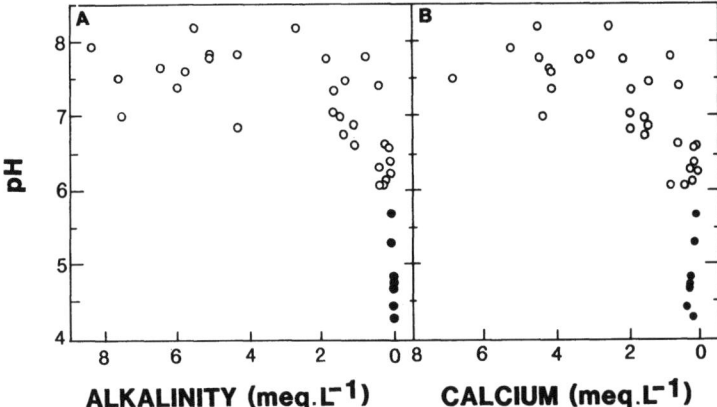

Fig. 2. The relationship of pH to alkalinity and calcium in the surface waters of small kettle-hole peatlands in northwestern Minnesota. Solid circles represent substantial cover by carpet-forming Sphagnum mosses.

Table 1. Characteristic components of vascular and bryophyte floras in the diverse landforms of large peatlands in northern Minnesota.[a] Data are from Glaser et al. 1981, Glaser 1983 a,b,c, Wheeler et al. 1983, Janssens, and Glaser, in press, and unpublished material.

	Vascular plants	Bryophytes
Forested Bog	Picea mariana Carex trisperma Vaccinium vitis-idaea var. minus Vaccinium myrtilloides Eriophorum spissum	Dicranum undulatum Pleurozium schreberi Dicranum polysetum
Unforested Bog	Carex oligosperma	Sphagnum fallax Sphagnum capillifolium Cladopodiella fluitans Sphagnum papillosum Sphagnum pulchrum Sphagnum flexuosum
Sphagnum Lawns (poor fen)	Carex oligosperma Carex aquatilis var. altior Carex rostrata var. utriculata Betula pumila var. glandulifera Carex pauciflora	Sphagnum papillosum Sphagnum capillifolium Sphagnum fallax Sphagnum majus
Fen Strings	Thelypteris palustris var. pubescens Betula pumila var. glandulifera Potentilla fruticosa	Campylium stellatum Sphagnum centrale
Fen Flarks (troughs)	Carex livida var. radicaulis Rhynchospora alba Carex limosa Carex lasiocarpa var. americana Triglochin maritima	Sphagnum contortum Aneura/Riccardia spp. Scorpidium scorpioides Calliergonella cuspidata Calliergon trifarium Drepanocladus lapponicus

continued/

Table 1 continued.

	Vascular plants	Bryophytes
Wooded Fen Islands	Larix laricina Carex disperma	Calliergon giganteum Sphagnum contortum Sphagnum teres Ptilidium pulcherrimum
Wooded Spring Fen	Picea mariana Larix laricina Carex disperma Carex gynocrates	Callicladium haldianum Dicranum undulatum Hylocomium splendens Myurella julacea Plagiomnium ellipticum Platygyrium repens Pleurozium schreberi Sphagnum warnstorfii Tomenthypnum nitens
Spring-fen Channels	Scirpus acutus S. cespitosus var. callosus S. hudsonianus Parnassia palustris var. neogaea	Aneura/Riccardia spp. Bryum pseudotriquetrum Campylium stellatum Cinclidium stygium Drepanocladus revolvens Fissidens adiantoides Scorpidium scorpioides Tomenthypnum nitens Moerkia hibernica

[a]Bog ericads (Ledum groenlandicum, Chamaedaphne calyculata, Vaccinium oxycoccus, etc.) are widespread throughout the peatlands, as are the bryophytes Aulacomnium palustre and Sphagnum angustifolium. Polytrichum strictum, Sphagnum magellanicum, and S. fuscum are common on bogs and on scattered ombrotrophic hummocks in fens.

surface waters from minerotrophic inputs as peat deposits accumulate and broaden leads to pronounced acidification once alkalinity is almost gone and calcium has dropped to a few hundred micro-equivalents per litre. In these peatlands pH usually declines below 6 as sedge meadows are invaded by carpet-forming species of Sphagnum, notably Sphagnum recurvum (s.l.) and S. papillosum. Scattered patches of Sphagnum species tolerant of circumneutral pH, such as S. teres, S. warnstorfii, species of Sphagnum Section Subsecunda, occur in many places above pH 6.

The transformation of circumneutral fen to strongly acid bog can be rapid, indicated by a bimodal pH frequency distribution for the vast peatlands of northern Minnesota (Fig. 3). Similar distributions characterize peatlands in northern Sweden (Gorham et al. 1984) and lakes in Norway (Wright and Henriksen 1977), Nova Scotia (Underwood et al. 1982) and Sudbury (E. Gorham and A.G. Gordon, unpublished data). As declining minerotrophic inputs of alkalinity are titrated away by organic acids from the decay process, acidity shifts quickly from a pH of around 6 to a pH of around 4, so that sites with intermediate pH values are relatively scarce.

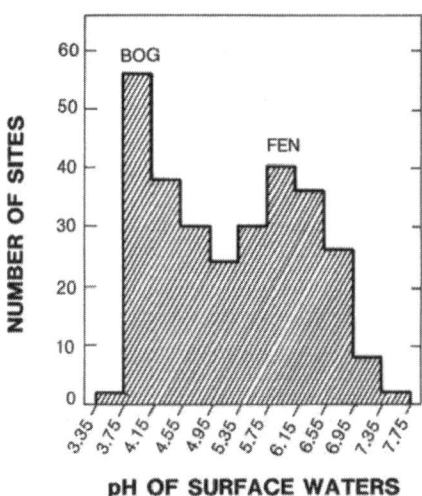

Fig. 3. The frequency distribution of pH in the peatlands of northern Minnesota (after Gorham et al. 1984).

There is also a shift in base-saturation of surface peats as bogs replace fens. Data from lakes, fens and bogs in the English Lake District (Fig. 4) indicate that as inputs of silt from marginal uplands decline and lake sediments and peats become increasingly organic, cation-exchange capacity increases greatly and reaches (on average) about 1 meq.g^{-1} dry wt in the most organic peats. This capacity is saturated predominantly by base-cations where inputs of silt keep the mineral ash content of peat above about 15% dry weight. However, as peats shift from minerotrophy to ombrotrophy, ash content declines to as low as 2% dry weight; at the same time adsorbed base-cations fall to very low concentrations and are replaced by adsorbed hydrogen ions (Gorham 1953, 1967). There is again a rapid transition from fen to bog, with a distinctly bimodal frequency distribution of percentage base saturation. All but two of forty fen peats range from 52 to 100% base saturation, whereas with but a single exception twenty-one bog peats range from 3 to 25% (Gorham 1953).

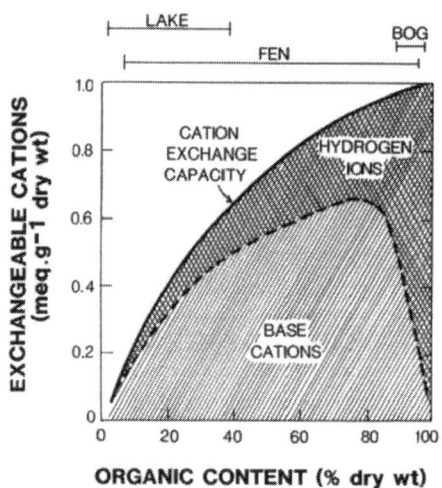

Fig. 4. Generalized relationships of exchangeable cations to organic content in surface lake sediments, fen peats and bog peats in the English Lake District (after Gorham 1953, 1967).

PLANT COMMUNITIES VULNERABLE TO ACIDIFICATION

Relationships of pH to alkalinity and to calcium in surface waters of peatlands suggest that once an ecosystem declines to the triple point marked by alkalinity (titrated to an end-point of pH 4.5) about 40 µeq.L^{-1} (Gorham et al. 1984), calcium about 150-250 µeq.L^{-1} (Gorham et al. 1985), and pH about 6, it is likely to change rapidly from fen to bog as minerotrophic inputs are progressively cut off. Although some plant species are affected much more than others, shifts in species assemblages that accompany this chemical alteration are very distinctive. In northern Minnesota (Fig. 5) typical fen species (e.g., Carex leptalea, Triglochin maritima, Scorpidium scorpioides, Sphagnum

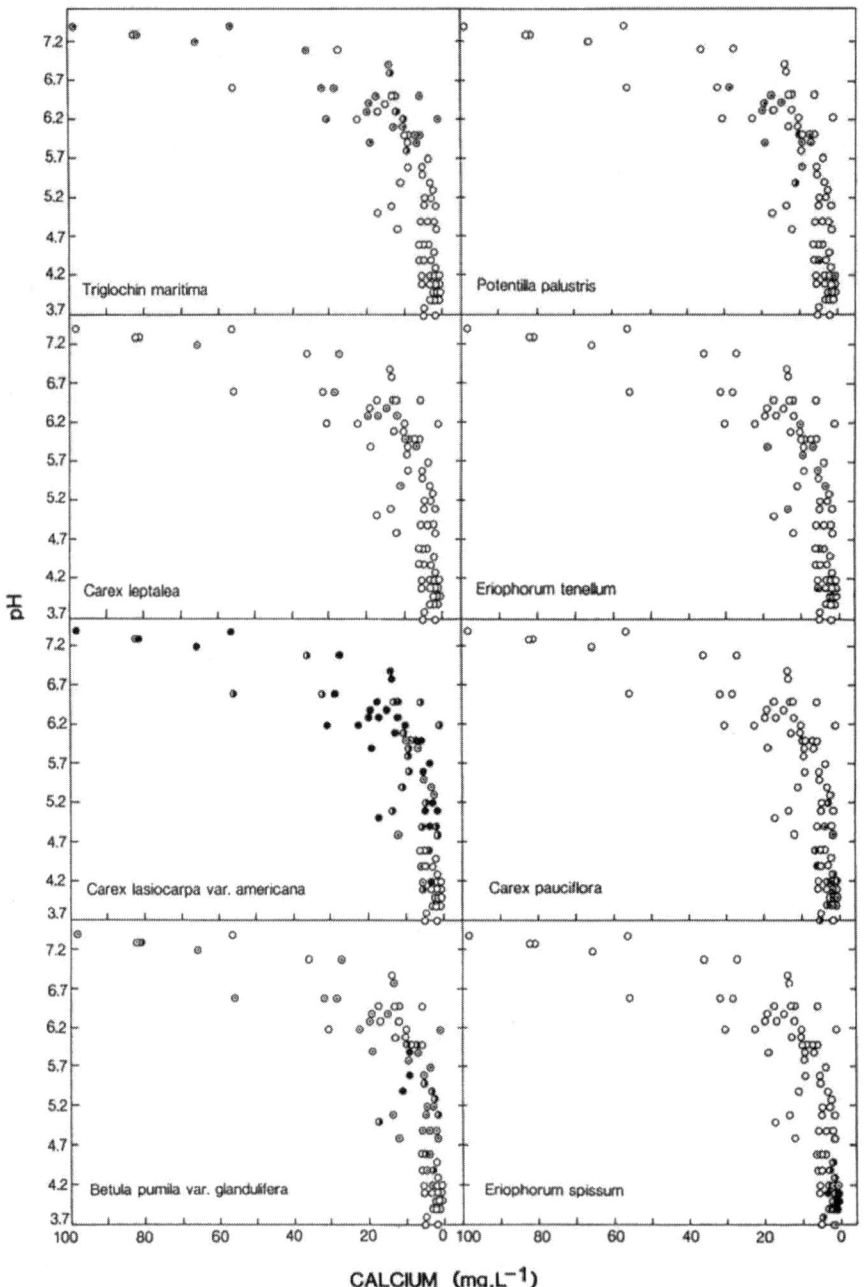

Fig. 5. Species shifts accompanying declines in the calcium concentration and pH of surface waters in northern Minnesota peatlands. Open circles indicate sites without the designated species. A) Vascular plants. The dotted circles represent a Braun-Blanquet cover value of rare or +, the half-filled circles cover 1 or 2, and the solid circles cover 3 or 4.

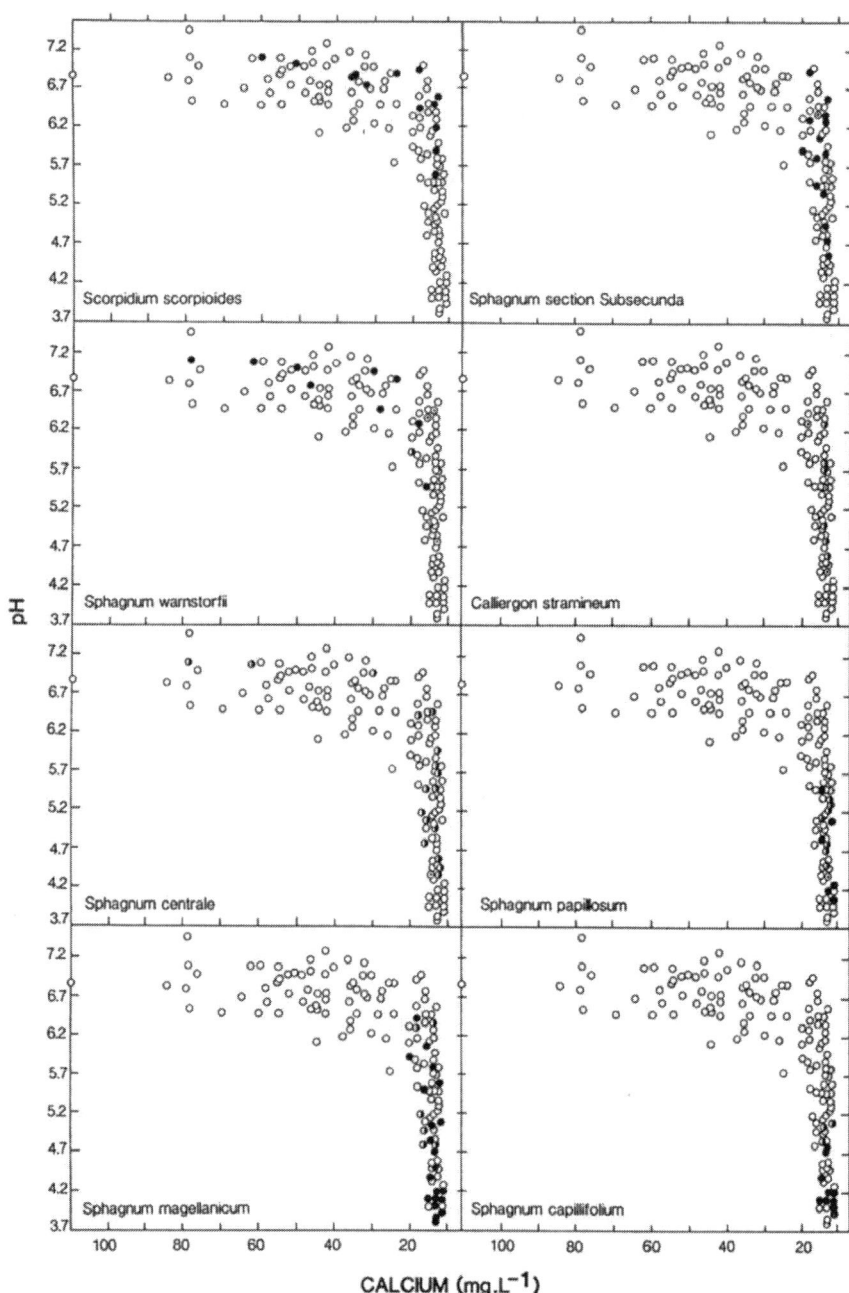

Fig. 5. B) Bryophytes. The dotted circles represent less than 33% frequency, the half-filled circles 33-67%, and the solid circles more than 67%.

warnstorfii) disappear. Other species appear and characterize intermediate conditions of pH and calcium concentration (e.g., Eriophorum tenellum, Potentilla palustris, Calliergon stramineum, Sphagnum Section Subsecunda). These are eventually replaced by species more characteristic of bogs (e.g., Eriophorum spissum, Carex pauciflora, Sphagnum capillifolium). A few species exhibit broad ranges of pH tolerance (e.g., Carex lasiocarpa, Betula pumila v. glandulifera) but are abundant over a much narrower range of conditions. (The distribution of Betula pumila var. glandulifera is related much more to low water-tables than to acidity.)

Species of intermediate pH and calcium concentrations, accompanied by more ubiquitous taxa, characterize the zone of vulnerability to acidification where pH drops rapidly below 6. This zone is the characteristic habitat of three plants rare in Minnesota (Juncus stygius, Xyris montana, Rhynchospora fusca) that occur in featureless or patterned water-tracks often categorized as poor fen (Fig. 6).

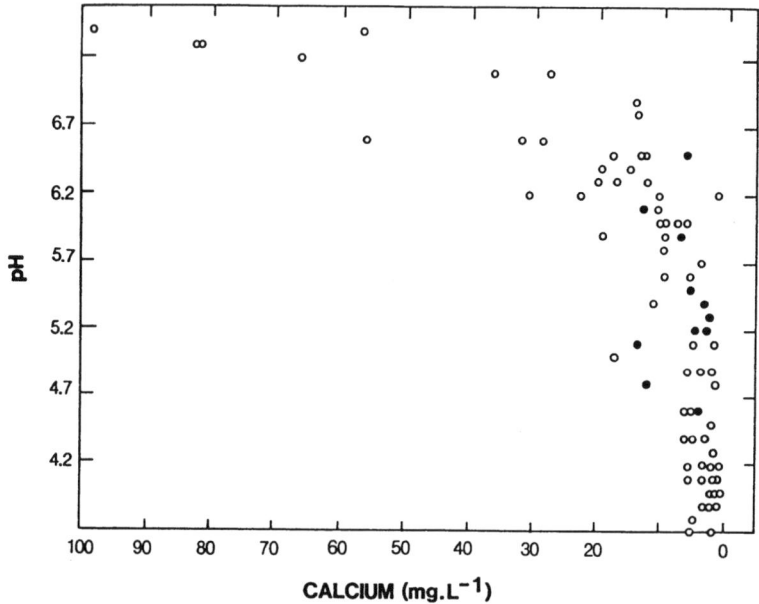

Fig. 6. The occurrence of three rare species - Juncus stygius var. americanus (n=5), Xyris montana (n=4), and Rhynchospora fusca (n=6) - in eleven sites along a gradient of calcium and pH in the surface waters of peatlands in northern Minnesota.

As fens are transformed into bogs the number of species declines. Fen relevé plots (10x10m) in northern Minnesota usually contain 12-26 species - tending toward the higher number at high pH and high calcium concentrations - whereas bog plots commonly contain 7-14 species.

RECONSTRUCTION OF PLANT SUCCESSION AND ASSOCIATED CHEMICAL CHANGES IN PEATLANDS

Recent advances in quantitative analysis of bryophyte stratigraphy in peat deposits (Janssens 1983) permit reconstruction in some detail of the transformation of circumneutral fen into acid bog - and _vice versa_ where hydrologic changes cause spatial oscillations of fen/bog ecotones (Janssens _et al._, in prep). Figure 7 illustrates such a reconstruction for the internal bog drain of a large, ovoid bog island with poorly developed black spruce in the west-central watershed of the Red Lake Peatland in northern Minnesota (Glaser _et al._ 1981 and unpublished). The sequence of vascular plants, reconstructed from local pollen and _Sphagnum_ spores, is shown in Fig. 7A. A sedge fen, grassy in its earlier stages, formed about 3300 years ago and changed rather suddenly into a _Sphagnum_-black spruce poor fen about 2500 years ago. The mosses show a similar shift in which a rich fen with _Calliergon trifarium_, _Scorpidium scorpioides_ and _Campylium stellatum_ was replaced by a poor fen with several species of _Sphagnum_, predominantly _S. magellanicum_, and ultimately by ombrotrophic bog with _Sphagnum fuscum_ and _S. capillifolium_ hummocks (Fig. 7B). With close-interval sampling of peat cores, the shift from fen to bog should be discernible on a time-scale of decades; bryophyte profiles are characterized by extremely sharp boundaries as hydrological and chemical changes affect the environments of local microsites. Fossil bryophyte assemblages include only a few taxa - mostly identifiable to species - at any given depth in a peat core. However, an entire core 5 cm in diameter may include throughout its length up to 20 species. A fossil bryophyte flora of 52 taxa, 45 species, and 15 species of _Sphagnum_ was recovered from a single watershed in the Red Lake Peatland; several of the fossil species are no longer found there.

Peat cores preserve a stratigraphic record of chemical change to some degree, although losses by decomposition and leaching occur even in deep peat over long periods (Clymo 1984b). For instance, a profile (Fig. 8) from Striber's Moss in the English Lake District (Gorham 1949) exhibited a pronounced decline in calcium and iron as minerotrophic inputs were lost and fen shifted to bog. A poor-fen community dominated by _Eriophorum vaginatum_ marked the transition (3.2 - 4.0 m) to raised bog, and is observable at the present time in a similar transitional position along the margin of a Swedish raised bog (Gorham 1952). A comparable decline of Si, Fe, Mn, and Ca was observed by Mornsjo (1968) in two Swedish bog profiles. However, both profiles exhibited a pronounced increase of these elements in the uppermost horizons owing to human interference: in one case by flooding, in the other by airborne inputs from cultivated soils.

Reconstruction of a stratigraphic profile of pH, to match profiles of plant fossils and of chemical elements, would be of great ecological value. By analogy with the use of diatom stratigraphy to reconstruct the acidification of lakes (Battarbee 1984, Davis and Anderson 1985), analysis of fossil plant remains should enable us to infer pH profiles in peat cores. There are

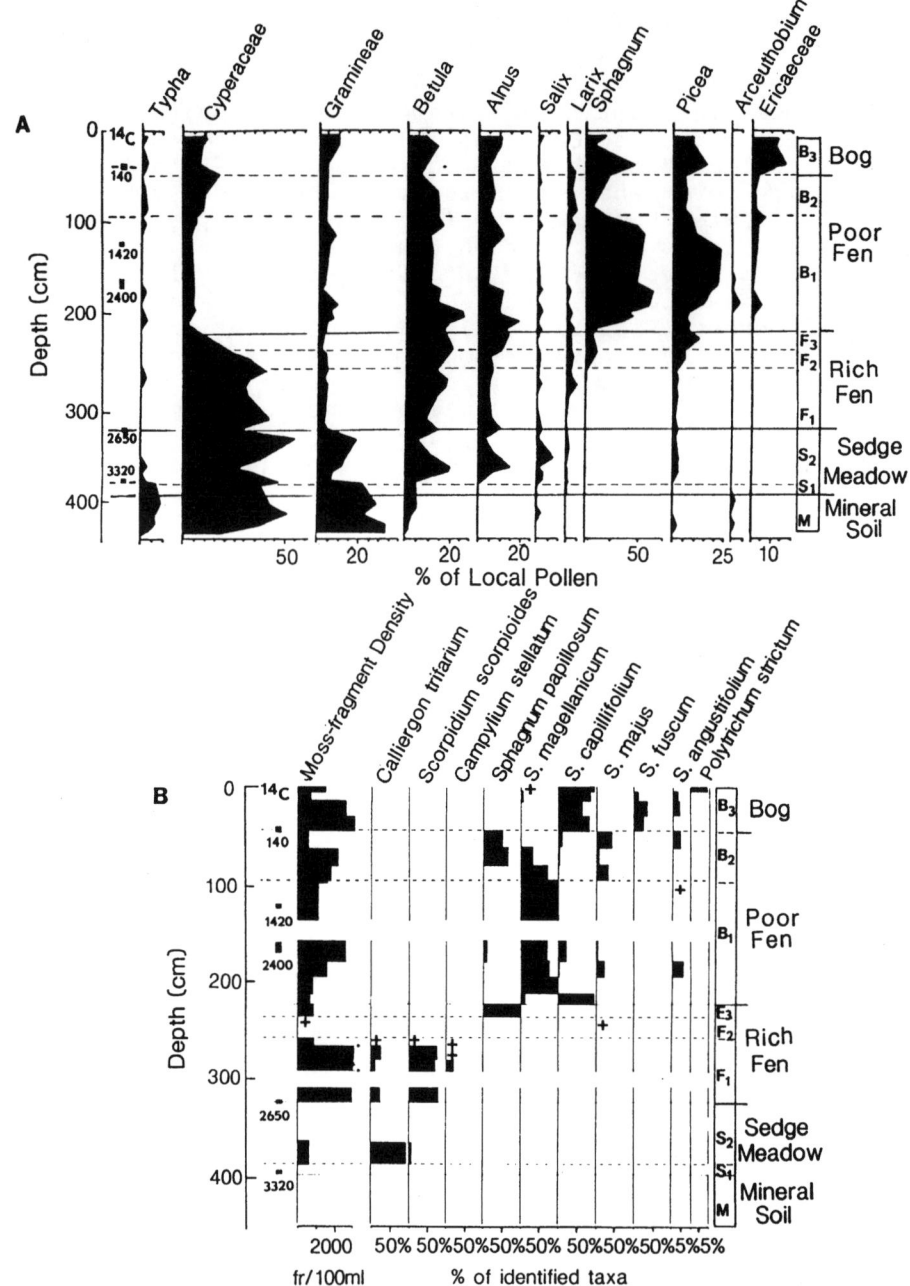

Fig. 7. Stratigraphic succession of plant fossils in a peat core (R8112) from a bog in the Red Lake Peatland of northern Minnesota. Pollen counts are as % of the sum of local pollen (excluding Sphagnum spores). Bryophyte counts are as % of the sum of identified species.

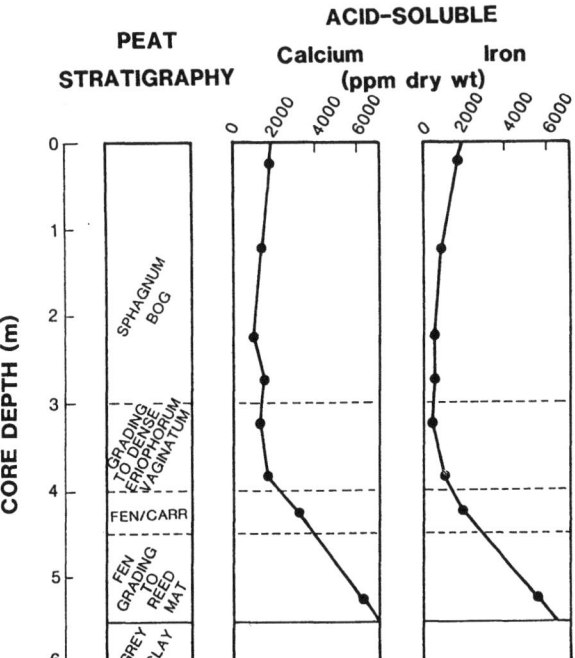

Fig. 8. Acid-soluble calcium and iron in a peat profile from a bog in the English Lake District (Gorham 1949). The dry ash was dissolved in 0.1N HCl.

many limitations to such techniques, but they offer the best hope of reconstructing the postglacial and postindustrial history of ecosystem acidification. We used data on the present distribution of 63 species of vascular plants in northern Minnesota peatlands as an initial test of the potentiality of this approach. First we separated the species of 70 relevés in northern Minnesota, exclusive of the Red Lake Peatland, into five groups of decreasing mean site (surface-water) pH (group 1, pH 6.52-7.01; group 2, pH 6.02-6.50; group 3, pH 5.36-5.94; group 4, pH 4.77-5.30; group 5, pH 4.03-4.68). Assigning the group number as the score applicable to each species within that group, we scored the 70 relevés for the presence of all vascular species within them. Finally we took averages of the scores for all species within each of the plots and regressed the actual pH values measured within the plots upon them. The correlation was extremely high ($r^2 = 0.92$). Then we used the regression of actual pH upon relevé score to compute the pH of 33 relevés in the Red Lake Peatland, scored according to the group-mean pH values established for the 70 relevés elsewhere in Minnesota. The correlation was equally high ($r^2 = 0.92$). Correspondence between computed and observed pH was close, except that computed bog values were about pH 4.0, whereas observed values centred around pH 3.8.

We next constructed an inferred pH profile, using bryophyte remains in the Red Lake peat core represented by Fig. 7 and a modification of the scoring method employed for vascular plants. Based on data for 48 common bryophyte species of Minnesota peatlands in 75 quadrats (Janssens and Glaser, to be published; Janssens, unpublished data), we calculated for each species its mean site (surface-water) pH and the 95% confidence interval around the mean (approximately 150 water samples were used, most obtained from depressions close to bryophyte populations). Then, assigning ($\overline{Y}_{pH} \times 100$) as the score applicable to each species, we scored 75 vegetation quadrats (50 x 50 cm) for all bryophyte species in them, taking averages - weighted for % cover - of the scores for all species within each quadrat. Finally, we calculated the correlation between pH measured in quadrats and quadrat scores (Fig. 9). It was sufficiently high ($\underline{r}^2 = 0.83$) that a stratigraphic application appeared worthwhile.

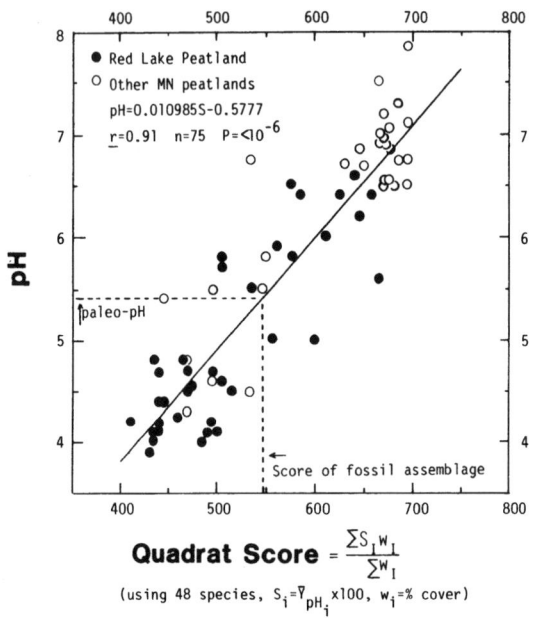

Fig. 9. Relationship between Quadrat Score (x) and measured pH (y) of 75 (50 x 50 cm) bryophyte plots. Individual species scores (S_i) equal their mean site pH times 100. Forty-eight common peatland taxa were collected in Minnesota with sufficient associated water chemistry to function as a preliminary data base. The Quadrat Score is calculated as a weighted average of the individual species scores. The weight is the % cover of the taxon in the quadrat. Measured pH is based on analyses of water samples collected in the plots. The linear regression (pH = 0.011 - 0.58) can be applied to infer pH from the weighted average score of a fossil assemblage (stippled line). In this instance the fossil assemblage score is based on the same 48 species, 39 of which are already found in the fossil record. The weight, however, is now based on fragment-density values resulting from quantitative fossil-bryophyte analysis (Fig. 7).

If scores predicted actual pH precisely, a score of 400 would correspond to pH 4.00 and a score of 700 to pH 7.00. However, presence of all the common bog species in poor fens (Janssens and Glaser, in press) results in higher than expected individual scores for these taxa and thus slightly high computed pH values for bog quadrats. A score of 400, for instance, corresponds to an actual pH of 3.82, whereas a score of 700 corresponds to a pH of 7.11. To correct inferred pH for this systematic error we used the linear regression equation, pH = 0.011 S - 0.58 (S: weighted average score of the quadrat in the test, or of the fossil assemblage in the application) as an intermediate corrective step to infer pH based on an incomplete data set not representative for all northern Minnesota peatlands. Inferences will be improved by additional surface sampling in the region.

We applied the inferred-pH regression formula (Fig. 9) to quantitative bryophyte analyses of Red Lake Peatland Core R8112 (Fig. 7) by calculating a weighted average score for each fossil assemblage. In this instance weight was based on moss-fragment density for each taxon in the assemblage (Fig. 7). The inferred-pH profile is plotted in Fig. 10. Major pH changes correspond to bryostratigraphical boundaries in Fig. 7. For example, the pH decline in zones F2 and F3 accompanies a shift from circumneutral rich fen with Calliergon trifarium and Scorpidium scorpioides to acid poor fen (pH about 4.5) with a variety of Sphagnum species. A distinct decline (to pH 4.1-4.2) occurs at the boundary between zones B2 and B3, where Sphagnum papillosum and S. majus disappear and bog hummock species (Sphagnum capillifolium, S. fuscum) become abundant. Inferred surface pH coincides with the measured pH of bog waters in the Red Lake Peatland (Glaser et al. 1981; Janssens and Glaser, in press).

PEATLAND RESPONSES TO ACID DEPOSITION

Surface bog waters from Manitoba to Newfoundland exhibit much lower concentrations of nitrate and sulphate than those in atmospheric precipitation (Table 2) despite the concentrating effect of evapotranspiration (cf. Urban et al., this volume). Presumably plant uptake is the primary mechanism for removal of nitrate, with microbial reduction the primary mechanism for removal of sulphate, although both processes affect each ion. These processes yield alkalinity that neutralizes hydrogen ions associated with sulphate and nitrate in acid deposition (cf. Hemond 1980). Under exceptionally severe acid deposition, as in industrial northern England (Fig. 1), microbes are unable to cope with the added load of sulphate, so that sulphuric acid from the atmosphere dominates coloured organic acids as a source of acidity in bog waters and sulphate concentrations remain high at 273 ± 30 μeq.L^{-1} (Gorham et al. 1985). Nitrate is undetectable in these English bog waters. It is from sites such as these, now dominated by the cottongrass Eriophorum vaginatum, that several acidophilous species of Sphagnum disappeared during the industrial age (Tallis 1964). Air pollution and associated acid deposition are likely to have been involved, the more so because experiments (Ferguson et al. 1978) indicated that these species

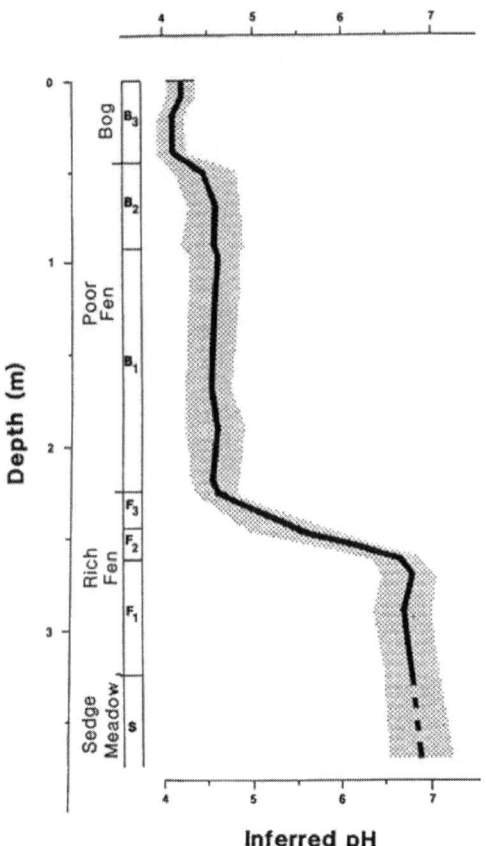

Fig. 10. The inferred pH profile for peat core R8112. See Fig. 9 for details of calculation and Fig. 7 for correspondence with zonation and fossil analysis. The lower and upper limits of the stippled band representing the error of the inferred pH are calculated as the weighted averages of the lower and upper 95% confidence limits of the individual scores for the taxa represented in the fossil assemblage.

are damaged by application of SO_2, SO_4^{2-}, and especially HSO_3^-, within the ranges of concentration observed in atmospheric deposition over northern England. In this region a substantial impact of acid deposition upon vulnerable sedge fens might be expected.

Even in peatlands where uptake and reduction lower nitrate and sulphate concentrations well below those in atmospheric precipitation, acid deposition can perhaps leach surface peats above the water-table (in fens vulnerable to acidification) to the point that Sphagnum invasion is either initiated or accelerated. If that happens it is likely to cause further autogenic acidification as Sphagnum generates polyuronic acids, peat rises above the groundwater-table, minerotrophic inputs to the peat surface are cut off, and the slowing of water flow allows the build-up of soluble, coloured organic acids produced during decomposition. A conceptual model of the process, which could be tested experimentally, is shown in Fig. 11.

Table 2. Nitrate and sulphate in the surface waters of bogs from Manitoba to Newfoundland, and in atmospheric precipitation at nearby stations (nd = not detected). Data from Munger and Eisenreich (1982) and Gorham et al. (1985).

		NO_3^-	SO_4^{2-}
		(μeq.L^{-1})	
Atmospheric precipitation		10-40	50-100
Bog waters	range	nd-3	nd-82
	mean	85% nd	18±16

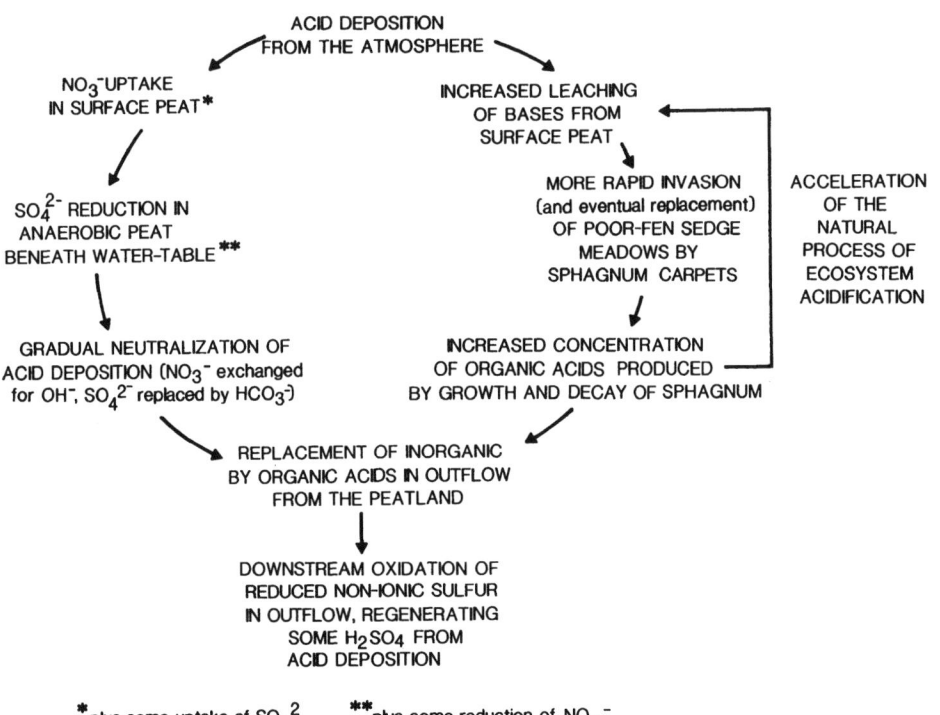

Fig. 11. A conceptual model of the chemical and biological effects of acid deposition upon a fen vulnerable to acidification.

THE ROLE OF PEAT BOGS IN LAKE ACIDIFICATION

Many areas on igneous and metamorphic rocks in eastern North America contain numerous streams and lakes with little buffering capacity (calcium about 50 µeq.L^{-1}) and many peat deposits that discharge acid, tea-coloured waters (DOC $\geqslant$ 10 mg.L^{-1}) into those streams and lakes. Lillie and Mason (1983) observed that 38% of a random set of Wisconsin lakes are brown in colour, mostly in the northern half of the state. Underwood (pers. comm.) found that colour (Hazen Pt/Co scale) ranged in 499 Nova Scotian lakes from less than 5 to 650 units, the antilog of the log$_{10}$ mean was 35 units (about 5.5 mg.L^{-1} DOC), and a quarter of the samples exceeded 75 units (about 10.2 mg.L^{-1} DOC). In Halifax County, Nova Scotia, lake acidification owes more to inputs of coloured organic acids from bogs than to acid deposition from the atmosphere (E. Gorham, J.K. Underwood, F.B. Martin, and J.G. Ogden III, in prep.). There is also evidence that atmospheric sulphate reduced to non-ionic form in bog waters (Gorham et al. 1985) and exported in appreciable amounts by bog outflow streams (Brown 1980; S.J. Eisenreich, pers. comm.) is re-oxidized after circulating in lakes downstream, restoring hydrogen ions that accompanied wet sulphate deposition initially (cf. Fig. 11). Because of reoxidation, peatland reduction processes will counter sulphuric acid in atmospheric deposition only to the extent that organic or pyritic sulphur is stored permanently in the peat deposit.

CONCLUSION

Long-term acidification is a natural characteristic of many peatland ecosystems, associated with pronounced changes in biotic communities. Although severe acid deposition has affected already acid Sphagnum bogs (Ferguson et al. 1978; Lee, this volume), its effects upon fens are conjectural. We suggest that weakly acid fens low in alkalinity are vulnerable to alteration by acid deposition that is likely to favour invasion by carpet-forming Sphagnum species, which are themselves capable of strongly acidifying their environment.

Acknowledgemnts

We thank the National Science Foundation (Grant DEB-7922142) for support, and R.S. Clymo and S.E. Bayley for helpful comments. Limnological Research Centre Contribution No. 318.

REFERENCES

Battarbee RW (1984) Diatom analysis and the acidification of lakes. Phil Trans Royal Soc London B 305: 451-477

Brown KA (1980) The distribution of sulphur compounds in a peat bog in relation to stream water chemistry. Laboratory Note No RD/L/N 150/80 (Job No VJ420), Central Electricity Generating Board, England

Clymo RS (1984a) Peat. In: Gore AJP (ed) Ecosystems of the world, Mires: swamp, bog, fen and moor, vol 4A, General Studies. Elsevier, New York, p 159

Clymo RS (1984b) The limits to peat bog growth. Phil Trans Royal Soc London B 303: 605-654

Clymo RS (this volume) Interactions of *Sphagnum* with water and air.

Davis RB, Anderson DS (1985) Methods of pH calibration of sedimentary diatom remains for reconstructing history of pH in lakes. Hydrobiologia 120: 69-87

Ferguson P, Lee JA, Bell JNB (1978) Effects of sulphur pollutants on the growth of *Sphagnum* species. Environ Pollut 16: 151-162

Glaser PH (1983a) *Eleocharis rostellata* and its relation to spring fens in Minnesota. Mich Botanist 22: 19-21

Glaser PH (1983b) Vegetation patterns in the North Black River peatland, northern Minnesota. Can J Bot 61: 2085-2104

Glaser PH (1983c) A patterned fen on the north shore of Lake Superior. Can Field Nat 97: 194-199

Glaser PH, Wheeler GA, Gorham E, Wright HE Jr (1981) The patterned mires of the Red Lake Peatland, northern Minnesota. J Ecol 69: 575-599

Gore AJP (1984) Introduction. In: Gore AJP (ed) Ecosystems of the world, Mires: swamp, bog, fen and moor, vol 4A, General Studies. Elsevier, New York, p 1

Gorham E (1949) Some chemical aspects of a peat profile. J Ecol 37: 24-27

Gorham E (1952) Variation in some chemical conditions along the borders of a *Carex lasiocarpa* fen community. Oikos 2: 217-240

Gorham E (1953) Chemical studies on the soils and vegetation of waterlogged habitats in the English Lake District. J Ecol 41: 345-360

Gorham E (1967) Some chemical aspects of wetland ecology. Tech Mem Assoc Comm Geotech Res, Nat Res Counc Can No 90, p 20

Gorham E, Bayley SE, Schindler DW (1984) Ecological effects of acid deposition upon peatlands: a neglected field in "acid-rain" research. Can J Fish Aquat Sci 41: 1256-1268

Gorham E, Eisenreich SJ, Ford J, Santelmann MV (1985) The chemistry of bog waters. In: Stumm W (ed) Chemical processes in lakes. Wiley, New York, p 339

Hemond HF (1980) Biogeochemistry of Thoreau's Bog, Concord, Massachusetts. Ecol Monogr 50: 507-526

Janssens JA (1983) A quantitative method for stratigraphic analysis of bryophytes in Holocene peat. J Ecol 71: 189-196

Janssens JA, Glaser PH (in press) The bryophyte flora and major peat-forming mosses at Red Lake Peatland, Minnesota. Can J Bot

Lee JA, Press MC, Woodin S, Ferguson P (this volume) Responses to acidic deposition in ombrotrophic mires in the U.K.

Lillie RA, Mason JW (1983) Limnological characteristics of Wisconsin lakes. Tech Bull Wisc Dep Nat Resour No 138
McKnight DM, Thurman EM, Wershaw RL, Hemond HF (1985) Biogeochemistry of aquatic humus substances in Thoreau's Bog, Concord, Massachusetts. Ecology 66: 1339-1352
Mornsjo T (1968) Stratigraphical and chemical studies on two peatlands in Scania, South Sweden. Bot Notiser (Lund) 121: 343-360
Munger JW, Eisenreich SJ (1982) Continental-scale variations in precipitation chemistry. Environ Sci Technol 17: 32A-42A
Oliver B, Thurman EM, Malcolm RL (1983) The contribution of humic substances to the acidity of coloured natural waters. Geochim Cosmochim Acta 47: 2031-2035
Schnoor JL, Stumm W (1985) Acidification of aquatic and terrestrial systems. In: Stumm W (ed) Chemical processes in lakes. Wiley, New York, p 311
Sjors H (1963) Bogs and fens on Attawapiskat River, Northern Ontario. Bull Nat Mus Canada 186: 45-133
Stumm W, Morgan JJ, Schnoor JL (1983) Sauer Regen, eine Folge der Storung hydrochemischer Kreislaufe. Naturwiss 70: 216-223
Tallis JH (1964) Studies on southern Pennine peats II. The patttern of erosion. J Ecol 52: 333-334
Underwood JK, Vaughan HH, Ogden JG III, Mann CG (1982) Acidification of Nova Scotia lakes II. Ionic balances in dilute waters. Manuscript, Nova Scotia Dep Environ, Halifax, N.S.
Urban NR, Eisenreich SJ, Gorham E (this volume) Proton cycling in bogs: geographic variation in northeastern North America.
Wheeler GA, Glaser PH, Gorham E, Wetmore CM, Bowers FD, Janssens JA (1983) Contributions to the flora of the Red Lake Peatland, northern Minnesota, with special reference to *Carex*. Amer Midl Nat 110: 62-96
Wright RF, Henriksen A (1977) Chemistry of small Norwegian lakes with special reference to acid precipitation. In: Rep. IR 33/77, SNSF Project, Oslo, p 1.1

INTERACTIONS OF SPHAGNUM WITH WATER AND AIR

R.S. Clymo

School of Biological Sciences, Queen Mary College, London E1 4NS

ABSTRACT

Peat-accumulating wetlands occupy 2-3% of the Earth's land surface. Sphagnum, an important constituent of much of the peatland vegetation, is responsible for initiating acid conditions in ombrotrophic bogs and, because it decays disproportionately slowly, becomes over-represented in peat. Several features of Sphagnum physiology are important: (1) the plant produces polyuronic acids which, by cation exchange, release H^+ into the bog water; (2) it is sensitive to the combination of high pH and high Ca^{2+} concentration together, though not to each separately; (3) it is sensitive to even moderate concentrations of o-phosphate, NO_3^- and NH_4^+; and (4) it is sensitive to moderate concentrations of HSO_3^-.

Cation exchange may be an important source of acidity in some bogs but is probably less important generally than was once thought. The role of coloured organic acids as primary sources of acid is not clear. Acid rain sensu stricto has not been shown to affect Sphagnum, but atmospheric pollution in the wide sense is responsible for its disappearance from badly polluted areas of the southern Pennines.

Since the last glaciation, peatlands have been a 'sink' for atmospheric carbon, but some bogs in Europe, at least, are becoming less effective as they approach the natural limit to their growth. Death of their vegetation, where it occurs, and mining of peat both contribute to increasing atmospheric CO_2 concentration, the extent of which can only be guessed. Nor do we know how peatlands would respond to increased concentrations of CO_2 in the atmosphere.

INTRODUCTION

Estimates of the area covered by peat-accumulating systems have tended to increase. A recent one (Kivinen and Pakarinen 1981) of 420 Mha - about 3% of the Earth's land surface - was obtained by summing national estimates by other authors who used a variety of criteria. Olson et al. (1983) used similar methods and recorded 90 Mha of bog and mire, and 380 of low-arctic tundra. Their central estimate of phytomass carbon in these two components was 2.0 and 3.8 Gt in a world non-ocean total of 560 Gt. Estimates of the phytomass of forests, which cover 4000 Mha, include that in trunks and branches of trees. However, the estimates for peatlands ignore the peat, which has almost as important a role in these ecosystems as trunks and branches do in forests. A crude estimate of peat mass can be made. Assuming that the mean bulk density of dry matter in peat, taken to be of average composition CH_2O, is 0.1 $g.cm^{-3}$ and that the mean depth of peat is 1 m, then the total carbon in peat and its surface vegetation (using cm for calculation) is about $420 \times 10^{14} \times 0.1 \times 100 \times (12/30) / 10^{15}$ = 170 Gt. The value of 1 m for peat depth is little better than a guess, but it seems possible that perhaps 1/4 to 1/3 of readily accessible vegetable carbon is in peatlands. These ecosystems have usually been considered as a sink for carbon, so a substantial change in the rate at which peat is accumulating might have noticeable effects of the concentration of CO_2 in the air.

Among peat-forming plants, one of the most important is Sphagnum - the bog moss. A general account of its ecology and physiology is given by Clymo and Hayward (1982). Of the 200-300 species about twenty are quantitatively important. Each species has its own range of tolerance of water supply and solute concentrations, but all are able to make their environment unusually acidic by cation exchange, and most have an unusually low rate of decay. Most Sphagnum-dominated peatlands are ombrotrophic, even if they rely to some extent on focussed drainage from a larger catchment to keep them in hydrological balance in dry periods. Two factors contribute to the vulnerability of peatlands to atmospheric pollutants when compared with many mineral sites: (1) the dependence of peatlands on precipitation, and (2) the sensitivity of Sphagnum to atmospheric pollutants because of its one-cell thick, uncuticularized leaves. The vulnerability of peatlands to inorganic compounds of S and N in areas with relatively high pollution has been convincingly shown; the evidence is reviewed by Lee et al. (this volume). They also give reasons for suspecting that there may be sub-lethal effects in less heavily polluted areas.

In this article I consider some general interactions between Sphagnum and water chemistry, and some possible effects on the peat-accumulation process.

EFFECTS OF SPHAGNUM ON WATER CHEMISTRY

One of the most obvious effects of Sphagnum plants is that they make the surrounding water acidic. Some, and probably most, of the initial effect is a consequence of cation exchange. Up to about 30% of the plant dry mass is uronic acid residues in long polymers. There seem to be two uronic acids in about equal amounts: galacturonic acid and 5-keto-D-mannuronic = D-lyxo-5-hexosulopyranuronic acid (Painter 1983a). The latter is unusual among naturally occurring sugars in its ability to crosslink glycan chains. The cation-exchange capacity of Sphagnum is close to that predicted from the measured polyuronic acid concentration (Clymo 1963; Spearing 1972). Both are related to the dryness of the microhabitat; hummock species have values about twice those of pool species. The same tendency is found among individuals of the same species in different microhabitats. Not all plant species of oligotrophic acid habitats have a large cation-exchange capacity. Roots of Eriophorum angustifolium, for example, have a capacity about 1/10 that of Sphagnum. There is no obvious explanation for these trends and differences.

Sphagnum is extraordinarily efficient at producing dry mass, which consists mostly of polysaccharides, with a small investment in cytoplasm. Most of the uronic acid is in the holocellulose fraction, and cation-exchange properties persist after death. It is likely, therefore, that most of the exchange capacity is in the cell walls. The whole plant may be considered as if it were in a cation exchange phase, which would come, if left for a short time into equilibrium with the surrounding solution. This does indeed occur (Fig. 1), showing that in the live Sphagnum plant most, if not all, the uronic acids are manufactured in the free acid form, i.e., as -COOH (or $-COO^- H^+$) rather than as $-COO^- R^+$, where R^+ represents some cation other than H^+.

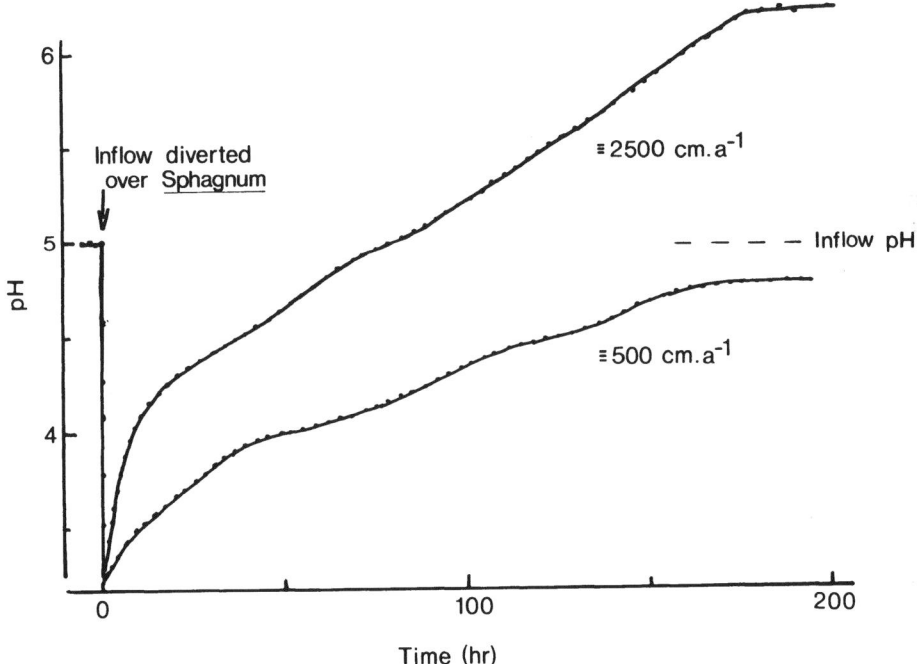

Fig. 1. pH of simulated rain flowing over Sphagnum recurvum at two rates. Both are much greater than normal precipitation averaged over a few days. The plants were in continuous light. At the arrow at t = 0, the solution, which had been bypassing the plants, was diverted over them.

The formation of a uronic acid rather than a sugar may be considered overall as:

$$-CH_2OH + O_2 \rightarrow -COOH + H_2O.$$

Given the presence of O_2, it is not necessary that H^+ be involved in this reaction. It is apparent from Fig. 1 that in the higher flow-rate the pH of the effluent rises as the H^+ on the exchange sites is slowly leached away. Eventually the pH of the outflow exceeds that of the inflow, perhaps as a consequence of photosynthesis (though curious results are found when light and dark alternate). The higher flow-rate, equivalent to 2500 $cm.y^{-1}$, is very much greater than natural rates of precipitation

averaged over periods of more than a day or so. Even though the
plants are living and, one assumes, producing new -COOH groups,
they cannot produce them fast enough to keep up with the monsoon
conditions of this part of the experiment. At the lower rate of
inflow (Fig. 1) the plants are able to maintain their pH below
that of the inflow. This steady-state pH clearly depends on the
plant growth rate and on the solution (precipitation) flow-rate.
It also depends on the concentration of solutes. A simple theory
(Clymo 1967) shows that with measured growth rates, effective
precipitation and solute concentrations, a mean pH of about 4.2
could be maintained in an unpolluted atmosphere. An extension of
the theory (Clymo 1984b) allows the effects of acidified rain,
evaporation, rain composition and temporal and spatial variations
of Sphagnum growth to be calculated. At Moor House (575 m altitude in the northern Pennines of England), during the year in
which detailed measurements of growth and rain chemistry were
available, Sphagnum contributed about half of the measured
acidity. The rest came in the rain, which had a volume-weighted
mean pH of 4.1.

The pH of Sphagnum hummocks at the Moor House site in the period
1968-70 showed autumn and spring values of from 4.2-4.5, with a
summer pH depression to 3.4, but with a sharp return to higher
values with the heavy autumn rains. At that time the pools
became, for a short time, more acid than the hummocks. The
values of pH calculated for the hummocks from the measured
properties of precipitation and from measured plant growth rates
agreed fairly well with the measured values except during
drought. This is assuming cation exchange to be the only process
contributing to further acidification of the bog water.

The concentration of other ions in the water around the plants
was also measured. During the summer, there were strong correlations in hummock samples within the group H^+, Na^+, K^+, Ca^{2+},
Mg^{2+}, Al^{3+}, Cl^- and SO_4^{2-}, implying that evaporative concentration may have been important. On the assumption that Cl^- is
relatively biologically inert, Cl^- concentration was used to
assess the extent of this process. The result was that the solution around the plants had, in the extreme cases, been concentrated 7.9-fold. Theoretically, a pH of 2.9 would be predicted
while the measured value was actually 3.4. Some of this discrepancy can be accounted for if it is allowed that the last rain
before the drought probably drained down the plants first, equilibrating with them as it did so, and then moved up again as
evaporation began. This was simulated as a 'batch' of rain (Fig.
2) moving down through ten layers of Sphagnum then back up again,
equilibrating with each layer as it moved. The simulation used
measured values of ion concentration in the rain and measured
values of Sphagnum growth. The immediate effect of the 'last'
rain, which had a lower concentration of cations than that in the
previous 3 months, was to allow the pH to rise by about 0.1 unit.
After evaporation to the point at which the sum of cation concentrations in the top layer was 7.9 times that in the 'last' rain,
and after replacement of solution from below, there was a very
steep concentration gradient at the surface. This sort of thing
occurs naturally as well; after long dry periods one may find the

surface of Sphagnum encrusted with brown tarry concentrates of what was once dissolved organic matter, or even (although rarely) with crystalline salts. The surface layer is very different from all the others, but the volume-weighted mean pH is 3.5, compared with the measured 3.4. (The choice of 10 layers was arbitrary, but the number of layers has only a tiny effect on the result because the 'last' rain is distributed among as many layers as are chosen.) This simulation is crude, but it does serve to show the sort and magnitude of effects that Sphagnum may have on the water around it. It is this water, of course, that eventually runs off into the surrounding streams.

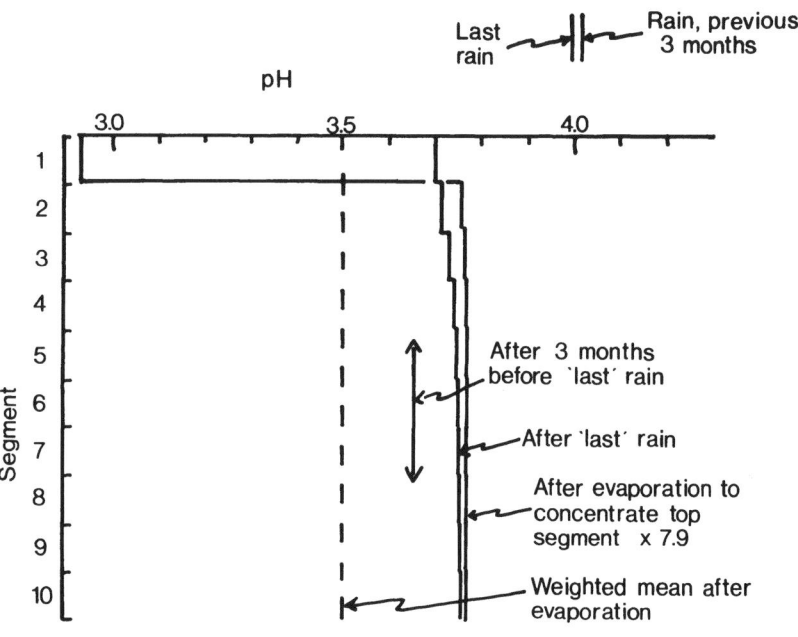

Fig. 2. Simulation of the consequences of cation exchange in a Sphagnum hummock. During April - June, the plants produced 2.4 g.dm^{-2}. This has been arbitrarily allocated to segments 2 - 9 and equilibrated with the measured 330 mm of rain of volume-weighted pH 4.01 and concentration of mono-, di- and tri-valent metallic cations 0.07, 0.16 and 0.04 mmol.L^{-1}. In the next three weeks, plants grew 0.27 g.dm^{-2} (segment 1) and received 39 mm of rain, pH 3.94, and cation concentrations 0.03, 0.06 and 0.5 mmol.L^{-1}. This rain was first assumed to flow down, displacing solution from the segment below and reaching a new equilibrium, in 10 equal aliquots. The following drought, which resulted in a 7.9-fold increase in the Cl$^-$ concentration, was simulated by allowing the solution to move upwards in aliquots, equilibrating at each stage with each segment. This, of course, produces a steep concentration gradient at the surface because solutes accumulate there.

While the concentrations of most ions in the solution around hummock-Sphagnum were strongly correlated with each other (and much more so than in rain) they were almost uncorrelated with those of NH_4^+ and NO_x^- ($NO_3^- + NO_2^-$), which were themselves correlated at r = 0.74, n = 38. The quotient NH_4^+/NO_x^- was 0.9 in rain, 7 in pools, and 9 in hummocks. It seems clear that there must be rapid and extensive interconversions of nitrogen-containing compounds - a conclusion supported by detailed studies reported i.e., by Hemond (1983). That these may be directly attributed to Sphagnum or to its associated microflora is shown by the work reviewed by Lee et al. (this volume).

Cation exchange is not the only possible cause of acidity in the surface water of Sphagnum bogs or, perhaps more importantly, in the runoff from them. From time to time it is claimed that high concentrations of CO_2 are found in bog water, and that these may cause the pH to fall to less than 4.0 (e.g., Villeret 1951). If CO_2 is bubbled through bog water the pH falls; conversely, for some samples, if the CO_2 is flushed out by bubbling N_2 gas through the sample then the pH rises to values of about 6.0. In only 5 of 29 samples of water around live Sphagnum from Moor House did the pH rise by more than 0.5 unit when treated in this way - a result similar to that recorded by Gorham (1956a). These five samples were all from places in which the water was probably static and may have been effectively subsurface drainage. They were the only samples that were visibly coloured yellow or brown. It may very well be that microbial activity in peat produces substantial amounts of CO_2; the concentration in peat 1-4 m below the surface was 1-10 mmol.L^{-1} (Claricoates, pers. comm.). This could contribute to the observed pH of about 4.0. But a peat-accumulating system must, on balance, be a sink for CO_2. If there are high concentrations in the catotelm (Ingram 1978) then there cannot also be high concentrations in the surface acrotelm. The CO_2 in the catotelm is thus a secondary, not a primary, source of acidity.

Another possible source of acidity is related to the supply and interconversion of S- and N-containing compounds. In Sphagnum bogs there are few of the neutralization reactions that exist in calcareous fen peat and mineral soils. Therefore, after the modifications imposed by cation exchange, most of the H^+ in precipitation contributes directly to the observed acidity. The anions may, however, undergo conversions which also affect the acidity. One example, simplistically:

$$2H^+ + SO_4^{2-} \text{ (anaerobic)} \rightarrow \leftarrow \text{ (aerobic) } H_2S + 2O_2,$$

is probably microbiologically mediated. Gorham (1956b) was the first to suggest that the reverse reaction might account for the observed approximately four-fold higher concentration of SO_4^{2-} in bog pools during dry weather. At the same time, the pH fell to below 4.0. The method used for analyses of SO_4^{2-} included organic anions (Gorham, pers. comm.). The concentration of other ions was also higher after dry weather: Na^+ and Cl^- by about two-fold; Ca^{2+} and Mg^{2+} by about four-fold. Evaporation may, therefore, have been partly responsible in the way already discussed. The mechanism of sulphide oxidation itself remains plausible. Hemond (1980, 1983) and Urban et al. (this volume) have made the most detailed attempts yet recorded to calculate the effects on acidity of assimilation of S and N into organic combination. In effect, when N in NH_4^+ is incorporated into neutral organic molecules, there is a concomitant acidification. The reverse is true of NO_3^- and SO_4^{2-}.

The last primary source of acidity may be loosely called excreted free organic acids. Ramaut (1955) extracted small amounts of an acid from Sphagnum recurvum and identified it as succinic acid or one of its polymers. He thought it might be excreted and might be at least partly responsible for the acidity of bog water. It is easy enough to identify a variety of organic acids in Sphagnum by chromatography, but in the only reported experiments in which C-14 was supplied to Sphagnum the rate of appearance of organic compounds in the water was tiny; over 1 day less than 0.1% of the C-14 fixed by the plants in soluble organic form appeared in the water, and most of that was in sugars (Clymo 1967). The rate of release of organic compounds is notoriously dependent on the environmental conditions, however, so perhaps this possibility should not be dismissed. More interesting is the recent demonstration (Hemond 1980; Gorham et al. 1985; Urban et al., this volume) that a very wide geographic range of North American bog waters contained organic anions at concentrations of 0.05-0.3 $mmol.L^{-1}$ - a result foreshadowed by Malmer (1963). This may be represented as pA 4.3-3.5 to allow comparison with pH. The concentration was strongly correlated with that of dissolved organic carbon and with the absorbance at 320 nm. On photooxidation the pH, in most cases, rose to 6.0 or above. This is a convincing demonstration that some of the H^+ is associated with these yellow or brown 'humic' or 'fulvic' acids of unknown chemical constitution. Painter (1983b) has recently shown that the coloured organic anions in one sample of peat water are probably derived from carbohydrates rather than from lignins. But whether or not these are the source of H^+ is a much more difficult question to answer. The dissolved organic anions are probably products of breakdown. Some might be colourless chunks of original plant material liberated from polymers in pieces small enough to be soluble. Small carbohydrates of this kind are usually rapidly attacked by microorganisms. Most of the yellow and brown anions are probably new productions by microorganisms. In the first case, the associated H^+ is scarcely new: the 'source' is the original plant. In the second case, it may be 'new', but one still has to account for the original H^+, both that in the water and that left in the exchange phase. This requires a clear model of the structure and processes in the surface layers - approximately the top 30-50 cm - of a peat bog.

Suppose the surface of a bog is Sphagnum-dominated. New material is added to the top few centimetres, while aerobic decay occurs down to the water table at perhaps 2-30 cm depth. Because the rate of diffusion of O_2 in water is only 10^{-4} that in air, the peat below the water table becomes anoxic as long as microorganisms are active, and the rate of decay then drops by several orders of magnitude. Aerobic decay in the acrotelm, and the accumulation of mass, eventually cause macroscopic structure at the base of the acrotelm to collapse. (The first snows of winter, at a time when the surface is not frozen solid, may be particularly important.) After collapse, the spaces between elements are very much smaller; therefore, hydraulic conductivity decreases by several orders of magnitude. Most excess water, therefore, flows <u>laterally</u> rather than through the newly collapsed structure and all that lies beneath it. It is this process that is the main cause of the high water table in bogs. Capillary forces are of secondary importance (Ingram 1983). This model is not universally applicable of course, but it will serve to direct attention to the fate of the organic matter.

Any particular piece of organic matter has to wait in aerobic conditions, during which the rate of decay is relatively great, until it collapses and passes into the catotelm. During this time perhaps 90% of the dry matter is lost (Clymo 1984a). Carboxylic acids may disappear by reactions that have the overall effect: $4COOH + O_2 \rightarrow 4CO_2 + 2H_2O$. There is an implied change in the concentration of H^+ insofar as the concentration of COOH was itself buffering H^+. But the acids associated with a metal cation, $-COO^-R^+$, cannot simply abandon the R^+ without a counter-ion. Their charge might be balanced if microorganisms use the $-COO^-$ as a substrate and release in its place equivalent amounts of alternative (yellow or brown) acid anions. Organic acids might also be produced using neutral plant carbohydrates (just as uronic acids are produced by plants). These would be <u>new</u> acids. The main possibilities are summarised in Fig. 3.

There appear to be differences in the bogs of England and those of North America with respect to organic anions. Hemond (1980), Gorham <u>et al</u>. (1984), and Urban <u>et al</u>. (this volume) all report substantial concentrations of yellow or brown organic anions in summer samples of bog water from North America. Hemond's samples were probably from the catotelm but the others were above or just below the water table. However, in similar midwinter samples from bogs in England Gorham <u>et al</u>. (1985) and Urban <u>et al</u>. (this volume) found no more than a very low concentration of organic anions perhaps because these pools had been flushed by recent rain (Gorham, pers. comm.). Clymo (1984b) reported that the concentration of cations and inorganic anions in an English blanket bog were close to equilibrium. The balance was maintained even in summer. It may be that the relatively high concentration of marine and pollution solutes masked the organic anions and that the relatively great excess of precipitation over evaporation prevented summer concentration of the organic anions in pools to the extent found in North America (Gorham, pers. comm.).

As Gorham <u>et al</u>. (1985) point out, the higher concentrations of organic anions are generally found in those areas in which precipitation is little more than evaporative, and perhaps below it in summer when the samples were collected. Not only is the lateral flux of water small, but solution around the lower part of <u>Sphagnum</u> plants may be drawn up to the apices by evaporation and concentrated there to the surprising extent already mentioned.

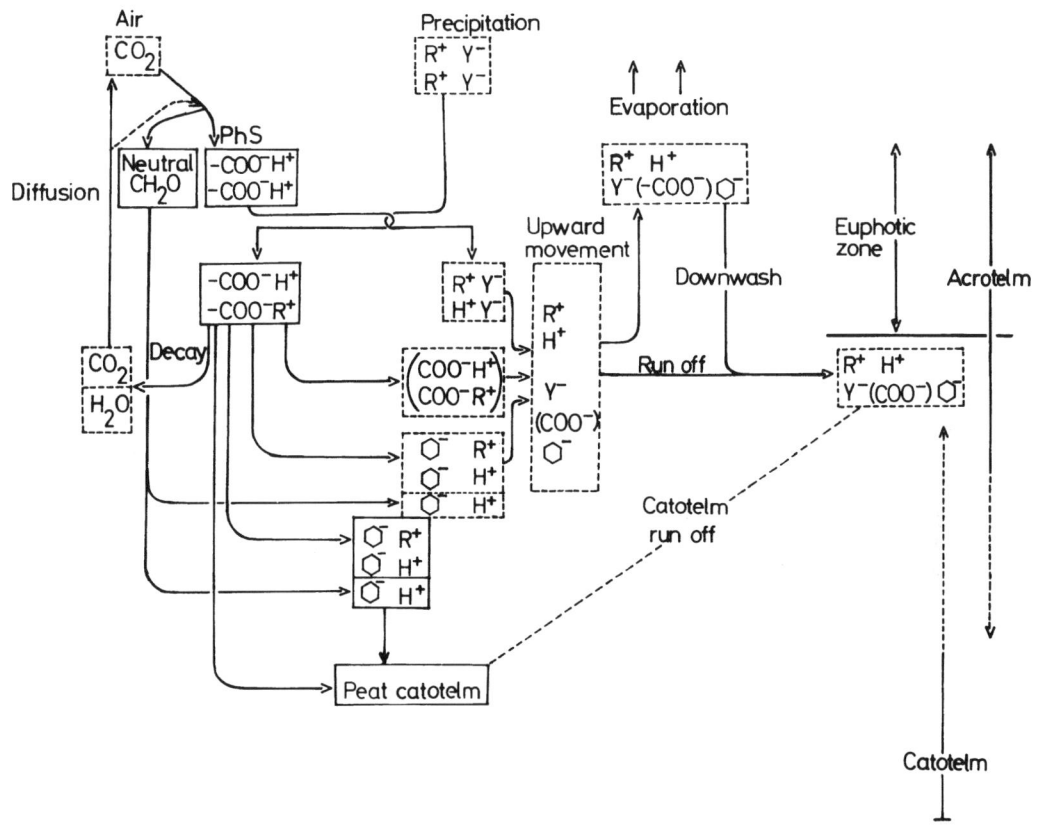

Fig. 3. Some of the processes involved in acid production in bog waters around Sphagnum. The effects of CO_2, of direct acid precipitation, of some forms of cation and anion uptake into plants, and of the interconversion of the various forms of N and S are omitted. Soluble forms are surrounded by broken lines, insoluble ones by a continous box. The symbol R^+ indicates a cation other than H^+; Y^- is an inorganic anion; $-COO^-$ represents uronic acid; ⌬⁻ represents yellow or brown coloured 'fulvic' or 'humic' acid, perhaps aromatic, but perhaps derived from carbohydrate (Painter 1983b) and shown in parentheses. Some of the CO_2 produced during decay may be taken up again at the plant apex without escaping to the air. Decay does continue in the catotelm and some water does flow through it. These three processes are shown by broken lines.

The existence of high concentrations of coloured organic anions serves to buffer the pH of the water at about 4.0, but it does not per se tell us about the primary origin of the H^+. We need to know to what extent these coloured anions are chemically equivalent to the $-COO^-R^+$ that have been destroyed, and to what extent they are new products in the H^+ form.

An approximate calculation of the chemical nature of the coloured anions can be made. Suppose Sphagnum productivity is 300 $g.m^{-2}.y^{-1}$ with 20% of dry mass as polyuronic acid. Suppose also that 50% of the H^+ is exchanged for other cations. This proportion will be greater the larger the value of P - E (precipitation - evaporation) and the greater the concentration of non-H^+ cations in the precipitation. Suppose further that 90% of the plant mass is lost during passage through the acrotelm (Clymo 1984a). The molecular mass of a uronic acid residue is 176; therefore the rate of destruction of those $-COO^-$ associated with non-H^+ cations is 300 x 0.2 x 0.9 x 0.5 / 176 = 150 $mmol.m^{-2}.y^{-1}$. For dissolved 'fulvic' and 'humic' acids the mean negative charge density, on a carbon basis, of many samples was 10 $mmol.g^{-1}$ (Oliver et al. 1983). If the lost non-H^+ uronic acids were all replaced by coloured organic anions in solution in this way then the mean concentrations would be:

P - E	$(cm.y^{-1})$	1	10	100
Coloured anions	$(mmol.L^{-1})$	15	1.5	0.15
Coloured anions as DOC	$(mg.L^{-1})$	1500	150	15

Most of the recorded concentrations lie toward the right hand side. On the other hand, as Gorham and Urban (pers. comm.) point out, for 14 samples in a transect across North America (Gorham et al. 1985) the mean concentration of H^+ was 107 $\mu mol.L^{-1}$, of other cations was 75, of $Cl^- + SO_4^{2-}$ was 21, (and of the anion deficit, probably mostly coloured organic anions, was 157 $\mu mol.L^{-1}$). This indicates a minimum of (107 - 21) / (107 - 75) = 0.47 $\mu mol.L^{-1}$ of 'new' coloured organic acids. However, this makes no allowance for undissociated acids.

In short, it is possible that the coloured organic anions are no more than microbially-produced replacements for non-N^+ associated carboxyl groups originally formed by the plants as polyuronic acids. One calculation indicates that they may all be replacements, but another indicates that half or more may be 'new'. We need more studies of the chemistry and microbiology of peat-forming ecosystems.

EFFECTS OF WATER CHEMISTRY ON SPHAGNUM

Most species of Sphagnum are unable to survive prolonged submergence in most, and particularly calcareous, groundwaters. From observations of their field behaviour as colonists in fens, it would seem that a few species, have a greater tolerance than most (e.g., S. squarrosum, S. fimbriatum). Is it the high ionic strength, more specifically Ca^{++}, or the pH to which the plants respond? Wilcox (1984) showed that S. recurvum continued to grow in NaCl at 40 mmol.L^{-1}, although at a reduced rate, and Paul (1908), recorded by Skene (1915), showed that several species of Sphagnum were able to grow in $CaSO_4$ solutions as concentrated as 15 mmol.L^{-1}, equivalent to a 1/2 Ca of 30 mmol.L^{-1}. Experimental work is bedevilled by the high cation-exchange capacity of the plants: any attempt to change the ambient concentration of cations is 'resisted' by the plants. It is necessary to flush them with large volumes of solution if approximate control concentrations in the solution are to be imposed. When this was done (Clymo 1973), it became clear that pH and Ca^{2+} concentration had independent effects, but also interacted. It was the combination of pH 7.5 and 1/2 Ca^{2+} concentration of 5 mmol.L^{-1} that reduced most species to bare survival. Two groups of species emerged in these experiments. The more tolerant were Sphagnum inundatum, S. squarrosum, S. subnitens, and S. recurvum. Less tolerant were the 'high moor' species S. papillosum, S. capillifolium, S. magellanicum, and S. cuspidatum.

Concentration gradients in the field can sometimes be very steep. For example, at Sunbiggin Tarn in northern England (National Grid reference NY 6808) 60-cm tall hummocks of S. fuscum rise apparently directly from mineral soil flushed by water so calcareous that tufa forms. The pH in water around the plants falls in adjacent 4-cm thick samples from 7.3 at the base to 4.1. The 1/2 Ca^{2+} concentration falls from 1.45 to 0.07 mmol.L^{-1} at the same time (Bellamy and Rieley 1966).

Rate of supply may be just as important as concentration (Pearsall 1950). Species such as Sphagnum inundatum and particularly S. recurvum may grow extremely rapidly (more than 50 cm.y^{-1}) in the flushed habitats of slow flowing ditches or drainage lines on hillsides. Flushing at rates up to the equivalent of precipitation at 30 cm.day^{-1} (3 L.dm^{-2}) increased growth by up to 15% (Clymo 1973). This increase is probably much less than that produced by favourable conditions in the field.

This comparison of rate and concentration was further investigated in an experiment with S. papillosum. Single plants, initially 4.0 cm long, were supported in a test tube with the capitulum top about 1 cm above the level of 17 mL of solution. The solution was either left in position and topped up with distilled water as necessary to compensate for loss by evaporation, or replaced (equivalent to flushing) every 2 or 3 days. The solutions contained one or more of $NaNO_3$, NH_4Cl, and NaH_2PO_4. No attempt was made to control the pH, but it was measured and varied between 4.1 and 5.2, tending to be lower in the unreplaced solutions. The most concentrated solutions of the two forms of N were 30 mmol.L^{-1}, and of ortho-P 6 mmol.L^{-1}. These were calculated to supply, in 17 ml, about 100 times the amount initially in the plants (i.e., x100 treatment). Dilutions to supply x50, x20, x10, x1, x0.1 and '0' (distilled water) were also used. Each treatment was duplicated and the tubes were randomized in racks, protected from rain, and put on a north-facing windowsill of a building in the relatively unpolluted Hampstead area of northwest London. The experiment ran from January to June 1981. As expected all plants in the x100 and x50 treatments of all three

solutes died within 2-3 weeks. Results for the rest are shown in Fig. 4. Separate '0' treatments were included for all three single-solute series, but these are in fact all the same treatment. The results of all six agreed well, as did most of the duplicates throughout the experiment. The results showed the following:
(1) In the unreplaced series all the treatments reduced growth, compared with the nominal '0', except perhaps that which was most dilute in NH_4^+.
(2) Replacing distilled water - the nominal '0' - increased growth, although in three of the six cases the plants died shortly before the end of the experiment in mid-summer. It may be that N and P had been removed to such an extent that the plants could not survive during active growth.
(3) Most of the other results are consistent with the hypothesis that, within the range of concentrations and rates of supply used, a higher supply reduces growth, although there may have been an initial stimulation, followed in spring by death.
(4) The x0.1 treatment with unreplaced NH_4^+ was less inhibitory than the similar one with NO_3^-.
(5) The mixtures of solutes allowed the plants to survive better than might have been predicted from the single-solute results, but the plants still grew poorly.

These results are consistent with field observations that dilute solutions of inorganic N and P compounds kill Sphagnum - at least those species in the 'high moor' group. But all these experiments lasted for a relatively short time. It is interesting, therefore, that the same effect is seen on hummocks where owls or grouse have perched and defecated: Sphagnum in the immediate vicinity dies, but that a little distance away may grow faster. A similar chance observation of increased growth of Sphagnum around ground rock phosphate applied in a fertilizer trial was made by McVean (1959). These two field observations extend the time scale over which effects may occur. Lastly, the growth of S. fuscum in the field is stimulated by balanced fertilization with N and P (Gardetto, pers. comm.).

The lowest concentrations supplied deliberately in the present experiments were 30, 30 and 6 $\mu mol.L^{-1}$ for NO_3^-, NH_4^+, and otho-P, respectively. The N concentrations are similar to those in rain in the UK and southern Norway (Barrett et al. 1983). No attempt was made to exclude gaseous dry deposition, although particulate deposition must have been much reduced if not completely prevented by the transparent covers. Even so, if unintended supplies had been important then one might not have expected the lowest deliberate rates of supply to have had such a depressing effect as did, for example, NO_3^-.

It seems clear that the growth of at least the 'high moor' species of Sphagnum is affected by concentrations or rates of supply of inorganic N that are commonly found nowadays in precipitation. Even present supplies may be beyond the nutrient and neutral ranges and into the toxic one. The results reported by Lee et al. (this volume) differ in detail but lead to similar conclusions.

525

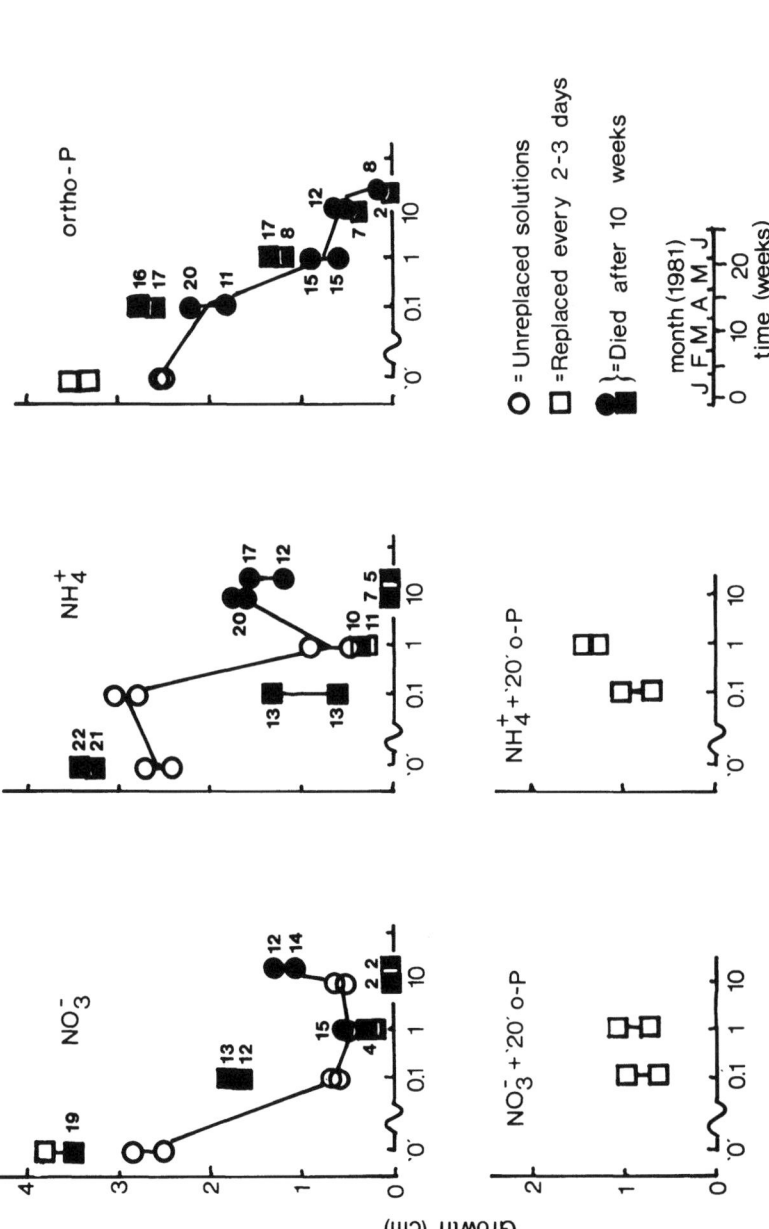

Fig. 4. Survival and growth of Sphagnum papillosum plants in relation to supply of NH_4^+, NO_3^- ortho-P, or a mixture. 'Concentration' is in units of the amount in the plants at the start of the experiment. In the replacement series (squares) the solution was replaced every 2 or 3 days. Filled symbols indicate plants that died after the number of weeks shown adjacent to the symbol. Other details are given in the text.

There is one curious exception to this. Boatman and Lark (1971) showed that the protonema of Sphagnum flourished in, and perhaps needed, concentrations of P of about 1 mmol.L^{-1}. In the field protonema is almost unknown, but it has recently been discovered (Clymo and Duckett, in press) that peat below Sphagnum contains abundant spores which can easily be stimulated to germinate and eventually to produce mature gametophytes, which are the plants one normally sees, in situ in low concentrations of solutes. High concentrations are not necessary.

This discussion has been selective, concentrating on pH, Ca, N and P. Other compounds, such as those containing S, have as great or greater effects (Lee et al., this volume.). Yet other compounds, such as those containing Pb and Cd, or AsO_4 and F have effects on the ultrastructure and growth of Sphagnum if the concentration is sufficiently high (Simola 1977). Doubtless others do too. Their importance in nature remains to be determined.

It is of some interest to know the cytoplasmic pH of Sphagnum. If it is similar to that of the environment in which the plants normally grow, then either it is unusually low or the proton gradient across the plasmalemma is unusually great. Preliminary results (Clymo and Hawkes, unpub.) with NMR spectroscopy of the ortho-P peak allow one to infer that the cytoplasmic pH of S. cuspidatum growing in a pool of pH 5.5 was also 5.5. This is an improbably low value for cytoplasm, and is being checked by an independent method. The value proved to be strongly 'buffered'; a change of pH of the solution by 2 units in either direction produced a change in the cytoplasm of no more than 0.1 - 0.2 units. These differences were maintained for several days at least, without impairing the ability to grow normally when returned to water of pH 5.5.

Change of pH in the water per se may be of much less importance to Sphagnum than quite small changes in the rate of supply of some inorganic compounds.

SPHAGNUM AND THE PEAT-ACCUMULATION PROCESS

The process by which Sphagnum formed at the surface is submerged by the rising acrotelm and eventually engulfed by the anoxic catotelm has been described. If the acrotelm maintains the same general character then it injects matter into the catotelm at an approximately constant rate, about 1/10 that at which matter is being added to the acrotelm by plant primary production. If this continues without change and decay ceases altogether in the catotelm then peat will accumulate without limit. But if decay continues, however slowly, then as time passes the integrated loss by decay throughout the growing peat column approaches more and more closely the (constant) rate at which it is injected into the top of the catotelm. The true rate of accumulation slows asymptotically towards zero even though the acrotelm surface is healthy and assimilating matter at the same rate that it always has done. There are two reasons for believing that this account is broadly correct. First, the age vs depth relation is curvilinear in many, but not all, European cases. This implies continued decay (or steadily accelerating productivity). Secondly, the concentrations of CH_4 and CO_2 in the peat increase steadily downwards (Claricoates, pers. comm.) to the base of the peat several metres deep. This is consistent only with continued production of CH_4, and hence of decay. The whole problem is considered in detail by Clymo (1984a). In the present context it is important because some, and perhaps many, peat-forming systems may be reaching the stage at which the true accumulation rate has become very small, so that they are no longer acting as effective sinks for atmospheric carbon. (In most North American bogs for which sufficient data are known (Gorham, pers. comm.) there seems to be, however, little evidence of a curvilinear age vs depth relation). Peat mining ('harvesting' is a euphemism) for burning or horticulture will make peat-bogs sources of atmospheric carbon. Draining may have the same effect as peat mining because it increases the effective depth of the acrotelm and hence the total mass exposed to aerobic decay.

Destruction of Sphagnum, or decrease in its growth rate caused by atmospheric pollution, may have complex effects. Sphagnum itself decays relatively slowly, probably as a direct result of its low N concentration (Coulson & Butterfield 1978). Not only does proportionally more of it survive into the catotelm, but it enables the catotelm to rise faster and hence it increases the fraction of other species that survive into the catotelm too. On the other hand, Sphagnum itself, excepting the aquatic species, usually forms a fairly thick acrotelm while a surface dominated by Eriophorum vaginatum, for example, may have a shallow acrotelm. Whether or not damage to a Sphagnum carpet, or its replacement by vascular plants, would turn peat bogs from sinks to sources of carbon cannot be confidently predicted at present.

The consequences of an increase in the concentration of CO_2 in the atmosphere are also unpredictable. Billings et al. (1983) showed that the direct increase of the rate of carbon storage in tundra was much smaller than the decreases, as a consequence of postulated increase in the temperature and lowering of the water table.

The links between those aspects of atmospheric pollution that most concern us now and those that seem likely to be occupying us by the end of the next decade remain unquantified. Sphagnum-dominated peat bogs may be useful systems on which to begin this task.

Acknowledgements

I thank E. Gorham and N. Urban for illuminating discussions during and after the meeting and Mrs. P. Ratnesar for technical help.

REFERENCES

Barrett CF, Atkins DHF, Cape JN, Fowler D, Irwin JG, Kallend AS, Martin A, Pitman JI, Scriven RA, Tuck AF (1983). Acid deposition in the United Kingdom. Warren Spring Laboratory, Stevenage, Herts SG1 2BX, UK

Bellamy DJ, Rieley J (1966) Some ecological statistics of a "miniature bog". Oikos 18: 33-40

Billings WD, Luken JO, Mortensen DA, Peterson KM (1983) Increasing atmospheric carbon dioxide: possible effects on arctic tundra. Oecologia (Berl) 58: 286-289

Boatman DJ, Lark PM (1971) Inorganic nutrition of Sphagnum papillosum Lindb., S. magellanicum Brid. and S. cuspidatum Ehrh. New Phytol 70: 1053-1059

Clymo RS (1963) Ion exchange in Sphagnum and its relation to bog ecology. Ann Bot (Lond) NS 27: 309-324

Clymo RS (1967) Control of cation concentrations, and in particular of pH, in Sphagnum dominated communities. In: Golterman HL and Clymo RS (eds) Chemical environment in the aquatic habitat. North Holland, Amsterdam, p 273-284

Clymo RS (1973) The growth of Sphagnum: some effects of environment. J Ecol 61: 849-869

Clymo RS (1984a) The limits to peat bog growth. Phil Trans R Soc Lond B 303: 605-654

Clymo RS (1984b) Sphagnum-dominated peat bog: a naturally acid ecosystem. Phil Trans R Soc Lond B 305: 487-499

Clymo RS, Duckett JG (in press) Regeneration of Sphagnum. New Phytol

Clymo RS, Hayward PM (1982) The ecology of Sphagnum. In: Smith AJE (ed) Bryophyte ecology. Chapman and Hall, London, p 229-289

Coulson JC, Butterfield JE (1978) An investigation of the biotic factors determining the rates of plant decomposition on blanket bog. J Ecol 66: 631-650

Gorham E (1956a) The ionic composition of some bog and fen waters in the English Lake District. J Ecol 44: 142-152

Gorham E (1956b) On the chemical composition of some bog waters from the Moor House nature reserve. J Ecol 44: 377-384

Gorham E, Eisenreich SJ, Ford J, Santelmann MV (1985) The chemistry of bog waters. In: Stumm W (ed) Chemical proceses in lakes. Wiley, New York, p 339-363

Hemond HE (1980) Biogeochemistry of Thoreau's Bog, Concord, Massachusetts. Ecol Monog 50: 507-526

Hemond HF (1983) The nitrogen budget of Thoreau's Bog. Ecol 64: 99-109

Ingram HAP (1978) Soil layers in mires: function and terminology. J Soil Sci 29: 224-227

Ingram HAP (1983) Hydrology. In: Gore AJP (ed) Mires: swamp, bog, fen and moor. Elsevier, Amsterdam, p 67-158

Kivinen E, Pakarinen P (1981) Geographical distribution of peat resources and major peatland complex types in the World. Annal Acad Sci Fenn A III Geol-Geogr 132: 1-28

Lee JA, Press MC, Woodin S, Ferguson P (this volume) Responses to acidic deposition in ombrotrophic mires in the U.K.

Malmer N (1963) Studies on mire vegetation in the Archaean area of southwestern Gotaland (south Sweden). III On the relation between specific conductivity and concentrations of ions in mire water. Bot Notiser 116: 249-256

McVean D (1959) Ecology of Alnus glutinosa (L.) Gaertn. VII Establishment of alder by direct seeding of shallow blanket bog. J Ecol 47: 615-618

Oliver BG, Thurman EM, Malcolm RL (1983) The contribution of humic substances to the acidity of colored natural waters. Geochim Cosmochim Acta 47: 2031-2035

Olson JS, Watts JA, Allison LJ (1983) Carbon in live vegetation of major world ecosystems. Oak Ridge National Laboratory publication 5862: 1-164

Painter TJ (1983a) Residues of D-lyxo-5-hexosulopyranuronic acid in Sphagnum holocellulose. Carbo Res 124: C18-C21

Painter TJ (1983b) Carbohydrate origin of aquatic humus from peat. Carbo Res 124: C22-C26.

Pearsall WH (1950) Mountains and moorlands. Collins, London

Ramaut, JL (1955) Extraction et purifiction de l'un des produits de l'acidité des eaux des hautes tourbièros et secreté par Sphagnum. Bull Acad R Belg (Cl Sci) Ser 5 41: 1168-1199

Simola LK (1977) The effect of lead, cadmium, arsenate, and fluoride ions on the growth and fine structure of Sphagnum nemoreum in aseptic culture. Can J Bot 55: 426-435

Skene M (1915) The acidity of Sphagnum, and it relation to chalk and mineral salts. Ann Bot (Lond) 29: 65-87

Spearing AM (1972) Cation-exchange capacity and galacturonic acid content of several species of Sphagnum in Sandy Ridge Bog, central New York State. Bryologist 75: 154-158

Urban NR, Eisenreich SJ, Gorham E (this volume) Proton cycling in bogs: geographic variation in northeastern North America.

Villeret S (1951) Recherches sur le role du CO_2 dans l'acidité des eaux tourbières a Sphaignes. C R Acad Sci Paris 232: 1583-1585

Wilcox DA (1984) The effects of NaCl deicing salts on Sphagnum recurvum P. Beauv. Env Exp Bot 24: 295-304

SOURCES OF ALKALINITY IN PRECAMBRIAN SHIELD WATERSHEDS UNDER NATURAL CONDITIONS AND AFTER FIRE OR ACIDIFICATION

S.E. Bayley and D.W. Schindler*

Department of Botany, University of Manitoba, Winnipeg, Manitoba, R3T 2N2
*Department of Fisheries and Oceans, Freshwater Institute, 501 University Cres., Winnipeg, Manitoba, R3T 2N6

ABSTRACT

Analysis of 13 years of chemical and hydrological records for three small watersheds in the Experimental Lakes Area of northwestern Ontario yielded the following conclusions:
1. The watersheds did not export significant alkalinity even though precipitation in the area has not been strongly acidic (long-term average = pH 4.9). Rates of alkalinity production in the terrestrial watersheds were a small fraction of those in lakes.
2. Forest fires did not cause an increase in the alkalinity yield of terrestrial watersheds. Instead, there was a slight tendency for alkalinity to decrease following fire due to higher releases of strong acid anions. This might have been due to drought rather than to fire per se.
3. Two years of experimental application of sulphuric and nitric acids to a small wetland at roughly 10 times normal rates deposited from the atmosphere caused little change in the acidity of outflow from the wetland. Sulphate and nitrate ions were almost completely retained.
4. In upland catchments with shallow soils, biological retention of sulphate, nitrate, and ammonium affected the alkalinity balance more than ion exchange and geochemical weathering.

INTRODUCTION

A number of studies have documented the yield of alkalinity from terrestrial watersheds in regions where precipitation has been strongly acidified by anthropogenic air pollutants (Galloway et al. 1983; Dillon et al. 1982; Wright 1983; April and Newton 1985). Typically, runoff from such areas possesses negative alkalinity[1], i.e., it has an acidifying effect on receiving waters. This is usually attributed to the exhaustion of buffering mechanisms in terrestrial ecosystems by acid precipitation (NRCC 1981; Galloway et al. 1983). None of the above studies included data on the alkalinity output from acid-vulnerable watersheds prior to acidification, or from areas where the acidity of deposition is low.

---

[1]Alkalinity is defined as: (1) $Alk = HCO_3^- - H^+$.
It may be also calculated as:
(2) $Alk = \Sigma$ base cations $- \Sigma$ strong acid anions. Negative alkalinity thus occurs when $H^+ > HCO_3^-$, or Σ strong acid anions $> \Sigma$ base cations.

The effect of acid precipitation on wetlands has been even less studied, although Gorham et al. (1984) pointed out that because alkalinity was low in "poor fen" type wetlands, they might acidify very rapidly when acid rain increases. Many of the buffering mechanisms important in lakes (Kelly et al. 1982; Schindler et al. 1986; Cook et al. 1986) or in upland watersheds (van Breemen et al. 1984) are believed not to operate in wetlands, due to the expected inhibition of microbiological activity by low pH, and to the high cation exchange capacity of bog "soils" (Puustjarvi 1959).

Finally, Brown (1984) suggested that the recent rapid increase in the acidity of waters in northeastern North America may have been caused by forest fire suppression rather than by acid precipitation. He suggested that prior to the turn of the century, periodic fires produced alkaline ash which provided buffering to downstream lakes and streams. He attributed the recent acidification of freshwaters to the reduced frequency of forest fires caused by fire suppression policies in the 20th century.

In the past 12 to 15 years, we have collected data which allow all of the above topics to be evaluated for extremely acid-sensitive ecosystems of the Precambrian Shield. This paper summarizes the amounts and sources of alkalinity in both upland and wetland watersheds in the Precambrian Shield of the Experimental Lakes Area of northwestern Ontario, reports preliminary results of acidification experiments in a Precambrian Shield wetland, and compares pre- and post-fire hydrology and stream chemistry to determine the effects of forest fire on alkalinity production.

Description of the Area

The Experimental Lakes Area (ELA) consists of 46 small lakes and their watersheds, plus several stream segments and wetlands in the Precambrian Shield, approximately 80 km east-southeast of Kenora, Ontario (Johnson and Vallentyne 1971; D.W. Schindler, unpublished data). There is no resident human population in the area except for inhabitants of the ELA research station. Human visitors to the experimental watersheds consist of occasional baitfishermen, trappers and canoeists, plus ELA researchers. Except for small parts of one or two recently-acquired watersheds, forests in the watersheds have never been logged.

Results reported here are for three small headwater streams in the drainage of Lake 239 (Rawson Lake, Fig. 1). Monitoring of precipitation and streamflow in this basin began in 1969, with methods improved several times during the period (Beaty 1981, and in prep.). Biases caused by changes in the hydrological methods are believed to be negligible in the context of this paper. Acidity of precipitation in the area has been low, with mean annual pH values between 4.9 and 5.0 prior to 1982, but with some tendency for lower pH since (Schindler, Linsey and Stainton, unpublished data.

Weekly samples for chemical analyses have been taken since 1970 in the Northwest (NW) Subbasin, and since 1971 in the Northeast (NE) and East (E) Subbasins. The period analysed here is for

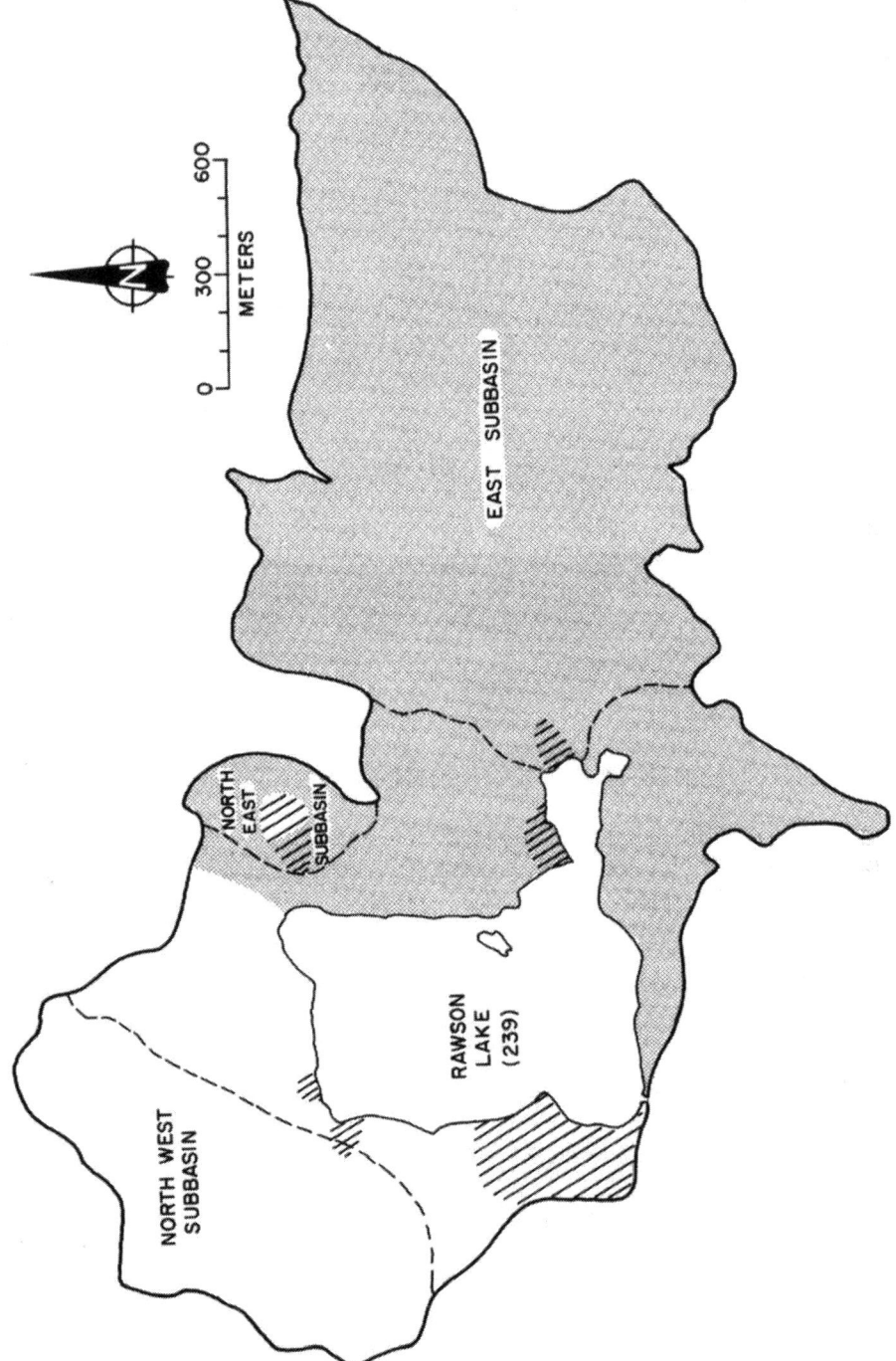

Fig. 1. A map of the Rawson Lake Watershed (Lake 239). Grey areas burned in 1974. All but the hatched areas burned in 1980. Modified from Schindler et al. (1980).

1971-1984, during which sampling and analytical methods were similar, and any changes in methods were carefully intercalibrated. Two of the watersheds, the NW and E Subbasins, drain largely upland areas with slopes of about 10%. The third (NE) Subbasin is about 1/3 wetland in area, with a much higher (16%) average slope (Table 1).

Table 1. Physical characteristics of the Lake 239 subbasins.

Characteristic	Basin		
	Northwest	Northeast	East
Area (ha)	56.4	10.6	170.3
Average slope (%)	11	16	10
Type of cover (%):			
Rock outcrop	21	3	10
Wetland	3	35	4
Upland with shallow mineral soil	72	62	80
Approx. depth of overburden (m) in the bottom of the basin	3	10	9

Two of the subbasins, the NE and E, were burned in 1974, one year after a severe windstorm caused widespread damage to the forests. The NW Subbasin was unburned, and little affected by the windstorm. Effects of the fire and windstorm on streamflow and yields of N, P, and K for the three years following the fire were reported by Schindler et al. (1980). In 1980, the NE and E Subbasins were re-burned in a second forest fire. The NW Subbasin was also burned, for the first time, in 1980. Here we shall concentrate on elements which significantly affect the consumption or production of alkalinity in the stream basins.

In 1983, an experimental acidification of the wetland in the NE Subbasin was begun using a 50-50 mixture of sulphuric and nitric acids to increase the annual hydrogen ion input to the wetland by about 10-fold. Results of the first 1.5 years of this experiment were reported by Bayley et al. (in press), and preliminary vegetation analyses by Vitt and Bayley (1984). Only results pertinent to the alkalinity budget will be summarized here.

MATERIALS AND METHODS

All three streams were gauged with V-notch weirs or flumes, with recorders providing a continuous flow record during the ice-free season. Weekly samples for chemical analyses were collected when streams were flowing, and perishable elements (pH, alkalinity, DIC, NO_3-N, and NH_4-N) were analysed the same day in the ELA field camp. Elements not affected by storage (Ca, Na, Mg, K, Fe, SO_4, and Cl) were analysed at the Freshwater Institute in Winnipeg. The methods of Stainton et al. (1977) were used for all ions except SO_4^{2-} which was measured using the gravimetric technique with barium chloride (1971, 1972) (APHA 1967) and the ion exchange technique (Stainton et al. 1977). These methods are now known to measure the sum of sulphate plus organic anions (M. Stainton, unpublished data). In 1981-1984, ion chromatography (Dionex) was used to measure SO_4^{2-}. Gran titrations for alkalinity were also performed from 1981-1984. Beaty (1981, and in prep.) gives flow records for all three streams during the period 1971-1984.

In what follows, the outflow of each chemical from a basin is referred to as the yield, while the production within a basin is the yield-precipitation input. Because Gran alkalinities were not performed before 1981, alkalinity was calculated for all years by ion balance, i.e., as the difference between nonprotolytic cations and anions (Fig. 2). Using this definition, alkalinity assumes negative values when strong acid anions exceed base cations. Negative alkalinity values were frequently found in stream samples with pH values below about 5.0. Gran titrations were also performed on waters with negative alkalinities by adding a known quantity of bicarbonate to the naturally acid samples, titrating with acid to the inflection point, and then subtracting the added bicarbonate from the value obtained. Results of Gran titrations and ion-balance alkalinities agreed reasonably well for upland catchments and precipitation (Schindler et al. 1986). Agreement for the wetland catchment was poorer, but the ion balance method was still useful to deduce changes in individual ions which could affect alkalinity.

All but a few of the precipitation events during the period of record were sampled for chemical analysis using an 0.25 m^2 "bulk" collector on an island in Lake 239 (summer) or on a low hill at the meteorological station just west of the lake (winter). All "bulk precipitation" samples were analyzed as for streamflow (Linsey et al., in press; Schindler et al., in prep.).

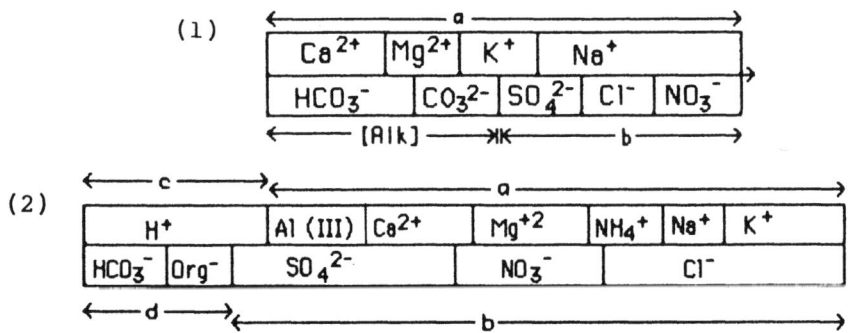

Fig. 2. A typical charge balance (in eq.L^{-1}) for (1) natural and (2) acid waters (from Schnoor and Stumm 1985). Alkalinity = a-b in the natural water, and a-b or d-c in the acidified water.

Interpreting the effects of forest fire was complicated by the occurrence of lower-than-average precipitation in 1979-81, the years just before and after the second forest fire (Fig. 3). Until 1979, it could be assumed that the effect of drought on the E and NE Subbasins, if they had not been burned, would have been similar to that on the NW (unburned) Subbasin. In order to simplify statistical comparisons, we divided data for all 3 Subbasins into three-year blocks as follows:

>   1971-73 - predisturbance;
>   1974-76 - post fire 1;
>   1977-79 - recovery;
> and 1980-82 - post fire 2.

These blocks and basins were then compared by analysis of variance (repeated measures design) followed by a multiple range test (Student-Newman-Keuls, and least significant difference).

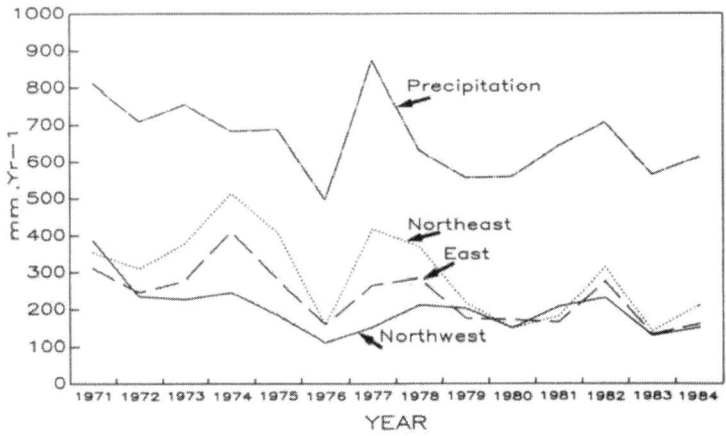

Fig. 3. Precipitation and annual runoff at the Experimental Lakes Area over the period of study. Data from Ken Beaty.

The division between the post fire 1 and recovery periods is somewhat arbitrary, designed in part to balance the distribution of data. However, Schindler et al. (1980) showed that water yield and yield of nutrients had returned to normal by three years after the 1974 fire, so that the assumption of a three-year recovery period is not unreasonable. As applied, the analysis is very conservative, i.e., any post-fire effects which were greater or less than three years in duration would tend to minimize differences between the post-fire 1 and recovery phases.

RESULTS

Gran Alkalinity Yield from ELA Watersheds, 1981-1984

Of the three watersheds, only the E Subbasin yielded significant alkalinity to Lake 239. Annual volume-weighted mean concentrations were 85-132 µeq.L^{-1} (Table 2). Alkalinity yield from the NW Subbasin was near zero, with slightly negative alkalinities in one year, and slightly positive values in the other two. The NE Subbasin yielded strongly negative alkalinities in all three years, due to the effects of the oligotrophic wetland. Of other watersheds sampled at ELA (Table 2), only one, the 661 Valley Inflow, had positive alkalinity. Both the 661 Valley Inflow and the E Subbasin contain deeper overburden than average for the area. Two streams entering Lake 114 and occasional samples from watershed 302W (unreported) had negative values. In summary, watersheds which either contain little overburden or those with high proportions of wetland supplied zero to negative alkalinity to downstream waters, despite the low acidity of deposition in the area.

Table 2. Typical Gran alkalinity from watersheds in the Experimental Lakes Area. NW, NE and E are volume-weighted mean annual concentrations. Values from other streams are averages from 5 to 12 samples per year.

Watershed	year	µeq.L^{-1}
239 NW stream	1981	- 1.2
	1982	3.5
	1983	7.2
239 E stream	1981	129
	1982	85
	1983	132
239 NE stream	1981	- 72
	1982	- 50
	1983	- 48
114 stream	1983	-4.0
	1984	-6.4
114 Cliff runoff	1983	- 15
	1984	- 11
661 Valley inflow	1983	41
	1984	45

Alkalinity Production in Watersheds

Alkalinity concentrations in outflows from both upland catchments were slightly higher than in precipitation, indicating that some buffering in these watersheds had occurred (Table 3). Of the two upland catchments studied in detail, the E Subbasin produced more alkalinity than the NW. In contrast, the NE Subbasin, with its large proportion of wetland, was always a net consumer of alkalinity.

Table 3. Annual alkalinity production (= yield - precipitation) in the three subbasins (based on Gran titrations) in eq.ha^{-1}.y^{-1}.

Year	Alkalinity of precipitation	NW	NE	East
1981	- 24	- 21	-107	237
1982	-153	161	- 3	389
1983	-111	120	43	282
1984	68	- 82	-140	NA*
1981-83, mean	-167	101	- 22	303
1981-84, mean	-108	55	- 52	NA*

*NA - not available

When the entire budget for Lake 239 is considered, the amount of alkalinity produced in the terrestrial catchment was small, with highest rates (in the E Subbasin), averaging about 20% of the alkalinity produced per unit area in situ in Lake 239 (Schindler et al., under review).

Alkalinity-Producing and Consuming Processes in the Watersheds, as Deduced from Ionic Budgets

By reasoning analogous to that shown in Fig. 2 for concentrations, the alkalinity yield from a watershed can be assumed to represent the balance between inputs and outputs of basic cations and acidic anions. The change in any ion affects this balance to either produce alkalinity (when output exceeds input for basic cations or input exceeds output for acid anions) or consume it (when input exceeds output for basic cations or output exceeds inputs for acid anions). It is therefore possible to deduce processes affecting the alkalinity of watersheds by calculating the balances of nonprotolytic cations and anions in input, output and storage. In addition, retention of organic anions (A-) in an ecosystem can produce alkalinity (Oliver et al. 1983), but we will not consider this aspect here. We assume that changes in storage of ions in the watersheds were negligible, i.e., that the outflows from watersheds were at steady-state, so that differences between inputs and outputs of a given ion represented the effect of that ion on alkalinity. Although the ion balance method gave slightly higher results, the ion balance and Gran-alkalinity methods showed good agreement for the E Subbasin, for precipitation and for Lake 239 (Schindler et al. 1986). As mentioned earlier, the agreement between the ion balance and Gran methods is poorer for other basins, probably because of unknown effects of dissolved organic matter on the charge assigned to different ions. As a result, our current conclusions for the NW and NE Subbasins should be regarded as less certain, but they would have little effect on calculations for the entire Lake 239 watershed.

Table 4. Production and yield of alkalinity from the three small catchments draining to Lake 239 for the three years 1981-83, due to reactions generating (+) or consuming (-) different ions. Data in keq. Sulphate measured by ion chromatography.

	Ca^{2+}	Mg^{2+}	Na^+	K^+	NH_4^+	SO_4^{2-}	Cl^-	NO_3^-
NW Subbasin								
precipitation input	27.3	11.7	9.2	2.4	26.9	39.9	12.0	26.4
outflow	37.6	25.4	18.9	7.9	0.5	65.8	7.2	2.9
alkalinity production (+) or consumption (-)	10.3	13.7	9.7	5.5	-26.4	-25.9	4.8	23.5
E Subbasin								
precipitation input	82.9	35.4	28.0	7.2	81.6	121.2	36.4	80.3
outflow (yield)	149.2	78.8	68.2	26.1	1.6	120.2	13.9	4.7
alkalinity production (+) or consumption (-)	66.3	43.4	40.2	18.9	-80.0	1.0	22.5	75.6
NE Subbasin								
precipitation input	5.1	2.2	1.7	0.4	5.0	7.5	2.2	4.9
outflow (yield)	6.2	3.7	3.3	1.0	0.1	3.5	0.6	0.0
alkalinity production (+) or consumption (-)	1.1	1.5	1.6	0.6	-4.9	4.0	1.6	4.9

In all of the watersheds, uptake of nitrate was the most important generator of alkalinity (Table 4). Calcium, magnesium, sodium and potassium exchange were also important in the two upland watersheds (E and NW). Sulphate reduction was important in the bog (NE) catchment. However, alkalinity-consuming processes were simultaneously at work. In both upland basins, NH_4^+ retention was most important, although SO_4^{2-} losses were a close second in the NW Subbasin (Table 4). In the wetland basin (NE) uptake of NH_4^+ the most important alkalinity-consuming process, although production of organic anions may also be of importance.

The Effects of Fire and Drought on Alkalinity Yield from the Watersheds

Two-way analysis of variance revealed that alkalinity yield from all three basins from 1974 onwards was less than or equal to that in 1971-72 (Fig. 4, Table 5). This coincides with an increase in export of sulphate (Fig. 5, Table 6), which may have been due to mineralization of organic material and reoxidation of reduced sulphur in the watersheds. The increased sulphate and decreased alkalinity from 1974 onward also occurred in the Northwest Subbasin, which was not burned until 1980, suggesting that the succession of drier-than-normal years in the mid- to late 1970s, rather than fire per se, was the cause. Reduced sulphur is deposited in the NE Subbasin as the result of microbial sulphate reduction during wetter years (Bayley et al., under review). Export of calcium and magnesium from the watersheds also increased after the fire (Fig. 6), counterbalancing to some extent the effects of sulphate export on alkalinity.

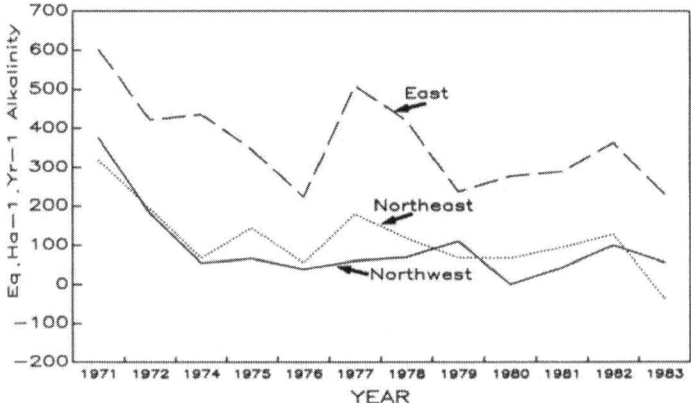

Fig. 4. Annual alkalinity yield in the three watersheds, 1971-83. There was no SO_4^{2-} or Cl^- analysed in 1973, so the year was omitted. Ion balance alkalinity was calculated as the difference between total non-protolytic cations minus total non-protolytic anions.

Table 5. Effects of fire on average alkalinity export from the three subbasins, as deduced by 2-way analysis of variance and the student Newman-Keuls' test (p = 0.05). Data in eq.ha^{-1}.ha^{-1}, calculated as the difference between basic cations and strong acid anions.

	Pre-disturbance 1971-72	Post-fire 1 1974-76	Recovery 1977-79	Post-fire 2 1980-82	mean
NW	278	53.3	79.7	46.7	99.5
NE	255	88.3	121.0	95.7	129.5
E	511	333.4	387.7	309.0	374.0
mean	348	158.6	196.1	150.4	

Differences between basins

Pre-disturbance	Post-fire 1	Recovery	Post-fire 2
E > NW = NE	E > NE = NW	E > NW = NE	E > NW = NE

Differences between periods

NW - Pre-disturbance > post-fire 1 = recovery = post-fire 2

NE - All periods equal

E - Pre-disturbance ≥ recovery ≥ post-fire 1 = post-fire 2

Interactions : none of significance

Table 6. Effects of fire on average export of sulphate from the 3 subbasins. Data in µeq.L^{-1}. Statistical procedures as in Table 5.

	Pre-disturbance 1971-72	Post-fire 1 1974-76	Recovery 1977-79	Post-fire 2 1980-82	mean
NW	80	117	138	228	146
NE	97	144	136	167	140
E	96	156	140	180	147
mean	91	139	138	192	

There were no significant differences between basins.

Differences between periods were as follows in the different subbasins:

NW - Post-fire 2 ≥ recovery ≥ post-fire 1 ≥ pre-disturbance

NE - Post-fire 2 > recovery = post-fire 1 ≥ pre-disturbance

E - Post-fire 2 = recovery ≥ post-fire 1 ≥ pre-disturbance

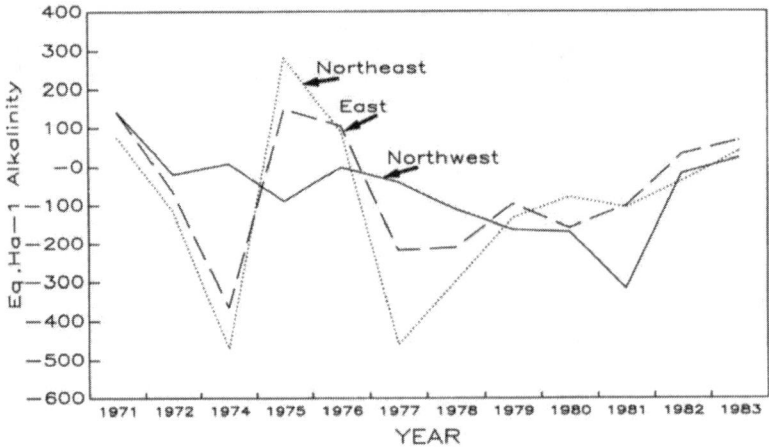

Fig. 5. Alkalinity production and consumption by reduction or reoxidation of sulphate in the three watersheds (calculated as export-precipitation of SO_4).

Effects of Experimental Acid Precipitation on the Wetland in the SE Subbasin

Chemical analysis of water samples from both the outflow and pools in the wetland revealed that sulphuric acid and nitric acids applied in monthly experiments were completely neutralized by the removal of the strong acid anions (Fig. 7, Bayley et al. 1986). Neutralization was very rapid, so that pH depressions were slight (usually undetectable) after each experiment, both in wetland pools and in the outflow. The efficiency of sulphate and nitrate retention by the wetland was not decreased by acidification (Table 7). Following reduction of sulphate, sulphides were stored in peat largely as organic sulphides or as iron sulphides (Behr 1985). Nitrate was removed primarily by plant assimilation.

Table 7. Sulphate and nitrate retention in the experimental wetland as % of total input. Data in parentheses are $eq.ha^{-1}.y^{-1}$.

	1981	1982	1983	1984
Nitrate retention	95 (80)	97 (148)	>99 (334)	>99 (457)
Sulphate retention	64 (405)	46 (441)	73 (510)	74 (802)

Prior to the acidification experiment, the wetland both produced alkalinity by reduction of sulphate and nitrate (Table 8), but unquestionably also consumed alkalinity due to production of organic anions. The sulphate and nitrate applied with the acid spray was removed in passage through the wetland, increasing alkalinity production and effectively neutralizing the added protons.

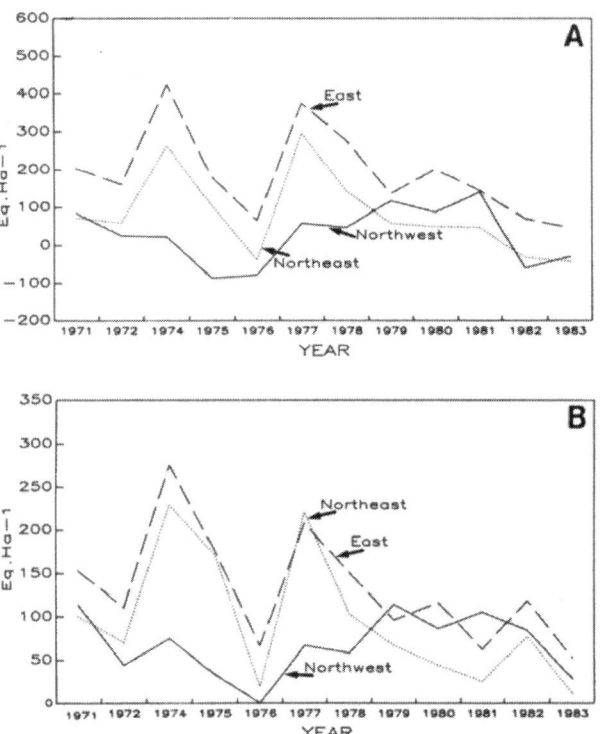

Fig. 6. Alkalinity production by A) calcium exchange (calculated as export-precipitation of Ca) and B) magnesium exchange (calculated as export-precipitation of Mg) from the watersheds.

DISCUSSION

A number of conclusions related to understanding the acidification process may be drawn from our work. Even when the acidity of precipitation is low, many watersheds in acid-sensitive areas do not yield positive alkalinity (Schindler, in press). The important alkalinity-producing and consuming mechanisms in such systems include a high proportion of biological processes, rather than being dominated by the geochemical weathering of rocks and soils. For lakes and wetlands, the production (though not necessarily the storage or yield) of alkalinity in ecosystems increases as the input of strong acids with biologically-reactive anions to the system increases. No equivalent experimental data are available for upland sites. In both lakes and wetlands, these processes are not totally efficient, so that while they

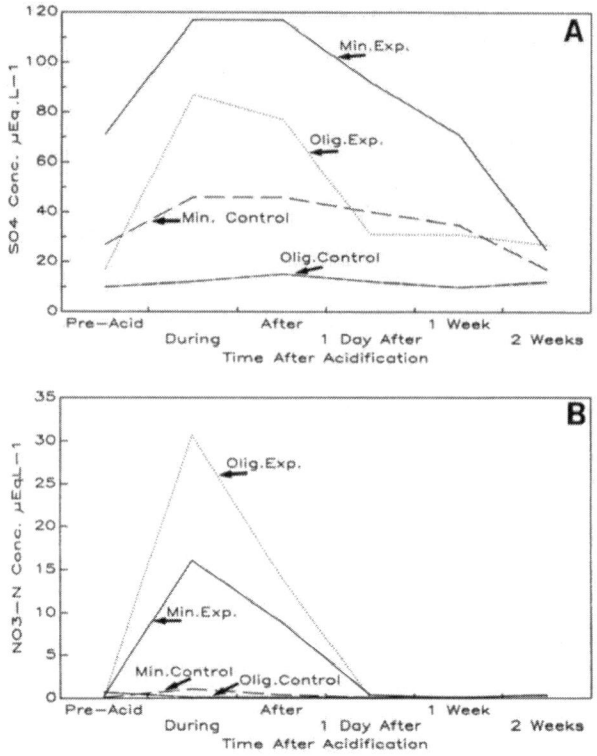

Fig. 7. Removal of (a) sulphate and (b) nitrate-nitrogen by the wetland after experimental acidification, June 1984. The mineral pool and central oligotrophic pool received acid while the mineral and oligotrophic control received lake water.

Table 8. Alkalinity budget for the experimental wetland (using Gran titration and assuming that upland runoff is proportional to the NW). Data in equivalents unless otherwise noted.

	1981	1982	1983	1984
Precipitation on wetland	- 87	-562	-407	248
Runoff from uplands to wetland	- 18	56	64	- 97
Roddy Lake water used in acid irrigation	--	--	418	564
Acid input in irrigation	--	--	-961	-2138
Total input	- 105	- 506	-886	-1423
Outflow	-1380	-1650	-715	- 765
Alkalinity produced (+) or consumed (-) by wetland	-1275	-1144	-171	- 658
eq.ha^{-1}	- 347	- 312	- 47	- 179

unquestionably retard the rate of ecosystem acidification, they do not prevent it.

The hypothesis of Brown (1984) that the buffering of watersheds by alkaline ash following forest fire prevented acidification of lakes prior to the 20th century appears to be incorrect, at least in northwestern Ontario. While the increased release of base cations from the watersheds would normally be expected to cause an increase in release of alkalinity, our results show that the droughts during which fires normally occur cause a corresponding reoxidation and export of sulphate, which consumes alkalinity. A strong pulse of nitrification also occurred in the year following fire (Schindler et al. 1980 and unpublished data) which must also have consumed significant alkalinity. Likens et al. (1969) also found high nitrate exports from a clear-cut watershed, with consequent consumption of significant alkalinity. Our results support the view that anthropogenically-caused acid precipitation, and not forest fire suppression, has caused the recent acidification of lakes and streams observed in this century.

The efficient removal of sulphate and nitrate was surprising at the low pH (≈ 4.0) of the wetland. Similar mechanisms slow the acidification of freshwater lakes (Schindler et al. 1980, 1986; Schindler and Turner 1982; Cook et al. 1986). Our wetland experiments and the results of others suggest that acid precipitation does not compromise the ability of wetlands to retain strong acid anions. Hemond (1980) showed that a bog in Massachusetts totally neutralized inputs of sulphuric and nitric acids, even though the pH of precipitation in the area is currently 4.1 and is believed to have been strongly acidic for at least 30 years. Urban et al. (this volume) found that from 40 to 90% of incoming hydrogen ions were neutralized in passing through wetlands in eastern North America.

However, while retention of sulphate and nitrate unquestionably protects ecosystems from acid precipitation to a degree, the reduction of sulphate and storage of sulphides may represent an ecological "time bomb" to be "exploded" upon lowering of the water table. Many decades of higher-than-normal sulphate input unquestionably causes higher-than-normal deposition of sulphides in wetlands and anoxic soils. If climatic conditions or human disturbances cause a lowering of the water table, these deposits may be re-oxidized. The result may be a rapid, pronounced acidification of waters flowing through the areas where sulphide has been deposited. Bayley et al. (under review) have shown this in the experimental wetland and Braekee (1981) found a high sulphate pulse after dry summers in Norwegian bog watersheds. Clymo (1984) observed similar events in the northern Pennines. Hultberg and Johansson (1981) observed the rapid acidification of aquifers in high-deposition areas of Sweden during periods of prolonged drought. Studies in the U.K. and in Finland have shown the acidification of streams draining wetlands when the latter are drained for agriculture or fuel-harvesting (Lehmusvuori 1981). All three of our catchments appeared to consume alkalinity by sulphate reoxidation during the dry years in the late 1970s and early 1980s (Fig. 5).

In the three watersheds, biological exchanges such as nitrate and ammonium uptake and/or sulphate reduction were at least as important as geochemical exchange or weathering reactions in determining the alkalinity balance. Furthermore, as the concentrations of sulphate, ammonium and nitrate in precipitation increase, we believe that alkalinity generated by these mechanisms will increase much more rapidly than that from geochemical exchange reactions. Henriksen (1982) estimated that acid precipitation would cause a maximum increase in base cation yield of 0.4 µeq per µeq of hydrogen ion added. In contrast, the efficiency of sulphate and nitrate retention in our watersheds and most others (see Schindler, in press) ranges from 60 to almost 100%.

In summary, even in areas as geologically simple as ELA, it is obvious that the proportions of bare bedrock, overburden, wetland and lake in a given catchment will greatly affect the alkalinity of the system because different mechanisms generate and consume alkalinity in the different areas. In order to predict the rates at which any catchment or the waters to which it drains will acidify due to acidic precipitation, or recover once the acidity of precipitation decreases, we must have a more comprehensive understanding of the mechanisms and rates of alkalinity generation in the different components of the catchment, rather than merely a "black box" treatment of the catchment as a whole.

Acknowledgements

Mike Stainton and Gary Linsey performed many of the analyses for this paper. Ken Beaty provided hydrological data. Mike Turner and Susan Kasian provided much of the data analysis. The work was supported by the Canadian Department of Fisheries and Oceans.

REFERENCES

APHA (American Public Health Association) (1967) Standard methods for the examination of water and wastewater. American Public Health Association Inc, Washington, DC, p 769
April R, Newton R (1985) Influence of geology on lake acidification in the ILWAS watersheds. Water Air Soi Pollut 26: 373-386
Bayley SE, Behr RS, Kelly CA (under review) Retention and release of sulfur from a freshwater wetland. Water Air Soil Pollut.
Bayley SE, Vitt DH, Newbury RW, Beaty KG, Behr R, Miller C (in press) Experimental acidification of a peatland: first year results. Can J Fish Aquat Sci
Beaty KG (1981) Hydrometeorological data for the Experimental Lakes Area, northwestern Ontario, 1969 through 1978 (In three parts), Can Data Rep Fish Aquat Sci 285: vi + 1-97 (Part I), v + 98-316 (Part II), iv + 317 (Part III)
Beaty KG (in prep) Water budget studies in the Experimental Lakes Area, northwestern Ontario. Can Tech Rep Fish Aquat Sci
Behr R (1985) Sulfur dynamics in an experimentally acidified mire in northwestern Ontario. MSc thesis, Univ. Manitoba, Winnipeg, p 105

Braekee FH (1981) Hydro chemistry of high altitude catchments in South Norway, 1. Effects of summer droughts with soil-vegetation characteristics. Rep Norw For Res Inst 36(8) 1-26

Brown W (1984) Maybe acid rain isn't the villain. Fortune 109: 170-174

Clymo RS (1985) Sphagnum-dominated peat bog: a naturally acid ecosystem. Phil Trans R Soc Lond B305: 487-499

Cook RB, Kelly CA, Schindler DW, Turner MA (in press) Mechanisms of hydrogen ion in an experimentally acidified lake. Limnol Oceanogr 30

Dillon PJ, Jeffries DS, Scheider WA (1982) The use of calibrated lakes and watersheds for estimating atmospheric deposition near a large point source. Water Air Soil Pollut 18: 241-258

EPRI (Electric Power Research Institute) (1983) The integrated lake-watershed acidification study: proceedings of the ILWAS annual review conference. EA-2827, Electric Power Research Institute, Palo Alto, CA, p 348

Galloway JN, Norton SA, Church MR (1983) Freshwater acidification from atmospheric deposition of sulfuric acid: a conceptual model. Environ Sci Technol 17: 541a-545a

Gorham E, Bayley SE, Schindler DW (1984) Ecological effects of acid deposition upon peatlands: a neglected field in "acid rain" research. Can J Fish Aquat Sci 41: 1256-1268

Hemond HF (1980) Biogeochemistry of Thoreau's Bog, Concord, Massachusetts. Ecol Monogr 50: 507-526

Henriksen A (1982) Preacidification pH values in Norwegian rivers and lakes. Acid Rain Report 3182. Norwegian Institute for Water Research, Oslo, Norway, p 24

Hultberg H, Jonasson S (1981) Acid groundwater. Nord Hydrol 12: 51-64

Johnson WE, Vallentyne JR (1971) Rationale, background, and development of experimental lakes studies in northwestern Ontario. J Fish Res Board Can 28: 123-128

Kelly CA, Rudd JWM, Cook RB, Schindler DW (1982) The potential importance of bacterial processes in regulating rate of lake acidification. Limnol Oceanogr 27: 868-882

Lehmusvuori M (1981) Runoff and leaching of nutrients from the drained and fertilized bog Laaviosuo in 1980. Suo 32: 134-137

Likens GE, Bromann FH, Johnson NM (1969) Nitrification: importance to nutrient losses from a cutover forested ecosystem. Science (Wash., DC) 163: 1205-1206

Linsey GA, Schindler DW, Stainton MP (in press) Atmospheric deposition of nutrients and major ions at the Experimental Lakes Area in northwestern Ontario, 1970-1982. Can J Fish Aquat Sci

NRCC (National Research Council of Canada) (1981) Acidification in the Canadian Aquatic Environment. NRCC No. 18475, Ottawa, p 369

Oliver BG, Thurman EM, Malcom RL (1983) The contribution of humic substances to the acidity of colored natural waters. Geochim Cosmochim Acta 47: 2031-2035

Puustjarvi V (1959) On the cation uptake mechanism of Sphagnum mosses. J Sci Agr Soc Finland 31: 103-119

Schindler DW (in press) The significance of in-lake production of alkalinity. Water Air Soil Pollut

Schindler DW, Turner MA (1982) Biological, chemical and physical responses of lakes to experimental acidification. Water Air Soil Pollut 18: 259-271

Schindler DW, Newbury RW, Beaty KG, Prokopowich J, Ruszczynski T, Dalton JA (1980) Effects of a windstorm and forest fire on chemical losses from forested watersheds and on the quality of receiving streams. Can J Fish Aquat Sci 37: 328-334

Schindler DW, Turner MA, Stainton MP, Linsey GA (in press) Natural sources of acid neutralizing capacity in low alkalinity lakes of the Precambrian Shield. Science (Wash., DC)

Schindler DW, Linsey GA, Stainton MP (in prep) Increasing acidity of precipitation in northwestern Ontario

Schnoor JL, Stumm W (1985) Acidification of aquatic and terrestrial systems. In: Stumm W (ed) Chemical processes in lakes. John Wiley & Sons, New York, p 311-338

Stainton MP, Capel MJ, Armstrong FAJ (1977) The chemical analysis of freshwater. 2nd Ed. Can Fish Mar Ser Misc Spec Publ 25: p 178

Urban NR, Eisenreich SJ, Gorham E (this volume) Proton cycling in bogs: geographic variations in northeastern North America.

van Breemen N, Driscoll CT, Mulder J (1984) Acidic deposition and internal proton sources in acidification of soils and waters. Nature 307: 599-604

Vitt DH, Bayley SE (1984) The vegetation and water chemistry of four oligotrophic basin mires in northwestern Ontario. Can J Bot 62: 1485-1500

Wright RF (1983) Input-output budgets at Langtjern, a small acidified lake in southern Norway. Hydrobiologia 101: 1-12

RESPONSES TO ACIDIC DEPOSITION IN OMBROTROPHIC MIRES IN THE U.K.

J.A. Lee, M.C. Press, S. Woodin and P. Ferguson
Department of Botany, The University, Manchester, U.K. M13 9PL

ABSTRACT

Ombrotrophic blanket mires are ecosystems of particular sensitivity to acidic deposition. The blanket mires of the southern Pennines of England have been extensively modified as the result of atmospheric pollution since the Industrial Revolution. Sphagnum species have virtually disappeared from the area probably as the result of high concentrations of sulphur dioxide or its solution products which were prevalent in the past. The poor growth of present-day transplants into the southern Pennine mires is associated with large increases in tissue nitrogen concentrations, suggesting that nitrogen deposition is of increasing importance, and this is supported by laboratory and field growth experiments. Further, in unpolluted environments, nitrate reductase activity of ombrotrophic Sphagnum species increases in response to nitrate deposition. However, this 'coupling' quickly disappears in the polluted southern Pennines, presumably as the result of the supra-optimal nitrogen supply, and nitrate is no longer retained by the moss. These observations are discussed in relation to the possible effects of increased nitrogen deposition on mires remote from industrial and urban regions.

INTRODUCTION

Peatlands account for 8.6% of the land surface area (2.7 Mha) in Great Britain and Ireland (Taylor 1983), and of these, ombrotrophic blanket mires are the most abundant. These mires are important from both hydrological and ecological viewpoints. In looking for effects of acidic deposition on terrestrial ecosystems, ombrotrophic blanket mires may be considered particularly susceptible for a combination of the following reasons: (i) they rely directly on the atmosphere for their supply of elements; (ii) this source of elements is usually at low supply; (iii) annual precipitation is often high (together with the second point accentuating the effects of any small increases in the concentration of supply); and (iv) many of these mires are dominated by bryophytes, principally Sphagnum species, which have leaves lacking a cuticle and are one cell thick. The dominant peat-forming plants are thus directly and continuously exposed to any changes in atmospheric element supply in the form of both wet and dry deposition resulting from atmospheric pollution.

HISTORICAL EVIDENCE FOR THE MODIFICATION OF MIRES BY ATMOSPHERIC POLLUTION

The Pennine range consists of flat-topped hills which are covered by ombrotrophic peat. The mires at the southern end of the range, which cover more than 52 Kha, are bounded by towns which were early centres of the Industrial Revolution. Of these towns Manchester is the largest and the most important, being directly upwind of a large part of the southern Pennines and being "the masterpiece of the Industrial Revolution" (Engels 1845). By the mid-nineteenth century it was said that the growth of Manchester (from 90,000 inhabitants in 1800 to 400,000 in 1840) had "defiled the atmosphere, polluted streams and destroyed the vegetation" (Webb and Webb 1922). It is no accident, therefore, that 'acid rain' was first described in Manchester (Smith 1852). Smith (1752) made extensive studies of the chemical composition of rain in towns, notably in Manchester where he lived, and concluded that "When the air has so much acid that two to three grains are found in a gallon of rain-water, or forty parts in a million, there is no hope for vegetation in a climate such as we have in the northern parts of the country." Correlations between the disappearance of plants from the region and atmospheric pollution began to be made by naturalists during the nineteenth century (see, for example, Press et al. 1983), but the best evidence comes from the ombrotrophic blanket peat itself. Tallis (1964) showed that the disappearance of Sphagnum remains from southern Pennine peat was associated with the appearance of soot particles in the peat profile, and further evidence of the pollution of the southern Pennine mires was provided by Gorham (1958 a,b) who showed the surface waters of these mires to be more acidic than others in the United Kingdom. The extensive peat deposits are largely formed of Sphagnum remains, but today Sphagnum species are almost completely absent from the surface of the mires.

Individual Sphagnum species grow over different ranges of chemical conditions. Before the Industrial Revolution, Pennine mires were similar to other ombrotrophic bogs in Great Britain and Ireland, characterized for example, by S. capillifolium, S. cuspidatum, S. fuscum, S. imbricatum, S. magellanicum, S. papillosum, S. pulchrum and S. tenellum. The only widespread and abundant species today, however, is S. recurvum, which has a wider ecological amplitude and is characteristic of sites with a richer solute supply. Experimental evidence that this widespread vegetation change resulted from atmospheric pollution was provided by Ferguson et al. (1978) and Ferguson and Lee (1980). These workers showed that Sphagnum species differed in their sensitivity to sulphur pollutants in both laboratory and field experiments, the minerotrophic S. recurvum being much more resistant than the ombrotrophic species. Fumigation at 131 µg $SO_2.m^{-3}$ caused a marked reduction in growth, and in some cases death, of these species, as compared to clean air controls (9 µg $SO_2.m^{-3}$). While it is uncertain what concentrations of SO_2 pervaded the southern Pennine blanket mires in the 19th century, it is likely that they were phytotoxic. Ferguson and Lee (1983a) showed that the present-day mean annual concentration of SO_2 at one bog surface was approximately half that in the surrounding towns. They also showed that as late as 1952 mean annual SO_2 concentrations in these towns were as high as c. 500 µg.m^{-3}. Smith (1872) recorded washings of Manchester air which gave a mean SO_2 concentration (from 11 measurements) of 1793 µg.m^{-3} (range 102-7338 µg.m^{-3}). Similar measurements for Buxton (a small town in the southern Pennine hills, to the south of Manchester), gave 1197 µg SO_2.m^{-3} (range 142 - 2961 µg.m^{-3}). Although these were almost instantaneous measurements of a small volume of air (1963 cm^3) and may perhaps have been contaminated by particulates, they probably indicate, together with the evidence of Ferguson and Lee (1983a), that concentrations of at least several hundred µg SO_2.m^{-3} were common in the upland region downwind of the industrial towns over a period of almost a century. There is little doubt that these concentrations would have had a dramatic effect on ombrotrophic Sphagnum species, and can explain the disappearance of these species from the southern Pennine mires.

RESPONSES TO PRESENT-DAY ACIDIC DEPOSITION IN SOUTHERN PENNINE MIRES

Mean annual concentrations of sulphur dioxide in the southern Pennines have been falling dramatically since at least 1952 (Ferguson and Lee 1983a) and are now unlikely on their own to account for the continuing absence of ombrotrophic Sphagnum species. The paucity of the plants in the southern Pennines today cannot, however, be interpreted in terms of an absence of disseminules. Transplants from a site remote from industrial influences to the southern Pennines showed poor growth in five out of six species. Only Sphagnum recurvum thrived (Ferguson and Lee 1983b).

Today evidence points to the importance of factors other than sulphur. First, there is strong historical evidence for a marked increase in nitrogen components of air (Salmon et al. 1978) and precipitation (Brimblecombe and Stedman 1982). Smith (1872) recorded an annual average ratio of sulphuric acid to nitric acid in Manchester rain for 1868 and 1869 of 43.4:1. However, Press (1983) recorded an annual average ratio of 3.4:1 in nearby southern Pennine bulk precipitation. This difference is not simply the result of a lower sulphuric acid concentration, but also of an increase in nitric acid by more than fourfold. Work at Manchester has also shown higher concentrations of combined nitrogen in southern Pennine air and rain as compared with a more remote mire site in North Wales (bulk deposition of nitrogen 1980-1982, 4.61 g.m^{-1}.y^{-1} and 2.40 g.m^{-2}.y^{-1}, respectively (Press and Lee 1982)). Second, Sphagnum transplanted to the southern Pennines showed marked changes in element concentration with a particularly large increase in total nitrogen within the first six months (Ferguson et al. 1984). These transplants were set into the peat surface, so some of the increase in element concentration could be attributed to release from the peat itself. However, even short-term transplants isolated from the bog surface so that they receive only wet and dry deposition from the atmosphere can show changes in total nitrogen in the capitula within 10 days (Fig. 1). Thus it would appear that the supply of combined nitrogen from the atmosphere is now supra-optimal for the growth of at least the ombrotrophic species. Laboratory experiments support this view and episodic nitrogen enrichment of a bog surface in North Wales (mimicking southern Pennine supply) reduced the growth of Sphagnum (Press 1983).

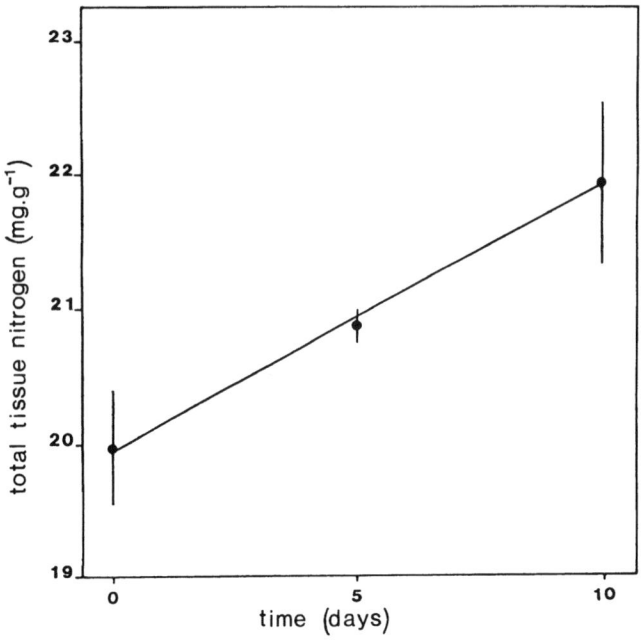

Fig. 1. Tissue nitrogen concentration ($mg.g^{-1}$ dry wt) in Sphagnum cuspidatum transplanted to the southern Pennines from an unpolluted region during a period in Nov. 1984. Vertical bars represent ± 1 standard error (n = 3). Regression equation y = 19.953 + .198x ($0.02 < p < 0.05$).

The extent to which bryophytes are closely coupled to their atmospheric nitrogen supply is perhaps most clearly demonstrated by the work of Woodin, Press & Lee (1985), utilising the substrate-inducible enzyme nitrate reductase. An ombrotrophic mire dominated by Sphagnum fuscum was studied in sub-arctic Sweden, at a site remote from urban and industrial sources of atmospheric pollutants. Previous work near this site had shown that although nitrate was readily measureable in precipitation, it was not detectable in the mire peat (Rosswall and Granhall 1980). Woodin et al. (1985) demonstrated that wet deposition of nitrate in each of several rain events over a three-week period induced nitrate reductase activity, and that the activity was proportional to the nitrate deposition. After each rain event, activity returned to near its constitutive level as nitrate was depleted. Thus, the constitutive enzyme activity of the moss was not sufficient to allow the assimilation of nitrate in precipitation.

A similar response of Sphagnum fuscum to rain events can be shown following transplantation to a more polluted atmosphere (Fig. 2), and it can be assumed that changes in the supply of solutes from the atmosphere will evoke rapid changes in the metabolism of bryophytes. This approach was used to examine the response of Sphagnum transplants to an exclusively atmospheric source of solutes in the southern Pennines. Plants were collected from the Berwyn Mountains, North Wales, a mire system similar to the southern Pennines (Tallis 1969) except that there are no major local sources of atmospheric pollutants and it apparently has not been significantly modified by atmospheric pollution. Plants of a small, possibly relict, southern Pennine population of S. cuspidatum were used for comparison. The plants were established on the bog surface in beakers at similar shoot densities for both populations. The plants were initially floated on distilled water. Figure 3 shows the response of Sphagnum cuspidatum to three rain events over a 10-day period. The Welsh transplants showed a rapid induction of nitrate reductase in response to the wet deposition of nitrate in the first rain event, a much smaller

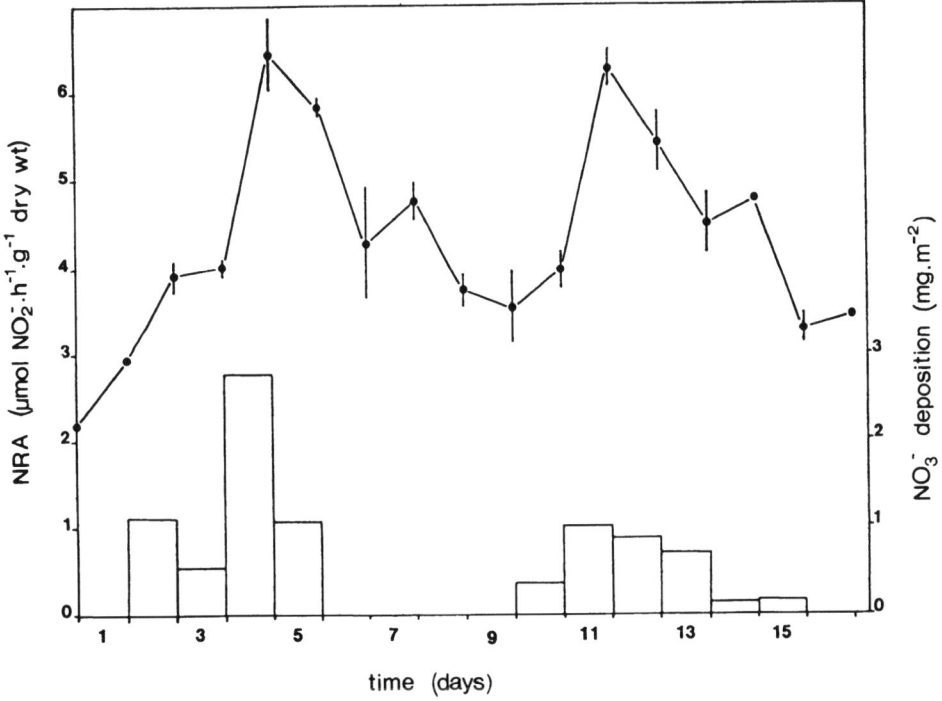

Fig. 2. Nitrate reductase activity (μmol $NO_2^-.h^{-1}.g^{-1}$ dry wt) (- • -) in Sphagnum fuscum transplanted to a polluted environment, in relation to NO_3^- -N deposition ($mg.m^{-1}$) during a period in Oct. 1984. Nitrate reductase activity figures represent means of 4 replicates, vertical bars represent ± 1 standard error.

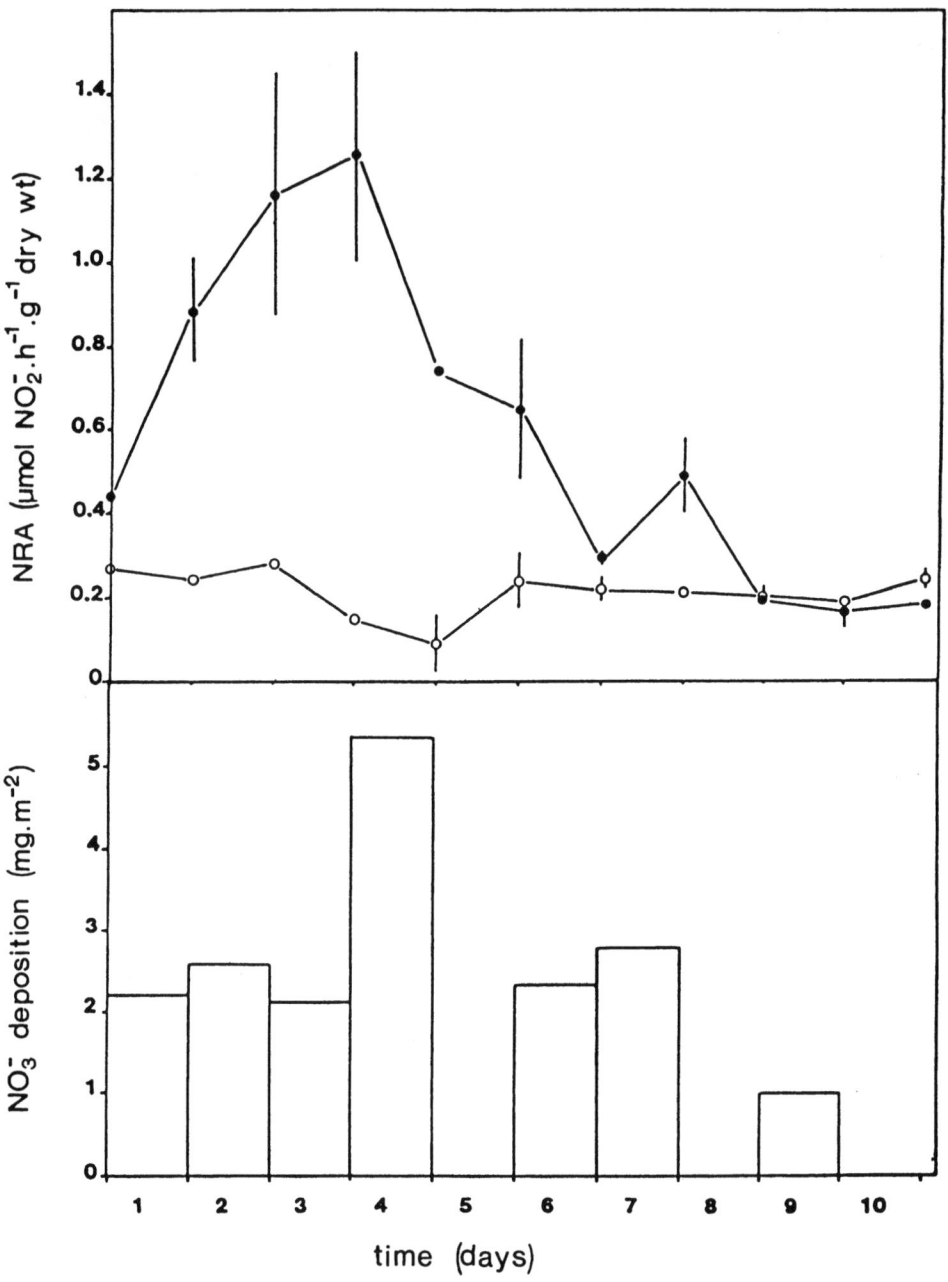

Fig. 3. Nitrate reductase activity (μmol $NO_2^-.h^{-1}.g^{-1}$ dry wt) in <u>Sphagnum cuspidatum</u> indigenous to the southern Pennines (-○-) and transplanted to the southern Pennines from an 'unpolluted' region (-●-) in relation to NO_3^- deposition ($mg.m^{-2}$) during a period in Nov. 1984. Nitrate reductase activity figures are means of 3 replicates, vertical bars represent ± 1 standard error.

response to the second, and no response to the third and smallest nitrate deposition event. In contrast, the southern Pennine population showed a low activity of nitrate reductase throughout, and no induction of the enzyme to any wet deposition event. Thus, it could be said that the southern Pennine population is 'uncoupled' from the normal response of Sphagnum species to 'unpolluted' wet deposition, and that the transplanted Welsh plants rapidly become 'uncoupled' when subjected to southern Pennine deposition. Indeed, uncoupling appeared to be initiated before the end of the first deposition event. This response may be reproduced even in remote sites by artifical episodic nitrate enrichment of the bog surface (Woodin et al. 1985).

Southern Pennine plants have a higher tissue ammonium-nitrogen concentration (338 $\mu g.g^{-1}$ dry wt.) compared with Welsh plants (98 $\mu g.g^{-1}$ dry wt.). The enhanced concentration of ammonium ions or of an amino acid probably inhibits the synthesis of the enzyme (Press and Lee 1982) and accounts for the 'uncoupling' process in the Welsh material. This 'uncoupling' is presumably an attempt to limit the assimilation of nitrogen or to prevent a nitrogenous solute from reaching toxic concentrations. Thus, in considering the response of these plants to nitrate deposition, it is essential to account for ammonium supply as well, since both contribute to a supra-optimal supply of combined nitrogen. (Rate of ammonium-nitrogen supply at the transplant site measured during 1980-1982 was 1.79 $g.m^{-2}.y^{-1}$ while that of nitrate-nitrogen was 1.43 $g.m^{-2}.y^{-1}$ (Press and Lee, 1982)). Little is known of the availability to plants of organic nitrogen in atmospheric deposition, but Press (1983) has shown that this can represent a large proportion of the total nitrogen deposition (as much as 36%), and perhaps should not be ignored since some amino acids can at least support the growth of Sphagnum nemoreum in axenic culture (Simola 1975). In the southern Pennines the poor growth of transplants, their rapid and massive accumulation of nitrogen, and their change in nitrogen metabolism when subjected to wet and dry deposition over a short period of time, point to the importance of present-day nitrogen supply from the atmosphere in affecting the ecology of mires in the region.

Evidence for other effects of present-day deposition on biological activity in ombrotrophic mires in the southern Pennines comes from the work of Press et al. (1985) on arylsulphatase activity of presumed microbial origin in peat. Arylsulphatase (aryl-sulphate sulphohydralase) catalyses the removal of the sulphate ion from a large number of arylsulphate esters. Dodgson and Rose (1975) report that microbial arylsulphatase activity may be affected by a range of inorganic and organic solutes including sulphur(IV) and sulphur(VI) ions. Press et al. (1985) showed that the most polluted region of the southern Pennine ombrotrophic mires had the lowest arylsulphatase activity in the surface peat, and that the further from industrial and urban areas, the higher the activity. (To aid comparison, all peat samples were taken from beneath Eriophorum vaginatum plants.) They also showed that surface peat transplanted from an 'unpolluted' mire surface into the southern Pennines showed a rapid loss in enzyme activity which suggests that some component (or components) of present-day acidic deposition was reducing enzyme activity. Laboratory experiments showed that the supply of nitrate, sulphur(IV) and sulphur(VI) ions could have been responsible for the low activity observed in the southern Pennines.

IMPLICATIONS OF INCREASED SOLUTE SUPPLY TO LESS GROSSLY POLLUTED MIRES

The southern Pennine mires are exceptional in that atmospheric deposition has removed the Sphagnum carpet, a process which may have accelerated peat erosion (Tallis 1965), leading to a modification of water-storage properties, a more rapid run-off and a partial silting of reservoirs. The low arylsulphatase activity in peat suggests that microbial processes are affected by acidic deposition. The removal of the dominant Sphagnum carpet would be expected to have a profound effect on the micro-flora by removing substrates and also by exposing the peat directly to the enriched solute supply of acidic deposition. But can acidic deposition have a marked effect on the ecology of ombrotrophic mires without the removal of the dominant Sphagnum species given their ability to affect the solute supply to other organisms? (See, for example, Clymo 1963; Clymo and Haywood 1982.)

There is some evidence to suggest that even at relatively remote mire sites in England and Wales, the supply of elements from the atmosphere is already optimal, or supra-optimal for the growth of Sphagnum species. Figure 4 shows the effects of increasing the supply of nitrate and ammonium ions on the growth of Sphagnum cuspidatum under laboratory conditions. The plants were collected from a 'remote' site in North Wales, together with a large volume of bog pool water. The plants were then grown in the bog pool water for 46 days, either unamended or enriched with nitrate and ammonium ions at concentrations of 0.01 and 0.10 mM. The solutions were changed every 2 days. (This rate of supply may be high, but the concentrations are within those observed in bulk deposition in Wales (Press 1983).) Increasing the concentration of these ions reduced the growth of Sphagnum cuspidatum, supporting the observations of Meade (1982) that no growth stimulation of Sphagnum species collected from the field occurs as the result of even small increases in nitrogen concentration. Rate of solute supply is also an important factor. Clymo (1973) demonstrated only very small increases in the growth of Sphagnum in an artificial rainwater solution by increasing the flushing rate 25 times. This work, together with the observations above, suggests that even modest increases in the concentration and rate of supply of atmospheric combined nitrogen may potentially reduce the growth of Sphagnum in the field, even when 'damage' is not readily visible.

However, more subtle changes may be associated with a supra-optimal nitrogen supply and these may perhaps be manifest before any effect on plant growth appears. Figure 5 shows the effects of daily deposition events under laboratory conditions on nitrate ion retention by Sphagnum capillifolium. The plants were grown

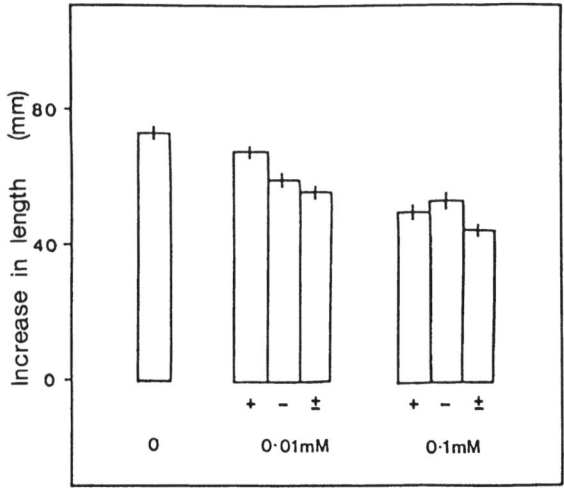

Fig. 4. Increase in length of <u>Sphagnum cusidatum</u> collected in North Wales and grown in the laboratory in bog pool water collected from the same site. The water was amended with either 0.01 or 0.1 mM nitrogen, supplied as either ammonium (+), nitrate (−) or as equimolar proportions of both (±). Plants were also grown in the unamended bog pool water. Vertical bars represent ± 1 standard error, based on at least 81 measurements.

in Buchner funnels and sprayed over a 30-minute period each day so that each funnel received 0.35 mg of NO_3-N in 50 cm^3 of solution (0.5 mM). The addition of ammonium ions markedly reduced the retention of nitrate and inhibited nitrate reductase activity. It is difficult to extrapolate from laboratory experiments to the field, but it is reasonable to assume that under 'normal' conditions of sub-optimal nitrogen supply, such as in the remote sub-arctic <u>Sphagnum fuscum</u> mires, little or no nitrate reaches the peat from atmospheric sources (Rosswall and Granhall 1980). As increases in the supply of nitrogen from the atmosphere occur progressively in more polluted regions, less nitrate is retained by the <u>Sphagnum</u> carpet and more enters the peat. The fate of this nitrate is far from certain. It may be taken up by the higher plant component of the mires. Most higher plants of ombrotrophic habitats are apparently poor utilisers of nitrate ions (Lee and Stewart 1978), but the increased supply may perhaps alter the competitive balance in favour of species with a higher potential to utilise the solute. Nitrate ions may have a profound effect on microbial ecology since, like <u>Sphagnum</u> species, microorganisms are directly and continuously exposed to any increase in solute supply. The evidence of Press <u>et al</u>. (1985) suggests that nitrate supply can affect microbial activity in peat, and may alter the composition of the microbial community. For example, nitrate ions will provide a substrate for denitrifying microorganisms, but denitrification is apparently low (Muller <u>et al</u>. 1980) or lacking in unpolluted ombrotrophic peat (Rosswall and Granhall 1980). Thus nitrate may in some circumstances reach the drainage water and play a role in the mobilization of ions such as trivalent aluminum ions which are potentially highly toxic to aquatic biota. The significance of this last-mentioned process is far from certain, but aluminum and nitrate ions are readily detectable in streams draining the southern Pennine uplands.

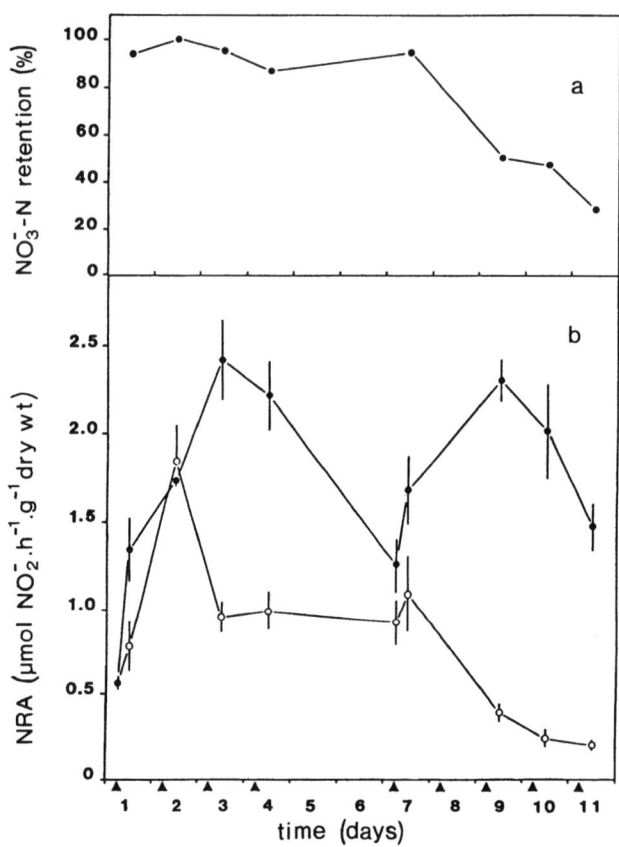

Fig. 5. The effect of nitrate and ammonium addition on (a) nitrate retention and (b) nitrate reductase activity in <u>Sphagnum capillifolium</u>. Plants were sprayed when indicated (▲) with 0.35 mg nitrate-N or 0.35 mg nitrate-N and 0.525 mg ammonium-N. (a) nitrate retention by ammonium plus nitrate treated plants expressed as a percentage of retention by nitrate-only treated plants. Figures are based on 3 replicates for each treatment; (b) nitrate reductase activity (μmol $NO_2^-.h^{-1}.g^{-1}$ dry wt) in nitrate-N treated plants (-●-) and in nitrate + ammonium treated plants (-○-). Vertical bars represent ± 1 standard error (n=3).

SUMMARY

The above discussion has been concerned almost exclusively with nitrogen deposition. But changes in the availability of other solutes may also be important in producing subtle effects on ombrotrophic mires. Availability of sulphate can influence microbial activity and thereby perhaps affect the production of hydrogen sulphide. Excess of sulphate may reduce growth of Sphagnum species and thereby reduce the nitrogen demand. The increased availability of sulphate could therefore change a given nitrogen deposition from being sub-optimal to being supra-optimal. Thus it is important to consider interactions between sulphate, nitrate and ammonium ions on the growth of Sphagnum species and the ecology of ombrotrophic mires. Such studies utilising field perturbation experiments at realistic concentrations of pollutants over long-term periods are at present lacking, but are essential to an understanding of the importance of acidic deposition on these widespread and sensitive ecosystems.

REFERENCES

Brimblecombe P, Stedman DH (1982) Historical evidence for a dramatic increase in the nitrate component of acid rain. Nature 298: 460-462

Clymo RS (1963) Ion exchange in Sphagnum and its relation to bog ecology. Ann Bot (Lond) 27: 308-324

Clymo RS (1973) The growth of Sphagnum: some effects of environment. J Ecol 61: 849-869

Clymo RS, Haywood PM (1982) The ecology of Sphagnum. In: Smith AJE (ed) Bryophyte ecology. Chapman and Hall, London, p 229-289

Dodgson KS, Rose FA (1976) Sulfohydrolases. In: Greenberg DM (ed) Metabolism of sulphur compounds, vol. 7, Metabolic pathways. Academic Press, New York, p 359-431

Engels F (1845) The condition of the working class in England. (Trans. Henderson WD and Chaloner WH). Oxford University Press, Oxford, p 50

Ferguson P, Lee JA (1980) Some effects of bisulphite and sulphate on the growth of Sphagnum species in the field. Environ Pollut (Ser A) 21: 59-71

Ferguson P, Lee JA (1983a) Past and present sulphur pollution in the southern Pennines. Atmos Environ 17: 1131-1137

Ferguson P, Lee JA (1983b) The growth of Sphagnum species in the southern Pennines. J Bryol 12: 579-586

Ferguson P, Lee JA, Bell JNB (1978) Effects of sulphur pollutants on the growth of Sphagnum species. Environ Pollut 16: 151-162

Ferguson P, Robinson RN, Press MC, Lee JA (1984) Element concentrations in five Sphagnum species in relation to atmospheric pollution. J Bryol 13: 107-114

Gorham E (1958a) The influence and importance of daily weather conditions in the supply of chloride, sulphate and other ions to fresh water from atmospheric precipitation. Phil Trans R Soc Lond B 241: 147-178

Gorham E (1958b) Free acid in British soils. Nature 181: 106

Lee JA, Stewart GR (1978) Ecological aspects of nitrogen assimilation. Adv Bot Res 6: 1-43

Meade R (1982) Adaptation of selected plants to ammonium as a source of nitrogen. Ph.D. Thesis, University of Manchester

Muller MH (1980) Denitrification in low pH spodosols and peats determined with the acetylene inhibition method. Appl Environ Microbiol 40: 235-239

Press MC (1983) Responses to acidic deposition in blanket bogs. Ph.D. Thesis, University of Manchester

Press MC, Lee JA (1982) Nitrate reductase activity in Sphagnum species in the South Pennines. New Phytol 92: 487-494

Press MC, Ferguson P, Lee JA (1983) Two hundred years of acid rain. Naturalist 108: 125-129

Press MC, Henderson J, Lee JA (1985) Arylsulphatase activity in peat in relation to acidic deposition. Soil Biol Biochem 17: 99-104

Rosswall T, Granhall U (1980) Nitrogen cycling in a subarctic ombrotrophic mire. In: Sonesson M (ed) Ecology of a subarctic mire. Ecol Bull 30: 209-234

Salmon L, Atkins DHS, Fisher EMR, Healy C, Law DV (1978) Retrospective trend analysis of the content of UK air particulate matter 1957-1973. Sci Total Environ 9: 161-200

Simola LK (1975) The effect of several protein amino acids and some inorganic nitrogen sources on the growth of Sphagnum nemoreum. Physiol Plant 35: 194-199

Smith RA (1852) On the air and rain of Manchester. Mem Lit Phil Soc Manchester (Second Series) 10: 207-217

Smith RA (1872) Air and rain. The beginnings of a chemical climatology. Longmans, Green and Co, London

Tallis JH (1964) Studies on southern Pennine peats. III. The behaviour of Sphagnum. J Ecol 52: 345-353

Tallis JH (1965) Studies on southern Pennine peats IV. Evidence of recent erosion. J Ecol 53: 509-520

Tallis JH (1969) The blanket bog vegetation of the Berwyn Mountains, North Wales. J Ecol 57: 765-787

Taylor JA (1983) The peatlands of Great Britain and Ireland. In: Gore AJP (ed) Mires: swamp, bog, fen and moor, B. Regional studies. Elsevier Scientific Publishing Co, Amsterdam, p 1-46

Webb B, Webb S (1922) English local government. Longmans, London, p 400-401

Woodin SJ, Press MC, Lee JA (1985) Nitrate reductase activity in Sphagnum fuscum in relation to atmospheric nitrate deposition. New Phytol 99: 381-388

THE STRATIGRAPHIC RECORD OF ATMOSPHERIC LOADING OF METALS AT THE OMBROTROPHIC BIG HEATH BOG, MT. DESERT ISLAND, MAINE, U.S.A.

S.A. Norton

Department of Geological Sciences, University of Maine at Orono, Orono, Maine 04469, U.S.A.

ABSTRACT

Ombrotrophic peat bogs receive all of their nutrients and a variety of non-essential substances from the atmosphere. Sources of inputs include marine aerosols, soil dust, products of air pollution, and normal atmospheric gases. Inputs may be deposited initially at or below the surface of the bog. Metals may be immobilized by adsorption, by incorporation into plant tissue, or by precipitation. Metals may be mobilized and migrate due to biological decay of organic matter, biological recycling, and changing chemical conditions below the living surface.

Cores from Big Heath, Mount Desert Island, Maine were dated using ^{210}Pb chronology (CRS model) and analyzed for bulk chemistry. Concentration profiles for major elements (Ca, Mg, Na, K, Fe, Mn, Al, and Ti) are consistent for replicate cores from either hollows or hummocks; however, profiles for hummocks and hollows are very different from each other. Neither site yielded concentration profiles related to atmospheric deposition, due to post-depositional processes. Net accumulation rates for major metals are consequently unrelated to atmospheric deposition rates, particularly for biophilic elements such as Ca, K, and Mn.

Net accumulation rates of Pb and Zn (largely attributable to atmospheric pollution) crudely mimic those observed in sediments of nearby lakes. However, considerable Zn moves out of the system. Integrated total Pb and ^{210}Pb values for hummock cores are about twice those for hollows, suggesting lateral migration of these components and, by implication, others.

The elemental composition of Sphagnum and vascular ("woody") plant tissue are quite different, even when growing side by side in the same atmospheric conditions. Conclusions based solely on the chemical composition of their remains in peat need to take this into account, else it is concluded that they have grown under quite different atmospheric conditions.

INTRODUCTION

Ombrotrophic peat bogs receive all of their nutrients and a variety of other substances from the atmosphere. The chemical inputs are: (1) marine-derived aerosols (largely Ca, Mg, Na, and K, plus the associated anions Cl^-, HCO_3^-, and SO_4^{2-}); (2) soil dust (largely Si, Al, Ca, Mg, Na, K, Fe, Mn, and Ti, combined in various minerals); (3) dry- and wet-fall components associated with fossil fuel consumption and related industrial activities (largely distinguished by the presence of excess Pb, Zn, Cu, Cd, V, S, and N [including NO_3^- and NH_4^+]); (4) atmospheric gases and normal precipitation (CO_2, N_2, and H_2O). These different materials may be deposited initially at different depths below the living surface of the bog.

As the bog surface grows upward and organic matter accumulates, metals accumulate in the stratigraphy by two major mechanisms. The organic matter incorporates various metals into tissue (e.g., Ca and K). Sphagnum, the dominant genus of the bog plant community, adsorbs large amounts of metals. Particularly after death of the plants, biological degradation results in the release of some of the various metals which then become available for biological recycling or downward migration. Downward migration of some elements (e.g., Mn, Fe, Zn) may be inhibited by gradients in the reduction/oxidation (redox) state of the water. Typically, aerobic conditions do not occur more than 5 to 10 cm below the groundwater table. Below that depth, biological decay is slow (Clymo 1978). However, the depth of the groundwater may vary up to 50 cm on a seasonal basis. Thus the site of the localization of some migrating metals may migrate considerably.

The input of soil dust has increased as a result of deforestation, agriculture, and various construction activities. However, the chronology and magnitude of these changes cannot be rigorously established. The chronology and magnitude of the deposition of atmospheric pollutants such as Pb and Zn have been established from the study of many lake sediment cores (reviewed in Norton, in press). However, although the deposition rate of atmospheric pollutants, such as Pb, into lake sediments parallels atmospheric deposition rates, the agreement is generally not on a 1:1 basis.

Lichens and bryophytes are known to accumulate most or all of their nutrients from the atmosphere. Consequently, they have been used to monitor atmospheric pollution. For example, Groet (1976) demonstrated regional pollution with Pb, Cu, Cd, Cr, Ni, and Zn in woodland sites in New York and New England. Concentrations of several elements were positively related to major regional and local pollution sources. Pakarinen (1981a,b) demonstrated regional pollution gradients utilizing chemical analysis of contemporary peat (1-5 year samples) from ombrotrophic bogs in Finland and northern continental Europe. Hvatum et al. (1983) found similar results in Norway. Time trends for deposition of heavy metals have been studied utilizing cores of peat from ombrotrophic bogs (e.g., Pakarinen et al. 1983 [Finland]; Livett et al. 1979 [Great Britain]). The cores, dated with increment dating and pollen, indicate increasing atmospheric deposition of these metals but the precise chronology (and thus deposition rates) is uncertain. Hemond (1980) demonstrated a sharp increase in the rate of deposition (net accumulation) of Pb over the last 100 years, based on a ^{210}Pb-dated core from Massachusetts. This time scale was consistent with that found in lake sediments although the accumulation rates found by Hemond for peat aged to be 100+ years old is probably much too high because of downward mobility of Pb in peat.

This study was undertaken to determine whether or not the chemical stratigraphy of peat bogs reflects atmospheric deposition rates for various energy-related pollutants.

METHODS

Big Heath is located east of Bass Harbor and south of Southwest Harbor on Mount Desert Island, Maine. The coring location is 350 meters from Maine Route #102a. Four cores (two from a hummock and two from an adjacent hollow) were retrieved using a stainless steel core tube 21 cm in diameter. The cores from hummocks were 60 and 72.5 cm long. The groundwater table on the date of collection (7/30/1982) was 35 cm below the top of the living surface of the Sphagnum. The pore water of the peat has a pH typically near 4. The cores from hollows were both 50 cm long; the groundwater table was 2.5 cm below the surface. Hummock to hollow relief is typically 50 cm. The hollow vegetation consisted dominantly of a Sphagnum-Carex lawn. Dominant species of Sphagnum include rubellum and fuscum. The hummock vegetation was typical of a Gaylussaccia shrub heath.

The cores were sectioned in the field in 2.5 cm increments except for the living surface (5 cm for the hummock; 10 cm for the hollow). Samples were sealed in plastic bags and transported immediately to the laboratory where they were weighed wet. They were dried at ca. 60°C to constant weight. From these data plus the calculated volume of each section of peat, we calculated water content of the peat and mass.cm^{-2}.$interval^{-1}$. Vascular woody plant material was physically separated from the Sphagnum and the components were processed separately. The peat and vascular components were crushed and homogenized in a Wiley Mill until they passed a 60 mesh sieve.

Approximately 1g of peat was used for determination of ^{210}Pb activity. Peat was digested and oxidized directly in boiling HNO_3; ^{210}Po was plated out according to methods developed by Eakins and Morrison (1978). Activity of ^{210}Po and ^{208}Po were measured on a Tracor Northern (TN-1700) alpha spectometer using counting times of 20,000 to 100,000 seconds. ^{210}Pb was calculated from ^{210}Po activity. ^{208}Po was used as an efficiency-of-recovery standard. Ages of intervals (Table 1) were calculated using a constant rate of supply model (CRS) developed by Appleby and Oldfield (1978). Lower in the core the CRS model fails mathematically. Data are shown for Core 2 (hummock) and Core 4 (hollow) (Figures 1 to 4).

Exactly 1.0000 g of dried peat was slowly brought to 600°C and oxidized for 3 hours. Weight loss was used to calculate ash content. The ash was transferred quantitatively into a linear polyethylene bottle for digestion using the method of Buckley and Cranston (1971). Resulting solutions were analyzed by atomic absorption spectrophotometry (Perkin-Elmer 703 flame plus HGA2200 graphite furnace).

RESULTS

Chronology of Cores

With steady state deposition of ^{210}Pb and steady state net accumulation of biomass, the ^{210}Pb activity profile (Figs. 1 and 2) should decrease exponentially with depth, decreasing to 50% of the surface activity at a depth corresponding to peat with an age of 22 years, the half life of ^{210}Pb. Such is clearly not the case for Core 2. There is no net decrease in activity at a depth of 40 cm. This phenomenon is the rule for all hummock cores from seven different ombrotrophic bogs in Maine. This relationship could be caused by: (1) non-steady state delivery of atmospherically-derived ^{210}Pb; (2) increasing productivity of the bog toward recent times, diluting the atmospheric ^{210}Pb flux; (3) biological decay of organic matter at depth, enriching immobile ^{210}Pb; (4) downward migration of ^{210}Pb during initial deposition and during degradation of biomass. Hypotheses (1) and (2) appear to be unlikely (Norton 1985). Hypothesis (3) is certainly true with respect to decay of organic matter (Clymo 1978) but the immobility of metals is undemonstrated. Hypothesis (4) has not been rigorously evaluated but appears to be probable to some unknown degree.

The CRS model was applied to the ^{210}Pb data for Cores 2 and 4. The resulting chronology is given in Table 1. No chronostratigraphic markers were present to check the validity of the ^{210}Pb chronology. Furthermore, for both cores, it is clear that the core was not quite long enough to reach background levels for ^{210}Pb. This will yield peat ages that are too old, the error increasing with increasing age. Another hummock core from Big Heath (Norton 1985) had a ^{210}Pb chronology which gave results consistently younger than "increment" dating (see El-Daoushy et al. 1982). A similar relationship was seen in three other peat deposits in Maine. The general validity of the "increment" dating has not been independently established for these bogs. If the chronology of increased Pb deposition from the atmosphere as derived from lake sediments in this area of Maine is accepted as correct (Kahl et al. 1985) and used to evaluate the ^{210}Pb peat chronology, peat from Core 2 ranging from 100 to 125 years old will be dated as 20 to 30 years too old. Dating of individual strata and thus temporal trends of concentration (dry and ignited bases), as well as accumulation rates, may be in error as much as 10 to 20%. There is close correspondence between the trends in total Pb concentration and chronology for Core 4 (hollow) and lake sediment data. This may be because peat in hollows is subjected to much less degradation once it is below the water table so that migration of Pb (and ^{210}Pb) may be inhibited.

Chemistry

The chemistry of Cores 2 and 4 are shown in Figs. 1 and 2. The traditional representation, concentration per unit dry weight, indicates independent behavior of various elements.

Ti, Al, Fe, Mn, and Al increase downward in concentration to a peak near the water table in Core 2. This is interpreted as the

Table 1. Chronology of a hummock (Core 2) and a hollow (Core 4) core from Big Heath, Mount Desert Island, Maine.

Depth	Age at bottom of Interval	
	Core 2	Core 4
0.0 - 5.0 cm	1980.2	
5.0 - 7.5	1979.5	
7.5 - 10.0	1978.9	1966.7
10.0 - 12.5	1978.3	1960.3
12.5 - 15.0	1977.6	1952.7
15.0 - 17.5	1976.9	1946.2
17.5 - 20.0	1976.2	1936.9
20.0 - 22.5	1975.5	1924.6
22.5 - 25.0	1974.4	1912.9
25.0 - 27.5	1972.8	1901.9
27.5 - 30.0	1970.3	1889.5
30.0 - 32.5	1967.8	1877.5
32.5 - 35.0	1964.2	1862.8
35.0 - 37.5	1956.6	1849.0
37.5 - 40.0	1945.3	1834.4
40.0 - 42.5	1930.8	1820.1
42.5 - 45.0	1912.2	1804.5
45.0 - 47.5	1893.1	1788.8
47.5 - 50.0	1870.9	1772.3
50.0 - 52.5	1859.8	
52.5 - 55.0	1845.1	
55.0 - 57.5	1836.4	
57.5 - 60.0	1827.4	

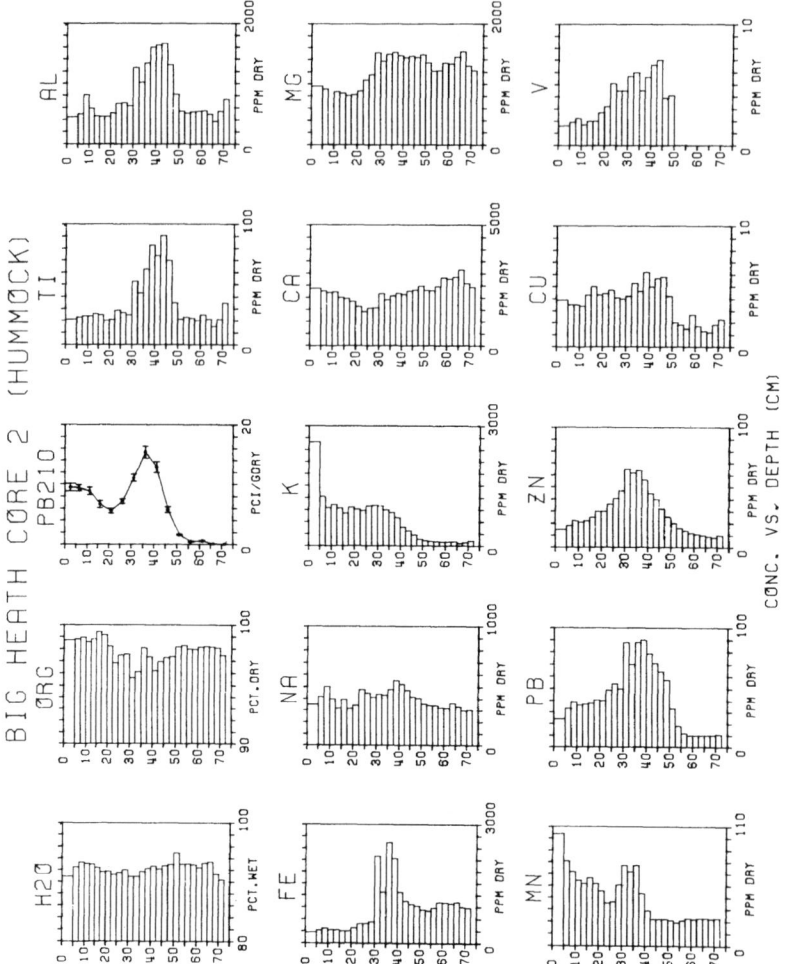

Fig. 1. Chemistry of hummock core (2) from Big Heath, Mount Desert Island, Maine. Depth in cm. Concentration units are for air-dried peat.

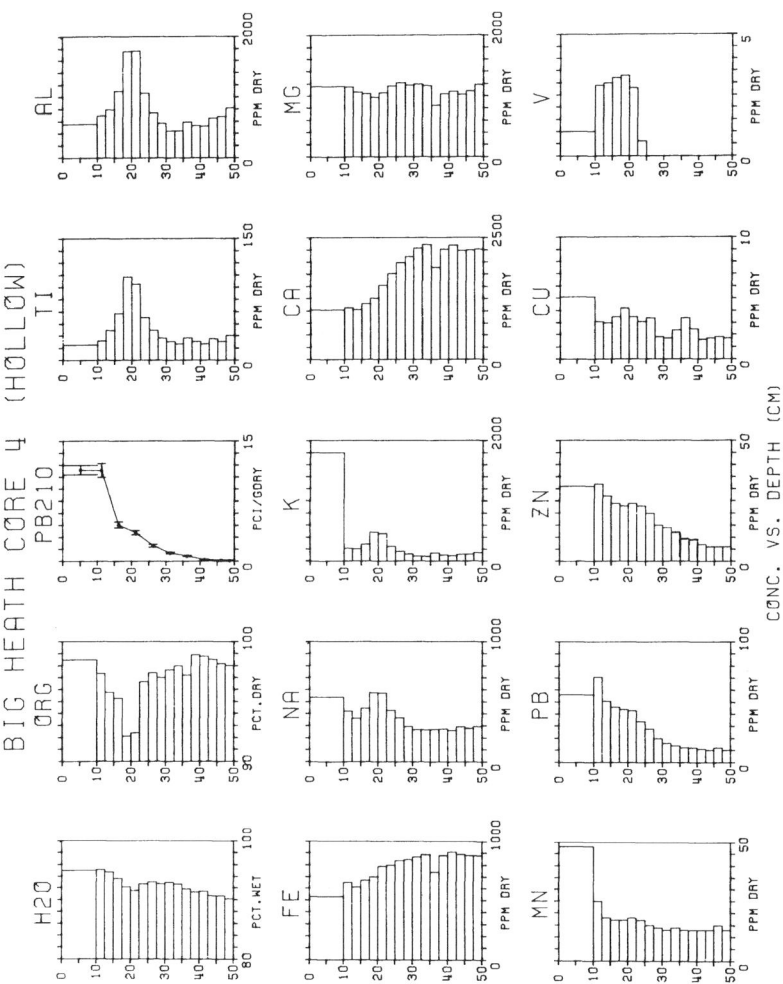

Fig. 2. Chemistry of hollow core (4) from Big Heath, Mount Desert Island, Maine. Depth in cm. Concentration units are for air-dried peat.

result of a long-term increase in atmospheric loading of atmospheric pollutants from fossil fuel emissions and soil dust starting more than 100 years ago, coupled with the concentration of these relatively refractory/immobile elements as a result of biological decay of organic matter above the water table. Mn, K, Ca, and Mg are concentrated in surface peats, relative to the aforementioned elements, presumably due to strong biological recycling. Mn is particularly concentrated in vascular plant material (Fig. 5).

If data are expressed in concentration per unit ignited (ashed) weight, different trends are apparent (Figs. 3 and 4). High concentrations of biophilic elements (Ca, Mg, K, Na) dilute other elements in surface peat. In Core 2, all reported elements (Fig. 3) have a minimum between about 20 and 35 cm. This is presumably due to dilution by abundant Si (not measured; calculated by difference). The sharp increase in all elements at 35 cm (the depth of the water table at collection) is due largely to a decrease in SiO_2. Very small errors in the determination of the ash content greatly magnify errors in chemistry, expressed on an ashed weight basis. The concentration of elements expressed as ppt (Figs. 3 and 4) is derived by multiplying the concentration in dry matter by the ratio (100)/(100 - % ash content). For Core 2, the ash content ranges from 95.61 to 99.44%. The multiplier thus ranges from 22.8 to 178.6.

The concentration trends in Core 4, for ashed peat, are shown in Fig. 4. From 17.5 to 22.5 cm there is a peak in ash content which dilutes the ^{210}Pb activity, the biologically important elements K, Ca, and Mg, and the redox-sensitive elements Fe and Mn. The atmospheric Pb and Zn inputs from pollution are also diluted by detritus, which is low in Pb. The immobile lithophilic elements Ti, Al, and Na are not diluted since they are important constituents of aerosol dust input (both soil dust and fossil fuel emissions) and not mobilized after deposition. Surface enrichments of K, Ca, Mg, and Mn (and possibly Cu) are probably related to biological recycling.

The chemistry of the twig fraction of Core 2 is shown in Fig. 5. The woody material typically represents 15 to 20% of the total biomass. Comparison of Figs. 1 and 5 reveals that although the chemical trends are crudely similar, certain metals partition strongly between Sphagnum and non-Sphagnum plant material. Most notably, Mn partitions strongly into the "twig" fraction whereas Al is preferentially found in the Sphagnum. Fe, under aerobic conditions, strongly prefers Sphagnum. The ash content of Sphagnum is typically twice that of the twigs. Therefore, the mass of Sphagnum is typically 90 to 93% of the total mass and accumulation rates based solely on Sphagnum are representative of the entire peat sample.

Accumulation Rates

The accumulation rate data depicted in Figs. 6 (a-d) and 7 are based only on the Sphagnum component and are therefore only about 90 to 93% of the total. The accumulation rate of lithophilic

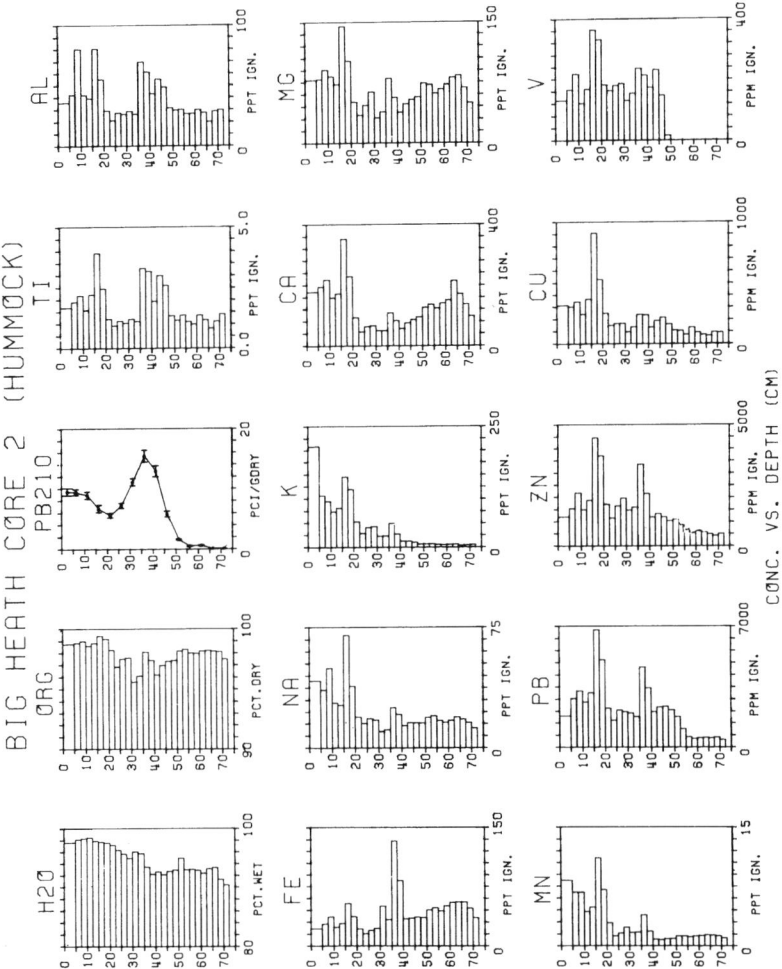

Fig. 3. Chemistry of hummock core (2) from Big Heath, Mount Desert Island, Maine. Depth in cm. Concentration units are for ashed peat.

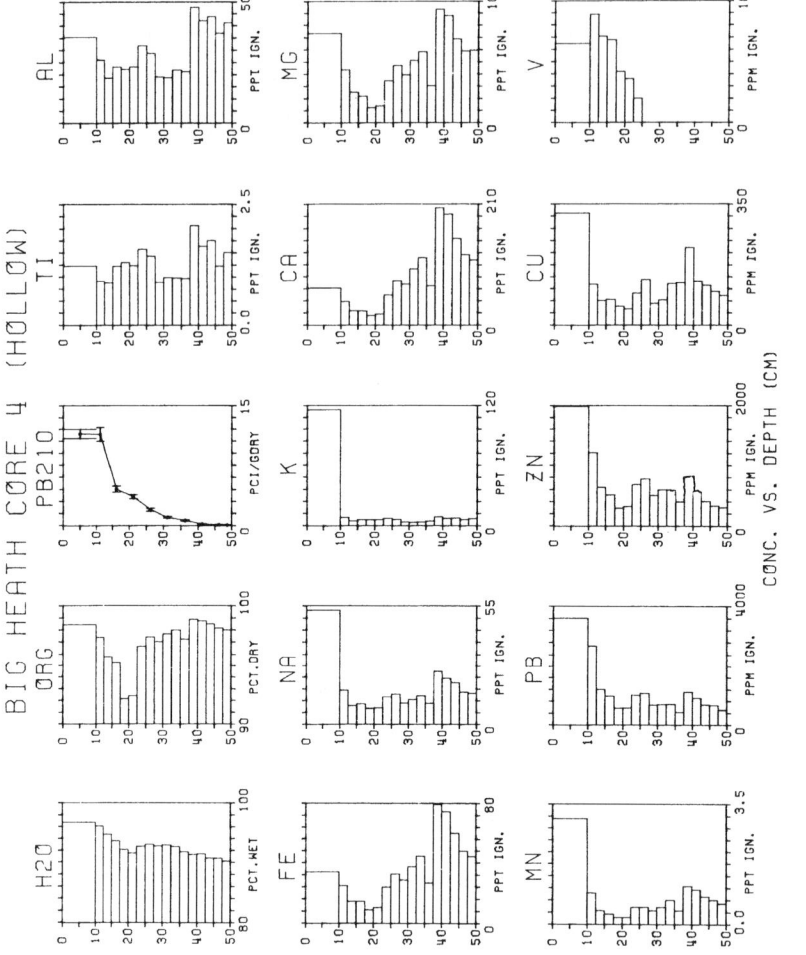

Fig. 4. Chemistry of hollow core (4) from Big Heath, Mount Desert Island, Maine. Depth in cm. Concentration units are for ashed peat.

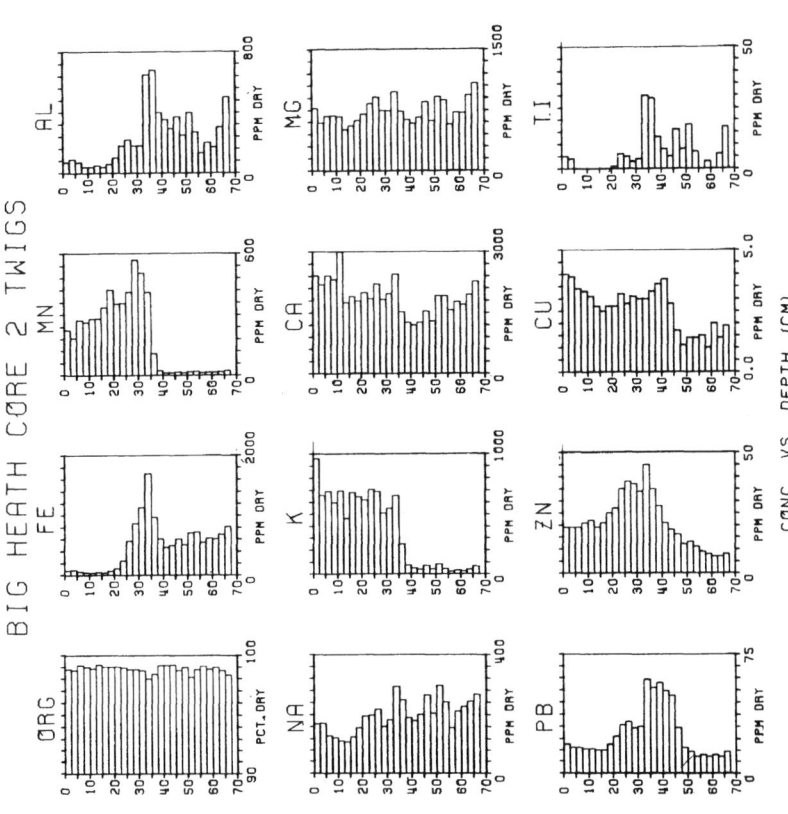

Fig. 5. Chemistry of "twigs" from hummock core (2) from Big Heath, Mount Desert Island, Maine. Depth in cm. Concentration units are for air-dried peat.

elements of relatively low mobility is represented by Ti (Fig. 6a). For Core 2, modern values are about four times 19th century values, with values increasing since about 1900. For the hollow core (4), a 1940 peak was followed by a return to background values of about 0.4 µg.cm^{-2}.y^{-1}. Background rates are comparable for hummocks and hollows. The persistent high modern rates for the hummock could be an artifact of the ^{210}Pb dating. Downward migration of ^{210}Pb would cause an apparent elevation of the rate. Alternatively, the net accumulation rate of Ti to the hummock could be greater. This seems less likely.

Biophilic elements such as Ca (Fig. 6b) and K show enormously elevated net accumulation rates in recent peat presumably due to biological recycling (Ca only for hummocks). These rates are unrelated to the deposition of dust (compare Figs. 6a and 6b).

Energy-related elements such as Pb have increased accumulation rates starting in the late 1800s (consistent with lake sediment data) but values differ between hummocks (Fig. 6c) and hollows (Fig. 6d). Core 2 reached a peak of about 4 µg.cm^{-2}.y^{-1} in the mid-1970s. The accumulation rate for Pb in Core 4 reached a plateau of about 1.2 µg.cm^{-2}.y^{-1} in the 1950s. The integrated Pb.cm^{-2} over the last 150 years for Core 2 is about twice that of Core 4, consistent with the accumulation rates. The accumulation rate for Pb and integrated anthropogenic Pb in the hollow core is slightly higher than those values for more remote bog sites in Maine. The higher hummock values may be due to local topography; more likely, it is possible that movement of water toward hummocks during dry seasons results in the transport of certain constituents to hummocks.

Normalization of the accumulation rate of Pb to that of TiO_2 for hollows (Fig. 7) or hummocks reveals that they covary until the last 30-40 years and that the accumulation rates are nearly equal. TiO_2 is emitted to the atmosphere during the burning of coal. Emission controls over the last few decades may have reduced TiO_2 emissions. Additionally, Pb has increased independently due to automotive Pb emissions.

The accumulation rate for Zn in the hummock generally parallels, but is consistently lower than, that of Pb. However, Zn is typically greater than Pb in precipitation. Therefore, one can deduce that Zn is lost from the system or deposited in much older peat. Copper behaves like zinc. The concentration profiles and accumulation rates for V increase later than for Pb, consistent with lake sediment profiles (Norton, in press). Accumulation rates for Pb, Zn, and Cu in hollows are about 0.3 to 0.5 time hummock rates, assuming correct chronology. The integrated Pb (the most conservative element) for peat deposited (dated) since 1850 is 151 µg.cm^{-2} and 82 µg.cm^{-2} for the hummock and hollow cores, respectively. The other adjacent replicate cores had 162 and 82 µg Pb.cm^{-2}, respectively. Integrated Zn and Cu are low with respect to total deposition due to the leakiness of the system for these elements.

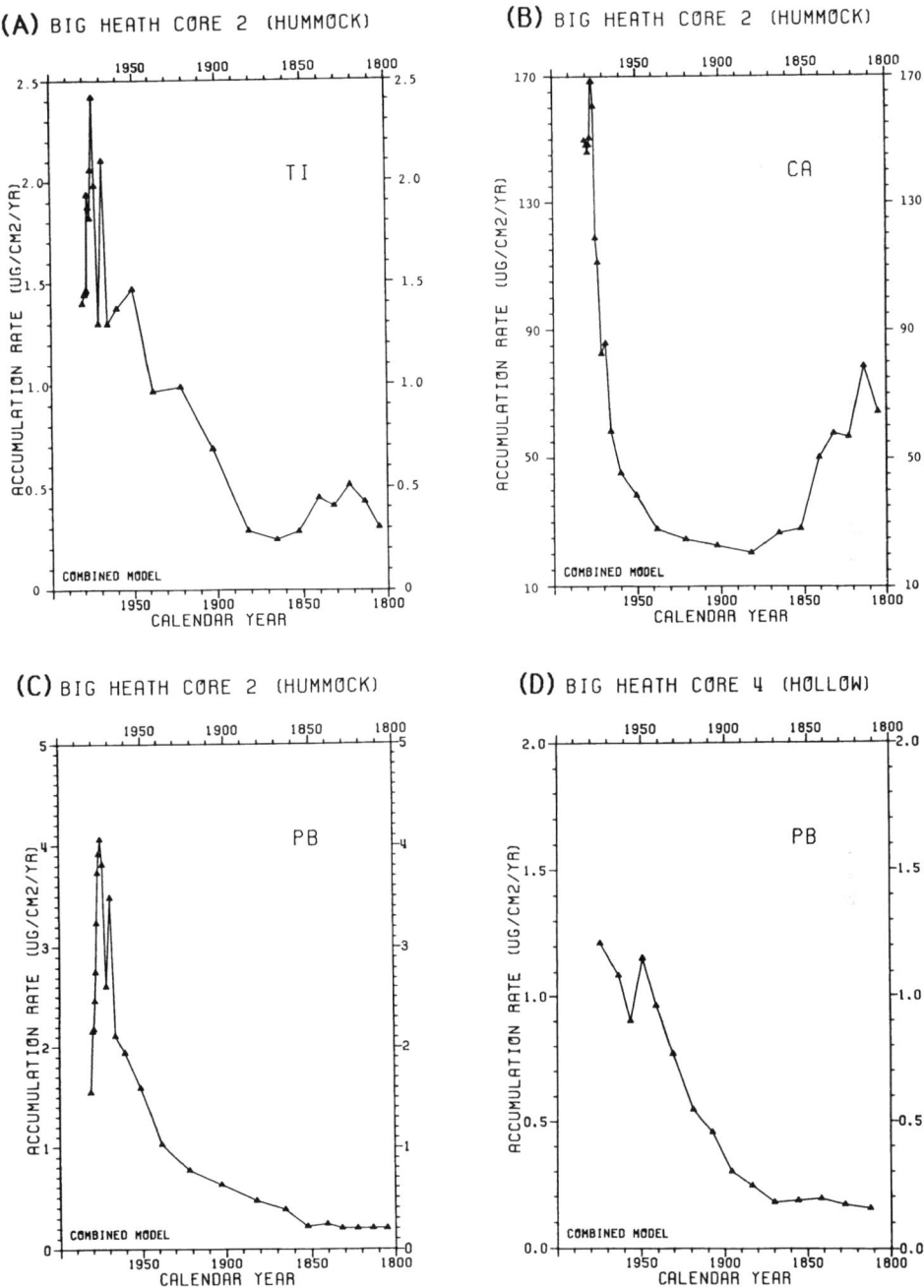

Fig. 6. Accumulation rate ($\mu g \cdot cm^{-2} \cdot y^{-1}$) of (a) Ti, (b) Ca, (c) Pb, for hummock core (2), and (d) Pb for hollow core (4) versus time. All cores were from Big Heath, Mount Desert Island, Maine.

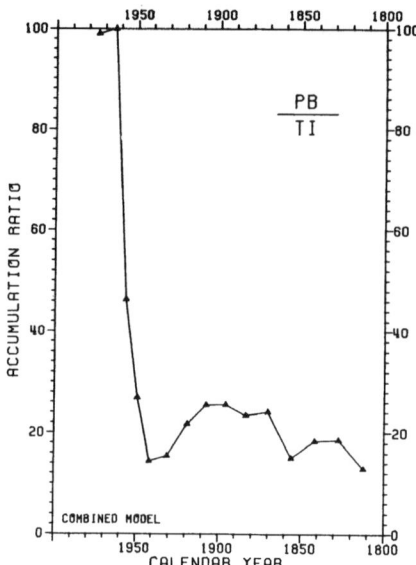

Fig. 7. Pb/Ti accumulation rate ratio for hollow core (4) from Big Heath, Mount Desert Island, Maine. The maximum ratio is set at 100.

CONCLUSIONS

The chemistry of peat cores from an ombrotrophic bog in Maine is characterized by strong concentration gradients. The bulk chemistry of any particular stratum is a combination of inputs which include: 1) aerosols from the ocean; 2) wind-blown soil dust; 3) material from polluted air masses; 4) gases (CO_2, N_2) and water, and internal fluxes of material due to: 1) mechanical migration, primarily downward; 2) chemical migration or concentration due to gradients in O_2, Eh, pH, etc.; and 3) biological recycling. The concentration of some immobile substances (e.g., Pb) may be incresed by biological decay. The activity of ^{210}Pb ($pCi.g^{-1}$) is a function of biological and radioactive decay.

Chemical profiles generally do not relate closely to trends in atmospheric deposition as interpreted from emission data and lake sediment cores. Net accumulation rates for most major elements are unrelated to atmospheric deposition rates. Titanium and Al probably most closely reflect the input of dust, some of which is derived from emissions related to fossil fuel consumption. Zinc and Cu accumulation rates crudely parallel but are lower than rates derived from lake sediments. Lead (and V) accumulation rates approximate average net accumulation rates derived from lake sediments.

Integrated ^{210}Pb (pCi.cm^{-2}) and energy-related pollutants (Pb, Zn, V, and Cu) are higher in hummocks, relative to hollows, suggesting migration of these constituents after deposition.

Although the general chronology of atmospheric pollution can be established with chemical stratigraphy in peat from Big Heath, Maine, the quantitative reconstruction of atmospheric deposition is not possible due to uncertainties in dating and chemical mobility.

Acknowledgements

Funding was provided by the Maine Geological Survey (using funds from the U.S. Department of Energy) and from the U.S. National Science Foundation (grant DEB-7922142 to E. Gorham). I was assisted in the conducting of field work and laboratory analyses by Geneva Blake and Marilyn Morrison. Dr. C.T. Hess assisted with many aspects of the ^{210}Pb dating. Denis Hanson developed the computer programs for data manipulation.

REFERENCES

Appleby PG, Oldfield F (1978) The calculation of lead-210 dates assuming a constant rate of supply of unsupported ^{210}Pb to the sediment. Catena 5: 1-8

Buckely DE, Cranston RE (1971) Atomic absorption analyses of 18 elements from a single decomposition of aluminosilicate. Chem Geol 7: 273-284

Clymo RS (1978) A model of peat bog growth. In: Heal OW, Perkins DF (eds) Ecological studies, Chapter 9, Vol 27. Springer-Verlag, Berlin, p 187-223

Eakins JD, Morrison RT (1978) A new procedure for the determination of lead-210 in lake and marine sediments. Intern J Applied Radiation Isotopes 29: 531-536

El-Daoushy F, Tolonen K, Rosenberg R (1982) Lead-210 and moss increment dating of two Finnish Sphagnum hummocks. Nature 296: 429-431

Groet, SS (1976) Regional and local variations in heavy metal concentrations of bryophytes in the northeastern United States. Oikos 27: 445-456

Hemond HF (1980) Biogeochemistry of Thoreau's Bog, Concord, Massachusetts. Ecol Mon 50: 507-526

Hvatum OO, Bolviken B, Steinnes E (1983) Heavy metals in ombrotrophic bogs. In: Hallberg R (ed) Environmental biogeochemistry. Ecol Bull 35: 351-356

Kahl JS, Anderson JL, Norton SA (1985) Water resource baseline data and assessment of impacts from acidic precipitation, Acadia National Park, Maine. U.S. Park Service, North Atlantic Region Water Resources Program Tech Rept 16

Livett EA, Lee JA, Tallis JH (1979) Lead, zinc, and copper analyses of British blanket peats. J Ecol 67: 865-891

Norton SA (1986) Geochemistry of selected Maine peat deposits. Maine Geol Survey Bull 34, p 38

Norton SA (1986) The chemical stratigraphy of lake sediments: monitoring and assessment of trends in acid deposition. In: Gibson J (ed) National Academy of Sciences, Environmental Studies Board

Pakarinen P (1981a) Metal content of ombrotrophic Sphagnum mosses in NW Europe. Ann Bot Fennici 18: 281-292

Pakarinen P (1981b) Nutrient and trace metal content and retention in reindeer lichen carpets of Finnish ombrotrophic bogs. Ann Bot Fennici 18: 265-274

Pakarinen P, Tolonen K, Heikkinen S, Nurmi A (1983) In: Hallberg R (ed) Accumulation of metals in Finnish raised bogs. Environmental biogeochemistry. Ecol Bull (Stockholm) 35: 377-382

PROTON CYCLING IN BOGS: GEOGRAPHIC VARIATION IN NORTHEASTERN NORTH AMERICA

N.R. Urban, S.J. Eisenreich, and E. Gorham[1]

Environmental Engineering Program, University of Minnesota, Minneapolis, Minnesota, U.S.A. 55455

[1]Dept. of Ecology and Behavioural Biology, University of Minnesota, Minneapolis, Minnesota, U.S.A. 55455

ABSTRACT

A detailed hydrogen ion budget has been constructed for the Marcell bog in north-central Minnesota based on a 5-year, intensive study of element cycles. Major features of the acidity balance for this site include the following: (1) production of organic acids (263 meq.m^{-2}.y^{-1}) is the dominant source of acidity and serves to buffer the bog water at pH 4; (2) sequestering of elements in peat is also a significant source of acidity (42.9 meq.m^{-2}.y^{-1}); (3) weathering of dustfall inputs is an important source of alkalinity (<76 meq.m^{-2}.y^{-1}) at this site which is situated near the major agricultural area of the plains; (4) nitrate and sulphate reduction contribute little alkalinity (<39.2 meq.m^{-2}.y^{-1}) because inputs (NO_3 and SO_4) to this bog are low. Analysis of peat and surface water from bogs across northeastern North America (Manitoba to Newfoundland) reveals the following: (1) production of organic acids across this region varies between 104 and 263 meq.m^{-2}.y^{-1}; (2) acidity-generation associated with net biological uptake (NBU, excluding nitrogen = 20-117 meq.m^{-2}.y^{-1}) varies in proportion to the rate of peat accumulation; (3) NBU-acidity exhibits high values in maritime bogs and lower values in mid-continental bogs; (4) bogs have a large capacity for sulphate reduction, and sulphate reduction becomes an increasingly important source of alkalinity as rates of sulphate deposition increase. From 60 to 93% of annual sulphate loadings are retained as reduced sulphur in bogs across eastern North America.

INTRODUCTION

The utility of acidity (alkalinity) budgets has been appreciated only recently, and few have been reported (e.g., Driscoll and Likens 1982; Kelley et al. 1982; Kilham 1982; Hemond 1980). There are three major benefits from construction of such budgets. First, all sources and sinks of acidity are placed in perspective, and the dominant processes are identified. The controversy over the importance of acid rain in ecosystem acidification has highlighted the need to compare the relative magnitudes of all sources and sinks (Rosenqvist 1978; Krug and Frink 1983; Stumm et al. 1983). Only by such comparison can the long-standing uncertainty over the cause of acidity in bogs be resolved (e.g., Ramaut 1954, Clymo 1964, 1967; Gorham 1967; Gorham et al. 1985).

Second, acidity budgets provide a holistic view of ecosystems. Whereas the same component processes of the hydrogen ion cycle (viz., nutrient uptake, decomposition, weathering, etc.) occur in all ecosystems, proton budgets provide a common framework for comparing widely different ecosystems. Subtle differences between successional stages (Gorham et al. 1979) and between different forest types (Alban 1982) are more readily apparent when viewed in the context of the proton cycle. Since many linkages between major element cycles are mediated by hydrogen ions (Likens et al. 1981; Schindler 1981; Wollast 1981), acidity budgets provide an overview of nutrient cycling. Theoretically it is even possible to express energy or electron flow in terms of proton transfers.

Finally, because many processes are integrated into the proton cycle, the ramifying effects of perturbations on ecosystems may be identified. The proton cycle is necessarily complex because it integrates both acid-base and oxidation-reduction reactions. Feedback exists on many levels as the intensity or concentration of hydrogen ions influences the capacity of the system to generate or consume acidity by such processes as nutrient uptake, decomposition, weathering, ion exchange, etc. Although no generalized model applicable to all systems has yet been formulated (see Schnoor et al. 1984, Chen et al. 1984), the acidity budget of a system can clarify the likely impact of such perturbations as timber harvesting, agricultural practices, and acid deposition (e.g., Driscoll and Likens 1982; Kilham 1982; Nilsson et al. 1982; Rosenqvist 1978). Acidity budgets also may clarify the probable consequences of acid deposition, nutrient (sewage) addition, and peat mining on wetlands (Gorham et al. 1984; Kelly and Harwell 1982).

The objectives of this paper are two-fold: (1) to present an overview of the proton budget of a small Minnesota bog to illustrate the major processes involved, and (2) to examine water and peat chemistry in bogs on a transect from Minnesota and Manitoba eastward to Nova Scotia and Newfoundland in order to assess geographic trends in the processes governing the acidity balance of bogs. The annual hydrogen ion budget ignores the internal dynamics of the cycle, which will be discussed in detail in a forthcoming paper. The geographic transect chosen exhibits large gradients in rates of precipitation, evapotranspiration, sulphate deposition, and cation input (Munger and Eisenreich 1983; Gorham et al. 1985). This paper will examine the extent to which the acidity balance of bogs changes in the face of changing inputs.

METHODS AND MATERIALS

Site Description

Marcell bog watershed S-2 is located in northcentral Minnesota (47° 32" N, 93° 28" W). The peat deposit (3.24 ha) occupies a kettle-hole and is surrounded by 6.48 ha of mineral-soil upland. A 5-7 m wide lagg surrounds the bog and funnels most of the upland run-off around it. The bog is forested with an even-aged (120 years old) stand of black spruce (Picea mariana) and an understory of ericaceous shrubs (Ledum groenlandicum, Chamaedaphne calyculata). Bryophytes (Sphagnum magellanicum, S. recurvum) dominate the ground-layer vegetation. A hummock-hollow microtopography exists (Verry 1984), but pools of standing water are absent. Limnic sediments seal the basin and have allowed the development of a perched bog water table (Verry and Timmons 1982).

Marcell experiences a continental climate with wide extremes in temperature (-40 to +40°C). Annual precipitation averages 760 mm, evapotranspiration 505 mm (Verry and Timmons 1982). Snow cover (November through April) prevents all but the top few centimeters from freezing.

Methods

Sources of acidity in bogs include atmospheric deposition, biological uptake of cations, and production of organic acids. Proton sinks include alkaline inputs, anion assimilation or reduction, decompositional release of cations, and weathering of dustfall inputs. Within this paper, the terms proton sink, alkalinity source, and negative acidity are used synonomously. A mass-balance approach was used to measure all processes. The system considered is the peat deposit (mire), i.e., the bog and the lagg. The upland is not included in the mass balance so upland run-off is considered an input to the mire. Rates of input and of all processes are expressed per unit area of the peat deposit. As stated above, this ignores internal gradients (e.g., bog to lagg). Details of sample collection and analysis and methods of calculation have been reported elsewhere (Verry 1975, 1983, 1984; Verry and Timmons 1977, 1982; Timmons et al. 1977; Buttleman 1982; Schurr 1983; Urban 1983; Grigal 1985; Grigal et al. 1985), and are briefly summarized below.

Wet deposition was collected weekly in a wet-only Aerochem Metrics sampler in a forest clearing one kilometer from the bog. This sampler was used during snow-free months in 1981-1984. Winter deposition was measured by sampling the snow pack once just before snowmelt. Earlier records of atmospheric deposition (Verry 1983; Verry and Timmons, 1977) have been incorporated into estimates of mean deposition. Estimates of dry deposition were based on dry bucket measurements (Verry 1983; Munger 1981) and aerosol measurements (Eisenreich et al. 1978).

Upland run-off was sampled from two run-off plots on the north and south slopes above the bog. Upland run-off has two components: surface run-off (occurring in the O horizon) and interflow (occurring in the A and B soil horizons). The volume and chemistry of surface run-off were measured directly in water collected from the run-off plots. Volume of interflow was calculated by hydrograph separation (Timmons et al. 1977). Composition of interflow was measured on samples from run-off plots. Chemical inputs from the upland were monitored for seven years.

Streamflow was monitored continuously at a v-notched weir. Water quality samples were taken biweekly except during spring periods of high flow when sampling was more frequent. The period of record for export of elements is 3 to 14 years, depending on the element.

Rates of element accumulation in peat were calculated as the differences between inputs and outputs and were also measured directly in six to ten peat cores. Peat cores (50 cm) were dated by assuming acid-insoluble ash (i.e., the residue after combustion at 550°C and dissolution in hot aqua regia) to be conservative and to have had an historically constant rate of input. This technique has been compared with other techniques (e.g., Pb-210, Cs-137, C-14, pollen, magnetic minerals; Rapaport et al. 1985). Decomposition is calculated as the difference between the rate of plant uptake (discussed below) and the rate of accumulation in peat.

Productivity of trees and shrubs (above- and below-ground), moss, and herbs was measured over a period of three years, and biomass and chemical composition were also measured (see Grigal 1985; Grigal et al. 1985).

Surface water samples and short peat cores (50 cm) were obtained from 28 ombrotrophic bogs along a transect stretching from Minnesota and Manitoba to Newfoundland (Fig. 1). Additional surface water samples were obtained from a coastal low-pocosin peatland in North Carolina, three Irish bogs, and three bogs in northern England. Locations of all sites and collection and analytical procedures have been reported previously (Gorham et al. 1985). Hummock cores collected from eight sites on the transect were analyzed for rates of element accumulation in the fashion described above.

All water samples were analyzed by the same protocol. Cations (Ca, Mg, Na, K) were measured by flame atomic absorption spectrophotometry (AA) and Fe and Al by flameless AA. Anions were measured by ion chromatography, ammonium by the method of Solorzano (1969). An Orion 701 pH meter was used to measure pH, alkalinity (Gran titration), and acidity. Acidity is defined here as the base-neutralizing capacity (BNC) to pH 7.0. For metal analyses, peat and plant tissues were combusted for 3 hours at 550°C and digested in hot aqua regia (1:1 $HCl:HNO_3$). Digestates were analyzed by flame AA or plasma emission spectroscopy. Organic nitrogen was measured by the semi-micro Kjeldahl method of Bremner (1965), and organic sulphur was analyzed with a LECO S-analyzer.

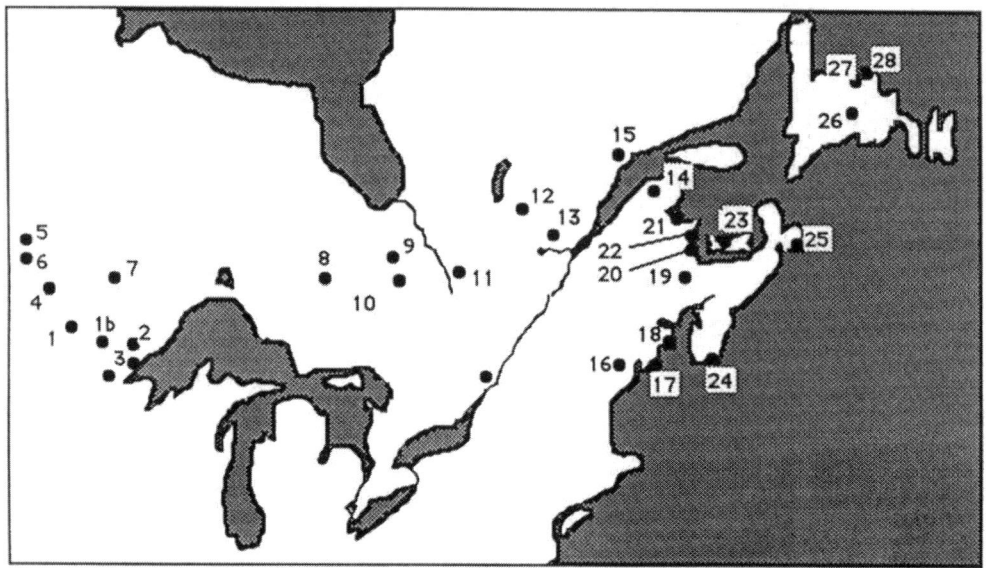

Fig. 1. Location of sites on the North American transect of bogs (after Gorham et al. 1985). Site numbers correspond to those given in Table 2. Site 1b is Marcell, Mn, the location of the site-intensive study.

RESULTS

Marcell: Inputs and Outputs

The Marcell bog acts as a large sink for alkalinity. Data collected earlier by Verry and coworkers (Verry 1975, 1983, 1984; Verry and Timmons 1977, 1982; Timmons et al. 1977) are averaged with data gathered in this study in order to base the mass balance on as long a time period as possible and thus compensate for large variations among years (see Likens et al. 1977). Atmospheric deposition provides a small input of alkalinity (1.6 meq.m^{-2}.y^{-1}) to this site because of its distance from both the prairie and major anthropogenic emission sources (Munger 1982; Thornton and Eisenreich 1982). The dominant input of alkalinity is upland run-off (44 meq.m^{-2}.y^{-1}), which occurs primarily in spring and fall and is channelled around the bog in the lagg. The bog is a large net source of acidity (142 meq.m^{-2}.y^{-1}). Acidity exported from the bog (Fig. 2) is chiefly in the form of fulvic acids produced by incomplete oxidation of organic matter (McKnight et al. 1985).

Examination of the mean annual export from the bog (Fig. 2) reveals that about half (121 meq.m^{-2}.y^{-1}) of the organic acids produced within the bog (263 meq.m^{-2}.y^{-1}) have been neutralized. Since alkalinity inputs total only 46 meq.m^{-2}.y^{-1}, there must be an additional net production of 96 meq.m^{-2}.y^{-1} of alkalinity.

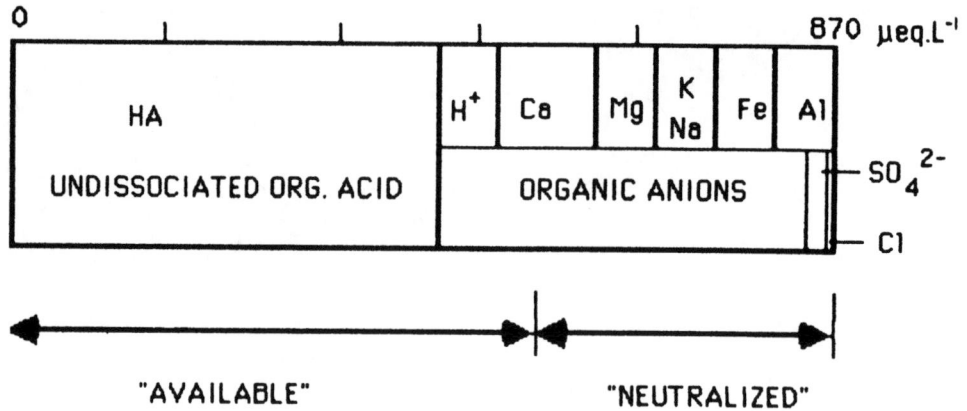

Fig. 2. Composition of mean annual export (streamflow) from the Marcell Bog. Bog waters typically have large concentrations of undissociated fulvic acids (measured by titration) in addition to the "anion deficit" or dissociated organic acid anions. That portion of the anion deficit balanced by cations other than H^+ represents "neutralized" acidity and indicates the magnitude of alkalinity inputs.

Weathering

The potential alkalinity contribution from weathering or dissolution of iron- and aluminum-containing minerals was calculated by assuming that all inputs of Fe and Al become organically bound within the mire. Neither the magnitude of dry deposition of Fe and Al (Eisenreich et al. 1978) nor the speciation of Fe and Al in upland run-off is well known. Consequently, alkalinity generated by weathering may be less than the maximum potential estimate (76 meq.m^{-2}.y^{-1}) presented here.

Denitrification

Denitrification potential (measured by laboratory assays with nitrate amendments) exceeds the annual supply of nitrate to the Marcell bog (Urban 1983). The actual denitrification loss depends on the relative rates of all processes competing for nitrate (plant uptake, dissimilatory reduction to ammonium, and denitrification). Some nitrate supplied in rainfall is taken up by the tree canopy (3.4 meq.m^{-2}.y^{-1} - Urban 1983; Verry and Timmons 1977). If all nitrate supplied to the bog surface in throughfall were lost to denitrification 12.2 meq.m^{-2}.y^{-1} of alkalinity would be generated.

Dissimilatory Reduction of Sulphate

Rates of dissimilatory sulphate reduction have not been measured directly. Dynamic cycles of sulphate reduction and reoxidation have been observed in other peatlands (R. Behr, pers. comm.; R.K. Wieder, pers. comm.). However, only about 10% of the sulphur stored within the Marcell bog occurs in reduced forms other than organic sulphur (primarily pyrite - Urban, unpub.). Thus, the net alkalinity generation from dissimilatory reduction is much less than 23.6 meq.m^{-2}.y^{-1}, the total sulphate input to the bog.

Plant Uptake

The elemental composition of plants in combination with productivity data (Grigal 1985; Grigal et al. 1985) permit calculation of the acidity flux associated with plant nutrient uptake. Acidity flux is defined as the excess of cation uptake over anion uptake (Nilsson et al. 1982). In this study sulphur, the dominant anionic nutrient, was measured only in Sphagnum. For all other plants, a N:S ratio of 3:1 was assumed based on work of Pastor and Bockheim (1984) and Driscoll and Likens (1982). With this assumption, sulphur uptake by Sphagnum amounts to 25 meq.m^{-2}.y^{-1}, and 63 meq.m^{-2}.y^{-1} are taken up by other plants. Nitrogen is taken up primarily as ammonium in bogs because nitrification is absent and atmospheric deposition of nitrate can supply only a small fraction of plant nitrogen requirements (Urban 1983; Hemond 1983; Rosswall and Granhall 1980; Martin and Holding 1978; Verhoeven et al. 1983). This generalization may not hold for all bogs in England (Press and Lee 1982; Woodin et al. 1985; Lee et al., this volume). At Marcell, nitrate deposition may supply at most 6% of the plant nitrogen requirement (Urban 1983). It is unclear at this time how much nitrate is taken up by plants and how much is denitrified. Since denitrification and nitrate assimilation both generate equivalent amounts of alkalinity, it is not essential to distinguish between them. Were all nitrate denitrified, plant uptake at Marcell would generate 827 meq.m^{-2}.y^{-1} of acidity (moss - 475 meq.m^{-2}.y^{-1}, above-ground trees - 177 meq.m^{-2}.y^{-1}, other plants - 175 meq.m^{-2}.y^{-1}).

Net Biological Uptake or Elemental Accumulation

Net biological uptake (NBU) is defined as the difference between annual rates of uptake and decomposition. Thus it is equal to the rate of storage of organic matter. There was good agreement among values for the 100-year element accumulation rates for replicate cores (including hummocks and hollows), and these values are in fair agreement with accumulation rates calculated by mass balance about the mire (Table 1). The biomass of the tree stand is in a steady state at present (Grigal et al. 1985) so element accumulations within the peat represent total net

Table 1. Element accumulation rates in peat: mass balance vs. direct measurements.

Element	Measured 100-year accum. rate[a] (meq.m^{-2}.y^{-1})	Mass balance[b] (meq.m^{-2}.y^{-1})
Ca	11.5 ± 2.8	37.4[c]
Mg	6.6 ± 1.5	7.4
Na	2.0 ± 0.1	7.0
K	2.8 ± 0.8	1.6
Fe	17.3 ± 4.3	-8[c]
Al	38.9 ± 5.0	10[c]
Cl	0.5 ± 0.1	5.7
NH_4		20.2
NO_3		15.6
SO_4		23.6
Total N	94.3 ± 7.1	83
Total S	14.4 ± 1.2	13.8
Net Biological Uptake	158 ± 10.3[d]	30.7[e]

[a] Mean ± 1 S.D.; n=6, 2 hummocks and 4 hollows.
[b] Sum of Inputs - Sum of Outputs.
[c] The discrepancy between accumulation rates of Ca is due to storage in wood, which is inadequately sampled in peat cores. Discrepancies for Fe and Al result from upward migration of these elements from fen peat which underlies the bog.
[d] NBU=Sum(Cations)-Sum(Anions). It is assumed that all N is taken up as ammonium. This value is incorrect because N transformations are not considered.
[e] NBU=Sum(Cations)-Sum(Anions). This incorporates assimilatory and dissimilatory anion reduction. Accounting for denitrification (12.2 meq.m^{-2}.y^{-1}) NBU = 42.9 meq.m^{-2}.y^{-1}.

biological uptake. As shown in Table 1, it is critically important to account for all nitrogen transformations to calculate the correct acidity flux. The mass balance for the bog indicates that the net acidity flux from nitrogen transformations is only 16.8 meq.m^{-2}.y^{-1} (i.e., ammonium retention - nitrate retention, where nitrate retention = plant NO_3 uptake + denitrification). Thus, acidity resulting from element <u>storage</u> in peat equals 42.9 meq.m^{-2}.y^{-1}, 30.7 (Table 1) - (-12.2) (acidity flux from denitrification, given in Table 3), not 158 meq.m^{-2}.y^{-1} as shown in column 1 of Table 1. Decomposition (784 meq.m^{-2}.y^{-1}) is calculated as the difference between plant uptake and accumulation. Data from Marcell suggest that, in the absence of an aggrading forest, calculation of NBU based on a single hummock core (as done for the transect sites) is a valid estimate for the whole bog.

North American Transect

Pertinent data for sites on the North American transect are summarized in Table 2. The tabulated values of sulphate concentration and loading in precipitation represent three- to five-year averages for sites in the Canadian Network for Sampling Precipitation (CANSAP - Barrie and Sirois 1982).

Uncertainty in estimated rates of organic acid production (OAP) and sulphate retention is difficult to assess. Large variations in sulphate concentrations and base neutralizing capacities of bog water samples from a single site were common (S.E. 1.8 to 21.6 µM SO_4). Pan evaporation rates represent an upper limit for evapotranspiration, but a lower limit is more difficult to establish. At Marcell, evapotranspiration calculated by mass balance equals that calculated by the Thornthwaite method (Verry and Timmons 1982). This latter technique will be treated as a lower limit to evapotranspiration in this discussion. Rates of total organic acid production were calculated from: OAP = measured concentration of available and neutralized acids x run-off, where run-off = precipitation - evapotranspiration. OAP may be overestimated because concentrations of organic acids are higher than the annual mean during summer due to evaporative concentration. Base neutralizing capacity of Marcell bog water varies three-fold over a year. Similarly, concentrations of organic anions vary four-fold annually in streams draining Nova Scotian bogs (Kerekes 1984). Export estimated in this fashion for Marcell (153 $meq.m^{-2}.y^{-1}$), however, shows good agreement with long-term measurements (142 $meq.m^{-2}.y^{-1}$). Thus, rates of OAP presented here are at least 10% too high owing to overestimation of the mean concentration and have an uncertainty of 25-50% arising from the uncertainty in actual evapotranspiration rates.

There are two difficulties in calculating accurate values of NBU-acidity for the transect sites. First, questions remain concerning the validity of all techniques for constructing dated profiles in recent strata of peat (Malmer and Holm 1984, Oldfield et al. 1979; El-Daoushy et al. 1982). Dates obtained with Pb-210 are generally younger than dates obtained with the acid-insoluble ash (AIA) technique used in this study. Consequently, fluxes of acidity from ion uptake (excluding N) based on dates obtained with Pb-210 for the transect cores (mean 81 ± 27 $meq.m^{-2}.y^{-1}$) are higher than those based on the AIA dating method (mean 56 ± 32 $meq.m^{-2}.y^{-1}$). The latter technique was chosen because it yielded dates consistent with measured profiles of chlorinated organic compounds (DDT, Toxaphene, PCB's) and the known historical input of these conservative compounds (Rapaport et al. 1985). Thus estimates of NBU-acidity may have a systematic bias from the dating technique, even though, as shown in Table 1, the precision of measurement is ± 7%.

The second and more intractable difficulty is determination of the acidity flux from nitrogen uptake. As discussed above for Marcell, it is impossible to deduce the acidity flux only from the nitrogen accumulation rate in the core. Nitrogen stored in peat may have entered the system as NO_3, NH_4, N_2, NO_x, HNO_3 or organic N, but it is only the net retention of charged species (nitrate and ammonium) which determines the acidity flux. Since inputs and outputs of nitrate and ammonium to and from the transect sites are not known, net retentions cannot be

Table 2. Summary of Data for North America Transect.

Site	Precipitation (cm)	Evapotranspiration Pan (cm)	Evapotranspiration Thornthwaite (cm)	SO$_4$ Conc. (μeq·L^{-1})	SO$_4$ Loading (meq·m^{-2})	Base Neutralizing Capacity (meq·L^{-1})	Organic Acid Production (meq·m^{-2}·y^{-1})	NBU-acidity (meq·m^{-2}·y^{-1})
Minnesota								
1b Marcell	76		50.5	29	18	0.60	251	138
3b Toivola						0.32		
Ontario								
7 Experimental Lake (ELA)	68.3	52.1	43	19	9.5	0.45	115	86
10a Diamond	88	53.3	37.5	48	42			115
10b Alfred								
Quebec								
11 Lac Parent						0.28		
12 Lac des Miserables	86.5	53.3	39.8	30	26			
13 Lac St. Jean						0.25	104	107
15 Sept Iles	132.9	56	36.8	25.2	33	0.19	183	
Maine								
16 Great Sydney Heath	81	67.3		42	34	0.48	188	
17 Bar Harbor	122	67.3	50.6	46	37	0.20	150	196
18 Carrying Place Cove						0.22		
New Brunswick								
20 Point Sapin	135	64	43.2	39	53	0.22	215	
21 Miscou Island						0.28		
22 Point Escuminac						0.13		

Table 2. Continued

Site	Precipitation (cm)	Evapotranspiration Pan (cm)	Evapotranspiration Thornthwaite (cm)	SO$_4$ Conc. (μeq.L^{-1})	SO$_4$ Loading (meq.m^{-2})	Base Neutralizing Capacity (meq.L^{-1})	Organic Acid Production (meq.m^{-2}.y^{-1})	NBU-acidity (meq.m^{-2}.y^{-1})
Prince Edward Island								
23 Foxley Moor						0.41		
Nova Scotia								
24 Cape Sable	157	66	50.6	63	99	0.22		
25 Fourchu						0.15		263
Newfoundland								
26a Gros Morne								
27 Gander Bay	135	55	39.5	17.5	24	0.16	186	240
North Carolina								
35 Croatan Marsh pocosin	132	130	87.2	30	40	0.66	104	

calculated. If wet deposition were the only input and if
retention of both species were 100%, acidity fluxes would range
from -25 to +7 $meq.m^{-2}.y^{-1}$. Dry deposition may be neglected from
charge-balance considerations, but retention of ammonium is not
100%. Ammonium retention at Marcell is only 90%, and Lee et al.
(this volume) observed nitrate in run-off from English bogs.
Nitrate deposition equals or exceeds ammonium deposition at all
sites from the Atlantic coast to a point in Ontario approximately
midway along the transect (Barrie and Sirois 1982; Munger and
Eisenreich 1983). Thus nitrogen uptake (including denitrification) will act to reduce the NBU-acidity flux from eastern
sites (by an amount <30 $meq.m^{-2}.y^{-1}$) and to increase the flux at
western sites (<10 $meq.m^{-2}.y^{-1}$). The range in NBU-acidity
fluxes may, therefore, be only 30 to 87 $meq.m^{-2}.y^{-1}$ (Table 2).

DISCUSSION

Marcell Hydrogen Ion Budget

Three major features stand out upon examination of the acidity
fluxes at Marcell (Table 3). First, the largest source of
acidity arises from nutrient uptake by the vegetation (827
$meq.m^{-2}.y^{-1}$). Based on measurements of exchangeable cations only
about 130 $meq.m^{-2}.y^{-1}$ or 15% of this represents ion exchange on
cell walls of Sphagnum; the remainder is assimilatory uptake
(some sulphate reduction may be dissimilatory). The majority of
this acidity, 784 $meq.m^{-2}.y^{-1}$, is neutralized upon the release of
cations during decomposition. The difference between uptake and
decomposition (42.9 $meq.m^{-2}.y^{-1}$), the net biological uptake (NBU
after Kilham 1983), represents acidity resulting from peat
accumulation (NBU-acidity). This mineral acidity is not exported
from the bog. Bog water at Marcell does not contain mineral
acidity (i.e., hydrogen ions in excess of organic anions). Thus
sources of alkalinity must exceed NBU-acidity.

The second salient feature is the magnitude of the production of
organic acids (263 $meq.m^{-2}.y^{-1}$). This process, generating twice
as much acidity as does ion exchange, represents the largest net
source of acidity for the bog (cf. Clymo, this volume). Other
investigators have shown the acids to be primarily fulvic acids
formed as a result of incomplete oxidation of organic matter
(McKnight et al. 1985; Perdue and Lytle 1983; Oliver et al.
1983). Fulvic acids from bogs are similar to fulvic acids from
other systems in having pKa's between 3 and 4 and in having 8-14
meq of acidic functional groups per mg DOC (Perdue and Lytle
1983; McKnight et al. 1985; Oliver et al. 1983). These acids are
responsible for buffering the pH of bogs throughout the world at
a value of about 4, and they represent the only acidity in the
outflow from the Marcell site.

Table 3. Acidity balance for Marcell Bog S-2.

	Acidity (meq.m^{-2}.y^{-1})	
Sources:		
Wet deposition	-0.05	(9)
Dry deposition	-1.5	(1.7)
Upland run-off	-44.3	(18.6)
Nutrient uptake	827	(248)
Organic acid production	263	(50)
Total	1044	(254)
Sinks:		
Denitrification	12.2	
Decomposition	784	
Weathering	76.0	
Outflow	142	(50)
Total	1014	

a Standard deviations given in parentheses. Uncertainty for plant uptake represents propagated errors from measurements of growth and element concentrations.

Finally, inputs of alkalinity to this mire and alkalinity generated in situ are sufficient not only to neutralize the acidity generated by NBU but also about half of the organic acids produced within the bog. Sources of alkalinity include atmospheric deposition (1.6 meq.m^{-2}), denitrification (12 meq.m^{-2}), dissimilatory sulphate reduction (<<23.6 meq.m^{-2}), weathering of Fe and Al inputs (<76 meq.m^{-2}), and run-off from the upland (44 meq.m^{-2} - mixed bicarbonate and organic anion alkalinity). Because most of the alkalinity input is derived from the upland, the lagg is much less acid (pH 5) than the highly acidic bog.

Comparison of the proton budget for Marcell with that for Thoreau's Bog in Massachusetts (Hemond 1980) reveals several differences.

```
              Inputs    +   Sources     =    Sinks     +    Outputs
Acid +   Run-off + Cation + Org.Acid = Anion + Weather- + Overflow
Dep.              Uptake   Formation   Uptake    ing      Mineral
                (NH4) (other)          (SO4) (NO3)        Organic

Thoreau's Bog:
  104 +     0    + 5    + 11    +  240  =  95   + 24    + 0   + 1.2 + 240

Marcell:
  -1.6 +  -44.3 + 20.2 + 55.4  +  263  = 23.6 + 15.6  + 76  + 0   + 142
```

All values above are in units of meq.m^{-2}.y^{-1}. Organic acids reported for Thoreau's Bog represent base neutralizing capacity to pH 11, not pH 7 as in this study. Sulphate uptake for Thoreau's Bog given above has been adjusted from that reported originally by Hemond to account for the inclusion of organic sulphur in the original measurement of sulphate. Hemond (1980) did not focus on specific processes within the bog but formulated a proton budget based on element mass balances about the entire system. At both sites, all terms in the acidity balance were measured; none was derived by difference. The slight imbalance in the Marcell budget may result from inaccurate estimates of dry deposition (especially of Fe, Al, SO_4, and NO_3) and some loss of water and ions through the bottom of the bog (deep seepage - Verry and Timmons 1982). The major differences between the two sites lie in the magnitude and quality of atmospheric deposition and upland inputs. Marcell receives a net input of alkalinity accompanied by a high cation loading and high inputs of weatherable dust particles. Thoreau's Bog receives much less soil dust but a high acid-sulphate loading. Consequently, half of the organic acids produced at Marcell are neutralized by alkalinity inputs. At Thoreau's Bog acid deposition is neutralized by anion reduction. Organic acid production at both sites is similar, but all other terms in the acid balance appear to be plastic responses to external inputs.

Geographic Variations in Acidity Production

Comparison of Thoreau's Bog and Marcell raises three questions concerning geographic trends in the proton balance of bogs. First, will acid deposition at current rates in North America be offset by nitrate and sulphate reduction or will it further acidify bogs? Gorham et al. (1985) have shown that English bogs have been acidified by acid deposition. Second, is organic acid production similar in all North American bogs? Third, is acidity from peat accumulation proportional to cation (nutrient) inputs to bogs? Surface water and peat chemistry data collected from the transect of North American bogs provide tentative answers to these questions.

1. Sulphate Retention Capacity: Were sulphate totally conserved in bogs then concentrations in bog waters would be low and unrelated to rates of deposition. Gorham et al. (1985) state that sulphate reduction obscures the effect of anthropogenic sulphate on bog water composition. However, sulphate concentrations increase from west to east across the transect (Fig. 3a), and this increase is related to the gradient in total sulphate deposition (Fig. 3b). Given that summer-time concentrations of sulphate in bog waters are proportional to deposition, comparison of concentrations in bog waters to concentrations in rain (corrected for the concentrating effect of evapotranspiration) should indicate the fraction of sulphate deposition retained within bogs. Such a comparison (Table 4) suggests that 60% to 93% of sulphate inputs are retained. The fraction retained decreases as the fraction of precipitation going to run-off increases (i.e., as the water retention time decreases). Because of this interplay between hydrology and sulphate reduction, it is impossible to define a fixed sulphate loading which will result in acidification of bogs.

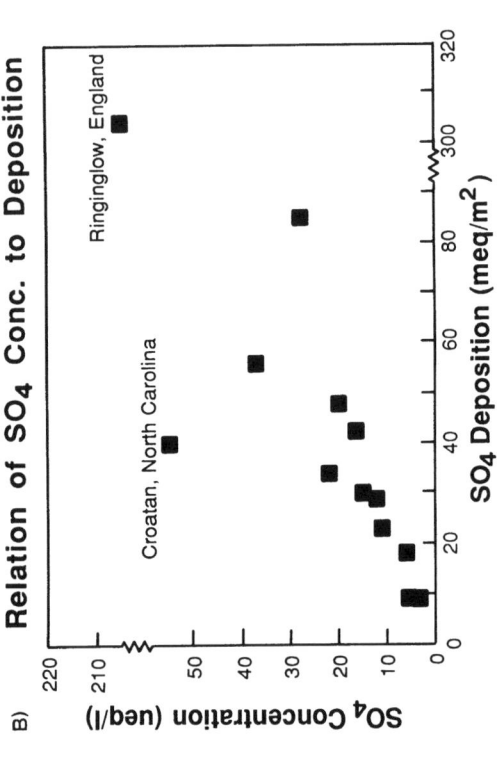
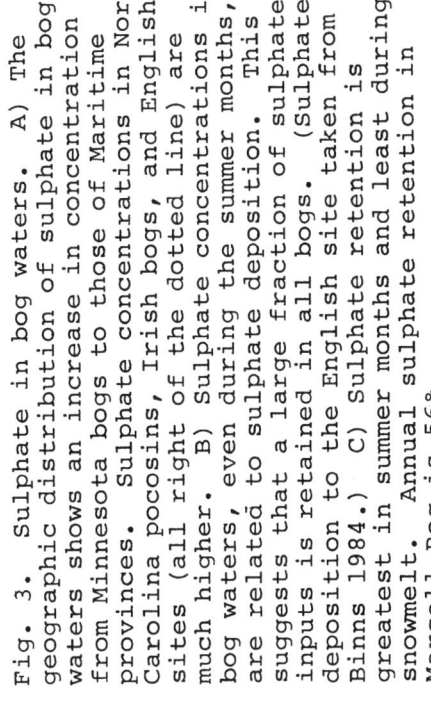
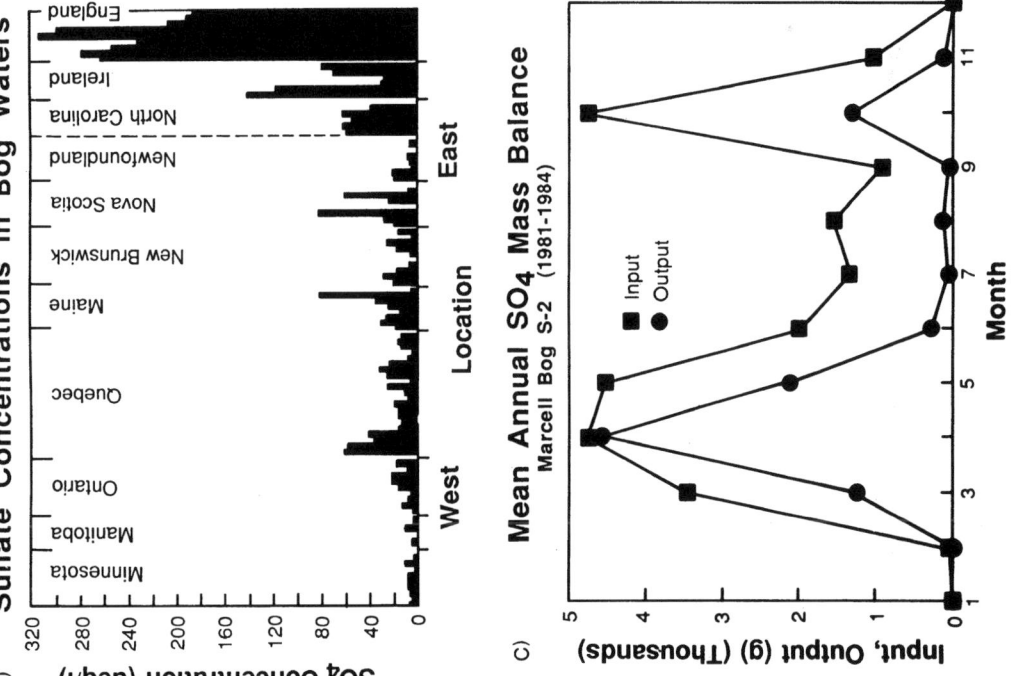

Fig. 3. Sulphate in bog waters. A) The geographic distribution of sulphate in bog waters shows an increase in concentration from Minnesota bogs to those of Maritime provinces. Sulphate concentrations in North Carolina pocosins, Irish bogs, and English sites (all right of the dotted line) are much higher. B) Sulphate concentrations in bog waters, even during the summer months, are related to sulphate deposition. This suggests that a large fraction of sulphate inputs is retained in all bogs. (Sulphate deposition to the English site taken from Binns 1984.) C) Sulphate retention is greatest in summer months and least during snowmelt. Annual sulphate retention in Marcell Bog is 56%.

Table 4. Estimated sulphate retention in bogs.

Site	Sulphate conc. ($\mu eq.L^{-1}$) Rain[a]	Bog	% Retention[b]
Marcell, Mn	85.7	6	93
ELA, Ontario	61.1	5.5	91
Diamond Lake, Ont.	102	16.3	84
Lac des Miserables, Quebec	65	15	77
Sept Iles, Que.	105	12	70
Great Sydney Heath, Maine	244	22	91
Bar Harbor, Maine	112	37	67
New Brunswick[c]	64.5	20	69
Cape Sable, N.S.	100	28	72
Gander Bay, Nfld.	27.5	11	60
Croatan pocosin, N. Carolina	172	55	68

[a] Rain concentration represents the volume-weighted mean annual concentration in precipitation corrected for concentration due to evapotranspiration (see Table 2).
[b] Retention = 1 - (Bog SO_4)/(corrected rain SO_4) x 100.
[c] Concentrations of sulphate in all bogs in New Brunswick were averaged as were the mean rainfall concentrations for all CANSAP stations in New Brunswick because bog and CANSAP sites were not near each other.

Annual sulphate retention may be less than 60-93% due to lower retention efficiency during snow-melt (Figure 3c). The Marcell bog exhibits about 93% sulphate retention during summer months but an annual retention of only 56%. Thus, small bogs may be susceptible to a spring flush of acidic sulphate, but larger systems which receive less snowmelt from surrounding uplands may exhibit an annual sulphate retention approaching 80%.

2. <u>Organic Acid Production</u>: There are no clear geographic trends in production rates of organic acids in bogs across the transect (Table 2, Fig. 4a). When base neutralizing capacity to pH 8 rather than pH 11 is used, production within Thoreau's Bog is only 195 meq.m^{-2}.y^{-1} (McKnight <u>et al</u>. 1985), similar to production in Maine bogs. Clear gradients of water run-off (precipitation - evapotranspiration) and BNC of bog waters do exist, but in opposite directions such that there is no clear gradient in the export of acidity. Additional sites would be required to determine if the apparent minimum in rates of production and export in central Canada is real.

3. <u>Net Biological Uptake</u>: There is an easterly increase in acid-generation by element sequestration in peat (Fig. 4b). As noted above, acidity fluxes from nitrogen uptake serve to offset this trend somewhat, but such fluxes are not large enough to

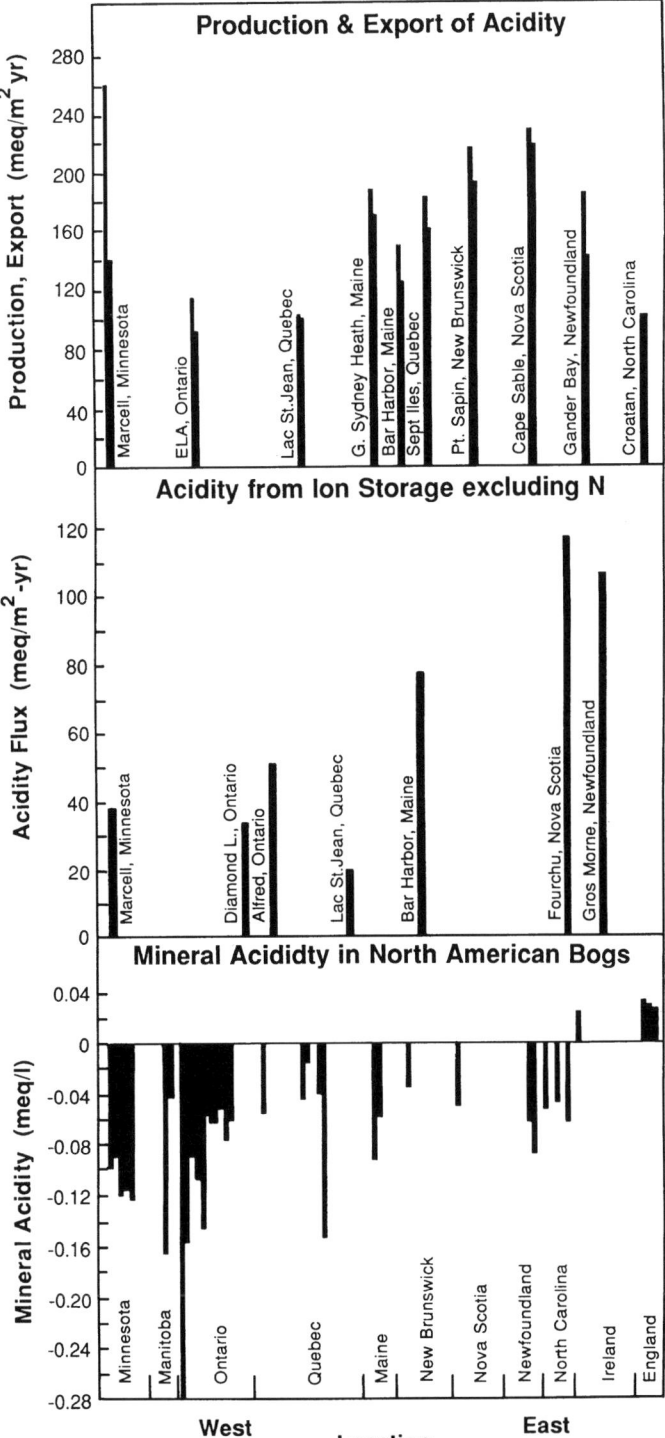

Fig. 4. Geographic trends in acid-producing processes. Total production of organic acids (left-hand bar) shows a minimum in Quebec and Ontario. Due to larger inputs of alkalinity, Minnesota bogs exhibit relatively low levels of organic acid export (right-hand bar). The pattern of acidity generation by element sequestering in peat shows an easterly increase. The combined effect of these processes is to create a gradient of increasing mineral acidity from west to east.

eliminate the gradient. The easterly increase in NBU-acidity is similar to the pattern of nutrient inputs (Barrie and Sirois 1982; Munger and Eisenreich 1983). Thus it appears that increased water flow rates, higher mean annual temperature and greater rates of nutrient input may enhance plant production, decomposition (recycling) and peat accumulation rates in bogs. These same conclusions were reached by Urban (1983) based on a comparison of nitrogen cycling in bogs and by Verhoeven et al. (1983).

4. Balance of Acidity: Not only do magnitudes of major processes generating and consuming acidity vary widely across North America, but there is also a subtle shift in the balance of acidity. Although the values of NBU-acidity, presented in Fig. 3b, may be overestimates, the bias in estimated NBU arising from the dating technique (discussed above) is similar in all sites so that trends in NBU are believed to be real. Estimated rates of organic acid production are also imprecise, and more intensive sampling is needed. The validity of the trend in NBU is supported by observation of a shift toward increasing importance of mineral acidity eastward across the transect (Fig. 4c). In no sites did mineral acidity actually exceed organic acidity. Nonetheless, eastern maritime bogs are more highly predisposed to further acidification by acid deposition due to two factors: lower water retention times in maritime bogs with correspondingly decreased sulphate reduction, and neutralization of low inputs of alkalinity by high rates of acidity production.

Acknowledgements

The authors thank E.S. Verry and A. Elling of the U.S. Forest Service who provided invaluable help at the Marcell study-site. We also thank D.F. Grigal, R.A. Rapaport, C.W. Gresham, and E.S. Verry for the use of unpublished data. The work of M.V. Santelmann in collecting samples and facilitating analysis of data from the N. American transect sites is greatly appreciated. Many thanks to R.S. Clymo for a careful review of this manuscript. This work was supported by the National Science Foundation (Grant No. DEB 7922142) and a Doctoral Dissertation Fellowship to N.U.

REFERENCES

Alban DH (1982) Effects of nutrient accumulation by aspen, spruce, and pine on soil properties. Soil Sci Soc Am J 46: 854-861
Barrie LA, Sirois A (1982) An analysis and assessment of precipitation chemistry measurements made by CANSAP, 1977-1980. Env Can Report AQRB-82-003-T
Binns WO (1984) Vegetation and soils, Sec 2. In: Acid rain, Report #14, Watt Committee on Energy, London
Bremner JM (1965) Total nitrogen. In: Black CA (ed) Methods of soils analysis, Am Soc Agron Inc, Madison, p 1149-1177

Buttleman CG (1982) Use of Rb/K ratio to define nutrient linkages between a perched bog and its surrounding upland. MSc Thesis, Uni of Minnesota, p 143

Canadian Climate Normals (1980) Vol 9, Env Can

Chen CW, Gherine SA, Dean JD, Hudson R, Goldstein RA (1984) Development and calibration of the integrated lake-watershed acidification study model. In: Schnoor JL (ed) Modeling of total acid precipitation impacts. Butterworth Publishers, Boston, p 175-204

Climate of Southern Ontario (1980) Climatological Studies #5, Env Can

Clymo RS (1964) The origin of acidity in Sphagnum bogs. Bryologist 67: 427-431

Clymo RS (1967) Control of cation concentrations and in particular of pH in Sphagnum dominated communities. In: Golterman HL, Clymo RS (eds) (1967) Chemical habitat Proc IBP Symp, Amsterdam and Nieuwersluis, 10-16 Oct 1966, NV Noord-Hollandsche Vitgeuers Maatschappij, Amsterdam, p 322

Clymo RS (this volume) Interactions of Sphagnum with water and air.

Driscoll CT, Likens GE (1982) Hydrogen ion budget of an aggrading forested ecosystem. Tellus 34: 283-292

Eisenreich SJ, Hollod GJ, Langevin S (1978) Precipitation chemistry and atmospheric deposition of trace elements in northeastern Minnesota. Minnesota Envir Qual Council

El-Daoushy F, Tolonen K, Rosenberg R (1982) Lead-210 and moss-increment dating of two Finnish Sphagnum hummocks. Nature 296: 429-431

Farnsworth RK, Thompson ES (1982) Mean monthly, seasonal, and annual pan evaporation for the United States. NOAA Tech Rep NWS 34

Gorham E (1967) Some chemical aspects of wetland ecology. Proc 12th Annual Muskeg Research Conf, Calgary, p 20-38

Gorham E, Vitousek PM, Reiners WA (1979) The regulation of chemical budgets over the course of terrestrial ecosystem succession. Ann Rev Ecol Syst 10: 53-84

Gorham E, Bayley SE, Schindler DW (1984) Ecological effects of acid deposition upon peatlands: a neglected field in acid-rain research. Can J Fish Aquat Sci 41: 1256-1268

Gorham E, Eisenreich SJ, Ford J, Santelmann MV (1985) The chemistry of bog waters. In: Stumm W (ed) Chemical processes in lakes. John Wiley and Sons, New York, p 339

Grigal DF (1985) Sphagnum production in forested bogs of northern Minnesota. Can J Bot 63: 1204-1207

Grigal DF, Buttleman CG, Kernik LK (1985) Biomass and productivity of the woody strata of forested bogs in northern Minnesota. Can J Bot 63: 2416-2424

Hemond HF (1980) Biogeochemistry of Thoreau's Bog, Concord, Massachusetts. Ecol Monogr 50: 507-526

Hemond HF (1983) The nitrogen budget of Thoreau's Bog. Ecology 64: 99-109

Kelly JR, Harwell MA (1982) Comparisons of the processing of elements by ecosystems: I. Nutrients. Ecosys Res Ctr Report No 21, Cornell University

Kelly CA, Rudd JW, Cook RB, Schindler DW (1982) The potential importance of bacterial processes in regulating rate of lake acidification. Limnol Oceanogr 27: 868-882

Kerekes J (1984) Review of the regional aquatic program. Proc of Atlantic Region LRTAP Monitoring and Effects Working Group, Bedford Inst Oceanogr, Dartmouth, NS

Kilham P (1982) Acid precipitation: its role in the alkalization of a lake in Michigan. Limnol Oceanogr 27: 856-867

Kilham P (1983) The biogeochemistry of bog ecosystems and the chemical ecology of Sphagnum. Mich Bot 21: 159-168

Krug EC, Frink CR (1983) Acid rain on acid soil: a new perspective. Science 221: 520-525

Lee JA, Press MC, Woodin S, Ferguson P (this volume) Responses to acidic deposition in ombrotrophic mires in the U.K.

Likens GE, Bormann FH, Johnson NM (1981) Interactions between major biogeochemical cycles in terrestrial ecosystems. In: Likens GE (ed) Some perspectives of the major biogeochemical cycles, John Wiley and Sons, New York, p 93

Likens GE, Bormann FH, Pierce RS, Eaton JS, Johnson NM (1977) Biogeochemistry of a forested ecosystem. Springer-Verlag, New York, p 146

Malmer N, Holm E (1984) Variation in the C/N-quotient of peat in relation to decomposition rate and age determination with Pb-210. Oikos 43: 171-182

Martin NJ, Holding AJ (1978) Nutrient availability and other factors limiting microbial activity in the blanket peat. Chap 6 In: Perkin DF and Heal OW (eds), British Moors and Montane Grasslands, Ecol Studies 27: 113-135

McKnight D, Thurman E, Wershaw R, Hemond H (1985) Biogeochemistry of aquatic humic substances in Thoreau's Bog, Concord, Mass., Ecology 66: 1339-1352.

Munger JW (1981) Environmental controls and ecological consequences of regional precipitation chemistry in Minnesota. MSc. Thesis, Uni of Minnesota, p 154

Munger JW (1982) Chemistry of atmospheric precipitation in the north-central US: influence of sulfate, nitrate, ammonia and calcareous soil particulates. Atmos Env 16: 1633-1645

Munger JW, Eisenreich SJ (1983) Continental-scale variations in precipitation chemistry. Envir Sci Tech 17: 32A-42A

Nilsson IS, Miller HG, Miller JD (1982) Forest growth as a possible cause of soil and water acidification: an examination of the concepts. Oikos 39: 40-49

Oldfield F, Appleby P, Cambray R, Eakins J, Barber K, Battarbee R, Pearson G, Williams J (1979) Pb-210, Cs-137, and Pu-239 profiles in ombrotrophic peat. Oikos 33: 40-45

Oliver BG, Thurman EM, Malcolm RL (1983) The contribution of humic substances to the acidity of colored natural waters. Geochim et Cosmochim Acta 47: 2031-2036

Pastor J, Bockheim JG (1984) Distribution and cycling of nutrients in an aspen-mixed hardwood spodosol ecosystem in Northern Wisconsin. Ecol 65: 339-353

Perdue EM, Lytle CR (1983) Distribution model for binding of protons and metal ions by humic substances. Envir Sci Tech 17: 654-661

Press CM, Lee JA (1982) Nitrate reductase activity of Sphagnum species in the south Pennines. New Phytol 92: 487-494

Rapaport RA, Urban NR, Capel PD, Baker JE, Looney BB, Eisenreich SJ, Gorham E (1985) New DDT inputs to North America: Atmospheric Deposition. Chemosphere 14: 1167-1173

Ramaut J (1954) Modificatons de pH apportees par la tourbe et le Sphagnum secs aux solutions salines et a l'eau bidistillee. Bulletin de l'Acad Roy de Belgique (Classe Des Sciences), Bruxelles, Ser 5 Vol 40, p 305-315

Rosenqvist IT (1978) Alternative sources for acidification of river water in Norway. Sci Total Envir 10: 39-49

Rosswall T, Granhall U (1980) Nitrogen cycling in a subarctic ombrotrophic mire. In: Sonesson M (ed), Ecology of a subarctic mire. Ecol Bull 30: 209-234

Ruffner JA (1978) Climates of the states. Gale Research Co, Detroit, p 1185

Schindler DW (1981) Interrelationships between the cycles of elements in freshwater ecosystems. In: Likens GE (ed) Some perspectives of the major biogeochemical cycles. John Wiley and Sons, New York, p 113

Schnoor JL, Palmer WD, Glass GE (1984) Modeling impacts of acid precipitation for Northeastern Minnesota. In: Schnoor JL (ed) Modeling of total acid precipitation impacts. Butterworth Publishers, Boston, p 155

Schurr KT (1983) Biogeochemistry of selected metals in a forested Sphagnum bog in Minnesota. MSc Thesis, Uni of Minnesota

Solorzano L (1969) Determination of ammonia in natural waters by the phenolhypochlorite method. Limnol Oceanogr 14: 799-801

Stumm W, Morgan JJ, Schnoor JL (1983) Sauer Regen, eine Folge der Storung hydrochemischer Kreislaufe. Naturewissenschaften 70: 216-223

Thomas MK (1953) Climatological atlas of Canada. Natural Res Council Can, No 3151, Ottawa

Thornton JD, Eisenreich SJ (1982) Impact of land-use on the acid and trace metal element composition of precipitation in the north central United States. Atmos Envir 16: 1945-1955

Timmons DR, Verry ES, Burwell RE, Holt RF (1977) Nutrient transport in surface runoff and interflow from an aspen-birch forest. J Envir Qual 6: 188-192

Urban NR (1983) The nitrogen cycle in a forested bog watershed in northern Minnesota. MSc Thesis, Uni of Minnesota, p 359

Verhoeven JT, van Beek S, Dekker M, Storm W (1983) Nutrient dynamics in small mesotrophic fens surrounded by cultivated land. Oecologia 60: 25-33

Verry ES (1975) Streamflow chemistry and nutrient yields from upland-peatland watersheds in Minnesota. Ecol 56: 1149-1157

Verry ES (1983) Precipitation chemistry at the Marcell experimental forest in north central Minnesota. Water Resources Res 19: 454-462

Verry ES (1984) Microtopography and water table fluctuation in a Sphagnum mire. In: Proc 7th Internat Peat Congress, Dublin

Verry ES, Timmons DR (1977) Precipitation nutrients in the open and under two forests in Minnesota. Can J For Res 7: 112-119

Verry ES, Timmons DR (1982) Water-borne nutrient flow through an uplant-peatland watershed in Minnesota. Ecol 63: 1456-1467

Wollast R (1981) Interactions between major biogeochemical cycles in marine ecosystems. In: Likens GE (ed) Some Perspectives of the Major Biogeochemical Cycles. John Wiley and Sons, New York, p 125

Woodin SJ, Press MC, Lee JA (1985) Nitrate reductase activity in *Sphagnum fuscum* in relation to wet deposition of nitrate from the atmosphere. New Phytol 99: 381-388

POTENTIAL SULPHUR GAS EMISSIONS FROM A TROPICAL RAINFOREST AND A SOUTHERN APPALACHIAN DECIDUOUS FOREST

B. Haines, M. Black*, J. Fail, Jr., L. McHargue† and G. Howell

Botany Department, University of Georgia, Athens, GA 30602, USA
* Environmental Health and Safety Lab., Georgia Tech Research Institute, Georgia Institute of Technology, Atlanta, GA 30332, USA
† Department of Biology, University of Miami, Coral Gables, FL 33124, USA

ABSTRACT

Potential emission rates of reduced sulphur gases were estimated for a tropical rainforest and a southern Appalachian deciduous forest. Potential emissions were sampled by using cuvettes placed on the forest floor, by incubating samples of leaf litter and soil in closed containers, by incubating living plant material in closed containers, and by pumping air from around plant canopies enclosed in transparent Tedlar bags. A gas chromatograph fitted with a sulphur specific detector identified and quantified sulphur gases from cuvette sampling and incubations. Quantification of potential H_2S emissions from plant canopies was attempted by Zn+Na acetate trapping followed by colorimetry. Sulphur emissions were not detected with cuvette sampling. Potential sulphur emission rates ($\pm$ standard deviation) from the litter of the tropical rainforest at La Selva, Costa Rica estimated by incubations were 5.9 (29.1), 8.4 (16.7) and 5.1 (12.9) g $S \cdot ha^{-1} \cdot y^{-1}$ for mature, secondary, and flooded stands, respectively. Potential rates from soil were 0.6 (3.2), 42 (184) and 1.2 (6.1) g $S \cdot ha^{-1} \cdot y^{-1}$ for the same stands. Relative to the 11.7 kg $S \cdot ha^{-1} \cdot y^{-1}$ SO_4-S input-output discrepancy that volatile sulphur loss was hypothesized to explain, the observed potential emission rates are at least 200 times too small. Potential emission rates from the leaf litter of the southern Appalachian deciduous forest at Coweeta ranged from 0.4 to 2.1 g $S \cdot ha^{-1} \cdot y^{-1}$ and were less than 1×10^{-5} g $S \cdot ha^{-1} \cdot y^{-1}$ for soil. These rates are too low to account for the SO_4-S input-output discrepancy at Coweeta. Potential emissions may be underestimated due to surface adsorption in sampling devices.

Some rainforest legumes emitted sulphur gases from seeds, wood samples, roots, or leaves, or from all these organs. The sulphur gases are ethyl mercaptan and carbon disulphide. These emissions, which are new to plant physiology research, may have community implications as anti-microbial or anti-herbivore agents and, as point sources in a rainforest, may create a sampling problem for ecosystem level studies. Sampling for H_2S emissions from plant canopies in the rainforest and at Coweeta has yet to detect this sulphur gas.

Quantifying the contribution of natural sulphur emissions to the atmospheric sulphur burden and to acid rain on a global scale is hampered by the great diversity of habitats, the temporal and

spatial variability of sulphur emissions within habitats, and by
analytical problems. Sulphur gases exist at low concentrations
in nature, consist of numerous chemical species, are transformed
from one species to another, and react with analytical surfaces.
Quantification of the contribution of natural sulphur emissions
to acid rain on a global basis is challenging.

INTRODUCTION

The proportional contributions of anthropogenic sulphur emissions
and natural sulphur emissions to the global atmospheric sulphur
burden (Ivanov 1983) and to the formation of acid rain are uncertain. Natural sulphur emissions come from geologic processes
such as volcanoes and degassing of the earth's crust and from
biological processes such as oxidation of organic sulphur compounds, and the reduction of sulphate by microbes, algae, and
higher plants. Uncertainties in estimating the magnitudes of
human-controlled and especially natural sulphur gas source
strengths (Ivanov 1983; Moller 1984) and magnitudes of sulphur
sink strengths make it difficult to balance the global sulphur
budget. Estimation of global sulphur biogenesis is challenging
both in biostatistical sampling and in analytical chemistry. The
biostatistical challenge results from the broad scale diversity
of terrestrial and aquatic habitats, and within habitats, from
spatial and temporal variability in microbial and plant-mediated
sulphur biogenic processes. The analytical challenge results
from: 1) the low concentrations of sulphur gases in nature, 2)
numerous sulphur gas species emitted by organisms, 3) variety of
transformation rates among sulphur gas species, and 4) reactivity
of sulphur gases with surfaces of sampling and analytical equipment.

Acidic precipitation is known for tropical rainforests in the
Amazon (Anon 1972; Clark et al. 1980; Haines et al. 1983;
Galloway et al. 1982) and in Costa Rica (Johnson et al. 1979;
Hendry et al. 1984). These sites are relatively remote from
industrial pollution sources. Acid precipitation has been reported from other remote sites including Amsterdam Island in the
South Indian Ocean (Galloway et al. 1982). Acid precipitation at
remote sites might result from: 1) long distance transport of
acid-forming materials from industrial regions, 2) from local
natural emissions, and 3) from some combination of the two
sources (Haines 1983).

Excesses of SO_4^{-2} inputs over SO_4^{-2} outputs are known for some
forested ecosystems (Haines 1983). Alternative hypotheses
(Haines 1983) to explain these input-output discrepancies include
1) sulphur accumulation in growing biomass, 2) sulphur adsorption
to soil, 3) conversion of SO_4^{-2} to organic sulphur compounds
which leave the system undetected by the SO_4^{-2} analytical
methods, 4) sulphur loss as gases, and 5) errors of estimation.
If the SO_4^{-2} input-output discrepancies result from sulphur biogenesis, the magnitude of the contribution to the atmospheric
sulphur burden would be sufficient to account for some of the
rainfall acidity reported for Amazonian and Costa Rican rainforest (Haines 1983). An investigation of hypothesis 4 was begun

relatively close to Athens, Georgia, in order to develop methodology and later extended to include sampling in a tropical rainforest. Progress with estimation of emissions from a southern Appalachian forest and a tropical rainforest is summarized here. Details of the several component studies will be reported elsewhere.

SITES AND METHODS

Potential emissions were estimated for a successional Robinia stand and a successional Pinus-mixed hardwood stand on watershed 6 and for a mature Quercus-Carya deciduous forested watershed 14 at the U.S. Forest Service Coweeta Hydrologic Laboratory, Otto, North Carolina in the Southern Appalachians (35° 04' N. Lat, 83° 26' W. Long, Elev 700-1500m) and for the tropical rainforest at the La Selva Biological Station of the Organization for Tropical Studies, Puerto Viejo de Sarapiqui, Heredia Prov., Costa Rica, Central America (10° 26'N. Lat, 84° 00' W. Long, Elev 35-150m).

Sulphur Quantification

Gas Chromatography: Gas samples were analyzed with a Perkin-Elmer Sigma 4B chromatograph (Perkin-Elmer, Norwalk, CT) fitted with a sulphur specific flame photometric detector, a 2 mm internal dia. teflon column packed with either a 4.5 or a 16.5 cm section of acetone washed (de Souza et al. 1975) Porapak QS (Supelco, Bellefonte, PA) supplied with N_2 carrier gas at 20 mL.min^{-1} and heated to between 60 and 100°C depending on the particular analysis. Reference sulphur dioxide, hydrogen sulphide, carbonyl sulphide, dimethyl sulphide, dimethyl disulphide, methyl mercaptan and ethyl mercaptan were supplied from permeation tubes (VICI Metronics, Santa Clara, CA) inserted into a Tracor model 432 Tri-Perm Permeation Calibration system (Tracor, Inc. Austin, TX). Standards were transferred from the permeation system into the chromatograph through multiposition zero dead volume sampling valves and a teflon sample loop using a sampling pump. Sample loop, valves and connecting teflon gas lines were heated to 65°C to minimize condensation of sulphur gases on surfaces. Identification of ethyl mercaptan and dimethyl sulphide in incubations of leaf litter and of ethyl mercaptan and carbon disulphide in incubations of legume roots was confirmed with a Finnigan OWA 3B gas chromatograph-mass spectrograph (Finnigan Corp., Sunnyvale, CA).

Colorimetry: Potential hydrogen sulphide emissions from the forest floor and from living plant leaves were quantified colorimetrically by aspirating air samples through pairs of impingers connected in series and containing 25 mL of Zn-Na acetate buffer (Johnson and Nishita 1952), through bubble flow meters, and through a battery powered vacuum pump. At the end of the sampling period p-aminodimethylaniline and ferric ammonium sulphate were added for colour development. Optical density was determined at 670 nm using a 2 cm cuvette in a Bausch and Lomb Mini Spectronic 20 (Bausch and Lomb, Rochester, NY) in Costa Rica or in a 10 cm cuvette in a Bausch and Lomb 700 Spectrophotometer at Coweeta. A standard curve was developed by aspirating H_2S

from a permeation standard through aliquots of the Zn-Na acetate trapping solution followed by colour development.

Sulphur Gas Samplings

<u>The Forest Floor</u>: Potential sulphur gas emissions from forest soil and forest floor litter were sampled with cuvettes placed on the forest floor and by the incubation of soil and litter samples in linear polypropylene bottles followed by analysis of head space gases.

Cuvettes, 50 x 50 x 3.5 cm, were constructed of stainless steel and lined with teflon film and fitted teflon bulkhead ports for removal of gas samples. Three cuvettes were placed on the rainforest floor in Costa Rica in two sampling episodes.

In Cuvette Sampling Episode 1, the gas in the head space was sampled to detect possible changes during a 30-day period. If the emission rates were of the magnitude of the 11.7 $kg.ha^{-1}.y^{-1}$ SO_4-S input-output discrepancy (Johnson <u>et al</u>. 1979), the emissions would change the concentration of sulphur in the head space by 2.7×10^{-6} $g.mL^{-1}.30$ $days^{-1}$, well above the 1.9×10^{-10} g dimethyl sulphide-S/mL detection limit of the gas chromatograph. Sulphur was not detected, thus the emissions are either less than the detection limit or emitted gases are rapidly converted to SO_4^{-2} in condensed moisture on the surfaces of the forest leaf litter and of the cuvettes.

In Cuvette Sampling Episode 2, cuvettes were modified to dynamic flux chambers similar to those described by Aneja (1984) where the air was stirred by two fans (Model 6000-16 LiCor, Inc., Lincoln, NB). Air was sampled from inside the cuvette (experimental stream) and from near the point of air entry port of the cuvette (control) by aspirating through two impingers in series, a bubble flow meter, to a 12 VDC vacuum pump.

<u>Incubations</u>: Litter and soil samples for incubations were taken from the deciduous forest of the Coweeta Hydrologic Lab., Otto, North Carolina and from a flooded forest, a secondary forest, and a mature forest stand in the rainforest at La Selva, Costa Rica. In each rainforest stand 12 sampling points were established at 10 m intervals on transects parallel to access trails. At each sampling point leaf litter was removed from a 0.0625 m^2 quadrat to a 250 mL Nalgene linear polypropylene wide mouth centrifuge jar (Nalge Co., Rochester, NY) fitted with a 5 x 9 mm rubber serum stopper. At the same point 2-4 soil cores 3 cm in dia. and 3 cm in depth were removed to other polypropylene jars. The time course of sulphur gas accumulation in the head space of the bottle was quantified by sampling with a gas syringe through the serum stopper followed by injection into the gas chromatograph.

<u>Sulphur Emissions from Plants</u>: During a study of the formation of nitrogen-fixing root nodules in legumes at La Selva, one of us (McHargue) found that some plants smelled of sulphur gases. Greenhouse- and field- grown plants of fourteen species were incubated in Nalgene polypropylene bottles and sampled to quantify sulphur gas emissions. Gas samples from incubation of

roots of Pithecellobium catenatum Donn. Smith grown in Miami, FL were analyzed with a gas chromatograph-mass spectrograph. Potential H_2S emissions from plant leaves were also investigated at La Selva (Haines and McHargue) and at Coweeta (Howell) by placing plant stems and leaves inside 30-liter capacity Tedlar gas sample bags (SKC, Eighty Four, PA). Air inside the bags was stirred with muffin fans at 980 $L.min^{-1}$ to minimize boundary layer resistance and leaf heating. Air was drawn from Tedlar bags through two impingers in series and through bubble flow meters at 0.5 to 1.5 $L.min^{-1}$. Control samples were obtained with an identical system of bags, fans, impingers, flow meters, and pumps but without an enclosed plant. Plants sampled at the Costa Rican site were seedlings of the early successional trees Heliocarpus appendiculatus Turcz. and Ochroma lagopus SW. grown in a 1:1 forest soil/river sand potting mix in plastic containers. Plants sampled at Coweeta were branches of stump sprouting trees exposed to full sunlight. Species sampled were Liriodendron tulipifera L., Carya glabra (Miller) Sweet, Quercus rubra L., Acer rubrum L., and Robinia pseudo-acacia L. The detection limits were less than $2 \times 10^{-10} g\ H_2S.m^{-2}.h^{-1}$ or $4 \times 10^{-12} g.g^{-1}\ dwt.h^{-1}$ for the two Costa Rican plants. Detection limits for Coweeta plants were less than 3×10^{-10} to $5 \times 10^{-9} g.m^{-2}.h^{-1}$ depending on volume of air sampled.

RESULTS AND DISCUSSION

Estimated sulphur emission rates from the forest floors of the tropical rainforest and the southern Appalachian deciduous forest are shown in Table 1. Sampling with cuvettes in Costa Rica did not detect sulphur gas. Incubations of litter and soil in bottles showed that dimethyl sulphide + ethyl mercaptan were the predominant species emitted at all sites. Methylmercaptan was occasionally detected. The analytical strategy was to screen some of the samples for all of the sulphur gases listed in the footnote of Table 1, then to quantify the predominant sulphur species in all of the samples. Thus, if sulphur gas appeared at low concentration or in just a few samples it would not have been quantified in all of the samples. Emissions per forest floor area were greater from the decomposing leaf litter than from the underlying soils in each of the forested ecosystems. Emissions from the soil may have resulted from incomplete removal of leaf litter from the soil surface.

Emission rates from the floors of the successional and mature southern Appalachian deciduous forests at Coweeta Hydrologic Laboratory are lower. The greater emission rates from the tropical rainforest than from the temperate forests is consistent with the hypothesis that forest floor emission rates increase towards the equator (Adams et al. 1981).

Sulphur emissions were detected by the flame photometric detector from excised roots of 10 out of 14 legume species tested (Table 2). Emissions from seeds, leaves, wood samples, and intact seedlings will be reported elsewhere.

Table 1. Potential DMS and C_2H_5SH sulphur gas emission rates, g S.ha^{-1}.y^{-1} (± standard deviation) of leaf litter and soils of three tropical and three temperate forest stands. Numbers of incubations are indicated as "n" and numbers of incubations in which sulphur was detected are indicated as "d". Incubations were screened for compounds listed in footnote[a] but DMS and C_2H_5SH were predominant forms detected.

Forest compartment and stand	n	d	DMS + C_2H_5SH	s.d.
Rainforest, La Selva, Costa Rica				
Litter, flooded	48	15	5.13	12.95
Litter, secondary	48	21	8.36	16.72
Litter, mature	48	11	5.93	29.14
Soil, flooded	24	1	1.23	6.05
Soil, secondary	24	2	41.8	184.0
Soil, mature	24	1	0.66	3.24
Temperate forest, Coweeta, North Carolina, USA				
Litter, Robinia	48	27	2.07	3.6
Litter, Pinus-hardwood	48	17	1.08	2.0
Litter, mature hardwood	48	8	0.43	1.2
Soil, Robinia	48	0	--	
Soil, Pinus-hardwood	48	0	--	
Soil, mature hardwood	48	0	--	

[a]

Abbreviation	Formula	Name
SO_2	SO_2	Sulphur dioxide
H_2S	H_2S	Hydrogen sulphide
COS	COS	Carbonyl sulphide
CS_2	CS_2	Carbon disulphide
DMS	CH_3SCH_3	Dimethyl sulphide
DMDS	CH_3SSCH_3	Dimethyl disulphide
CH_3SH	CH_3SH	Methyl mercaptan (methanethiol)
C_2H_5SH	CH_3CH_2SH	Ethyl mercaptan (ethanethiol)

Table 2.— Emission rates of ethyl mercaptan + carbon disulphide sulphur (g S.g^{-1} dwt tissue.h^{-1} ± standard deviation (S.D.)) for incubation times less than and greater than 24h for roots of 14 legume species. Numbers of samples with detectable sulphur/total numbers of samples analyzed are indicated in parentheses. Mean ($\bar{x}$) rates calculated only for samples with detectable emissions. NT = not tested.

	< 24 h			> 24 h		
	$\bar{x}$ ($\times 10^{-7}$)	S.D.		$\bar{x}$ ($\times 10^{-7}$)	S.D.	
Acacia mangium Willd.	NT			NT		
Gliricidia sepium (Jacq.) Stend.	3.5	1.9	(2/3)	3.28	1.17	(3/3)
Enterolobium cyclocarpum (Jacq.) Griseb.	3.9		(1/1)	0.41		(1/1)
Leucaena multicapitula Schery	0.9	0.5	(3/3)	0.46	0.15	(3/3)
Mimosa pigra L.	128	100	(2/3)	45.5	3.54	(3/3)
Mimosa pudica L.	1.9		(1/1)	2.2		(1/1)
Mimosa sp.	16.7	3.04	(2/2)	0.94	0.45	(2/2)
Pentaclethra macroloba (Willd.) Kuntze	0		(0/3)	0		(0/3)
Pithecellobium arboretum (L.) Urban	225	65	(2/2)	0.7	0.32	(2/2)
P. catenatum Donn. Smith	203		(1/1)	8.76		(1/1)
P. gigantifolium (Schery) J. Leon	NT			0		(0/1)
P. longifolium (H. & B.) Standl.	NT			0		(0/1)
P. pedicellare (DC.) Benth.	39.1	14.5	(3/3)	21.7	14.6	(3/3)
Stryphnodendron excelsum Harms	717	428	(3/3)	16.2	7.9	(3/3)

Since the retention times of ethyl mercaptan and carbon disulphide were identical the proportions of the two compounds are unknown, and therefore, the results are expressed as total sulphur. A gas chromatographic-mass spectrograph analysis found ethyl mercaptan and carbon disulphide in the proportions of 1:7 in gas samples from incubation of roots of Pithecellobium catenatum grown in Miami, Fla. Ethyl mercaptan + carbon disulphide emissions are potentially significant at three ecological levels. First, ethyl mercaptan emissions are new to plant physiology and carbon disulphide emissions are known for only four plant species. In a review of the literature on natural plant products, sulphur chemistry, and plant pathology we have found no mention of the occurrence of ethyl mercaptan among the many sulphur compounds known for plants (Daly and Deverall 1983; Lewis and Papavizas 1971; Misaghi 1982; Nicholas 1973; Richmond 1973; Schönbeck and Schlösser 1976). Carbon disulphide emissions are known from the minced leaves of Brassica oleracea capitata L. (Bailey et al. 1961) and from Medicago sativa L., Zea mays L. and Quercus lobata Nee (Westberg and Lamb 1984). Second, the material may have community implications because it may confer anti-microbial, anti-nematode, or anti-insect properties upon the producing plant. Third, at the ecosystem level some plant species may constitute point sources of sulphur emissions to the atmosphere from within tropical forests.

Potential emission rates of H_2S from live leaves were less than 2×10^{-10}g and 5×10^{-9}g $H_2S.m^{-2}.h^{-1}$, the detection limits of the method for Costa Rica and Coweeta, respectively. Sampling was performed because H_2S emissions have been reported from the leaves of Avena sativa L., Beta vulgaris L., Coleus blumei Benth., Cucumis sativus L., C. melo L., Cucurbita pepo L., Glycine max L. Merr., Gossypium hirsutum L., Hordeum vulgarae L., Medicago sativa L., Nicotiana tabacum L., Phaseolus vulgaris L., Pinus sylvestris L., Picea abies (L). Karsten and Zea mays L. (Kinraide and Staley 1985; Sekiya et al. 1982; Wilson et al. 1978; Winner et al. 1981; Hallgren and Fredriksson 1982; Filner et al. 1984; Rennenberg 1984; Spaleny 1977). Subsequent to our sampling efforts, the emissions of various combinations of the sulphur gases COS, DMS, CS_2, and DMDS (see footnote of Table 1 for definitions) have been reported from Medicago sativa L., Zea mays L., Phaseolus vulgaris L., Lycopericon, Pisum, Quercus lobata Nee and Pinus ponderosa Dougl. by Westberg and Lamb (1984).

These studies were designed to quantify the contribution of sulphur emissions from forests to the atmospheric sulphur burden and to acid rain. If, following additional sampling, the CH_3SCH_3 and C_2H_5SH emission rates fall between the 1.0×10^{-5}g $S.ha^{-1}.y^{-1}$ detection limit and the highest reliable value of 8.0 g $S.ha^{-1}.y^{-1}$ (Table 1) and if the emissions of other sulphur compounds are not detected, two conclusions can be drawn. First, the rates of $CH_3SCH_3 + C_2H_5SH$ emissions reported for the rainforest at La Selva fall within the range of 1-70 g $S.ha^{-1}.y^{-1}$ and are close to the weighted average of 20 g $S.ha^{-1}.y^{-1}$ which can be calculated from summary statistics reported for forest and swamp soils of the United States by Adams et al. (1979). Second, the magnitude of the sulphur emissions from the forest floors of the two forests are too small to account for either the SO_4-S input-output discrepancy or for the rainfall acidity reported earlier for the rainforest at La Selva (Johnson et al. 1979) or at Coweeta (Swank and Douglass 1977).

Possible alternative explanations of the SO_4-S input-output discrepancy which have yet to be excluded are sulphur accumulation in biomass (Hypothesis 1), sulphur accumulation in soils (Hypothesis 2), loss of organic sulphur compounds in streamwater (Hypothesis 3), or volatile emissions from live plant leaves (variation of Hypothesis 4).

FUTURE RESEARCH

Emission rates estimated here must be considered minimum estimates for the forest floor. The magnitude of underestimation due to adsorption of sulphur compounds to surfaces of sampling cuvettes, incubation bottles, and gas sampling bags is presently unknown. Problems of underestimating natural sulphur emissions have recently been explored by Westberg and Lamb (1984) and by Adams and Farwell (1984) and will not be discussed here. Balancing the global sulphur budget will be more challenging than balancing the global carbon budget. The exchange of sulphur gases between the atmosphere and the surfaces of the land and water involves numerous sulphur gas species each at relatively low concentrations. The exchange of carbon along the same pathways is dominated by CO_2 at relatively high concentrations, making quantification far easier. After decades of research, the global carbon budget still is not understood well enough to balance (Bolin et al. 1979) and is the subject of continuing active research (Woodwell 1984; Lemon 1983). We hope that a cooperative international effort can be organized in the 1990s as part of the International Geosphere-Biosphere Program (National Research Council 1983) to sample sulphur exchange processes between the atmosphere and biosphere at selected sites around the globe to refine our understanding of the global sulphur budget.

Future efforts of our group will focus on three general problems: improved trapping, lighter weight analytical systems, and more extensive sampling. Improved sampling efforts will involve: 1) continued development of sample pre-concentration by solid adsorbent trapping (Black et al. 1978; Steudler and Kijowski 1984) for use in remote areas where cryogenic trapping is not possible; 2) learning cryogenic trapping methodology for use in regions where liquid N_2 is available; 3) quantification and subsequent correction for surface adsorption of sulphur gases to incubation jars, teflon lined cuvettes, and transparent Tedlar gas sampling bags. A lighter weight analytical gas chromatograph having a flame photometric detector, which can be dismantled into pieces weighing no more than 32 kg each for transport as passenger baggage on international air carriers, needs to be developed.

More extensive sampling is envisaged to include collaboration with local investigators both in North America and in rainforests

around the globe. Comparisons are planned among sites of the US
National Science Foundation supported Long Term Ecological
Research (LTER) program (Callahan 1984). These include desert,
prairie, coniferous forest, deciduous forest, salt marsh and
freshwater swamp ecosystems. Rainforest sampling will be expanded to include Australian, Malaysian, Mexican, and West African
sites. The goals are both to perform comparative ecological
studies among systems as well as to quantify the contributions of
these systems to the global sulphur balance. Collaboration with
persons using aerodynamic approaches to estimate CO_2 fluxes between plant communities and the atmosphere would be useful. If we
measured sulphur gas concentration profiles while they measure
CO_2 profiles, we would calculate simultaneous CO_2 and sulphur
exchange, thereby learning about the interactions of the two gas
exchange processes while at the same time contributing toward the
refining of both global carbon and sulphur budgets.

Acknowledgements

Research supported by U.S. National Science Foundation grants
BSR8104700, BSR8012093 and by U.S.D.A. Forest Service,
Southeastern Forest Experiment Station, Coweeta Hydrologic Lab.,
Otto, North Carolina to the University of Georgia. We thank
personnel of the U.S. Forest Service Coweeta Hydrologic
Laboratory and the Organization for Tropical Studies for
logistical support without which this research could not have
been accomplished.

REFERENCES

Adams, DF, Farwell SO (1984) A study of error sources in natural
  sulfur emission measurements - a preliminary examination. In:
  Aneja VP (ed) Environmental impact of natural emissions. Air
  Pollution Control Association, Pittsburgh, PA, p 54
Adams DF, Farwell SO, Pack MR, Bamesberger WL (1979) Preliminary
  measurement of biogenic sulfur-containing gas emissions from
  soils. J Air Pollut Control Assoc 29(4): 380-383
Adams DF, Farwell SO, Pack MR, Robinson E (1981) Biogenic sulfur
  gas emissions from soils in eastern and southeastern United
  States. J Air Pollut Control Assoc 31(10): 1083-1089
Aneja VP (1984) The role of tidal and diurnal variations on the
  release of biogenic sulfur compounds from Coastal Marine
  sediments. In: Aneja VP (ed). Environmental impact of natural
  emissions. Air Pollution Control Association, Pittsburgh, PA,
  p 1-20
Anon (1972) Regenwasseranalysen aus Zentralamazonien ausgeführt
  in Manaus, Amazonas, Brasilien, von Dr. Harald Ungemach.
  Amazoniana (Kiel) 3: 186-198
Bailey SD, Bazinet ML, Driscoll JL, McCarthy AJ (1961) The
  volatile sulfur components of Cabbage. J Food Sci 26: 163-170
Black MW, Herbst RP, Hitchcock DR (1978) Solid adsorbent
  preconcentration and gas chromatographic analysis of sulfur
  gases. Anal Chem 50: 848-851
Bolin B, Degens ET, Kempe S, Ketner P (eds) (1979) The global
  carbon cycle, Scope 13. John Wiley and Sons, New York, p 491

Callahan JT (1984) Long-term ecological research. Bioscience 34(6): 363-367
Clark HL, Clark KE, Haines BL (1980) Acid rain in Venezuelan Amazon. In: Furtado JI (ed). Tropical Ecology and Development. Proc of the Vth Int Symp of Tropical Ecology, International Society of Tropical Ecology, Kuala Lumpur, p 633
Daly JM, Deverall BJ (eds) (1983) Toxins and plant pathogenesis. Academic Press, New York, p 181
Filner P, Rennenberg H, Sekiya J, Bressan RA, Wilson LG, Le Cureux , Shimei T (1984) Biosynthesis and emission of hydrogen sulfide by higher plants. In: Koziol MJ, Whatley FR (ed) Gaseous air pollutants and plant metabolism. Butterworths, London, p 291
Galloway JN, Likens GE, Keene WC, Miller JM (1982) The composition of precipitation in remote areas of the world. J Geophys Res 87(11): 8771-8786
Haines BL (1983) Forest ecosystem SO_4-S input-output discrepancies and acid rain: are they related? Oikos 41: 139-143
Haines BL, Jordan C, Clark H, Clark K (1983) Acid rain in an Amazon rainforest. Tellus 35B: 77-80
Hallgren J, Fredriksson S (1982) Emission of hydrogen sulfide from sulfur dioxide-fumigated pine trees. Plant Physiol 70: 456-459
Hendry CD, Berish CW, Edgerton ES (1984) Precipitation chemistry at Turrialba, Costa Rica. Water Resources Research 20: 1677-1684
Ivanov MV (1983) Major fluxes of the global biogeochemical cycle of sulphur. In: Ivanov MV, Freney JR (eds) The global biogeochemcal sulphur cycle. John Wiley and Sons, New York, p 449
Johnson CM, Nishita H (1952) Microestimation of sulfur in plant materials, soil and irrigation waters. Anal Chem 24(4): 736-742
Johnson DW, Cole DW, Gessel SP (1979) Acid precipitation and soil sulfate absorption properties in a tropical and in a temperate forest soil. Biotropica 11: 38-42
Kinraide TB, Staley TE (1985) Cysteine-induced H_2S emission, ATP depletion, and membrane electrical responses in oat coleoptiles. Physiol Plant 64: 217-222
Lemon ER (ed) (1983) CO_2 and plants: The response of plants to rising levels of atmospsheric carbon dioxide. Amer Assoc Adv Science. Selected Symposia Series, Westview Press, Inc., Boulder, Colorado
Lewis JA, Papavizas GC (1971) Effects of sulfur-containing volatile compounds and vapors from cabbage decomposition on Aphanomyces euteiches. Phytopath 61: 208-214
Misaghi IJ (1982) Physiology and biochemistry of plant-pathogen interactions. Plenum Press, New York, p 287
Moller D (1984) On the global natural sulphur emission. Atmos Environ 18(1): 29-39
National Research Council (1983) Toward an international geosphere-biosphere program: a study of global change. National Academy Press, Washington, DC
Nicholas HJ (1973) Miscellaneous volatile plant products. In: Miller LP (ed) Phytochemistry, vol II. Van Nostrand Reinhold Co., New York, p 381

Rennenberg H (1984) The fate of excess sulfur in higher plants. Ann Rev Plant Physiol 35: 121-153

Richmond DV (1973) Sulfur compounds. In: Miller LP (ed) Phytochemistry vol III. Van Nostrand Reinhold Co., New York p 41

Schönbeck F, Schlösser E (1976) Preformed substances as potential protectants. In: Heitefuss R, Williams PH (eds) Encyclopedia of plant physiology, New Series vol 4. Physiological Plant Pathology, Springer-Verlag, Berlin p 653

Sekiya J, Schmidt A, Wilson LG, Filner P (1982) Emission of hydrogen sulfide by leaf tissue in response to L-cysteine. Plant Physiol 70: 430-436

Souza de TLC, Lane DC, Bhatia SP (1975) Analysis of sulfur-containing gases by gas-solid chromatography on a specially treated Porapak QS Column packing. Anal Chem 47(3): 543-545

Spaleny J (1977) Sulphate transformation to hydrogen sulphide in spruce seedlings. Plant and Soil 48: 557-563

Steudler PA, Kijowski W (1984) Determination of reduced sulfur gases in air by solid adsorbent preconcentration and gas chromatography. Anal Chem 56: 1432-1436

Swank WT, Douglass JE (1977) Nutrient budgets for undisturbed and manipulated hardwood forest ecosystems in the mountains of North Carolina. In: Correll DL (ed) Watershed research in Eastern North America: A workshop to compare results. Chesapeake Bay Center for Environmental Studies, Smithsonian Institution, Edgewater, Maryland, p 343

Westberg H, Lamb B (1984) Estimation of biogenic sulfur emissions from the continental U.S. In: Aneja VP (ed) Environmental impact of natural emissions. Air Pollution Control Association, Pittsburgh, PA, p 41

Wilson LG, Bressan RA, Filner P (1978) Light-dependent emission of hydrogen sulfide fom plants. Plant Physiol 61: 184-189

Winner WE, Smith CL, Koch GW, Mooney HA, Bewley JD, Krouse HR (1981) Rates of emission of H_2S from plants and patterns of stable sulphur isotope fractionation. Nature 289(5799): 672-673

Woodwell GM (ed) (1984) The role of terrestrial vegetation in the global carbon cycle: Measurement by remote sensing, Scope 23. John Wiley and Sons, New York, p 247

PERSPECTIVES ON ESTABLISHING THE RELATIONSHIP BETWEEN ACIDIC
DEPOSITION AND VEGETATION RESPONSES*

L.S. Evans* and K.F. Lewin

Terrestrial and Aquatic Ecology Division, Department of Applied
Science, Brookhaven National Laboratory, Upton, NY 11973
* also: Laboratory of Plant Morphogenesis, Manhattan College, The
Bronx, NY 10471

ABSTRACT

In order to understand the response of vegetation to an air
pollutant, definite cause and effect relationships must be established. Survey methods which document only the mutual occurrence
of two events will not provide adequate information on which to
base a cause and effect relationship. To administer experimental
treatments, knowledge of the spatial, temporal, and other characteristics of ambient precipitation must be understood and technical equipment must be available to administer such treatments.
For acidic deposition studies, exclusion of ambient acidic deposition must be accomplished so that the effects of administered
treatments can be separated from the effects of ambient precipitation. Previous experience has demonstrated that the exclusion
of ambient deposition and administration of experimental treatments should be accomplished under conditions which most closely
approximate natural environments because vegetation responses
have been shown to differ under controlled environmental conditions compared with field conditions. In experiments performed
under natural field conditions, many plants per replicate and a
large number of treatment replicates are necessary in order to
separate treatment effects from random environmental variables.
The following factors may make a linkage between acidic deposition and forest decline difficult to establish. The chemical,
physiological, and episodic nature of acidic deposition exposures
at either low or high elevation sites are not well documented.
Visible symptoms exhibited by needles or trees in decline are not
well characterized. The decline in forest growth does not appear
to be of recent origin. Visible foliar symptoms which are
directly attributable to ambient acidic deposition have rarely
been observed. Environmental factors other than acidic deposition may be concurrently influencing these tree species. Consideration of these aspects is necessary to establish a relationship between acidic deposition and forest decline in the northern
Appalachians.

INTRODUCTION

In order to understand the impacts of ambient air pollution on
terrestrial vegetation, well-designed field experiments capable
of documenting small changes in plant response are needed (Évans
and Thompson 1984). Information from ecological, agronomic, and
air-pollution effects research provides the perspective necessary
to determine if there is a relationship between acidic deposition
and vegetation effects. The purpose of this paper is to provide

---

* Copyright of this paper: U. S. Government

perspective for present and planned research that seeks to establish such relationships between acidic deposition and vegetation responses. The paper can be divided into three sections: (1) experimental approaches necessary to establish a relationship between acidic deposition and vegetation responses, (2) perspectives obtained from an established air pollutant (ozone) - vegetation (ponderosa pine) response relationship and (3) current obstacles to overcome in order to establish a relationship between acidic deposition and forest decline in the northern Appalachians.

EXPERIMENTAL APPROACHES NECESSARY TO ESTABLISH CAUSE AND EFFECT RELATIONSHIPS

Cause and Effect Relationships

The word <u>cause</u> is defined as "something that produces an effect, result, or consequence" or "the person, event, or condition for an action or result." The word <u>effect</u> is defined as "something brought about by a cause or agent; result" (Morris 1982). To determine if an air pollutant such as acidic deposition causes a particular effect such as a change in plant growth, development and/or reproduction, a relationship between cause and effect must be established. For an adequate cause and effect relationship, the effect should occur when the causative event occurs, on a repetitive basis. For example, in experiments aimed at determining the effects of acidic deposition on vegetation, specific plant responses should occur repeatedly after exposure to acidic deposition.

The Role of Associative Surveys

The purpose of survey approaches is to document the mutual occurrence of two or more events. Such approaches are helpful in that they tend to exclude certain factors from further consideration, and facilitate hypothesis construction. However, such associative surveys, which document only the co-occurrence of events, do not establish cause and effect relationships.

Need for Manipulative Experiments

In order to establish if cause and effect relationships are present, manipulative experiments are required. An experimental protocol should be established whereby known quantities and intensities of the hypothesized causal agent are applied to the target organism which is then monitored to see if the expected effect occurs. Care must be taken in designing and performing these experiments to insure that: 1) The actual quantities and intensities of the suspected causal agent are known and are in amounts similar to those found under ambient conditions where the noted effect occurs. 2) The experimental protocol does not alter the environment around the target species to such an extent that the target would not respond in the same way as it would under ambient conditions. 3) The experimentally controlled treatment is either the only variable, or else other sources of variation, either random or systematic have been accounted for by the statistical design of the experiment or by other experiments.
4) Uncontrolled ambient exposures of the hypothesized causal

agent are either eliminated or incorporated into the experimental design. These factors are discussed in more detail in later sections of this paper. Failure to consider these factors when designing an experiment can cause the experimental result obtained to be of questionable value in the formulation of a quantitative cause and effect relationship.

Characterization of Ambient Deposition

To administer experimental treatments, the characteristics of the independent variable (the various experimental treatments imposed) under natural conditions must be considered. To administer acidic deposition treatments, knowledge of the spatial, temporal, and other characteristics of ambient acidic deposition (rain, snow, cloud moisture, rime ice) must be understood (Evans et al. 1982c; Evans 1984) and suitable technical equipment must be available to administer such treatments. The interpretation of experimental data to establish cause and effect relationships must, at least, consider both the natural variabilities of acidic deposition and the degree to which experimental protocols can implement suitable treatment regimens.

Exclusion of Ambient Deposition

In order to administer experimental treatments adequately, it may be necessary to isolate the system, at least partially, from the natural environment. For example, automatically moveable rainfall exclusion shelters have been employed in acidic deposition experiments conducted out-of-doors (Lewin and Evans 1984) to cover vegetation when natural precipitation occurs. Exclusion of ambient precipitation is necessary in order to characterize accurately the effects of the imposed treatments.

Minimal Perturbation of Ambient Environment

In order to establish cause and effect relationships occurring in natural environments, manipulative experiments should be performed under conditions which most closely approximate the natural environment. This situation is particularly germane in acidic deposition-vegetation response research since there is sufficient evidence that the effects of simulated rain applications on vegetation under controlled-environmental conditions may differ markedly from effects obtained under field conditions. Two examples illustrate this situation. Root (hypocotyl) yields of radishes were reduced after exposure to simulated acidic rainfalls under controlled environmental conditions (Lee et al. 1981; Evans et al. 1982a). In contrast to these results, Troiano et al. (1982) demonstrated higher yields at high acidity rainfalls compared with controls with radishes grown under field conditions and no significant yield effects were demonstrated in another experiment which utilized 45 plots per treatment (Evans et al. 1982b). In addition, seed yields of Amsoy soybeans showed no consistent effects due to acidic deposition under controlled environment conditions (Evans and Lewin 1980). However, experiments with field-grown Amsoy soybeans have shown a negative effect of acidic deposition on yield (Evans et al. 1983, 1984a, 1985; Banwart et al. 1984). This information restricts the advisability of using results of experiments performed under controlled environmental conditions to predict effects of acidic deposition under ambient conditions.

Differences in vegetation responses to acidic deposition between controlled environments and more natural conditions may arise because of differences in macroclimate or microclimate. Important plant physiological functions which may be affected in this way are: photosynthesis, respiration, translocation, water and nutrient uptake.

Moveable rainfall exclusion shelters have been shown to minimize changes in a plant's microclimate (Dugas and Upchurch 1984). Such moveable shelters have been used to assess the effects of acidic deposition on field-grown soybeans over four growing seasons (Evans et al. 1983, 1984a, 1985). Such structures have a minimum impact on overall plant microclimate because they cover the crop only when ambient rainfalls occur or when simulated rainfalls are applied.

Crop coverage by the shelter will decrease wind velocity, and increase relative humidity. If the test plants are covered during the daytime, the amount of light reaching the canopy will be reduced and the air temperature under the shelter will be elevated. The actual magnitude of these effects will be minimized when the crop is being covered during ambient rainfalls or for short periods at night such as when treatments are applied. Since the shelters cover test plants only at these times, which normally comprise only a small fraction of the growing season, the overall effect of the shelter on their growth would be expected to be minimal.

In contrast, open-top chambers (OTC) which have also been used to study acidic deposition effects (Jacobson et al. 1980; Troiano et al. 1983) have been shown to alter yields of crops grown within them compared with crops grown in ambient conditions (Heagle et al. 1973). The microclimate within such chambers differs from the ambient conditions outside the chambers (Olszyk et al. 1980), which may influence plant growth and productivity (Clarke et al. 1983). Recently, it has been suggested that the "experimental conditions (in such chambers) are probably closer to greenhouse than to field experimentation" (Troiano and McCune 1984), although they also stated that this applied only to a particular experiment of theirs. More recently, Shriner (1985) has demonstrated that open-top chambers can have very marked effects on crop yields. In their experiment, soybean seed yields were about 20% lower in chambers than in ambient conditions.

Need for Adequate Treatment Replication

In addition to perturbing the environment of experimental vegetation minimally, adequate replication of experimental plots is absolutely necessary in such experiments to overcome plant-to-plant and plot-to-plot variations (Nelson and Rawlings 1983; Hurlbert 1984; Evans et al. 1984b). Adequate replication is necessary because plant microclimates, local soil fertilization levels, soil moisture gradients, and localized soil texture heterogeneities may vary significantly within a field. The occurrence of significant variation within a small area is supported by the fact that statistically significant differences in yield have usually occurred both among and within latin squares in our field experiments with soybeans (Evans and Thompson 1984).

PERSPECTIVES OBTAINED FROM OZONE-PONDEROSA PINE RESEARCH: AN ESTABLISHED LINK

Knowledge gained from the established link between ozone and ponderosa pine can provide perspective for present and planned studies to determine if there if a linkage between acidic deposition and spruce dieback in the northern Appalachians. In the last few decades the presence of ozone has been linked to dieback of ponderosa pine in the San Bernardino National Forest of California. The decline of ponderosa pine has been explained by some rather easily performed experiments in which the visual, histological, and physiological effects of ozone under controlled environmental conditions were very similar to those obtained in the forest with comparable ambient ozone concentrations.

Several conditions prevailed which were conducive to establishing the link between ozone and injury to ponderosa pine. They were:

1) The chemical, physical, and episodic characteristics of ozone exposures were well documented.
2) The visual and histological symptoms exhibited by needles in the forest were of relatively recent origin and became well characterized.
3) Visual, histological, and physiological symptoms which developed in nature, could be accurately mimicked under controlled conditions.
4) These studies proceeded with funding over a relatively long time period (approximately ten years).

This knowledge linkage is substantiated by visible injury symptoms such as needle chlorosis, terminal dieback of needles, premature abscission (Miller et al. 1963), loss of chlorophyll (Richards et al. 1968), and both visible and histological evidence (Evans and Miller 1972, 1975; Miller and Evans 1974).

PERSPECTIVES FOR ACIDIC DEPOSITION-RED SPRUCE DECLINE: OBSTACLES TO ESTABLISHING A RELATIONSHIP

This situation with ozone and ponderosa pine is important for perspective in present and planned acidic rain-forestry studies. From present information, establishing a link between acidic deposition (cause) and forest tree decline (effect) in the northeastern United States will be much more difficult than the above mentioned ozone-ponderosa pine linkage because:

1) The chemical, physiological, and episodic nature of acidic deposition exposures at low and high elevations are not well documented and are relatively unpredictable.
2) Visible symptoms which represent needles or trees in

decline are not well characterized.
3) Decline in forest growth does not appear to be of recent origin.
4) Visible foliar symptoms which are attributable directly to ambient acidic deposition have been observed rarely.
5) From available evidence, it is obvious that dendrological results show a decline in some forest species (particularly red spruce and balsam fir) in the northern Appalachians. However, environmental factors may significantly alter the sensitivity of the declining species to acidic deposition. This would make the establishment of a quantitative forest growth-air pollutant linkage more difficult.
6) Researchers in the acidic deposition-forest decline field may be asked to assess if there is a relationship after a relatively short time period. If so, results of preliminary experiments may not be evaluated adequately in time to design more appropriate long-range experiments which, in turn, could establish such linkages.

These differences in knowledge linkages between the acidic deposition-forest decline and ozone-ponderosa pine decline are quite significant and should not be forgotten in research planning.

SUMMARY

In summary, the above observations have been drawn from previous research efforts. Adequate realization of the benefits and limitations of various research approaches must be maintained so that present and planned experiments can demonstrate valid conclusions about the information linkages between acidic deposition and decline of red spruce and balsam fir in the northern Appalachians.

Acknowledgements

Although the research described in this article has been funded wholly or in part by the United States Environmental Protection Agency through Interagency Agreement DW930196-01-1 to Brookhaven National Laboratory, it has not been subjected to the Agency's required peer and policy review and therefore does not necessarily reflect the views of the Agency and no official endorsement should be inferred. Support was also obtained from the United States Department of Energy, under Contract No. DE-AC01-76CH0016 and from Associated Universities, Inc., under Contract No. 550925-S.

REFERENCES

Banwart WL, Hassett JJ, Vasalis BL (1984) Acid rain and its effect on corn and soybean yields. Proc Illinois Fertilizer and Chemical Dealers' Conference, p 19-21
Clarke BB, Henninger MR, Brennan E (1983) An assessment of potato losses caused by oxident air pollution in New Jersey. Phytopath 73: 104-108

Dugas WA, JR., Upchurch DR (1984) Microclimate of a rainfall shelter. Agron J 76: 867-871

Evans LS (1984) Acidic precipitation effects on terrestrial vegetation. Ann Rev Phytopath 22: 397-420

Evans LS, Gmur NF and Mancini D (1982b) Effects of simulated acidic rain on yields of Raphanus sativus, Lactuca sativa, Triticum aestivum and Medicago sativa. Environ Exptl Bot 22: 445-453

Evans LS, Hendrey GR, Thompson KH (1984b) Comparison of statistical designs and experimental protocols used to evaluate rain acidity effects on field-grown soybeans. J Air Pollut Control Assoc 34: 1107-1114

Evans LS, Lewin KF (1980) Growth, development and yield responses of pinto beans and soybeans to hydrogen ion concentrations of simulated acid rain. Environ and Exptl Bot 21: 103-113

Evans LS, Thompson KH (1984) Comparison of experimental designs to detect yields of crops exposed to acidic precipitation. Agron J 76: 81-84

Evans LS, Lewin KF, Conway CA, Patti MJ (1981) Seed yields (quantity and quality) of field-grown soybeans exposed to simulated acidic rain. New Phytol 89: 459-470

Evans LS, Lewin KF, Cunningham EA, Patti MJ (1982a) Effects of simulated acidic rain on field-grown crops. New Phytol 91: 429-441

Evans LS, Lewin KF, Patti MJ, Cunningham EA (1983) Productivity of field-grown soybeans exposed to simulated rain. New Phytol 93: 377-388

Evans LS, Lewin KF, Patti MJ (1984a) Effects of simulated acidic rain on yields of field-grown soybeans. New Phytol 96: 207-213

Evans LS, Lewin KF, Santucci KA, Patti MJ (1985) Effects of frequency and duration of simulated acidic rainfalls on soybean yields. New Phytol 100: 191-198

Evans LS, Miller PR (1972) Ozone damage to ponderosa pine: a histological and histochemical appraisal. Am J Bot 59: 297-304

Evans LS, Miller PR (1975) Histological comparison of single and additive O_3 and SO_2 injuries to elongating ponderosa pine needles. Amer J Bot 62: 416-421

Evans LS, Raynor GS, Jones DM (1984c) Frequency distribution for durations and volumes of rainfalls in the eastern United States in relation to acidic precipitation. Water Air Soil Pollut 23: 187-195

Evans LS, Thompson KH (1984) Comparison of experimental designs to detect yields of crops exposed to acidic precipitation. Agron J 76: 81-84

Heagle AS, Body DE, Heck WW (1973) An open-top field chamber to assess the impact of air pollution on plants. J Environ Qual 2: 365

Hurlbert S (1984) Pseudo-replication in the design of ecological field experiments. Ecol Monogr 54: 187-

Jacobson JS, Troiano J, Colavito L, Heller LL, McCune DC (1980) Polluted rain and plant growth. In: Toribara TY, Miller MW, Morrow PE (eds) Polluted Rain, Plenum Press, New York, p 291-299

Lee JJ, Neely GE, Perrigan SC, Grothaus LS (1981) Effects of simulated acid rain on yield, growth and foliar injury of several crops. Environ Exp Bot 21: 171-185

Lewin KF, Evans LS (1984) Design of an experimental system to determine the effects of rainfall acidity on vegetation. Brookhaven National Laboratory Report 34649

Miller PR, Evans LS (1974) Histopathology of oxidant injury and winter fleck injury on needles of western pines. Phytopath 64: 801-806

Miller PR, Parmeter JR, Jr., Taylor OC, Cardiff EA (1963) Ozone injury to the foliage of ponderosa pine. Phytopath 53: 1072-1076

Morris WI (1982) The American Heritage Dictionary, 2nd ed. Houghton Mifflin Co., Boston, MA

Nelson LA, Rawlings JO (1983) Ten common misuses of statistics in Agronomic Research and Reporting. J Agron Education 12: 100-105

Olszyk DM, Tibbitts TM, Hertsberg WM (1980) Environment in open-top field chambers for air pollution studies. J Environ Qual 9: 610-615

Richards BL, Taylor OC, Edmunds GF, Jr. (1968) Ozone needle mottle of pines in southern California. J Air Pollut Control Assoc 18: 73-77

Shriner DS (1985) Acidic deposition: effects on agricultural crops. Final Report to the Electric Power Research Institute (Project 1908-2)

Troiano J, Colavito L, Heller L, McCune DC, Jacobson JS (1983) Effects of acidity of simulated rain and its joint action with ambient ozone on measures of biomass and yield in soybean. Environ Exp Bot 23: 113-119

Troiano J, Heller L, Jacobson JS (1982) Effect of added water and acidity of simulated rain on growth of field-grown radish. Environ Pollut 29: 1-11

Troiano J, McCune D (1984) Commentary. J Air Pollut Control Assoc 34: 1115-1116

Group Summary Report: FOREST DECLINE

P. Manion (Chairman)
A. Johnson (Rapporteur)
W.O. Binns
E. Bondietti
J.B. Bucher
E.R. Cook
D. Godbold
B.L. Haines
T. Ingestad
D. Johnson
S.E. Lindberg
S.N. Linzon

E. Matzner
J. McBride
S.B. McLaughlin
L. Pitelka
B. Prinz
D.J. Raynal
M.R.D. Seaward
G.H. Tomlinson
B. Tveite
G. Tyler
M. Tyree
H.W. Zoettl

A number of phenomena involving the deterioration of trees in forests of Europe and North America have attracted considerable public and scientific attention during the past five years. The level of concern caused the undertaking of national forest surveys in Europe in which crown density was measured on a subjective scale.

Substantial agreement exists that two of the cases which have been reviewed in this conference represent major phenomena which are not easily or satisfactorily explained by obvious factors known to cause major disturbances in forests. Those cases are described below, and a set of recommendations for research follow. These recommendations reflect not only our interest in discerning the causes of the current decline phenomenon, but also our need to better understand the response of forest trees to stress from natural causes, from air pollution stress, and from their various combinations. An increased understanding of whole tree responses to stress will provide a better capability for early detection of serious problems in the forests for better prediction of under what conditions of air pollution serious forest problems might occur.

Norway spruce decline on acid, nutrient-poor soils in the Federal Republic of Germany

Although diseases involving chlorosis of older needles have long been known, the present phenomenon can be considered unusual because the rapid increase in the geographical extent of yellowing of older needles during the past ten years has not been reported previously, at least during the past century. It almost certainly would not have gone unnoticed had it occurred. This decline phenomenon occurs on nutrient-poor soils and to the greatest extent on high elevation sites.

The readily identifiable characteristics of this disease type are:

- The most recent year class of needles remains green.
- The yellowing of older needles begins at the tip of the needle and moves toward the base.

- The yellowing is associated with cation nutrient deficiencies, most notably magnesium, but sometimes potassium or zinc.
- Chlorosis is most pronounced in sun-exposed branches.
- Chlorosis occurs in trees of all ages.
- Shedding of the chlorotic foliage generally begins with the oldest needles.
- The disease occurs in areas which have not been investigated previously as areas subject to pollutant stress because of their distance from industrial and urban sources.

Other disease types are also present in the German forests, adding to the general perception of widespread forest decline. These disease types have not been well characterized by our group and are not treated in detail here. Several species showing a large variety of symptoms, often in varying combinations, add to the complex picture that has emerged from the Federal Republic of Germany.

The hypotheses which have been advanced to explain the decline of Norway spruce on acid sites have evolved during the past few years to the extent that there are now many common components. The major hypotheses reflect the consensus that air pollution is involved. The hypotheses which have been most extensively researched propose that there are nutrient imbalances involved, that these imbalances are exacerbated by acid, nutrient-poor soils and ozone, and that above- and below-ground functions are impaired.

One hypothesis proposes that a major predisposing stress is soil acidification with the accompanying loss of nutrient cations and mobilization of aluminum. Aluminum is manifested through antagonisms with calcium and magnesium and disturbance of root growth. Ozone is considered an inciting stress in this hypothesis. Another hypothesis suggests that ozone is an important driving force. It is proposed that ozone damage, accompanied by leaching by acidic mist or precipitation, leaches nutrients from foliage, and alters the allocation of assimilates so as to inhibit root growth and normal nutrient uptake. This further reduces vigor and reinforces nutrient deficiencies above ground. The low nutrient status of the soil is proposed to be a critical predisposing factor which allows the development of cation nutrient deficiencies.

The extent to which the current major hypotheses can satisfactorily explain the disease type is a matter for further research. There is disagreement over the role of heavy metals in the disease type. Laboratory studies indicate that elevated metal levels in soil water could reduce root development. However, it is not known how the well-controlled laboratory studies, which used reasonable levels of metals, equate with the availability of metals in the forest floor. Overall, the long term effects of accumulating heavy metals in forest ecosystems must be fully evaluated now.

It should also be noted that forest delines of a type similar in symptoms and of recent onset are now described by Austria,

Switzerland and France, while forests in Poland, Czechoslovakia, East Germany and some other parts of the Federal Republic show clear signs of damage by pollutants of industrial origin.

Red spruce decline in high elevation forests of the Adirondacks and Northern Appalachians

The decline of red spruce has few similarities with the disease type described above. Although, as in Germany, deterioration is most pronounced at higher elevations, the symptoms observed are substantially different than those reported in Europe. The most readily observed characteristics are:

- substantial spruce mortality on many mid-to-upper slopes of the Adirondack, Green and White Mountains.
- the death of twigs and needles at the top of the crown and ends of the branches.
- a reduction in radial increment which is often abrupt. It is synchronous across northern New York, Vermont and New Hampshire and is apparent across all age classes.

The phenomenon is not ubiquitous, and is less conspicuous at low elevations. Stands of widely different biomass, age and disturbance history are affected.

The hypotheses advanced to explain this phenomenon are tentative, owing to a lack of data. The elevations at which red spruce deterioration is most readily observed are expected to receive relatively high levels of deposition of acid substances, heavy metals (and probably other substances) due in large part to the interception of cloud water.

Concentrations of ozone and other gaseous pollution also should be better defined. The changes observed in tree rings and field reports suggest that the phenomenon started or intensified during the 1960's, a time of two regional climatic anomalies - prolonged and severe summer drought, and a succession of colder than normal winters. Major unknowns in the attempt to identify the key factors involved are the lack of a refined analysis of relevant climatic conditions, a sound understanding of possible biotic influences, and the lack of data on the doses of pollutants delivered to the trees. Mechanistic models which might account for the symptoms in light of natural and/or anthropogenic factors are lacking. Decline is also reported in red spruce in the southern Appalachians, although the symptoms appear to be somewhat different.

RESEARCH NEEDS

(These are not listed in any order of priority)

1. Establish a series of reference plots for long-term monitoring of forest decline; measure both above- and below-ground parameters.

2. Measure total atmospheric deposition inputs, including gaseous, particulate, aerosol, and dissolved substances; attempt to quantify, on a site basis, the most significant deposition components and processes.

3. Increase the precision of forest decline surveys to determine objectively changes in tree and stand vitality. Evaluate root, as well as crown and foliar injury, as a basis for identification of causal factors and agents.

4. Characterize the role of forest stand management, development and disturbance history to generate a chronology of conditions influencing present decline. Physical and biological factors (e.g., pathogens, insects) should be assessed.

5. Evaluate the significance of climatic factors which influence past and present forest growth. Dendroclimatological assessment is needed.

6. Study nutrient cycling in the forest ecosystem including inputs, storage, biocycling, and outputs. Investigate interactions between atmospheric and forest canopies.

7. Investigate the allocation and translocation of carbon and nutrients in root and shoot systems under varying conditions of nutrient deficiency and physical stress.

8. Characterize changes in tree growth allometry as a function of stand conditions and decline status. Detailed stem analysis methodology may be required. The question of factors regulating root growth should be addressed in laboratory and field studies. Emphasis should be given especially to quantitative root analysis.

9. Design field experiments to assess multiple pollutant stresses. By nature, these must be long-term investigations and be performed in areas large enough for suitable replication on uniform sites. A large area experiment aimed at optimum nutrition of the ecosystem should be included. The use of open-top chambers and ambient deposition exclusion approaches to measure individually and interactively the influences of substances, such as sulfur and nitrogen oxides and ozone, is appropriate. Use of realistic loadings and dosages is important. Ideally, experiments should also be replicated in areas with potentially important differences in soil, climate or pollutant regimes.

10. Consider the significance of genetic diversity and variability in interpreting injury or damage symptoms under field conditions. Consider genetic diversity in designing experimental treatments.

11. Utilize findings from studies of point source pollution effects in evaluating atmospheric deposition effects of forest ecosystems.

12. Conduct forest fertilization studies as diagnostic tests of soil nutrient deficiencies. Utilize foliar analysis and foliar application of nutrients in studying effects of nutrients on vegetation.

13. Encourage experimental studies in labs in clarifying the mechanisms of action both of soil toxicants, such as aluminum, on root systems, and of air pollutants, such as ozone, on cell membrane systems and cuticles.

Group Summary Report: FOREST SOILS

N. van Breemen (Chairman)
N. Foster (Rapporteur)
G. Abrahamsen
D.W. Johnson

I. Morrison
L. Robitaille
G. Tyler

STATE OF KNOWLEDGE

Acid deposition has probably increased the rate of base cation leaching in most soils to which it is exposed. The degree to which this increased leaching is manifested as a change in soil acidity will vary with soil base cation reserves, weathering rate and the amount and duration of the acid deposition impact. Due to variations in all of these factors, broad generalizations are hazardous, but classification of soils relative to potential change (e.g., CEC, base saturation) in affected areas may be useful in identifying those soils most likely to change.

Reports based on scientific information indicate that as a result of acid deposition, soil pH and base content have decreased over the past decades in parts of central Europe and Scandinavia. Increased base leaching has been reported in North America, but soil pH values still remain well within those considered normal. The evidence for soil acidification comes from comparisons among old and new soil investigations, hydrogen budgets, deposition models, and experiments with artificial acidification. There are indications that acid deposition has also acidified deeper soil horizons below the root zone. Soil acidity can also be increased by biotic uptake of base cations (K^+, Ca^{2+}, Mg^{2+}, Na^+), by leaching of base cations in association with the transport of organic and inorganic anions (HCO_3^-, NO_3^-, SO_4^{2-}, Cl^-) produced naturally within the soil, and by humus accumulation. In central Europe and southern Sweden, atmospherically derived sulfuric and nitric acid has led to increased aluminum concentrations in soil solution in very acid forest soils. This situation would probably be reversed with a reduction in air pollution.

FUTURE RESEARCH

A great difficulty is the uncertainty related to determining the relative contribution of current levels of acid deposition from the atmosphere and the other processes to the change in soil pH. In general, therefore, better quantification of the processes that contribute to soil acidification is needed, including the role of atmospheric deposition. Specifically we recommend that:

1. Better base line information be collected in both polluted and non-polluted areas. Reference areas where the soil and vegetation are described and analyzed be established. Some of the sites be used for examining element cycling on a long-term basis. Transects from point sources of pollutants and coastal areas be examined.

2. There is a need to study how well and how fast soils recover after pollution is abated (reference to smelter-damaged areas, previous acid irrigation experiments with parallel computer simulation modelling of the processes could be helpful here).

3. More attention be paid to quantifying the rates of base cation supply by mineral weathering and seeing whether the rates are accelerated by acid deposition.

4. Deficiencies in measuring atmospheric deposition, the acidity, base and sulfur status of soil, and solute fluxes in soil be addressed. Seasonal variation in soil properties be considered. Better quantification of the hydrologic processes in soil is needed.

5. The implications of soil acidification on soil fauna and flora (particularly mycorrhizal fungi) and microbiological processes (decomposition, nitrification, etc.) be given more attention.

6. The implication of soil acidification on soil fertility, including the generation of toxic compounds, be given more attention. In areas where forest decline is hypothesized to be due to soil-mediated effects, long-term fertilization trials should be established to test this hypothesis.

Group Summary Report: PERSPECTIVES OF AIR POLLUTION EFFECTS ON
CROP PLANTS

T.A. Mansfield (Chairman)   D.M. Reid
L.S. Evans (Rapporteur)   D.S. Shriner
V.S. Berg   M.H. Unsworth
E.J. Pell   W.E. Winner

Deposition of materials from the atmosphere onto the earth's surface involves many complicated processes. Gases, liquids, and solids are deposited over time periods of minutes to months after they are placed into the atmosphere. Once airborne, matter can change its chemical and physical characteristics. Constituents in rain that confer acidity such as H^+, NO_3^-, and SO_4^{2-}, as well as gaseous pollutants such as O_3, SO_2, and NO_x are of concern because they may have affected ecosystems in North America and Europe.

The purpose of this summary is to highlight areas of uncertainty, and also areas for research in the near future. There are many gaps in our knowledge of the impact that air pollution has on vegetation, and it will not be possible to discuss all these in detail. It is important to emphasize that air pollutants represent an <u>additional</u> environmental stress imposed on plants. Plants are continually exposed to a range of natural stresses, and they have morphological adaptations and physiological and biochemical mechanisms to enable them to survive in a highly variable environment.

It has been recognized for many years that gaseous pollutants such as O_3, SO_2, and oxides of nitrogen, both individually and in combination, can significantly affect plant growth (Reinert, 1984). At the present time, O_3 is considered to be a major air pollutant which significantly reduces the growth, development and survival of vegetation in the United States (Heck <u>et al</u>. 1982). Specific effects of acidic rain have been described (Evans, 1984). Recently, the constituents in acidic rain have been shown to have economic consequences (Medeiros <u>et al</u>. 1984), but the impact of acidic rain is thought to be many times lower than that of ozone (Medeiros <u>et al</u>. 1984). In western Europe, acidic rain and ozone were, from a historical perspective, thought to be of minor importance compared with sulfur dioxide. However, with slightly lower SO_2 concentrations in recent times coupled with newly perceived problems caused by O_3 (Skarby and Sellden, 1984) and acidic rain (Ulrich <u>et al</u>. 1981), the current relative importance of these pollutants in imposing stress on vegetation is unclear.

Over the last few decades, numerous experiments have been performed to support field observations of the impact of air pollutants on crops and natural ecosystems. Experimental techniques have been continually refined to enable responses to be studied under controlled environmental conditions, and under conditions close to those in the field. Although there are still differences of opinion among those using different methodologies, there is, nevertheless, now enough agreement for some general statements of principle to be made. There is a wealth of information on herbaceous plants which may be helpful in interpreting some of the current problems in forests. The methodology developed for herbs can be adapted for trees, and more critical experiments to solve urgent problems may be designed using the large body of information on the pollution responses of herbs.

Many factors can limit plant growth, for example, the availability of water, carbon dioxide, mineral nutrients, extremes of temperature (including duration of frost-free period), irradiance, photoperiod, herbivory and attack by microbial pathogens. Among these many hazards experienced by plants, air pollutants are probably a minor stress factor; nevertheless, they appear to be ecologically and economically significant. An additional stress which is of small importance in otherwise favorable conditions may be of overriding significance if a plant is already greatly stressed by its natural environment and, therefore, close to its limits of survival. Forest trees at high altitude are more likely to be in this situation than a well-managed field crop, as are natural species or crops at the climatic limits of their geographical range.

Plants are integrators of their total environment and the response to a single stress factor is not always simple. For example, water stress usually results in stomatal closing which, in turn, usually results in reduced CO_2 uptake and slower growth. This effect will be superimposed on any direct effect of water stress on the biochemical mechanisms of CO_2 fixation. Another result of stomatal closure could be an increased resistance to some diseases, reduced exposure to gaseous air pollutants, and reduced herbivory since the foliage may be less palatable.

Another example is the complex responses to increased nitrogen. Increased input of nitrogen often stimulates growth, but increased nitrogen fertilization in late summer can deleteriously affect the process of hardening (i.e., acclimation) that is essential for survival over winter. It is possible that dry deposition of NO_x, or wet deposition of nitrate, may act in this way.

Relatively little information is available to evaluate the inventory of air pollutants that can affect plant growth. New compounds are being emitted into the atmosphere each year. There are few data to assess the impacts of these new pollutants. Moreover, an inadequate amount of information is available regarding interactions of known air pollutants with other stress factors that may limit growth.

Although we have mentioned the possible transfer of information from studies of one plant type to another, it must be emphasized that relatively few relationships have been investigated extensively enough to be viewed as models. Also, we do not yet have enough data to provide a secure base to enable information from one pollutant-species response relationship to be applied to others. Many of the species that have been tested were chosen because they are of economic importance; therefore, some agricultural crops and a few other species have received great attention. The majority of these species have had some genetic manipulation by plant breeders. Since air pollutants have been present for many decades it may be likely that cultivated plants selected in variety trials are more resistant to air pollutants than vegetation in unmanaged ecosystems.

Although cultivated plants may be more resistant to air pollutants because of selection procedures, it should also be recognized that air pollutants near point sources can be major stress factors that alter ecosystem structure and function. Since relatively few studies have been performed on ecosystems affected by air pollutants, it is not prudent to over-generalize from effects on one species to effects on another species, or from one location to another. With the present level of knowledge it is difficult to determine the level of generalization from species to species and from ecosystem to ecosystem that is permissable scientifically. Moreover, because of the constraints of the Heisenberg uncertainty principle, the level or degree of inference from controlled experiments with imposed treatments to real world situations is always of concern.

In addition to over-generalizations of the types mentioned above, there is always the risk of applying knowledge of short-term effects to predict what may happen over longer time periods. Many experiments used to assess the effects of air pollutants have durations of hours, days, or weeks. Results of such studies have been used to predict effects over longer time periods, such as entire growing seasons and entire life cycles of organisms. It should be recognized that compensatory mechanisms exist so that a short-term reduction in growth may not result in a long-term reduction. The opposite can also be true. Namely, a short term increase in growth may not result in a beneficial effect on growth in the long term. It must be remembered that species viability results from successful reproduction and adaptation, parameters rarely measured in research over the short term.

This short synopsis has sought to highlight areas of uncertainly and areas of research needed to pin-point gaps in knowledge. It has been recognized for many decades that air pollutants are stress factors among a large number of other chemical and physical components that can limit plant productivity under natural and agricultural situations. A complete evaluation of the impacts of air pollutants on plant growth is impossible to achieve with the data presently available. Overall assessments are difficult because so few species have been tested, the experimental procedures used usually measure only short-term effects on growth, few air pollutants have been adequately examined, and the results of many experiments may fall under the criticism that the experimental exposure conditions may significantly affect the degree and manner of response. Much additional research must be undertaken in order to understand how air pollutants affect growth and development, agricultural crop productivity, terrestrial ecosystem dynamics, and particularly plants exposed to severe natural stresses from the environments.

REFERENCES

Evans LS (1984) Acidic precipitation effects on terrestrial vegetation. Ann Rev Phytopath 22: 397-420

Heck WW, Taylor OC, Adams R, Bingham G, Miller J (1982) Assessment of crop loss from ozone. J Air Pollut Control Assoc 32: 353-361

Medeiros WH, Moskowitz PD, Coveney EA, Thode Jr HC, Oden NL (1984) Oxidant and acid precipitation effects on soybean yield: cross-sectional model development. Environment International 10: 27-33

Reinert RA (1984) Plant responses to air pollutant mixtures. Ann Rev Phytopath 22: 421-442

Skarby L, Sellden G (1984) The effects of ozone on crops and forests. Ambio 13: 68-72

Ulrich B, Mayer R, Khanna PK (1980) Chemical changes due to acid precipitation in a loss-derived soils in central Europe. Soil Sci 130: 193-199

Group Summary Report: WETLANDS

            E. Gorham (Chairman)   M. Havas
            J.A. Lee (Rapporteur)   J. Jeglum
            J. Anderson   S.A. Norton
            S. E. Bayley   D.W. Schindler
            R. S. Clymo   N.R. Urban

INTRODUCTION

Wetlands include ecosystems on predominantly inorganic sediments and those that accumulate organic peat. Peat-accumulating ecosystems cover about 460 Mha - approximately 3% of the Earth's land surface - and contain about 150 Gt of carbon - a value that may be compared with an estimate of 560 Gt of carbon in the Earth's non-marine biomass. On the whole, these ecosystems are not of great economic importance, although they are locally exploited for agriculture (drained fens, cranberry bogs), forestry, and mining of peat. They do, however, have substantial effects on the chemistry of the water that runs from them to streams and lakes. Damage to peat-forming ecosystems is likely to be quickly reflected in water quality and possibly in the ability to act as absorbents of aerial pollutants. In spite of the actual and potential importance of peat-accumulating ecosystems, the money spent on assessing the consequences of damaging them by aerial pollution is probably less than 1% of that spent for a similar purpose on forests or on agriculture. This is short-sighted.

Peat mining, draining, or general damage to surface vegetation by air pollution can result in the release of substantial amounts of CO_2 to the atmosphere. If the rate of efflux from a damaged surface were 500 $g.m^{-2}.yr^{-1}$ - a not implausible value - the concentration of CO_2 above the peat surface to a height of 5 km could be increased by 100 ppm. Conversely, the effects of a continued increase in atmospheric CO_2 concentration, with concomitant increase in temperature and perhaps a fall in water-table, is unpredictable. Some work on responses to CO_2 enrichment by peat-accumulating ecosystems in the arctic tundra has begun (Billings et al. 1983). We need to know what would happen to a much broader range of such systems.

ISSUES

Because there has been little research on the effects of air pollution and acid deposition upon wetlands, we have sought chiefly to identify problems deserving study. Wetlands with inorganic soils (e.g., marshes) that possess large reserves of base cations and alkalinity capable of buffering acid deposition have not been considered. A major factor complicating our examination of various problems has been the fact that peatland ecosystems undergo a natural acidification process that has transformed - during the postglacial period - many circumneutral, minerotrophic fens receiving a part of their supply of minerals from adjacent soils into strongly acid (pH 3.6 - 4.5), ombrotrophic bogs receiving their supply of minerals solely from the atmosphere.

Five major problems have been identified, as follows:

1. What kinds of alteration (damage) might be expected from aerial pollution and acid deposition?

In the case of gaseous pollutants, the effects of sulphur dioxide have probably been severe on the cryptogams of northern English bogs over the past century or more. Effects of nitrogen on bryophytes are uncertain, but should be sought.

Acid deposition would be expected to favour the invasion and spread of bog mosses (Sphagnum spp.) in fens with waters low in calcium and bicarbonate alkalinity. This may be occurring in central England and elsewhere, and should be looked for. The accompanying nitrogen may be important as a nutrient, favouring Sphagnum recurvum s.l. In bogs this nitrogen probably increases the growth of bog mosses such as Sphagnum fuscum and of trees such as black spruce (Picea mariana) and tamarack (Larix americana). Other plant nutrients may also accompany acid deposition, for instance phosphorus and potassium.

Over the long term sulphur may be stored in amounts greater than normal in peat deposits subjected to acid deposition. This sulphur could be mobilized as an acid pulse by drawdown of the water-table during a drought and the consequent oxidation of reduced sulphur. Effects upon the biota could be severe both locally and downstream. It is also possible that acidification of shallow ground-waters (and wells) could ensue.

Metals such as lead, copper, zinc, mercury, etc. could also be enriched in peat deposits as a consequence of acid deposition, and be released by drawdowns during drought as well as by drainage and burning of peat deposits. There may have been metal damage to bog bryophytes in England under the severe atmospheric loadings of the past century, and perhaps to micro-organisms; the possibility deserves investigation. In the Sudbury smelting area of Canada, Sphagnum bogs have been severely affected by a combination of high SO_2 levels and toxic metal inputs. Use of peat as a medium for the growth of economic plants, either on-site or in nurseries, greenhouses and gardens, may also be compromised by high metal concentrations in areas of severe aerial pollution and acid deposition. The mobility of metals and its relationship to both mineral and organic acidity needs much more study.

2. By what means can peatland ecosystems counter acid deposition and associated aerial pollutants?

The uptake of plant nutrients can generate alkalinity, but the balance between cation and anion retention as plants decay to form peat is uncertain and requires study. We do not know the degree to which amounts of nitrogen, sulphur, or base cations trapped in deep peats will be greater in areas of high acid deposition than in relatively pristine regions.

Export of metals as organic complexes in outflows may – like trapping in deep peats – be viewed as a detoxication mechanism for peatland ecosystems. Again, we do not have data to compare such export under different deposition regimes.

3. What sorts of peatlands might be most susceptible to damage by acid deposition and associated aerial pollutants?

Fens with waters low in calcium and bicarbonate alkalinity but relatively high pH (about 6.0) may be transformed by acid deposition, which should favour their invasion by carpet-forming Sphagnum mosses capable of acidifying such sites down to pH 3.6 - 4.5. An experimental test of the hypothesis should be carried out. The effect of differing rates of water flow in determining vulnerability to acidification should also be examined. At low to moderate rates of acid deposition, nutrient effects may complicate and perhaps offset the toxic effects expected of acidification. Moreover, acidification from pH 6.0 to pH 4.5 could favour development of commercial conifers on continental peatlands, because Sphagnum is a good seedbed for them.

Where aerial pollution is unusually severe, there can be marked toxic effects of gaseous pollutants upon plants that lack a cuticle, such as the bryophytes (and lichens) that dominate the surfaces of many acid bogs.

The relationship of metal toxicity to site pH is uncertain, and should be investigated further. The conventional wisdom of greater toxicity at low pH is suspect, especially where organic matter is abundant in both solid and dissolved phases. Moreover, although many metals may be less toxic in the organic form, in some cases (e.g., mercury) the organic compounds may be more toxic.

4. Is peatland acidification, whether natural or owing to acid deposition, likely to have effects upon other ecosystems?

There will certainly be effects upon receiving waters if drawdown of peatland water-tables releases acids owing to sulphur oxidation and perhaps mobilizes metals as well. Conversion of fens to bogs is also likely to increase export of coloured organic acids by outflow streams.

5. Are there early-warning indicators of damage to peatlands by acid deposition and associated aerial pollutants?

Loss of diversity in lichen and bryophyte species is probably the earliest visible sign of damage to the biota by gaseous aerial pollutants. Loss of older needles by conifers is also an early consequence of such pollution, and should be sought in the black spruce that is abundant on many continental peatlands.

Among animals, disappearance of molluscs with calcareous shells (or perhaps initially, shell thinning) should be looked for in fen waters that are becoming acidified. A loss of alkalinity by such fen waters is an early chemical sign, but both chemical and biological changes will be hard to distinguish from those due to the natural processes of peatland acidification mentioned earlier. Paleoecological and paleogeochemical analyses of dated peat cores may help to establish associations between the onset of acidification and the spread of severe urban/industrial aerial pollution.

RESEARCH RECOMMENDATIONS

Our group made nine recommendations for research on the effects of acid deposition and associated aerial pollutants upon peatland ecosystems, as follows:

1. More analyses of acidity/alkalinity budgets are needed.

Such budgets should be constructed, where possible, in closed basins (such as those of the Experimental Lakes Area of northwestern Ontario) where ground-water inputs from outside the watershed are not appreciable. Budgets are needed particularly in fens with little remaining alkalinity and a pH of about 6.0, but a budget for an unforested oceanic bog would also be valuable for comparison with those already constructed at forested sites. Further studies are also needed to assess techniques for measuring acidity and alkalinity in strongly coloured waters rich in dissolved organic matter.

2. Experimental acidification of peatland ecosystems should be carried out in a variety of sites.

Such experiments should focus in the first instance upon vulnerable fens with waters low in alkalinity. The experiments should be continued long enough to demonstrate any long-term biotic responses, and to examine at least one drought cycle - looking for the effects of drawing down the water-table upon sulphur oxidation, acidification, and the biotic responses to it. The export of coloured organic acids and metals should be followed in the peatland outflows, again through a drought cycle and with special attention to metal speciation. Microbiological studies should be carried out in conjunction with the biogeochemical analyses of the effects of acidification upon the cycles of carbon, nitrogen and sulphur.

3. Further work is needed on trace metal speciation and mobility.

Research should focus on the influence of acidity and redox potential on speciation and mobility in pore waters, surface waters and outflows. Metals should include Al, Fe, Pb, Cd, Hg, Cu, Zn, etc. As a particular case, the magnetic minerals that are emitted in large amounts to the atmosphere from industrial sources should also be studied, because if not substantially reduced below the water-table they may provide a marker useful in dating the onset of local industrialization.

4. Stable and radioactive isotopes should be used more widely.

Such techniques can be useful, for instance, in studying groundwater fluxes, sulphur sources, nitrogen metabolism, metal mobility, etc.

5. Further experimentation is needed on the responses of bryophytes to mixtures of sulphur dioxide and oxides of nitrogen.

Not enough is known about the interactions of these gaseous pollutants at low concentrations, or their interactions with acid deposition in affecting moss metabolism, growth and reproduction. Ozone might also be included; there are continental peatlands in areas where relatively high atmospheric concentrations have been observed.

6. More attention should be paid to inputs of ammonium ions and gaseous ammonia to peatlands.

Increased deposition of ammonium ions may affect the uptake of nitrate ions and nitrogen metabolism in general. There is also some question as to whether acid bogs are sinks (as expected) or sources for gaseous ammonia.

7. Experimental deposition of acids in clean, remote areas should compare the responses of peatland bryophytes and tree seedlings with those of tree seedlings in upland forests and of crop plants in agricultural soils.

Bryophytes might well prove to be the plants most sensitive to moderate levels of acid deposition.

8. Paleoecological and paleogeochemical studies are needed to assess the course of peatland acidification in areas of severe acid deposition as well as in relatively pristine regions.

Studies of bryophyte remains in dated peat cores, coupled with a variety of chemical analyses, could help to assess the possible influence of acid deposition upon peatland ecosystems. Stratigraphic pH profiles can be inferred using assemblages of bryophyte fossils in the same way that limnologists use death assemblages of diatoms. They should be calibrated where acidification experiments are being carried out, and conversely, can be of use in establishing the stability (or instability) of sites being considered for such experiments. In connection with such stratigraphic studies, there needs to be a great deal more research to compare a variety of dating techniques applicable to strata deposited over the last 200 years.

9. Studies are needed concerning the effects of peatland damage by acid deposition and associated aerial pollutants upon the cycles of carbon dioxide and methane in peatlands.

Release of carbon dioxide might be stimulated, and its uptake retarded, by any damage that increased the rate of decomposition and decreased the rate of photosynthesis. Methane release would also be affected by damage to the plant cover, particularly if it had effects upon the ability of the peatland to maintain a high water-table. Conversely, the role of peatlands - damaged and undamaged - in trapping carbon dioxide and releasing methane should be studied in relation to currently increasing concentrations of atmospheric carbon dioxide, the likelihood of warming temperatures, and the possibility of falling water-tables.

REFERENCE

Billings WD, Luken JO, Mortensen DA, Peterson KM (1983) Oecologia (Berl.) 58: 286-289.

AUTHOR INDEX

Abrahamsen, G., 321
Adamson, N., 201
Allen, O.B., 451
Atkinson, C.J., 427

Baes III, C.F., 307
Bayley, S.E., 531
Beckerson, D.W., 451
Berg, V.S., 145
Berggren, D., 347
Bergkvist, B., 347
Binns, W.O., 69
Black, M., 599
Bucher, J.B., 43

Clymo, R.S., 513
Cook, E.R., 277
Cox, R.M., 155
Crossley, A., 171

Deveau, J.L., 451
Dimma, D.E., 101
Dixon, M., 411

Eisenreich, S.J., 577
Eldhuset, T., 401
Evans, L.S., 611, 627

Fail, J. Jr., 599
Falkengren-Grerup, U., 347
Ferguson, P., 549
Flanagan, L.B., 201
Folkeson, L., 347
Foster, N.W., 377, 625
Freer-Smith, P.H., 131

Glaser, P.H., 493
Godbold, D.L., 387
Göransson, A., 401
Gorham, E., 493, 577, 631

Haines, B., 599
Howell, G., 599
Hutchinson, T.C., 411
Hüttermann, A., 387

Ingestad, T., 401

Janssens, J.A., 493
Johnson, A.H., 83, 619
Johnson, D.W., 333
Johnson, J.W. Jr., 481
Jung, K.-D., 1

Krause, G.H.M., 1
LeBlanc, D.C., 291
Lee, J.A., 549
Lewin, K.F., 611
Lindberg, S.E. 117
Linzon, S.N., 101
Lovett, G.M., 117

Manion, P.D., 267
Mansfield, T.A., 131
Matzner, E., 25
McBride, J.R., 217
McHargue, L., 599
McIlveen, W.D., 101
McLaughlin, D.L., 101
McLaughlin, S.B., 307
Meiwes, K.-J., 117
Miller, P.R., 217
Morrison, I.K., 377
Mulder, J., 361

Norton, S.A., 561

Ormrod, D.P., 451

Pande, P.C., 131
Pell, E.J., 229
Press, M.C., 549
Prinz, B., 1

Raynal, D.J., 291
Redfern, D.B., 69
Reid, D.M., 241
Rennolls, K., 69
Rühling, Å., 347

Saxe, H., 463
Schindler, 531
Scott, M., 411
Seaward, M.R.D., 439
Shriner, D.S., 481
Soto, C., 411

Tischner, R., 387
Tomlinson, G.H., 189
Tveite, B., 59
Tyler, G., 347
Tyree, M.T., 201

Ulrich B., 25
Unsworth, M.H., 171
Urban, N.R., 577

van Breemen, N., 361

Wheeler, G.A., 493

White, E.H., 291

Whitmore, M.E., 131
Winner, W.E., 427
Woodin, S., 549

Zoettl, H.W., 255

SPECIES INDEX

Abies sp., 31, 411
Abies alba, 6, 8
Abies balsamea, 18, 84, 204-205
Abies concolor, 224
Abies fraseri, 311
Abies grandis, 79
Abies procera, 79
Abies religiosa, 225
Acacia mangium, 605
Acer sp., 446
Acer rubrum, 103, 603
Acer saccharum, 84, 103, 158-161, 207, 224, 363
Alnus sp., 504
Aneura sp., 496-497
Arceuthoblum, 504
Armillaria mellea, 48, 86, 101, 112, 115, 271-272
Artemisia tilesii, 152
Aulacomnium palustre, 497

Berberis aquifolium, 148
Beta vulgaris, 605
Betula sp., 358, 411, 504
Betula alleghaniensis, 103, 159-161, 363, 407
Betula papyrifera, 86, 103, 158-160, 407
Betula pendula, 140-141, 402
Betula pubescens, 140
Betula pumila, 496, 500, 502
Blastophagus sp., 49
Brassica oleracea, 151, 606
Bryoria fuscescens, 443
Bryum pseudotriquetrum, 497
Buellia punctata, 443

Callicladium haldianum, 497
Calliergon stramineum, 502
Calliergon trifarium, 496-497, 503-504, 507
Calliergonella cuspidata, 496
Calluna vulgaris, 348, 363
Calocedrus decurrens, 224
Campylium stellatum, 496-497, 503-504
Carex aquatilis, 496
Carex disperma, 496
Carex gynocrates, 497
Carex lasiocarpa, 496, 500, 502
Carex leptalea, 499-500
Carex limosa, 496
Carex livida, 496
Carex oligosperma, 496

Carex pauciflora, 496
Carex rostrata, 496
Carex trisperma, 496
Carpinus betulus, 348
Carya sp., 601
Carya glabra, 603
Camellia sp., 157
Campylium stellatum, 496
Ceratocystis fagacearum, 271
Chamaedaphne calyculata, 497, 579
Chrysomyxa sp., 48
Cinclidium stygium, 497
Cladina sp., 412-413, 419
Cladina mitis, 412, 414, 420-421
Cladina rangiferina, 412, 414, 420-424
Cladina stellaris, 412, 414, 419-420, 422-424
Cladopodiella fluitans, 496
Climacium dendroides, 430
Coleus blumei, 606
Cucumis melo, 606
Cucumis sativus, 606
Cucurbita pepo, 606
Cytospora kunzii, 86

Dactylis glomerata, 133, 141
Dendroctonous sp., 92
Dicranum polysetum, 496
Dicranum undulatum, 496-497
Dieffenbachia maculata, 465, 469, 471, 474-477
Diervilla lonicera, 157-158, 160-161
Drepanocladus lapponicus, 496
Drepanocladus revolvens, 497

Elatobium abietinum, 77
Enterolobium cyclocarpum, 605
Epinotia pygmaeana, 48
Eriophorum angustifolium, 514
Eriophorum spissum, 496, 600
Eriophorum tenellum, 502
Eriophorum vaginatum, 503, 507, 527, 555

Fagus grandifolia, 103, 363
Fagus sylvatica, 6, 20, 348, 358
Ficus benjamina, 464, 468-470, 472-477
Ficus elastica 464, 469-470, 473-477
Fissidens adiantoides, 497
Fomes pini, 86
Fraxinus sp., 207, 443, 446
Fraxinus americanus, 103, 224
Fraxinus excelsior, 447

Geranium carolinianum, 157
Gliricidia sepium, 605
Glycine max, 132, 236, 484, 606
Gossypium hirsutum, 606
Gremeniella abientina, 65

Hedera canariensis, 464, 469-470, 472-477
Hedera helix, 464, 469-470, 473-477
Helianthus sp., 210
Heliocarpus appendiculatus, 603
Heterobasidion annosum, 272
Hibiscus rosa-sinensis, 464-465, 469-471, 473-477
Hordeum vulgarae, 606
Hylocomium splendens, 412, 497
Hypogymnia physodes, 9

Ips typographus, 59

Juncus stygius, 502

Larix sp., 504
Larix americana, 632
Larix laricina, 497
Lecanora conizaeoides, 443, 444
Lecanora muralis, 441, 443
Ledum groenlandicum, 497, 579
Lepidium virginicum, 156
Leucaena multicapitula, 605
Liriodendron tulipifera, 603
Lobaria pulmonaria, 446-447
Lolium multiflorum, 133
Lolium perenne, 141
Lophodermium macrosporum, 48
Lophodermium piceae, 48, 20
Lophodermium pinastri, 49
Lycopericon sp., 606

Maianthemum canadense, 157-158
Medicago sativa, 606
Mimosa sp., 605
Mimosa pigra, 605
Mimosa pudica, 605
Mindarus abietinus, 48
Moerkia hibernica, 497
Myurella julacea, 497

Nectria coccinea, 272
Nephrolepis exaltata, 465, 469, 471-477
Nicotiana tabacum, 606

Ochroma lagopus, 603
Oenothera parviflora, 157-158, 160-161, 163-165

Parmelia incurva, 443, 445
Parmelia sulcata, 443, 445
Parmeliopsis ambigua, 443, 445, 447
Parnassia palustris, 497
Pentaclethra macroloba, 605
Phaeophysica orbicularis, 443
Phaseolus vulgaris, 132, 482, 606
Phleum pratense, 133, 141
Picea sp., 411, 504
Picea abies, 6-8, 16, 18-19, 70, 224, 294, 348, 358, 387-398, 401, 407, 606
Picea lambertiana, 224
Picea mariana, 436, 496-497, 579, 632
Picea rubens, 83, 311, 407
Picea stichensis, 72
Pinus sp., 411, 601, 604
Pinus banksiana, 157-161, 377-378, 412
Pinus contorta, 79
Pinus echinata, 292, 311
Pinus eliotii, 17
Pinus hartwegii, 225
Pinus jeffreyi, 221
Pinus montezumae, 225
Pinus nigra, 21
Pinus ponderosa, 221, 279, 606
Pinus resinosa, 158-161, 294
Pinus rigida, 311
Pinus strobus, 18, 156-166, 223, 311
Pinus sylvestris, 72, 294, 363, 402, 407, 606
Pisum sp., 606
Pithecellobium arboretum, 605
Pithecellobium catenatum, 603, 605-606
Pithecellobium gigantifolium, 605
Pithecellobium longifolium, 605
Pithecellobium pedicellare, 605
Plagiomnium ellipticum, 497
Platygyrium repens, 497
Pleurozium shreberi, 412-419, 496-497
Poa pratensis, 133-142
Polytrichum strictum, 497, 504
Populus deltoides, 156, 158
Populus tremuloides, 159-161
Potentilla fruticosa, 496
Potentilla palustris, 500, 502
Prunus serotina, 103
Pseudevernia furfuracea, 443
Pseudotsuga menziesii, 79
Ptilidium pulcherrimum, 497

Quercus sp., 407, 447, 601
Quercus alba, 119
Quercus kelloggii, 224
Quercus lobata, 606
Quercus prinus, 119
Quercus robur, 348
Quercus rubra, 603

Ramilina sp., 447
Ramalina farinacea, 443
Rhizosphaera kalkhoffii, 20, 48
Rhynchospora alba, 496
Rhynchospora fusca, 502
Riccardia sp., 496-497
Robinia sp., 601, 604
Robinia pseudo-acacia, 603

Salix sp., 443, 504
Scirpus acutus, 497
Scirpus cespitosus, 497
Scirpus hudsonianus, 497
Scolisciosporum chlorococcum, 443
Scorpidium scorpioides, 496-497, 499, 501, 503-504, 507
Solanum tuberosum, 238
Sphagnum sp., 493-495, 507-508, 510, 513-527, 549-559, 561-563, 568, 588, 632-633
Sphagnum angustifolium, 497, 504
Sphagnum capillifolium, 496, 501-504, 507, 523, 550, 556, 558
Sphagnum contortum, 496-497
Sphagnum cuspidatum, 523, 526, 553-554, 556-557
Sphagnum fallax, 496
Sphagnum fimbriatum, 522
Sphagnum flexuosum, 496
Sphagnum fuscum, 494, 497, 503-504, 507, 523-524, 550, 553, 557, 563, 632
Sphagnum imbricatum, 550
Sphagnum inundatum, 523
Sphagnum magellanicum, 497, 501, 503-504, 523, 550, 579
Sphagnum majus, 496, 504, 507
Sphagnum nemoreum, 555
Sphagnum papillosum, 496, 498, 501, 504, 507, 523, 525, 550
Sphagnum pulchrum, 496, 550

Sphagnum recurvum, 498, 514, 519, 522-523, 550-551, 579, 632
Sphagnum rubellum, 563
Sphagnum section, 501-502
Sphagnum squarrosum, 522-523
Sphagnum stramineum, 501
Sphagnum subnitens, 523
Sphagnum tenellum, 550
Sphagnum teres, 497-498
Sphagnum warnstorfii, 497-499
Stryphnodendron excelsum, 605
Suillus bovinus, 401, 403, 406

Thelypteris palustris, 496
Thuja occidentalis, 204, 207
Tilia cordata, 140
Tomenthypnum nitens, 497
Trebouxia sp., 424-425
Triglochin maritima, 496, 499-500
Trillium grandiflorum, 158-161, 164-165
Tsuga canadensis, 103, 157-161, 207, 279, 311
Typha sp., 504

Ulmus sp., 446
Umbilicaria vellea, 424
Usnea sp., 443, 447
Usnea subfloridana, 443-444

Vaccinium oxycoccus, 497
Vaccinium vitis-idaea, 496
Vaccinium myrtilloides, 496
Viscum album, 49

Xanthium, 211
Xanthoria candelaria, 443
Xanthoria elegans, 443, 446
Xyris montana, 502

Zea mays, 606

SUBJECT INDEX

abscission (leaf), 244, 246-248
acid deposition/precipitation, see also acidification, acid rain, 83-84, 103, 164, 165, 213, 241, 270, 291-304, 321, 440, 451, 531, 600, 627
  effects on peatlands, 507-509, 532, 549-559, 631-634
  experimental approaches, 611-616
  impact on soils, 107, 322-328, 333-343, 361-374
  research needs, 634-636
acid fog, 10-13, 19, 256
acidification
  artificial, 66
  of lakes, 545
  of peatlands, 493-510, 514, 516, 518
  of soil, 27, 30-31, 35, 38-39, 66, 71, 107, 321-328, 347-359, 620, 625-626
  stromal, 212
acidity, 177, 516
  budgets, 577, 581, 588-594
  effects on pollen, 157, 161
acid precipitation - see acid deposition
acid rain, 10-13, 18-19, 83, 145-152, 155-167, 267, 532, 600
  effects on vegetation, 183
  effects on wetlands, 542-543
  simulated acid rain studies, 14, 224, 324-328, 411-425, 448, 481-490, 517
acoustic emissions, 206
acrotelm, 518-520, 526-527
active rooting zone, 107
air pollution, see also NO_2, O_3, SO_2, and other specific pollutants, 35, 38-39, 212, 230, 267, 271, 321-328, 427-428, 481-490, 514
  and lichens, 439-448
  as cause of forest damage, 6, 9, 10-12, 50-54
  in Europe, 51
  in Norway, 59-60
  indicators, 155, 440-441
alcohol, 147
aldehyde, 147
alder, 3, 323
alfalfa, 231, 238, 483-484, 487
alkalinity, 494-495, 498-499, 633-634
  sources in watersheds, 531-546, 579, 581, 589
aluminum, 267, 311, 313-316, 361-374, 516, 588
  in bog waters, 580, 582-583
  in foliage, 107-109, 258-261, 263, 401-407, 418-419
  in peat cores, 566-571, 580
  in roots, 30, 107, 110-111, 387-398
  in soil, 27-30, 33-34, 104-107, 110, 128, 259, 261, 322-323, 334-335, 338-339, 348, 353, 359, 379-380, 402, 425, 561, 620, 625
aluminum oxide, 334
aluminum sulphate, 371
Amazon, 600
ammonium (NH_4^+), 15, 326, 535, 539, 546, 555-556, 558, 580, 583-584, 588, 635
angiosperms, 244
anion(s), 29, 31, 322, 333-335, 353, 531, 538, 625
  in bog waters, 520-522
apothecia, 422-423
arylsulphatase, 555-556
ash tree, 45, 267, 273
  mountain, 3
  white, 103, 224
AsO_4, 526
aspect ratio, 176
aspen, 336
  trembling, 158-162
ATP (adenosine triphosphate), 210-212
ATPase(es), 233
Austria, 46-48, 51, 53-54, 620
automobile exhaust, 52, 117
auxin, 209

β-carotene, 17
β-diketone, 147
basaluminite ($Al_4(SO_4)(OH)_{10} \cdot 5H_2O$), 371
barley, 135-136, 245
base neutralizing capacity, 586-587, 589
base saturation, 27, 105-107, 259, 261, 323, 326-328, 350, 498-499, 625
bean(s), 245
  bush, 482-484, 487
  green, 231, 233
beech, 3-4, 6, 20, 27-29, 45-46, 50,

267, 272, 315, 338, 348, 358-359
American, 103
beetle(s), 20, 39, 272
   Japanese, 484
   spruce, 92, 270
birch, 88-89, 140-141, 158-161, 267, 358-359, 402
   European, 403-407
   white, 86, 88, 103
   yellow, 103
bog(s), 494-499, 502-507, 509-510, 545, 549-559
boron, 107-111
broad-leaved trees, 45, 50, 72, 74, 139
brome grass, 245
brunisols, 106-107
bryophyte(s), 549-553, 562, 579, 635
buffering capacity, 7, 26-27, 31, 166, 510

cadmium, 313-314, 316, 359, 526, 561
   effects on pollen, 159, 164-165
   effects on roots, 389-390, 393-395
   in foliage, 8, 107-109
   in roots, 107, 110-111
   in soil, 107, 110, 310, 353, 355, 389-380
calcium, 314, 494-495, 498-503, 505, 516, 522, 535, 539-540, 543, 561
   in foliage, 6-8, 15, 107-109, 247, 256, 258, 260-262, 418-419
   in bog waters, 580, 582, 633
   in groundwater, 326
   in peat cores, 566-573, 580, 583
   in roots, 30, 107, 110-111
   in soil, 66, 104, 106-107, 110, 259, 261, 323, 336, 338-339, 353, 365, 378-381, 383-384, 620
calcium/aluminum ratio, 27, 30, 313, 396, 401, 407
calcium/hydrogen ion ratio, 30
cambium, 207-208, 296, 313-314
Canabinaceae, 243
Canada, 85, 92, 101-115, 160, 166, 218-219, 222, 224, 244-245, 271, 339, 377-378, 464, 493, 498, 507-510, 532, 545, 578, 580, 585-587, 592-593
carbohydrates, 209, 211, 519-521
carbon, 18, 128, 180, 366
   fixation, 209
carbon-14, 519
carbonates, 335
carbonic acid, 321
carbon dioxide (CO_2), 210-212, 428, 464-465, 518, 527, 562, 608, 631
   effect on photosynthesis, 468-471
   effect on respiration, 472
   effect on transpiration, 473
carbonyl sulphide (COS), 601-606
carotenoid, 243
cation(s), 26, 29, 31, 149-151, 322, 324, 348, 361, 364, 373, 351, 538, 625
   in bog waters, 516, 520-522, 579
   in vegetation, 336
   leaching, 377-384
cation exchange capacity (CEC), 104-107, 327, 334, 337, 339, 378-383, 499, 523, 532, 625
   of <u>Sphagnum</u>, 514-515, 517
catotelm, 518, 520, 526
cavitation, 205, 207, 212
cedar
   Alaska, 271
cell membrane, 233
cellulase, 146
cellulose, 146
cell wall, 149, 243
Central America, 222
cereal(s), 135-137
cherry
   black, 103
chloramphenicol, 159
chloride, 516-517, 535, 539, 561, 582-583
   deposition, 51
   in foliage, 8, 15, 107-109
   in groundwater, 326
   in roots, 107, 110-111
   in soil, 107, 110
chlorine, 53
chlorophyll, 15, 17-18, 418
chloroplasts, 17, 210, 212, 234, 236, 422-424, 429
chlorosis, 15, 17, 223, 244, 247, 256
chromium, 107-111, 310
cloudwater, 117
   chemical composition, 84, 177-180
   collection, 177-180
   deposition, 171-185
   particulates, 180-181
coffee, 315-316
competition, 442, 445
condensation, 184
condensation nuclei, 177
conductance (K), 203
conifers, 13, 18, 45, 50, 72, 74, 78-79, 88, 139, 164, 176, 180, 224, 292, 313, 315-316, 337, 387
copper, 311, 314, 316, 356, 561
   effects on pollen tubes, 159-162,

165
  in foliage, 107-109
  in peat cores, 566-572, 632
  in roots, 107, 110-111
  in soils, 107, 110, 310
cores
  peat, 503-506
corn, 483-487
Costa Rica, 600-608
cotton, 246, 315
cottongrass, 507
crops
  air pollution effects on, 222, 451-463, 481-490, 627-630
crown density, 64, 72, 77
crown dieback, 87-88, 102, 268
cryptogam(s), 412, 421, 425, 427, 632
cubic smoothing splines, 283
cucumber, 245
Cucurbitaceae, 243
culture media, 160
cuticle, 145-152, 412, 428, 430
cutin, 146-147, 149
Cyperaceae, 504
Czechoslovakia, 3, 185, 621

deciduous trees, 337
  sensitivity to ozone, 222-223
decline, see forest decline
decline index, 103-106
decomposition, 26, 503, 519, 589, 594
defoliation, 113
dendrochronology, 277-304, 622
denitrification, 582-583, 589
deposition, 84-85
  bulk, 119, 122-123
  cloud, 124-125, 127
  dry, 51, 61, 119-122, 125-127, 131-142, 159, 165, 172, 178, 184, 279, 326, 411, 549, 551, 579, 588-589
  nitrogen, 62
  sulphur, 62
  velocities, 119, 173-174
    dry (V_d), 125-127
  wet, 51, 61, 118-120, 122, 159, 165, 178, 184, 279, 326, 411, 549, 551, 579, 588-589
desert, 431-436
desiccation, 428, 433-434
dieback, see also forest damage, forest decline, 25, 269
dimethyl disulphide ((CH_3)$_2S_2$), 601
dimethyl sulphide ((CH_3)$_2S$), 601, 603-606
disequilibrium index (I_p), 363, 372
dissolved inorganic carbon (DIC), 534
dissolved organic carbon (DOC), 368-369, 494-495, 510, 522, 588
dissolved organic mattter (DOM), 356-357
dormancy, 243
dose response, 137-139, 140-141, 243, 246
drag coefficient, 175
drought, 39, 301, 304, 540
  response of trees, 201-213

ED_{10}, 165
electron transport, 210-211
electron dense bodies, 429
emissions, 117
  nitrogen oxide, 51-52, 118, 447, 463-477, 518
  sulphur, 599-608
endodermis, 33, 396
endoplasmic reticulum, 234
English Lake District (UK), 498-499
enzymes, 242
epicuticular wax, 146-148, 150
epidermal cells, 146, 151
epinasty, 247
epiphytes, 121
Ericaeceae, 504
ester, 147
ethylene, 241-249
  sensitivity of vegetation, 244, 246
  symptoms of exposure, 241, 246
ethyl mercaptan, 601, 603-606
Euphorbiaceae, 243
Europe, also see individual countries, 25, 27, 31, 54, 62, 131, 255, 272, 323, 347, 411, 427, 493, 562, 625
  air pollution, 51, 60
evaporation, 184, 203, 518, 520
evapotranspiration, 361
Experimental Lakes Area (Canada), 532-546

fatty acids, 230
feather moss(es), 411-425, 430
fen(s), 494, 496-499, 502-503, 507, 509, 518, 532
fertilization
  experimental, 255-263, 623
Finland, 545, 562
fir, 3, 20, 45, 47-48, 50, 89, 124-125, 225, 272, 315
  balsam, 18, 84-86, 88-90, 315, 616
  Douglas, 3, 209

Fraser, 311-312, 314
Grand, 79
Noble, 79
silver, 6, 8, 31, 43, 46-47, 70, 257, 263
fire, 534, 536, 545
  effects on alkalinity yield, 540-541
fluorescein diacetate (FDA), 238
fluorescence, 210
fluoride, 8, 12, 272, 526
fluorine, 53, 107-111
flux density, 175-176
fog, 10-12, 52, 173
fog deposition, 125, 127
fog water
  collector, 178
foliage damage, 132, 151, 155, 222, 224, 235
foliage loss, 87-88, 277
  relative foliage loss, 44-49, 54
foliar leaching, 112, 125, 147, 149-152
  rate, 145
forest, 175, 184
  boreal, 411-425
  coniferous, 46, 119, 348-349, 387
  deciduous, 46, 119, 127, 151, 348-350, 363, 432-435, 599-608
  evergreen, 432-435
  mixed deciduous-coniferous, 46
forest damage
  causes, 2, 87, 90, 92-96, 388
    air pollution, 3, 9-10, 12-14, 50-54
    aging, 63-64
    climate, 64-65, 77, 79, 284-285
    drought, 59, 63, 65, 112-113, 115, 202, 284
    fire, 279
    forest insects, 59, 77, 112, 279
    frost, 63, 279
    fungus, 59, 112
    mistletoe, 49
    needle blight fungus, 49
    ozone, 224-228
    pine shoot beetle, 49
    snow, 59
    spruce bark beetle, 59, 63, 65
    wind, 59, 63, 279
  classification/inventories of damage, 44, 47, 52, 54, 63, 71-79, 622
    in Austria, 46-48
  in Britain, 72-79
  in France, 49-50
  in Germany, 3-5, 77
  in Italy, 47-49
  in Norway, 63-64
  in Sweden, 78
  in Switzerland, 44-47
  damage threshold, 54
  site characteristics, 4, 12, 45-46, 172, 256
  symptoms, 2, 6-9, 59, 63, 72, 79, 86, 102-103, 113, 117, 124, 256, 277, 619-621
    chlorosis, 6, 71
    foliage loss, 6, 44-49, 54
    needle yellowing, 6, 72, 78
forest decline, see also forest damage, 86-87, 223, 267-273, 619-621
  fir, 3, 272, 616
  in Austria, 46-47, 53-54
  in Britain, 69-80
  in Canada, 101-115
  in Europe, 62, 66, 74, 224, 401
  in France, 49-50, 54
  in Germany, 1-21, 25-39, 43-44, 71, 123, 172
  in Italy, 47, 49, 54
  in Norway, 59-66
  in Switzerland, 44-54
  in United States, 123, 172, 292
  Norway spruce, 70, 621
  red spruce 83, 99, 83-96, 288, 616, 621
  research needs, 621-623
  silver fir, 70
  sugar maple, 101-115, 225
formic acid, 218
France, 49-50, 621
frost, 17, 39, 128, 185
fructose bisphosphatase (FBP), 212
fulvic acid, 397-398, 494, 519, 521-522, 581-582, 588
fungus, 20, 39, 48-49, 65, 86, 272

gas exchange, 150
gas exposure system, 465-467
genetic studies
  of ozone toxicity, 231-232
Germany (FDR), 10-11
Germany (FRG), 3-11, 18, 25-39, 43, 70-71, 74, 79, 117-118, 123-125, 185, 257, 272, 310, 316, 338, 347, 388, 397-398, 401, 494, 619
gibbsite, 371-373
glucanase, 246
Gramineae, 504

grana, 236
gran alkalinity titration, 535-538, 544, 580
grasses, 133-139, 141, 185
grassland, 175, 184
green spruce aphid, 77
guard cells, 150, 210, 231
gymnosperms, 244

hardiness, 247
heather, 348, 363-364
heavy metals, 9, 25, 267, 387-398, 620
hemlock, 103, 157-162, 279-280, 311-312, 314
   western, 175
hexokinase, 396
honey fungus, 48
honeysuckle
   bush, 157-158, 160-162
hormone(s)
   plant growth, 242
hornbeam, 348
Hubbard Brook Experimental Forest, 84-85, 363-373
humic acid, 398, 519, 521, 522
humus, 387-388, 397, 625
   formation, 336-337, 341
hydathodes, 150
hydrocarbons, 25
   unsaturated, 218
hydrogen ion, 30, 33, 156-157, 322, 334-336, 338, 348, 353, 365, 489, 495, 534, 578
   budget, 27, 358-359, 588-594
   deposition, 11, 256
   in cloudwater, 178, 180-181
   in peatlands, 516-519, 522
   wet deposition, 51
hydrogen chloride (HCl), 9
hydrogen fluoride (HF), 9
hydrogen peroxide (H_2O_2), 218, 230, 235
hydrogen sulphide (HS), 601, 606
hyponasty, 247

Indian Ocean, 600
indicator species, 412, 440-441
immissionsrate, 12
insects, 20, 39, 48, 112
Instituto Nacional de Investigaciones Forestales, 225
International Geosphere-Biosphere Program (IGBP), 607
ion exchange, 149, 151, 233, 352, 588
ions, see also anions, cations, 31

in cloudwater, 177
iron, 247, 311-314, 316, 503, 535
   in bog waters, 580, 582
   in foliage, 53, 107-109, 258-259, 261, 418-419
   in peat cores, 566-571, 580
   in roots, 107, 110-111
   in soil, 105, 107, 110, 259, 261, 361-362, 364, 379-381, 383, 561
Italy, 47-49

jurbanite ($Al(SO_4)(OH) \cdot 5H_2O$), 30, 371-373

kaolinite, 371-372
ketone, 147

larch, 45, 294
La Selva Biological Station of the Organization for Tropical Studies, 601
LD_{50}, 161-162, 164, 166
leaching
   of soils, 128, 184, 324, 333-334, 341, 365, 377-384, 412, 419, 625
   foliar, 12-13, 18, 112, 125, 145, 147, 149-152, 262, 336, 418-419
lead, 164, 236, 356, 526, 561-562
   ^{210}Pb, 562-564, 566-570, 572, 574-575
   in foliage, 8, 107-109, 263
   in peat cores, 566-575
   in roots, 107, 110-111, 389-392
   in sediments, 307-308, 310
   in soil, 85, 105, 107, 110, 397
   staining, 234
leaf conductance, 210, 233
leaf specific conductivity (LSC), 203-205
lichens, 9, 411-425, 427, 439-448, 562
lignens, 519
lime, 334
lime potential (K_L), 337
limestone, 7, 46
liming, 263
linear regression coefficients, 454
lipids, 148, 233
litter analysis, 602-604
lysimeter, 28, 378, 383, 401-402

magnesium, 164, 256, 397, 516, 535, 543, 561
   deficiency, 6, 18, 30, 79, 224
   deposition, 51
   in bog waters, 582

in foliage, 6, 7, 14-16, 18-19,
30, 78, 107-109, 256, 258-261,
263, 418-419
 in groundwater, 326
 in peat cores, 566-571, 580, 583
 in roots, 107, 110-111
 in soil, 19, 66, 104, 107, 110,
259, 261, 323, 336, 338-339,
353-354, 365, 378-384
maize, 398, 452-461
manganese, 310-311, 313-314, 316,
503, 539, 561
 in foliage, 107-109, 256, 258-
261, 418-419
 in peat cores, 566-571
 in roots, 30, 107, 110-111
 in soil, 107, 110, 259, 261, 338,
353
maple, 45, 101-115, 267
 sugar, 84, 103, 158-162, 224
  decline, 225
  dieback, 101-115, 225
mercury, 247, 632
 effects on roots, 389-395, 397
 in soils, 397
mesophyll, 223, 233
metal(s), see also heavy metals,
trace metals, individual elements
 atmospheric loading, 561-575
 solubility, 347-359
 speciation, 356-357, 634
metaloprotien sites, 242
methylmercaptan, 601, 603
Mexico, 219, 225
microfibril, 243
microorganisms, 31
microtubule, 243
mineralization, 31, 336
 of sulphate, 383
mining, 66
mire, see bog(s)
mitochondria, 234-236, 422-424, 429
model(s), 285
 constant rate of supply model
(CRS), 563-564
 decline, 267, 270
 distance-independent diameter
model of competition intensity,
296
 growth-climate regression model,
297-300, 302
 linear aggregate for ringwidth
series, 278
 stepwise regression models, 283-
288, 301
 unit pipe model, 202-203

 Weibull dose-reponse model, 459
molluscs, 633
molybdenum, 107-111
moss(es), 411-425, 427-436, 494-510,
513-527
mRNA, 242
mycorrhizae, 6, 33, 402-403
mycostatin, 159

n-alkane, 147
NADPH, 210
needle loss, 6, 38, 48, 258, 633
 causes, 48
needle yellowing, 6, 72, 78, 256,
258, 615
net biological uptake (NBU), 583,
586-588, 592
Netherlands, 361-374
"neuartige Waldschaden", 1-21, 255
neutralization
 of soils, 350
NH_4^+, see ammonium
nickel, 311, 313-314, 316
 in foliage, 107-109
 in roots, 107, 110-111
 in soil, 105, 107, 110
nitrate reductase, 142, 557-558
nitrates, 9, 11, 15-16, 31-34, 322,
334-335, 489, 507, 509, 539-540,
542, 544-545, 552, 556, 557-558,
582, 585, 588
 deposition, 52, 117-128
nitric acid (HNO_3), 178, 417-418,
551, 625
nitric acid vapour, 119-121, 127,
172
nitric oxide (NO), 451
 effects on photosynthesis, 468-
471, 475-476
 effects on respiration, 472
 effects on transpiration, 472-477
 responses of herbaceous and woody
plants, 131-144
 sensitivity of vegetation, 476-
477
nitrification, 31, 33, 323
nitrite reductase, 121, 142
nitrogen, 31, 121-122, 142, 209,
336, 518, 521, 527, 534, 551-
552, 555, 561-562, 583-585, 588,
635
 deficiency, 78, 121, 125
 deposition, 51, 124, 127, 184-
185, 256, 324-328
 in foliage, 107-109, 128, 258-
261, 263, 418, 522
 in roots, 107, 110-111

in soil, 105, 107, 110, 128, 259, 261, 323
inorganic, 524-525
mineralization, 321, 336
nitrogen dioxide (NO_2), 150, 218, 256, 451
  biochemical effects of, 141
  effects on growth, 133-134, 137-141
  effects on photosynthesis, 468-471, 475-476
  effects on respiration, 472
  effects on transpiration, 132, 472-476
  effects on vegetation, 132-142, 627
  effects on yield, 135-137
  sensitivity of vegetation, 476-477
nitrogen oxides (NO_x), 9, 25, 61, 117, 120, 127, 131-132, 150, 177, 220, 585
  emissions, 51-52, 118, 447, 463-477, 518
North America, see also Canada, Mexico, United States, 85, 87, 92, 117, 161, 166, 217, 222, 411, 427, 493, 510, 520, 590, 625
Norway, 59-66, 324-328, 338, 417, 498, 524, 562
nucleus, 422-423, 429
nutrient(s), 7, 9, 13, 18, 25, 209, 267
nutrient cycling, 119, 127, 310, 622

oak, 4, 45, 50, 119, 125, 267, 268, 271, 348, 352
oat, 245
occult deposition, 180, 184-185
occult precipitation, 172, 173, 175-176
onion, 231
ontogeny, 291
organic acids, 361-362, 494, 509, 519, 579, 581-582, 585, 590, 592-594
organic colloids, 107
orthoganol polynomials, 283
osmoregulation, 208
osmotic potential, 429
oxidants (also see specific oxidants)
  photochemical, 217-225
  photooxidants, 256-261
oxidation, 230, 233, 518
ozone (O_3), 10-12, 25, 71, 172, 217, 229-239, 262, 615, 620-621, 635
  concentration in air, 10, 256
  effects on vegetation, 13-14, 17-21, 132, 155-156, 166, 217-225, 247, 451-461, 482-483, 487-489, 627
  emission, 51-52, 62
  monitoring, 218

palisade cells, 151
paludification, 493
PAN (peroxyacetyl nitrate), 52, 218-219
PPN (peroxypropionyl nitrate), 218
parenchyma cell, 236
pathogens, 128, 142, 152, 267-273
peach, 243, 315
peas, 243
peat, 563
  accumulation, 594
  mining, 527, 631
peatlands, see also bog(s), 493-510, 513-529, 549, 561-575, 577-594
pectin, 146, 313
pectinase, 146
pesticide, 150
petunia, 155-164, 231, 245
pH, 51-52, 61, 145, 150-152, 178, 211-212, 440-441, 447, 515-517, 534
  cytoplasmic, 526
  effect on pollen germination, 156-166
  effect on vegetation, 16, 183
  peatlands, 498, 500-502
  seepage water, 27
  soil, 34, 35, 104, 106-107, 324, 326-328, 342, 347-355, 358, 625
phloem, 208
  and trace metals, 311, 313-314
phosphate, 15
phosphoribulokinase, 212
phosphorus, 313, 402, 524, 534
  in foliage, 107-109, 258-259, 261, 263
  in roots, 107, 110-111, 396
  in soils, 107, 110, 336, 339
photosynthesis, 15, 17, 38, 209-212, 419-421, 430, 463-477
  effects of ozone on, 223
phytohormones, 19
pigmentation, 244
pine, 4, 17, 20, 45, 47-48, 50, 127, 225, 294, 363, 368
  Austrian, 21
  jack, 157-162, 336, 378, 412
  Jeffrey, 221

loblolly, 243
lodgepole, 79
pitch, 311-312, 314
ponderosa, 221-224, 279-280, 615
red, 158-162, 294, 297-301
Scots, 63, 72, 74, 76-79, 262, 294, 297-301, 402, 404-406, 417
short-leaf, 292, 311-312, 314, 316
white, 18, 156-166, 224, 311-312, 314
pineapple, 243
pK, 588
plasmalemma, 526
plasma membrane, 233-236, 238
plastrochron index, 458
podetium, 422-423
podzolization, 364
podzols, 105-107, 324, 351, 356, 402
Poland, 494, 621
pollen, 155-167, 244, 504
pollen tube, 155, 159
pollination, 155
polonium, 563
polypeptides, 242
polyuronic acid, 494, 508, 522
poplar, 151, 315, 398
potassium, 212, 397, 516, 534, 535, 539, 561
    deficiency, 78, 294
    deposition, 51, 85
    in bog waters, 580, 582
    in foliage, 15, 107-109, 256, 258-263
    in groundwater, 326
    in peat cores, 566-572, 580, 583
    in roots, 107, 110-111
    in soil, 104, 106, 107, 110, 259, 261, 323, 336, 341, 365, 378, 561
potassium chloride (KCl), 366
    in soils
potato, 231-233, 235, 247
power plants, 292-311
precipitation, 121, 172
    acidity, 62
    chemistry, 60
    collection, 119, 122, 127
    in Britain, 75
primary producers, 25
principal component analysis (PCA), 297
proteins, 234-235
protein-lipid bilayer, 233
proton, 31, 33, 38
    cycling in bogs, 577-594
    movement through cuticle, 147-150

protonema, 526
protoplasts, 235
pulverised fuel ash (P.F.A.), 180
pyrenoid body, 422-424

radish, 247, 483-484, 487, 618
rain detector, 484
rainfall exculsion shelter, 614
rainforest, 599-608
red blood cells, 235
relative decline index, 103-106
relative growth rate (R_G), 405
respiration, 463-477
rhizosphere, 336
ribose, 425
ribosome, 429
ribulose bisphosphate carboxylase (Rubisco), 235, 238
rice, 238
rime ice, 84
root rot, 271
roots, 13, 17-18, 26, 32-33, 35-37, 107, 110, 112, 128
    effects of metals on physiology, 387-398
Rothamsten Experimental Station, 339
RUBP carboxylase, 212

Scandinavia, 323, 347, 353, 464, 493, 625
senescence, 244, 248, 279
sesquioxides, 362, 383
sex ratios, 244
Shenandoah National Park (USA), 221-222
silica, 368, 370, 373, 503, 561
silicate, 26, 367-369, 384
"silver tinsel effect", 6
Sky Forest (USA), 221
smelting, 307, 310-312, 427
sodium (Na), 516, 535, 539
    deposition, 51, 85
    in bog waters, 580, 582
    in foliage, 107-109
    in groundwater, 326
    in peat cores, 566-571, 580, 583
    in roots, 107, 110-111
    in soil, 107-110, 365, 378-381, 384
soil, 26, 37, 104-107, 244, 602-604, 607, 625
    acidification, 19, 27, 30-31, 35, 38-39, 61, 71, 267, 316, 321-328, 333-343, 361, 412, 620
    chemistry, 32, 259-262, 354-358, 364-373, 380
    classification, 104, 106
    horizons/profiles, 106, 324,

327, 347, 353-355, 380-381
  leaching, 321, 322, 333-335, 341, 365, 377-384, 625
  nutrient deficiency, 256
  pH, 104, 31-33, 35, 106-107, 162, 323, 327, 339, 347-355, 358, 366, 378
  pH profile, 364
  research needs, 626
  sensitivity, 31, 104-107, 378
  temperature, 31
  toxicity, 35, 38
  weathering, 26, 338
Soviet Union, 494
soybean, 234-235, 247, 452-461, 483-484, 487, 489, 613
spidermite(s), 484
spinach, 212, 234
Spodosols, 361-374
sporophytes, 162
spruce, 3-4, 8, 15, 17, 20, 27-31, 33, 36, 43, 45, 46-48, 50, 52, 65, 89, 125, 185, 272, 294, 310, 353-355, 357-359, 615
  black, 579, 632
  Norway, 6, 8, 15-16, 18-19, 27, 33, 36-37, 63, 72, 74, 76-79, 175, 185, 256-257, 259, 261-263, 294, 297-305, 315, 388-398, 402, 404-407, 619
  red, 83-96, 269, 277, 279-282, 284-290, 311, 314-315, 336, 616
  sitka, 72, 74, 76-77, 79
  white, 18, 336
spruce needle miner, 48
starch
  in roots, 17, 112
stemflow, 119-121, 123-124, 358
sterility, 244
stigma, 156-157, 163-164, 166-167
stomata, 132, 150, 210, 223, 235, 243-244, 428, 430, 433, 463-477
stomatal conductance, 210, 428, 431
Storage and Retrieval of Aerometric Data (SAROAD), 218
stress(es), 18, 35, 38, 87, 112, 128, 134, 172, 184, 247-249, 256, 268-270, 294, 412, 620, 627
  water, 201-213
succession, 503-507
sucrose, 159, 160, 165
sulphate, 11, 15, 322, 377, 389, 451, 508, 535, 539, 541-542, 544-545
  in bog waters, 507, 509, 516, 559, 582-587, 590-592
  in soil, 105, 333, 334-336, 339,

363, 368-370, 373, 379, 381-383
  retention capacity, 590-592
sulphur, 12, 120, 256, 313, 321, 368, 521, 526, 556, 584
  deposition, 51, 60, 66, 74-75, 78, 185, 324-328
  emission, 599-608
  in foliage, 6-8, 52, 107-109
  in roots, 107, 110-111, 396
  in soil, 29-30, 107, 110, 336
  quantification, 601
sulphur dioxide, 3, 9, 19, 25, 60, 66, 71, 131, 132, 150, 172, 256, 439, 441-445
  absorption rate, 147, 428, 430-436
  biochemical effects of, 141
  effects on growth, 133-134, 137-141
  effects on pollen, 156
  effects on transpiration, 132
  effects on vegetation, 21, 132-142, 166, 185, 247, 249, 451-461, 508, 550-551, 627, 635
  effects on yield, 135
  emission, 9, 51, 313, 601
sulphuric acid (H_2SO_4), 417-418, 551, 625
sunflower, 245, 247
surfactants, 150
Sweden, 62, 122, 316, 323-325, 338-339, 347-349, 351-353, 357, 359, 402, 498, 503, 545, 552
Switzerland, 44-46, 51-54, 621
sycamore, 3

tamarack, also see larch, 632
Tannensterben, 70
tent caterpillar, 102, 112-115
thin sectioning, 234
throughfall, 119-124, 358
titanium, 561
  in peat cores, 566-574
tobacco, 155, 231, 233
tomato, 142, 156, 242, 452-461, 483-484, 487
trace metals (see also individual metals)
  in forests, 85, 120, 307-317
  responses of pollen, 157, 159
translocation of assimilates, 18
transpiration, 210, 463-477
tree ring analyses, 65, 86, 90-92, 113-114, 277-304
  chronologies, 293
  indices, 113
  standardization, 283

trichomes, 148, 150-151
trillium, 158-161
tropics, 431-436
tundra, 431-435, 513

ultra-violet irradiation, 247
United Kingdom, 69-80, 132, 180, 182-183, 310, 339, 441-447, 464, 493, 495, 507, 516, 520, 523-524, 545, 549-559, 580
United States, 118, 122, 124-125, 218-219, 224, 244, 270, 272, 284-285, 292-293, 307, 315, 482, 607, 627
  Adirondack Mountains, 84, 86-89, 92-95, 118, 177, 270, 304, 316, 621
  Alaska, 271
  Appalachian Mountains, 83-96, 615, 621
  California, 218-222, 224
  Camel's Hump Mountain, 87, 89, 93, 311, 313-314, 316
  Colorado River Valley, 218
  Georgia, 339, 601
  Great Smoky Mountains National Park, 311-314
  Hubbard Brook, 84-85
  Maine, 85, 92, 307, 493, 561-575, 587, 592
  Massachusetts, 545, 562, 589
  Minnesota, 494-507, 580, 587, 591-593
  Michigan, 102
  Mount Moosilauke, 84-85, 122-124, 127
  New England, 96, 277, 284, 307, 562
  New Hampshire, 84-86, 88-91, 94, 122, 279, 316, 362-363, 368
  New Jersey, 292
  New Mexico, 279
  New York, 86-89, 92-95, 177, 273, 277, 294, 362, 368
  North Carolina, 87, 580, 591-592
  Ohio River Valley, 224
  Pennsylvania, 279
  Shenandoah National Park, 221-222
  Tennessee, 18, 119, 310, 314, 316, 339-342, 483
  Texas, 218, 271
  Vermont, 84-91, 93-95, 307, 310, 314, 316
  West Virginia, 92, 270
uronic acid, 514-515, 521-522

vacuoles, 396, 429
vanadium, 561
  in foliage, 107-109
  in peat cores, 566-572
  in roots, 107, 110-111
  in soil, 107, 110

Waldsterben, see also forest decline, 25, 39, 44, 272
Walker Branch Watershed, 119-125, 127, 339-342
water potential, 205-206, 211
water relations, 233
wax layer, 146-150
weathering, 582, 589
wetlands, 534-535, 537, 540, 543-545, 631-636
wettability (leaf), 147-149
wheat, 135-137, 185, 483-484, 487-488
willow, 243
wind tunnel, 175
wood formation, 208

x-ray probe, 30
xylem, 201, 204-206, 208-209, 311, 313-314
xylogenesis, 243

zinc, 307, 311-314, 356, 561-562, 632
  effect on roots, 389
  in foliage, 8, 15, 107-109, 256, 258-259, 261, 418-419
  in peat cores, 566-572
  in roots, 107, 110-111
  in soil, 105, 107, 110, 259, 261, 310, 353, 397